GEOMETRIC FORMULAS

TRIANGLE

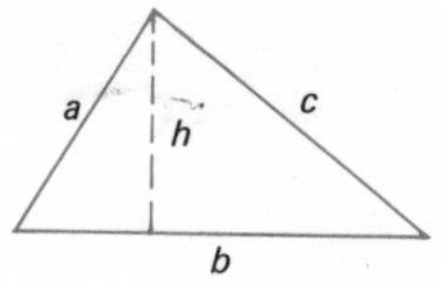

Area $= \frac{1}{2} bh$
Perimeter $= a + b + c$

RIGHT TRIANGLE

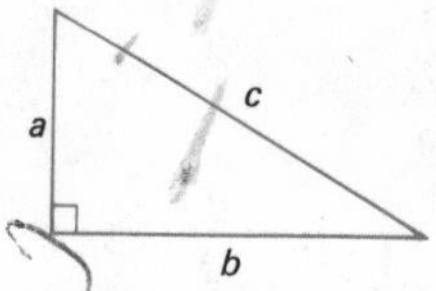

Pythagorean theorem
$a^2 + b^2 = c^2$

SIMILAR TRIANGLES

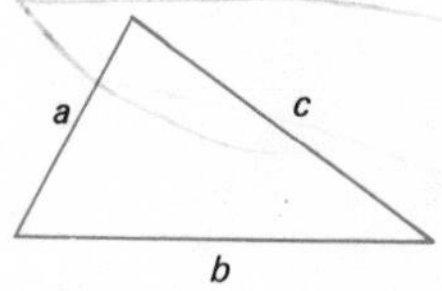

$\frac{a}{a'} = \frac{b}{b'} = \frac{c}{c'}$

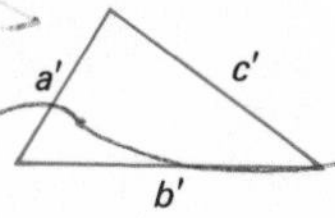

SQUARE

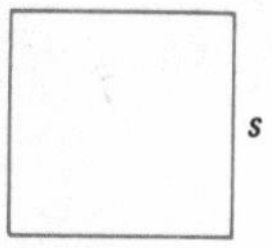

Area $= s^2$
Perimeter $= 4s$

RECTANGLE

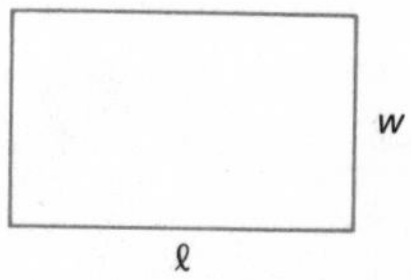

Area $= \ell w$
Perimeter $= 2\ell + 2w$

PARALLELOGRAM

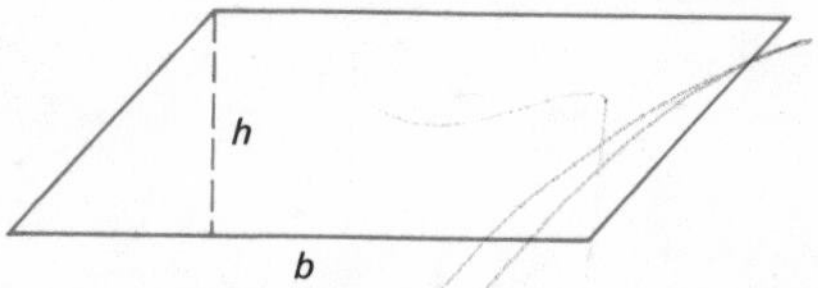

Area $= bh$

TRAPEZOID

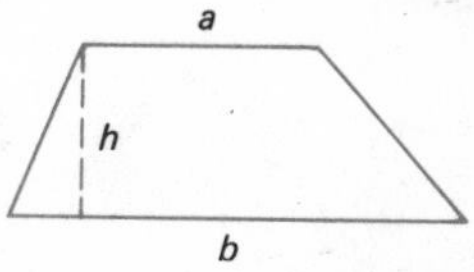

Area $= \frac{1}{2} h\ (a + b)$

CIRCLE

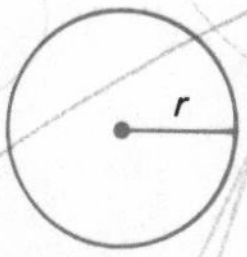

Area $= \pi r^2$
Circumference $= 2\pi r$

SPHERE

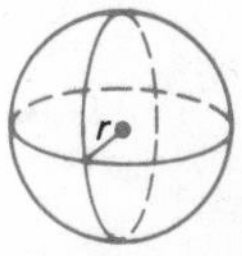

Volume $= \frac{4}{3} \pi r^3$
Surface area $= 4\pi r^2$

RIGHT CIRCULAR CYLINDER

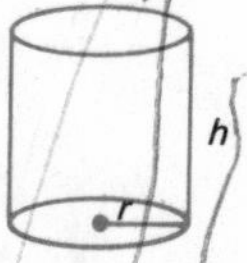

Volume $= \pi r^2 h$
Lateral surface area $= 2\pi r h$

RIGHT CIRCULAR CONE

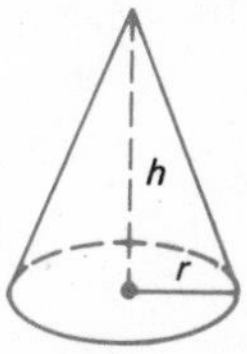

Volume $= \frac{1}{3} \pi r^2 h$
Lateral surface area $= \pi r \sqrt{r^2 + h^2}$

RECTANGULAR SOLID

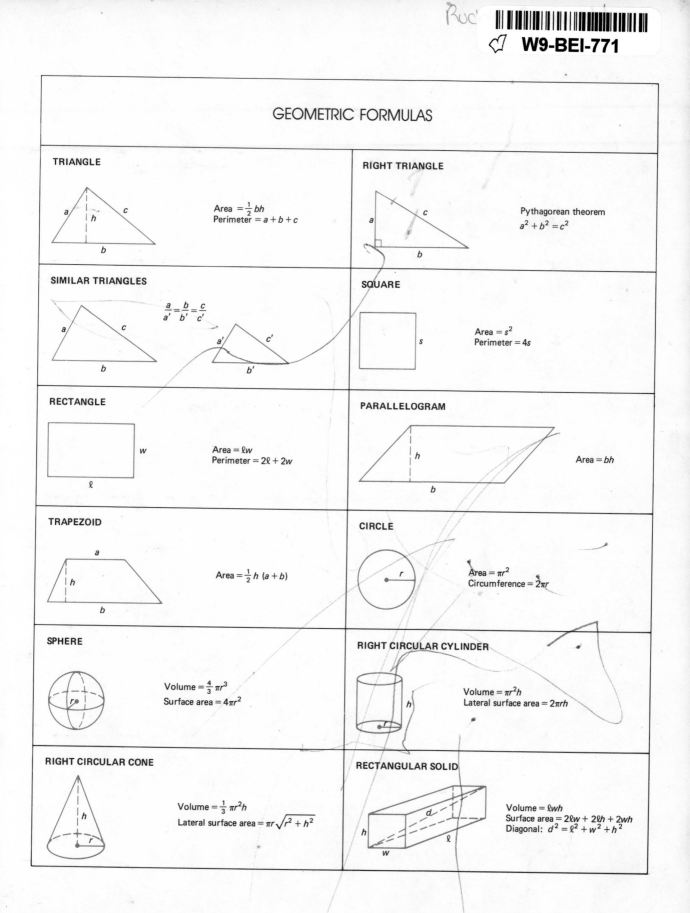

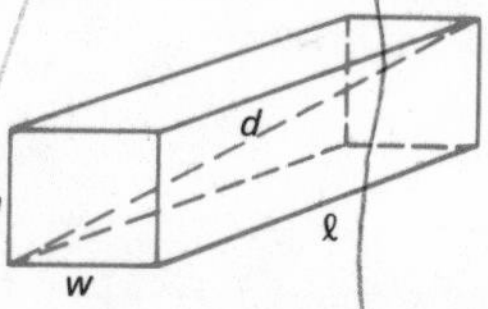

Volume $= \ell w h$
Surface area $= 2\ell w + 2\ell h + 2wh$
Diagonal: $d^2 = \ell^2 + w^2 + h^2$

ALGEBRA FOR COLLEGE STUDENTS

Fourth Edition

ALGEBRA FOR COLLEGE STUDENTS

Max A. Sobel

Montclair State College

Norbert Lerner

State University of New York at Cortland

PRENTICE HALL, Englewood Cliffs, NJ 07632

Library of Congress Cataloging-in-Publication Data

Sobel, Max A.
Algebra for college students / Max A. Sobel,
Norbert Lerner. – 4th ed.
p. cm.
Includes index.
ISBN 0-13-025933-0
1. Algebra. I. Lerner, Norbert. II. Title.
QA154.2.S6 1991
512.9–dc20 90-40847
CIP

Editorial/production supervision: Rachel J. Witty, Letter Perfect, Inc.
Acquisitions editor: Priscilla McGeehon
Interior design: Meryl Poweski
Cover design: Meryl Poweski
Prepress buyer: Paula Massenaro
Manufacturing buyer: Lori Bulwin

Printed in the United States of America
10 9 8 7 6 5 4 3 2 1

ISBN 0-13-025933-0

Prentice-Hall International (UK) Limited, *London*
Prentice-Hall of Australia Pty. Limited, *Sydney*
Prentice-Hall Canada Inc., *Toronto*
Prentice-Hall Hispanoamericana, S.A., *Mexico*
Prentice-Hall of India Private Limited, *New Delhi*
Prentice-Hall of Japan, Inc., *Tokyo*
Simon & Schuster Asia Pte. Ltd., *Singapore*
Editora Prentice-Hall do Brasil, Ltda., *Rio de Janeiro*

CONTENTS

PREFACE

TO THE INSTRUCTOR

This fourth edition of *Algebra for College Students* has been written to provide beginning college students with the fundamental algebraic concepts and skills necessary for further study of mathematics. It should also assist students in understanding the applications of mathematics to other subjects.

Many sections have been rewritten to improve the exposition. This includes a variety of new applications and additional display examples. Most exercise sets have been enhanced with the introduction of additional questions of various degrees of difficulty.

Emphasis has been given throughout the text to the use of a scientific calculator, especially in the chapter on exponential and logarithmic functions. Calculator displays have been included, and exercises that should be solved by a calculator have been designated by the logo, although the instructor still maintains the option of not allowing their use in class or on tests.

Many users of previous editions have contributed valuable suggestions that have been incorporated into this fourth edition. Please feel free to communicate with the authors and provide your input for future revisions. Some of the changes that have been incorporated into this fourth edition are included in the following summary:

Chapter 1. Section 1.5, Operations with Real Numbers, is new. This section extends the fundamental operations to include rational and irrational numbers. The calculator is used to obtain approximations when using irrational numbers.

Chapter 2. This chapter reviews the fundamental algebraic concepts and skills required for the remaining course of study. Negative exponents are developed in Section 2.2 (rather than in Chapter 5, as in the third edition). Scientific notation is now covered in this section. The work on Pascal's triangle has been extended, but has been placed in Chapter 12. However, this work can be extracted from Chapter 12 and covered earlier, if needed.

Chapter 3. The introductory work on quadratic equations is included as Section 3.6. This is a more natural placement than in the preceding editions, where it was located in the chapter on fundamental operations. More extensive work on quadratic equations is done in Chapter 5.

Chapter 4. Synthetic division has been moved into Chapter 8, preceding the work on the factor, remainder, and rational root theorems.

Chapter 5. This chapter contains the same subject matter as in the third edition. Also, the algebraic work dealing with the quadratic formula and applications of quadratic equations has been included here.

Chapter 6. This chapter is similar to Chapter 6 in the third edition.

Chapter 7. The work from Chapter 8 of the third edition on quadratic functions and inequalities is covered here. In addition a more extensive treatment of the conic section is included. Each conic section is covered in an individual section.

Chapter 8. This is essentially a new chapter. It includes sections on graphing polynomial, rational, and radical functions. The method of shifting (translating) fundamental curves to obtain graphs of more complicated functions is a unifying concept. The work on synthetic division and the factor, remainder and rational root theorems are also included.

Chapter 9. This chapter has been completely rewritten. After some introductory work on inverse functions, exponential functions of the form $y = b^x$ are introduced. Then the logarithmic functions are developed as the inverses of the exponential functions. The work with the laws of logarithms follows, as do applications of exponential growth and decay. Common logarithms are included as an optional section.

Chapter 10. The subject matter from Chapter 7 in the third edition on systems of linear equations is covered here. In addition, some work on systems of linear inequalities, followed by an introduction to linear programming, is included. There is also a section on solving nonlinear systems of equations.

Chapter 11. This chapter is similar to Chapter 10 in the third edition on Sequences and Series.

Chapter 12. This is a new chapter and introduces the student to permutations, combinations, and probability. A section on the binomial expansion is also included. The binomial formula is developed in two ways, one of which uses binomial coefficients. However, the work on this formula, without using the binomial coefficients, can be extraced and used earlier in the course, if needed.

It is important to note that this book was written with the expectation that students would read it, and not just refer to it as a source of exercises. Thus, we have worked carefully to make the exposition clear. To this end, we have used an informal, yet precise, approach in presenting basic mathematical concepts throughout.

You may find it helpful to note the following pedagogical features, which have been included in this edition to assist your students in learning the fundamentals of algebra. Some have been successfully used in past editions, and others have been added for this current edition, as noted.

Margin Notes

Throughout the text margin notes have been inserted to enhance the exposition, raise questions, point out interesting facts, show alternate procedures, give references and reminders, and caution the student to avoid errors.

Test Your Understanding

These short sets of exercises are found within most sections of the text and encourage students to *think carefully*. Students can use them to test their knowledge of new material, before attempting to solve the section exercises. Answers to all of these are given at the end of each chapter, providing an excellent means of self-study.

Caution Items

Where appropriate, caution items appear in the text or in the margin notes to alert students to typical errors that are to be avoided. Students can see both the correct and the incorrect solution to an exercise.

Illustrative Examples

The text contains more than 500 illustrative examples with detailed solutions. These appear in every section and allow students to see sample solutions prior to working on the end-of-section exercises. They can refer to these while working on the exercises, if they experience any difficulties. These sample solutions are especially helpful as review material, when students have been absent from class.

Challenges

This is a new feature. In each chapter, we have included challenging problems that encourage students to *think creatively*. These can be used to test the students' skill in solving a more difficult problem or one that has an unusual twist to it.

Explorations

A set of Exploration exercises in each chapter encourages students to *think creatively*. These demand higher-order thinking skills and do not depend solely on applying routine skills.

Written Assignments

This is a new feature. Throughout the book, written exercises are featured at the end of many of the standard sets of exercises. These generally ask for a written explanation or description rather than only an algebraic solution of an exercise. It is generally agreed that we should encourage practice in writing in mathematics courses, and it is hoped that this experience will help students with their other college courses as well.

Summary

This is a new feature. Each chapter contains a concise chapter review, which includes key terms and basic concepts and formulas.

Review Exercises

At the end of each chapter, there is a set of review exercises identical to some of the illustrative examples developed in the text. Students can use them as a self-study review of the chapter by comparing their answers with the worked-out solutions found in the body of the text.

Chapter Tests

Each chapter concludes with two forms of a chapter test, standard answer and multiple choice. Answers to both tests are provided at the end of the book. These enable students to test their knowledge of the work of the chapter in preparation for an instructor's test on the material.

Cumulative Reviews

This is a new feature. At the end of every three chapters cumulative review questions test the work of the course to date. More than any other subject, the study of mathematics is cumulative, and students need to be certain that they have not forgotten previously learned skills. The answers are included in the back of the book.

Inside Covers

Inside the front and back covers, summaries of useful information, including basic graphs and important algebraic and geometric formulas are presented for easy reference.

SUPPLEMENTARY MATERIALS

The following supplementary materials are available for this book.

1. *Instructor's Solution Manual* This manual provides the insructor with completely worked-out solutions for every even-numbered exercise in the text.
2. *Student Solutions Manual* This manual contains completely worked-out solutions for the odd-numbered exercises found at the end of each section, as well as for each of the chapter tests.
3. *Instructor's Manual with Tests* Includes testing material for classroom use.
4. *Interactive Algebra Tutor Software* Provides tutorial assistance and drill problems for students.

ACKNOWLEDGMENTS

The preparation of this fourth edition of *Algebra for College Students* has been guided by many people. First, we wish to thank the many students and instructors who used the first three editions and contributed their comments and suggestions.

For their detailed reviews, criticisms, and suggestions we thank

Martin Brown
Jefferson Community College, Louisville, KY

Laurene V. Fause
Florida Institute of Technology, Melbourne, FL

Roy Fraser
Skyline College, San Bruno, CA

Julie D. Horne
University of Georgia, Athens, GA

Merwin J. Lyng
Mayville State College, Mayville, ND

Thomas A. McCready
California State University, Chico, CA

Chantal Shafroth
North Carolina Central University, Durham, NC

Nancy Joe Shaw
Elizabethtown College, Elizabethtown, PA

Lee-Ing Tong
Southeastern Massachusetts University,
North Dartmouth, MA

For her thorough and careful writing of the manuscript for the *Student Solutions Manual,* as well as for her many helpful comments and contributions, we thank Margaret Hall Babcock.

For their assistance in the preparation of supplementary materials, we also acknowledge:

Joe May
North Hennepin Community College, Brooklyn Park, MN

William Radulovich
Florida Community College

Deborah White
College of the Redwoods, Mendocino Coast, CA

For their valuable comments we are grateful to

Andrew N. Aheart
West Virginia State College, Institute, WV

Sonya M. Oetzel
Madison Area Technical College, Madison WI

Roger Riveland
Bismarck Jr. College, Bismarck, ND

For her outstanding editing, constant patience, and cooperation, we express our sincere appreciation to our production editor, Rachel J. Witty, Letter Perfect, Inc.

For providing us with guidance and direction, and the occasional prodding we needed throughout the development of the text, we thank our editor at Prentice Hall, Priscilla McGeehon.

For their patience and endurance throughout the many months that we devoted to manuscripts, galleys, page proofs, and endless telephone conversations, we offer our thanks to our wives Manya and Karin.

We hope that you and your students will enjoy using this book. We invite your reactions and welcome your suggestions.

Max A. Sobel
Norbert Lemer

SUGGESTIONS FOR THE STUDENT

We have prepared this book for you to read and to enjoy as well. Every effort has been made to make each topic meaningful, with clear exposition and numerous worked-out examples to serve as models for the exercises that you will attempt. Nevertheless, we recognize that many students enter this course with weak backgrounds, and often with a distaste for mathematics due to prior experiences.

We urge you to be patient, and suggest that if you are willing to devote the necessary time and effort to the course, then you should become successful. Based on the authors' many years of teaching experience, we would like to offer a short list of suggestions that should help to ensure this success:

1. Make every possible effort to keep up to date. Set aside regular time periods for the work in this course, and stick to your plan.
2. Be patient! Always try to reason things out on your own first, despite any difficulties that you may encounter. Even a modest effort along these lines will prove to be rewarding in the long run.
3. Read each section with paper and pencil at hand. Try to solve the illustrative examples before looking at the worked-out solutions provided.
4. Attempt each of the Test Your Understanding exercises, whether or not they are assigned. Reread the section if you have difficulty with any of these.
5. Try as many exercises as possible at the end of each section. Complete the odd-numbered ones first and check your answers with those given at the back of the book. Keep in mind that sometimes your answer may be correct, even though it is not in the same form as the given answer. If you miss very many, reread and study the section again.
6. Prior to a test, make use of the review exercises at the end of each chapter. You can check your results by referring to the designated section from which these are taken where you will find the completely worked-out solution for each one.
7. Complete each of the chapter tests under testing conditions. That is, do not refer to the text as you complete these, and set a fixed period of time for your work, usually an hour. The answers are given at the back of the book.
8. If convenient, find time to study and solve problems cooperatively with a classmate. Such efforts can be beneficial, as you explain ideas to one another.

We are convinced that even if you have had a negative attitude towards mathematics in the past, an honest attempt to learn it properly will not only result in greater success, but will also lead to a self-awareness that you have more ability and talent than you ever gave yourself credit for! Please feel free to let us know, if you have any comments, criticisms, or suggestions. We look forward to hearing from you, as you continue with your study of *Algebra for College Students*. Good luck!

Max A. Sobel
Norbert Lerner

CHAPTER

1 THE SET OF REAL NUMBERS

1.1 SETS OF NUMBERS

Welcome to the *set* of students who will spend the semester using this text. The word **set** is an important one in mathematics and, as in its everyday usage, it means a collection of things. Many words are commonly used to denote sets:

A *flock* of geese
A *swarm* of bees
A *squadron* of planes
A *team* of players
A *herd* of cattle

Can you find other such words that denote sets?

In mathematics we are often concerned with different *sets of numbers*. In most of your past work you have been dealing with the **set of real numbers.** Here are some examples of real numbers:

$$5 \qquad -3 \qquad 0 \qquad \tfrac{3}{4} \qquad \sqrt{2} \qquad -4.75 \qquad -1\tfrac{1}{2} \qquad \pi$$

We can illustrate these numbers on a **number line,** where each number is the **coordinate** of some point on the line.

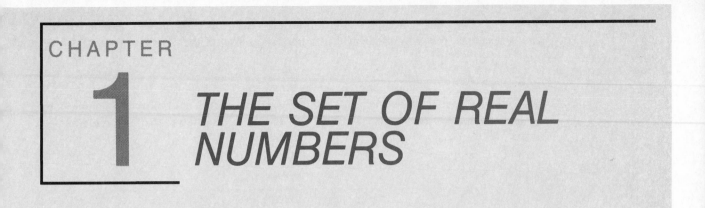

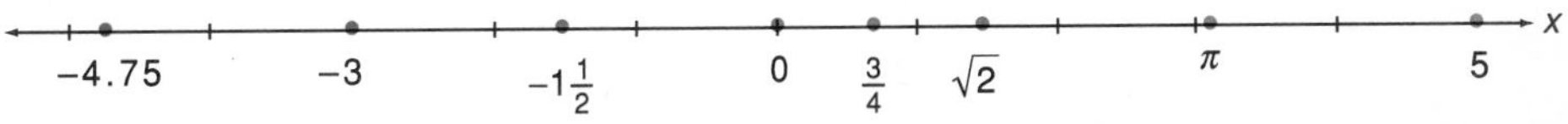

WATCH THE MARGINS! We will use these for special notes, explanations, challenges, and hints throughout the book.

Sometimes we will only need to use a **subset,** or part, of the real numbers. For example, the set N of **natural numbers** is a subset of the set of real numbers:

$$N = \{1, 2, 3, \ldots\}$$

Every natural number is a real number, but not every real number is a natural number. Each of the numbers 1, 2, 3, . . . is said to be a **member** of or an **element** of set N.

According to this definition, every set is a subset of itself.

> **DEFINITION OF SUBSET**
>
> Set A is a **subset** of set B if every element of A is also an element of B. To state this in symbols, we write
>
> $A \subseteq B$ Set A is a subset of set B.

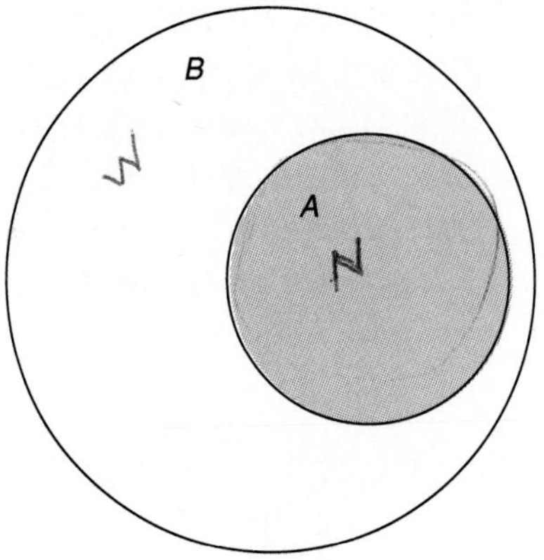

$A \subseteq B$

The set of natural numbers is also referred to as the set of **counting numbers.** We name a set with a capital letter and include the members of the set within a pair of braces. The three dots are used here to indicate that the set goes on without end; that is, N is an example of an **infinite set.**

Note that 0 is *not* a member of the set of natural numbers. However, if we add the number 0 to the set of natural numbers, we produce another infinite set called the set of **whole numbers,** W:

$$W = \{0, 1, 2, 3, \ldots\}$$

The symbol $\nsubseteq$ is read "is not a subset of."

Every natural number is also a whole number, but not every whole number is a natural number. Thus $N \subseteq W$, but $W \nsubseteq N$ because 0 is an element in W that is not in N.

Both sets that we have just introduced are examples of infinite sets. However, some of the sets we will use will be **finite sets.** The members of a finite set can be listed and counted, and there is an end to this counting. For example, the set of whole numbers that are less than 5 is an example of a finite set:

$$\{0, 1, 2, 3, 4\}$$

At times a set can be classified as finite even though it has so many elements that no one would really want to list them all. Thus the set of counting numbers from 1 through 1,000,000 is a large set, but nevertheless finite, inasmuch as it does have a last member. We adopt the convention of using three dots to indicate that some members of a set are not listed and can write this as

Note: In Exercise 8 the word "between" means that 1 and 10 are not included.

$$\{1, 2, 3, \ldots, 1{,}000{,}000\}$$

TEST YOUR UNDERSTANDING
Think Carefully

Throughout this book we shall occasionally pause for you to test your understanding of the ideas just presented. If you have difficulty with these brief sets of exercises, then you should reread the material of the section before going ahead. Answers will be found at the end of the chapter.

List the elements in each of the following sets.

1. The set of natural numbers less than 4. {1,2,3}
2. The set of whole numbers less than 5. {0,1,2,3,4}
3. The set of counting numbers from 1 through 1000.
4. The set of whole numbers greater than 10.
5. The set of odd natural numbers.
6. The set of even natural numbers less than 10.
7. The set of even natural numbers.
8. The set of whole numbers *between* 1 and 10.
9. The set of whole numbers greater than 100.
10. The set of counting numbers between 5 and 6.

(Answers: Page 31)

You may have been puzzled by Exercise 10 because there is obviously no counting number between 5 and 6. Such a set is then said to be empty. Thus the **empty set,** or **null set,** is the set that contains no elements. We will need this language later to describe the solution of certain types of equations. For example, consider the set of whole numbers x that make this equation true:

$$x + 2 = x$$

There is no whole number (or any other type of number) that will make this equation true. We say that the solution is the empty set, and denote this by the symbol $\varnothing$.

Another set that we shall use often is the set of **integers,** I:

$$I = \{\ldots, -3, -2, -1, 0, 1, 2, 3, \ldots\}$$

The set of integers can be thought of as consisting of three subsets:

It should be noted that the set of positive integers is just another way of describing the set of counting numbers. Unless the plus signs (+) are needed for special emphasis, we generally omit them.

The set of negative integers:	$\{\ldots, -3, -2, -1\}$
The set consisting of zero:	$\{0\}$
The set of positive integers:	$\{+1, +2, +3, \ldots\}$

Having introduced the set of integers we are now able to solve such equations as $x + 2 = 0$. The solution is $x = -2$, a number that is a member of the set of integers but that was not included in the set of whole numbers.

Note that every integer can be written in fractional form. For example:

$$3 = \frac{3}{1} \qquad -2 = \frac{-2}{1} \qquad 0 = \frac{0}{1}$$

However there are fractions that cannot be written as integers, such as $\frac{2}{3}$ and $-\frac{3}{4}$. The collection of all such numbers is called the set of **rational numbers.**

For any integer a we have $a = \frac{a}{1}$. Therefore, every integer is a rational number.

DEFINITION OF RATIONAL NUMBER

A **rational number** is one that can be written in the form $\frac{a}{b}$, where a and b are integers, $b \neq 0$.

Note that a number in decimal form, such as 2.75, is a rational number because it can be written in the form $\frac{a}{b}$:

$$2.75 = 2\frac{75}{100} = \frac{275}{100}$$

On the other hand there are numbers that cannot be expressed as the *ratio* of two integers, such as $\sqrt{2}$. This fact so surprised the ancient Pythagoreans that it is said they kept the fact secret for some time! Such numbers that cannot be written as the quotient of integers are called **irrational numbers.** Some other examples of irrational numbers are

$$\sqrt{5} \qquad \sqrt{12} \qquad \sqrt[3]{4} \qquad -\sqrt{17} \qquad \pi$$

When the set of rational numbers and the set of irrational numbers are combined, we have the set of **real numbers.** The following diagram shows how the set of real numbers may be subdivided into various subsets, as we have just discussed.

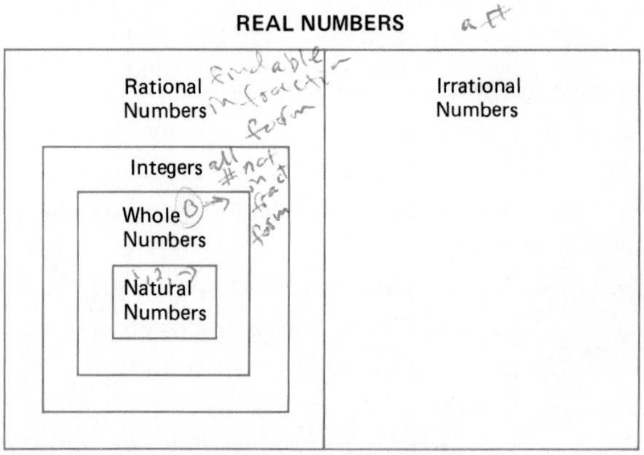

EXAMPLE 1 To which subsets of the real numbers do each of the following belong?

(a) 5 **(b)** $\frac{2}{3}$ **(c)** $\sqrt{7}$ **(d)** -14

Solution

(a) 5 is a natural number, a whole number, an integer, a rational number, and a real number.

(b) $\frac{2}{3}$ is a rational number and a real number.

(c) $\sqrt{7}$ is an irrational number and a real number.

(d) -14 is an integer, a rational number, and a real number. ■

EXAMPLE 2 Classify as true or false: Every whole number is a natural number.

Solution In order for a statement to be true, it must be true for all possible cases; otherwise it is false. Since 0 is a whole number but is *not* a natural number, the statement is false. ■

EXERCISES 1.1

List the elements in each of the sets described in Exercises 1 through 10.

1. The set of natural numbers less than 2.
2. The set of odd natural numbers less than 10.
3. The set of even natural numbers.
4. The set of whole numbers less than 3.
5. The set of integers greater than 100.
6. The set of natural numbers from 10 through 10,000.
7. The set of odd whole numbers.
8. The set of natural numbers that are divisible by 5.
9. The set of whole numbers that are not natural numbers.
10. The set of natural numbers that are not whole numbers.

For Exercises 11 through 20, answer true or false. If false, explain your answer.

11. Every natural number is a whole number.
12. Every whole number is a natural number.
13. The set of natural numbers is a subset of the set of integers; that is, $N \subseteq I$.
14. The set of integers is a subset of the set of natural numbers; that is, $I \subseteq N$.
15. Every integer is a rational number.
16. Every real number is a rational number.
17. Every rational number is the coordinate of some point on the number line.
18. Some rational numbers are irrational.
19. Between any two integers there is always another integer.
20. The set of whole numbers contains the negative of each of its members.

Two sets A and B are said to be in **one-to-one correspondence** *if each element of A corresponds to one and only one element of B and each element of B corresponds to one and only one element of A. For example, here is one way to show such a correspondence between set* $V = \{a, e, i, o, u\}$ *and set* $F = \{1, 2, 3, 4, 5\}$:

$$\begin{array}{ccccc} a & e & i & o & u \\ \updownarrow & \updownarrow & \updownarrow & \updownarrow & \updownarrow \\ 1 & 2 & 3 & 4 & 5 \end{array}$$

21. Show a one-to-one correspondence between the elements of the set $A = \{a, b, c\}$ and the elements of the set $B = \{m, n, p\}$.
22. Show five other ways of establishing the correspondence for Exercise 21.
23. Show a one-to-one correspondence between the set of whole numbers less than 5 and the set of integers between -3 and 3.

*24. Demonstrate a one-to-one correspondence between the set of natural numbers and the set of integers.

For Exercises 25 through 28, note that the empty set, $\varnothing$, *is considered to be a subset of every set. Furthermore, a set is also considered to be a subset of itself.*

25. List all the possible subsets for the set $M = \{1, 2\}$.
26. Repeat Exercise 25 for set $N = \{1, 2, 3\}$.
27. How many subsets are there for a set consisting of (a) 2 elements, (b) 3 elements, (c) 4 elements, and (d) 5 elements?

*28. In general, how many subsets are there for a set consisting of n elements?

29. The set of real numbers has the **density property:** between any two real numbers there is always another real number. Show, with an example, that this property is *not* true for the set of integers.
30. The set of rational numbers also has the density property (see Exercise 29). For example, we can find a rational number between $\frac{13}{15}$ and $\frac{14}{15}$ by finding the average of the two, or by changing each denominator to 30:

$$\left.\begin{array}{l} \frac{13}{15} = \frac{26}{30} \\ \frac{14}{15} = \frac{28}{30} \end{array}\right\} \quad \text{Clearly, } \frac{27}{30} \text{ lies between the two given numbers.}$$

Find a rational number between $\frac{26}{30}$ and $\frac{27}{30}$.

31. We can locate a point with irrational coordinate $\sqrt{2}$ on the number line. At the point with coordinate 1 construct a 1-unit segment perpendicular to the number line. Connect the endpoint of this segment to the point labeled 0. This becomes the hypotenuse of a right triangle, and by the Pythagorean theorem is equal to the square root of 2, written as $\sqrt{2}$. Using a compass, this length can then be transferred to the number line, thus locating a point with coordinate $\sqrt{2}$. This demonstrates that irrational numbers correspond to points on the number line.

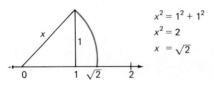

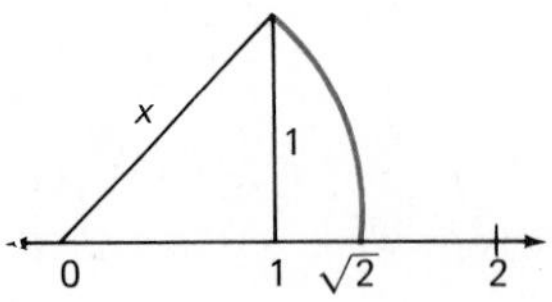

$$x^2 = 1^2 + 1^2$$
$$x^2 = 2$$
$$x = \sqrt{2}$$

Show how to locate a point on the number line with coordinate $\sqrt{5}$.

*Throughout the text, an asterisk will be used to indicate that an exercise is more difficult than usual or involves some unusual aspect.

CHALLENGE
Think Creatively

Throughout this book we will pose occasional challenges for you to consider. Some will be in the form of explorations that may lead to the discovery of an interesting mathematical pattern, some may be especially difficult but interesting problems, and some may be presented just for fun. In this latter category, consider the following set of numbers:

$$A = \{8, 5, 4, 9, 1, 7, 6, 3, 2, 0\}$$

Find a rule that describes the *order* in which the members of the set are listed.

1.2 PROPERTIES OF THE REAL NUMBERS

Throughout your work in mathematics you have been dealing with binary operations. **A binary operation** combines any two members of a set, such as the operations of addition and multiplication. In this section we shall consider the system of real numbers together with the basic operations of addition and multiplication, and will review properties of this system.

Note that the product $a \times b$ may also be written in a number of other ways, such as $a \cdot b$, $(a)(b)$, or just ab.

We assume that the variables a, b, and c represent any real numbers.

FUNDAMENTAL PROPERTIES OF ADDITION AND MULTIPLICATION

Closure Properties	For addition: For multiplication:	$a + b$ is a real number $a \times b$ is a real number
Commutative Properties	For addition: For multiplication:	$a + b = b + a$ $a \times b = b \times a$
Associative Properties	For addition: For multiplication:	$(a + b) + c = a + (b + c)$ $(a \times b) \times c = a \times (b \times c)$
Distributive Property		$a \times (b + c) = (a \times b) + (a \times c)$

The last property stated in the list is a very important one that combines the operations of addition and multiplication. It allows us, for example, to evaluate the expression $5(3 + 9)$ in two different ways.

(a) Add first, then multiply.

$$5(3 + 9) = 5(12) = 60$$

(b) Multiply first, then add.

$$5(3 + 9) = (5)(3) + (5)(9) = 15 + 45 = 60$$

The distributive property can also be written in this form: $(b + c)a = ba + ca$

The same result is obtained either way. This example illustrates the use of the **distributive property of multiplication over addition: $a(b + c) = ab + ac$.**

This example shows how the distributive property can be used to do mental arithmetic.

EXAMPLE 1 Use the distributive property to find the product 7×61.

Solution Think of 61 as 60 + 1.

$$7 \times 61 = 7 \times (60 + 1)$$
$$= (7 \times 60) + (7 \times 1)$$
$$= 420 + 7 = 427$$ ■

To disprove a statement, it is only necessary to find one specific example where the statement does not hold. Such an example is called a **counterexample.**

TEST YOUR UNDERSTANDING
Think Carefully

Name the property of real numbers being illustrated.

1. $3 + (\frac{1}{2} + 5) = (3 + \frac{1}{2}) + 5$ **2.** $3 + (\frac{1}{2} + 5) = (\frac{1}{2} + 5) + 3$
3. $6 + 4(2) = 4(2) + 6$ **4.** $8(-6 + \frac{2}{3}) = 8(-6) + 8(\frac{2}{3})$
5. $(17 \cdot 23)59 = (23 \cdot 17)59$ **6.** $(17 \cdot 23)59 = 17(23 \cdot 59)$
7. Does $3 - 5 = 5 - 3$? Is there a commutative property for subtraction?
8. Give a *counterexample* to show that the set of real numbers is not commutative with respect to division. (That is, use a specific example to show that $a \div b \neq b \div a$.)
9. Does $(8 - 5) - 2 = 8 - (5 - 2)$? Is there an associative property for subtraction?
10. Give a counterexample to show that the set of real numbers is not associative with respect to division.
11. Are $3 + \sqrt{7}$ and $3\sqrt{7}$ real numbers? Explain.

Use the distributive property to find each product mentally.

12. 9×51 **13.** 8×72 **14.** 12×31 **15.** 6×205

(Answers: Page 31)

The set of real numbers contains two very special numbers, 0 and 1, that are called **identity elements.** We call 0 the identity element with respect to addition, or the **additive identity,** because the sum of any number and 0 is that number. In other words, when 0 is added to another number, the "identity" of the other number is not changed. Similarly, 1 is the **multiplicative identity** because the product of any real number and 1 is that real number. In general, we have the following additional properties.

Identity Properties	For addition:	$a + 0 = 0 + a = a$ This property is also known as the **addition property of zero.**
	For multiplication:	$a \cdot 1 = 1 \cdot a = a$ This property is also known as the **multiplication property of one.**
Inverse Properties	For addition:	For each real number a there exists another real number $-a$ such that $a + (-a) = (-a) + a = 0$ We call $-a$ the **additive inverse** or **opposite** of a.
	For multiplication:	For each real number a, $a \neq 0$, there exists a real number $\frac{1}{a}$ such that

The symbol ≠ is real "is not equal to."

$$a \cdot \frac{1}{a} = \frac{1}{a} \cdot a = 1$$

We call $\frac{1}{a}$ the **multiplicative inverse** or **reciprocal** of a.

EXAMPLE 2 Find the additive and multiplicative inverses of **(a)** 3, **(b)** $\frac{2}{5}$.

Solution

(a) The additive inverse of 3 is -3; $3 + (-3) = 0$.
The multiplicative inverse of 3 is $\frac{1}{3}$; $3 \times \frac{1}{3} = 1$.

(b) The additive inverse of $\frac{2}{5}$ is $-\frac{2}{5}$; $\frac{2}{5} + (-\frac{2}{5}) = 0$.
The multiplicative inverse of $\frac{2}{5}$ is $\frac{5}{2}$; $\frac{2}{5} \times \frac{5}{2} = 1$. ■

EXAMPLE 3 What basic property of the real numbers is illustrated by each of the following?

(a) $6 + (17 + 4) = (17 + 4) + 6$

(b) $\frac{3}{4} + (-\frac{3}{4}) = 0$

(c) $57 \times 1 = 57$

(d) $\frac{2}{3}(12 + 36) = \frac{2}{3}(12) + \frac{2}{3}(36)$

(e) $(-43)\left(\frac{1}{-43}\right) = 1$

(f) $5\sqrt{3}$ is a real number

Solution

(a) Commutative, for addition

(b) Inverse, for addition

(c) Identity, for multiplication

(d) Distributive

(e) Inverse, for multiplication

(f) Closure for multiplication ■

There are other important properties of 0 that will be needed throughout our course of study.

Multiplication Property of Zero	Any number times 0 is equal to 0. $a \cdot 0 = 0 \cdot a = 0$
Zero-product Property	If the product of two numbers is 0, then at least one of them must be 0. If $ab = 0$, then $a = 0$ or $b = 0$ or both equal 0.

Finally, we list the following properties of opposites that will often be used in algebraic work.

Opposite of an Opposite	The opposite of the opposite of a number a is the number a. $-(-a) = a$
Opposite of a Sum	The opposite of a sum is the sum of the opposites. $-(a + b) = (-a) + (-b)$
Opposite of a Product	The opposite of a product of two numbers is the product of one number times the opposite of the other. $-(ab) = (-a)b = a(-b)$

EXERCISES 1.2

Name the property illustrated by each of the following.

1. $5 + 7$ is a real number.
2. $8 + \sqrt{7} = \sqrt{7} + 8$
3. $5 + (-5) = 0$
4. $9 + (7 + 6) = (9 + 7) + 6$
5. $(5 \times 7) \times 8 = (7 \times 5) \times 8$
6. $(5 \times 7) \times 8 = 5 \times (7 \times 8)$
7. $\frac{1}{4} + \frac{1}{2} = \frac{1}{2} + \frac{1}{4}$
8. $(4 \times 5) + (4 \times 8) = 4(5 + 8)$
9. $-13 + 0 = -13$
10. $1 \times \frac{1}{9} = \frac{1}{9}$
11. $\frac{1}{2} + (-\frac{1}{2}) = 0$
12. $3 - 7 = 3 + (-7)$
13. $0(\sqrt{2} + \sqrt{3}) = 0$
14. $\sqrt{2} \times \pi$ is a real number.
15. $(3 + 9)(7) = (3)(7) + (9)(7)$
16. $\dfrac{1}{\sqrt{2}} \cdot \sqrt{2} = 1$

Use a property to replace the variable n by a real number to make each statement true.

17. $7 + n = 3 + 7$
18. $\sqrt{5} \times 6 = 6 \times n$
19. $(3 + 7) + n = 3 + (7 + 5)$
20. $6 \times (5 \times 4) = (6 \times n) \times 4$
21. $5(8 + n) = (5 \times 8) + (5 \times 7)$
22. $(3 \times 7) + (3 \times n) = 3(7 + 5)$
23. $-(n) = 9$
24. $-(n + 4) = 12 + (-4)$
25. If $7(x + 2) = 0$, then $x = -2$. Why?
26. If $x(x - 6) = 0$, then $x = 0$ or $x = 6$. Why?

Give the basic property that justifies each of the numbered steps.

27. $ab + 2b + a \cdot 3 + 6 = ab + 2b + a \cdot 3 + 3 \cdot 2$

$$\begin{aligned} &= ab + 2b + 3a + 3 \cdot 2 && \text{(i)} \\ &= (a + 2)b + 3(a + 2) && \text{(ii)} \\ &= (a + 2)b + (a + 2)3 && \text{(iii)} \\ &= (a + 2)(b + 3) && \text{(iv)} \end{aligned}$$

28. $d[(a + b) + c] = d(a + b) + dc$ (i)

$$\begin{aligned} &= (da + db) + dc && \text{(ii)} \\ &= da + (db + dc) && \text{(iii)} \\ &= da + (bd + cd) && \text{(iv)} \\ &= (bd + cd) + da && \text{(v)} \\ &= (b + c)d + da && \text{(vi)} \\ &= da + (b + c)d && \text{(vii)} \end{aligned}$$

Answer true or false to each statement. If false, give a specific counterexample to justify your answer. (Recall that a true statement must be true for all possible cases; otherwise it is false.)

29. The set of whole numbers is closed with respect to multiplication.
30. The set of natural numbers is closed with respect to subtraction.
31. The set of integers is closed with respect to division.
32. Except for 0, the set of rational numbers is closed with respect to division.
33. The set of integers is commutative with respect to subtraction.
34. The set of rational numbers is associative with respect to multiplication.
35. The set of rational numbers contains the additive inverse for each of its members.
36. Except for 0, the set of rational numbers contains the multiplicative inverse for each of its members.

37. The product of any two real numbers is a real number.
38. The quotient of any two real numbers is a real number.

39. Note that $2^4 = 4^2$. Is the operation of raising a number to a power a commutative operation? Justify your answer.
40. Give an example to show that the set of irrational numbers is not closed with respect to addition.
41. Repeat Exercise 40 for multiplication.
42. Give an example showing that addition does *not* distribute over multiplication; that is, $a + (b \cdot c) \neq (a + b) \cdot (a + c)$.

Written Assignment: Explain why the set of real numbers have the closure property for both subtraction and division, excluding division by 0.

1.3 OPERATIONS WITH INTEGERS: ADDITION AND SUBTRACTION

In this section we will review the rules of signs for addition and subtraction of integers. These same rules, however, apply to *any* real numbers. The number line is useful for explaining the addition of two integers. Consider, for example, the sum $5 + (-7)$. On a number line represent 5 by drawing an arrow that starts at the **origin,** 0, and extends 5 units *to the right*. From this point draw a second arrow that is 7 units of length extended *to the left*. This second arrow terminates at the point whose coordinate is -2; $5 + (-7) = -2$.

At times a plus sign is used for emphasis with a positive number. Thus $5 + (-7)$ *may be written as* $(+5) + (-7)$. *Similarly,* $2 + 3 = 5$ *may be written as* $(+2) + (+3) = +5$.

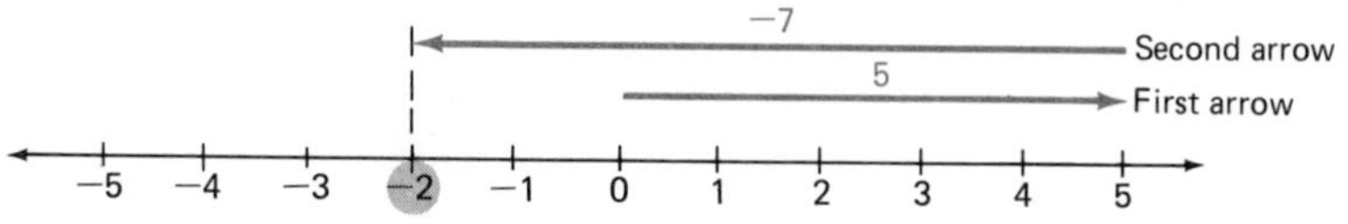

EXAMPLE 1 Compute: $(-2) + (-5)$

Solution Start at 0 and draw an arrow that is 2 units in length to the left. Then draw another arrow to the left that is 5 units in length. The sum is -7; $(-2) + (-5) = -7$.

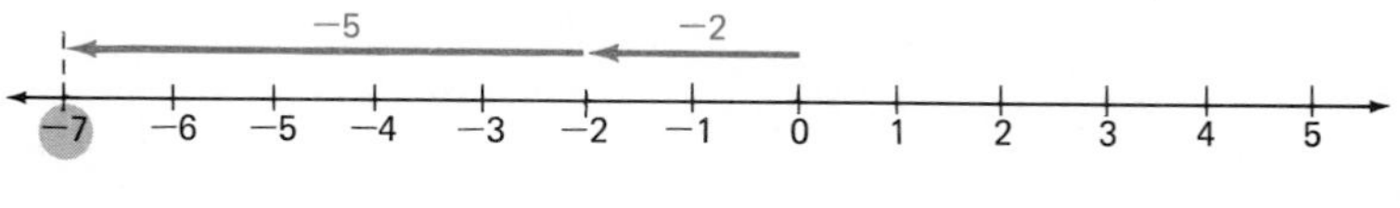

Drawing a number line each time we wish to add two integers soon becomes cumbersome. Often we may substitute a mental picture of a number line instead. With a little practice you will soon be able to do such problems quickly by inspection, and extend the process to include more than two numbers.

$$\begin{aligned} &\underbrace{(+4) + (-7)} + (-5) \\ = \ &\underbrace{\quad (-3) \qquad + (-5)} \\ = \ &\qquad -8 \end{aligned}$$

Here we group $(+4)$ with (-7). According to the associative property, we could just as well group (-7) with (-5). The final result will be the same. ■

TEST YOUR UNDERSTANDING
Think Carefully

Find each sum.

1. $(+5) + (+6)$
2. $(+6) + (-3)$
3. $(-2) + (-3)$
4. $(-4) + (+6)$
5. $(-5) + (+1)$
6. $(-1) + (+3)$
7. $(-13) + (-9)$
8. $(-9) + (-13)$
9. $(-2) + (-5) + (-3)$
10. $(+3) + (+7) + (-5)$
11. $(-2) + (-3) + (-5)$
12. $(-3) + (-7) + (+12)$
13. $(-5) + (+7) + (+5)$
14. $(+4) + (-8) + (+6)$
15. $(-1) + (+6) + (-8)$
16. $(-7) + (-5) + (+12)$

(Answers: Page 31)

Once you know how to add with positive and negative numbers, often called **signed numbers,** you should be able to subtract without difficulty. Here are arithmetic examples that suggest a procedure for finding the difference of any two real numbers.

$15 - 8 = 7$ is the same as $15 + (-8) = 7$
$23 - 6 = 17$ is the same as $23 + (-6) = 17$

In other words, to subtract one number from another, *add the opposite* of the number to be subtracted. In symbols we may write:

DEFINITION OF SUBTRACTION

For all real numbers a and b, $a - b = a + (-b)$.

We can use this definition to prove that multiplication distributes over subtraction. (See Exercise 61.)

This rule enables us to subtract any two real numbers by changing from a subtraction problem to an addition problem. Thus the set of real numbers is also closed with respect to subtraction.

EXAMPLE 2 Find: **(a)** $(-7) - (+5)$ **(b)** $14 - 19$ **(c)** $(-2) - (-6)$

Solution

change to addition

(a) $(-7) - (+5) = (-7) + (-5) = -12$

use the opposite

With practice, you will soon be able to omit the middle step in completing a subtraction problem.

(b) $14 - 19 = 14 + (-19) = -5$
(c) $(-2) - (-6) = (-2) + (+6) = +4$ (or 4) ■

An alternative way to define subtraction is to say that $a - b = c$ provided that $b + c = a$. In other words, the difference c when added to b must equal a. Thus $12 - 7 = 5$ because $7 + 5 = 12$, also note that $(-2) - (-6) = +4$ because $(-6) + (+4) = -2$.

TEST YOUR UNDERSTANDING
Think Carefully

Find each difference.

1. $(+12) - (+5)$	**2.** $(+8) - (+3)$	**3.** $(-6) - (+3)$
4. $(+5) - (-4)$	**5.** $(+7) - (-7)$	**6.** $(-7) - (-7)$
7. $(-5) - (-9)$	**8.** $(-9) - (-7)$	**9.** $(-8) - (+6)$
10. $6 - 9$	**11.** $7 - (-10)$	**12.** $(-3) - (-5)$
13. $(-13) - 7$	**14.** $(-13) - (-7)$	**15.** $0 - (-9)$

(Answers: Page 31)

EXAMPLE 3 Compute: $[(-8) + (+5)] - (-20)$

Solution First add within the brackets and then subtract.

$$\underbrace{[(-8) + (+5)]} - (-20)$$
$$\underbrace{(-3) \quad - (-20)}$$
$$\underbrace{(-3) \quad + (+20)}$$
$$+17$$

■

Using the number line is not very practical for an addition example such as $426 + (-758)$. More general rules are needed. Such rules can be stated by using the concept of **absolute value,** which is the distance that a number is from the origin without regard to direction.

What do the numbers -5 and $+5$ have in common? Obviously, they are different numbers and are the coordinates of two distinct points on the number line. However, they are both the same distance from 0, the **origin,** on the number line.

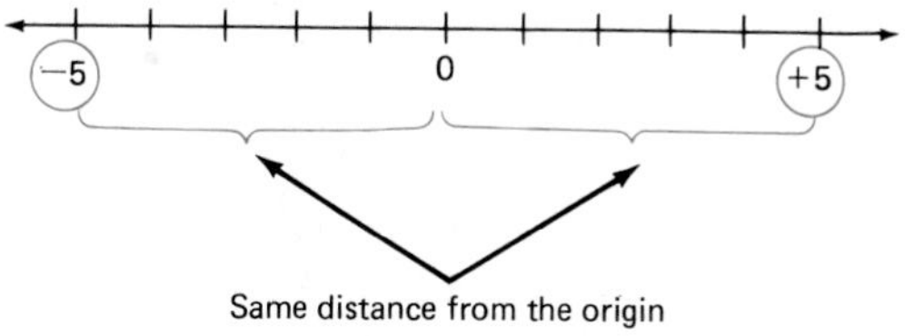

In order words, -5 is as far to the left of 0 as $+5$ is to the right of 0. We show this fact by using **absolute value notation** as follows:

$|-5| = 5$ read as "The *absolute value* of -5 is 5."

$|+5| = 5$ read as "The *absolute value* of $+5$ is 5."

Geometrically, for any real number x, $|x|$ is the distance (without regard to direction) that x is from the origin. Note that for a positive number, $|x| = x$; $|+5| = 5$. For a negative number, $|x| = -x$; that is, $|-5| = -(-5) = 5$. Also, since 0 is the origin, it is natural to have $|0| = 0$.

We can summarize this by the following definition.

In words, if x is positive or zero, the absolute value of x is x. If x is negative, the absolute value is the opposite of x. For example:

$$|+3| = 3$$
$$|-3| = -(-3) = 3$$
$$|0| = 0$$

DEFINITION OF ABSOLUTE VALUE

For any real number x,

$$|x| = \begin{cases} x & \text{if } x \geq 0 \\ -x & \text{if } x < 0 \end{cases}$$

CAUTION
This is **incorrect:**

$$|(-4) + 9| = |-4| + |9| = 4 + 9 = 13$$

The mistake is the assumption that the absolute value of a sum is the same as the sum of the absolute values of each term. You must simplify within the absolute value symbols first. Here is the **correct** *simplification:*

$$|(-4) + 9| = |5| = 5$$

EXAMPLE 4 Evaluate: **(a)** $|-3| + |+7| + |-2|$ **(b)** $|-13| - |19|$

Solution
(a) $|-3| + |+7| + |-2| = 3 + 7 + 2 = 12$
(b) $|-13| - |19| = 13 - 19 = -6$ ■

Our earlier experience of finding sums using the number line can now be used to write the general rules for adding signed numbers. First recall that the sum of two positive numbers is positive: $(+5) + (+6) = +11$. Also, the sum of two negative numbers is negative: $(-2) + (-5) = -7$. These observations can be generalized as follows:

To add two numbers of like sign (both positive or both negative):
add their absolute values and prefix their common sign.

EXAMPLE 5 Evaluate: $(-18) + (-21)$

Solution $|-18| = 18$ and $|-21| = 21$. Since both numbers are negative, the sum will be negative. Thus

$$(-18) + (-21) = -(18 + 21) = -39$$ ■

From the number line we see that $(+9) + (-5) = +4$ and also that $(+5) + (-7) = -2$. These results suggest the following general rule:

To add two numbers of unlike sign:
subtract their absolute values (the smaller from the larger) and prefix the sign of the number with the larger absolute value.

EXAMPLE 6 Evaluate: **(a)** $(+37) + (-25)$ **(b)** $426 + (-758)$

Solution
(a) Since $|+37| = 37$ is greater than $|-25| = 25$, we prefix the plus sign.

$$(+37) + (-25) = +(37 - 25)$$
$$= +12$$

(b) Since $|-758| = 758$ is greater than $|426| = 426$, we prefix the minus sign.

$$\begin{aligned} 426 + (-758) &= -(758 - 426) \\ &= -332 \end{aligned}$$

■

EXERCISES 1.3

Find each sum.

1. $(+5) + (-9)$
2. $(+8) + (-6)$
3. $(-6) + (+6)$
4. $(-9) + (-3)$
5. $(-5) + (+12)$
6. $(+3) + (-8)$
7. $(+8) + (-10)$
8. $(-3) + (-7)$
9. $(-12) + 2$
10. $(-65) + (-95)$
11. $(-65) + (95)$
12. $(+65) + (-95)$
13. $(+2) + (-3) + (-7)$
14. $(-3) + (-5) + (-9)$
15. $(-7) + (+12) + (-5)$
16. $(-1) + 6 + (-12)$
17. $(-52) + (-84) + (+175)$
18. $183 + (135) + (-174)$

Find each difference.

19. $(+6) - (+2)$
20. $(-5) - (-9)$
21. $(-2) - (+8)$
22. $(+7) - (-9)$
23. $(+8) - (-5)$
24. $(-6) - (-12)$
25. $(-12) - 15$
26. $17 - (-13)$
27. $(-15) - (-17)$
28. $(-81) - (+132)$
29. $(+132) - (-81)$
30. $(-132) - (-81)$
31. $0 - (+7)$
32. $0 - (-7)$
33. $(-175) - 252$
34. $(-1932) - (-1701)$
35. $164 - (-2309)$
36. $(-135) - (-180)$
37. $180 - 135$
38. $180 - (-135)$

Compute the following; work within the brackets first in each case.

39. $[(-3) + (-7)] - (+5)$
40. $[(-3) - (-7)] - (+5)$
41. $(+7) + [(-5) - (+2)]$
42. $(+7) - [(-5) - (+2)]$
43. $(-9) + [(-6) - (-3)]$
44. $[(-9) - (-6)] - (-3)$

Evaluate.

45. $|-12|$
46. $-|+7|$
47. $-|-5|$
48. $-|0|$
49. $|(-3) + (-2)|$
50. $|(-3) - (-2)|$
51. $|(-2) - (-3)|$
52. $|0 - (-9)|$
53. $|(+7) - (-12)|$
54. $|(-5) - (+12)|$
55. $-|(-3) - (-7)|$
56. $-|(-7) - (-3)|$
57. $-|(-7) + (-3)|$
58. $-(|-2| + |-3|)$
59. $-(|-2| - |-3|)$
60. $-(|+2| - |+3|)$
61. Give a reason for each step in this proof that multiplication distributes over subtraction.

$$\begin{aligned} a(b - c) &= a[b + (-c)] && \text{(i)} \\ &= ab + a(-c) && \text{(ii)} \\ &= ab + [-(ac)] && \text{(iii)} \\ &= ab - ac && \text{(iv)} \end{aligned}$$

In Exercises 62 and 63, describe the conditions on the real numbers x and y to obtain the stated result.

*62. $x + y$ is positive

*63. $x + y$ is negative

*64. By making use of absolute values, write the rule for adding two negative numbers x and y by completing the equation

$$x + y = \underline{\hspace{4cm}}$$

*65. Follow the instructions of Exercise 64 for adding numbers of unlike signs. Assume that x is positive and y is negative.
(a) $x + y =$ ____________, if $|x|$ is greater than $|y|$.
(b) $x + y =$ ____________, if $|y|$ is greater than $|x|$.

Written Assignment: Explain why $x^2 = |x^2| = |x|^2$ for any real number x.

1.4 OPERATIONS WITH INTEGERS: MULTIPLICATION AND DIVISION

Once again, for the sake of clarity and simplicity, we will develop rules of signs for multiplication and division of integers. But these same rules of sign apply to *all* real numbers.

The rules for multiplying signed numbers can be based upon our experience of multiplication in arithmetic. For example, $3 \times 4 = 4 + 4 + 4 = 12$. But as *signed* numbers, we have $3 = +3$, $4 = +4$, and $12 = +12$. Thus

$$(+3) \times (+4) = 3 \times 4 = 12 = +12$$

This suggests the following:

The product of two positive numbers is positive.

Next consider the product of a positive and a negative number such as $(+3) \times (-4)$. It is reasonable to write

$$(+3) \times (-4) = 3 \times (-4) = (-4) + (-4) + (-4) = -12$$

Furthermore, the commutative property of multiplication now gives

$$(-4) \times (+3) = (+3) \times (-4) = -12$$

The preceding illustrates this rule:

The product of a positive number and a negative number is negative.

Most students of mathematics have difficulty justifying the rules for multiplication of two negative numbers, and generally learn by rote the that product is a positive number. The distributive property may be used to justify the fact that the product of two negative numbers is positive. Consider the problem of evaluating $(-3)[(+4) + (-4)]$. It can be done two different ways. Adding within the brackets first, we obtain the following:

$$\begin{aligned}(-3)[(+4) + (-4)] &= (-3)(0)\\ &= 0\end{aligned}$$

On the other hand, multiplying first and then adding, we have

$$(-3)[(+4) + (-4)] = (-3)(+4) + (-3)(-4)$$
$$= (-12) \quad + (?)$$

In the first approach the answer is seen to be 0. Therefore, in the second method shown, we need to find a number that must be added to -12 to obtain a sum of 0. The only possibility is $+12$, and thus $(-3) \times (-4) = +12$ in order that the distributive property will hold for all real numbers.

The preceding observations lead to these rules for multiplying signed numbers:

RULES OF SIGNS FOR MULTIPLICATION

The product of two positive numbers is positive.
The product of two negative numbers is positive.
The product of a positive number and a negative number is negative.

It is interesting to note the result when a number is multiplied by -1. Here are two specific illustrations:

$$(-1)(+5) = -5 \qquad (-1)(-5) = +5$$

The product of -1 and any number results in the additive inverse of that number.

For all real numbers x, $\quad -1 \cdot x = -x$.

Note how this result, in conjunction with the distributive property, can be used to demonstrate the property of the opposite of a sum (See page 9.) Thus

$$-(a + b) = -1(a + b) = (-1 \cdot a) + (-1 \cdot b) = (-a) + (-b)$$

EXAMPLE 1 Find the product: $(-3) \times (-4) \times (+5)$

Solution

$$(-3) \times (-4) \times (+5) = [(-3) \times (-4)] \times (+5)$$
$$= (+12) \times (+5)$$
$$= +60$$

■

EXAMPLE 2 Evaluate: $(-7)[(+5) + (-2)]$

Solution It is customary to work within the brackets first.

$$(-7)[(+5) + (-2)] = (-7)(+3) = -21$$

We can check this result by using the distributive property:

$$(-7)[(+5) + (-2)] = (-7)(+5) + (-7)(-2)$$
$$= (-35) + (+14) = -21$$

■

TEST YOUR UNDERSTANDING
Think Carefully

Evaluate.

1. $(+4)(+5)$
2. $(+4)(-5)$
3. $(-4)(+5)$
4. $(-4)(-5)$
5. $(-7)(-12)$
6. $(-6)(+11)$
7. $(-2)(-3)(-4)$
8. $(+3)(-5)(-10)$
9. $(-3)(+2)(-6)$
10. $(+3)[(-5) + (-2)]$
11. $(-2)[(-7) + (+5)]$
12. $(-5)[(+6) + (-9)]$

(Answers: Page 31)

To divide real numbers we use the concept that division is the inverse operation of multiplication. That is, for example, $15 \div 3 = 5$ since $3 \times 5 = 15$. In general:

DEFINITION OF DIVISION

For real numbers a and b, with $b \neq 0$, the quotient $a \div b$ is defined to equal q, provided that $bq = a$. That is,

$$a \div b = q \quad \text{provided that } bq = a$$

Alternatively, $a \div b = a \times \dfrac{1}{b} = \dfrac{a}{b} \quad (b \neq 0)$

This definition implies that the set of real numbers is closed with respect to division, provided we do not allow division by 0. By this definition,

See the discussion following Example 5 as to why division by 0 is undefined.

$$(-12) \div (+4) = -3 \quad \text{because we know } (+4)(-3) = -12$$
$$(-12) \div (-4) = +3 \quad \text{because we know } (-4)(+3) = -12$$

These results lead to the following general results for dividing signed numbers.

RULES OF SIGNS FOR DIVISION

The quotient of two positive numbers is positive.
The quotient of two negative numbers is positive.
The quotient of a positive number and a negative number is negative.

EXAMPLE 3 Evaluate: $(-2)[(-12) \div (-3)]$

Solution Within the brackets note that $(-12) \div (-3) = (+4)$. Therefore, we may write:

$$(-2)[(-12) \div (-3)] = (-2)[+4] = -8$$ ■

Let us use the expression in Example 3 to see if there is a distributive property for multiplication over division. Suppose we try to do the evaluation as follows:

$$(-2)[(-12) \div (-3)] = [(-2)(-12)] \div [(-2)(-3)]$$
$$= (+24) \div (+6)$$
$$= +4$$

This is not the same result as in Example 3, and thus the distributive property does *not* apply. It is therefore important that for such problems we work within the grouping symbols first.

At times we may encounter a problem where no parentheses are used. For example, consider the expression $3 + 5 \times 7$. Actually there are two ways to evaluate this expression:

(a) Multiply, then add.

$$3 + 5 \times 7 = 3 + 35$$
$$= 38$$

(b) Add, then multiply.

$$3 + 5 \times 7 = 8 \times 7$$
$$= 56$$

According to this agreement, $3 + 5 \times 7 = 38$

In the absence of grouping symbols we agree to the convention that all multiplications and divisions will be done first, in order from left to right. Then all additions and subtractions will be completed, in order from left to right.

EXAMPLE 4 Evaluate: $7 + 3 \times 8 \div 4 - (-15)$

Solution

$$7 + 3 \times 8 \div 4 - (-15) = 7 + 24 \div 4 - (-15)$$
$$= 7 + 6 - (-15)$$
$$= 13 - (-15)$$
$$= 13 + 15$$
$$= 28$$

■

The preceding convention about the order of operations is used when evaluating expressions that contain grouping symbols (parentheses, brackets, braces) according to the following agreement:

Order of Operations

1. Begin with the innermost grouping symbols and work outward.
2. Complete all multiplications and divisions, in order, from left to right.
3. Complete all additions and subtractions, in order, from left to right.

These procedures are illustrated in the following example.

EXAMPLE 5 Evaluate: $30 - (8 + 60 \div 5[9 - 2 \cdot 3])$

Solution

$$
\begin{aligned}
30 - (8 + 60 \div 5[9 - 2\cdot 3]) &= 30 - (8 + 60 \div 5[9 - 6])\\
&= 30 - (8 + 60 \div 5\cdot 3)\\
&= 30 - (8 + 12\cdot 3)\\
&= 30 - (8 + 36)\\
&= 30 - 44\\
&= -14
\end{aligned}
$$

■

This is very important: DIVISION BY ZERO IS NOT POSSIBLE.

Division by 0 is not possible. This can be explained using the definition of division. Suppose, for example, we assumed that $7 \div 0 = q$ is some real number. Then the definition implies that $0 \cdot q = 7$. But $0 \cdot q = 0$, which leads to the contradiction that $7 = 0$. Thus the assumption that $7 \div 0$ is a real number is false and we conclude that $7 \div 0$ is undefined.

EXERCISES 1.4

Find each product.

1. $(-5)(+8)$
2. $(+7)(-7)$
3. $(-8)(-8)$
4. $(-9)(+6)$
5. $(-5)(-8)$
6. $(-9)(0)$
7. $(-4)(-13)$
8. $(+7)(-19)$
9. $(-12)(-8)$
10. $(-3)(-6)(+5)$
11. $(+5)(-2)(+2)$
12. $(+5)(-5)(-5)$
13. $(+2)(+8)(-8)$
14. $(-2)(-8)(+8)$
15. $(-2)(-8)(-8)$
16. $(-11)(+6)(-3)$
17. $(-9)(-12)(+12)$
18. $(-6)(+21)(-10)$

Find each quotient.

19. $(-24) \div (-6)$
20. $(+24) \div (-6)$
21. $(-24) \div (+6)$
22. $(+24) \div (+6)$
23. $(-21) \div (-3)$
24. $(-21) \div (+3)$
25. $(0) \div (-5)$
26. $(-100) \div (-10)$
27. $(+42) \div (-7)$
28. $(-52) \div (+4)$
29. $(-48) \div (-6)$
30. $(-72) \div (+8)$
31. $(+144) \div (-12)$
32. $(-144) \div (+12)$
33. $(-144) \div (-12)$
34. $(-51) \div (+3)$
35. $(+55) \div (-5)$
36. $(-256) \div (-16)$

Evaluate.

37. $(-3)[(+7) + (-9)]$
38. $(-8)[(-5) - (-2)]$
39. $(+6)[(-21) \div (3)]$
40. $(-8)[(-12) \div (+4)]$
41. $(-24) - [(-6) \div (-2)]$
42. $[(-24) \div (-6)] \div (-2)$
43. $(+10)[(-8) - (-2)]$
44. $(-10)[(-12) + (+3)]$
45. $[(-12) \div (-3)] \div (+2)$
46. $[(-48) \div (+3)] \div (16)$
47. $|-3|\cdot|-5|$
48. $|3|\cdot|-5|$
49. $|-3|\cdot|5|$
50. $|-2|\cdot|-3|\cdot|+2|$
51. $|-3|\cdot|-5|\cdot|0|$
52. $-(|+2|\cdot|-3|\cdot|-5|)$
53. $|-20| \div |-4|$
54. $|+20| \div |-4|$
55. $|-20| \div |+4|$
56. $-(|-32| \div |-4|)$
57. $-|(-32) \div (-4)|$
58. $-|(+32) \div (+4)|$
59. $(-7) + 16 \div 2 - (-1)$
60. $(-21) \div 3 + 2 \times (-4)$
61. $9(4 - 3 - 7)$
62. $9(4 - [3 - 7])$
63. $(8 - 12)(7 - 3)$
64. $8 - 12(7 - 3)$
65. $14 - 5\cdot 2 \div 2 - 6 + 8$
66. $(14 - 5\cdot 2) \div 2 - 6 + 8$
67. $20 - [5 + 3(6[(-4) + 2])]$
68. $24 \div 3 - 3 + 18[(-4)\cdot 2 - 1] \div 9$
69. $8 - [(9 - 3)4 - 8] \div 2$
70. $(8 - [(9 - 3)4 - 8]) \div 2$

In Exercises 71 through 74 a **mistake** *has been made. Find the mistake and give the correct answer.*

71. $|\frac{3}{4} - 2| = \frac{3}{4} - 2 = -\frac{5}{4}$ *wrong answer*
72. $|-7 - 5| = |-7| - |5| = 7 - 5 = 2$ *wrong answer*
73. $(9 - 4)5 - 2 = (9 - 4)3 = 5 \cdot 3 = 15$ *wrong answer*
74. $8 + 12 \div 4 \times 3 = 20 \div 4 \times 3 = 5 \times 3 = 15$ *wrong answer*

*75. Is there a distributive property for division over multiplication? Justify your answer with examples.

*76. Is there a distributive property of subtraction over multiplication? Justify your answer with examples.

Written Assignment: Explain why $\frac{0}{0}$ must remain undefined. (*Hint:* If $\frac{0}{0}$ is to be some value, it must be a unique value.)

EXPLORATIONS
Think Critically

1. Show a one-to-one correspondence between the set of whole numbers and the set of integers. Does this mean that there are just as many integers as there are whole numbers? Explain your answer.
2. Does the set of real numbers contain a multiplicative inverse for each member of the set? Explain your answer.
3. For each pair of numbers below evaluate both $|x + y|$ and $|x| + |y|$.
   (a) $x = 3, y = 7$ **(b)** $x = -3, y = 7$
   (c) $x = 3, y = -7$ **(d)** $x = -3, y = -7$
   Now make a conjecture for showing the relationship between $|x + y|$ and $|x| + |y|$ for *any* pair of real numbers x and y. You may use either symbols or words to describe the relationship.
4. Use geometry to locate the point on the number line whose coordinate is the irrational number $\sqrt{29}$.
5. Does division distribute over multiplication? That is, does $a \div b \cdot c = (a \div b) \cdot (a \div c)$ for all real numbers a, $b \neq 0$, and $c \neq 0$? Discuss.
6. Does multiplication distribute over division? That is, does $a \cdot b \div c = (a \cdot b) \div (a \cdot c)$ for all real numbers $a \neq 0$, b, and $c \neq 0$? Discuss.

1.5 OPERATIONS WITH REAL NUMBERS

For simplicity, we used the set of integers to illustrate the rules for computation with signed numbers in the preceding sections. However, as noted earlier, the same rules apply for the set of real numbers. Thus, for example, we can operate with fractions and decimals and apply the same rules. To illustrate, recall that we are able to find a difference such as $(-8) - (+17)$ in this way:

$$(-8) - (+17) = (-8) + (-17) \leftarrow a - b = a + (-b)$$
$$= -25$$

In the same way, consider these differences:

$$(-8.23) - (+7.45) = (-8.23) + (-7.45) = -15.68$$
$$(-2\tfrac{1}{4}) - (+\tfrac{2}{3}) = (-2\tfrac{3}{12}) + (-\tfrac{8}{12}) = -2\tfrac{11}{12}$$

Since computations with fractions frequently cause difficulties in algebra, we will review briefly the rules of operations with rational numbers.

More detailed work with fractions as used in algebra will be discussed in Chapter 4.

OPERATIONS WITH FRACTIONS

	Example	Rule
Reducing	$\frac{30}{45} = \frac{2 \times 15}{3 \times 15} = \frac{2}{3}$	$\frac{ac}{bc} = \frac{a}{b}$
Multiplication	$\frac{2}{3} \times \frac{4}{5} = \frac{2 \times 4}{3 \times 5} = \frac{8}{15}$	$\frac{a}{b} \times \frac{c}{d} = \frac{a \times c}{b \times d} = \frac{ac}{bd}$
Division	$\frac{3}{4} \div \frac{5}{7} = \frac{3}{4} \times \frac{7}{5} = \frac{21}{20}$	$\frac{a}{b} \div \frac{c}{d} = \frac{a}{b} \times \frac{d}{c} = \frac{ad}{bc}$
Addition and Subtraction (Like and unlike denominators)	$\frac{7}{10} - \frac{4}{10} = \frac{7-4}{10} = \frac{3}{10}$ $\frac{3}{4} + \frac{2}{3} = \frac{3}{4} \cdot \frac{3}{3} + \frac{2}{3} \cdot \frac{4}{4}$ $= \frac{9}{12} + \frac{8}{12} = \frac{17}{12}$	$\frac{a}{c} \pm \frac{b}{c} = \frac{a \pm b}{c}$ $\frac{a}{b} \pm \frac{c}{d} = \frac{a}{b} \cdot \frac{d}{d} \pm \frac{c}{d} \cdot \frac{b}{b}$ $= \frac{ad}{bd} \pm \frac{bc}{bd} = \frac{ad \pm bc}{bd}$

EXAMPLE 1 Evaluate: **(a)** $(+3.5)(-2.5)$ **(b)** $(-\frac{3}{4}) \div (-1\frac{1}{2})$

Solution

(a) The product of a positive number and a negative number is negative. Thus multiply the absolute values of the two numbers and prefix a minus sign.

$$\text{Since } 3.5 \times 2.5 = 8.75, \text{ then } (+3.5)(-2.5) = -8.75$$

(b) The quotient of two negative numbers is positive. Thus use the rules for division of fractions in arithmetic and divide the absolute values of the two numbers. Then prefix a plus sign to the answer.

$$\tfrac{3}{4} \div 1\tfrac{1}{2} = \tfrac{3}{4} \div \tfrac{3}{2} = \tfrac{3}{4} \times \tfrac{2}{3} = \tfrac{6}{12} = \tfrac{1}{2}$$

Thus $(-\frac{3}{4}) \div (-1\frac{1}{2}) = +\frac{1}{2}$.
Since the result is positive, it can be given just as $\frac{1}{2}$, without the plus sign. ■

Sometimes we need to approximate with the indicated sum of two real numbers when one, or both, of the numbers is irrational, such as $5 + \sqrt{2}$. As we shall see in Chapter 5, such a *sum* is often left in this exact or *simplified form*. However, in many applied problems, it is necessary to obtain a rational *approximation* for this sum.

Using a calculator, and considering the first three decimal places only, we find

$\sqrt{2} \approx 1.414$ The symbol $\approx$ is read *"is approximately equal to."*

The result 1.414 is not *exactly* equal to the square root of 2. For example, if you now use your calculator to *square* this approximation (multiply it by itself), your calculator will show the following result:

This point warrants emphasis: 1.414 *is a rational approximation for the irrational number* $\sqrt{2}$.

$$1.414 \times 1.414 = 1.999396$$

However, note that the symbol $\sqrt{2}$ represents an exact number; it is the number whose square is 2. That is,

$$\sqrt{2} \times \sqrt{2} = (\sqrt{2})^2 = 2$$

Normally, we will give results rounded to a specified number of decimal places. When using such approximations we often agree to use the equal (=) sign for simplicity. For example, we will state results in this form:

Rounded to three decimal places, $\sqrt{2} = 1.414$
Rounded to seven decimal places, $\sqrt{2} = 1.4142136$

Let us now return to the original number under discussion, $5 + \sqrt{2}$:

$5 + \sqrt{2} = 5 + 1.414 = 6.414$, rounded to three decimal places.
$5 + \sqrt{2} = 5 + 1.41421 = 6.41421$, rounded to five decimal places.

EXAMPLE 2 Use a calculator to find the following rounded to three decimal places:
(a) $(+\sqrt{89}) \div (-2)$ **(b)** $\sqrt{2} - \sqrt{3}$

Note that 9.434 *is a* ***rational approximation*** *of* $\sqrt{89}$.

Solution

(a) From a calculator we find that $\sqrt{89} = 9.43398$, rounded to five decimal places. Rounded to three places, we have $\sqrt{89} = 9.434$. Also, we know that the quotient of a positive number and a negative number is negative. Thus:

$$(+\sqrt{89}) \div (-2) = (+9.434) \div (-2) = -4.717 \text{ to three decimal places.}$$

(b) To three decimal places we have $\sqrt{2} - \sqrt{3} = 1.414 - 1.732 = -0.318$. ■

EXAMPLE 3 Use a calculator to evaluate $(3\sqrt{28} - 9.4) \div (-7.3)$, rounded to three decimal places.

Solution The following sequence of calculator steps assumes that the calculator has keys for parentheses, square roots, and changing of signs.

$(3\sqrt{28} - 9.4) \div (-7.3)$

=	(	28	√	×	3	−	9.4	)	÷	7.3	+/−	=	−0.8869189
	↑	↑	↑	↑	↑	↑	↑	↑	↑	↑	↑	↑	↑
	press	enter	press	press	enter	press	enter	press	press	enter	press	press	Final Display

The calculator sequences here apply to many but not all calculators. Thus it may be necessary to refer to your calculator instruction manual for an appropriate sequence.

Then to three decimal places the evaluation is −0.887. This can also be done without using parentheses using this sequence.

28 [√] [×] 3 [−] 9.4 [=] [÷] 7.3 [+/−] [=] −0.8869189

↑ (at the first [=]) The calculator display after this step should be 6.4745079 ■

Throughout this text we use the logo [calculator] to indicate that a problem or exercise is to be completed through the use of a calculator. In place of a calculator, where appropriate, consider the use of Table I, which gives the square roots of numbers from 1 through 200 to three decimal places.

Observe that we use circles when entering numbers. Such entries themselves may need a sequence of keys. Thus to enter 9.4 requires the sequence ⑨ [.] ④. We use rectangles for all other steps in the sequence.

As mentioned earlier, real numbers are either rational or irrational. It is possible to distinguish one type from the other by recognizing their decimal forms. Rational numbers have two types of decimal representations. Sometimes the result will be a *terminating decimal,* as in the following examples:

$$\tfrac{3}{4} = 0.75 \qquad \tfrac{7}{8} = 0.875 \qquad \tfrac{23}{10} = 2.3$$

A method for converting a repeating decimal into fraction form is considered in Exercises 61 and 62.

Other rational numbers will produce a *repeating decimal:*

$$\tfrac{2}{3} = 0.666\ldots \qquad \tfrac{19}{22} = 0.86363\ldots \qquad \tfrac{3}{7} = 0.428571428571\ldots$$

Usually, a bar is placed over the set of digits that repeat, so that the preceding illustrations can be written in this way:

$$\tfrac{2}{3} = 0.\overline{6} \qquad \tfrac{19}{22} = 0.8\overline{63} \qquad \tfrac{3}{7} = 0.\overline{428571}$$

The decimal form of a rational number can be found by division. For example:

$$\tfrac{7}{8} = 7 \div 8 = 0.875 \qquad \tfrac{19}{22} = 19 \div 22 = 0.86363\ldots = 0.8\overline{63}$$

```
    .875            .86363 . . .
 8|7.000        22|19.00000
   6 4             17 6
     60             1 40
     56             1 32
      40               80
      40               66
       0              140
                      132
                        80
                        66
                        14
```

Note that the division continues endlessly because the remainders 14 and 80 keep on repeating in succession, giving the repeating cycle 63 in the quotient.

In 1989, two Columbia University mathematicians used a supercomputer to establish a world record by computing π to over one billion digits! If these digits were printed on a straight line, the line would stretch out for over 1200 miles. Computer experts use such computations to test the speed and accuracy of new computers.

Some decimals neither terminate nor repeat. You are probably familiar with the number π from your earlier study of geometry. This real number is *not* a rational number; it cannot be expressed as the quotient of two integers. The decimal representation for π goes on endlessly without repetition. Here are the first 100 places:

π = 3.14159 26535 89793 23846 26433 83279 50288 41971 69399
37510 58209 74944 59230 78164 06286 20899 86280 34825
34211 70679 . . .

In general, every irrational number can be represented as a nonterminating, nonrepeating decimal. The following decimals represent irrational numbers, but there is a *pattern* to the digits in these particular numbers that allows you to extend each one endlessly without repetition. Can you find the pattern and write the next 10 decimal places for each?

Note that most irrational numbers have no pattern in their decimal forms, such as those for π, $\sqrt{2}$, $\sqrt[3]{17}$, etc.

$$0.23223222322223\ldots$$

$$0.05055055505555\ldots$$

In summary, we have the following:

> Every real number can be represented by a decimal. If the decimal terminates or repeats, the number is a rational number; otherwise, it is an irrational number.

EXERCISES 1.5

Evaluate.

1. $(-5.72) + (-8.63)$
2. $(+8.76) + (-3.59)$
3. $(-12.72) + (-8.62)$
4. $(+0.76) + (+0.85)$
5. $(+0.93) + (-0.72)$
6. $(-0.87) - (+0.69)$
7. $(+9.56) - (+2.72)$
8. $(-5.63) - (-7.25)$
9. $(-2.83) - (+5.83)$
10. $(+6.79) - (-5.63)$
11. $(-8.73) - (+2.68)$
12. $(+6.89) - (-9.23)$
13. $(-0.85)(+0.75)$
14. $(-6.2)(-7.9)$
15. $(-2.34)(-7.62)$
16. $(+8.32)(+6.59)$
17. $(-7.62)(+0.39)$
18. $(+6.28)(-0.99)$
19. $(-7.25) \div (-0.05)$
20. $(+8.65) \div (-0.5)$
21. $(-10.35) \div (+0.23)$
22. $(+4.23) \div (+4.7)$
23. $(+56.25) \div (-7.5)$
24. $(-120) \div (-0.048)$

Complete the indicated operations. Do not use decimal forms.

25. $(-\frac{3}{4}) + (-\frac{2}{3})$
26. $(-\frac{1}{4}) - (-\frac{1}{3})$
27. $(+\frac{5}{6}) - (-\frac{2}{3})$
28. $(-5\frac{7}{8}) + (-2\frac{1}{4})$
29. $(-3\frac{2}{3}) - (-2\frac{3}{4})$
30. $(+1\frac{5}{9}) + (-4\frac{2}{3})$
31. $(+\frac{2}{3})(-\frac{1}{4})$
32. $(-\frac{1}{2}) \div (-\frac{3}{4})$
33. $(+\frac{3}{8}) \div (-\frac{3}{4})$
34. $(-7\frac{1}{3})(+\frac{6}{11})$
35. $(-5\frac{1}{4}) \div (+1\frac{1}{2})$
36. $(-8\frac{2}{5})(-\frac{5}{7})$
37. $[(-\frac{1}{2}) + (-\frac{2}{3})] \div (+\frac{1}{4})$
38. $(+\frac{3}{5}) - [(-\frac{1}{3}) + (-\frac{1}{5})]$
39. $(-2\frac{1}{7})[(-7\frac{1}{2}) + (-8\frac{5}{6})]$
40. $[(-3\frac{1}{4}) - (+2\frac{7}{8})] \div (-1\frac{3}{4})$

Use a calculator to approximate each of the following, rounded to three decimal places.

41. $\sqrt{7} + 5$
42. $3 - \sqrt{5}$
43. $12 - \sqrt{150}$
44. $7\sqrt{7} + 2$
45. $-5 + 2\sqrt{3}$
46. $8\sqrt{6} - 3$
47. $\sqrt{38} \div 2$
48. $-\sqrt{75} + \sqrt{82}$
49. $\sqrt{137} \div (-8)$
50. $18 \div \sqrt{7}$
51. $(3 + \sqrt{15}) \div (-7)$
52. $(2\sqrt{13} - 12) \div (-0.6)$

For Exercises 53 and 54, assume that the decimals given continue the same pattern endlessly.

53. Write the next 10 digits for each decimal: **(a)** 0.454454445 . . . **(b)** 0.121221222 . . .
54. Consider this decimal: 0.35335333533335 . . . . **(a)** How many 3's will there be between the 49th and the 50th digit 5? **(b)** How many 3's will there be altogether before the 50th digit 5?

55. List the numbers of each set in order, from smallest to largest:
(a) $\{0.23,\ 0.232332333\ldots,\ 0.\overline{23},\ 0.23233\}$
(b) $\{0.070070007\ldots,\ 0.\overline{07},\ 0.07007,\ 0.07\}$

Use division to express each fraction as a terminating or repeating decimal.

56. $\frac{27}{16}$ **57.** $\frac{5}{7}$ **58.** $\frac{45}{123}$ **59.** $\frac{13}{64}$ **60.** $\frac{9}{13}$

61. Every repeating decimal can be expressed as a rational number in the form a/b. Consider, for example, the decimal 0.727272 . . . and the following process:

Let $n = 0.727272\ldots$ Then $100n = 72.727272\ldots$

$$\begin{aligned} & 100n = 72.727272\ldots \\ \text{Subtract:}\quad & \underline{\quad n = \ \ 0.727272\ldots} \\ & 99n = 72 \\ \text{Solve for } n\text{:}\quad & n = \frac{72}{99} = \frac{8}{11} \end{aligned}$$

Use this method to express each decimal as the quotient of integers and check by use of a calculator.

(a) 0.454545 . . . $(0.\overline{45})$ **(b)** 0.373737 . . . $(0.\overline{37})$ **(c)** 0.234234 . . . $(0.\overline{234})$ (*Hint:* Let $n = 0.234234\ldots$; multiply by 1000.)

62. Study the illustration given below for $n = 0.2737373\ldots$ and then convert each repeating decimal into a quotient of integers.

Multiply n by 1000:	$1000n = 273.\overline{73}$	The decimal point is *behind* the first cycle.
Multiply n by 10:	$10n = 2.\overline{73}$	The decimal point is in *front* of the first cycle.
Subtract:	$990n = 271$	
Divide:	$n = \frac{271}{990}$	

(a) 0.4585858 . . . **(b)** 3.21444 . . . **(c)** $2.0\overline{146}$ **(d)** $0.00\overline{123}$

Written Assignment: Write the repeating decimals for $\frac{1}{7}$, $\frac{2}{7}$, $\frac{3}{7}$, $\frac{4}{7}$, $\frac{5}{7}$, and $\frac{6}{7}$. Describe a pattern for the digits that appear. Then repeat this experiment for the decimal representations of $\frac{1}{13}, \frac{2}{13}, \ldots, \frac{12}{13}$.

CHAPTER 1 SUMMARY

Review these key terms and concepts so that you are able to define or describe them. A clear understanding of these will be very helpful when reviewing the developments of this chapter.

The important subsets of the set of real numbers:

Natural (or **counting**) **Numbers:**	{1, 2, 3, . . .}
Whole Numbers:	{0, 1, 2, 3, . . .}
Integers:	{. . . , −3, −2, −1, 0, 1, 2, 3, . . .}
Rational Numbers:	Numbers that can be written as the ratio of two integers, $\frac{a}{b}$, $b \neq 0$.
Irrational Numbers:	Numbers that are not rational, such as $\sqrt{2}$.
Real Numbers:	The collection of the set of rational numbers and the set of irrational numbers.

Properties of the real numbers:

Closure Properties:	$a + b$ is a real number. $a \times b$ is a real number.
Commutative Properties:	$a + b = b + a$ $a \times b = b \times a$
Associative Properties:	$(a + b) + c = a + (b + c)$ $(a \times b) \times c = a \times (b \times c)$
Distributive Property:	$a \times (b + c) = (a \times b) + (a \times c)$
Identity Properties:	$a + 0 = 0 + a = a$ $a \cdot 1 = 1 \cdot a = a$
Inverse Properties:	For each real number a there exists another number $-a$ such that $a + (-a) = (-a) + a = 0$. For real number a, $a \neq 0$, there exists a number $\frac{1}{a}$ such that $a \cdot \frac{1}{a} = \frac{1}{a} \cdot a = 1$.
Multiplication Property of Zero:	$a \cdot 0 = 0 \cdot a = 0$
Zero-Product Property:	If $ab = 0$, then $a = 0$, or $b = 0$, or both equal 0.

Properties of opposites:

$$-(-a) = a \qquad -(a + b) = (-a) + (-b) \qquad -(ab) = (-a)b = a(-b)$$

Definition of **absolute value:**

$$|x| = x \quad \text{if } x \text{ is positive or } x = 0$$

$$|x| = -x \quad \text{if } x \text{ is negative}$$

Rules for operating with positive and negative numbers:

1. To add two numbers of like sign, add their absolute values and prefix their common sign.
2. To add two numbers of unlike sign, subtract their absolute values and prefix the sign of the number with the larger absolute value.
3. To subtract two numbers make use of the following definition, and then proceed as in addition: $a - b = a + (-b)$
4. For multiplication or division of signed numbers:
   (a) The product or quotient of two positive numbers is positive.
   (b) The product or quotient of two negative numbers is positive.
   (c) The product or quotient of a positive and a negative number is negative.

REVIEW EXERCISES

The solutions to the following exercises can be found within the text of Chapter 1. Try to answer each question before referring to the text.

Section 1.1

1. Use set notation to show the set of counting numbers from 1 through 1,000,000.
2. To which subsets of the real numbers do each of the following belong?
   (a) 5 **(b)** $\frac{2}{3}$ **(c)** $\sqrt{7}$ **(d)** -14
3. Classify as true or false, and explain your answer: Every whole number is a natural number.

Section 1.2

State each property.

4. The closure properties for addition and multiplication.
5. The commutative properties for addition and multiplication.
6. The associative properties for addition and multiplication.
7. The distributive property.
8. The identity properties for addition and multiplication.
9. Use the distributive property to find the product 7×61.
10. Find the additive and multiplicative inverses of **(a)** 3, **(b)** $\frac{2}{5}$.
11. What basic property of the real numbers is illustrated by each of the following?
    (a) $6 + (17 + 4) = (17 + 4) + 6$ **(b)** $\frac{3}{4} + (-\frac{3}{4}) = 0$
    (c) $57 \times 1 = 57$ **(d)** $\frac{2}{3}(12 + 36) = \frac{2}{3}(12) + \frac{2}{3}(36)$
    (e) $(-43)(\frac{1}{-43}) = 1$ **(f)** $5\sqrt{3}$ is a real number

Section 1.3

12. Show the sum with a number line picture: $(-2) + (-5)$.
13. Complete this definition of subtraction: For all real numbers a and b, $a - b = ?$
14. Find: **(a)** $(-7) - (+5)$ **(b)** $(-2) - (-6)$
15. Compute: $[(-8) + (+5)] - (-20)$
16. State the definition of absolute value.
17. Evaluate: **(a)** $|-3| + |+7| + |-2|$ **(b)** $|-13| - |19|$
18. State the rule of signs for adding two numbers of unlike signs.
19. Evaluate: **(a)** $(+37) + (-25)$ **(b)** $426 + (-758)$

Section 1.4

20. State the rules of signs for multiplying two real numbers.
21. Find the product: $(-3) \times (-4) \times (+5)$
22. State the definition of division.

Evaluate.

23. $(-7)[(+5) + (-2)]$
24. $(-2)[(-12) \div (-3)]$
25. $7 + 3 \times 8 \div 4 - (-15)$
26. $30 - [8 + 60 \div 5(9 - 2 \cdot 3)]$

Section 1.5

27. Evaluate: **(a)** $(+3.5)(-2.5)$ **(b)** $(-2\frac{3}{4}) \div (-1\frac{1}{2})$
28. Use a calculator to find the quotient, rounded to three decimal places:
    (a) $(+\sqrt{89}) \div (-2)$ **(b)** $\sqrt{2} - \sqrt{3}$
29. Describe the manner in which real numbers can be classified according to their decimal representations.
30. Use a calculator to evaluate $(3\sqrt{28} - 9.4) \div (-7.3)$, rounded to three decimal places.
31. Find the decimal forms of $\frac{7}{8}$ and $\frac{19}{22}$.

CHAPTER 1 TEST: STANDARD ANSWER

Use these questions to test your knowledge of the basic skills and concepts of Chapter 1. Then check your answers with those given at the back of the book.

1. Check the boxes in the table to indicate the set to which each number belongs. The first row is completed for you.

	Whole numbers	Integers	Rational numbers	Irrational numbers	Negative integers
-6		✓	✓		✓
0.231					
$\sqrt{5}$					
$\frac{2}{3}$					
1991					

List the elements in each set.

2. The set of whole numbers less than 5.
3. The set of natural numbers between 3 and 8.
4. The set of opposites of the first five natural numbers.
5. The set of integers between -3 and $+2$.
6. The set of whole numbers greater than 3.

Classify each statement as true or false.

7. Every real number is a rational number.
8. Every rational number is a real number.
9. Negative irrational numbers are not real numbers.
10. The set of integers is a subset of the set of rational numbers.
11. Some irrational numbers are integers.
12. Zero is a rational number.

Name the property illustrated by each statement.

13. $5\frac{1}{2} \times 7 = 7 \times 5\frac{1}{2}$
14. $(8 + 3) + \sqrt{7} = (3 + 8) + \sqrt{7}$
15. $5(9 + 6) = (5 \times 9) + (5 \times 6)$
16. $(3 \times 7) \times 8 = 3 \times (7 \times 8)$
17. $(-6) + 0 = -6$
18. $\frac{3}{7} \times \frac{7}{3} = 1$

Compute.

19. **(a)** $(+3) + (-7)$ **(b)** $(-12) + (+17)$
20. **(a)** $(-9) - (-3)$ **(b)** $(-7) - (+8)$
21. **(a)** $(-3) \times (+7)$ **(b)** $(-9) \times (-6)$
22. **(a)** $(-24) \div (+3)$ **(b)** $(-72) \div (-8)$
23. **(a)** $(-7)[(-5) + (+3)]$ **(b)** $(-3)[(+2) - (-5)]$
24. **(a)** $-|(-3) - (-7)|$ **(b)** $-(|-3| - |-7|)$
25. **(a)** $(-18) - 12 \div 2 \times 5$ **(b)** $[2 - 9] \times [3 - (-5)] \div (4 - 9 - 3)$

CHAPTER 1 TEST: MULTIPLE CHOICE

1. To which sets of numbers does the number -20 belong?
   I. Natural numbers **II.** Whole Numbers **III.** Integers **IV.** Rational numbers **V.** Real numbers
   (a) Only I **(b)** Only III **(c)** Only II, III, and IV **(d)** Only III, IV, and V **(e)** None of the preceding
2. Which of these statements are true?
   I. Every integer is the coordinate of some point on the number line.
   II. Every rational number is a real number.
   III. Every point on the number line can be named by a rational number.
   (a) Only I **(b)** Only II **(c)** Only I and II **(d)** I, II, and III **(e)** None of the preceding
3. Which of the following is an illustration of the commutative property for addition?
   I. $(7 + 3) + 8 = 7 + (3 + 8)$
   II. $7 + (3 + 8) = (3 + 8) + 7$
   III. $7(3 + 8) = (3 + 8)7$
   (a) Only I **(b)** Only II **(c)** Only III **(d)** I, II, and III **(e)** None of the preceding
4. Which of these statements are true?
   I. The set of integers contains the additive inverse for each of its members.
   II. The set of integers contains the multiplicative inverse for each of its members.
   III. Every rational number can be represented by a terminating decimal or by a repeating decimal.
   (a) Only I **(b)** Only II **(c)** Only III **(d)** I, II, and III **(e)** None of the preceding
5. What is the multiplicative inverse of $\frac{2}{3}$?
   (a) $-\frac{2}{3}$ **(b)** $\frac{3}{2}$ **(c)** $-\frac{3}{2}$ **(d)** 1 **(e)** None of the preceding
6. What is the additive inverse of -5?
   (a) 0 **(b)** 1 **(c)** $-(-5)$ **(d)** -5 **(e)** None of the preceding
7. Which property is illustrated by $7(8 + 9) = (7)(8) + (7)(9)$?
   (a) Commutative **(b)** Associative **(c)** Distributive
   (d) Inverse **(e)** None of the preceding
8. Which of the following represents an irrational number?
   (a) $0.\overline{73}$ **(b)** 0.73 **(c)** 0.737373 **(d)** 0.737337333
   (e) None of the preceding
9. Which of the following rational numbers cannot be represented by a terminating decimal?
   (a) $\frac{3}{4}$ **(b)** $\frac{7}{8}$ **(c)** $\frac{5}{9}$ **(d)** $\frac{13}{25}$ **(e)** None of the preceding
10. For negative numbers a and b, which of the following are true?
    I. $|a + b| = -(a + b)$
    II. $|a + b| = |a| + |b|$
    III. $|a + b| = a + b$
    (a) Only I **(b)** Only II **(c)** Only III **(d)** Only I and II **(e)** I, II, and III
11. Which of the following is false?
    (a) $|8 - 2| = 8 - 2$ **(b)** $|6 + 8| = |-6| + |-8|$
    (c) $|\pi - 5| = -(\pi - 5)$ **(d)** $|2 - 2.1| = 2 - 2.1$ **(e)** $|(-7) - (-2)| = |-7| - |-2|$
12. Which of the following sets of numbers is a subset of the set of rational numbers?
    I. Natural numbers **II.** Whole numbers **III.** Integers
    (a) Only I **(b)** Only II **(c)** Only III **(d)** Only I and II **(e)** I, II, and III
13. $-60 \div 2[4 - 3(4 \cdot 2 - 10)] =$
    (a) -300 **(b)** -3 **(c)** 15 **(d)** 60 **(e)** None of the preceding

14. Which of the following sets is an example of the empty set?

I. The set of natural numbers between 5 and 6.

II. The set of rational numbers between $\frac{1}{3}$ and $\frac{2}{3}$.

III. The set of integers between -1 and $+1$.

(a) Only I **(b)** Only II **(c)** Only III **(d)** Only I and II **(e)** I, II, and III

15. Which of the following must be true about two numbers a and b, if a is positive and b is negative?

I. The product ab is negative.

II. The difference $a - b$ is positive.

III. The sum $a + b$ is positive.

(a) Only I **(b)** Only II **(c)** Only III **(d)** Only I and II **(e)** I, II, and III

ANSWERS TO THE TEST YOUR UNDERSTANDING EXERCISES

Page 3

1. {1, 2, 3}
2. {0, 1, 2, 3, 4}
3. {1, 2, 3, . . . , 1000}
4. {11, 12, 13, . . .}
5. {1, 3, 5, . . .}
6. {2, 4, 6, 8}
7. {2, 4, 6, . . .}
8. {2, 3, 4, 5, 6, 7, 8, 9}
9. {101, 102, 103, . . .}
10. There are none

Page 8

1. Associative property for addition
2. Commutative property for addition
3. Commutative property for addition
4. Distributive property
5. Commutative property for multiplication
6. Associative property for multiplication
7. No; no
8. $12 \div 3 \neq 3 \div 12$
9. No; no
10. $(8 \div 4) \div 2 \neq 8 \div (4 \div 2)$
11. Yes; the closure properties for addition and multiplication of real numbers, respectively.
12. $9(50 + 1) = 450 + 9 = 459$
13. $8(70 + 2) = 560 + 16 = 576$
14. $12(30 + 1) = 360 + 12 = 372$
15. $6(200 + 5) = 1200 + 30 = 1230$

Page 12

1. +11 2. +3 3. −5 4. +2 5. −4 6. +2 7. −22
8. −22 9. −10 10. +5 11. −10 12. +2 13. +7 14. +2
15. −3 16. 0

Page 13

1. +7 2. +5 3. −9 4. +9 5. +14 6. 0 7. +4 8. −2
9. −14 10. −3 11. +17 12. +2 13. −20 14. −6 15. +9

Page 18

1. +20 2. −20 3. −20 4. +20 5. +84 6. −66 7. −24
8. +150 9. +36 10. −21 11. +4 12. +15

CHAPTER

2 FUNDAMENTAL OPERATIONS

2.1 INTEGRAL EXPONENTS

Much of mathematical notation can be viewed as efficient abbreviations of lengthier statements. For example:

$$4^9 = 4 \times 4 \times 4 \times 4 \times 4 \times 4 \times 4 \times 4 \times 4$$

This illustration makes use of a *positive integral exponent*. In this section we shall explore the use of integers as exponents.

The most common ways of referring to b^n are "b to the nth power," "b to the nth," or "the nth power of b."

DEFINITION OF POSITIVE INTEGRAL EXPONENT

If n is a positive integer and b is any real number, then

$$b^n = \underbrace{b \cdot b \cdot \cdots \cdot b}_{n \text{ factors}}$$

The number b is called the **base** and n is called the **exponent.**

Some illustrations of the definition follow:

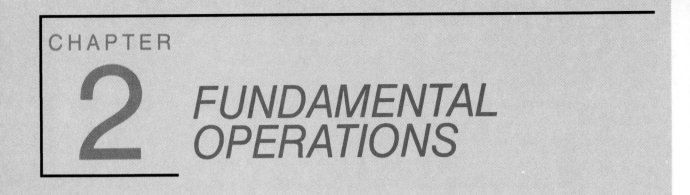

CAUTION
Be careful when you work with parentheses. Note that
$(-3)^2 \neq -3^2$.
$(-3)^2 = (-3)(-3) = 9$
$-3^2 = -(3 \times 3) = -9$
Under what conditions does $(-a)^n = -a^n$*?*

$$b^1 = b$$
$$(a + b)^2 = (a + b)(a + b)$$
$$(-2)^3 = (-2)(-2)(-2) = -8$$
$$(\tfrac{1}{3})^4 = \tfrac{1}{3} \cdot \tfrac{1}{3} \cdot \tfrac{1}{3} \cdot \tfrac{1}{3} = \tfrac{1}{81}$$
$$(-1)^5 = (-1)(-1)(-1)(-1)(-1) = -1$$
$$10^6 = 10 \cdot 10 \cdot 10 \cdot 10 \cdot 10 \cdot 10 = 1{,}000{,}000$$

A number of important rules concerning positive integral exponents can be established on the basis of the preceding definition. Example 1 illustrates the manner in which a specific example can be used to help lead to the generalization of a rule.

EXAMPLE 1 Simplify: $b^3 \cdot b^4$

Solution

$$b^3 \cdot b^4 = \underbrace{\overbrace{(b \cdot b \cdot b)}^{3\text{ factors}} \cdot \overbrace{(b \cdot b \cdot b \cdot b)}^{4\text{ factors}}}_{3 + 4 = 7\text{ factors}} = b^7$$

In brief,

$$b^3 \cdot b^4 = b^{3+4} = b^7$$

■

The preceding example can be generalized where b is any real number, and m and n are any positive integers.

$$b^m \cdot b^n = \underbrace{\overbrace{(b \cdot b \cdots b)}^{m\text{ factors}}\overbrace{(b \cdot b \cdots b)}^{n\text{ factors}}}_{m + n\text{ factors}} = b^{m+n}$$

We restate this generalization as Rule 1 in the following list. In a similar manner, the other rules can also be motivated through the use of specific cases and justified, in general, as in the preceding case. In the list of rules that follow, m and n are any positive integers, a and b are any real numbers, and there is the usual understanding that denominators cannot be zero.

When multiplying powers with a common base, add the powers and use the same base.

RULE 1. $b^m b^n = b^{m+n}$

Illustrations:

$$2^3 \cdot 2^4 = 2^{3+4} = 2^7$$
$$x^3 \cdot x^4 = x^7$$

RULE 2. $$\frac{b^m}{b^n} = \begin{cases} b^{m-n} & \text{if } m > n \\ 1 & \text{if } m = n \\ \dfrac{1}{b^{n-m}} & \text{if } m < n \end{cases}$$

These illustrate the rule for $\frac{b^m}{b^n}$.

Illustrations:

$(m > n)$	$(m = n)$	$(m < n)$
$\frac{2^5}{2^2} = 2^{5-2} = 2^3$	$\frac{5^2}{5^2} = 1$	$\frac{2^2}{2^5} = \frac{1}{2^{5-2}} = \frac{1}{2^3}$
$\frac{x^5}{x^2} = x^{5-2} = x^3$	$\frac{x^2}{x^2} = 1$	$\frac{x^2}{x^5} = \frac{1}{x^{5-2}} = \frac{1}{x^3}$

The power of a power is the product of the powers with the same base.

RULE 3. $(b^m)^n = b^{mn}$

Illustrations:

$$(2^3)^2 = 2^{3 \cdot 2} = 2^6$$
$$(x^3)^2 = x^{3 \cdot 2} = x^6$$

The power of a product is the product of the powers.

RULE 4. $(ab)^m = a^m b^m$

Illustrations:

$$(2 \cdot 3)^5 = 2^5 \cdot 3^5$$
$$(xy)^5 = x^5 y^5$$

The power of a quotient is the quotient of the powers.

RULE 5. $\left(\frac{a}{b}\right)^m = \frac{a^m}{b^m}$

Illustrations:

$$\left(\frac{3}{2}\right)^5 = \frac{3^5}{2^5}$$
$$\left(\frac{x}{y}\right)^5 = \frac{x^5}{y^5}$$

Here is a proof of Rule 4. You can try proving the other rules by using similar arguments.

Eventually, the rules for exponents will be extended to include all the real numbers (not just positive integers) as exponents. We will find that all the rules stated here will still apply.

$(ab)^m = (ab)(ab) \cdots (ab)$ (by definition)

$= (a \cdot a \cdot \cdots \cdot a)(b \cdot b \cdot \cdots \cdot b)$ (by repeated use of the commutative and associative laws for multiplication)

$= a^m b^m$ (by definition)

The proper use of these rules can simplify computations, as in the following example.

In this chapter occasional use is made of some basic properties of fractions with which you should be familiar.

EXAMPLE 2 Evaluate $12^3(\frac{1}{6})^3$

Solution This expression can be evaluated without use of any of the rules of exponents.

$$\begin{aligned} 12^3(\tfrac{1}{6})^3 &= 1728(\tfrac{1}{216}) \\ &= \tfrac{1728}{216} \\ &= 8 \end{aligned}$$

Using Rule 4, the work proves to be much easier.

$$\begin{aligned} 12^3(\tfrac{1}{6})^3 &= (12 \cdot \tfrac{1}{6})^3 \\ &= 2^3 \\ &= 8 \end{aligned}$$

■

Example 3 (b) illustrates the use of more than one rule of exponents when simplifying. Note that the last step can be done showing more detail. Thus

$$\frac{x^6y^5}{x^4y^6} = \frac{x^6}{x^4} \cdot \frac{y^5}{y^6} = x^2 \cdot \frac{1}{y} = \frac{x^2}{y}$$

EXAMPLE 3 Simplify: (a) $\left(-\frac{2}{3}\right)^5\left(\frac{9}{4}\right)^5$ (b) $\frac{(x^3y)^2y^3}{x^4y^6}$

Solution

(a) $$\left(-\frac{2}{3}\right)^5\left(\frac{9}{4}\right)^5 = \left(-\frac{2}{3} \cdot \frac{9}{4}\right)^5 = \left(-\frac{3}{2}\right)^5 = -\frac{243}{32}$$

(b) $$\frac{(x^3y)^2y^3}{x^4y^6} = \frac{(x^3)^2y^2y^3}{x^4y^6} = \frac{x^2}{y}$$

■

Care must be taken with both the multiplication and the division rules when the bases are not the same, as in the following example.

EXAMPLE 4 Simplify: $\frac{4^5}{8^3}$

CAUTION
The following is **incorrect:**

$$\frac{4^5}{8^3} = \left(\frac{4}{8}\right)^2$$

Rule 2 does not apply here because the bases are different, and Rule 5 does not apply since the exponents are different.

Solution The bases are not the same, so that Rule 2 does not apply. However, rather than finding 4^5 and 8^3, the problem can be simplified in this way:

$$\frac{4^5}{8^3} = \frac{(2^2)^5}{(2^3)^3} = \frac{2^{10}}{2^9} = 2$$

■

Here is a summary of the rules for exponents that have just been stated. In each case m and n are positive integers and a and b are any real numbers, $b \neq 0$.

FUNDAMENTAL RULES FOR EXPONENTS

Rule	Illustration
1. $b^m \cdot b^n = b^{m+n}$	$x^3 \cdot x^5 = x^8$
2. $\frac{b^m}{b^n} = \begin{cases} b^{m-n} & \text{if } m > n \\ 1 & \text{if } m = n \\ \frac{1}{b^{n-m}} & \text{if } m < n \end{cases}$	$\frac{x^7}{x^3} = x^4$; $\frac{x^7}{x^7} = 1$; $\frac{x^3}{x^7} = \frac{1}{x^4}$
3. $(b^m)^n = b^{mn}$	$(x^3)^4 = x^{12}$
4. $a^m b^m = (ab)^m$	$x^3 y^3 = (xy)^3$
5. $\frac{a^m}{b^m} = \left(\frac{a}{b}\right)^m$	$\frac{x^3}{y^3} = \left(\frac{x}{y}\right)^3$

TEST YOUR UNDERSTANDING
Think Carefully

Evaluate each of the following.

1. 5^3 **2.** $(-\frac{1}{2})^5$ **3.** $(-\frac{2}{3})^3 + \frac{8}{27}$ **4.** $(10^3)^2$

5. $2^3(-2)^3$ **6.** $(\frac{1}{2})^3 8^3$ **7.** $\frac{17^8}{17^9}$ **8.** $\frac{(-2)^3 + 3^2}{3^3 - 2^2}$

9. $\frac{(-12)^4}{4^4}$ **10.** $(ab^2)^3(a^2b)^4$ **11.** $\frac{2^2 \cdot 16^3}{(-2)^8}$ **12.** $\frac{(2x^3)^2(3x)^2}{6x^4}$

(Answers: Page 80)

Thus far our discussion of exponents has been restricted to the use of positive integers only. Now let us consider the meaning of 0 as an exponent. In particular, what is the meaning of 5^0? We know that 5^3 means that we are to use 5 three times as a factor. But certainly it makes no sense to use 5 zero times. The rules of exponents will help to resolve this dilemma.

This discussion provides meaning for the use of 0 as an exponent. That is, an expression like 5^0 will now be defined.

We would like these laws for exponents to hold even if one of the exponents happens to be zero. That is, we would *like* Rule 2 to give

$$\frac{5^2}{5^2} = 5^{2-2} = 5^0$$

But it is already known that

$$\frac{5^2}{5^2} = \frac{25}{25} = 1$$

Thus 5^0 ought to be assigned the value 1. Consequently, in order to *preserve* the rules of exponents, we decide to let $5^0 = 1$. And, from now on, we will agree to the following:

Notice that the definition calls for b to be a real number different from 0. That is, we do not define an expression such as 0^0; this is said to be undefined.

DEFINITION OF ZERO EXPONENT

If b is a real number different from 0, then

$$b^0 = 1$$

EXAMPLE 5 Simplify: $\left(\frac{x^3}{y^0}\right)\left(\frac{x}{y}\right)^0$

Solution: $\left(\frac{x^3}{y^0}\right)\left(\frac{x}{y}\right)^0 = \left(\frac{x^3}{1}\right)(1) = x^3$ ■

Exponential notation is used in a variety of situations. Example 6 shows the use of exponents to analyze a situation in which a substance is decreasing exponentially.

EXAMPLE 6 Suppose that a radioactive substance decays so that $\frac{1}{2}$ the amount remains after each hour. If at a certain time there were 320 grams of the substance, how much will remain after 8 hours? How much after n hours?

Solution Since the amount remaining after each hour is $\frac{1}{2}$ of the grams at the end of the preceding hour, we find the remaining amount by multiplying the preceding number of grams by $\frac{1}{2}$.

	Grams remaining
Start: 0 hours	$320(\frac{1}{2})^0 = 320$
After 1 hour	$320(\frac{1}{2})^1 = 160$
After 2 hours	$320(\frac{1}{2})^2 = 80$
After 3 hours	$320(\frac{1}{2})^3 = 40$
⋮	⋮
After 8 hours	$320(\frac{1}{2})^8 = \frac{5}{4}$

Observe that the power of $\frac{1}{2}$ is the same as the number of hours that the substance has been decaying. Assuming that the same pattern will continue, we conclude that there are $320(\frac{1}{2})^n = \frac{320}{2^n}$ grams remaining after n hours. ■

Many errors are made when working with exponents because of misuses of the basic rules and definitions. This list shows some of the common errors that you should try to avoid.

CAUTION: Learn to Avoid These Mistakes

WRONG	RIGHT
$5^2 \cdot 5^4 = 5^8$ (Do not multiply exponents.) $5^2 \cdot 5^4 = 25^6$ (Do not multiply the base numbers.)	$5^2 \cdot 5^4 = 5^6$ (Rule 1)
$\frac{5^6}{5^2} = 5^3$ (Do not divide the exponents.) $\frac{5^6}{5^2} = 1^4$ (Do not divide the base numbers.)	$\frac{5^6}{5^2} = 5^4$ (Rule 2)

CAUTION: Learn to Avoid These Mistakes	
WRONG	RIGHT
$(5^2)^6 = 5^8$ (Do not add the exponents.)	$(5^2)^6 = 5^{12}$ (Rule 3)
$(-2)^4 = -2^4$ (Misreading the parentheses)	$(-2)^4 = (-1)^4 2^4 = 2^4$ (Rule 4)
$(-5)^0 = -1$ (Misreading definition of b^0)	$(-5)^0 = 1$ (Definition of b^0)
$5^3 + 5^3 = 5^6$ (Adding exponents does not apply because of plus sign.)	$5^3 + 5^3 = (1 + 1)5^3 = 2 \cdot 5^3$ (Distributive)

EXERCISES 2.1

Classify each statement as true or false. If it is false, correct the right side of the equality to obtain a true statement.

1. $3^4 \cdot 3^2 = 3^8$
2. $(2^2)^3 = 2^8$
3. $2^5 \cdot 2^2 = 4^7$
4. $\dfrac{9^3}{9^3} = 1$
5. $\dfrac{10^4}{5^4} = 2^4$
6. $\left(\dfrac{2}{3}\right)^4 = \dfrac{2^4}{3}$
7. $(-27)^0 = 1$
8. $(-2)^4 = -2^4$
9. $3^4 + 3^4 = 3^8$
10. $(a^2b)^3 = a^2b^3$
11. $(a + b)^0 = a + 1$
12. $a^2 + a^2 = 2a^2$

Evaluate.

13. 10^5
14. $2^0 + 2^1 + 2^2$
15. $(-3)^2(-2)^3$
16. $\left(\dfrac{2}{3}\right)^0 + \left(\dfrac{2}{3}\right)^1$
17. $\left[\left(\dfrac{1}{2}\right)^3\right]^2$
18. $\left(\dfrac{1}{2}\right)^4(-2)^4$
19. $\dfrac{3^2}{3^0}$
20. $\dfrac{(-2)^5}{(-2)^3}$
21. $\left(-\dfrac{3}{4}\right)^3$
22. $\dfrac{2^3 \cdot 3^4 \cdot 4^5}{2^2 \cdot 3^3 \cdot 4^4}$
23. $\dfrac{8^3}{16^2}$
24. $\dfrac{2^{10}}{2^5 \cdot 2^3}$

Simplify.

25. $(x^3)^2$
26. $x^3 \cdot x^9$
27. $(x^2y^3)^2$
28. $\dfrac{x^9}{x^3}$
29. $\left(\dfrac{x^2}{y^3}\right)^2$
30. $\dfrac{x^3y}{y^4}$
31. $a^3 \cdot a^2 \cdot a$
32. $(ab^0)^3$
33. $(ab^3)^0$
34. $(2a)^3(3a)^2$
35. $(2x^3y^2)^0$
36. $(-2a^2b^0)^4$
37. $\dfrac{(x^2y)^4}{(xy)^2}$
38. $\left(\dfrac{x^2y}{xy^2}\right)^4$
39. $\left(\dfrac{x^3}{y^2}\right)^4\left(\dfrac{-y}{x^2}\right)^2$
40. $\dfrac{5x^0y^3}{3xy}$
41. $\left(\dfrac{2x^4y^2}{3xy^4}\right)^2$
42. $\left(\dfrac{3a^2}{b}\right)^2\left(\dfrac{-2a}{3b}\right)^3$
43. $\dfrac{(x^3y^3)^4}{(x^2y)^2}$
44. $\left(\dfrac{3xy^2}{2x^2y^3}\right)^2$
45. $\dfrac{(x-2y)^6}{(x-2y)^4}$
46. $\dfrac{(x^2 \cdot x^3)^2}{x^4}$
47. $\dfrac{(x^a \cdot x^{2a})^2}{x^{2a}}$
48. $\left(\dfrac{x^a}{y^b}\right)\left(\dfrac{x^{2a}}{y^b}\right)^2$

49. Evaluate $\dfrac{(1{,}000{,}000)(10{,}000)}{10{,}000{,}000}$ by first converting into powers of 10.

50. Evaluate $\dfrac{16^3 \cdot 8^5}{128 \cdot 32}$ by first converting into powers of 2.

In Exercises 51 through 56, evaluate each expression in two ways. First, substitute the given values directly into the expression and then simplify; and second, substitute into the expression after it has been simplified algebraically.

Example: Substituting $x = -2$, $y = 3$ into $\dfrac{(xy^2)^3}{x^4y^3}$ gives

$$\frac{(-2 \cdot 3^2)^3}{(-2)^4 3^3} = \frac{(-2)^3 3^6}{(-2)^4 3^3} = \frac{3^3}{-2} = -\frac{27}{2}$$

Also, since $\dfrac{y^3}{x}$ is the simplified form, substitution gives $\dfrac{3^3}{-2} = -\dfrac{27}{2}$.

51. Use $x = -2$, $y = 3$ for the expression in Exercise 30.
52. Use $a = \frac{1}{6}$ for the expression in Exercise 34.
53. Use $x = -\frac{1}{2}$, $y = 4$ for the expression in Exercise 37.
54. Use $a = 2$, $b = -6$ for the expression in Exercise 42.
55. Use $x = 2.3$, $y = 1.5$ for the expression in Exercise 44. (Compute to three decimal places.)
56. Use $x = 0.25$, $y = 0.13$ for the expression in Exercise 40. (Compute to three decimal places.)
57. Assume that a substance decays such that $\frac{1}{2}$ the amount remains after each hour. If there were 640 grams at the start, how much remains after 7 hours? How much after n hours?
58. If a rope is 243 feet long and you successively cut off $\frac{2}{3}$ of the rope, how much remains after five cuts? How much after n cuts?
59. For the rope in Exercise 58, how much remains after five cuts if each time you cut off $\frac{1}{3}$? How much is left after n cuts?
60. A company has a 4-year plan to increase its work force by $\frac{1}{4}$ for each of the 4 years. If the current work force is 2560, how big will it be at the end of the 4-year plan? Write an exponential expression that gives the work force after n years.
61. When an investment of P dollars earns $i\%$ interest per year and the interest is compounded annually, the formula for the final amount A after n years is $A = P(1 + i)^n$, where i is in decimal form. Find the amount A if \$1000 is invested at 10% compounded annually for 3 years.
62. Using the formula in Exercise 61, approximate the number of years it would take for a \$1000 investment to double when it is invested at 10% interest compounded annually.
63. "Raising to a power" is not an associative property. Verify this by showing that $(2^3)^4 \neq 2^{(3^4)}$.
*64. The intensity of earthquakes is normally given as a number between 1 and 10 on the *Richter scale*. The major San Francisco earthquake of 1906 measured 8.3 on this scale, the one on northern California in 1989 measured 6.9. An earthquake that measures 6 is 10 times as intense as one that measures 5 on the Richter scale, one that measures 5 is 10 times as intense as one that measures 4, etc. How much more intense is an earthquake that measures 8 than one that measures 1 on this scale?

Written Assignment: We have said that 0^0 is undefined. The following shows why we have not defined this to be equal to 1. *Suppose* that $0^0 = 1$. Then $1 = \dfrac{1}{1} = \dfrac{1^0}{0^0} = \left(\dfrac{1}{0}\right)^0$

(a) What rule for exponents is being used in the last step?
(b) What went wrong?

2.2 NEGATIVE EXPONENTS AND SCIENTIFIC NOTATION

Our next objective is to give meaning to negative integer exponents. For example, we want to decide the meaning of x^{-3}. Our guideline in making this decision will be that *the preceding rules of exponents are to apply for all kinds of integer exponents*. That is, we want to *preserve* the structure of the basic rules. With this in mind, observe the effect of Rule 1 when a negative exponent is involved.

We use a variable, such as x, to represent any allowable real number.

$$(x^3)(x^{-3}) = x^{3+(-3)} = x^0 = 1$$

Dividing both sides of $(x^3)(x^{-3}) = 1$ first by x^3 and then by x^{-3} produces these two statements

$$x^{-3} = \frac{1}{x^3} \quad \text{and} \quad x^3 = \frac{1}{x^{-3}}$$

We are now ready to make the following definiton:

Note that the exponents $-n$ and n are opposites, and that $-n$ may be negative or positive.

DEFINITION OF b^{-n}

If n is an integer and $b \neq 0$, then

$$b^{-n} = \frac{1}{b^n}$$

EXAMPLE 1 Use the definition $b^{-n} = \dfrac{1}{b^n}$ to simplify:

(a) 3^{-2} **(b)** $\dfrac{1}{3^{-2}}$ **(c)** $(-3)^{-3}$ **(d)** $\left(\dfrac{1}{3}\right)^{-2}$

Solution

(a) $3^{-2} = \dfrac{1}{3^2} = \dfrac{1}{9}$ **(b)** $\dfrac{1}{3^{-2}} = 3^2 = 9$

(c) $(-3)^{-3} = \dfrac{1}{(-3)^3} = \dfrac{1}{-27} = -\dfrac{1}{27}$ **(d)** $\left(\dfrac{1}{3}\right)^{-2} = \dfrac{1}{(\frac{1}{3})^2} = \dfrac{1}{\frac{1}{9}} = 1 \cdot \dfrac{9}{1} = 9$ ■

In view of the definition of b^{-n} we are now able to condense Rule 2 of exponents from three separate cases into just one case.

Another form of this rule is $\dfrac{b^m}{b^n} = \dfrac{1}{b^{n-m}}$

RULE 2. (revised): $\dfrac{b^m}{b^n} = b^{m-n}$

Illustrations:

$$\frac{3^4}{3^2} = 3^2 \qquad \frac{3^2}{3^4} = 3^{-2} \qquad \frac{3^2}{3^2} = 3^0 = 1$$

$$\frac{x^8}{x^2} = x^6 \qquad \frac{x^2}{x^8} = x^{-6} \qquad \frac{x^2}{x^2} = x^0 = 1$$

In the following examples, the basic rules of exponents are applied to situations involving negative exponents. We use the instruction "simplify" to mean that the results should be written using only positive exponents.

EXAMPLE 2 Simplify: **(a)** x^3y^{-5} **(b)** $\dfrac{x^{-3}}{y^{-2}}$

Solution

(a) $x^3y^{-5} = x^3 \cdot \dfrac{1}{y^5} = \dfrac{x^3}{y^5}$ **(b)** $\dfrac{x^{-3}}{y^{-2}} = \dfrac{x^{-3}}{1} \cdot \dfrac{1}{y^{-2}} = \dfrac{1}{x^3} \cdot \dfrac{y^2}{1} = \dfrac{y^2}{x^3}$ ■

The result of Example 2(b) demonstrates that an exponential *factor* in the numerator (or denominator) becomes an exponential *factor* in the denominator (or numerator) provided that the exponent has the opposite sign. This observation is used in the next example.

EXAMPLE 3 Simplify: $\dfrac{a^4b^{-3}}{a^{-2}b^2}$

CAUTION
Be sure to apply this procedure only to ***factors****. For example,* $\dfrac{1+2^{-1}}{4}$ *is* **not** *equal to* $\dfrac{1}{4+2} = \dfrac{1}{6}$*. Here is the correct simplification for this example:*
$\dfrac{1+2^{-1}}{4} = \dfrac{1+\frac{1}{2}}{4} = \dfrac{\frac{3}{2}}{4} = \dfrac{3}{8}$

Solution The factor b^{-3} in the numerator becomes the factor b^3 in the denominator, and the factor a^{-2} in the denominator becomes the factor a^2 in the numerator. Thus

$$\frac{a^4b^{-3}}{a^{-2}b^2} = \frac{a^2a^4}{b^2b^3}$$
$$= \frac{a^6}{b^5} \quad \text{(Rule 1)}$$

As another approach, use the revised Rule 2 and subtract exponents. Then rewrite the result, using positive exponents:

Another option here is to write $\dfrac{a^4b^{-3}}{a^{-2}b^2} = \dfrac{a^{4-(-2)}}{b^{2-(-3)}} = \dfrac{a^6}{b^5}$

$$\frac{a^4b^{-3}}{a^{-2}b^2} = a^{4-(-2)}b^{-3-2}$$
$$= a^6b^{-5}$$
$$= \frac{a^6}{b^5}$$ ■

EXAMPLE 4 Simplify: $\left(\dfrac{a^{-2}b^3}{a^3b^{-2}}\right)^5$

Solution There are several ways to proceed; here are two.

(a) $\left(\dfrac{a^{-2}b^3}{a^3b^{-2}}\right)^5 = (a^{-5}b^5)^5 = (a^{-5})^5(b^5)^5 = a^{-25}b^{25} = \dfrac{b^{25}}{a^{25}}$

(b) $\left(\dfrac{a^{-2}b^3}{a^3b^{-2}}\right)^5 = \left(\dfrac{b^5}{a^5}\right)^5 = \dfrac{(b^5)^5}{(a^5)^5} = \dfrac{b^{25}}{a^{25}}$ ■

The revised definition can be used to simplify a fraction raised to a negative power. Thus it follows from the definition that

$$\left(\frac{a}{b}\right)^{-1} = \frac{b}{a}, \text{ since } \left(\frac{a}{b}\right)^{-1} = \frac{1}{\frac{a}{b}} = \frac{b}{a}$$

In other words, a fraction to the -1 power is the *reciprocal* of the fraction. Can you show that $\left(\frac{a}{b}\right)^{-2} = \left(\frac{b}{a}\right)^{2}$?

TEST YOUR UNDERSTANDING
Think Carefully

*Each of the following statements is **incorrect** and represents an error commonly made. Find the correct simplification for each expression at the left.*

1. $2^{-3} = -\frac{1}{2^3}$

2. $\frac{2^3}{2^{-4}} = 2^{3-4}$

3. $(-2)^{-4} = 2^4$

4. $\left(\frac{1}{2}\right)^{-3} = \left(-\frac{1}{2}\right)\left(-\frac{1}{2}\right)\left(-\frac{1}{2}\right)$

5. $(4^{-3})^5 = 4^{-3+5}$

6. $(5^{-3})(4^{-3}) = 20^9$

7. $(a + b)^{-1} = a^{-1} + b^{-1}$

8. $\frac{8^{-3}}{4^{-1}} = 2^{-3-(-1)}$

(Answers: Page 80)

Exponents are often used by scientists to help write very large or very small numbers in a form called **scientific notation.** Such usage is also helpful in simplifying certain computations. Here are some illustrations of scientific notation:

$$623{,}000 = 6.23 \times 10^5 \qquad 0.00623 = 6.23 \times 10^{-3}$$

$$6230 = 6.23 \times 10^3 \qquad 0.0000623 = 6.23 \times 10^{-5}$$

It is easy to verify that these are correct. For example:

$$6.23 \times 10^5 = 6.23 \times 100{,}000 = 623{,}000$$

$$6.23 \times 10^{-3} = 6.23 \times \frac{1}{10^3} = \frac{6.23}{1000} = 0.00623$$

The preceding illustrations indicate that *a number N has been put into scientific notation when it has been expressed as the product of a number between 1 and 10 and an integral power of 10.* Thus:

$$N = x(10^c)$$

where $1 \le x < 10$ and c is an integer.

WRITING A NUMBER IN SCIENTIFIC NOTATION

Place the decimal point behind the first nonzero digit. (This produces the number between 1 and 10.) Then determine the power of 10 by counting the number of places you moved the decimal point. If you moved the decimal point to the left, then the power is positive; and if you moved it to the right, it is negative.

Illustrations:

$$2{,}070{,}000. = 2.07 \times 10^6$$

six places left

$$0.00000084 = 8.4 \times 10^{-7}$$

seven places right

To convert a number given in scientific notation back into standard notation, all you need do is move the decimal point as many places as indicated by the exponent of 10. Move the decimal point to the right if the exponent is positive and to the left if it is negative.

EXAMPLE 5 Write 1.21×10^4 in standard notation.

Solution Move the decimal point in 1.21 four places to the right.

$$1.21 \times 10^4 = 1.2100. = 12{,}100$$ ■

EXAMPLE 6 Write 1.21×10^{-2} in standard notation.

Solution Move the decimal point in 1.21 two places to the left.

$$1.21 \times 10^{-2} = .01.21 = 0.0121$$ ■

TEST YOUR UNDERSTANDING Think Carefully

Convert into scientific notation.

1. 739 **2.** 73,900 **3.** 0.00739

4. 0.739 **5.** 73.9 **6.** 7.39

Convert into standard notation.

7. 4.01×10^3 **8.** 4.01×10^{-3} **9.** 1.11×10^{-2}

10. 1.11×10^5 **11.** 9.2×10^{-4} **12.** 4.27×10^0

(Answers: Page 80)

Sometimes scientific notation can help simplify arithmetic computations, as in the following examples.

EXAMPLE 7 Use scientific notation to compute $\dfrac{1}{800{,}000}$.

Solution

In scientific notation the solution to Example 7 is written 1.25×10^{-6}.

$$\frac{1}{800{,}000} = \frac{1}{8 \times 10^5} = \frac{1}{8} \times \frac{1}{10^5} = 0.125 \times 10^{-5}$$

$$= 0.00000125$$ ■

EXAMPLE 8 Use scientific notation to evaluate

$$\frac{(2{,}310{,}000)^2}{(11{,}200{,}000)(0.000825)}$$

Solution

$$\frac{(2{,}310{,}000)^2}{(11{,}200{,}000)(0.000825)} = \frac{(2.31 \times 10^6)^2}{(1.12 \times 10^7)(8.25 \times 10^{-4})}$$

$$= \frac{(2.31)^2 \times (10^6)^2}{(1.12 \times 10^7)(8.25 \times 10^{-4})} \qquad [(ab)^n = a^n b^n]$$

Can you verify this result on your calculator?

$$= \frac{(2.31)^2}{(1.12)(8.25)} \times \frac{10^{12}}{(10^7)(10^{-4})} \qquad [(a^m)^n = a^{mn}]$$

$$= 0.5775 \times 10^9$$

$$= 577{,}500{,}000$$

■

EXERCISES 2.2

Evaluate.

1. 2^{-1}
2. 2^{-2}
3. 2^{-3}
4. $(-2)^3$
5. $(-2)^{-3}$
6. $(-2)^{-4}$
7. $(-2)^0$
8. 1^{-9}
9. $\dfrac{2^{-3}}{2^3}$
10. $\left(\dfrac{1}{2}\right)^{-1}$
11. $\dfrac{10^0}{10^{-2}}$
12. $\dfrac{10^{-2}}{10^{-3}}$
13. $\left(\dfrac{3}{4}\right)^{-2}$
14. $\dfrac{(-8)^{-3}}{(-4)^{-3}}$
15. $3^2 \cdot 4^2 \cdot 12^{-3}$

Simplify. Express all answers using positive exponents.

16. x^{-7}
17. a^2b^{-3}
18. $x^{-4}x^5$
19. $x^{-4}x$
20. $2x^{-3}$
21. $(2x)^{-3}$
22. $2^{-3}x$
23. $\dfrac{1}{x^{-2}}$
24. $\left(\dfrac{a^{-2}}{b^5}\right)^0$
25. $\dfrac{x^2}{x^{-3}}$
26. $\left(\dfrac{a}{b}\right)^{-2}$
27. $\dfrac{a^{-2}}{b^{-2}}$
28. $\left(\dfrac{1}{x}\right)^{-1}$
29. $\left(\dfrac{1}{x}\right)^{-2}$
30. $\left(\dfrac{1}{x^{-1}}\right)^{-2}$
31. $(x^3)^5$
32. $\dfrac{x^{12}}{x^{-3}}$
33. $x^5 \cdot x^7$
34. $(2x)^3(3x)^{-4}$
35. $(x^3y^2)(x^2y^4)$
36. $(-2x^3y)^2(-3x^2y^2)^3$
37. $\dfrac{(x^4y^2)^2}{x^2y}$
38. $\left(\dfrac{x^3}{y^2}\right)^4\left(\dfrac{-y}{x^2}\right)^{-2}$
39. $\dfrac{(x-2y)^6}{(x-2y)^2}$
40. $(2x^3y^2)^0$
41. $x^5 \cdot x^{-2}$
42. $x^2 \cdot x^{-5}$
43. $(x^{-3})^2$
44. $(x^3)^{-2}$
45. $(3x^{-2}y^2)^3$
46. $\dfrac{x^{-2}y^3}{x^3y^{-4}}$
47. $\dfrac{2x^{-2}y^{-3}}{x^2y^{-6}}$
48. $\dfrac{5x^0y^{-2}}{x^{-1}y^{-2}}$
49. $\dfrac{(x^{-2}y^2)^3}{(x^3y^{-2})^2}$
50. $\dfrac{(xy)^{-2}}{(xy)^3}$
51. $\dfrac{(2x^3y^{-2})^2}{8x^{-3}y^2}$
52. $\dfrac{(-3a)^{-2}}{a^{-2}b^{-2}}$
53. $\dfrac{3a^{-3}b^2}{2^{-1}c^2d^{-4}}$
54. $\dfrac{(a+b)^{-2}}{(a+b)^{-8}}$
55. $\dfrac{8x^{-8}y^{-12}}{2x^{-2}y^{-6}}$
56. $\dfrac{-12x^{-9}y^{10}}{4x^{-12}y^7}$
57. $\dfrac{(2x^2y^{-1})^6}{(4x^{-6}y^{-5})^2}$
58. $\dfrac{(a+3b)^{-12}}{(a+3b)^{10}}$
59. $\dfrac{(-a^{-5}b^6)^3}{(a^8b^4)^2}$
60. $\dfrac{(2a^{12}b^{-6})^2}{(4a^{-6}b^4)^3}$
61. $\dfrac{(s+t)^{-2}}{(s+t)^{-8}}$
62. $\dfrac{(r+s)^{-5}}{(r-s)^{-2}}$

Find a value of x to make each statement true.

63. $2^x \cdot 2^3 = 2^{12}$ 64. $2^{-3} \cdot 2^x = 2^6$ 65. $2^x \cdot 2^x = 2^{16}$

66. $2^x \cdot 2^{x-1} = 2^7$ 67. $\frac{2^x}{2^2} = 2^{-5}$ 68. $\frac{2^{-3}}{2^x} = 2^4$

Write each number in scientific notation.

69. 4680 70. 0.0092 71. 0.92 72. 0.9 73. 7,583,000
74. 93,000,000 75. 25 76. 36.09 77. 0.000000555 78. 0.57721
79. 202.4 80. 7.93

Write each number in standard notation.

81. 7.89×10^4 82. 7.89×10^{-4} 83. 3.0×10^3 84. 3.0×10^{-3} 85. 1.74×10^{-1}
86. 1.74×10^0 87. 1.74×10^1 88. 2.25×10^5 89. 9.06×10^{-2}

Express each of the following as a single power of 10.

90. $\frac{10^{-3} \times 10^5}{10}$ 91. $\frac{10^8 \times 10^4 \times 10^{-5}}{10^2 \times 10^3}$ 92. $\frac{10^{-3}}{10^{-5}}$

93. $\frac{10^1 \times 10^2 \times 10^3 \times 10^4}{10^{10}}$ 94. $\frac{10^9 \times 10^{-2}}{10^6 \times 10^{-9}}$ 95. $\frac{(10^2)^3 \times 10^{-1}}{(10^{-3})^4}$

Compute, using scientific notation. Use a calculator to verify your result.

96. $\frac{1}{5000}$ 97. $\frac{1}{0.0005}$ 98. $\frac{2}{80{,}000}$ 99. $\frac{(0.000025)}{(0.0625)(0.02)}$

100. $\frac{(240)(0.000032)}{(0.008)(12{,}000)}$ 101. $(0.0006)^3$ 102. $\frac{(40)^4(0.015)^3}{24{,}000}$

103. Light travels at a rate of about 186,000 miles per second. The average distance from the sun to the earth is 93,000,000 miles. Use scientific notation to find how long it takes light to reach the earth from the sun.

104. Based on information given in Exercise 103, use scientific notation to show that 1 light-year (the distance light travels in 1 year) is approximately $5.87 \times 10^{12} =$ 5,870,000,000,000 miles.

Written Assignment: Explain, in words and examples, how to write a number in scientific notation and how to convert a number given in scientific notation back into standard form.

2.3 ADDITION AND SUBTRACTION OF POLYNOMIALS

Algebraic expressions often contain variables designated by letters of the alphabet. For example, the expression

$$3x^2 - 4x + 5$$

Note that we can use other letters as well in such expressions. Thus, this can also be written as $3n^2 - 4n + 5$, $3a^2 - 4a + 5$, etc.

contains the *variable* x, which represents any real number. When $x = -2$ the value of the expression is

$$3(-2)^2 - 4(-2) + 5 = 25$$

Similarly, when $x = 0$, the value of $3x^2 - 4x + 5$ is

$$3(0)^2 - 4(0) + 5 = 5$$

*Note that $x^3 + x^{1/2}$ and $x^{-2} + 3x + 1$ are **not** polynomials because of the fractional and negative exponents in the expressions.*

The expression $3x^2 - 4x + 5$ is an example of a **polynomial in x;** all the exponents of x are non-negative integers. Its **degree** is 2 because 2 is the largest power of the variable x. The **coefficients** of the polynomial are 3, -4, and 5 ($5 = 5x^0$). Also, 5 is referred to as the **constant term** of the polynomial.

A polynomial is said to be in *standard form* if its terms are arranged so that the powers of the variable are in descending or ascending order. Thus the polynomial $x^2 + 3 + 7x + 5x^3$ can be rewritten in standard form as follows:

$$5x^3 + x^2 + 7x + 3$$

***Like terms** have exactly the same power of x; the numerical coefficients need not be the same.*

*$9x^2$, $-6x^2$, x^2 are **like terms.***

*$4x^5$, $4x^3$, are **unlike terms.***

Since polynomials always represent real numbers, we can add and subtract them according to the rules of addition and subtraction of real numbers. For example, to add $5x^3 + x^2 + 7x + 3$ and $3x^2 - 4x + 5$, we proceed as follows, collecting **like terms:**

$$\begin{aligned}
&(5x^3 + x^2 + 7x + 3) + (3x^2 - 4x + 5)\\
&\quad = 5x^3 + (x^2 + 3x^2) + (7x - 4x) + (3 + 5)\\
&\quad = 5x^3 + (1 + 3)x^2 + (7 - 4)x + 8\\
&\quad = 5x^3 + 4x^2 + 3x + 8
\end{aligned}$$

Notice that each polynomial was written in terms of *decreasing order* of the exponents. This is convenient, but not at all necessary.

As an alternative method, the polynomials can be written one below the other, putting like terms in the same column, and adding:

$$\begin{array}{r}
5x^3 + x^2 + 7x + 3\\
3x^2 - 4x + 5\\
\hline
5x^3 + 4x^2 + 3x + 8
\end{array}$$

The word "add" or "subtract" is often replaced by the command to "simplify" as in Example 1.

EXAMPLE 1 Simplify and check: $(a^2 - 2a) + (3a^2 + 3a + 7) + (5a^2 - 1)$

Solution

$$\begin{aligned}
&(a^2 - 2a) + (3a^2 + 3a + 7) + (5a^2 - 1)\\
&\quad = (a^2 + 3a^2 + 5a^2) + (3a - 2a) + (7 - 1)\\
&\quad = (1 + 3 + 5)a^2 + (3 - 2)a + 6\\
&\quad = 9a^2 + a + 6
\end{aligned}$$

Check: If the answer is correct, then the result obtained is true for all real numbers a. Therefore, we can check by choosing a *convenient* value for a and substituting into both the original expression and the final result.

Let $a = 3$ and substitute.

$$(a^2 - 2a) + (3a^2 + 3a + 7) + (5a^2 - 1)$$
$$= (9 - 6) + (27 + 9 + 7) + (45 - 1)$$
$$= 3 + 43 + 44 = 90$$

Now let $a = 3$ in the sum.

$$9a^2 + a + 6 = 81 + 3 + 6 = 90$$ ■

It is important to note that the manner of checking illustrated in Example 1 is not completely foolproof. For example, suppose you had arrived at the *incorrect* answer of $7a^2 + 5a + 6$. Then by substituting $a = 0$ in the original problem and in the answer you would obtain the same result of 6! So, to increase the reliability of the checking process, you might try a second number as a replacement for a, or just redo the entire problem.

Expressions may contain more than one variable. Here is an expression in the two variables x and y:

$$-2x^4 + 7x^2y + xy^3 - y$$

This is a **polynomial in x and y** because each variable only has positive integers as exponents. The rules for adding such polynomials are the same as before.

Example 2 demonstrates how the column method for addition can be used to find the sum of more than two polynomials.

EXAMPLE 2 Find the sum of the following three polynomials.

$$-2x^4 + 7x^2y + xy^3 - y$$
$$2x^4 + 17x^2y - 2xy^3 - 5y + 4$$
$$4x^4 + 5xy^3 + 2y - 10$$

Solution Place like terms in the same columns. Like terms have the same variables with the same powers; the coefficients need not be the same.

$$\begin{array}{rrrrr}
-2x^4 & +\ 7x^2y & +\ xy^3 & -\ y & \\
2x^4 & +\ 17x^2y & -\ 2xy^3 & -\ 5y & +\ 4 \\
4x^4 & & +\ 5xy^3 & +\ 2y & -\ 10 \\
\hline
4x^4 & +\ 24x^2y & +\ 4xy^3 & -\ 4y & -\ 6
\end{array}$$ ■

TEST YOUR UNDERSTANDING
Think Carefully

1. Add $3x^2 - 2x + 7$ and $x^2 + 3x - 6$ and check for $x = 2$.
2. Add $5x^4 - 4x^3 + 3x^2 - 2x + 1$ and $-5x^4 + 6x^3 + 10x$.
3. Add $x + 2y$ and $3x - 5y - 9$.
4. Add $x^2y + xy + xy^2$ and $3x^2y - 2xy - xy^2$.
5. Add $\frac{1}{2}x + 5$, $\frac{3}{2}x - 9$, and $4x + 1$.
6. Add $x^2 + x + 1$, $-3x - 4$, $6x - 5$, and $x^3 + x^2$.
7. Add $xy + x - y$, $7xy + \frac{1}{3}x - \frac{1}{2}y$, and $-8xy + \frac{2}{3}x - \frac{3}{2}y$.
8. Simplify: $(6x^5 + 2x^3 - 13x + 13) + (2x^5 + x^4 - x^3 + 12)$
9. Simplify: $(a^2 + 5a) + (a^2 - 4a + 3) + (5a^2 + 5a + 5)$
10. Simplify: $(r + s) + (r - 2s) + (2r + s) + (-3r - 3s)$

(Answers: Page 80)

To subtract polynomials we use the definition $a - b = a + (-b)$. Thus to subtract $3x^2 - 4x + 5$ from $5x^3 + x^2 + 7x + 3$, we first note that the opposite of $3x^2 - 4x + 5$ is $-3x^2 + 4x - 5$. Then, following the rules for subtraction, we add this opposite as in the following:

$$
\begin{aligned}
&(5x^3 + x^2 + 7x + 3) - (3x^2 - 4x + 5) \\
&\quad = (5x^3 + x^2 + 7x + 3) + (-3x^2 + 4x - 5) \\
&\quad = 5x^3 + (x^2 - 3x^2) + (7x + 4x) + (3 - 5) \\
&\quad = 5x^3 - 2x^2 + 11x - 2 && \text{(combining terms)}
\end{aligned}
$$

The preceding subtraction can also be completed by using the column method. Subtract each term in the bottom row from the like term in the top row to find the difference:

To check this result, add the second and third polynomials to obtain the first, as is done in arithmetic.

$$
\begin{array}{rrrr}
5x^3 + & x^2 + & 7x + & 3 \\
 & 3x^2 - & 4x + & 5 \\
\hline
5x^3 - & 2x^2 + & 11x - & 2
\end{array}
$$

$3 - (5)$

$7x - (-4x)$

$x^2 - (3x^2)$

$5x^2 - (0)$

EXAMPLE 3 Subtract $a^2 - 5a - 2$ from $2a + 3$.

Solution

As a short-cut, we can simply change all the signs in $a^2 - 5a - 2$ and then proceed as in addition.

$$
\begin{aligned}
(2a + 3) - (a^2 - 5a - 2) &= (2a + 3) + (-a^2 + 5a + 2) \\
&= -a^2 + (2a + 5a) + (3 + 2) \\
&= -a^2 + 7a + 5
\end{aligned}
$$

■

Note that we can also complete a subtraction problem by using the property $-b = (-1)b$. Thus Example 3 can be completed in this way:

$$
\begin{aligned}
(2a + 3) - (a^2 - 5a - 2) &= (2a + 3) - 1(a^2 - 5a - 2) \\
&= 2a + 3 - a^2 + 5a + 2 \\
&= -a^2 + 7a + 5
\end{aligned}
$$

EXAMPLE 4 Simplify:

(a) $(x - 3y) - (5x + y) - (3y - 2x)$

(b) $[(2x - x^2) - (x^3 - 5x^2 + 8x)] - [(9 - 2x^3 + 4x^2) - (6x - x^2)]$

Solution

(a) $(x - 3y) - (5x + y) - (3y - 2x)$

$$= (x - 3y) + (-5x - y) + (-3y + 2x)$$
$$= [x + (-5x) + 2x] + [-3y + (-y) + (-3y)]$$
$$= -2x + (-7y)$$
$$= -2x - 7y$$

(b) Just as we did with numerical expressions, the simplification is completed by working outward from the inner parentheses.

Note some of the shortcuts used in this solution. For instance, in going from the second to the third line we have combined like terms and then arranged them in decreasing powers of x, within each bracket all at once.

$$[(2x - x^2) - (x^3 - 5x^2 + 8x)] - [(9 - 2x^3 + 4x^2) - (6x - x^2)]$$
$$= [2x - x^2 - x^3 + 5x^2 - 8x] - [9 - 2x^3 + 4x^2 - 6x + x^2]$$
$$= [-x^3 + 4x^2 - 6x] - [-2x^3 + 5x^2 - 6x + 9]$$
$$= -x^3 + 4x^2 - 6x + 2x^3 - 5x^2 + 6x - 9$$
$$= x^3 - x^2 - 9$$
■

Sometimes, for convenience, we will name a polynomial in this way:

*The symbol $p(x)$ is read as "p of x" or "p at x." It does **not** mean p times x.*

$$p(x) = 3x^2 - 4x + 5$$

The notation $\boldsymbol{p(x)}$ tells us that we are considering a polynomial in x. However we can use any other letter as well to name the polynomial, such as:

$$f(x) = 3x^2 - 4x + 5$$
$$s(x) = 3x^2 - 4x + 5$$

Using this notation, the symbol $p(-2)$ represents the *value of the polynomial* when the variable is set equal to -2; it does *not* mean p times -2.

EXAMPLE 5 Find $p(3)$ for the polynomial $p(x) = -2x^2 + 3x + 4$.

Solution $p(3) = -2(3)^2 + 3(3) + 4 = -18 + 9 + 4 = -5$
That is, when $x = 3$, $-2x^2 + 3x + 4 = -5$ ■

A calculator can be used to avoid tedious computations when evaluating a polynomial $p(x)$. For example, to find $p(12.17)$ for $p(x) = -2x^2 + 3x + 4$ we may use this calculator sequence:

$p(12.17) =$ (2) [+/−] [×] (12.17) [x^2] [+] (3) [×] 12.17 [+] (4) [=] − 255.7078

EXERCISES 2.3

Add and check for $x = 2$.

1. $3x^2 + 5x - 2$
   $\underline{5x^2 - 7x + 9}$

2. $3x^3 - 7x^2 - 8x + 12$
   $\underline{x^3 - 2x^2 + 8x - 9}$

3. $x^3 - 3x^2 + 2x - 5$
   $\underline{5x^2 - x + 9}$

4. $4x^2 + 9x - 17$
   $\underline{2x^3 - 3x^2 + 2x - 11}$

Subtract and check.

5. $\begin{array}{r} 5x^2 - 9x - 1 \\ \underline{2x^2 + 2x + 7} \end{array}$

6. $\begin{array}{r} 3x^3 - 2x^2 - 8x + 9 \\ \underline{2x^3 + 5x^2 + 2x + 1} \end{array}$

7. $\begin{array}{r} x^3 - 2x^2 + 6x + 1 \\ \underline{-\ \ x^2 - 6x - 1} \end{array}$

8. $\begin{array}{r} 4x^3 + x^2 - 2x - 13 \\ \underline{2x^2 + 3x + 9} \end{array}$

Simplify by performing the indicated operations.

9. $(3x^2 + 8x + 7) + (x^2 + 9x + 1)$
10. $(x^3 + 5x^2 - 2x - 3) + (2x^3 - x^2 + 7x + 2)$
11. $(3x^3 - 5x + 9) + (x^2 - 5x - 3)$
12. $(7x^2 + 5x + 8) - (3x^2 + 2x - 5)$
13. $(3x^3 - 2x^2 + 7x - 5) - (x^3 + 6x^2 - 7x + 1)$
14. $(2x^2 - 9x + 11) - (x^3 + 6x^2 - 2x - 5)$
15. $(3x + 5y) + (8x - 2y)$
16. $(-4x^3 - 6x^2 + 11x - 15) + (5x^3 + 5x^2 - 5x - 5)$
17. $(2xy + 3y + 4) + (3xy - 3y + 1)$
18. $5x + (y - 2x)$
19. $(x + y) + 3y$
20. $(x^5 - 6x^3 + x^2 + x + 9) + (2x^4 + x^3 + 3x - 7)$
21. $(x + 2y) + (2x + y) + (3x + 3y)$
22. $(a^2 - 2ab + b^2) + 4ab$
23. $(7x + 5) - (2x + 3)$
24. $(x^3 + 3x^2 + 3x + 1) - (x^2 + 2x + 1)$
25. $(x^2y + xy^2) - (2xy^2 - xy)$
26. $(6a^2 - 3a) - (a^3 - 4a^2 + 7a)$
27. $x - (x + 2)$
28. $(x + y) - (y - x)$
29. $(2a + 3b - 6c) - (a - 2b + 3c)$
30. $(x + 3y) + (2x - 2y) - (6x + 10y)$
31. $(a - b) - (2a + 3b) - (b - a)$
32. $(x^2 + 2y^2 + 6) - (3x^2 - 2y^2 - 6) - (x^2 + 4y^2 + 1)$

33. $7x - (x - 3y) - 2y$
34. $10 - [8 - (x + 2)] - x$
35. $[(2d + 3c) + (2d - c)] + (-4d + c + 5)$
36. $5a - [a - (3a + 8)]$
37. $(x^4 - 3x^3 + 4x^2 - x + 1) + (2x^4 + 2x^3 - x^2 + 3x + 7) + (-5x^4 - x^3 + 2x^2 + x + 2)$
38. $(2x - 3y) - [(x - y) - (4x - y)]$
39. $[(-x^2 + 5x + 4) - (x^3 - 2x)] - [(x^3 + 2x^2 + 2) - (1 + x + x^2)]$

Find the sum of the polynomials.

40. $x^3 - 3x^2 + 7x - 1,\ x^4 + 2x^3 - 6x - 3,\ 3x^4 + 6x^2 + 9$
41. $x + 3y - z,\ 2y + 3z + x,\ 4z + 3x - 2y$
42. $x^2 + 2x + 1,\ -3x^2 + 3x + 3,\ -3x^2 - 7x + 2,\ 10x^2 - 2x + 5$
43. $a^2 - ab + 2b,\ 2ab + b^2,\ -a^2 + ab,\ a^2 + 2b^2 - 3ab$
44. $5x^3y + 4x^2y^2 - xy^3 + a^2,\ 3xy^3 - x^2y^2 + 5a^2 - x^3y,\ 2a^2 - 9x^2y$

In each of the following, subtract the first polynomial from the second.

45. $x^2 - 3x + 4,\ 5x^2 + 2x + 1$
46. $3x^3 + 2x^2 + x + 1,\ -x^3 - x^2 - x - 1$
47. $a + b - 2,\ 3b + 4$
48. $2s^2 - 2s,\ s^2 - s + 3$
49. $4a^2 + 3b^2 - 5a + 3b,\ 2a - b^2 + 4a^2 - 6b$
50. $xy - 2xz + 3yz,\ 5xy - 4xz - 6yz$
51. $a - b + c - d,\ -a - b - c - d$
52. $5 - xy,\ x^2y + xy - 3x^2$

53. Let $p(x) = 3x^2 + x - 5$. Find **(a)** $p(-2)$; **(b)** $p(2)$.
54. Let $f(x) = -x^2 - 3x + 10$. Find **(a)** $f(-1)$; **(b)** $f(5)$.
55. Let $s(x) = -2x^2 + 8x - 3$. Find **(a)** $s(0)$; **(b)** $s(-5)$.
56. Let $p(x) = 8x^3 - 4x^2 + 6x - 7$. Find **(a)** $p(\frac{1}{2})$; **(b)** $p(-4)$.
57. Let $h(x) = -5x^2 + 3x - 17$. Find **(a)** $h(2.9)$; **(b)** $h(-0.57)$.
58. Let $f(x) = 2x^3 - 7x^2 + 5x + 1$. Find **(a)** $f(1.5)$; **(b)** $f(-3.8)$.
59. Let $p(x) = x^2 - 3x + 7$ and $r(x) = 2x^2 + 5x - 8$. **(a)** Find $p(2.53) + r(2.53)$. **(b)** Find the sum $f(x) = p(x) + r(x)$ and evaluate for $f(2.53)$.

Written Assignment: To subtract one polynomial, $p(x)$, from another one, $f(x)$, we can change the signs of $p(x)$ and proceed as in addition. Explain what this rule really means.

CHALLENGE
Think Creatively

Find a replacement for x so that the sum of the numbers in every row, column, and diagonal is the same; that is, so as to form a **magic square.**

$x + 3$	$x - 4$	$-3(3 - x)$
$x - 2$	$x(x - 4)$	$x + 2$
$4(x - 4)$	$3(x - 2)$	$x - 3$

2.4 MULTIPLICATION OF POLYNOMIALS

In the preceding section we learned how to add and subtract polynomials. Next we turn our attention to finding the products of polynomials.

The simplest type of polynomial is one that consists of a single term, and is called a **monomial.** Each of the following is an example of a monomial:

$$3x \qquad 2y^3 \qquad -5xy \qquad -2x^2y^3$$

In the expression $-2x^2y^3$, -2 is called the (numerical) **coefficient** of the monomial; -2, x^2, and y^3 are some of the *factors* of the monomial. We can use the rules of exponents to multiply two monomials, as follows:

$$\begin{aligned}(-2x^2y^3)(3x^4y) &= (-2)(3)(x^2x^4)(y^3y)\\ &= -6x^{2+4}y^{3+1} \qquad (b^m \cdot b^n = b^{m+n})\\ &= -6x^6y^4\end{aligned}$$

To find the product of a monomial and a polynomial we make use of the distributive property, as follows:

$$\begin{aligned}3x^2(4x^7 - 3x^4 - x^2 + 15) &= 3x^2(4x^7) - 3x^2(3x^4) - 3x^2(x^2) + 3x^2(15)\\ &= 12x^9 - 9x^6 - 3x^4 + 45x^2\end{aligned}$$

In the first line we used an *extended version of the distributive property*, namely,

$$a(b - c - d + e) = ab - ac - ad + ae$$

where a plays the role of $3x^2$ in the preceding illustration.

TEST YOUR UNDERSTANDING
Think Carefully

(Answers: Page 80)

Find each product.

1. $(2x)(3x^3)$ **2.** $(-3x)(-9x^4)$ **3.** $(5x^2y)(-2x^3y^2)$
4. $(-3a^2b^2)(-4ab^3)$ **5.** $a(-2ab)(3ab^2)$ **6.** $2x(x^3 + x)$
7. $-3x^2(2x^2 - 5x + 7)$ **8.** $2x^3(x^2 - 9x + 2)$ **9.** $-3a^2b(3a^2b - 7ab^2 + 1)$

Next let us find the product of two polynomials:

$$(x + 2)(x^3 + 7x^2 - 4)$$

You can think of $x + 2$ as a and visualize the product in this way:

$a(x^3 + 7x^2 - 4) =$
$a \cdot x^3 + a \cdot 7x^2 - a \cdot 4$

We will analyze the process first, and then search for an efficient way to obtain the product. Use the distributive property to write

$$(x + 2)(x^3 + 7x^2 - 4) = (x + 2)(x^3) + (x + 2)(7x^2) - (x + 2)(4)$$

Use the distributive property three more times to find these products:

$$(x + 2)x^3 = x^4 + 2x^3$$
$$(x + 2)7x^2 = 7x^3 + 14x^2$$
$$(x + 2)4 = 4x + 8$$

Thus we have

$$\begin{aligned}(x + 2)(x^3) + (x + 2)(7x^2) - (x + 2)(4)\\ = (x^4 + 2x^3) + (7x^3 + 14x^2) - (4x + 8)\end{aligned}$$

The complete solution is as follows:

$$\begin{aligned}(x + 2)(x^3 + 7x^2 - 4) &= (x + 2)x^3 + (x + 2)7x^2 - (x + 2)4\\ &= (x^4 + 2x^3) + (7x^3 + 14x^2) - (4x + 8)\\ &= (x^4 + 2x^3) + (7x^3 + 14x^2) + (-4x - 8)\\ &= x^4 + (2x^3 + 7x^3) + 14x^2 + (-4x) + (-8)\\ &= x^4 + 9x^3 + 14x^2 - 4x - 8\end{aligned}$$

A study of the preceding problem shows that each term in $x^3 + 7x^2 - 4$ is eventually multiplied by each term in $x + 2$. Thus, as a shortcut, multiply each term within the second set of parentheses by each term in the first set and combine like terms.

$$\begin{aligned}(x + 2)(x^3 + 7x^2 - 4) &= \underbrace{(x)(x^3) + (x)(7x^2) - (x)(4)}_{\text{Multiply each term by } x.} + \underbrace{(2)(x^3) + (2)(7x^2) - (2)(4)}_{\text{Multiply each term by 2.}}\\ &= x^4 + 7x^3 - 4x + 2x^3 + 14x^2 - 8\\ &= x^4 + 9x^3 + 14x^2 - 4x - 8\end{aligned}$$

The following scheme shows another effective way to compute this product.

This technique is merely a rearrangement of what was done earlier. Therefore, we are able to use this schematic way with the assurance that it produces correct results. It is particularly useful when each of the polynomials contains three or more terms, as in Example 1.

$$\begin{array}{rrrrrl} & x^3 & +\ 7x^2 & & -\ 4 & \\ & & & x & +\ 2 & \\ \hline & +\ 2x^3 & +\ 14x^2 & & -\ 8 & (2 \text{ times } x^3 + 7x^2 - 4)\\ x^4 & +\ 7x^3 & & -\ 4x & & (x \text{ times } x^3 + 7x^2 - 4)\\ \hline x^4 & +\ 9x^3 & +\ 14x^2 & -\ 4x & -\ 8 & \end{array}$$

Arrange the given polynomials as shown. The first line shows the product of $x^3 + 7x^2 - 4$ and 2. The second line shows the product by x. These *partial products* are written so that like terms are in the same columns. The final product is the sum of these partial products.

Note that the column method is a convenient way to organize your work. Be certain to keep like terms in the same column. Let $x = 2$ and check the solution.

EXAMPLE 1 Multiply $3x^3 - 8x + 4$ by $2x^2 + 5x - 1$

Solution

$$
\begin{array}{rrrrrrl}
 & & 3x^3 & - & 8x & + 4 & \\
 & & 2x^2 & + & 5x & - 1 & \\
\hline
 & & -3x^3 & & +8x & -4 & (-1 \text{ times } 3x^3 - 8x + 4) \\
\text{(add)} & 15x^4 & & -40x^2 & +20x & & (5x \text{ times } 3x^3 - 8x + 4) \\
6x^5 & & -16x^3 & +8x^2 & & & (2x^2 \text{ times } 3x^3 - 8x + 4) \\
\hline
6x^5 & +15x^4 & -19x^3 & -32x^2 & +28x & -4 &
\end{array}
$$

■

Try to discover a pattern in these three examples. Given the products, could you write the original two factors?

EXAMPLE 2 Multiply $x - a$ by each of the following:

(a) $x^2 + ax + a^2$

(b) $x^3 + ax^2 + a^2x + a^3$

(c) $x^4 + ax^3 + a^2x^2 + a^3x + a^4$

Solution

(a) $(x - a)(x^2 + ax + a^2) = x^3 + ax^2 + a^2x - ax^2 - a^2x - a^3$
$= x^3 - a^3$

(b) $(x - a)(x^3 + ax^2 + a^2x + a^3) = x^4 + ax^3 + a^2x^2 + a^3x - ax^3$
$-a^2x^2 - a^3x - a^4$
$= x^4 - a^4$

(c) $(x - a)(x^4 + ax^3 + a^2x^2 + a^3x + a^4) = x^5 - a^5$ ■

EXAMPLE 3 Simplify: $3x[(2x - 1)(x^2 + 5x - 4) - 2(x - 3)(x^2 + 2x + 2)]$

Solution To simplify, work outward from the inner grouping symbols.

$$
\begin{aligned}
&3x[(2x - 1)(x^2 + 5x - 4) - 2(x - 3)(x^2 + 2x + 2)] \\
&\quad = 3x[2x^3 + 9x^2 - 13x + 4 - 2(x^3 - x^2 - 4x - 6)] \\
&\quad = 3x[2x^3 + 9x^2 - 13x + 4 - 2x^3 + 2x^2 + 8x + 12] \\
&\quad = 3x[11x^2 - 5x + 16] \\
&\quad = 33x^3 - 15x^2 + 48x
\end{aligned}
$$

■

EXERCISES 2.4

Find each product. Be certain to combine all like terms.

1. $(5x^2)(-4x^3)$
2. $(-3x)(7x^5)$
3. $(4x^3)(-4x^3)$
4. $(x^2)(5x^2)$
5. $(-2x^2)(-2x^2)$
6. $(x^4)(-3x^3)$
7. $(ab^3)(6a^3b)$
8. $(-3a^2b)(-2a^4b^2)$
9. $(2a^3b^3)(-2ab^4)$
10. $(-2x)(5x^2)(3x)$
11. $(5x)(x^3)(-3x^3)$
12. $(-xy)(5x^2)(6y^4)$
13. $(2ax)(-3a^2x)(ax^2)$
14. $(-xy^3)(-8x^3y)(x^2y)$
15. $x(-3x^2y^3)(-3x^3y^2)$
16. $3x(2x^2 - 3x)$
17. $(-2a)(a^3 - 3)$
18. $x(3x^2 - 2)$
19. $(-2x)(x^3 + 5x^2 - 1)$
20. $(-3x^2)(x^2 + 2x - 4)$
21. $x^3(2x^2 - 5x + 3)$
22. $x^2(x^3 - 3x^2 + 7x + 2)$
23. $2a^2b(a^2 - 3ab + b^2)$
24. $(-2x^2y)(x^2 + 3xy + y^2)$
25. $(x + 1)(x^2 + 3x + 4)$
26. $(x + 2)(x^2 - 2x + 4)$
27. $(x - 1)(x^2 + 5x - 2)$
28. $(x - 2)(x^2 + 2x + 4)$
29. $(x + 1)(x + 1)$
30. $(2x + 3)(2x + 3)$
31. $(2x + 3)(3x + 2)$
32. $(5x - 4)(4x + 6)$
33. $(2x - 1)(3x^2 - 2x + 5)$
34. $(3x - 2)(2x^3 + 5x - 1)$
35. $(x^2 + x + 1)(x^2 - 3x - 2)$
36. $(x^2 - 2x + 3)(x^3 + x^2 - 2x + 1)$
37. $(x - 2)(x^5 + 2x^4 + 4x^3 + 8x^2 + 16x + 32)$
38. $(x + 1)(x + 1)(x + 1)$
39. $4x[(x^2 + 2x - 1)(x - 1) - (2x + 1)(x^3 - 4x - 5)]$
40. $(x^2 - x + 1)(x^2 - x + 1)(x^2 - x + 1)$
41. Let $p(x) = 3x + 5$ and $r(x) = x^2 + 2x - 7$.
   (a) Find the product $p(0.85) \times r(0.85)$.
   (b) Find the product $f(x) = p(x) \cdot r(x)$ and evaluate for $f(0.85)$.
42. Let $r(x) = x^2 - 2x + 5$ and $s(x) = 5x + 3$.
   (a) Find the product $r(0.13) \times s(0.13)$.
   (b) Find the product $p(x) = r(x) \cdot s(x)$ and evaluate for $p(0.13)$.
43. **(a)** For $r(x)$ and $s(x)$ in Exercise 42, find $f(x) = 2xr(x) - x^2s(x) - 10x$.
   (b) Evaluate $f(-5.27)$.

2.5 THE PRODUCT OF BINOMIALS

A **binomial** is a polynomial with two terms. Each of the following is an example of a binomial:

$$3x + 2 \qquad 8x^2 + 7x \qquad 8a^3 - 27b^3$$

Next, observe how the distributive property is used to multiply two **binomials** (polynomials having two terms).

$$\begin{aligned}(2x + 3)(4x + 5) &= (2x + 3)4x + (2x + 3)5 \\ &= (2x)(4x) + (3)(4x) + (2x)(5) + (3)(5) \\ &= 8x^2 + 12x + 10x + 15 \\ &= 8x^2 + 22x + 15\end{aligned}$$

Here is a shortcut that can be used to multiply two binomials.

$(2x + 3)(4x + 5)$:

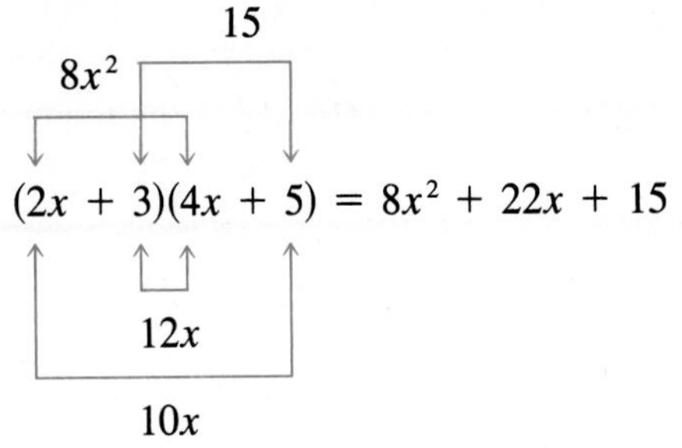

$8x^2$ is the product of the *first* terms in the binomials.

$10x$ and $12x$ are the products of the *outer* and *inner* terms;

$$10x + 12x = 22x$$

15 is the product of the *last* terms in the binomials.

In general, we may write the product $(a + b)(c + d)$ in this way:

Keep this diagram in mind as an aid to finding the product of two binomials mentally. Some students find it helpful to remember the first letter of each step, FOIL.

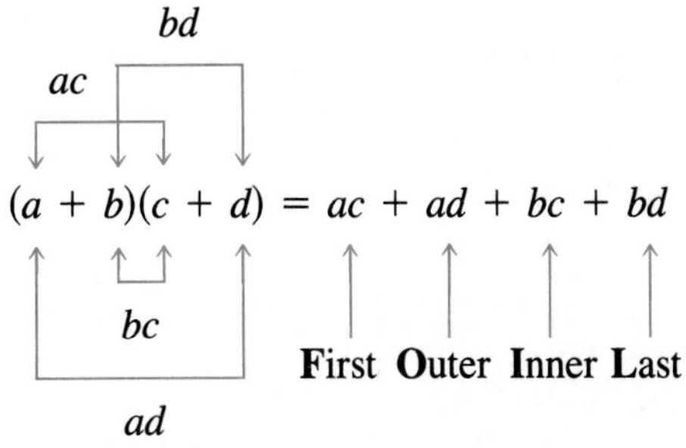

EXAMPLE 1 Find the product: $(5x - 9)(6x + 2)$

Solution

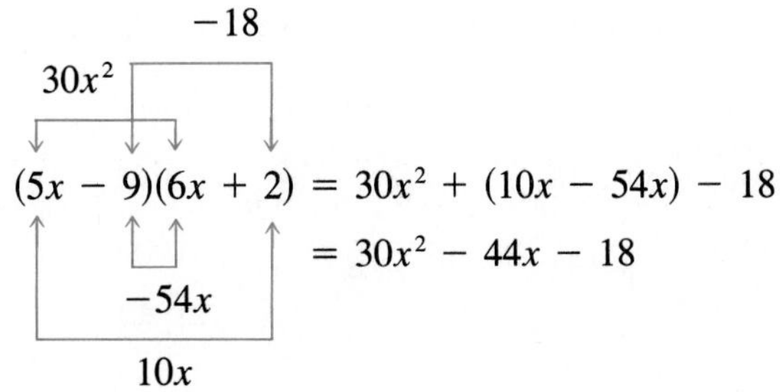

■

With a little practice you should be able to find the products of binomials without having to draw the arrows shown in the preceding example. Thus, by inspection, you should be able to write

You may also be able to combine the inner terms mentally and immediately write the final trinomial.

$$\begin{aligned}(5x - 9)(6x + 2) &= 30x^2 + 10x - 54x - 18\\ &= 30x^2 - 44x - 18\end{aligned}$$

EXAMPLE 2 Find the product of $ax + b$ and $cx + d$ by:

(a) Making use of the distributive property.
(b) Using the method of writing the binomials in column form.
(c) Using the visual inspection method (FOIL).

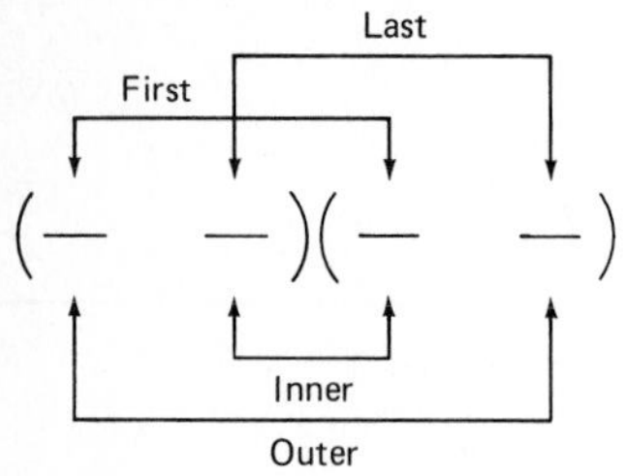

Solution

(a) $(ax + b)(cx + d) = (ax + b)cx + (ax + b)d$
$$= acx^2 + bcx + adx + bd$$
$$= acx^2 + (bc + ad)x + bd$$

(b)
$$
\begin{array}{r}
ax + b \\
cx + d \\
\hline
adx + bd \\
acx^2 + bcx \\
\hline
acx^2 + (bc + ad)x + bd
\end{array}
$$

(c)

$$(ax + b)(cx + d) = acx^2 + (bc + ad)x + bd$$

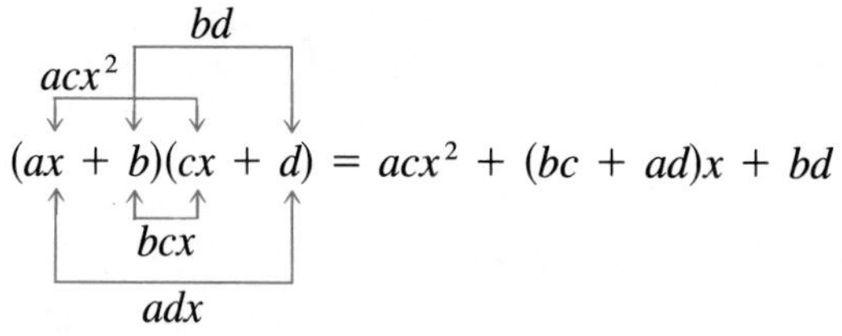

EXAMPLE 3 Find the products: **(a)** $(a - b)(a + b)$ **(b)** $(a - b)(a - b)$

Solution

(a) $(a - b)(a + b) = a^2 + ab - ab - b^2 = a^2 - b^2$
(b) $(a - b)(a - b) = a^2 - ab - ab + b^2 = a^2 - 2ab + b^2$

We can use the result of Example 3(a) to perform certain computations mentally. For example, to find the product 38×42 mentally, first convert the factors into the form $a - b$ and $a + b$ and evaluate $a^2 - b^2$ as follows:

The trick here is to have two factors, one as much above a number as the other is below, where the number is easily squared, such as 40.

$$38 \times 42 = (40 - 2)(40 + 2)$$
$$= (40)^2 - (4)^2 \qquad (a - b)(a + b) = a^2 - b^2$$
$$= 1600 - 16$$
$$= 1584$$

TEST YOUR UNDERSTANDING
Think Carefully

Find each product.

1. $(2x + 3)(3x + 2)$ **2.** $(5x + 1)(2x + 1)$ **3.** $(3x - 2)(2x - 3)$
4. $(5a - 1)(2a - 1)$ **5.** $(4x + 3)(2x - 3)$ **6.** $(3n - 8)(n + 5)$
7. $(3x + 1)(3x + 1)$ **8.** $(5y + 3)(5y + 3)$ **9.** $(7x - 2)(7x - 2)$

Find each product mentally.

(Answers: Page 81) **10.** 15×25 **11.** 33×27 **12.** 96×104

More than two polynomials may be involved in a product. For example, here is a product of three binomials:

$$\begin{aligned}(x + 2)(x + 3)(x + 4) &= [(x + 2)(x + 3)](x + 4) \\ &= (x^2 + 5x + 6)(x + 4) \\ &= (x^3 + 5x^2 + 6x) + (4x^2 + 20x + 24) \\ &= x^3 + 9x^2 + 26x + 24\end{aligned}$$

Sometimes more than one operation is involved, as demonstrated in the next example.

EXAMPLE 4 Simplify by performing the indicated operations:

$$(x^2 - 5x)(3x^2) + (x^3 - 1)(2x - 5)$$

Solution Multiply first and then combine like terms.

$$\begin{aligned}(x^2 - 5x)(3x^2) + (x^3 - 1)(2x - 5) &= (3x^4 - 15x^3) + (2x^4 - 5x^3 - 2x + 5) \\ &= 5x^4 - 20x^3 - 2x + 5\end{aligned}$$ ■

Compare the areas of the two congruent squares in terms of the segments a and b.

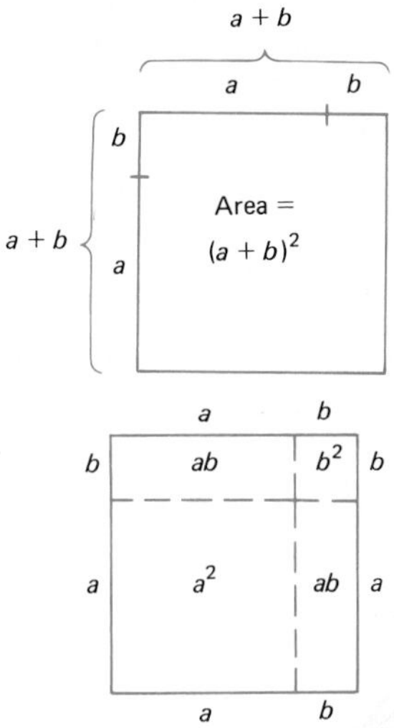

Explain how these figures provide a geometric intepretation for the expansion of $(a + b)^2$.

The product of $a + b$ times itself is given by

$$(a + b)(a + b) = a^2 + 2ab + b^2$$

Using exponents, we have

$$(a + b)(a + b) = (a + b)^2$$

Thus

$$(a + b)^2 = a^2 + 2ab + b^2$$

We say that $a^2 + 2ab + b^2$ is the *expansion,* or *expanded form,* of $(a + b)^2$.

Here are several special forms of products of binomials that are worth remembering for future use.

CAUTION
$(a + b)^2 \neq a^2 + b^2$
$(a - b)^2 \neq a^2 - b^2$

SPECIAL PRODUCTS

$$(a - b)(a + b) = a^2 - b^2$$
$$(a - b)(a - b) = (a - b)^2 = a^2 - 2ab + b^2$$
$$(a + b)(a + b) = (a + b)^2 = a^2 + 2ab + b^2$$

EXAMPLE 5 Expand: $(a + b)^3$

Solution First write

$$(a + b)^3 = (a + b)(a + b)^2$$

The result of Example 5 provides a formula for the cube of a binomial, that is, for all expansions of the form $(a + b)^3$. Try to use it to expand $(x + 3)^3$ mentally.

Then, use the expansion of $(a + b)^2$:

$$(a + b)^3 = (a + b)(a^2 + 2ab + b^2)$$
$$= a^3 + 2a^2b + ab^2 + a^2b + 2ab^2 + b^3$$
$$= a^3 + 3a^2b + 3ab^2 + b^3$$

■

EXERCISES 2.5

Find each product.

1. $(x + 1)(2x + 3)$
2. $(x + 2)(x + 5)$
3. $(x + 6)(x + 4)$
4. $(y + 6)(y - 4)$
5. $(x - 6)(x + 4)$
6. $(b - 6)(b - 4)$
7. $(x - 3)(x - 3)$
8. $(n - 3)(n + 3)$
9. $(2x + 1)(3x + 1)$
10. $(4y - 2)(y + 7)$
11. $(8x + 10)(2x - 5)$
12. $(5a + 3)(4a + 6)$
13. $(5x - 3)(4x - 6)$
14. $(5y + 3)(4y - 6)$
15. $(5x - 3)(4x + 6)$
16. $(x + 2)(x + 2)$
17. $(x - 2)(x - 2)$
18. $(2a + b)(2a + b)$
19. $(2a - b)(2a - b)$
20. $(12x - 8)(7x + 4)$
21. $(x + \frac{1}{2})(x + \frac{1}{2})$
22. $(x - \frac{1}{2})(x + 4)$
23. $(x + \frac{1}{2})(x - 4)$
24. $(x - 0.1)(x - 0.1)$
25. $(15x + 30)(18x - 27)$
26. $(10x - 15)(20x - 25)$
27. $(-2x + 3)(3x + 6)$
28. $(-2x - 3)(3x + 6)$
29. $(-2x - 3)(3x - 6)$
30. $(-2x - 3)(-3x - 6)$
31. $(\frac{1}{2}x + 4)(\frac{1}{2}x - 4)$
32. $(\frac{2}{3}x + 6)(\frac{2}{3}x + 6)$
33. $(\frac{1}{2}x - \frac{1}{3})(\frac{1}{2}x - \frac{1}{3})$
34. $(\frac{1}{2}x + \frac{1}{3})(\frac{1}{2}x - \frac{1}{3})$
35. $(7 + 3x)(9 - 4x)$
36. $(7 - 3x)(4x - 9)$
37. $(ax + b)(ax + b)$
38. $(ax + b)(ax - b)$

Square each binomial.

39. $a - 2b$
40. $3x - 5$
41. $-3x - 4y$
42. $xy - 1$
43. $2x - \frac{1}{2}$
44. $\frac{2}{3}x + 6y$

Expand each of the following and combine like terms.

45. $(a - b)^2$
46. $(x - 1)^3$
47. $(x + 1)^4$
48. $(a + b)^4$
49. $(a - b)^4$
50. $(2x + 3)^3$
51. $(\frac{1}{2}x^2 - 4)^2$
52. $(\frac{1}{3}x + 3)^3$
53. $(\frac{1}{2}x - 1)^3$

Perform the indicated operations and combine like terms.

54. $(x + y)^2 - (x - y)^2$
55. $(a - 3b)(a + 3b) - (2a + b)(2a + b)$
56. $(x + 1)(3x + 1) + (x + 1)(3x + 2) + (x + 1)(3x + 3)$
57. $(x^2 + 8)(2x - 9) + (2x - 1)(3x + 2)(x - 1)$
58. $5(x + 5)(x - 5)(x + 5)$
59. $3x(1 - x)(1 - x)$
60. $10x(2x + 1)(4x - 1) - 3(x + 9)(x - 9)$

61. $1000(\frac{1}{10}x + 1)(\frac{1}{10}x + 1)(\frac{1}{10}x - 1)$
62. $64(\frac{1}{2}x + \frac{1}{4})(\frac{1}{8}x + \frac{1}{16})$
63. $(c^2 + 3d^2)(c^2 - 3d^2)$
64. $(x^2 + 1)(x^2 - 1)$
65. $(x^3 + 2)(x^3 + 2)$
66. $(a^3 - b^2)(a^3 + b^2)$
67. $(p^3 - q^2)(p^3 - q^2)$
68. $[(a + b) - c][(a + b) + c]$
69. $(a^n - 4)(a^n + 4)$
70. $(a^{2n} + b^n)(a^{2n} - 2b^n)$

71. Recall that $(ab)^n = a^n b^n$. Use this idea to compute $(x + 1)^2(x - 1)^2$ two different ways and compare.

72. Use the idea given in Exercise 71 in conjunction with $a(b + c) = ab + ac$ to evaluate $(4x + 12)^3$ in two different ways.

Written Assignment: The "FOIL" method for multiplying two binomials is a shortcut of a mathematical procedure. Give an explanation as to why it works.

CHALLENGE
Think Creatively

Here is a shortcut for squaring a number whose units digit is 5:

$35^2 = 1225$ (Multiply 3×4 and append 25)

$65^2 = 4225$ (Multiply 6×7 and append 25)

Thus, to square 85, multiply 8 by the next higher integer ($8 \times 9 = 72$) and append 25; $85^2 = 7225$. *Prove* that this will always work by representing a two-digit number as $10n + 5$ and finding the expansion of $(10n + 5)^2$.

2.6 INTRODUCTION TO FACTORING

When a number is written as the product of several other numbers, each is called a **factor** of the given number. For example:

$30 = 2 \times 3 \times 5$ 2, 3, and 5 are factors of 30.

$(x - 2)(x + 3) = x^2 + x - 6$ $x - 2$ and $x + 3$ are factors of $x^2 + x - 6$

Loosely speaking, factoring is "unmultiplying." In general, we will be looking at certain types of expressions and rewriting them as products of other expressions.

One of the skills we will need to have is that of finding the **greatest common factor (GCF)** of the terms of a polynomial. When dealing with natural numbers, the GCF is the largest number that is a factor of each of the numbers under consideration. For example:

Numbers	GCF
12, 15	3
18, 30	6
18, 24, 42	6
15, 19, 23	1

Note: *There is no natural number, other than 1, by which 15, 19, and 23 are divisible.*

A prime number is a positive integer greater than 1 whose only positive integer factors are 1 and itself. The first 10 primes are 2, 3, 5, 7, 11, 13, 17, 19, 23, 29.

The **prime factored form** is sometimes useful to find the GCF of a set of integers. For example, to find the GCF of 36, 54, and 90 first write their prime factored forms:

$$36 = 2 \cdot 18 = 2 \cdot 2 \cdot 3 \cdot 3 = 2^2 \cdot 3^2$$

$$54 = 2 \cdot 27 = 2 \cdot 3 \cdot 3 \cdot 3 = 2 \cdot 3^3$$

$$90 = 2 \cdot 45 = 2 \cdot 3 \cdot 15 = 2 \cdot 3 \cdot 3 \cdot 5 = 2 \cdot 3^2 \cdot 5$$

Now take each prime that appears in *all* of the factored forms to the *lowest* power that it appears in any of them. In this case we use 2 and 3^2. Their product $2 \cdot 3^2 = 18$ is the GCF. It is the largest integer that is a factor of each of the given integers 36, 54, 90.

When factoring a polynomial, begin by determining the greatest common factor of the terms of the polynomial. When the terms have no common variable factors, the GCF is the largest integer that is a factor of all of the coefficients. Thus 3 is the GCF of $3x + 15$ because 3 is the largest integer factor of both 3 and 15. Now use the distributive property to obtain

$$\begin{aligned} 3x + 15 &= 3(x) + 3(5) \\ &= 3(x + 5) \end{aligned}$$

EXAMPLE 1 Factor out the GCF: $36x^2 - 54x + 90$

Solution Since there are no variable factors common to all the terms of the polynomial, we use 18, the GCF of 36, 54, and 90 that was found earlier. Thus

$$\begin{aligned} 36x^2 - 54x + 90 &= 18(2x^2) - 18(3x) + 18(5) \\ &= 18(2x^2 - 3x + 5) \end{aligned}$$ ■

When the terms have common variable factors, the GCF is the product of the greatest common factor of the coefficients times the *lowest* powers of those variables appearing in all terms, as shown in the following example.

EXAMPLE 2 Factor out the GCF: $4b^2x^3 + 12b^3x^4 - 6bx^2$

Solution The greatest common factor of the coefficients is 2. The variable b is in all terms and $b^1 = b$ is the lowest power that appears. The variable x is in all terms and x^2 is the lowest power. Therefore, GCF $= 2bx^2$.

Now use the distributive property.

$$\begin{aligned} 4b^2x^3 + 12b^3x^4 - 6bx^2 &= 2bx^2(2bx) + 2bx^2(6b^2x^2) - 2bx^2(3) \\ &= 2bx^2(2bx + 6b^2x^2 - 3) \end{aligned}$$ ■

TEST YOUR UNDERSTANDING
Think Carefully

Find the greatest common factor for each set of integers.

1. 8, 12
2. 15, 21
3. 18, 42
4. 10, 15, 25
5. 12, 24, 36
6. 30, 54, 84
7. 42, 56, 84
8. 48, 96, 120, 144

Factor out the GCF.

9. $6x + 8$
10. $3x + 36$
11. $5ax + 5ay$
12. $2x^2 + 4x$
13. $3ab^2 - 3a^2b$
14. $18a^3b^2 - 27a^2b^3$
15. $8x^4 + 12xy + 16xy^2$
16. $35a - 21a^2x^2 + 14ax$
17. $32x^2y - 24xy^2 - 12xy$

(Answers: Page 81)

From our work of the preceding section we know that the following product is correct:

$$(a - b)(a + b) = a^2 - b^2$$

With this result, it now becomes possible to take a binomial that happens to have the form $a^2 - b^2$ and write it in the *factored form* $(a + b)(a - b)$.

THE DIFFERENCE OF TWO SQUARES

$$a^2 - b^2 = (a - b)(a + b)$$

EXAMPLE 3 Factor: $x^2 - 25$

Solution

$$x^2 - 25 = x^2 - 5^2 = (x + 5)(x - 5)$$ ■

This example demonstrates that factoring the difference of squares requires ***recognition*** *that the given terms can be written as squares.*

EXAMPLE 4 Factor: $4a^2 - 9b^2$

Solution Think of $4a^2$ as $(2a)^2$ and $9b^2$ as $(3b)^2$. Then use $x^2 - y^2 = (x + y)(x - y)$, with $x = 2a$ and $y = 3b$.

$$4a^2 - 9b^2 = (2a + 3b)(2a - 3b)$$ ■

In all cases of factoring, unless indicated otherwise, we will want the **complete factored form.** That is, each of the factors is to be in a form such that no further factorizations are possible. To accomplish this, it is best to *look first for common monomial factors* before attempting any other procedure. This is illustrated in the following example.

EXAMPLE 5 Factor: $18x^4 - 8x^2y^2$

Solution

Note that the GCF, $2x^2$, is factored out first. Then $9x^2 - 4y^2$ is factored as the difference of two squares.

$$\begin{aligned} 18x^4 - 8x^2y^2 &= 2x^2(9x^2 - 4y^2) \\ &= 2x^2(3x + 2y)(3x - 2y) \end{aligned}$$ ■

In the preceding examples we factored the *difference* of two squares. What about the *sum* of two squares? How can we factor $x^2 + 25$? Verify that each of the following is correct:

$$(x + 5)(x + 5) = x^2 + 10x + 25$$
$$(x - 5)(x - 5) = x^2 - 10x + 25$$
$$(x + 5)(x - 5) = x^2 - 25$$

There does not seem to be any pair of factors whose product is $x^2 + 25$. Actually *we cannot factor the sum of two squares using coefficients that are real numbers.*

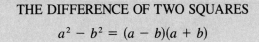

The factored forms in the preceding examples had integers as their numerical coefficients. Such coefficients are largely dependent upon the coefficients of the expression that is being factored. Consider the binomial $x^2 - 2y^2$. It is also possible to factor this as the difference of two squares, but not with integral coefficients. In order to do this, first observe that

$$2y^2 = (\sqrt{2}y)^2$$

Then

$$\begin{aligned} x^2 - 2y^2 &= x^2 - (\sqrt{2}y)^2 \\ &= (x + \sqrt{2}y)(x - \sqrt{2}y) \end{aligned}$$

Unless otherwise indicated, we will assume that only integral coefficients are to be used in factoring. Without this agreement about integer coefficients there would be an endless variety of factored forms possible. Here are just a few for $2x - 6y$.

$$\begin{aligned} 2x - 6y &= \frac{1}{2}(4x - 12y) \\ &= 4\left(\frac{1}{2}x - \frac{3}{2}y\right) \\ &= -\frac{5}{3}\left(-\frac{6}{5}x + \frac{18}{5}y\right) \end{aligned}$$

EXAMPLE 6 Factor: $a^4 - 16$

Solution Think of a^4 as $(a^2)^2$, and 16 as 4^2. Then

$$a^4 - 16 = (a^2 - 4)(a^2 + 4)$$

Since $a^2 - 4 = (a + 2)(a - 2)$, the final result is

$$a^4 - 16 = (a + 2)(a - 2)(a^2 + 4)$$ ■

TEST YOUR UNDERSTANDING Think Carefully

(Answers: Page 81)

Factor completely.

1. $x^2 - 49$	**2.** $18x^2 - 2$	**3.** $4a^2 - 25b^2$
4. $ax^2 - 81a$	**5.** $100a^2x - 81x$	**6.** $32a^2b - 50b$
7. $x^4 - 81$	**8.** $16 - x^4$	**9.** $81b^4 - 16$

Verify, by multiplication, that each of the following products is correct:

$$(x - y)(x^2 + xy + y^2) = x^3 - y^3$$
$$(x + y)(x^2 - xy + y^2) = x^3 + y^3$$

These tell us how to factor the sum or difference of two cubes:

CAUTION
Show that the following is ***in-correct.***

$a^3 - b^3 = (a - b)(a^2 + b^2)$

THE SUM AND DIFFERENCE OF TWO CUBES

$$a^3 + b^3 = (a + b)(a^2 - ab + b^2)$$
$$a^3 - b^3 = (a - b)(a^2 + ab + b^2)$$

Check the results in Examples 7 and 8 by multiplication.

EXAMPLE 7 Factor: $8a^3 - 27$

Solution Write $8a^3 = (2a)^3$ and $27 = 3^3$. Then, using the preceding form,

$$\begin{aligned} 8a^3 - 27 &= (2a)^3 - 3^3 \\ &= (2a - 3)[(2a)^2 + (2a)(3) + 3^2] \\ &= (2a - 3)(4a^2 + 6a + 9) \end{aligned}$$

■

It is very important that whenever you attempt to factor, you first look for common factors in the given terms.

EXAMPLE 8 Factor: $x^3y + 64y^4$

Solution

$$\begin{aligned} x^3y + 64y^4 &= y[x^3 + 64y^3] \\ &= y[x^3 + (4y)^3] \\ &= y(x + 4y)(x^2 - 4xy + 16y^2) \end{aligned}$$

■

EXAMPLE 9 Factor: $x^6 - 1$

Solution There are different ways that this can be done.

(a) Using the formula for the difference of two cubes, we have

$$\begin{aligned} x^6 - 1 = (x^2)^3 - 1^3 &= (x^2 - 1)(x^4 + x^2 + 1) \\ &= (x - 1)(x + 1)(x^4 + x^2 + 1) \end{aligned}$$

(b) Using the formula for the difference of two squares, we have

$$\begin{aligned} x^6 - 1 = (x^3)^2 - 1^2 &= (x^3 - 1)(x^3 + 1) \\ &= (x - 1)(x^2 + x + 1)(x + 1)(x^2 - x + 1) \end{aligned}$$

Notice that in (b) we continued to factor difference and sum of two cubes. This gives the complete factorization for $x^6 - 1$ since neither of the quadratic factors is factorable. ■

CAUTION: Learn to Avoid These Mistakes

WRONG	RIGHT
$3x + 1 = 3(x + 1)$	$3x + 1$ is not factorable by using integers.
$x^3 - y^3 = (x - y)(x^2 + y^2)$	$x^3 - y^3 = (x - y)(x^2 + xy + y^2)$
$x^3 + 8$ is not factorable.	$x^3 + 8 = (x + 2)(x^2 - 2x + 4)$
$x^2 + y^2 = (x + y)(x + y)$	$x^2 + y^2$ is not factorable by using real numbers.
$4x^2 - 9y^2 = (2x - 3y)(2x - 3y)$	$4x^2 - 9y^2 = (2x - 3y)(2x + 3y)$; factor as the difference of two squares.

EXERCISES 2.6

Factor out the GCF.

1. $5x - 5$
2. $3a - 3b$
3. $7x + 14$
4. $x^2 + 2x$
5. $ax + ay$
6. $ax^2 + bx^2$
7. $2x^2 + 4x$
8. $2x - 4x^2$
9. $2x^2 - 4x^3$
10. $x^2y + xy^2$
11. $2ab^2 - 4a^2b$
12. $3a^2b^2 + 6a^2b^3$
13. $8x^3 + 4x^2 - 4x$
14. $5a^5 - 10a^3 + 15a^2$
15. $-12x^3y + 9x^2y^2 - 6xy^3$
16. $4ab^3 - 8ab^4 + 12ab^5$
17. $ax + bx - cx$
18. $a^{10} - a^7 + a^4 + a^3$

Factor each binomial by using the formula for the difference of two squares; check your answers by multiplying.

19. $x^2 - 9$
20. $x^2 - 16$
21. $4 - x^2$
22. $49 - x^2$
23. $x^2 - 100$
24. $x^2 - 10{,}000$
25. $4x^2 - 9$
26. $x^2 - y^2$
27. $64a^2 - b^2$
28. $25x^2 - 144y^2$
29. $a^2 - 121b^2$
30. $169a^2 - 625b^2$

Use the formula for the difference of two squares to find each product.

Example: $(42)(38) = (40 + 2)(40 - 2) = 1600 - 4 = 1596$

31. (41)(39)
32. (52)(48)
33. (83)(77)
34. (57)(63)
35. (96)(104)
36. (65)(75)
37. (123)(117)
38. (252)(248)

Factor each binomial by using the formulas for the sum or diference of two cubes.

39. $x^3 - 8$
40. $x^3 + 64$
41. $x^3 - 125$
42. $27 - x^3$
43. $216 - a^3$
44. $1000 + x^3$
45. $8x^3 + 1$
46. $27a^3 + 8$
47. $27a^3 - 8$
48. $125x^3 - 64$
49. $8 - 27a^3$
50. $1331x^3 - y^3$
51. $8x^3 + 343y^3$
52. $125a^3 + 343$
53. $1 - 729b^3$

Factor (completely) each of the following.

54. $x^4 - y^4$
55. $16a^4 - b^4$
56. $81x^4 - 256y^4$
57. $a^8 - b^8$
58. $256x^8 - 1$
59. $a^{16} - 1$

Factor completely by first factoring out the GCF.

60. $8a^2 - 2b^2$
61. $ax^3 + ay^3$
62. $4ax^2 - 4a$
63. $81x^3 - 3y^3$
64. $x^3y - xy^3$
65. $5a^3 - 125a$
66. $-ab^3 + a^3b$
67. $5 - 80x^4$
68. $27a^3x^3 - 45axy^2$
69. $250c^3d - 2d^4$
70. $3x^5 - 48x$
71. $128p^5q + 2p^2q^4$

72. **(a)** Factor $a^6 - b^6$ completely by first using the formula for the difference of two squares.
 (b) Write $a^6 - b^6 = (a^2)^3 - (b^2)^3$ and factor using the formula for the difference of two cubes. Compare your result here to part (a).
 (c) Use the result observed in part (b) to factor $x^4 + 4x^2 + 16$.

73. **(a)** Factor $x^4 - y^4$ completely and then use your answer to get

$$x^4 - y^4 = (x - y)(x^3 + x^2y + xy^2 + y^3)$$

 (b) Use Exercise 72 to get a similar factored form for $x^6 - y^6$.
 (c) Record the results from parts (a) and (b) in the appropriate spaces below and try to discover the pattern that will give you a similar factored form for the remaining entries.

$$x^2 - y^2 = (x - y)(x + y)$$
$$x^3 - y^3 = (x - y)(x^2 + xy + y^2)$$
$$x^4 - y^4 =$$
$$x^5 - y^5 =$$
$$x^6 - y^6 =$$
$$x^7 - y^7 =$$
$$x^8 - y^8 =$$

Check these three answers by multiplication.

*74. Use your observation from Exercise 73 to write a general formula for a factored form for $x^n - y^n$, where n is a positive integer.

*75. Find a formula (similar in structure to the one found in Exercise 74) for $x^n + y^n$, where n is an odd positive integer.

Use rational number coefficients to factor as the difference of squares.

Example: $x^2 - \frac{9}{4} = x^2 - (\frac{3}{2})^2 = (x - \frac{3}{2})(x + \frac{3}{2})$

76. $x^2 - \dfrac{1}{25}$ 77. $\dfrac{1}{9}x^2 - 16$ 78. $\dfrac{a^2}{4} - \dfrac{b^2}{100}$ 79. $\dfrac{c^2}{36} - \dfrac{49}{d^2}$

80. The large circle has radius R and each small circle has radius a. Write the area of the shaded portion in factored form and use the result to find this area when radius $R = 15.7$ and radius $a = 3.1$.

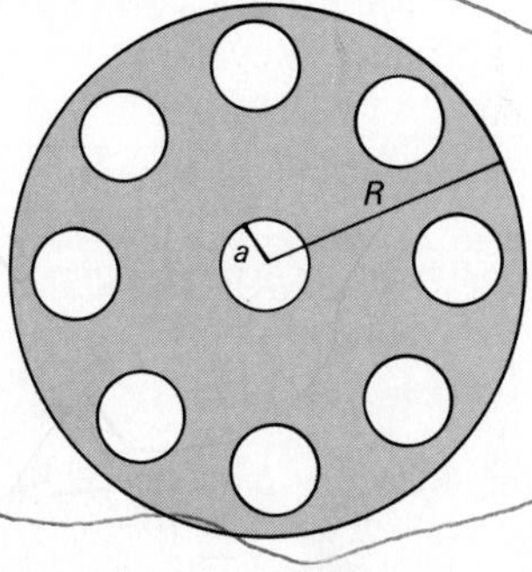

81. **(a)** If the four congruent square corners are cut from the large square, write the area of the resulting figure in factored form. Use this result to find this area when $y = 12.8$ and $x = 2.4$.

(b) Explain why the expression $(y - 2x)^2 + 4x(y - 2x)$ is also the area of the remaining part and show that it is equivalent to the result in part (a).

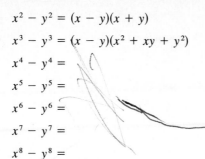

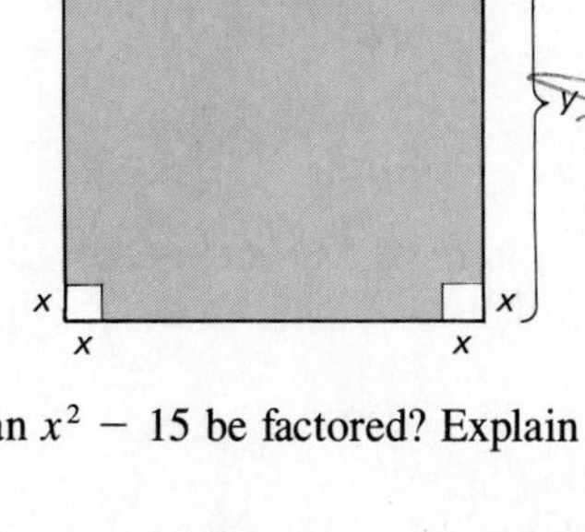

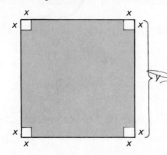

Written Assignment: Can $x^2 - 15$ be factored? Explain your answer.

CHALLENGE
Think Creatively

A *Fibonnaci sequence* is one where each term, beginning with the third term, is the sum of the two preceding terms. For example:

$$1, 1, 2, 3, 5, 8, 13, 21, 34, 55, 89, 144, \ldots$$

$$4, 6, 10, 16, 26, 42, 68, 110, 178, 288, 466, \ldots$$

Now for the sequences above, or for any that you may choose to write, the sum of the first 10 terms is equal to 11 times the seventh term. (Try it.) Thus you can *predict* the sum of the first 10 terms after having written only 7 of the terms! Now try to *prove* that this will always be the case by writing a general Fibonnaci sequence of 10 terms:

$$a, b, a + b, a + 2b, 2a + 3b, \ldots, 21a + 34b$$

Find the sum, and show that it is equal to 11 times the seventh term.

2.7 FACTORING TRINOMIALS

By multiplication we have already established the following products:

$$(a + b)^2 = (a + b)(a + b) = a^2 + 2ab + b^2$$

$$(a - b)^2 = (a - b)(a - b) = a^2 - 2ab + b^2$$

*A **trinomial** is a polynomial of three terms.*

Each product is called a *perfect square trinomial,* that is, each is the square of a binomial. In each case the first and last terms are squares of a and b, respectively, and the middle term is twice their product. Reversing the procedure gives us two more general factoring forms.

Observe that in $a^2 \pm 2ab + b^2$ the middle term (ignoring signs) is twice the product of the square roots of the end terms. Hence the factored form is the square of the sum (or difference) of these square roots.

PERFECT SQUARE TRINOMIALS

$$a^2 + 2ab + b^2 = (a + b)^2$$

$$a^2 - 2ab + b^2 = (a - b)^2$$

EXAMPLE 1 Factor: $a^2 + 2a + 1$

Solution Since a^2 is the square of a, 1 is the square of 1, and $2a$ is twice the product of a and 1, $a^2 + 2a + 1$ is a perfect square trinomial. Thus

$$a^2 + 2a + 1 = a^2 + 2(a)(1) + (1)^2$$

$$= (a + 1)^2$$

■

CAUTION
*Some students will try to factor a trinomial such as $4x^2 - 6xy + 9y^2$ as a perfect square trinomial. This won't work because $6xy \neq 2(2x)(3y)$. So $4x^2 - 6xy + 9y^2$ is **not** a perfect square trinomial.*

EXAMPLE 2 Factor: $25s^2 - 40st + 16t^2$

Solution Since $25s^2$ is the square of $5s$, $16t^2$ is the square of $4t$, and $40st$ is twice the product of $5s$ and $4t$, we have

$$25s^2 - 40st + 16t^2 = (5s)^2 - 2(5s)(4t) + (4t)^2$$
$$= (5s - 4t)^2$$

■

Remember always to search for a common monomial factor first. In this case the GCF = x. After factoring this out, the other factor is a perfect square trinomial.

EXAMPLE 3 Factor: $x^3 + 14x^2 + 49x$

Solution

$$x^3 + 14x^2 + 49x = x(x^2 + 14x + 49)$$
$$= x(x + 7)^2$$

The result can be checked by working backward and multiplying to obtain the original trinomial. ■

Another factoring technique that we will consider deals with trinomials that are not necessarily perfect squares. For example, let us factor $x^2 + 7x + 12$. From our experience with multiplying binomials we can anticipate that the factors will be of this form:

$$x^2 + 7x + 12 = (x + \underline{\ ?\ })(x + \underline{\ ?\ })$$

We need to fill in the blanks with two integers whose product is 12. Furthermore, the middle term of the product must be $+7x$. The possible choices for the two integers are

12 and 1   6 and 2   4 and 3

To find the correct pair is now a matter of trial and error. These are the possible factorizations:

$$(x + 12)(x + 1)$$
$$(x + 6)(x + 2)$$
$$(x + 4)(x + 3)$$

Only the last form gives the correct middle term of $7x$. Therefore, we conclude that $x^2 + 7x + 12 = (x + 4)(x + 3)$.

EXAMPLE 4 Factor: $x^2 - 10x + 24$

Solution The final term, +24 must be the product of two positive numbers or two negative numbers. (Why?) Since the middle term, $-10x$, has a minus sign, the factorization must be of this form:

$$x^2 - 10x + 24 = (x - \underline{\ ?\ })(x - \underline{\ ?\ })$$

Now try all the pairs of integers whose product is 24:

1 and 24   2 and 12   3 and 8   4 and 6

Since the sum of the pair must equal 10, the integers are 4 and 6. Thus

$$x^2 - 10x + 24 = (x - 4)(x - 6)$$

Check this result by multiplication. ■

EXAMPLE 5 Factor: $18x^2 + 11x + 1$

Solution Since 1 and 1 is the only pair of positive integers whose product is 1, we begin with this form:

$$18x^2 + 11x + 1 = (\_\_\_\_x + 1)(\_\_\_\_x + 1)$$

The blanks must be filled by two positive integers whose product is 18 and whose sum is 11. Those pairs whose product is 18 are:

1 and 18   2 and 9   3 and 6

Therefore, the required integers are 2 and 9 because their sum is 11. Thus

$$18x^2 + 11x + 1 = (2x + 1)(9x + 1)$$ ■

TEST YOUR UNDERSTANDING
Think Carefully

Factor the following trinomial squares.

1. $a^2 + 6a + 9$
2. $x^2 - 10x + 25$
3. $4x^2 + 12xy + 9y^2$
4. $9a^2 - 6ab + b^2$
5. $25a^2 - 70ab + 49b^2$
6. $49h^2 + 28hk + 4k^2$

Find positive integers p and q having the given product and sum.

7. $pq = 16 \quad p + q = 10$
8. $pq = 16 \quad p + q = 17$
9. $pq = 24 \quad p + q = 14$
10. $pq = 42 \quad p + q = 17$

Factor each trinomial.

11. $a^2 + 8a + 15$
12. $x^2 - 12x + 20$
13. $x^2 - 25x + 100$
14. $36 + 15y + y^2$
15. $45a^2 - 18a + 1$
16. $1 + 16b + 28b^2$

(Answers: Page 81)

Let us now consider a more complicated factoring problem. If we wish to factor the trinomial $15x^2 + 43x + 8$, we need to consider possible factors both of 15 and of 8. Because of the "+" signs in the trinomial, the factorization will be of this form:

$$15x^2 + 43x + 8 = (\underline{\ ?\ }x + \underline{\ ?\ })(\underline{\ ?\ }x + \underline{\ ?\ })$$

Here are the different possibilities for factoring 15 and 8:

$$15 = 15 \cdot 1 \qquad 15 = 5 \cdot 3$$
$$8 = 8 \cdot 1 \qquad 8 = 4 \cdot 2$$

Using $15 \cdot 1$, write the form

$$(15x + \underline{\ ?\ })(x + \underline{\ ?\ })$$

Try 8 and 1 in the blanks, *both ways,* namely:

$$(15x + 8)(x + 1) \longrightarrow \text{middle term: } 23x$$
$$(15x + 1)(x + 8) \longrightarrow \text{middle term: } 121x$$

*With a little luck, and much more experience, you can often avoid exhausting **all** the possibilities before finding the correct factors. You will then find that such work can often be shortened significantly.*

Neither gives a middle term of $43x$; so now try 4 and 2 in the blanks both ways. Again you see that neither case works. Next consider the form

$$(5x + \underline{\ ?\ })(3x + \underline{\ ?\ })$$

Once again try 4 and 2 both ways, and 8 with 1 both ways. Here is the correct answer:

$$15x^2 + 43x + 8 = (5x + 1)(3x + 8)$$

Factoring $15x^2 - 43x + 8$ is very similar to factoring $15x^2 + 43x + 8$. The only difference is that both binomial factors have minus signs instead of plus signs; that is,

$$15x^2 - 43x + 8 = (5x - 1)(3x - 8)$$

Example 6 shows us that not every trinomial can be factored.

EXAMPLE 6 Factor: $12x^2 - 9x + 2$

Solution Consider the forms

$$(12x - \underline{\ ?\ })(x - \underline{\ ?\ })$$
$$(6x - \underline{\ ?\ })(2x - \underline{\ ?\ })$$
$$(4x - \underline{\ ?\ })(3x - \underline{\ ?\ })$$

In each form try 2 with 1 both ways. None of these produces a middle term of $-9x$; hence we say that $12x^2 - 9x + 2$ is *not factorable* with integral coefficients. ■

In all of the examples we have studied thus far, the constant term has been positive. When this term is negative, you may find it easier to leave out the signs in the two binomial forms as you try various cases. Thus, to factor $x^2 - 8x - 20$ we begin with this form:

$$x^2 - 8x - 20 = (x \quad \underline{\ ?\ })(x \quad \underline{\ ?\ })$$

Next fill in the blanks with the two positive integers 10 and 2 whose product is 20 and whose difference is 8. Thus

$$x^2 - 8x - 20 = (x \quad 10)(x \quad 2)$$

Finally, we fill in the signs such that the middle term of the product will be $-8x$.

$$x^2 - 8x - 20 = (x - 10)(x + 2)$$

EXAMPLE 7 Factor: $5x^2 - 13x - 6$

Solution Since 5 and 1 is the only pair of positive integer factors of 5, begin with this form:

$$5x^2 - 13x - 6 = (5x \quad \_\_\_\_)(x \quad \_\_\_\_)$$

The blanks must be filled with pairs of positive integers whose product is 6. The possibilities are:

1 and 6 2 and 3

But because the *x*-terms in the binomials are not the same, these possibilities need to be tried *both* ways. Here is a list of the resulting forms; next to each is the *difference* of the outer and inner products without regard to sign

CAUTION
Be careful when inserting the signs. In this problem the form $(5x - 2)(x + 3)$ *is wrong because the middle term obtained is* $+13x$ *rather than* $-13x$.

		Middle term without regard to sign
$(5x \quad 1)(x \quad 6)$	$\longrightarrow$	$29x$
$(5x \quad 6)(x \quad 1)$	$\longrightarrow$	x
$(5x \quad 2)(x \quad 3)$	$\longrightarrow$	$13x$
$(5x \quad 3)(x \quad 2)$	$\longrightarrow$	$7x$

Since the difference of the outer and inner products is $13x$ in the third case, we insert the signs to obtain $-13x$ as the middle term. Thus

$$(5x + 2)(x - 3) = 5x^2 - 13x - 6$$ ■

EXAMPLE 8 Factor: $15x^2 + 7x - 8$

Solution Try these forms:

$$(5x \quad \underline{\ ?\ })(3x \quad \underline{\ ?\ })$$
$$(15x \quad \underline{\ ?\ })(x \quad \underline{\ ?\ })$$

In each form try 4 with 2 and 8 with 1 both ways. The factored form is $(15x - 8)(x + 1)$. ■

EXAMPLE 9 Factor: $21x^2 + 183x - 54$

Solution First factor out the GCF, and then factor the resulting trinomial.

$$\begin{aligned} 21x^2 + 183x - 54 &= 3(7x^2 + 61x - 18) \\ &= 3(7x - 2)(x + 9) \end{aligned}$$ ■

EXERCISES 2.7

Factor each of the following perfect square trinomials.

1. $x^2 + 4x + 4$
2. $x^2 - 8x + 16$
3. $a^2 - 14a + 49$
4. $r^2 - 2r + 1$
5. $1 + 2b + b^2$
6. $100 - 20x + x^2$
7. $4a^2 + 8a + 4$
8. $4a^2 - 8a + 4$
9. $9x^2 - 18xy + 9y^2$
10. $64a^2 + 64a + 16$

Factor each trinomial.

11. $x^2 + 5x + 6$
12. $x^2 - 7x + 10$
13. $x^2 + 20x + 51$
14. $12a^2 - 13a + 1$
15. $20a^2 - 9a + 1$
16. $4 - 5b + b^2$
17. $x^2 + 20x + 36$
18. $a^2 - 24a + 63$
19. $9x^2 + 6x + 1$
20. $x^2 - 20x + 64$
21. $25a^2 - 10a + 1$
22. $8x^2 + 14x + 3$
23. $3x^2 + 20x + 12$
24. $5x^2 + 31x + 6$
25. $14x^2 + 37x + 5$
26. $9x^2 - 18x + 5$
27. $8x^2 - 9x + 1$
28. $30a^2 - 17a + 1$
29. $4a^2 + 20a + 25$
30. $a^2 - 9a + 18$
31. $b^2 + 18b + 45$
32. $6x^2 + 12x + 6$
33. $8x^2 - 16x + 6$
34. $12x^2 + 92x + 15$
35. $18t^2 - 67t + 14$
36. $12a^2 - 25a + 12$
37. $15x^2 - 7x - 2$
38. $15x^2 + 7x - 2$
39. $6a^2 + 5a - 21$
40. $6a^2 - 5a - 21$
41. $4x^2 + 4x - 3$
42. $15x^2 + 19x - 56$
43. $12b^2 - 34b - 6$
44. $10x^2 - 7xy - 12y^2$
45. $24a^2 + 25ab + 6b^2$

Factor each trinomial when possible. When appropriate, first factor out the common monomial.

46. $3a^2 + 6a + 3$
47. $5x^2 + 25x + 20$
48. $18x^2 - 24x + 8$
49. $4ax^2 + 4ax + a$
50. $x^2 + x + 1$
51. $a^2 - 2a + 2$
52. $49r^2s - 42rs + 9s$
53. $6x^2 + 2x - 20$
54. $50a^2 - 440a - 90$
55. $6a^2 + 4a - 9$
56. $36x^2 - 96x + 64$
57. $2x^2 - 2x - 112$
58. $15 + 5y - 10y^2$
59. $2b^2 + 12b + 16$
60. $4a^2x^2 - 4abx^2 + b^2x^2$
61. $a^3b - 2a^2b^2 + ab^3$
62. $12x^2y + 22xy^2 - 60y^3$
63. $16x^2 - 24x + 8$
64. $16x^2 - 24x - 8$
65. $25a^2 + 50ab + 25b^2$

Factor each trinomial by first factoring out the negative of the GCF.

Example: $-8x^2 + 2x + 3 = -1(8x^2 - 2x - 3)$
$= -(4x - 3)(2x + 1)$

66. $-x^2 + 9x - 18$
67. $-30x^2 + x + 3$
68. $-4a^3 - 23a^2 + 6a$
69. $-12x^3y + 12x^2y^2 - 3xy^3$

Factor out a fraction so that the terms of the resulting trinomial factor have integer coefficients. Then complete the factoring.

Example: $x^2 - \frac{7}{6}x + \frac{1}{3} = \frac{1}{6}(6x^2 - 7x + 2)$
$= \frac{1}{6}(3x - 2)(2x - 1)$

70. $\frac{1}{2}x^2 - 3x - 8$
71. $\frac{1}{4}x^2 + x - 8$
72. $x^2 - \frac{3}{4}x - \frac{1}{4}$
73. $\frac{1}{3}x^2 + \frac{1}{3}x - \frac{1}{4}$

CHALLENGE
Think Creatively

Here is a "proof" that 1 = 2! See if you can find the error.

Let $a = b$

Then $a \cdot a = a \cdot b$ (Multiply each side by a.)

So $a^2 = ab$

Then $a^2 - b^2 = ab - b^2$ (Subtract b^2 from each side.)

$(a - b)(a + b) = b(a - b)$ (By factoring)

Thus $a + b = b$ (Divide each side by $a - b$.)

And $b + b = b$ (Since $a = b$)

$2b = b$

$2 = 1$ (Divide each side by b.)

EXPLORATIONS
Think Critically

1. Explain why for any real number b, $b \neq 0$, b^0 is *defined* to equal 1.
2. When factoring polynomials it is useful to first factor out the GCF. What is the advantage in doing this?
3. The least common multiple (LCM) of two positive integers is the smallest number that is a multiple of each number. Show that the product GCF × LCM for two numbers, such as 24 and 40, is equal to the product of the two numbers. Explain why this is so.
4. A *prime triplet* is a set of three prime numbers P_1, P_2, and P_3 such that P_2 is 2 more than P_1 and P_3 is 2 more than P_2. Find a prime triplet and explain why there is only one such triplet.
5. A geometric interpretation for the expansion of $(a + b)^2$ is shown on page 57. Find a geometric interpretation for the expansion of $(a - b)^2$.
6. The trinomial $x^4 + x^2 + 1$ can be written as the difference of two squares. This is done by first writing $x^4 + x^2 + 1 = (x^4 + 2x^2 + 1) - x^2$. Convert this form into the difference of two squares and obtain a factored form consisting of two trinomial factors. Use this technique to factor each of the following:
(a) $4x^4 + 11x^2 + 9$ **(b)** $x^4 - 3x^2y^2 + y^4$

2.8 USING VARIOUS FACTORING METHODS

Sometimes an expression containing four terms does not appear to be factorable, such as the following:

$$a^2 - 6b + ab - 6a$$

However, this expression can be factored by first **regrouping** the terms. Since this expression has no common monomial factor in all of its terms, we try to regroup the four terms into two pairs of binomials such that each binomial has its own common factor. With this in mind, group a^2 and $-6a$ together, as well as ab and $-6b$.

$$\begin{aligned} a^2 - 6b + ab - 6a &= (a^2 - 6a) + (ab - 6b) \\ &= a(a - 6) + b(a - 6) \end{aligned}$$

At this point, note that $(a - 6)$ is a common factor of both terms. Thus we may use the distributive property to factor out this binomial:

$$a(a - 6) + b(a - 6) = (a + b)(a - 6)$$

In brief, the work condenses to:

$$\begin{aligned} a^2 - 6b + ab - 6a &= a^2 - 6a + ab - 6b \\ &= a(a - 6) + b(a - 6) \\ &= (a + b)(a - 6) \end{aligned}$$

A temporary substitution may be helpful when factoring out a binomial. In this case let $x = a - 6$; then

$$\begin{aligned} a(a - 6) + b(a - 6) &= ax + bx \\ &= (a + b)x \\ &= (a + b)(a - 6) \end{aligned}$$

EXAMPLE 1 Factor $a^2 - 6b + ab - 6a$ using a different grouping than that shown above.

Solution

$$\begin{aligned} a^2 - 6b + ab - 6a &= a^2 + ab - 6a - 6b \\ &= a(a + b) - 6(a + b) \\ &= (a - 6)(a + b) \end{aligned}$$ ■

CAUTION
*When factoring by grouping do not stop prematurely. Note that in this Example, $4x(2y - x) + 5(2y - x)$ is **not** a factored form because of the plus sign between the terms.*

EXAMPLE 2 Factor $8xy - 4x^2 - 5x + 10y$ by grouping.

Solution

$$\begin{aligned} 8xy - 4x^2 - 5x + 10y &= 4x(2y - x) - 5(x - 2y) \\ &= 4x(2y - x) + 5(2y - x) \\ &= (4x + 5)(2y - x) \end{aligned}$$ ■

TEST YOUR UNDERSTANDING
Think Carefully

Factor out the common binomial.

1. $2x(y - 8) + y(y - 8)$ **2.** $(a^2 + b)a - (a^2 + b)b$

Factor by grouping.

3. $xy - 4y + 7x - 28$ **4.** $5x - y^2 + xy - 5y$
5. $2a^2 + 3ab + 18a + 27b$ **6.** $35pq - 7q - 20p + 4$
7. $x^3 - 12y - 2x^2y + 6x$ **8.** $2r^2 - 2rs^2 + r - s^2$

(Answers: Page 81)

At times we can factor by grouping, and then factor again to obtain the complete factored form. This is illustrated in Example 3.

Not all groupings are productive. Thus the grouping

$$(x^3 - 18) + (2x^2 - 9)$$

does not lead to the factored form.

EXAMPLE 3 Factor: $x^3 + 2x^2 - 9x - 18$

Solution Since this is not a trinomial, we attempt to factor by grouping.

$$\begin{aligned} x^3 + 2x^2 - 9x - 18 &= x^2(x + 2) - 9(x + 2) \\ &= (x + 2)(x^2 - 9) \end{aligned}$$

Now observe that $x^2 - 9$ can be factored as the difference of squares. Therefore, the complete factored form is

$$x^3 + 2x^2 - 9x - 18 = (x + 2)(x + 3)(x - 3)$$

■

As you studied the lessons on factoring you were usually directed to factor expressions by using certain methods. However, in future mathematical work you will encounter problems in which factoring is needed, but there will not be any directions given as to what methods to use. You will then need to decide for yourself which methods are appropriate. It is therefore worthwhile to review the basic procedures that we have used in this chapter.

This table illustrates the various techniques for factoring that have been presented in this chapter.

Procedure	Illustration
Greatest common factor	$6x^3y - 8x^2y^3 = 2x^2y(3x - 4y^2)$
Difference of squares	$16x^2 - 25y^2 = (4x - 5y)(4x + 5y)$
Difference of cubes	$x^3 - 27 = (x - 3)(x^2 + 3x + 9)$
Sum of cubes	$x^3 + 27 = (x + 3)(x^2 - 3x + 9)$
Perfect square trinomials	$4x^2 - 12x + 9 = (2x - 3)^2$
Trinomials	$4x^2 - 5x - 6 = (4x + 3)(x - 2)$
Grouping	$x^3 - 2x^2 + 3x - 6 = (x^2 + 3)(x - 2)$

EXAMPLE 4 Factor: $50a^3 - 40a^2b + 8ab^2$

Solution First look for common monomial factors.

$$50a^3 - 40a^2b + 8ab^2 = 2a(25a^2 - 20ab + 4b^2)$$

Now observe that the second factor is a trinomial and test to see if it can be factored. Since it proves to be a perfect square trinomial, the complete factored form is

$$50a^3 - 40a^2b + 8ab^2 = 2a(5a - 2b)(5a - 2b) \quad \text{or} \quad 2a(5a - 2b)^2$$

■

EXAMPLE 5 Factor: $2x^4 - x^3y + 54xy^3 - 27y^4$

Solution First look for common monomial factors. There are none, so try grouping.

$$2x^4 - x^3y + 54xy^3 - 27y^4 = x^3(2x - y) + 27y^3(2x - y)$$
$$= (x^3 + 27y^3)(2x - y)$$

The first factor is the sum of two cubes. Thus

$$2x^4 - x^3y + 54xy^3 - 27y^4 = (x + 3y)(x^2 - 3xy + 9y^2)(2x - y)$$ ■

EXAMPLE 6 Factor: $x^4 - x^2 - 12$

Solution There are no common factors. Since the degree is more than 2, it does not appear to be a factorable trinomial. However, noting that $x^4 = (x^2)^2$ makes it possible to rewrite the expression by letting $u = x^2$ and substituting to obtain $u^2 - u - 12$. Since this is a factorable trinomial in u, we proceed by factoring in terms of u. Then replace u by x^2, and factor further if possible. Here is the complete solution:

Remember that any factoring problem can be checked by multiplying. Check this answer.

$$\begin{aligned} x^4 - x^2 - 12 &= u^2 - u - 12 \qquad (u = x^2) \\ &= (u - 4)(u + 3) \\ &= (x^2 - 4)(x^2 + 3) \\ &= (x - 2)(x + 2)(x^2 + 3) \end{aligned}$$ ■

EXERCISES 2.8

Factor out the common binomial factor.

1. $x(y - 2) + 3(y - 2)$
2. $a^3(b + 5) - b(b + 5)$
3. $(2h - 3)7 - (2h - 3)3h$
4. $(x + 3y)x^2 + (x + 3y)y^3$

Factor completely by first factoring out the common factors.

5. $(2a + b)a^2 - (2a + b)b^2$
6. $3(a + 1)x^3 + 24(a + 1)$
7. $(x - 1)a^3 - (x - 1)ab^2$
8. $x^2(x + y) + 4x(x + y) + 4(x + y)$

Factor each of the following by grouping.

9. $a^2 - 2b + 2a - ab$
10. $x^2 - y - x + xy$
11. $x + 1 + y + xy$
12. $-y - x + 1 + xy$
13. $ax + by + ay + bx$
14. $a^2b + 3a + ab^2 + 3b$
15. $2 - y^2 + 2x - xy^2$
16. $y^2 + 2y^3 + 2 + 4y$
17. $ax^3 - by^3 + axy^2 - bx^2y$
18. $21x^2 + 20y + 15xy + 28x$

Factor.

19. $16x^3y + 4x^2y - 42xy$
20. $3s^2 - 30s^2t + 75s^2t^2$
21. $40ab^3 - 5a^4$
22. $x^6 + x^2y^4 - x^4y^2 - y^6$
23. $a^3x - b^3y + b^3x - a^3y$
24. $a^4 + a^3 - ab^3 - b^3$
25. $81 - 24b^3$
26. $10x^4 - 100{,}000$
27. $7a^2 - 35b + 35a - 7ab$
28. $x^5 - 16xy^4 - 2x^4y + 32y^5$
29. $a^6 - 2a^3 + 1$
30. $x^4 - 2x^2 + 1$
31. $6x^5y - 3x^3y^2 - 30xy^3$
32. $2x^4 + 8x^2 - 42$
33. $a^6 - 2a^3b^3 + b^6$
34. $3x^4 + 6x^2 + 3$
35. $2c^4 - 8d^7 + 16cd^6 - c^3d$
36. $a^3b^3 + b^3 - a^3 - 1$
37. $(x + 5)^2 + 5(x + 5) - 14$
38. $(a^2 - 2ab + b^2) - 9(a - b) + 18$

*39. $a^4 + 3a^3 + 3a^2 + a$

*40. $5x^3y - 15x^2y^2 + 15xy^3 - 5y^4$

CHAPTER 2 SUMMARY

Review these key terms and concepts so that you are able to define or describe them. A clear understanding of these will be very helpful when reviewing the developments of this chapter.

Fundamental Rules for Exponents:

$$b^m \cdot b^n = b^{m+n} \qquad (b^m)^n = b^{mn} \qquad \frac{a^m}{b^m} = \left(\frac{a}{b}\right)^m$$

$$\frac{b^m}{b^n} = b^{m-n} \qquad a^m b^m = (ab)^m$$

A **zero exponent** is defined as $b^0 = 1$.

Also, $b^{-n} = \dfrac{1}{b^n}$ defines a **negative exponent.**

A number N is in **scientific notation** if it expressed as the product of a number between 1 and 10 and an integral power of 10:

$$2{,}070{,}000 = 2.07 \times 10^6 \qquad 0.00000084 = 8.4 \times 10^{-7}$$

Polynomials can be added and subtracted by adding or subtracting the *like terms* of each. To multiply two binomials, we can think of the letters of the word *FOIL:*

$$(5x - 9)(6x + 2) = 30x^2 - 44x - 18$$

First terms: $(5x)(6x) = 30x^2$
Outer terms: $(5x)(2) = 10x$
Inner terms: $(-9)(6x) = -54x$ $\quad$ $(10x - 54x = -44x)$
Last terms: $(-9)(2) = -18$

Special Products and Factoring Forms:

$$a^2 - b^2 = (a - b)(a + b)$$

$$a^3 + b^3 = (a + b)(a^2 - ab + b^2)$$

$$a^3 - b^3 = (a - b)(a^2 + ab + b^2)$$

$$a^2 + 2ab + b^2 = (a + b)^2$$

$$a^2 - 2ab + b^2 = (a - b)^2$$

Always search for a *common monomial factor* first. Also, consider *factoring by grouping.*

$$\begin{aligned} 2ax^3 + 4ax^2 - 18ax - 36a &= 2a(x^3 + 2x^2 - 9x - 18) \\ &= 2a[x^2(x + 2) - 9(x + 2)] \\ &= 2a(x + 2)(x^2 - 9) \\ &= 2a(x + 2)(x - 3)(x + 3) \end{aligned}$$

REVIEW EXERCISES

The solutions to the following exercises can be found within the text of Chapter 2. Try to answer each question before referring to the text.

Section 2.1

Simplify.

1. $12^3(\frac{1}{6})^3$ 2. $\left(-\frac{2}{3}\right)^5\left(\frac{9}{4}\right)^5$ 3. $\frac{(x^3y)^2y^3}{x^4y^6}$ 4. $\frac{4^5}{8^3}$

5. Simplify: $\left(\frac{x^3}{y^0}\right)\left(\frac{x}{y}\right)^0$

6. Suppose that a radioactive substance decays so that $\frac{1}{2}$ the amount remains after each hour. If at a certain time there were 320 grams of the substance, how much will remain after 8 hours? How much after n hours?

7. Explain the motivation behind the definition $b^0 = 1$.

Section 2.2

Simplify.

8. 3^{-2} 9. $\frac{1}{3^{-2}}$ 10. $(-3)^{-3}$ 11. $\left(\frac{1}{3}\right)^{-2}$

12. x^3y^{-5} 13. $\frac{a^4b^{-3}}{a^{-2}b^2}$ 14. $\left(\frac{a^{-2}b^3}{a^3b^{-2}}\right)^5$

15. Write in scientific notation: **(a)** 2,070,000 **(b)** 0.00000084

16. Write in standard notation: **(a)** 1.21×10^4 **(b)** 1.21×10^{-2}

17. Use scientific notation to compute: $\frac{1}{800{,}000}$

18. Use scientific notation to evaluate: $\frac{(2{,}310{,}000)^2}{(11{,}200{,}000)(0.000825)}$

Section 2.3

19. Add $5x^3 + x^2 + 7x + 3$ and $3x^2 - 4x + 5$.
20. Simplify: $(a^2 - 2a) + (3a^2 + 3a + 7) + (5a^2 - 1)$
21. Add these three polynomials:

$$-2x^4 + 7x^2y + xy^3 - y; \qquad 2x^4 + 17x^2y - 2xy^3 - 5y + 4; \qquad 4x^4 + 5xy^3 + 2y - 10$$

22. Subtract $3x^2 - 4x + 5$ from $5x^3 + x^2 + 7x + 3$.
23. Subtract $a^2 - 5a - 2$ from $2a + 3$.
24. Simplify: $(x - 3y) - (5x + y) - (3y - 2x)$
25. Simplify: $[(2x - x^2) - (x^3 - 5x^2 + 8x)] - [(9 - 2x^3 + 4x^2) - (6x - x^2)]$
26. Find $p(3)$ for the polynomial $p(x) = -2x^2 + 3x + 4$.

Section 2.4

27. Multiply: $(-2x^2y^3)(3x^4y)$
28. Multiply: $3x^2(4x^7 - 3x^4 - x^2 + 15)$
29. Multiply: $(x + 2)(x^3 + 7x^2 - 4)$
30. Multiply $3x^3 - 8x + 4$ by $2x^2 + 5x - 1$.

31. Multiply $x^3 + ax^2 + a^2x + a^3$ by $x - a$.
32. Simplify: $3x[(2x - 1)(x^2 + 5x - 4) - 2(x - 3)(x^2 + 2x + 2)]$

Section 2.5

33. Find the product: $(5x - 9)(6x + 2)$
34. Multiply $ax + b$ by $cx + d$.
35. Find the product: **(a)** $(a - b)(a + b)$ **(b)** $(a - b)(a - b)$
36. Find the product: $(x + 2)(x + 3)(x + 4)$
37. Perform the indicated operations and simplify: $(x^2 - 5x)(3x^2) + (x^3 - 1)(2x - 5)$
38. Expand: $(a + b)^3$

Section 2.6

Factor out the GCF.

39. $36x^2 - 54x + 90$
40. $4b^2x^3 + 12b^3x^4 - 6bx^2$

Factor.

41. $x^2 - 25$
42. $18x^4 - 8x^2y^2$
43. $a^4 - 16$
44. $8a^3 - 27$
45. $x^3y + 64y^4$
46. $x^6 - 1$

Section 2.7

Factor, if possible.

47. $a^2 + 2a + 1$
48. $25s^2 - 40st + 16t^2$
49. $x^3 + 14x^2 + 49x$
50. $x^2 - 10x + 24$
51. $18x^2 + 11x + 1$
52. $15x^2 + 43x + 8$
53. $12x^2 - 9x + 2$
54. $x^2 - 8x - 20$
55. $5x^2 - 13x - 6$
56. $15x^2 + 7x - 8$
57. $21x^2 + 183x - 54$

Section 2.8

Factor.

58. $a^2 - 6b + ab - 6a$
59. $8xy - 4x^2 - 5x + 10y$
60. $x^3 + 2x^2 - 9x - 18$
61. $50a^3 - 40a^2b + 8b^2$
62. $2x^4 - x^3y + 54xy^3 - 27y^4$
63. $x^4 - x^2 - 12$

CHAPTER 2 TEST: STANDARD ANSWER

Use these questions to test your knowledge of the basic skills and concepts of Chapter 2. Then check your answers with those given at the back of the book.

1. Classify each statement as true or false.
   (a) $3^5 \cdot 3^2 = 3^{10}$ **(b)** $(3^5)^2 = 3^{10}$ **(c)** $(3^0)^2 = 1$
   (d) $\dfrac{x^3(-x)^2}{x^5} = x$ **(e)** $\left(\dfrac{x^6}{x^2}\right)^2 = x^9$

Evaluate.

2. $(-2)^3(-5)^2$
3. $(\frac{3}{4})^0 + (\frac{3}{4})^2$
4. $\dfrac{(-2)^3(-2)^4}{(-2)^2}$

Simplify.

5. $(-3a^3b^0)^2$
6. $(2a)^5(\frac{1}{2}a)^3$
7. $\left(\dfrac{a^3b}{ab^3}\right)^2$

Simplify. Express the answers using positive exponents.

8. $\dfrac{a^{-3}b^2}{a^{-4}b^{-3}}$ 9. $(3x^{-2}y^3)^0$ 10. $\dfrac{(2x^3y^{-2})^2}{x^{-2}y^3}$ 11. $\dfrac{(3x^2y^{-3})^{-1}}{(2x^{-2}y^2)^{-2}}$

12. Write each number in scientific notation: **(a)** 23,700 **(b)** 0.00508

13. Write each number in standard form: **(a)** 5.63×10^5 **(b)** 8.06×10^{-7}

Simplify by performing the indicated operations.

14. $(2x^2 - 5x + 3) + (x^2 + 7x - 5)$
15. $(5x^2 + 2x - 7) - (x^2 - 3x + 9)$
16. $(-2x^2)(x^3 + x^2 - 3x - 2)$
17. $(3x - 2)(5x + 3)$
18. $(2x + 3)(x^2 - 3x + 5)$
19. $(x + 2)(2x - 1) - (x + 3)(x - 3)$

Factor.

20. $49x^2 - 4y^2$
21. $8x^3 + 27$
22. $15x^3y^2 - 10x^2y^4 + 5xy^3$
23. $a^3 + 7a - a^2b - 7b$
24. $12a^2 - 2ab - 4b^2$
25. $128x^4 - 2xy^3$

CHAPTER 2 TEST: MULTIPLE CHOICE

1. Which of the following are true?

   I. $(-3)^2 = -3^2$ **II.** $\dfrac{x^5}{y^3} = \left(\dfrac{x}{y}\right)^2$ **III.** $\dfrac{5^{-2}}{12^{-2}} = \left(\dfrac{5}{12}\right)^{-2}$

   (a) Only I **(b)** Only II **(c)** Only III **(d)** I, II, and III **(e)** None of the preceding

2. Which of the following are true?

   I. $x^2 \cdot x^4 = x^8$ **II.** $\dfrac{x^{12}}{x^3} = x^4$ **III.** $x^{-5} = -\dfrac{1}{x^5}$

   (a) Only I **(b)** Only II **(c)** Only III **(d)** Only I and II **(e)** None of the preceding

3. Which of the following is equivalent to $(a + b)^{-1}$?

   (a) $a^{-1} + b^{-1}$ **(b)** $\dfrac{1}{a + b}$ **(c)** $(-a) + (-b)$ **(d)** $\dfrac{1}{a} + \dfrac{1}{b}$ **(e)** $\dfrac{1}{a - b}$

4. Simplify: $\dfrac{(x^3y^{-2})^2}{(x^{-2}y^0)^{-3}}$

   (a) $\dfrac{1}{y^4}$ **(b)** x^{10} **(c)** y^3 **(d)** $\dfrac{x^{12}}{y^4}$ **(e)** None of the preceding

5. Using scientific notation the expression $\dfrac{(0.00054)(9200)}{370}$ can be rewritten as

   (a) $\dfrac{(5.4)(9.2)}{3.7} \times 10^3$ **(b)** $\dfrac{(5.4)(9.2)}{3.7} \times 10^{-3}$ **(c)** $\dfrac{(5.4)(9.2)}{3.7} \times 10^2$

   (d) $\dfrac{(5.4)(9.2)}{3.7} \times 10^{-2}$ **(e)** None of the preceding

6. Simplify: $12 - [7 - (x - 5)] - x$

   (a) $-2x$ **(b)** $10 - 2x$ **(c)** $10 + 2x$ **(d)** 0 **(e)** 10

7. Multiply: $(2x - 3)(3x - 2)$

   (a) $6x^2 - 6$ **(b)** $6x^2 + 6$ **(c)** $6x^2 - 13x + 6$ **(d)** $6x^2 - 13x - 6$ **(e)** None of the preceding

8. Which of the following are correct?

   I. $(x + a)(x + a) = x^2 + a^2$
   II. $(x + a)(x + a) = x^2 + 2ax + a^2$
   III. $(x + a)(x + a) = (x + a)^2$

   (a) Only I **(b)** Only II **(c)** Only III **(d)** Only II and III **(e)** I, II, and III

9. Expand: $(a + b)^3$

(a) $a^3 + b^3$ (b) $a^3 + a^2b + ab^2 + b^3$ (c) $a^3 + 3a^2b + 3ab^2 + b^3$ (d) $a^3 - 3a^2b + 3ab^2 - b^3$
(e) $a^3 + 3a^2b^2 + 3ab + b^3$

10. The factored form of $(3x + 2)^2 - (2x - 1)^2$ is:

(a) $(9x^2 + 12x + 4) - (4x^2 - 4x + 1)$ (b) $(x + 3)^2$
(c) $5(x^2 + 1)$ (d) $(5x + 1)(x + 3)$ (e) $5x^2 + 3$

11. Find the GCF of $24x^2y$, $36xy^3$, and $40x^3y^2$.

(a) $4xy$ (b) $4x^3y^3$ (c) $2x^2y^2$ (d) $2x^3y^3$ (e) None of the preceding

12. Which of the following is the factorization of $27a^3 - 8$?

(a) $(3a - 2)^3$ (b) $(3a - 2)(9a^2 + 6a + 4)$ (c) $(3a - 2)(9a^2 - 12a + 4)$ (d) $(3a - 2)(a^2 - 4)$
(e) This cannot be factored

13. Which of the following is the factorization of $4x^2 - 6xy + 9y^2$?

(a) $(2x - 3y)(2x - 3y)$ (b) $(2x + 3y)(2x + 3y)$ (c) $(2x - 3y)(2x + 3y)$ (d) $(2x)^2 - (3y)^2$
(e) This cannot be factored

14. When $8xy - 4x^2 - 5x + 10y$ is factored completely, one of the factors is:

(a) $4x$ (b) $4x + 5$ (c) $4x - 5$ (d) $2x + y$ (e) $2x - y$

15. When $x^{16} - y^{16}$ is factored completely, the number of factors is:

(a) 2 (b) 4 (c) 5 (d) 8 (e) This cannot be factored

ANSWERS TO THE TEST YOUR UNDERSTANDING EXERCISES

Page 36

1. 125 2. $-\frac{1}{32}$ 3. 0 4. 1,000,000 5. -64 6. 64
7. $\frac{1}{17}$ 8. $\frac{1}{23}$ 9. 81 10. $a^{11}b^{10}$ 11. 64 12. $6x^4$

Page 42

1. $\frac{1}{2^3} = \frac{1}{8}$ 2. $2^7 = 128$ 3. $\frac{1}{(-2)^4} = \frac{1}{16}$ 4. 8
5. $4^{-15} = \frac{1}{4^{15}}$ 6. $(20)^{-3} = \frac{1}{8000}$ 7. $\frac{1}{a + b}$ 8. $\frac{4}{8^3} = \frac{1}{128}$

Page 43

1. 7.39×10^2 2. 7.39×10^4 3. 7.39×10^{-3} 4. 7.39×10^{-1}
5. 7.39×10^1 6. 7.39×10^0 7. 4010 8. 0.00401
9. 0.0111 10. 111,000 11. 0.00092 12. 4.27

Page 47

1. $4x^2 + x + 1$ 2. $2x^3 + 3x^2 + 8x + 1$ 3. $4x - 3y - 9$
4. $4x^2y - xy$ 5. $6x - 3$ 6. $x^3 + 2x^2 + 4x - 8$
7. $2x - 3y$ 8. $8x^5 + x^4 + x^3 - 13x + 25$ 9. $7a^2 + 6a + 8$
10. $r - 3s$

Page 51

1. $6x^4$ 2. $27x^5$ 3. $-10xy^5y^3$ 4. $12a^3b^5$
5. $-6a^3b^3$ 6. $2x^4 + 2x^2$
7. $-6x^4 + 15x^3 - 21x^2$ 8. $2x^5 - 18x^4 + 4x^3$ 9. $-9a^4b^2 + 21a^3b^3 - 3a^2b$

Page 56

1. $6x^2 + 13x + 6$
2. $10x^2 + 7x + 1$
3. $6x^2 - 13x + 6$
4. $10a^2 - 7a + 1$
5. $8x^2 - 6x - 9$
6. $3n^2 + 7n - 40$
7. $9x^2 + 6x + 1$
8. $25y^2 + 30y + 9$
9. $49x^2 - 28x + 4$
10. $(20 - 5)(20 + 5) = 20^2 - 5^2 = 375$
11. $(30 + 3)(30 - 3) = 30^2 - 3^2 = 891$
12. $(100 - 4)(100 + 4) = 100^2 - 4^2 = 9984$

Page 60

1. 4
2. 3
3. 6
4. 5
5. 12
6. 6
7. 14
8. 24
9. $2(3x + 4)$
10. $3(x + 12)$
11. $5a(x + y)$
12. $2x(x + 2)$
13. $3ab(b - a)$
14. $9a^2b^2(2a - 3b)$
15. $4x(2x^3 + 3y + 4y^2)$
16. $7a(5 - 3ax^2 + 2x)$
17. $4xy(8x - 6y - 3)$

Page 62

1. $(x + 7)(x - 7)$
2. $2(3x + 1)(3x - 1)$
3. $(2a + 5b)(2a - 5b)$
4. $a(x + 9)(x - 9)$
5. $x(10a + 9)(10a - 9)$
6. $2b(4a + 5)(4a - 5)$
7. $(x^2 + 9)(x + 3)(x - 3)$
8. $(4 + x^2)(2 + x)(2 - x)$
9. $(9b^2 + 4)(3b + 2)(3b - 2)$

Page 68

1. $(a + 3)^2$
2. $(x - 5)^2$
3. $(2x + 3y)^2$
4. $(3a - b)^2$
5. $(5a - 7b)^2$
6. $(7h + 2k)^2$
7. $p = 2, q = 8$
8. $p = 1, q = 16$
9. $p = 2, q = 12$
10. $p = 3, q = 14$
11. $(a + 3)(a + 5)$
12. $(x - 10)(x - 2)$
13. $(x - 20)(x - 5)$
14. $(12 + y)(3 + y)$
15. $(15a - 1)(3a - 1)$
16. $(1 + 2b)(1 + 14b)$

Page 73

1. $(2x + y)(y - 8)$
2. $(a^2 + b)(a - b)$
3. $(x - 4)(y + 7)$
4. $(x - y)(y + 5)$
5. $(2a + 3b)(a + 9)$
6. $(5p - 1)(7q - 4)$
7. $(x^2 + 6)(x - 2y)$
8. $(2r + 1)(r - s^2)$

CHAPTER

3 INTRODUCTION TO EQUATIONS AND INEQUALITIES

3.1 INTRODUCTION TO EQUATIONS

A statement such as $2x - 3 = 7$ is said to be a **conditional equation.** It is true for some replacements of the variable x, but not true for others. For example, $2x - 3 = 7$ is a true statement for $x = 5$ but is false for $x = 7$. On the other hand, an equation such as $3(x + 2) = 3x + 6$ is called an **identity** because it is true for *all* real numbers x.

To *solve* an equation means to find the numbers x for which the given equation is true; these are called the **solutions** or **roots** of the given equation. Let us solve the equation $2x - 3 = 7$, showing the important steps.

$$2x - 3 = 7 \qquad \text{(Add 3 to each side of the equation.)}$$
$$(2x - 3) + 3 = 7 + 3$$
$$2x = 10 \qquad \text{(Multiply each side by } \tfrac{1}{2}.\text{)}$$
$$\tfrac{1}{2}(2x) = \tfrac{1}{2}(10)$$
$$x = 5$$

This solution can be checked by substituting 5 for x in the original equation.

$$2(5) - 3 = 10 - 3 = 7$$

In the preceding solution we made use of the following two basic **properties of equality.**

The strength of these two properties is that they produce ***equivalent equations,*** *equations having the same roots. Thus the addition property converts $2x - 3 = 7$ into the equivalent form $2x = 10$.*

ADDITION PROPERTY OF EQUALITY

For all real numbers a, b, c, if $a = b$, then $a + c = b + c$.

MULTIPLICATION PROPERTY OF EQUALITY

For all real numbers a, b, c, if $a = b$, then $ac = bc$.

The properties of equality can also be used to solve more complicated equations. The procedure is to *collect all terms with the variable on one side of the equation and all constants on the other side*. Although most of the steps shown can be done mentally, the next two examples will include the essential details that comprise a formal solution.

Since subtraction is the inverse operation of addition, we could just as well subtract 5 instead of adding -5. Thus the second step would be

$$3x + 5 - 5 = 2x + 1 - 5$$

In general, if $a = b$, then

$$a - c = b - c$$

EXAMPLE 1 Solve for x: $3x + 5 = 2x + 1$

Solution Note the use of the additive inverse property in the steps below.

$$3x + 5 = 2x + 1$$

$$3x + 5 + (-5) = 2x + 1 + (-5) \qquad \text{(Add } -5 \text{ to each side.)}$$

$$3x = 2x + (-4) \qquad \text{(Combine terms.)}$$

$$3x + (-2x) = 2x + (-4) + (-2x) \qquad \text{(Add } -2x \text{ to each side.)}$$

$$1x = -4 \qquad \text{(Combine terms.)}$$

$$x = -4$$

■

The **solution set** for an equation is the set of replacements for the variable that will make the equation true. The solution set for the equation in Example 1 is $\{-4\}$. However, for simplicity we usually write the solution in the form $x = -4$.

There are three more basic properties of equality that underlie the work when solving equations.

PROPERTIES OF EQUALITY

Reflexive Property	$a = a$
Symmetric Property	If $a = b$, then $b = a$
Transitive Property	If $a = b$ and $b = c$, then $a = c$

The reflexive property states that any number is equal to itself. The symmetric property allows the interchanging of the sides of an equation, so that in Example 1 we could just as well begin with $2x + 1 = 3x + 5$ and end with $-4 = x$. The transitive property is used in the next example and explained in the margin.

EXAMPLE 2 Solve for n: $2(n + 3) = n + 5$

Solution Here we have an additional step because of the parentheses, which can be eliminated by applying the distributive property. Try to give a reason for each step in the work that follows.

The second equation is justified by two properties. First, $2n + 6 = 2(n + 3)$ by the distributive property. Also, since $2(n + 3) = n + 5$, the transitive property gives $2n + 6 = n + 5$. For the sake of efficiency, all of this is usually done in a single step.

$$2(n + 3) = n + 5$$
$$2n + 6 = n + 5$$
$$2n + 6 + (-6) = n + 5 + (-6)$$
$$2n = n + (-1)$$
$$2n + (-n) = n + (-1) + (-n)$$
$$n = -1$$

This solution can be checked by substituting the value found for n in the original equation. Does $2[(-1) + 3] = (-1) + 5$? ■

The equations that we are dealing with here are called **linear equations.** In each case the exponent of the variable is understood to be 1; that is, $x = x^1$.

A linear equation can be written in the general form

$$ax + b = 0$$

where a and b are real numbers, with $a \neq 0$.

Frequently, linear equations are not given in this stated form but can be rewritten in this form if necessary. However, the essential feature of a linear equation is that the degree of the variable is 1. The following examples show the use of both properties of equality to solve a linear equation.

EXAMPLE 3 Find the solution set: $5x - 3 = 3x + 1$

Solution

Since division is the inverse operation of multiplication, the multiplication step could be replaced by $\frac{2x}{2} = \frac{4}{2}$. In general, if $a = b$, then $\frac{a}{c} = \frac{b}{c}$ $c \neq 0$

$$5x - 3 = 3x + 1$$
$$5x - 3 + 3 = 3x + 1 + 3 \quad \text{(Addition property of equality)}$$
$$5x = 3x + 4 \quad \text{(Combining terms)}$$
$$5x + (-3x) = 3x + 4 + (-3x) \quad \text{(Addition property of equality)}$$
$$2x = 4 \quad \text{(Combining terms)}$$
$$(\tfrac{1}{2})(2x) = (\tfrac{1}{2})(4) \quad \text{(Multiplication property of equality)}$$
$$x = 2$$

The solution set is {2}. You can check this by substituting 2 for x in the original equation. ■

EXAMPLE 4 Solve for x: $\frac{3}{4}x - 2 = \frac{1}{3}x - 12$

Solution

An alternative way to solve $\frac{5}{12}x = -10$ is to multiply first by 12 and then divide by 5:

$$5x = -120$$
$$x = -24$$

$$\frac{3}{4}x - 2 = \frac{1}{3}x - 12$$
$$\frac{3}{4}x - 2 + 2 = \frac{1}{3}x - 12 + 2$$
$$\frac{3}{4}x = \frac{1}{3}x - 10$$
$$\frac{3}{4}x + (-\tfrac{1}{3}x) = \frac{1}{3}x - 10 + (-\tfrac{1}{3}x) \qquad (\tfrac{3}{4} - \tfrac{1}{3} = \tfrac{5}{12})$$
$$\frac{5}{12}x = -10$$
$$(\tfrac{12}{5})(\tfrac{5}{12}x) = (\tfrac{12}{5})(-10) \qquad (\tfrac{12}{5} \times \tfrac{5}{12} = 1)$$
$$x = -24$$

■

TEST YOUR UNDERSTANDING Think Carefully	*Solve each linear equation for x.*		
	1. $x + 3 = 9$	**2.** $x - 5 = 12$	**3.** $x - 3 = -7$
	4. $2x + 5 = x + 11$	**5.** $3x - 7 = 2x + 6$	**6.** $3(x - 1) = 2x + 7$
	7. $3x + 2 = 5$	**8.** $5x - 3 = 3x + 1$	**9.** $x + 3 = 13 - x$
	10. $2(x + 2) = x - 5$	**11.** $2x - 7 = 5x + 2$	**12.** $4(x + 2) = 3(x - 1)$
(Answers: Page 124)	**13.** $\frac{5}{8}x = -40$	**14.** $\frac{5}{6}x + 2 = \frac{1}{3}x - 4$	**15.** $\frac{7}{8}x - 15 = \frac{1}{2}x + 9$

The next two examples show the use of properties of equality to solve a *formula* for one of the variables in terms of the others.

Here is an example of the use of this formula: When C = 20, then F = $\frac{9}{5}(20) + 32 = 36 + 32 = 68$. Thus 20° Celsius = 68° Fahrenheit.

EXAMPLE 5 The formula relating degrees Fahrenheit and degrees Celsius is $\frac{5F - 160}{9} = C$. Solve for F in terms of C.

Solution Try to explain each step shown.

$$\frac{5F - 160}{9} = C$$
$$5F - 160 = 9C$$
$$5F = 9C + 160$$
$$F = (\tfrac{1}{5})(9C + 160)$$
$$F = \tfrac{9}{5}C + 32$$

■

EXAMPLE 6 A formula that gives an approximate relationship between a person's weight, W, in pounds and height, H, in inches is the following:

$$W = \tfrac{11}{2}H - 220$$

(a) Solve for H in terms of W.
(b) Use your result to approximate the height of someone who weighs 154 pounds.

Solution

(a)

$$W = \tfrac{11}{2}H - 220$$

$$W + 220 = \tfrac{11}{2}H$$

$$2(W + 220) = 11H$$

$$\frac{2(W + 220)}{11} = H$$

(b) For $W = 154$:

$$H = \frac{2(154 + 220)}{11} = \frac{2(374)}{11} = \frac{748}{11} = 68$$

The person's height is approximately 68 inches, or 5 ft, 8 in. ■

EXERCISES 3.1

Solve for x and check each result.

1. $3x - 2 = 10$
2. $5x + 1 = 21$
3. $-2x + 1 = 9$
4. $-3x - 2 = 10$
5. $-3x - 5 = 7$
6. $3x + 2 = -13$
7. $2x - 1 = -17$
8. $-2x + 3 = -12$
9. $2(x + 1) = 11$
10. $3(x - 2) = 15$
11. $3x + 7 = 2x - 2$
12. $2.5x - 8 = x + 3$
13. $\frac{2}{5}x = -8$
14. $\frac{2}{3}x = 12$
15. $-\frac{3}{4}x = 24$
16. $\frac{3}{4}x = \frac{1}{4}x + 20$
17. $\frac{4}{5}x = \frac{1}{5}x - 15$
18. $\frac{2}{3}x - 7 = 5$
19. $\frac{1}{2}x + 7 = 2x - 3$
20. $\frac{5}{2}x - 5 = 3x + 7$
21. $\frac{4}{3}x - 7 = \frac{1}{3}x + 8$
22. $5x - 1 = 5x + 1$
23. $\frac{3}{5}(x - 5) = x + 1$
24. $5(x + 4) = \frac{5}{2}x - 5$
25. $\frac{7}{2}x + 5 + \frac{1}{2}x = \frac{5}{2}x - 6$
26. $2(x + 3) - x = 2x + 8$
27. $-3(x + 2) + 1 = x - 25$
28. $\frac{4}{3}(x + 8) = \frac{3}{4}(2x + 12)$
29. $1 - 12x = 7(1 - 2x)$
30. $2(3x - 7) - 4x = -2$

Find the solution set.

31. $3x + 5 + x = 2x - 6$
32. $6x - 2x - 3 + 2 = 2x + 1$
33. $(3y - 9) - y = -(3y - [y + 3])$
34. $5(x + 4) = 2x - 7$
35. $-3(n + 2) + 1 = n - 25$
36. $2(n + 3) - n = 2n + 8$
37. $5(2z - 1) = -3(z + 8) - 7$
38. $4(y + 8) = 3(2y + 12)$
39. $1 - 12u = 7(1 - 2u)$
40. $2(3x - 7) - 4x = -2$
41. $x + 2(x + 2) = 6x + 16$
42. $5 - 3(2 - b) + 10(12 - 2b) = 0$
43. $4(1 - \frac{1}{2}x) + 3x = \frac{1}{2}x - \frac{13}{2}$
44. $\frac{3}{4}x - 14 = -\frac{1}{4}(24 + x)$
45. $2(t - 3) + 4(t + 7) = 3(6t + 4) - 4(10t + 1)$
46. $5[2x - 4(x - 5)] = 5(x - 1)$

Solve for the indicated variable.

47. $P = 2\ell + 2w$ for w
48. $P = 4s$ for s
49. $N = 10t + u$ for t
50. $F = \frac{9}{5}C + 32$ for C
51. $C = 2\pi r$ for r
52. $7a - 3b = c$ for b
53. $6 + 4v = w - 1$ for v
54. $2(r - 3s) = 6t$ for s

55. A formula for the area of a *trapezoid* in terms of its altitude h and its bases a and b is $A = \frac{1}{2}h(a + b)$.

(a) Solve this formula for b in terms of the other variables.

(b) Solve for b, rounded to two decimal places, if $A = 12.566$ sq. cm, $h = 3.05$ cm, and $a = 2.76$ cm.

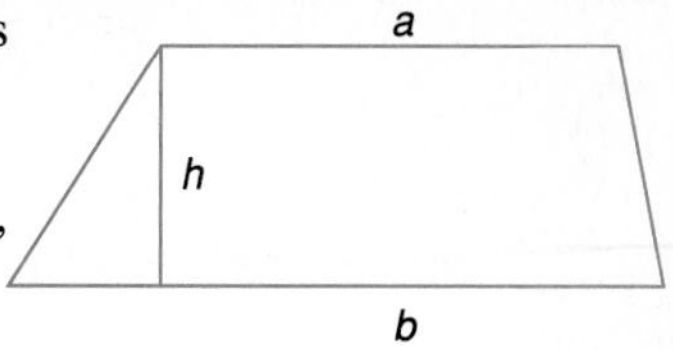

56. The area of a triangle is found as the product of one-half the base b, by the altitude h.

(a) Solve the formula $A = \frac{1}{2}bh$ for b in terms of A and h.

(b) Solve for b, rounded to two decimal places, if $A = 3.76$ and $h = 2.18$.

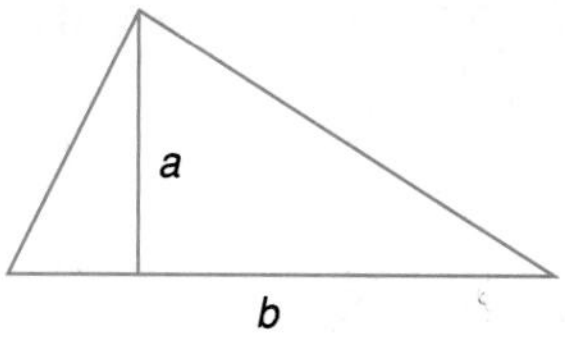

Using the formulas in Example 5, make these conversions.

57. 10° Celsius to Fahrenheit
58. 0° Celsius to Fahrenheit
59. 1° Celsius to Fahrenheit
*60. −4° Fahrenheit to Celsius
61. 32° Fahrenheit to Celsius
62. 77° Fahrenheit to Celsius

In Exercises 63 through 68, decide whether the equation is consistent, inconsistent, or an identity and give the solution. For an ***inconsistent equation*** *use ∅, the null set. You can recognize such an equation if it produces a false result such as $x + 2 = x + 3$, producing the false result $2 = 3$. An* ***identity*** *is identified by arriving at a true statement such as $2(x + 3) = 2x + 6$, producing $6 = 6$. A* ***consistent equation*** has a unique solution.

63. $5x - 1 = 5x + 1$
64. $4x - (2 - x) = 5x - 2$
65. $4x - (2 - x) = 5x - (x - 2)$
66. $2(3x - 2) + 10 = -3[2 - (1 + 2x)] - 16$
67. $3(3x - 2) - 2x = 6 + (x - 6)2 + 5x$
68. $4[2(3x + 1) - 3x] = -2(2 - 15x)$

Two equations are ***equivalent*** *provided that they have the same solution sets. Find the value of c that will make the two equations equivalent.*

*69. $3(x + 7) = -2(x + 1) + 3;\quad 5x + 14 = c$

*70. $4x - c = 13;\quad 2x - 3 = c$

71. Let ℓ, w, and h be the dimensions of a rectangular box. If the volume is 200 cubic feet, write the formula for the height in terms of ℓ and w.

*72. If the total surface area of the box in Exercise 71 is 210 square feet, write the formula of h in terms of ℓ and w. (*Hint:* There are six sides.)

3.2 APPLICATIONS: PROBLEM SOLVING

In this section we shall explore the solution of problems that are expressed in words. Our task will be to translate the English sentences of a problem into an equation that we can solve. Before attempting to solve specific problems, it will be helpful first to gain some practice in translating verbal phrases into mathematical expressions. Study the following list of phrases that involve a number x.

ENGLISH PHRASE	MATHEMATICAL PHRASE
three times a number	$3x$
a number increased by 5	$x + 5$
a number decreased by 7	$x - 7$

ENGLISH PHRASE	MATHEMATICAL PHRASE
a number divided by 3	$\frac{x}{3}$, or $\frac{1}{3}x$
two consecutive integers	x and $x + 1$
two consecutive even integers	$2x$ and $2x + 2$
three more than twice a number	$2x + 3$
five less than three times a number	$3x - 5$
the sum of two consecutive integers	$x + (x + 1)$
the product of two consecutive integers	$x(x + 1)$
the number of cents in x nickels	$5x$
two numbers whose sum is 10	x and $10 - x$
the number of inches in x feet	$12x$
an 8% commission earned on sales of x dollars	$.08x$
hours to travel x miles at 40 miles per hour	$\frac{x}{40}$

Although it is traditional to use x to represent a number, any other letter can be used as well. Thus two consecutive integers can also be represented by n and n + 1.

TEST YOUR UNDERSTANDING
Think Carefully

Translate each verbal phrase into a mathematical phrase.

1. Seven times a number.
2. A number decreased by 10.
3. The sum of two consecutive even integers.
4. The sum of four times a number and five times the same number.

Write an equation for each of the following. Do not solve. Use x to represent each number to be found.

5. Two less than five times a number is 33.
6. The sum of two consecutive integers is 63.
7. The product of two consecutive integers is 90.
8. A number is increased by 5. The sum is then multiplied by 3 to give a product of 39.
9. A number is decreased by 3. The difference is then multiplied by 8 to give a product of 48.
10. Three-fourths of a number, decreased by 2, is 28.

(Answers: Page 124)

Solving verbal problems often creates difficulties for many students. To become a good problem solver you need some patience and much practice. Study the solution to the following sample problem in detail, as well as the examples that follow.

Problem: The length of a reactangle is 1 centimeter less than twice the width. The perimeter is 28 centimeters. Find the dimensions of the rectangle.

1. Reread the problem and try to picture the situation given. Make note of all the information stated in the problem.
   The length is one less than twice the width.
   The perimeter is 28.
2. Determine what it is you are asked to find. Introduce a suitable variable, usually to represent the quantity to be found. When appropriate, draw a figure.

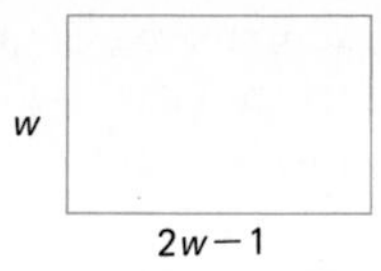

Let w represent the width.
Then $2w - 1$ represents the length.

3. Use the available information to compose an equation that involves the variable.

   The perimeter is the distance around the rectangle. This provides the necessary information to write an equation.

$$w + (2w - 1) + w + (2w - 1) = 28$$

4. Solve the equation.

$$\begin{aligned} w + (2w - 1) + w + (2w - 1) &= 28 \\ 6w - 2 &= 28 \\ 6w &= 30 \\ w &= 5 \end{aligned}$$

5. Return to the original problem to see whether the answer obtained makes sense. Does it appear to be a reasonable solution? Have you answered the question posed in the problem?

   The original problem asked for both dimensions. If the width w is 5 centimeters, then the length $2w - 1$ must be 9 centimeters.

6. Check the solution by direct substitution of the answer into the original statement of the problem.

   As a check, note that the length of the rectangle, 9 centimeters, is 1 centimeter less than twice the width, 5, as given in the problem. Also, the perimeter is 28 centimeters.

7. Finally, state the solution in terms of the appropriate units of measure.

   The dimensions are 5 centimeters by 9 centimeters.

EXAMPLE 1 Find two numbers whose sum is 42 if the larger number is three less than twice the smaller number.

Solution Two numbers, a smaller and a larger, need to be found. Since the statement of the problem describes the larger in terms of the smaller, we begin by using a variable for the smaller.

This problem can also be solved by choosing x to represent the larger number. However, this would be more difficult. Try it.

Let x represent the smaller number.
Then $2x - 3$ represents the larger because it is three less than twice the smaller.

The fact that the sum of the numbers must equal 42 can now be used to write an equation that can be solved for x.

$$\begin{aligned} x + (2x - 3) &= 42 \\ 3x - 3 &= 42 \\ 3x &= 45 \\ x &= 15 \\ 2x - 3 = 2(15) - 3 &= 27 \end{aligned}$$

The two numbers are 15 and 27. Check to see that this solution is correct. ■

EXAMPLE 2 Each year a student uses the interest earned on two separate investments to help pay for college costs. The annual interest rates are 11% and 13%. If the total amount invested is \$15,200 and the total interest earned per year is \$1840, how much is invested at each rate?

Solution

Let x be the amount invested at 11%.

Then $15{,}200 - x$ is the amount invested at 13%.

This formula is a special case of the simple interest formula $I = prt$, where t is the time of the investment. When r is an annual rate, then t is expressed in years. In this example, $t = 1$.

Now use the formula $I = pr$, where I is the interest earned on the principal of p dollars invested at the rate r (in decimal form) per year. Then

$0.11x$ = yearly interest earned at 11%

$0.13(15{,}200 - x)$ = yearly interest earned at 13%

Since total annual interest is \$1840, we have

Interest at 11%	+	Interest at 13%	=	Total interest
↓		↓		↓

$$0.11x + 0.13(15{,}200 - x) = 1840$$

$$11x + 13(15{,}200 - x) = 184{,}000 \longleftarrow \text{(Multiply by 100 to avoid decimals.)}$$

$$11x + 197{,}600 - 13x = 184{,}000$$

$$-2x = -13{,}600$$

$$2x = 13{,}600$$

$$x = 6800$$

$$\text{and} \quad 15{,}200 - x = 8400$$

Guidelines for problem solving:

Read the problem. List the given information.
↓
What is to be found? Introduce a variable and state what the variable represents. Draw a figure or use a table if appropriate.
↓
Write an equation.
↓
Solve the equation.
↓
Does the answer seem reasonable? Have you answered the question stated in the problem?
↓
Check your answer to the equation in the original problem.
↓
State the solution to the problem.

Thus \$6800 is invested at 11% and \$8400 at 13%. Use a calculator to check that $0.11(6800)$ plus $0.13(8400)$ equals 1840. ■

EXAMPLE 3 A car leaves a certain town at noon, traveling due east at 40 miles per hour. At 1:00 P.M. a second car leaves the town traveling in the same direction at a rate of 50 miles per hour. In how many hours will the second car overtake the first car?

Solution Problems of motion of this type often prove difficult to students of algebra, but need not be. The basic relationship to remember is this:

$$\text{Rate} \times \text{Time} = \text{Distance} \qquad (r \times t = d)$$

For example, a car traveling at a rate of 60 miles per hour for 5 hours will travel 60×5 or 300 miles.

Now we need to explore the problem to see what part of the information given will help form an equation. The two cars travel at different rates, and for different amounts of time, but both travel the same distance from the point of departure until they meet. This is the clue: Represent the distance each travels and equate these quantities.

This diagram shows what the situation looks like at 1:00 P.M.:

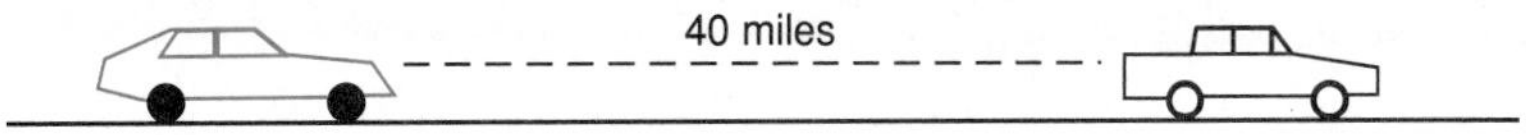

Let us use x to represent the number of hours it will take the second car to overtake the first. Then the first car, having started an hour earlier, travels $x + 1$ hours before they meet. You may find it helpful to summarize this information in tabular form.

	Rate	Time	Distance
First car	40	$x + 1$	$40(x + 1)$
Second car	50	x	$50x$

Equating the distances, we have an equation that can be solved for x:

$$50x = 40(x + 1)$$
$$50x = 40x + 40$$
$$10x = 40$$
$$x = 4$$

The second car overtakes the first in 4 hours. Does this solution seem reasonable? Let us check the solution. The first car travels 5 hours at 40 miles per hour for a total of 200 miles. The second car travels 4 hours at 50 miles per hour for the same total of 200 miles.

Solution The second car takes 4 hours to overtake the first. ■

EXAMPLE 4 David has a total of \$4.10 in nickels, dimes, and quarters. He has twice as many nickels as dimes, and two more quarters than dimes. How many dimes does he have?

Although some of these problems may not seem to be very practical to you, they will help you develop basic skills at problem solving that will be helpful later when you encounter more realistic applications.

Solution Begin by letting x represent the number of dimes. Then from the statement of the problem we can let $2x$ represent the number of nickels, and $x + 2$ the number of quarters.

Next, observe that whenever you have a certain number of a particular coin, then the total value is the number of coins times the value of that coin. For example, if you have 8 dimes, then you have 10×8 or 80¢. If you have 9 nickels, then you have 5×9 or 45¢. For this problem:

$10x$ represents the value of the dimes, in cents.
$5(2x)$ represents the value of the nickels, in cents.
$25(x + 2)$ represents the value of the quarters, in cents.

This information is summarized in the following table.

	Value of each coin (¢)	Number of coins	Value of coins (¢)
Nickels	5	$2x$	$5(2x)$
Dimes	10	x	$10x$
Quarters	25	$x + 2$	$25(x + 2)$

Since the total value of these coins is \$4.10, or 410¢, we write and solve the following equation:

$$\underset{\downarrow}{\text{Value of dimes}} + \underset{\downarrow}{\text{Value of nickels}} + \underset{\downarrow}{\text{Value of quarters}} = \underset{\downarrow}{\text{Total value}}$$

$$10x + 5(2x) + 25(x + 2) = 410$$
$$10x + 10x + 25x + 50 = 410$$
$$45x + 50 = 410$$
$$45x = 360$$
$$x = 8$$

Always check by returning to the original statement of the problem. If you just check in the constructed equation, and the equation you formed was incorrect, you will not detect the error.

Check: If David has 8 dimes, then he must have $2x$ or 16 nickels, and $x + 2$ or 10 quarters. The total amount is

$$8(10) + 16(5) + 10(25) = 80 + 80 + 250 = 410, \quad \text{that is,} \quad \$4.10$$

Solution David has 8 dimes. ■

EXAMPLE 5 Marcella has a base salary of \$250 per week. In addition, she receives a commission of 12% of her sales. Last week her total earnings were \$520. What were her total sales for the week?

Solution Let x represent her total sales. Then make use of this relationship:

$$\underset{\downarrow}{\text{Salary}} + \underset{\downarrow}{\text{Commission}} = \underset{\downarrow}{\text{Total earnings}}$$

$$250 + 0.12x = 520 \qquad \text{(Recall: } 12\% = 0.12\text{)}$$
$$0.12x = 270$$
$$x = \frac{270}{0.12}$$
$$x = 2250$$

Check this result in the given problem.

Marcella's total sales for the week were \$2250.
Use a calculator to show that $250 + 0.12(2250) = 520$. ■

Now it's your turn! Use the guidelines suggested earlier in this section, and try to solve as many of the problems as you can. Don't become discouraged if you have difficulty; be assured that most mathematics students have trouble with word problems. Time and practice will most certainly help you develop your skill at problem solving.

EXERCISES 3.2

For each question assign a variable, write an equation, and solve the problem. See the guidelines used in this section as well as the worked-out verbal problems.

1. Five more than three times a certain number is 29. Find the number.
2. Three less than five times a certain number is 37. Find the number.
3. Find the dimensions of a rectangle whose perimeter is 56 inches if the length is three times the width.
4. Find two consecutive integers whose sum is 73. (Consecutive integers can be represented as x and $x + 1$.)
5. Find two consecutive even integers whose sum is 74.
6. Find two numbers whose sum is 27 if the larger is one less than three times the smaller.
7. Ellen's age is two more than five times Robert's age. The sum of their ages is 14. How old is each?
8. Each of the two equal sides of an isosceles triangle is 3 inches longer than the base of the triangle. The perimeter is 21 inches. Find the length of each side.

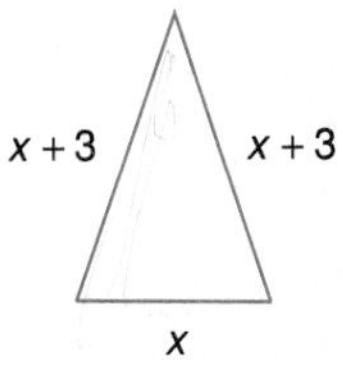

9. Find a number such that two-thirds of the number increased by one is 13.
10. How much money should be invested at an annual interest rate of 9% in order to earn $450 interest for the year?
11. $1500 is invested at an annual rate of 12%. Use the formula $I = prt$ to find the interest earned over each period of time.
    (a) $\frac{1}{2}$ year **(b)** 18 months **(c)** 40 months
    (See the marginal note, page 90.)
12. Maria has $169 in ones, fives, and tens. She has twice as many one-dollar bills as she has five-dollar bills, and five more ten-dollar bills than five-dollar bills. How many of each type bill does she have?
13. Carlos spent $5.35 on stamps, in denominations of 15¢, 20¢, and 25¢. He bought one-half as many 25¢ stamps as 15¢ stamps, and three more 20¢ stamps than 15¢ stamps. How many of each type did he buy?
14. Robert goes for a walk at a speed of 3 miles per hour (mph). Two hours later Tony attempts to overtake him by jogging at the rate of 7 mph. How long will it take him to reach Robert?
15. Two cars leave a town and travel in opposite directions. One car travels at the rate of 45 mph, and the other at 55 mph. In how many hours will the two cars be 350 miles apart?
16. Sara is 5 years older than Jan. Darsi is twice as old as Jan. The sum of the three ages is 65. How old is each?
17. Find a number such that three-fourths of the number increased by five is 23.
18. The width of a painting is 4 inches less than the length. The frame that surrounds the painting is 2 inches wide and has an area of 240 square inches. What are the dimensions of the painting? (*Hint:* The total area minus the area of the painting alone is equal to the area of the frame.)
19. Bob is 20 years older than Ben. In 10 years, Bob will be twice as old as Ben. How old is each now? (*Hint:* If Ben is x years old now, he will be $x + 10$ years old in 10 years.)
20. Joanne's age is five more than twice Kim's age. In 5 years Joanne will be twice as old as Kim. How old is each now? Explain your answer.
21. A financial advisor invested an amount of money at an annual rate of interest of 9%. She invested $2700 more than this amount at 12% annually. The total yearly income from these investments was $1794. How much did she invest at each rate? (*Note:* Use the formula $I = Prt$, where I is the interest earned on the principal of P dollars invested at the rate r (in decimal form) per year. In this case the time, t, is 1 year.)

22. A senior class earns 10.5% interest per year on one investment. The class also earns 11.5% on another investment that is \$400 more than the first investment. If the total yearly interest is \$959, find the amount of each investment.

23. Luis earns a monthly salary of \$1225, plus a commission of 8% of his total sales for the month. Last month his total earnings were \$1750. What were his total sales for the month?

24. Andrew paid \$9010 for a car, which included a 6% sales tax on the base cost of the car. What was the cost of the car, without the sales tax?

25. The total cost of two certificates of deposit is \$12,800. The annual interest rates on these certificates are 8% and 9%. The yearly interest on the 9% certificate is \$217 more than that on the 8% certificate. What is the cost of each certificate?

26. Find three consecutive integers whose sum is 114.

27. Find five consecutive even integers such that twice the sum of the first three is 46 more than the sum of the last two.

28. Amy travels 27.5 miles to get to work by car. The first part of her trip is along a country road on which she averages 35 miles per hour, and the second part is on a highway where she averages 48 miles per hour. If the time she travels on the highway is 5 times the amount of time she travels on the country road, what is the total time for the trip?

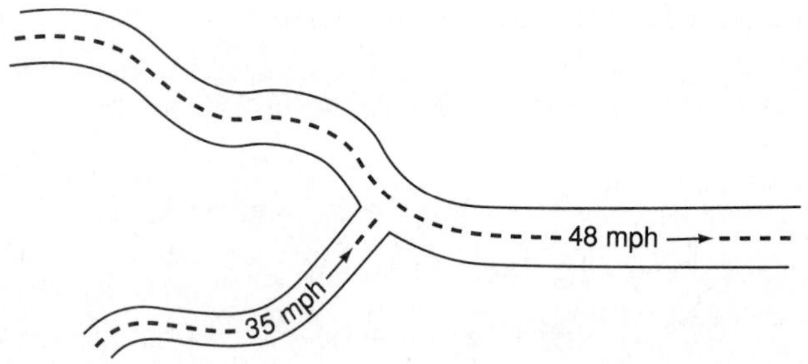

29. Prove that the measures of the angles of a triangle cannot be represented by consecutive odd integers. (*Hint:* The sum of the measures is 180.)

30. The unit's digit of a two-digit number is three more than the ten's digit. The number is equal to four times the sum of the digits. Find the number. (*Hint:* We can represent a two-digit number as $10t + u$.)

31. The length of a rectangle is 1 inch less than three times the width. If the length is increased by 6 inches and the width is increased by 5 inches, then the length will be twice the width. Find the dimensions of the rectangle.

32. The length of a rectangle is 1 inch less than twice the width. If the length is increased by 11 inches and the width is increased by 5 inches, then the length will be twice the width. What can you conclude about the data for this problem?

33. Following is a set of directions for a mathematical trick. First try it. Then use algebraic representations of each phrase (direction) to show why the trick works.

Think of a number.
Add 2.
Multiply by 3.
Add 9.
Multiply by 2.
Divide by 6.
Subtract the number with which you started.
The result is 5.

*34. A taxi charges 80 cents for the first $\frac{1}{6}$ mile and 20 cents for each additional $\frac{1}{6}$ mile. If a passenger paid \$6.00, how far did the taxi travel?

Written Assignment: Explain why the solution to a problem always must be checked by returning to the original statement of the problem.

CHALLENGE
Think Creatively

Here is a challenging problem that can be solved algebraically. However, first draw a diagram and try solving the problem by simple arithmetic and logical thinking!

Two boats begin their journeys back and forth across a river at the same time, but from opposite sides of the river. The first time that they pass each other they are 700 feet from one of the shores of the river. After they each make one turn, they pass each other again at a distance of 400 feet from the other shore. How wide is the river? Assume each boat travels at a constant speed and that there is no loss of time in making a turn.

3.3 INTRODUCTION TO STATEMENTS OF INEQUALITY

As you continue your study of mathematics you will find a great deal of attention given to *inequalities*. We begin our discussion of this topic by considering the ordering of the real numbers on the number line. In the following figure we say that *a is less than b* because a lies to the left of b. In symbols, we write $a < b$.

Also note that b lies to the right of a. That is, $b > a$; this is read "*b is greater than a*." Two inequalities, one using the symbol < and the other >, are said to have the *opposite sense*.

It may help to note that the inequality symbol "points" to the smaller of the two numbers and "opens wide" to the larger.

Note that $a < 0$ means a is to the left of the origin, which means that a is negative. Likewise, $a > 0$ means a is positive. Here are two examples, of the use of these **symbols of inequality.**

$3 < 7$ *3 is less than 7* $5 > -2$ *5 is greater than −2*

Algebraically, we define $a < b$ as follows:

$$a < b \text{ provided that } b - a \text{ is positive } (b - a > 0)$$

For example, $-7 < -4$ because $-4 - (-7) = +3$, a positive number.

A fundamental property of inequalities states that for any two real numbers a and b, either a is less than b, or a equals b, or a is greater than b. In symbols:

TRICHOTOMY PROPERTY

For any real numbers a and b, exactly one of the following is true:

$$a < b \qquad a = b \qquad a > b$$

Just as there is a transitive property of equality, so is there one for inequalities.

TRANSITIVE PROPERTY

For all real numbers a, b, and c, if $a < b$ and $b < c$, then $a < c$.

The transitive property can be illustrated on a number line in this way:

Draw a diagram to show that this is also true: If $a > b$ and $b > c$, then $a > c$.

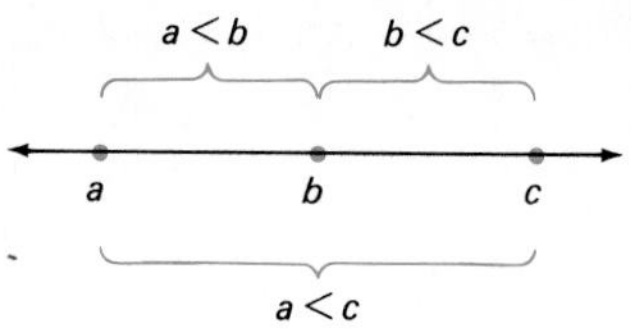

When an inequality involves a variable, we are usually interested in finding the values of the variables for which the given inequality is true. This collection of values is called the **solution set** of the inequality. For example, intuitively we see that the solution set for the inequality $x + 2 < 7$ consists of all real numbers less than 5. That is, the solution set consists of all x where $x < 5$. However, we want to have more formal methods of solution that can be applied to this and to other more complicated problems. Study the following examples. They show that *the same number can be added to each side of an inequality and the sense will be maintained.*

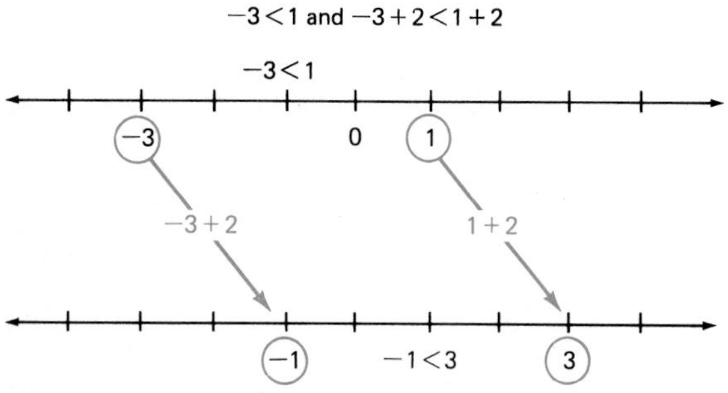

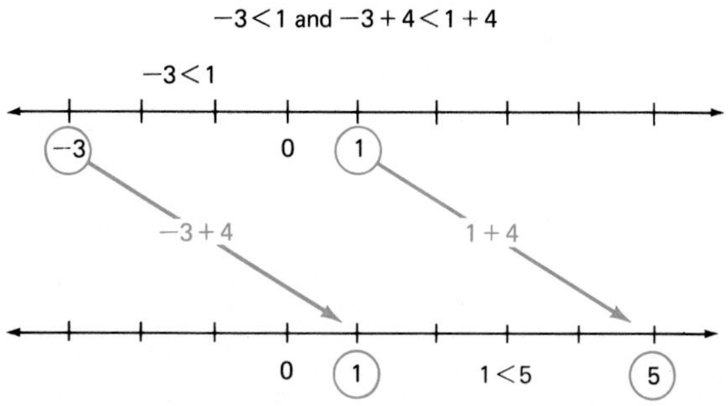

This gives rise to the following important property for inequalities:

ADDITION PROPERTY OF ORDER
For all real numbers, a, b, and c:
If $a < b$, then $a + c < b + c$.
If $a > b$, then $a + c > b + c$.

This property can now be used to solve an inequality such as $x + 2 < 7$.

Instead of adding -2 we can just as well subtract 2. Then the second step becomes $x + 2 - 2 < 7 - 2$. In general, if $a < b$, then $a - c < b - c$; if $a > b$, then $a - c > b - c$.

$$x + 2 < 7$$
$$x + 2 + (-2) < 7 + (-2) \qquad \text{(Addition property of order)}$$
$$x + 0 < 5$$
$$x < 5$$

When nothing is said to the contrary it will always be assumed that we are using the set of real numbers. Therefore, the solution set here consists of all real numbers that are less than 5. Rather than use a verbal description, we can use braces and write this solution set in **set-builder notation** as follows:

$$\{x \mid x < 5\}$$

This is read as "the set of all x such that x is less than 5."

$$\{ \; x \; | \; x \; < \; 5 \; \}$$

The set of all x such that x is less than 5.

The graph of this set is shown by the heavily shaded arrow drawn to the left of 5 to indicate all real numbers less than 5. The open circle at 5 indicates that 5 *is not* included in the solution set.

EXAMPLE 1 Solve and graph the inequality: $-4n - (3 - 5n) > 8$

Solution First simplify the left side.

$$-4n - (3 - 5n) > 8$$
$$-4n - 3 + 5n > 8$$
$$n - 3 > 8$$
$$n - 3 + \textcircled{3} > 8 + \textcircled{3}$$

The solution set is $\{n \mid n > 11\}$. The graph of the solution set has an open circle at 11 and points to the right.

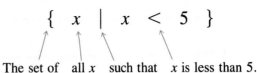

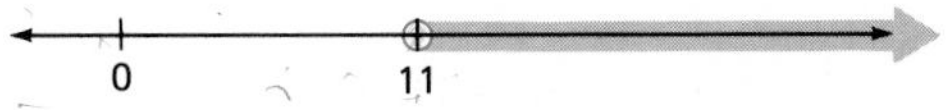

■

These additional symbols of inequality are also used quite often:

The addition property of order also applies for inequalities using the symbols ≤ and ≥. For example, if $a \le b$, then $a + c \le b + c$.

$a \le b$ means *a is less than or is equal to b;* that is, $a < b$ or $a = b$.
$a \ge b$ means *a is greater than or is equal to b;* that is, $a > b$ or $a = b$.

For example, if $x \le 3$ and x is a whole number, then x may be replaced by any whole number that is *less than or equal to* 3, namely, 0, 1, 2, or 3.

EXAMPLE 2 Find and graph the solution set: $3x + 2 \le 2x - 1$

Solution We apply the addition property of order twice.

$$3x + 2 \le 2x - 1$$
$$3x + 2 + (-2) \le 2x - 1 + (-2)$$
$$3x \le 2x - 3$$
$$3x + (-2x) \le 2x - 3 + (-2x)$$
$$x \le -3$$

The graph of the solution set is shown on the following page.

For $3x + 2 \le 2x - 1$, the solution set consists of all real numbers that are less than or equal to -3, that is, $\{x \mid x \le -3\}$.

The graph of the solution set has a heavy dot at -3 to show that -3 is in the solution set. The heavily shaded arrow to the left of -3 shows that all real numbers less than -3 are also in the solution set.

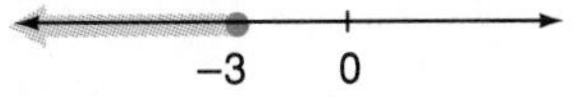

■

TEST YOUR UNDERSTANDING
Think Carefully

Solve the inequality and graph the solution set.

1. $x + 3 < 12$ **2.** $x - 5 < 13$ **3.** $x - 1 > 8$

4. $x + 7 > 2$ **5.** $x + (-5) \le 9$ **6.** $x + (-3) \ge -5$

7. $3x + 8 < 2x + 12$ **8.** $2x - 7 \ge x + 8$ **9.** $3(x + 5) > 2(x + 3)$

10. $5(x + 7) \le 4x - 8$

(Answers: Page 124)

When the addition property of order is applied, **equivalent inequalities** are produced, which means that they have the same solution set. Thus, in Example 2, $3x + 2 \le 2x - 1$ and $x \le -3$ are equivalent; they have the same solution set, $\{x \mid x \le -3\}$.

Equivalent inequalities can also be obtained through a multiplication process. There are two cases. First observe that

$$1 < 3 \quad \text{and} \quad 3(1) < 3(3)$$

$1 < 3$
$3(1) < 3(3)$
$3 < 9$

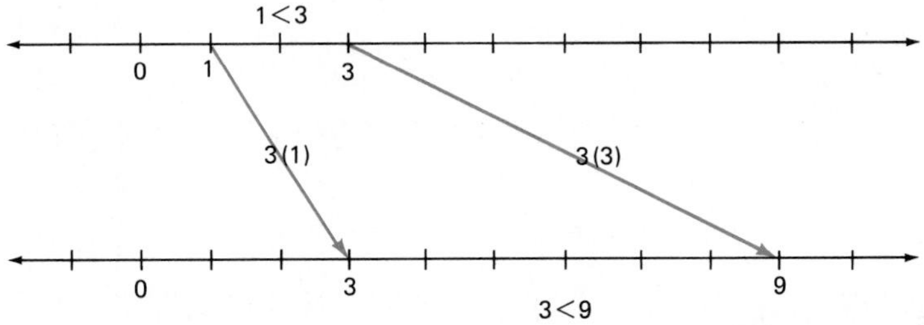

This suggests that when each side of an inequality is multiplied by a *positive* number, the resulting inequality has the same sense as the given inequality; that is, *the sense is preserved.*

Next observe what happens when the multiplier is negative.

$$1 < 3 \quad \text{and} \quad -3(1) > -3(3)$$

$1 < 3$
$-3(1) > -3(3)$
$-3 > -9$

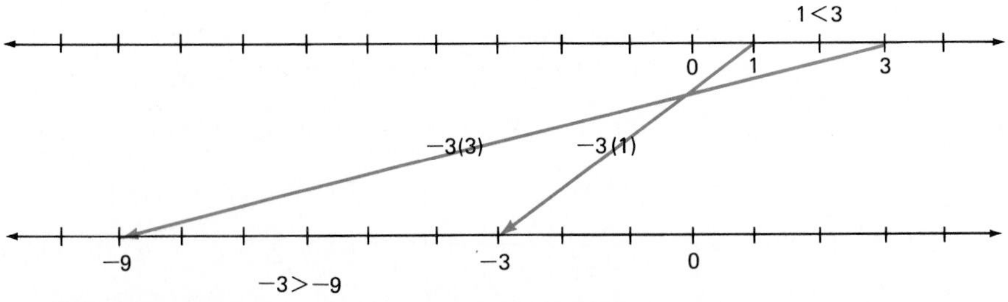

This suggests that when each side of an inequality is multiplied by a *negative* number, the resulting inequality has the opposite sense as the given inequality; that is, *the sense is reversed.*

In general we have

Note that similar properties hold for the inequality when $a > b$. Also, note that when $c = 0$, then $ac = bc$; that is, both sides will be equal to 0.

MULTIPLICATION PROPERTY OF ORDER

For all real numbers a, b, and c:

If $a < b$ and c is positive, then $ac < bc$.

If $a < b$ and c negative, then $ac > bc$.

EXAMPLE 3 Solve the inequality: $-2x > 8$

Solution Multiply each side by $-\frac{1}{2}$, the multiplicative inverse of -2. According to the multiplication property of order, this will reverse the sense of the inequality.

Instead of multiplying by $-\frac{1}{2}$, we can just as well divide by -2. Then the second step becomes

$$\frac{-2x}{-2} < \frac{8}{-2}$$

$$-2x > 8$$
$$(-\tfrac{1}{2})(-2x) < (-\tfrac{1}{2})(8)$$
$$x < -4$$

The solution set is $\{x \mid x < -4\}$. ■

In the following example both of the properties of order are used to solve for x. Although the detailed solution is shown here, with practice some of the steps can be done mentally.

EXAMPLE 4 Find the solution set: $5(3 - 2x) \geq 10$

Solution Multiply each side by $\frac{1}{5}$ (or divide by 5).

$$\tfrac{1}{5} \cdot 5(3 - 2x) \geq \tfrac{1}{5}(10) \qquad \text{(Multiplication property of order)}$$
$$3 - 2x \geq 2$$

Add -3 to each side.

$$3 - 2x + (-3) \geq 2 + (-3) \qquad \text{(Addition property of order)}$$
$$-2x \geq -1$$

Note the change in the sense of the resulting inequality because we are multiplying (or dividing) by a negative number.

Multiply by $-\frac{1}{2}$ (or divide by -2).

$$-\tfrac{1}{2}(-2x) \leq -\tfrac{1}{2}(-1) \qquad \text{(Multiplication property of order)}$$
$$x \leq \tfrac{1}{2}$$

The solution set is $\{x \mid x \leq \frac{1}{2}\}$. ■

The following examples illustrate how statements of inequality can be used to solve applied problems.

EXAMPLE 5 A department store wishes to purchase a number of computers prior to running a sale. Each computer costs \$600 wholesale, plus a 5% sales tax. In addition, they must pay a \$50 delivery charge for the entire shipment. The manager of the store has allocated *at most* \$10,000 for the initial purchase. How many computers should be ordered?

Solution Let x = the number of computers to be purchased. The cost of each computer is

$$\$600 + 0.05(\$600) = 1.05(\$600) = \$630$$

For x computers, the total cost will be $630x$, in dollars. Thus:

Note: *At most \$10,000 means less than or equal to \$10,000.*

Cost of computers	+	Delivery charges	≤	Total amount allocated
↓		↓		↓
$603x$	+	50	≤	$10{,}000$

$$603x + 50 - 50 \le 10{,}000 - 50$$

$$630x \le 9950$$

$$\frac{630x}{630} \le \frac{9950}{630}$$

$$x \le 15.79 \quad \text{(to two decimal places)}$$

Thus there is enough money allocated to order 15 computers. (There is not quite enough available to order 16. How much more money would be needed to be able to order 16 computers?) ■

EXAMPLE 6 A car traveling at 54 mph passes another car that is traveling in the same direction at 46 mph. How long will it take until the faster car is more than 10 miles ahead of the slower car?

Solution Let t = the time in hours the cars must travel. Now use the formula $d = rt$.

$$54t = \text{distance traveled by the faster car}$$

$$46t = \text{distance traveled by the slower car}$$

Since they are traveling in the same direction the distance between the cars is the difference of the distances traveled.

$$54t - 46t = \text{distance between the cars}$$

But the distance between must be *more than* 10 miles, which translates into an inequality that we can solve.

distance between (points to $54t - 46t$) more than (points to $>$)

$$54t - 46t > 10$$
$$8t > 10$$
$$t > \tfrac{5}{4}$$

The time must be more than $1\frac{1}{4}$ hours, or 1 hour and 15 minutes. ■

EXERCISES 3.3

Use the addition property of order to solve each inequality.

1. $x + 5 > 17$
2. $x - 3 > 15$
3. $x - 8 < 5$
4. $x + 12 < 2$
5. $x - 7 \geq -3$
6. $x + 6 \leq -7$
7. $x + (-2) \leq -5$
8. $x + (-7) \geq -11$
9. $5x - 4 < 6 + 4x$
10. $7 - 3x < 10 - 4x$
11. $3x + 12 > 2x - 5$
12. $2x - 7 < x + 8$
13. $x - 17 \leq 2x + 1$
14. $12 - 6x \geq 3 - 5x$

15. Solve the inequality $11 - 2x \leq 16 - 3x$ for **(a)** the set of whole numbers and **(b)** the set of natural numbers.
16. Solve the inequality $2(7 - 4x) > 1 - 3(3x - 2)$ for **(a)** the set of negative integers and **(b)** the set of real numbers.

Use the multiplication property of order to find the solution set.

17. $7x > 35$
18. $8x < 56$
19. $3x \geq -21$
20. $9x \leq -45$
21. $-5x < 50$
22. $-3x > 27$
23. $\frac{1}{2}x > 18$
24. $\frac{1}{4}x > -7$
25. $\frac{4}{5}x \leq -8$
26. $\frac{1}{6}x \geq \frac{1}{2}$
27. $-\frac{2}{3}x < 2$
28. $-\frac{3}{4}x < 20$
29. $-6x \geq -24$
30. $-8x \leq -40$
31. $\frac{7}{3}x > \frac{5}{6}$
32. $\frac{9}{5} \geq -\frac{4}{15}x$

In Exercises 33 through 50, solve the inequality and graph the solution set.

33. $3x + 5 \geq 17$
34. $5x - 3 \leq 22$
35. $-2x + 1 \leq 19$
36. $-3x + 2 \geq 20$
37. $3x + 7 < x - 9$
38. $5x - 6 < 2x + 9$
39. $3(x + 5) > x - 7$
40. $2(x + 1) > 3(x - 2)$
41. $3x + 5 + x > 2(x - 1)$
42. $5x - 3 - 2x > 3x + 8$
43. $\frac{1}{2}x + 5 < 3$
44. $\frac{1}{3}(x + 4) \geq 2$
45. $\frac{1}{2}x - 5 > \frac{1}{4}x + 3$
46. $\frac{3}{4}x + 2 < \frac{5}{8}x - 3$
47. $-\frac{3}{5}x - 6 < -\frac{2}{5}x + 7$
48. $\frac{7}{8}x - 7 > \frac{1}{8}x + 5$
49. $0.5(x + 6) > 0.3x - 2$
50. $x - 7 < 0.6x + 3$

Solve each inequality for x.

51. $ax + b > c,\ a > 0$
52. $3x - 2b < P$
53. $3x - 2 \not< 1$ ($\not<$ means is not less than)
54. $2x + 1 \not> 5$ ($\not>$ means is not greater than)
55. $-8 \not\leq -\frac{2}{3}x - 2$ ($\not\leq$ means is not less than or equal to)
56. $\frac{2}{9}(3x + 9) \not\geq 1 - \frac{4}{3}x$ ($\not\geq$ means is not greater than or equal to)

In Exercises 57 through 64, find all possible real numbers to satisfy each of the following.

57. Two more than five times a number is greater than 17.
58. Six minus four times a number is less than 12.

59. Two more than three times a number is more than nine minus four times the number.

60. If a number is decreased by two, then three times that difference is greater than five times the number.

61. Cne hundred times a number is less than fifty more than seventy-five times the number.

62. Two-thirds of a number is more than half the number minus one.

63. A number decreased by 3 is less than five-eighths of the number.

64. Three times the sum of twice a number plus seven is more than forty-eight times the number.

Solve each word problem using an inequality.

65. Two cars traveling in opposite directions on a highway pass each other and continue at the rates of 51 mph and 48 mph. When will they be more than 66 miles apart?

66. Two trains start from the same station at the same time and travel in opposite directions. One train travels at 72 mph and the other at 64 mph. How long will the trains be less than 170 miles apart?

67. The length of a rectangle is 27 inches. What is the largest width possible if the area is less than 63 square inches?

68. A college student plans to earn at least 36 credit hours distributed among English, mathematics, and computer science. She intends to take twice as much computer science as English, and three times as much mathematics as English. What is the minimum number of credits that she will have completed in each subject?

69. A student's grades on three tests were 81, 87, and 78. What is the lowest score the student can get on the fourth test to have an average of at least 85? at least 80?

70. Mr. Jackson intends to invest part of $15,000 in bonds that earn 11% interest per year, and the rest in savings certificates that pay 13% interest per year. What is the maximum amount that he can invest in bonds and still earn a total yearly interest of at least $1840?

71. Each week a family allows itself to spend at most $153 for entertainment, transportation, and food. They also plan to spend one and one-half times as much for transportation as for entertainment, and four times as much for food as for transportation. What is the most they allow themselves to spend on each item?

72. How many television sets can a department store order for $15,000 if each set costs $250 wholesale, plus 6% sales tax. There is also a $100 delivery charge for the total shipment.

73. A salesman who sells computers earns $1200 a month plus a 5% commission on each sale. If each computer sells for $750, how many must he sell in a month in order to earn at least $2000 total for the month?

*74. Assume that $p < q$. For what kind of real numbers x is each statement true? Your answer should be one of $x > 0$ or $x < 0$ or $x = 0$.

(a) $px < qx$ **(b)** $xp = xq$ **(c)** $qx < px$ **(d)** $\frac{p}{x} > \frac{q}{x}$

Use the multiplication property of order to prove each statement.

*75. If a is positive and $a < b$, then $\frac{1}{a} > \frac{1}{b}$.

*76. If a, b, c, and d are positive such that $a < b$ and $c < d$, then $ac < bd$.

Written Assignment: In your own words, state the addition and the multiplication properties of equality and of order.

3.4 COMPOUND INEQUALITIES

Intervals of real numbers and their graphs will be useful in the study of inequalities. For example, the set of real numbers *between* -2 and 3 may be written in these forms:

Note that the word "between" means that the boundary points −2 and 3 are not included.

$$-2 < x < 3 \qquad \text{In interval notation: } (-2, 3)$$

$$\text{In set-builder notation: } \{x \mid -2 < x < 3\}$$

If both boundary points are to be included, then the inequality can be written as

$$-2 \le x \le 3 \qquad \text{In interval notation: } [-2, 3]$$

$$\text{In set-builder notation: } \{x \mid -2 \le x \le 3\}$$

The following summary shows various types of intervals. The solid dot in the graphs means that the boundary point is included in the interval. The open circle indicates that this point is *not* included.

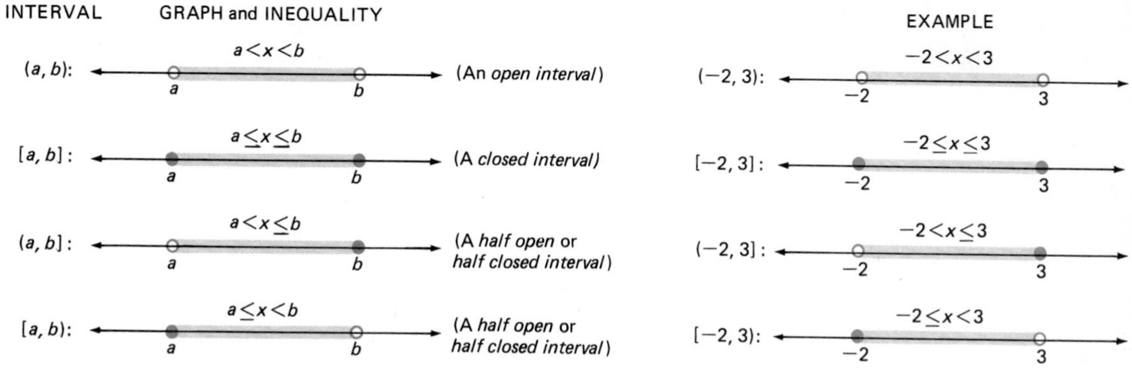

There are also *unbounded intervals*. For example, the set of all $x > 5$ is denoted by $(5, \infty)$. Similarly, $(-\infty, 5]$ represents all $x \le 5$. The symbols ∞ and $-\infty$ are read "plus infinity" and "minus infinity" but do *not* represent numbers. They are symbolic devices used to indicate that *all* x in a given direction, without end, are included, as in the following figure.

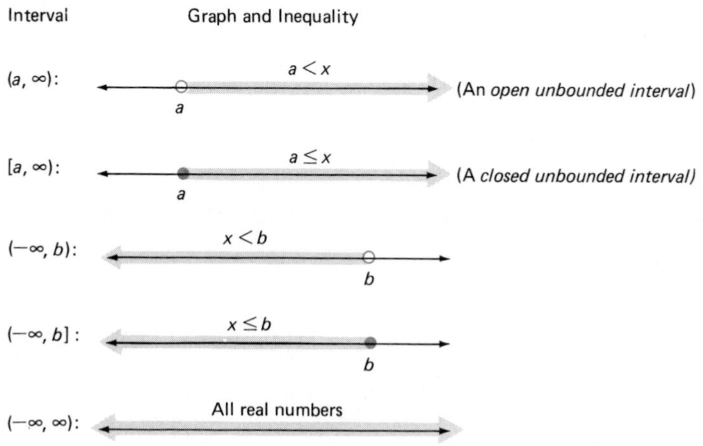

TEST YOUR UNDERSTANDING
Think Carefully

Show each of the following intervals as a graph on a number line.

1. $(-4, 1]$ **2.** $[0, 3]$ **3.** $[-1, 2)$ **4.** $(-2, \infty)$

Express each inequality in interval notation.

5. $-1 \le x \le 2$ **6.** $0 < x < 3$ **7.** $x \le 0$
8. $1 \le x < 4$ **9.** $-3 \le x$ **10.** $-5 < x < -2$

(Answers: Page 124)

Suppose that the conditions in a problem require that *2 more than x* be less than or equal to 3, and also that *1 more than x* be greater than or equal to -2. These two conditions can be expressed as a **compound inequality** that uses the word "and" as a *connective:*

$$x + 2 \le 3 \quad and \quad x + 1 \ge -2$$

A compound statement that uses the word "and" is true if and only if *both* parts are true. Thus, to solve the preceding compound we need to solve each of the inequalities $x + 2 \le 3$ and $x + 1 \ge -2$ separately, and then find all the common values in the two solution sets.

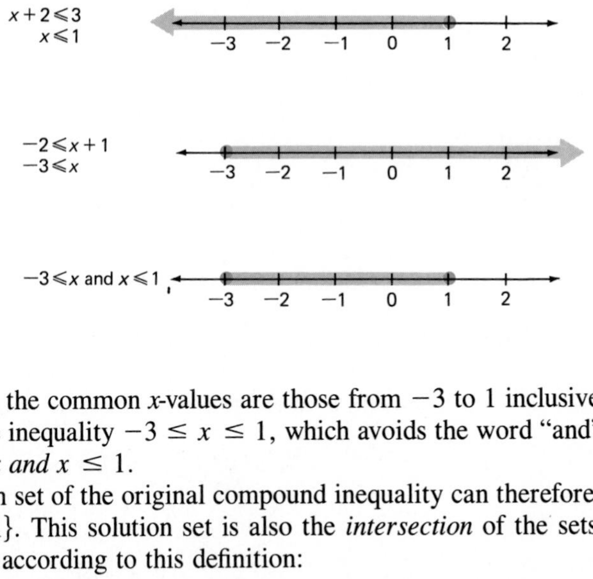

You can see that the common *x*-values are those from -3 to 1 inclusive. This can be expressed by the inequality $-3 \le x \le 1$, which avoids the word "and" but is equivalent to $-3 \le x$ *and* $x \le 1$.

The solution set of the original compound inequality can therefore be written as $\{x \mid -3 \le x \le 1\}$. This solution set is also the *intersection* of the sets $\{x \mid -3 \le x\}$ and $\{x \mid x < 1\}$, according to this definition:

> The **intersection** of two sets of numbers A and B, written $A \cap B$, is the set of numbers that are in both set A and B.

Since the interval notation $[-3, 1]$ is an abbreviation for the set of all x between and including the boundary points, it is also correct to write $[-3, 1] = \{x \mid -3 \le x \le 1\}$.

Thus,

$$\{x \mid -3 \le x\} \cap \{x \mid x \le 1\} = \{x \mid -3 \le x \le 1\}$$

EXAMPLE 1 Solve the compound inequality and graph the solution set: $2x + 4 \ge x + 1$ *and* $3x - 20 > -17$.

Solution Solve and graph the individual inequalities and find their intersection.

$2x + 4 \geqslant x + 1$
$x \geqslant -3$

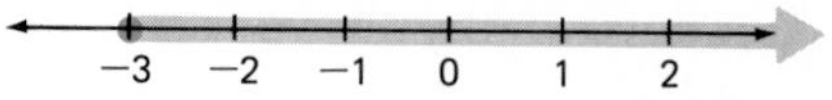

$3x - 20 > -17$
$3x > 3$
$x > 1$

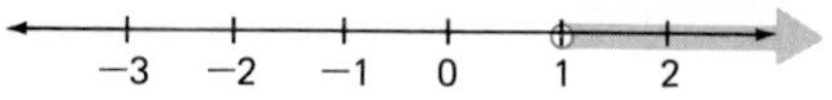

Form the intersection of the individual sets. Note that the common x-values are $x > 1$.

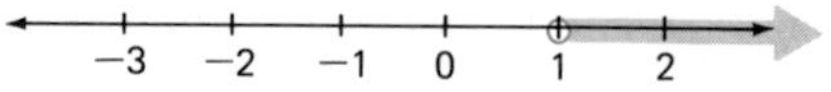

Solution set: $\{x \mid x > 1\}$ ■

EXAMPLE 2 Solve the compound inequality $3 < 2x + 7 < 15$ and graph the solution set.

Solution The given inequality is equivalent to

$$3 < 2x + 7 \quad and \quad 2x + 7 < 15$$

but it is not necessary to solve these individually. Rather, we can work with the given form as follows:

The adjacent graph is the ***open interval*** *(−2, 4) because it contains all numbers between the boundaries −2 and 4, but does not include these boundaries themselves.*

$$3 < 2x + 7 < 15$$

$$-4 < 2x < 8 \qquad \text{(Add } -7 \text{ to each part.)}$$

$$-2 < x < 4 \qquad \text{(Multiply each part by } \tfrac{1}{2}.\text{)}$$

The solution set is $\{x \mid -2 < x < 4\}$, and has the following graph:

■

EXAMPLE 3 Solve: $x + 2 > 3$ *and* $x + 3 < 2$

Solution First solve and graph each part separately.

$x + 2 > 3$
$x > 1$

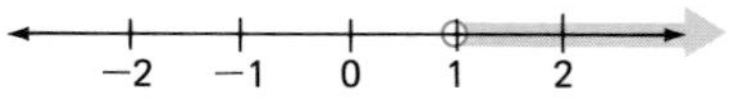

$x + 3 < 2$
$x < -1$

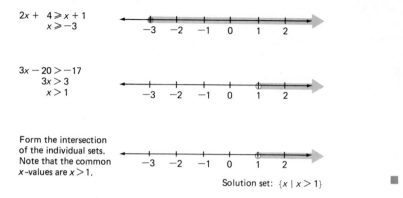

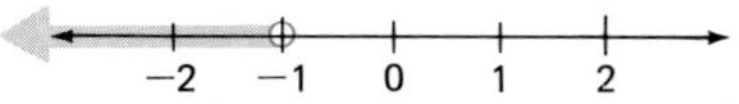

It should be clear that there are no points *common* to the two graphs; their intersection is the empty set. In other words, there are no real numbers that are both greater than 1 and less than −1 at the same time. The solution set is the empty set, $\varnothing$. ■

Another type of compound statement makes use of the connective "or." A statement that uses the word "or" is true if either part is true or if both parts are true. For example, let us graph the solution set for this compound statement:

$$2x + 3 < 1 \quad or \quad x - 5 > -4$$

First solve and graph each part separately.

$$2x + 3 < 1$$
$$2x < -2$$
$$x < -1$$

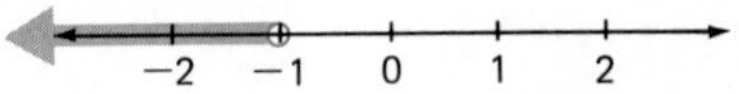

$$x - 5 > -4$$
$$x > 1$$

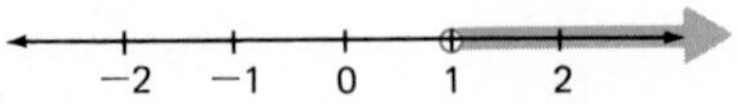

$x < -1$ or $x > 1$

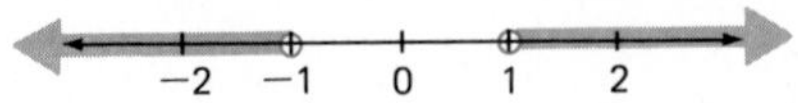

Note that the final graph consists of all points less than -1 as well as those that are greater than 1. The solution set is

$$\{x \mid x < -1 \quad or \quad x > 1\}$$

The set is also the *union* of the following two sets

$$\{x \mid x < -1\} \qquad \{x \mid x > 1\}$$

according to this definition:

> The **union** of two sets of numbers A and B, written $A \cup B$, is the set of numbers that are in set A or in set B or in both.

*Note that $\{x \mid 1 < x < -1\}$ is an **incorrect** way to write the solution. Can you explain why?*

Thus

$$\{x \mid x < -1\} \cup \{x \mid x > 1\} = \{x \mid x < -1 \quad \text{or} \quad x > 1\}$$

EXAMPLE 4 Graph the solution set: $2x - 1 \geq 5$ *or* $-2x + 1 \geq 3$

Solution First solve and graph each part separately.

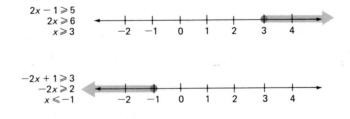

$$2x - 1 \geq 5$$
$$2x \geq 6$$
$$x \geq 3$$

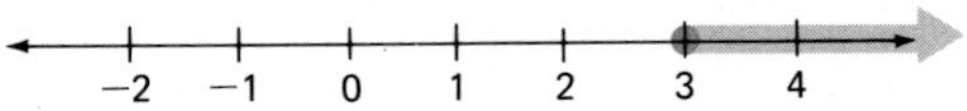

$$-2x + 1 \geq 3$$
$$-2x \geq 2$$
$$x \leq -1$$

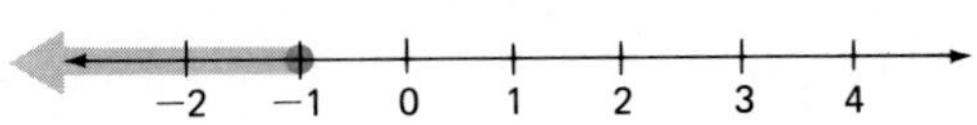

Now form the union of the two parts to give the graph of the solution set:

$$\{x \mid x \leq -1 \quad or \quad x \geq 3\}$$

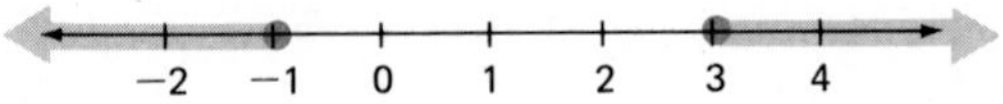

For the sake of simplicity we often will omit the set-builder notation. Thus the solution to this example can be written as $x \leq -1$ *or* $x \geq 3$. ■

TEST YOUR UNDERSTANDING
Think Carefully

Find the solution set.

1. $x - 2 < 7$ and $x + 3 > 5$
2. $3x + 2 < 5$ and $x + 1 > -2$
3. $x + 1 < 2$ or $x - 2 > 3$
4. $3x - 2 < 7$ or $4x + 1 > 17$
5. $2x - 3 \leq 5$ and $-3x + 1 \leq 7$
6. $2x + 1 \geq 5$ or $-2x - 1 \geq -3$

Graph the solution set.

7. $x + 1 < 3$ and $x - 1 > -3$
8. $x + 5 < 3$ or $x + 3 > 5$

(Answers: Page 125)

This example illustrates how inequalities can be used in an applied problem.

EXAMPLE 5 A small family business employs two part-time workers per week. The total amount of wages they pay to these employees ranges from \$128 to \$146 per week. If one employee will earn \$18 more than the other, what are the possible amounts earned by each per week?

Solution Let x = wages paid to the employee that earns the smaller amount. Then $x + 18$ = wages for the other employee. Since the sum of the wages is at least \$128 but no more than \$146, the sum $x + (x + 18)$ satisfies this compound inequality:

$$128 \leq x + (x + 18) \leq 146$$

Now simplify to get the possibilities for x.

Note: −18 is added to each part of the compound inequality, and then each part is divided by 2.

$$128 \leq 2x + 18 \leq 146$$
$$110 \leq 2x \leq 128$$
$$55 \leq x \leq 64$$

To get the result for the other employee, add 18 to the preceding inequality and simplify.

$$55 + 18 \leq x + 18 \leq 64 + 18$$
$$73 \leq x + 18 \leq 82$$

One part-time employee earns from \$55 to \$64 per week and the other earns from \$73 to \$82 per week. ■

EXERCISES 3.4

Show each of the following intervals as a graph on a number line.

1. $(-3, -1)$
2. $(-3, -1]$
3. $[-3, -1)$
4. $[-3, -1]$
5. $[0, 5]$
6. $(-1, 3)$
7. $(-\infty, 0]$
8. $[2, \infty)$

Express each inequality in interval notation.

9. $-5 \leq x \leq 2$
10. $0 < x < 7$
11. $-6 \leq x < 0$
12. $-2 < x \leq 4$

13. $-10 < x < 10$ 14. $3 \le x \le 7$ 15. $x < 5$ 16. $x \le -2$
17. $-2 \le x$ 18. $2 < x$ 19. $x \le -1$ 20. $x < 3$

Write an inequality for each of the following graphs. Also express each as an interval of real numbers.

For example: 

$-1 \le x \le 3; [-1, 3]$

21. 

22. 

23. 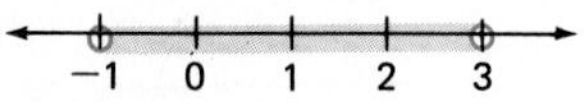

24. 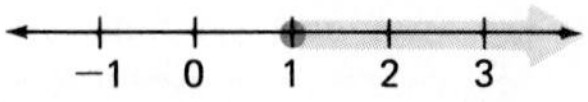

25. 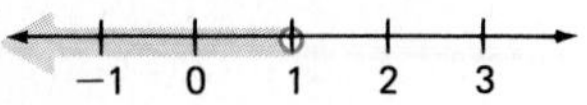

26. 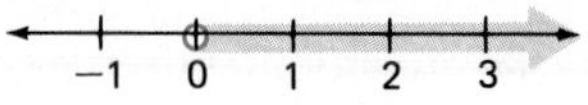

Graph each set.

27. $\{x \mid -1 < x < 2\}$ 28. $\{x \mid -2 \le x \le 1\}$
29. $\{x \mid x > 2 \text{ or } x < -1\}$ 30. $\{x \mid x \le -1 \text{ or } x \ge 2\}$
31. $\{x \mid x < -1 \text{ or } x < 1\}$ 32. $\{x \mid x < -1 \text{ and } x \le 1\}$

Solve the compound inequality and graph the solution set.

33. $x + 2 \le 5$ *and* $x + 3 \ge 1$ 34. $x + 2 < 1$ *or* $x - 1 > 1$
35. $0 < x + 2 < 7$ 36. $x - 1 > 4$ *or* $x + 2 < 3$
37. $2x + 3 > 5$ *and* $3x + 1 < 7$ 38. $2x + 3 \le 1$ *or* $3x + 1 \ge 10$
39. $-2x + 3 < 5$ *or* $-x + 1 > 4$ 40. $-7 < -3x - 1 \le 8$
41. $4(2x + 1) < 2(5 + x)$ *and* $3x - 1 > 2x - 3$
42. $-3x - 1 \ge 5 - x$ *and* $1 - x < -(3x - 2)$

Solve the compound inequality.

43. $2 < n + 3 < 7$ 44. $n + 1 \le 5$ *and* $n - 2 > 1$
45. $2n + 3 \le 2$ *or* $2n - 5 \ge -4$ 46. $n + 3 < 1$ *or* $n - 2 \le 1 - 3n$
47. $-2 \le 3n + 2 \le 7$ 48. $3n + 1 \le 9$ *and* $4n + 1 \le 8$
49. $\frac{5}{4}n - \frac{1}{2} \ge \frac{13}{4}$ *or* $\frac{n}{2} + \frac{1}{6} \le -\frac{7}{3}$ 50. $\frac{4}{5}n - \frac{7}{10} \ge \frac{1}{5}$ *or* $\frac{10}{12} - \frac{2}{3}n > 5$
51. $4 - \frac{3}{2}n \le 5\left(1 - \frac{1}{2}n\right)$ *and* $1 - \frac{n}{3} \ge -\frac{7}{12}n$
52. $\frac{1}{2}(5n - 2) \le \frac{3}{4}(n - 3)$ *or* $2(n + \frac{3}{4}) + 4 \ge \frac{11}{2}$

Graph the solution set.

*53. $(x < -1$ *or* $x > 1)$ *and* $x = 2$ *54. $x < -1$ *or* $(x > 1$ *and* $x = 2)$
*55. $(x \ge -3$ *and* $x \le 2)$ *and* $x > 0$ *56. $(x \le -3$ *or* $x \ge 1)$ *or* $x = -1$
*57. $(x < 2$ *and* $x > -2)$ *or* $x \ge 3$ *58. $(x \le -1$ *or* $x \ge 3)$ *and* $x < 0$

59. Write the compound statement having this graph.

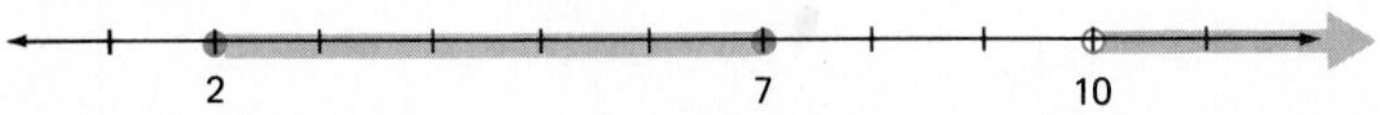

60. In order to get a grade of B+ in an algebra course, a student must have a test average of at least 86% but less than 90%. If the student's grades on the first three tests were 85%, 86%, and 93%, what grades on the fourth test would guarantee a grade of B+?

61. In Exercise 60, what grades on the fourth test would guarantee a B+ if the fourth test counts twice as much as each of the other tests?

62. If x satisfies $\frac{7}{4} < x < \frac{9}{4}$, then what are the possible values of y where $y = 4x - 8$? (*Hint:* Apply the multiplication and addition properties of inequality to the given inequality to obtain $4x - 8$ as the middle part.)

63. For a given time period, the temperature in degrees Celsius varied between 25° and 30°. What was the range in degrees Fahrenheit for this time period? (*Hint:* Begin with $25 < C < 30$ and apply the idea in Exercise 62 using $F = \frac{9}{5}C + 32$.)

64. Suppose that a machine is programmed to produce rectangular metal plates such that the length of the plate will be 1 more than twice the width w. When the entry $w = 2$ cm is made, the design of the machine will only guarantee that the width is within a one-tenth tolerance of 2. That is, $2 - 0.1 < w < 2 + 0.1$.

(a) Within what tolerance of 5 centimeters is the length?
(b) Find the range of values for the area.

65. A delivery service will accept a package only if the sum of its length ℓ and its girth g is no more than 110 inches. It also requires that each of the three dimensions be at least 2 inches.

(a) If $\ell = 42$ inches, then what are the allowable values for the girth g?
(b) If $\ell = 42$ inches and $w = 18$ inches, then what are the allowable values of h?

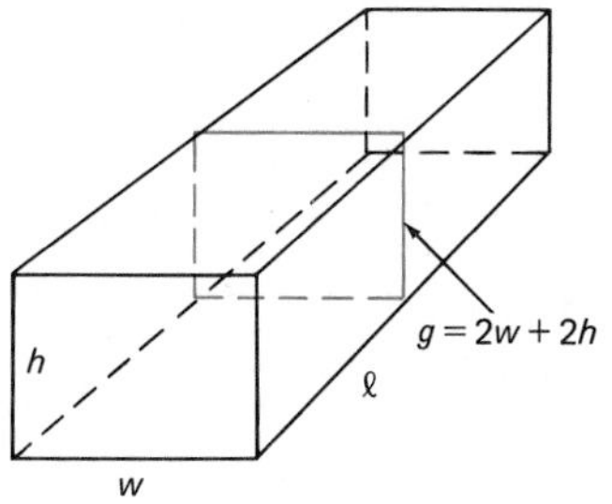

66. The sum of an integer and 5 less than three times this integer is between 34 and 54. Find all possible pairs of such integers.

67. A store has two part-time employees who together are paid a weekly total from \$150 to \$180. If one of the employees earns \$15 more than the other, what are the possible amounts earned by each per week?

68. A store has three part-time employees who together are paid a weekly total from \$210 to \$252. Two of them earn the same amount and the third earns \$12 less than the others. Find the possible amounts earned by each per week.

69. A supermarket has 20 part-time employees who together are paid a weekly total from \$1544 to \$1984. Twelve of them earn the same amount and the remaining eight each earn \$22 less. Find the possible amounts earned by each employee per week.

3.5 INEQUALITIES AND ABSOLUTE VALUE

Some equations and inequalities involve the absolute value of a number. For example, an equation such as $|x| = 2$ is really a compound statement in disguise. It is just another way of saying that $x = 2$ or $x = -2$. Thus the graph consists of the two points with coordinates 2 and -2; each of these points is 2 units from the origin.

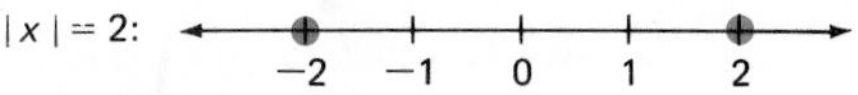

Now consider the inequality $|x| < 2$. Here we are considering all the real numbers whose absolute value *is less than* 2. On the number line, we can think of the solution as consisting of all points whose distance from the origin is less than 2 units, that is, all of the real numbers between -2 and 2. The set is:

Note that this can also be written as $x > -2$ and $x < 2$.

$$\{x \mid x > -2\} \cap \{x \mid x < 2\} = \{x \mid -2 < x < 2\}$$

The graph of $|x| < 2$ is the interval $(-2, 2)$.

$|x| < 2$:
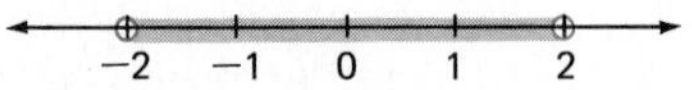

Next, let us draw the graph of $|x| > 2$. We need to find the set of real numbers whose absolute value *is greater than* 2. On the number line this calls for all the points whose distance from the origin is more than 2 units. We can express the solution set in this way:

Using interval notation the solution set can be written as

$$(-\infty, -2) \cup (2, \infty)$$

$$\{x \mid x < -2\} \cup \{x \mid x > 2\} = \{x \mid x < -2 \quad or \quad x > 2\}$$

$|x| > 2$:
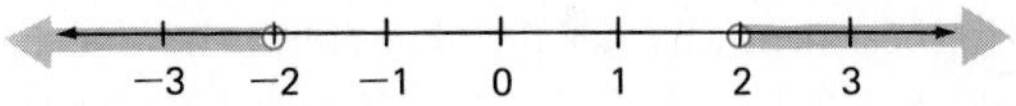

The preceding observations can be generalized as follows:

Properties (ii) and (iii) can be modified by replacing the inequality symbols, $<, >$ by $\leq, \geq$. For example, $|x| \leq k$ is equivalent to $-k \leq x \leq k$.

	Absolute Value Statement	Equivalent Compound Form
(i)	$\lvert x \rvert = k$	$x = k$ or $x = -k$
(ii)	$\lvert x \rvert < k$	$-k < x < k$
(iii)	$\lvert x \rvert > k$	$x < -k$ or $x > k$

EXAMPLE 1 Solve for x: $|x - 3| = 5$

Solution Let $x - 3$ play the role of x in Property (i). Therefore,

$$x - 3 = 5 \quad or \quad x - 3 = -5$$
$$x = 8 \quad or \quad x = -2$$

The solution set may also be written in the form $\{-2, 8\}$.

Check: $|8 - 3| = |5| = 5$; $|-2 - 3| = |-5| = 5$ ■

TEST YOUR UNDERSTANDING Think Carefully

(Answers: Page 125)

Solve.

1. $\lvert x \rvert = 5$	**2.** $\lvert x - 2 \rvert = 5$	**3.** $\lvert x + 2 \rvert = 5$
4. $\lvert x + 2 \rvert = 0$	**5.** $\lvert x + 2 \rvert = -1$	**6.** $\lvert x - 1 \rvert < 0$
7. $\lvert x - 1 \rvert \geq 0$	**8.** $\lvert 2x - 3 \rvert = 2$	**9.** $\lvert 2x + 3 \rvert = 2$

Note that the answer to each of the following equations is given as $\varnothing$.

$$|x + 2| = -1$$
$$|x - 1| < 0$$

In each case the solution set is the empty set because the absolute value of a number can never be less than 0. On the other hand, the solution to $|x - 1| \geq 0$ is the set of all real numbers because the absolute value of every number is always greater than or equal to 0.

Here are some other examples that deal with the graphs of statements of inequality involving absolute value.

EXAMPLE 2 Solve: $|x - 2| \leq 3$.

Solution Let $x - 2$ play the role of x in Property (ii). Consequently, $|x - 2| \leq 3$ is equivalent to $-3 \leq x - 2 \leq 3$. Now add 2 to each part to isolate x in the middle.

$$\begin{array}{rcl} -3 \leq & x - 2 & \leq 3 \\ +2 & +2 & +2 \\ \hline -1 \leq & x + 0 & \leq 5 \end{array} \qquad \text{Thus } -1 \leq x \leq 5.$$

■

In Example 2 we could also have solved the inequality by writing and solving two separate inequalities in this way:

$$|x - 2| \leq 3 \quad \text{means} \quad x - 2 \geq -3 \quad \textit{and} \quad x - 2 \leq 3$$

Thus the solution is $x \geq -1$ and $x \leq 5$. The word "and" indicates that we are to consider the values for x that satisfy *both* conditions. Therefore, the solution set consists of those numbers that are both greater than or equal to -1 and less than or equal to 5: that is, $\{x \mid -1 \leq x \leq 5\}$.

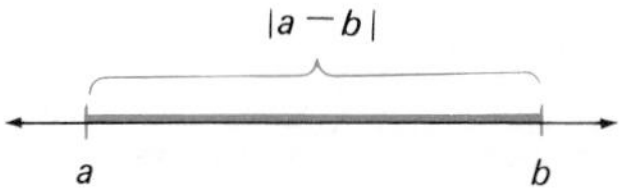

The quantity $|a - b|$ also represents the distance between the points a and b on the number line. For example, the distance between 8 and 3 on the number line is $5 = |8 - 3| = |3 - 8|$. The idea of distance between points on a number line can be used to give an alternative way to solve Example 2. Think of the expression $|x - 2|$ as the distance between x and 2 on the number line. Then consider all the points x whose distance from 2 *is less than or equal to* 3 units.

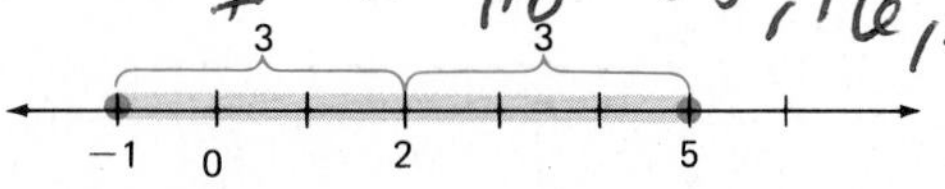

EXAMPLE 3 Graph the solution set of $|x + 1| > 2$.

Solution Rewrite as a compound statement using Property (iii).

$$x + 1 > 2 \quad \textit{or} \quad x + 1 < -2$$

Solving, we find that $x > 1$ *or* $x < -3$.

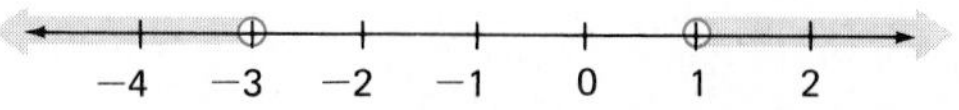

The solution set is $\{x \mid x < -3 \textit{ or } x > 1\}$. ■

As an alternative solution to Example 3, think of $|x + 1|$ as $|x - (-1)|$ so as to have it in the form $|a - b|$. This now represents the distance between x and -1, and we wish to find all the points that are more than 2 units from the point -1. The graph consists of all the points to the left of -3 and to the right of 1.

In summary, we can generalize the preceding properties as follows:

> If $|x - a| = k$, then $x - a = k$ or $x - a = -k$.
>
> If $|x - a| < k$, then $x - a > -k$ and $x - a < k$; that is, $-k < x - a < k$.
>
> If $|x - a| > k$, then $x - a < -k$ or $x - a > k$.

EXAMPLE 4 Convert each compound inequality into absolute value form.

(a) $3 < x < 7$ **(b)** $x < 3$ *or* $x > 7$

Solution

(a) $3 < x < 7$. First draw the graph.

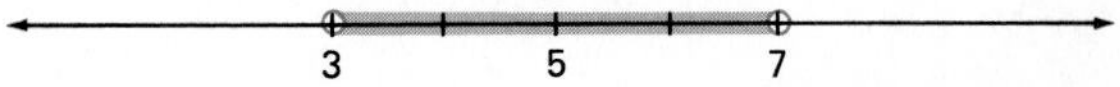

Observe that 5 is the *midpoint* of the segment. Therefore, if x is between 3 and 5, the distance from x to 5 is less than 2, as is the case if x is between 5 and 7. Thus the absolute value inequality is $|x - 5| < 2$.

(b) $x < 3$ or $x > 7$. First draw the graph.

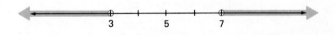

For each $x < 3$, the distance from 5 is more than 2, as is the case for each $x > 7$. Therefore, the absolute value inequality is $|x - 5| > 2$. ■

Recall that you will find sets of CAUTION items throughout the text. These illustrate errors that are often made by students. Study them carefully so that you understand the errors shown and can avoid such mistakes.

CAUTION: Learn to Avoid These Mistakes

WRONG	RIGHT
$|x| = -2$ has the solution $x = -2$.	There is no solution; the absolute value of a number can never be negative.
$|x - 1| < 3$ if and only if $x < 4$.	$|x - 1| < 3$ if and only if $-3 < x - 1 < 3$; that is, $-2 < x < 4$.
If $|x| > 6$, then $-6 < x$ *or* $x < 6$.	If $|x| > 6$, then $x < -6$ *or* $x > 6$.
If $|x - 7| < 4$, then the distance of x from 4 is less than 7 units.	If $|x - 7| < 4$, then the distance of x from 7 is less than 4 units.

EXERCISES 3.5

Solve for x and check.

1. $|x| = 3$
2. $|x + 1| = 3$
3. $|x - 1| = 3$
4. $|x + 1| = 0$
5. $|x - 3| = 1$
6. $|x + 3| = 1$
7. $|5 - x| = 8$
8. $|12 - x| = 4$
9. $|2x - 3| = 7$
10. $|3x + 4| = 16$

Graph the solution set.

11. $|x + 1| = 3$
12. $|x - 1| < 3$
13. $|x - 1| > 3$
14. $|x + 2| = 3$
15. $|x + 2| \le 3$
16. $|x + 2| > 3$
17. $|-x| = 5$
18. $|x| \le 5$
19. $|x| \ge 5$
20. $|x - 5| = \frac{1}{2}$
21. $|x - 5| \le 3$
22. $|x - 5| \ge \frac{1}{2}$
23. $|x - 3| < 1$
24. $|2 - x| < 3$
25. $|2x - 1| < 7$
26. $|3x - 6| < 9$
27. $|4 - x| < 2$
28. $|1 + 5x| < 1$
29. $|2x - 1| < 2$
30. $|4x - 1| > 1$
31. $2 - |x - 1| > 0$

Graph the solution set for Exercises 32 through 45.

32. $|x| \le 5$ *and* $|x| \ge 2$
33. $|x| \le 2$ *or* $|x| \ge 5$
34. $|x + 3| = |x| + 3$
35. $|x + 3| = x + 3$
36. $-2|x| + 5 \le |x| - 1$
37. $3|x| + 5 \le |x| + 7$
38. $|3x - 1| + 2 \ge 7$
39. $|2x + 3| + 5 \le 8$
40. $|x| \ge 3$ *or* $|x| \le 1$
41. $|x| \ge 1$ *and* $|x| \le 3$
42. $|x - 2| < 1$ *or* $|x + 2| > 6$
43. $|x + 1| > 2$ *and* $|x - 1| < 3$
44. $3|x + 1| < 2|x + 1| + 5$
45. $5|x - 1| < 3|x - 1| + 6$

46. If $|x - h| = k$, x represents the set of numbers whose distance from h on the number line is equal to k. Use this interpretation to solve the following for x:
    (a) $|x - 3| = 7$ **(b)** $|x + 4| = 5$ (*Hint:* Think of $x + 4$ as $x - (-4)$).
47. We can interpret $|x - h| \le k$ as the set of values of x for which the distance from h on the number line is less than or equal to k. Use this interpretation to solve the following for x:
    (a) $|x - 4| \le 8$ **(b)** $|x + 3| \le 1$
48. Following the manner of Exercises 46 and 47, interpret $|x - h| \ge k$ in terms of distance on the number line. Use this intepretation to solve the following for x:
    (a) $|x - 5| \ge 2$ **(b)** $|x + 5| \ge 3$

*49. Draw and compare the graphs of $|x| \le 2$ and $|x - 3| \le 2$. Do the same for $|x| \ge 2$ and $|x - 3| \ge 2$. What relationship can you state between each pair of graphs?

Write each inequality in terms of absolute value.

50. **(a)** $2 \le x \le 6$ **(b)** $x \le 2$ *or* $x \ge 6$
51. **(a)** $-2 < x < 4$ **(b)** $x \le -2$ *or* $x \ge 4$
52. **(a)** $-5 \le x \le 3$ **(b)** $x < -5$ *or* $x > 3$
53. **(a)** $2 < x < 5$ **(b)** $x < 2$ *or* $x > 5$

Use Property (ii) to prove that the first inequality implies the second.

Example: If $|x - 2| < \frac{1}{3}$, then

$$-\tfrac{1}{3} < x - 2 < \tfrac{1}{3} \qquad \text{(Property ii)}$$

$$-1 < 3x - 6 < 1$$

$$|3x - 6| < 1 \qquad \text{(Property ii)}$$

54. $|x - 4| < \frac{1}{2}$; $|2x - 8| < 1$
55. $|x - 6| < \frac{2}{5}$; $|\frac{1}{2}x - 3| < \frac{1}{5}$
56. $|x - \frac{1}{2}| < \frac{1}{12}$; $|4x - 2| < \frac{1}{3}$

57. **(a)** Confirm that $|x| \cdot |y| = |xy|$ using these pairs of values: $x = 7$, $y = 5$; $x = -7$, $y = 5$; $x = -7$, $y = -5$.

*(b) Here is a proof that $|x| \cdot |y| = |xy|$ when $x < 0$ and $y > 0$. Give a reason for each step. (*Hint:* See the definition of $|x|$ on page 14.)

$$
\begin{aligned}
|x| \cdot |y| &= (-x)y \qquad \text{(i)} \underline{\hspace{3cm}} \\
&= -(xy) \qquad \text{(ii)} \underline{\hspace{3cm}} \\
&= |xy| \qquad \text{(iii)} \underline{\hspace{3cm}}
\end{aligned}
$$

*58. Write a proof that $|x| \cdot |y| = |xy|$ when $x < 0$ and $y < 0$. (*Hint:* See Exercise 57(b).)

59. Cite at least four different examples to confirm this inequality:

$$|x - y| \geq ||x| - |y||$$

Be certain to illustrate each of these cases:
(a) $x > 0, y > 0$; **(b)** $x > 0, y < 0$;
(c) $x < 0, y > 0$; **(d)** $x < 0, y < 0$

60. Repeat Exercise 59 for the following inequality, known as the *triangle inequality:*

$$|x + y| \leq |x| + |y|$$

Solve for x.

*61. $\dfrac{|x + 2|}{x + 2} = 1$ *62. $\dfrac{|x - 2|}{x - 2} = 1$ 63. $\dfrac{1}{|x - 3|} > 0$ 64. $\dfrac{3}{|x - 2|} < 0$

Written Assignment: Explain how to obtain the graph of $|x - 2| < 3$ by using the graph of $|x| < 3$.

EXPLORATIONS Think Critically

1. If $a < b$ and $c < d$, is it true that $a - c < b - d$? Explain your answer.
2. What absolute value inequality represents the set of real numbers that are less than three units from the point -2 on the number line?
3. Describe a relationship in everyday life to satisfy each of the following:
   (a) symmetric and transitive, but not reflexive
   (b) transitive, but neither reflexive nor symmetric
   (*Hint:* Consider such relations as "is younger than," "is a sibling of," etc.)

*The **arithmetic mean** or average of two numbers a and b is* $\dfrac{a + b}{2}$. *The **geometric mean** or average of two positive numbers a and b is* $\sqrt{ab}$. *(Questions 4 and 5 deal with these averages.)*

4. Prove that for any real numbers a and b, with $a < b$, that $a < \dfrac{a + b}{2} < b$.
5. Prove that for any positive numbers a and b, the arithmetic mean is greater than or equal to the geometric mean: that is $\dfrac{a + b}{2} \geq \sqrt{ab}$ (*Hint:* Consider $(a - b)^2$)

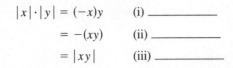

3.6 INTRODUCTION TO QUADRATIC EQUATIONS

Quadratic inequalities will be studied in Chapter 7.

Thus far we have dealt with linear equations and inequalities. In this section we shall introduce *quadratic equations*. The variable in a quadratic equation has an exponent of 2, as in the following:

$$x^2 - 8x - 3 = 0 \qquad x^2 - 4 = 0 \qquad 12x^2 - 25x = 7$$

In general:

Can you explain why we must say $a \neq 0$?

> A **quadratic equation** can be written in the form
>
> $$ax^2 + bx + c = 0$$
>
> where a, b, and c are real numbers and $a \neq 0$.

In each of the examples of quadratic equations shown above the variable appears to the *second power,* but not higher. Note also that the variable may appear to the first power as well. If the **quadratic polynomial** $ax^2 + bx + c$ can be factored, then the equation can be solved by making use of the **zero-product property.**

If the product $ab = 0$, then $a = 0$ or $b = 0$, or both a and b equal 0.

Using this property, here is the solution for $x^2 - 4 = 0$.

$$x^2 - 4 = 0$$

$$(x - 2)(x + 2) = 0 \qquad \text{(by factoring)}$$

$$x - 2 = 0 \quad \text{or} \quad x + 2 = 0 \qquad \text{(by the zero-product property)}$$

$$x = 2 \quad \text{or} \quad x = -2 \qquad \text{(by solving each equation for } x\text{)}$$

Note that the zero-product property reduces a quadratic equation to two linear equations that we already know how to solve.

The solution for $12x^2 - 25x - 7 = 0$ is found by using the same basic technique. See if you can explain each step in the following solution.

$$12x^2 - 25x - 7 = 0$$

$$(4x + 1)(3x - 7) = 0$$

$$4x + 1 = 0 \quad \text{or} \quad 3x - 7 = 0$$

$$x = -\tfrac{1}{4} \quad \text{or} \quad x = \tfrac{7}{3}$$

Check:

$$x = -\tfrac{1}{4}: \qquad 12(-\tfrac{1}{4})^2 - 25(-\tfrac{1}{4}) - 7 = \tfrac{3}{4} + \tfrac{25}{4} - 7 = 0$$

$$x = \tfrac{7}{3}: \qquad 12(\tfrac{7}{3})^2 - 25(\tfrac{7}{3}) - 7 = \tfrac{196}{3} - \tfrac{175}{3} - 7 = 0$$

The *roots* of the equation are $-\frac{1}{4}$ and $\frac{7}{3}$.

EXAMPLE 1 Solve: $4x^2 + 8x = -4$

Solution First divide through by 4.

$$x^2 + 2x = -1$$

Then write in the form $ax^2 + bx + c = 0$, factor, and use the zero-product property.

*In Example 1 the only solution to the equation is $x = -1$. This is called a **double root** of the equation.*

$$x^2 + 2x + 1 = 0$$
$$(x + 1)(x + 1) = 0$$
$$x + 1 = 0$$
$$x = -1$$

Check: $4(-1)^2 + 8(-1) = 4 - 8 = -4$ ■

EXAMPLE 2 Solve: $x^2 + 3x = 0$

CAUTION
A common error is to divide an equation like $x^2 + 3x = 0$ by x to obtain $x + 3 = 0$. This produces the root $x = -3$, but loses the other root, $x = 0$.

Solution First factor out the common x.

$$x(x + 3) = 0$$

Then use the zero-product property.

$$x = 0 \quad \text{or} \quad x + 3 = 0$$
$$x = 0 \quad \text{or} \quad x = -3$$

Check:

$$0^2 + 3(0) = 0: \qquad (-3)^2 + 3(-3) = 9 - 9 = 0$$ ■

TEST YOUR UNDERSTANDING
Think Carefully

Solve for x and check.

1. $x^2 + 2x - 15 = 0$
2. $x^2 - 16x + 64 = 0$
3. $2x^2 - x = 0$
4. $4x^2 + 12x = -9$
5. $6x^2 + 9x - 6 = 0$
6. $9x^2 - 1 = 0$
7. $6x^2 = -2x$
8. $4x^2 = 4$
9. $x^2 - x = 0$
10. $2x^2 - 7x + 6 = 0$
11. $9x^2 - 12x + 4 = 0$
12. $2x^2 + 5x = 3$

(Answers: Page 125)

There is a variety of word problems that gives rise to quadratic equations. If these equations can be factored, then they can be solved by the method just shown. For example, consider this problem:

> Find the number(s) such that, if 24 is added to its square, the result is 14 times the number.

In symbols, we are looking for x such that $x^2 + 24 = 14x$. To solve this equation, first add $-14x$ to both sides. This places the equation in the general form $ax^2 + bx + c = 0$. Then factor and use the zero-product property.

$$x^2 - 14x + 24 = 0$$

$$(x - 12)(x - 2) = 0$$

$$x - 12 = 0 \quad \text{or} \quad x - 2 = 0$$

$$x = 12 \quad \text{or} \quad x = 2$$

Therefore, the only possible solutions are 2 and 12. Both of these check in the original statement.

Review the guidelines for problem solving on page 90. Notice the importance here of drawing a suitable diagram.

EXAMPLE 3 A rectangular garden has dimensions 8 yards by 14 yards. Inside this rectangle a path (border) of uniform width is going to be constructed, leaving 55 square yards for grass and flowers. How wide should the path be?

Solution Using x as the width of the path, the dimensions of the inside rectangle become $8 - 2x$ by $14 - 2x$. Then, according to the problem,

$$(8 - 2x)(14 - 2x) = 55$$

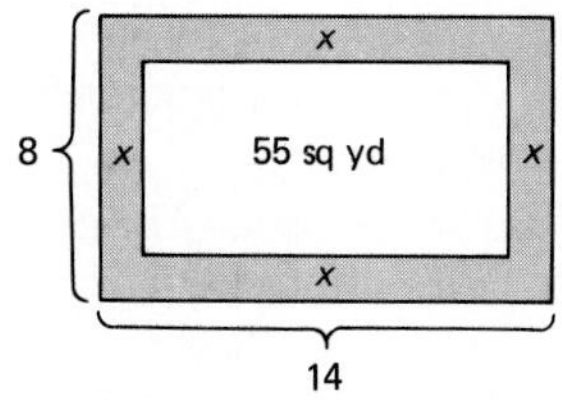

CAUTION
The zero-product property is sometimes used ***incorrectly****, as in the following:*

$$(8 - 2x)(14 - 2x) = 55$$

$$8 - 2x = 55 \quad \textit{or}$$
$$14 - 2x = 55$$

Explain why this is incorrect.

Since the right-hand side of this equation is not 0, we cannot set the factors equal to zero. It is first necessary to change this equation into the form $ab = 0$. To do this, first multiply the factors on the left:

$$112 - 44x + 4x^2 = 55$$

Add -55 to each side:

$$4x^2 - 44x + 57 = 0$$

Now factor:

$$(2x - 3)(2x - 19) = 0$$

Then

$$2x - 3 = 0 \quad \textit{or} \quad 2x - 19 = 0$$

$$x = \tfrac{3}{2} \quad \textit{or} \quad x = \tfrac{19}{2}$$

Hence the only possibilities for x are $\frac{3}{2}$ and $\frac{19}{2}$. Both of these check in the initial equation $(8 - 2x)(14 - 2x) = 55$. However, *we must go back to the original statement of the problem to check*. There it becomes clear that a path $\frac{19}{2}$ yards wide is not a possibility. (Why not?) With $x = \frac{3}{2}$, the inside length is $14 - 2(\frac{3}{2}) = 11$ and the width is $8 - 2(\frac{3}{2}) = 5$, leaving an area of $11 \times 5 = 55$ square yards, as required. The answer is $\frac{3}{2}$ yards. ■

The quadratic equations used in this section were carefully selected so that they could be solved by factoring. There are, however, many quadratics that cannot be solved this way. For example, the equation

$$x^2 - 8x - 3 = 0$$

cannot be solved by the factoring method, because $x^2 - 8x - 3$ is not factorable with integers. You will learn a general method that will allow you to solve this or any other quadratic equation in Chapter 5.

EXERCISES 3.6

Solve for x and check.

1. $x^2 - 4x + 4 = 0$
2. $x^2 + 14x + 49 = 0$
3. $x^2 - 10x = 0$
4. $x^2 = 100$
5. $10x^2 - 13x - 3 = 0$
6. $x = 2x^2$
7. $4x^2 - 32x = -64$
8. $20x^2 - 60x = -45$
9. $x^2 - 16x - 30 = 6$
10. $9x^2 = 64$
11. $(x + 1)(x + 2) = 30$
12. $3x^2 - 14x = 10x$
13. $2x(x + 6) = 22x$
14. $4x^2 = 40x - 100$
15. If 16 is added to the square of a number, the result is 25. Find all such numbers.
16. Find the number(s) that, if 12 is added to its square, the result is 13 times the number.
17. If 64 is added to the difference between the square of a number and 16 times the number, the result is zero. Find the number.
18. A rectangular garden is 10 yards by 20 yards. The garden consists of an inner rectangle that has an area of 144 square yards surrounded by a path of uniform width. How wide is the path?
19. A square piece of tin is 10 inches on a side. Squares of equal size are cut out of the four corners. If the resulting figure has an area of 64 square inches, how big is each of the squares?

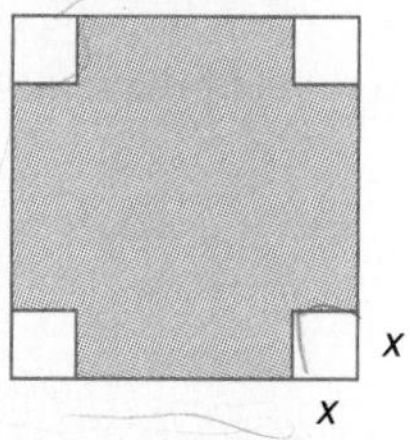

20. Find two positive numbers that differ by 7 and whose product is 144.
21. A number multiplied by twice itself is 242. Find all such numbers.
22. The unit's digit of a two-digit number is 3 more than the ten's digit. The square of the unit's digit is 6 more than the number. Find the number. (*Hint:* The number can be represented as $10t + (t + 3)$, where t is the ten's digit.)
23. Explain why the equation $x^2 + 1 = 0$ has no solutions.
24. Explain why the equation $(x + 1)^2 = -1$ has no solutions.
25. The length of a rectangle is 9 inches more than the width. If the length is decreased by 10 inches and the width is doubled, the area of the new rectangle is 42 square inches more than the original area. Find the dimensions of the original rectangle.
26. The base of a triangle is 7 inches longer than its altitude. If the altitude is increased by 3 inches and if the base is cut in half, the resulting triangle has an area that is 24 square inches less than the area of the original triangle. Find the base and altitude of the given triangle.
27. Factor and solve for x in the equation $x^3 + x^2 - 6x = 0$.
28. A rectangular patio with one of its longer sides against a house has dimensions 15 feet by 21 feet. The other three sides are surrounded by a path of uniform width. If the combined area of the patio and the path is 396 square feet, how wide is the path?
29. The area of a rectangle is 112 square centimeters. If the perimeter is 46 centimeters, what are the dimensions of the rectangle?

30. A ball that is thrown straight up from ground level reaches the height h feet in t seconds according to the formula $h = 144t - 16t^2$. Find how long it takes for the ball to be 320 feet high on its way up, and again on its way down.

31. A supermarket receives a shipment of boxes containing canned peaches. The number of boxes is 18 less than the number of cans in each box. Find the number of boxes and the number of cans in each box if there were 144 cans of peaches in the shipment.

32. Find four consecutive integers such that the sum of the squares of the first two is 83 more than the product of the last two.

33. Find three consecutive odd integers such that the product of the second and third is 27 less than twice the product of the first and second.

34. Factor by grouping and solve for x in the equation $x^3 + x^2 - 4x - 4 = 0$.

35. A rectangular box has a volume of 96 cubic inches. If the length is 8 inches more than the height, and the width is 2 inches less than the height, find the dimensions of the box. $(V = L \times W \times H)$

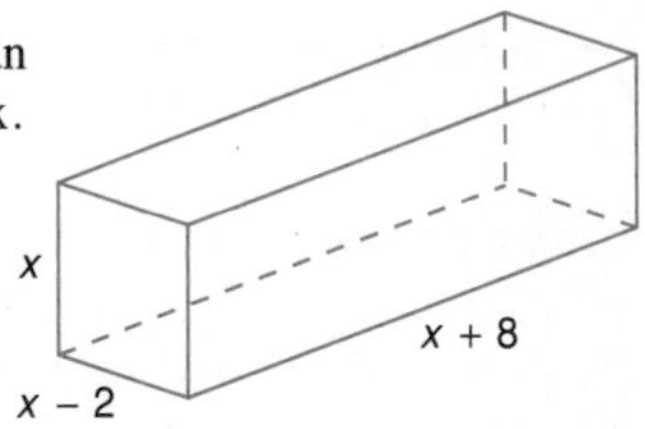

36. (a) Verify that $x^2 + x + 1 = (x + \frac{1}{2})^2 + \frac{3}{4}$.
(b) Use part (a) to explain why $x^2 + x + 1 = 0$ has no solutions.
(c) Use part (b) to explain why $x^3 - 1 = 0$ has only one solution.

CHAPTER 3 SUMMARY

Review these key terms and concepts so that you are able to define or describe them. A clear understanding of these will be very helpful when reviewing the developments of this chapter.

Properties of Equality:

Addition:	If $a = b$, then $a + c = b + c$.
Multiplication:	If $a = b$, then $ac = bc$.
Reflexive:	$a = a$
Symmetric:	If $a = b$, then $b = a$.
Transitive:	If $a = b$ and $b = c$, then $a = c$.

General Form of a Linear Equation: $ax + b = 0$ where $a \neq 0$.

Definition of $a < b$:

Geometric: a is to the left of b on the number line.

Algebraic: $b - a$ is positive.

Properties of Inequality:

Trichotomy Property:	For real numbers a and b, exactly one of the following is true: $a < b$ $\quad a = b$ $\quad a > b$
Transitive Property:	If $a < b$ and $b < c$, then $a < c$.
Addition Property of Order:	If $a < b$, then $a + c < b + c$. If $a > b$, then $a + c > b + c$.
Multiplication Property of Order:	If $a < b$ and $c > 0$, then $ac < bc$. If $a < b$ and $c < 0$, then $ac > bc$.

Interval Notation:

Bounded intervals		Unbounded intervals	
(a, b):	$a < x < b$	(a, ∞):	$x > a$
$(a, b]$:	$a < x \leq b$	$[a, \infty)$:	$x \geq a$
$[a, b)$:	$a \leq x < b$	$(-\infty, b)$:	$x < b$
$[a, b]$:	$a \leq x \leq b$	$(-\infty, b]$:	$x \leq b$

The **intersection** of two sets A and B, written $A \cap B$, is the set of numbers that are in both set A and set B.

The **union** of two sets A and B, written $A \cup B$, is the set of numbers that are in set A or in set B or in both.

Equations and Inequalities with Absolute Value:

1. $|x| = k$ is equivalent to $x = k$ or $x = -k$.
2. $|x| < k$ is equivalent to $x > -k$ and $x < k$; that is $-k < x < k$.
3. $|x| > k$ is equivalent to $x < -k$ or $x > k$.

General Form of a Quadratic Equation: $ax^2 + bx + c = 0$ where $a \neq 0$.
A quadratic equation that can be factored can be solved by using the **zero-product property:**

If $ab = 0$, then $a = 0$ or $b = 0$, or both a and b equal 0.

REVIEW EXERCISES

The solutions for the following exercises can be found within the text of Chapter 3. Try to answer each question before referring to the text.

Section 3.1

Solve.

1. $3x + 5 = 2x + 1$
2. $2(n + 3) = n + 5$
3. $5x - 3 = 3x + 1$
4. $\frac{3}{4}x - 2 = \frac{1}{3}x - 12$
5. $2x - 3 = 7$
6. $\dfrac{5F - 160}{9} = C$ for F
7. A formula that relates a person's weight, W, in pounds and height, H, in inches follows below. Solve for H in terms of W and use the result to approximate the height of someone who weighs 154 pounds.

$$W = \frac{11}{2}H - 220$$

Section 3.2

8. The length of a rectangle is one centimeter less than twice the width. The perimeter is 28 centimeters. Find the dimensions.
9. Find two numbers whose sum is 42 if the larger number is three less than twice the smaller number.
10. Each year a student uses the interest earned on two separate investments to help pay for college costs. The annual interest rates are 11% and 13%. If the total amount invested is \$15,200 and the total interest earned per year is \$1840, how much is invested at each rate?

11. A car leaves a certain town at noon, traveling due east at 40 miles per hour. At 1:00 P.M. a second car leaves the town traveling in the same direction at a rate of 50 miles per hour. In how many hours will the second car overtake the first car?

12. David has a total of \$4.10 in nickels, dimes, and quarters. He has twice as many nickels as dimes, and two more quarters than dimes. How many dimes does he have?

13. Marcella has a base salary of \$250 per week. In addition, she receives a commission of 12% of her sales. Last week her total earnings were \$520. What were her total sales for the week?

Section 3.3

Solve and graph.

14. $x + 2 < 7$

15. $-4n - (3 - 5n) > 8$

16. $3x + 2 \leq 2x - 1$

17. $5(3 - 2x) \geq 10$

18. A department store wishes to purchase a number of computers prior to running a sale. Each computer costs \$600 wholesale, plus a 5% sales tax. In addition, they must pay a \$50 delivery charge. The manager of the store has allocated at most \$10,000 for the initial purchase. How many computers should be ordered?

19. A car traveling at 54 mph passes another car that is traveling in the same direction at 46 mph. How long will it take for the faster car to be more than 10 miles ahead of the slower car?

Section 3.4

Solve the compound inequality and graph the solution set.

20. $x + 2 \leq 3$ and $x + 1 \geq -2$

21. $2x + 4 \geq x + 1$ and $3x - 20 \geq -17$

22. $3 < 2x + 7 < 15$

23. $x + 2 > 3$ and $x + 3 < 2$

24. $2x + 3 < 1$ or $x - 5 > -4$

25. $2x - 1 \geq 5$ or $-2x + 1 \geq 3$

26. A small family business employs two part-time workers per week. The total amount of wages they pay to these employees ranges from \$128 to \$146 per week. If one employee will earn \$18 more than the other, what are the possible amounts earned by each per week?

Section 3.5

27. Graph: $|x| < 2$

28. Solve: $|x - 3| = 5$

29. Solve and graph: $|x - 2| \leq 3$

30. Solve and graph: $|x + 1| > 2$

31. Convert each compound inequality into absolute value form.

(a) $3 < x < 7$ **(b)** $x < 3$ or $x > 7$

Section 3.6

Solve for x and check:

32. $x^2 - 4 = 0$

33. $12x^2 - 25x - 7 = 0$

34. $4x^2 + 8x = -4$

35. $x^2 + 3x = 0$

36. Find the number(s) such that, if 24 is added to its square, the result is 14 times the number.

37. A rectangular garden has dimensions 8 yards by 14 yards. Inside this rectangle a path (border) of uniform width is going to be constructed, leaving 55 square yards for grass and flowers. How wide should the path be?

CHAPTER 3 TEST: STANDARD ANSWER

Use these questions to test your knowledge of the basic skills and concepts of Chapter 3. Then check your answers with those given at the back of the book.

Solve for x.

1. $3x - 5 = 22$ **2.** $-2x + 3 = 11$ **3.** $4(x + 2) = 2x + 1$

4. $3(x - 1) = x + 5$ **5.** $7(x + 2) > 3(x - 2)$ **6.** $3(x + 2) < x + 10$

7. $3x + 3 \leq 20 + 5x$ **8.** $-5x + 1 \geq -2x + 12$

Graph on a number line.

9. $x + 2 > 7$ **10.** $-2x + 1 \geq 3$ **11.** $-2 < x < 1$

12. $3x - 2 \leq 10$ **13.** $|x + 2| \geq 1$ **14.** $|x - 3| < 2$

Solve and graph.

15. $x + 3 \leq 5$ *and* $x + 4 \geq 1$ **16.** $2x + 5 < 3$ *or* $3x - 2 > 4$

17. Show each interval on a number line: **(a)** $(-2, 0)$ **(b)** $[-1, 1]$ **(c)** $(-\infty, \frac{3}{2}]$

18. Solve for x and check.
(a) $3x^2 + 13x - 10 = 0$ **(b)** $(x - 1)(x + 1) = 8$

19. A formula for the surface area S of a cylinder with height h and radius of the base r is $S = 2\pi r^2 + 2\pi rh$. Solve this formula for h in terms of the other variables.

20. Convert each compound inequality into absolute value form.
(a) $1 < x < 5$ **(b)** $x < 3$ *or* $x > 7$

21. Find the dimensions of a rectangle whose perimeter is 52 inches if the length is 5 inches more than twice the width.

22. A total of $13,000 is invested in two separate accounts that pay 8% and 9% annual interest, respectively. If the 9% investment earns $258 less interest in a year than the 8% investment, then how much is invested at the 9% rate?

23. Two airplanes pass each other and continue in opposite directions at 240 mph and 320 mph, respectively. When will they be more than 140 miles apart?

24. A rectangular picture, including its frame, measures 12 inches by 16 inches. If the picture itself has an area of 96 square inches, how wide is the frame? (Assume the frame is of uniform width.)

25. **(a)** How many calculators can a department store order for $2500 if each calculator costs $18 wholesale, plus a 5% sales tax? There is also a $25 delivery charge for the entire shipment.
(b) Assuming the same sales tax and delivery charge, how many cameras can be bought if the first 100 cameras cost $18 each and those above 100 cost $12 each?

CHAPTER 3 TEST: MULTIPLE CHOICE

1. Which of the following makes use of the addition property of equality?
(a) If $3(x + 5) = 2x + 7$, then $3x + 15 = 2x + 7$.
(b) If $2x + 3 = 7$, then $2x + 3 + (-3) = 7 + (-3)$.
(c) If $\frac{2}{3}x = 8$, then $\frac{3}{2}(\frac{2}{3}x) = \frac{3}{2}(8)$.
(d) If $2x + 5 + x = x + 3$, then $3x + 5 = x + 3$.
(e) If $5 = x + 2$, then $x + 2 = 5$.

2. If $-3(x + 4) + x = 2(2x - 3) - x + 4$, then $x =$
(a) 10 **(b)** 2 **(c)** -2 **(d)** -10 **(e)** None of the preceding

3. If $3x - y = 3y + 4$, then $y =$

(a) $\frac{3}{4}x - 1$ (b) $\frac{4}{3}x + 1$ (c) $3x - 4$ (d) $3y - 4$ (e) None of the preceding

4. The sum of three consecutive integers is 48. Which of the following equations can be used to solve for the integers?

I. $n + (n + 1) + (n + 2) = 48$
II. $(n - 1) + n + (n + 1) = 48$
III. $(n + 1) + (n + 2) + (n + 3) = 48$

(a) Only I (b) Only II (c) Only III (d) Only I and II (e) I, II, and III

5. Manya has 20 coins consisting of dimes and quarters. The total value of the coins is \$3.80. Which equation can be used to find the number of dimes, d?

(a) $10d + 25(20d) = 380$ (b) $25d + 10(20 - d) = 380$
(c) $10d + 25(20 - d) = 380$ (d) $10d + (25 - d) = 380$ (e) $.10d + .25(d - 20) = 3.80$

6. The sum of an integer n and 4 more than twice itself is less than 91. Which inequality can be used to find the largest such integer?

(a) $n + (2n + 4) \geq 91$ (b) $3n + 4 < 91$ (c) $n + (4n + 2) < 91$
(d) $2(n + 4) < 91$ (e) None of the preceding

7. If $2n - 7 \leq 5n + 2$, then

(a) $n \leq -3$ (b) $n \leq 3$ (c) $n \geq -3$ (d) $n \geq 3$ (e) None of the preceding

8. Which of the following is the graph for $-2(n - 1) \geq 5 + n$?

(a) 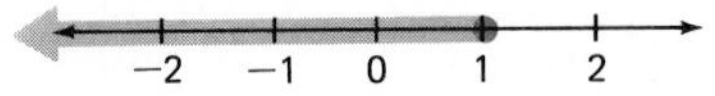

(b) 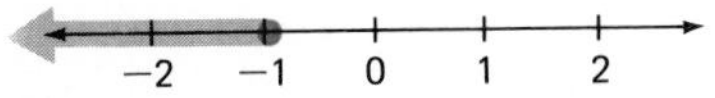

(c) 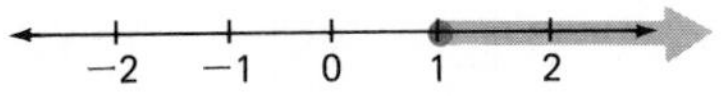

(d) 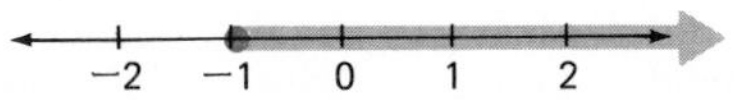

(e) None of the preceding

Which of the following is *not* true for all real numbers x?

9. (a) $x + 3 = 3 + x$ (b) $x + 2 > x$ (c) $x - 1 < x$
(d) $x + 3 = x$ (e) $|x| \geq x$

10. Which of the following is *not* represented by the graph at the right?

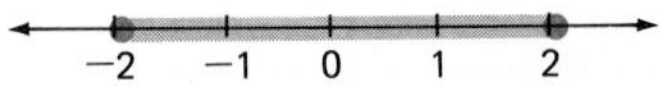

(a) $x + 3 \leq 1$ or $x - 1 \geq 1$ (b) $x + 3 \geq 1$ and $x - 1 \leq 1$
(c) $-2 \leq x \leq 2$ (d) $|x| \leq 2$ (e) $x \leq 2$ and $x \geq -2$

11. Assume that $a > b$ and x is negative. Which of the following is false?

(a) $a + x > b + x$ (b) $a - x > b - x$ (c) $\frac{a}{x} < \frac{b}{x}$ (d) $ax < bx$ (e) $2ax > 2bx$

12. Which of the following is the solution set for $|2x - 3| < 7$?

(a) $\{x | x < 5\}$ (b) $\{x | -5 < x < 5\}$ (c) $\{x | -2 < x < 5\}$
(d) $\{x | x > -2 \text{ or } x < 5\}$ (e) None of the preceding

13. Which of the following has as its graph all points within 3 units of -2 on the number line?

(a) $|x + 2| < 3$ (b) $|x + 2| > 3$ (c) $|x - 2| < 3$ (d) $|x - 2| > 3$ (e) None of the preceding

14. Twice the square of a number is 27 more than three times the number. Which of the following equations can be used to solve for all such numbers?

(a) $2x^2 = 3x - 27$ (b) $2x^2 + 3x = 27$ (c) $2x^2 + 3x - 27 = 0$
(d) $2x^2 - 27 = 3x$ (e) $(2x)^2 = 3x + 27$

15. The solution(s) for the equation $(x - 2)(x - 4) = 3$ is:

(a) $x = 2, x = 4$ (b) $x = 5, x = 7$ (c) $x = 1, x = 5$
(d) $x = 3$ (e) None of the preceding

ANSWERS TO THE TEST YOUR UNDERSTANDING EXERCISES

Page 85

1. $x = 6$ 2. $x = 17$ 3. $x = -4$ 4. $x = 6$ 5. $x = 13$
6. $x = 10$ 7. $x = 1$ 8. $x = 2$ 9. $x = 5$ 10. $x = -9$
11. $x = -3$ 12. $x = -11$ 13. $x = -64$ 14. $x = -12$ 15. $x = 64$

Page 88

1. $7x$ 2. $x - 10$ 3. $2x + (2x + 2)$ 4. $4x + 5x$
5. $5x - 2 = 33$ 6. $x + (x + 1) = 63$ 7. $x(x + 1) = 90$ 8. $3(x + 5) = 39$
9. $8(x - 3) = 48$ 10. $\frac{3}{4}x - 2 = 28$

Page 98

1. $\{x \mid x < 9\}$

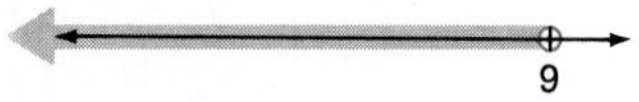

2. $\{x \mid x < 18\}$

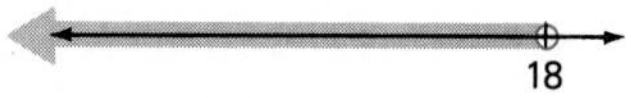

3. $\{x \mid x > 9\}$

4. $\{x \mid x > -5\}$

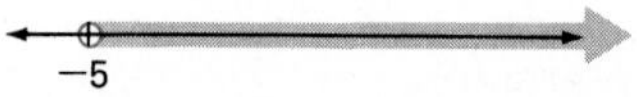

5. $\{x \mid x \leq 14\}$

6. $\{x \mid x \geq -2\}$

7. $\{x \mid x < 4\}$

8. $\{x \mid x \geq 15\}$

9. $\{x \mid x > -9\}$

10. $\{x \mid x \leq -43\}$

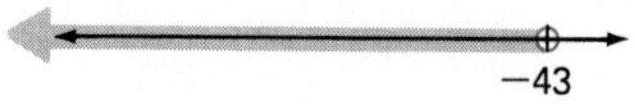

Page 104

1.
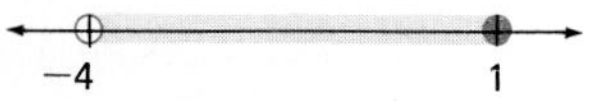

2.
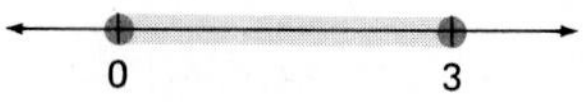

3.

4.
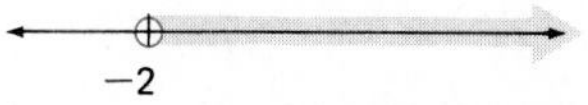

5. $[-1, 2]$ 6. $(0, 3)$ 7. $(-\infty, 0]$ 8. $[1, 4)$ 9. $[-3, \infty)$ 10. $(-5, -2)$

Page 107

1. $\{x \mid 2 < x < 9\}$
2. $\{x \mid -3 < x < 1\}$
3. $\{x \mid x < 1 \text{ or } x > 5\}$
4. $\{x \mid x < 3 \text{ or } x > 4\}$
5. $\{x \mid -2 \le x \le 4\}$
6. $\{x \mid x \le 1 \text{ or } x \ge 2\}$
7. 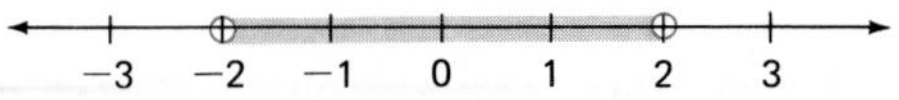

8. 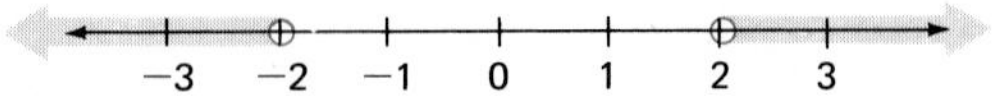

Page 110

1. $\{-5, 5\}$
2. $\{-3, 7\}$
3. $\{-7, 3\}$
4. $\{-2\}$
5. $\varnothing$
6. $\varnothing$
7. All real numbers
8. $\{\frac{1}{2}, \frac{5}{2}\}$
9. $\{-\frac{1}{2}, -\frac{5}{2}\}$

Page 116

1. $-5, 3$
2. 8
3. $0, \frac{1}{2}$
4. $-\frac{3}{2}$
5. $-2, \frac{1}{2}$
6. $-\frac{1}{3}, \frac{1}{3}$
7. $-\frac{1}{3}, 0$
8. $-1, 1$
9. $0, 1$
10. $\frac{3}{2}, 2$
11. $\frac{2}{3}$
12. $-3, \frac{1}{2}$

CUMULATIVE REVIEW QUESTIONS: CHAPTERS 1–3

Evaluate.

1. $(-8)[(+3) + (-7)]$
2. $(-5)[(-15) \div (+3)]$
3. $(-7)(-8) \div (-2)(+4)$
4. $8 + 6 \div 2 - 5 - (-1)$
5. $-|(-7) - (-2)|$
6. $(-2)(|-8| - |-5|)$

Which basic property of the real numbers is illustrated by each of the following?

7. $(-7) + (8 + 9) = (8 + 9) + (-7)$
8. $(-5)[2 + (-3)] = [2 + (-3)](-5)$
9. $(-6)[(-8) + 5)] = (-6)(-8) + (-6)(5)$
10. $(-12) + [5 + (-3)] = [(-12) + 5] + (-3)$

Classify each statement as true or false.

11. Every irrational number is a real number.
12. The set of rational numbers is a subset of the set of real numbers.
13. Every real number can be represented by either a terminating or a repeating decimal.
14. Every irrational number can be represented by a repeating decimal.
15. Every positive integer is a whole number.

Simplify. (Express all answers using positive exponents only.)

16. $\dfrac{(x^2y^3)^2y^3}{x^4y^4}$
17. $\dfrac{x^5y^0}{(5x + 1)^0}$
18. $\dfrac{(a^{-3}b^2)^3}{a^2b^{-3}}$
19. $\left(\dfrac{2x^2}{y}\right)^2\left(\dfrac{-2y}{x}\right)^3$
20. $\dfrac{x^{-5}y^{-2}}{x^{-2}y^{-5}}$
21. $\left(\dfrac{x^3y^0}{x^{-1}}\right)\left(\dfrac{y^{-2}}{x^{-2}y^{-1}}\right)^{-1}$

22. State, in symbols, the zero-product property.
23. Write in scientific notation: **(a)** 3,050,000 **(b)** 0.000057
24. Write in standard notation: **(a)** 4.72×10^5 **(b)** 1.08×10^{-4}

Perform the indicated operation and simplify.

25. $(7x^3 + 5x^2 - 7x + 8) + (x^3 - 8x^2 + 2x - 7)$
26. $(x^3 - 7x^2 + 9x - 2) - (2x^3 + x^2 - 5x + 7)$
27. $5x^2(3x^3 - x^2 + 5)$
28. $(x - 2)(2x^2 - x + 3)$
29. $(5x - 2y)(3x + 7y)$
30. $(2a^2 + b^2)(2a^2 + b^2)$

Factor, if possible.

31. $2x^3 - 50x$
32. $x^2 + 6xy + y^2$
33. $2x^2 + 5x + 3$
34. $x^4y - 64xy$
35. $3x^3 - 13x^2 + 14x$
36. $x^4 - 2x^3 + 3x - 6$

Solve for x.

37. $3(x + 7) = x - 5$
38. $\frac{2}{3}x + 1 = \frac{1}{2}x - 8$
39. $ax - b^2 = 2x$

40. The length of a rectangle is 3 centimeters more than twice the width. The perimeter is 36 centimeters. Find the dimensions.
41. Allison receives a base salary of \$225 per week. In addition, she receives a commission of 8% of her sales. Last week her total earnings were \$505. What were her total sales for the week?

Solve and graph.

42. $3x + 5 \le 2x + 3$
43. $|x + 1| \le 3$
44. $|3x - 2| > 1$
45. $2x + 3 \le x + 1$ and $-5x + 2 < 22$

Solve for x.

46. $x^2 - 5x = 0$
47. $4x^2 + 13x = 12$
48. $2x^2 - x - 3 = 0$
49. Show each interval of real numbers on a number line: **(a)** $(-3, 7]$ **(b)** $(-\infty, 2)$
50. How many VCRs can a store order for \$12,000 if each costs \$240 wholesale, plus a 5% sales tax? There is also a \$50 delivery charge for the total shipment.

CHAPTER 4 FUNDAMENTAL OPERATIONS WITH RATIONAL EXPRESSIONS

4.1 SIMPLIFYING RATIONAL EXPRESSIONS

A rational expression is the quotient of two polynomials, such as the following:

$$\frac{x+3}{x^2-2} \qquad \frac{n^2-5n+1}{2n+3} \qquad \frac{3a-5b}{2a+3b}$$

Since division by zero is not defined, we assume throughout our work that the denominators are *not* zero. This will save us the trouble of specifically stating this each time that we work with a fraction that has variables in the denominator.

EXAMPLE 1 For what values is the following rational expression undefined?

$$\frac{x^2+3x}{x^2-x-2}$$

Solution The expression is undefined when the denominator is equal to 0. Thus we set the denominator equal to 0 and solve for x:

$$x^2 - x - 2 = 0$$
$$(x+1)(x-2) = 0$$
$$x+1=0 \quad \text{or} \quad x-2=0$$
$$x=-1 \quad \text{or} \quad x=2$$

The given expression is undefined when $x = -1$ or when $x = 2$ since these values cause the denominator to be equal to 0. ■

A rational expression is in *simplest form,* or *reduced to lowest terms,* if the numerator and denominator have no common factor other than 1. This is the same idea as reducing a fraction in arithmetic to *lowest terms*. For example, $\frac{3}{4}$ is in its simplest form, but $\frac{9}{15}$ can be further simplified by dividing the numerator and denominator by their greatest common factor (GCF).

$$\frac{9}{15} = \frac{\overset{1}{\cancel{3}} \cdot 3}{\underset{1}{\cancel{3}} \cdot 5} = \frac{3}{5} \qquad \text{(The GCF of 9 and 15 is 3.)}$$

This example illustrates the *fundamental principle of fractions*.

CAUTION
Be careful to divide ***common factors*** *only. Observe that*

$$\frac{3+7}{5+7} = \frac{10}{12} = \frac{5}{6}$$

But this is ***wrong:***

$$\frac{3+\overset{1}{\cancel{7}}}{5+\underset{1}{\cancel{7}}} = \frac{4}{6}$$

7 is ***not*** *a common factor.*

FUNDAMENTAL PRINCIPLE OF FRACTIONS

For real numbers a, b, and c where $b \neq 0$ and $c \neq 0$,

$$\frac{ac}{bc} = \frac{a}{b}$$

According to this principle, both the numerator and denominator of a fraction may be multiplied by or divided by the same nonzero number without changing the value of the fraction.

In the following the instruction to *simplify* means to write the given fraction in lowest terms. This is done by using the fundamental principle to divide the numerator and denominator by their GCF. The result should be a fraction for which the GCF of the numerator and denominator is 1.

EXAMPLE 2 Simplify: $\dfrac{8x^2y^2}{12x^3y}$

Solution Write the numerator and denominator in a factored form that shows their GCF.

Note that the fundamental principle $\frac{ac}{bc} = \frac{a}{b}$ has been applied here using $a = 2y$, $b = 3x$, and $c = 4x^2y$, which is the GCF.

$$\frac{8x^2y^2}{12x^3y} = \frac{2y \cdot 4x^2y}{3x \cdot 4x^2y} = \frac{2y}{3x}$$ ■

Instead of identifying the GCF of the numerator and denominator, it is usually easier to reduce a fraction to lowest terms by dividing the numerator and denominator by their common factors. Thus the fraction in Example 2 can be simplified as in the following way:

$$\frac{8x^2y^2}{12x^3y} = \frac{\overset{2}{\cancel{8}}\overset{1}{\cancel{x^2}}\overset{y}{\cancel{y^2}}}{\underset{3}{\cancel{12}}\underset{x}{\cancel{x^3}}\underset{1}{\cancel{y}}} = \frac{2y}{3x}$$

As an alternative, the fraction can be simplified in this way:

$$\frac{8x^2y^2}{12x^3y} = \frac{2 \cdot 4 \cdot x \cdot x \cdot y \cdot y}{3 \cdot 4 \cdot x \cdot x \cdot x \cdot y} = \frac{2y}{3x}$$

EXAMPLE 3 Simplify: $\dfrac{x^2 + 5x + 6}{x^2 + 2x}$

Note that $x^2 + 2x \neq 0$ and thus $x \neq 0$ and $x \neq -2$.

Solution Factor the numerator and denominator. Then use the fundamental principle to divide the numerator and denominator by their greatest common factor.

$$\frac{x^2 + 5x + 6}{x^2 + 2x} = \frac{\cancel{(x + 2)}(x + 3)}{\cancel{(x + 2)}x} = \frac{x + 3}{x}$$

■

TEST YOUR UNDERSTANDING
Think Carefully

Indicate the values for which each rational expression is undefined.

1. $\dfrac{x^2 + 3x}{x^2 - 3x}$
2. $\dfrac{x^2 + 5}{x^2 - 4x - 5}$
3. $\dfrac{x^2 + 4}{x^2 - 4}$

Simplify.

4. $\dfrac{15xyz}{25x}$
5. $\dfrac{(2x^2y)^2}{8xy^3}$
6. $\dfrac{8x^3y^2}{-4x^2y}$
7. $\dfrac{-12a^4b^3}{-18a^2b^2}$
8. $\dfrac{x^2 + 2x}{x}$
9. $\dfrac{x^2}{x^2 + 2x}$
10. $\dfrac{x^2 - 9}{x^2 - 5x + 6}$
11. $\dfrac{x^2 + 6x + 5}{x^2 - x - 2}$
12. $\dfrac{x^2 - 4}{x^4 - 16}$
13. $\dfrac{3x^2 + x - 10}{-5x + 3x^2}$
14. $\dfrac{4x^2 + 4x - 3}{4x^2 + 12x + 9}$
15. $\dfrac{x^2 + x - 20}{30 + 16x + 2x^2}$

(Answers: Page 176)

EXAMPLE 4 Simplify: $\dfrac{5a - 3b}{3b - 5a}$

Solution At first glance it seems that no further simplification is possible. However, we have here a situation where the numerator is the negative (opposite) of the denominator. You can see this by factoring -1 out of the numerator.

$$\frac{5a - 3b}{3b - 5a} = \frac{(-1)(-5a + 3b)}{(1)(3b - 5a)} = \frac{(-1)\cancel{(3b - 5a)}}{(1)\cancel{(3b - 5a)}}$$

$$= \frac{(-1)}{(1)}$$

$$= -1$$

■

The work in Example 3 can be shortened by dividing the numerator and denominator by $3b - 5a$:

Note that any quotient of the form $\dfrac{x - y}{y - x}$ is equal to -1.

$$\frac{5a - 3b}{3b - 5a} = \frac{\overset{-1}{\cancel{5a - 3b}}}{\underset{1}{\cancel{3b - 5a}}} = \frac{-1}{1} = -1$$

Example 4 demonstrates that we need to be careful with the signs involved in a fraction. Recall from our work with integers that the quotient of a positive and a negative number is negative. For example:

$$\frac{-12}{4} = -3 \quad \text{and} \quad \frac{12}{-4} = -3$$

Thus

$$\frac{-12}{4} = \frac{12}{-4} = -\frac{12}{4} = -3$$

The same relationship holds for rational expressions, so that we may assign a negative sign (meaning -1) to either the numerator or the denominator. Generally, however, we place the sign so as to apply to the entire fraction:

CAUTION
What is **wrong** *with this?*
$\frac{-x+3}{5} = -\frac{x+3}{5}$

$$\frac{-a}{b} = \frac{a}{-b} = -\frac{a}{b}$$

Also note that $\frac{-a}{-b} = \frac{a}{b}$.

EXAMPLE 5 Simplify: $\frac{x^2 - 4}{2 - x}$

Solution

$$\frac{x^2 - 4}{2 - x} = \frac{\overset{-1}{\cancel{(x-2)}}(x+2)}{\underset{1}{\cancel{2-x}}}$$

$$= \frac{-1(x+2)}{1}$$

$$= \frac{-(x+2)}{1} \qquad (-1 \cdot n = -n, \text{ see page 17})$$

$$= -(x+2) \qquad \left(\frac{a}{1} = a\right)$$

EXAMPLE 6 Simplify: $\frac{x^2 - 6x + 9}{3x^2 - x^3}$

Solution

$$\frac{x^2 - 6x + 9}{3x^2 - x^3} = \frac{\overset{-1}{\cancel{(x-3)}}(x-3)}{\underset{1}{\cancel{(3-x)}}x^2}$$

$$= \frac{-(x-3)}{x^2}$$

$$= -\frac{x-3}{x^2} \qquad \left(\frac{-a}{b} = -\frac{a}{b}\right)$$

EXERCISES 4.1

Indicate the values, if any, for which each rational expression is undefined.

1. $\frac{x + 3y}{x}$
2. $\frac{x^2 + 5x}{x^2 - 5x}$
3. $\frac{2x + 3y}{x + y}$
4. $\frac{x^2 - 5x + 2}{x^2 + 2}$
5. $\frac{x^2 + 6x + 9}{x^2 - 9}$
6. $\frac{x^2 + 4x + 4}{x^2 - 4x + 4}$
7. $\frac{x^3 + x^2 + x}{x^2 + 4}$
8. $\frac{2x^2 + x - 3}{2x^2 - x - 3}$

Simplify, if possible.

9. $\frac{24}{40}$
10. $\frac{36}{60}$
11. $\frac{72}{48}$
12. $\frac{80}{120}$
13. $\frac{90}{54}$
14. $\frac{32}{160}$
15. $\frac{8xy}{12yz}$
16. $\frac{-3abc}{6ac}$
17. $\frac{24mn}{-3m}$
18. $\frac{3ax}{5by}$
19. $\frac{15x^2y}{20xy}$
20. $\frac{9ab}{3b^2}$
21. $\frac{-24abc^2}{36bc^2d}$
22. $\frac{-8x^2yz}{4xy^2z}$
23. $\frac{7a^3b^3}{14a^2b^5}$
24. $\frac{x + 2}{2 + x}$
25. $\frac{x - 2}{2 - x}$
26. $\frac{-(x - 2)}{2 - x}$
27. $\frac{2x - 3}{3 - 2x}$
28. $\frac{2x + 3}{3x + 2}$
29. $\frac{2x - 3}{3x - 2}$
30. $\frac{x^2 + 5x}{5x}$
31. $\frac{x^2 + 5x}{x + 5}$
32. $\frac{x^2 - 5x}{5 - x}$
33. $\frac{n + 1}{n^2 - 1}$
34. $\frac{n - 1}{n^2 - 1}$
35. $\frac{n + 1}{n^2 + 1}$
36. $\frac{x + 1}{(x + 1)^2}$
37. $\frac{(x + 1)^2}{x^2 - 1}$
38. $\frac{x - 2}{2 + x}$
39. $\frac{a^2 - a - 2}{(a + 1)^2}$
40. $\frac{b^2 + b - 2}{b^2 - 1}$
41. $\frac{(-a)^3}{a^5}$
42. $\frac{(x - 5)^2}{(5 - x)^2}$
43. $\frac{(x - 5)^3}{(5 - x)^3}$
44. $\frac{(x - 5)^4}{(5 - x)^2}$

*In Exercises 45 through 50, evaluate the fraction, if possible, for the given values of the variables. If not possible, say **undefined**.*

45. $\frac{15x}{6x^2y}$
    (a) $x = 4, y = 5$
    (b) $x = -2, y = 3$
46. $\frac{x^2 - 1}{2x - 4}$
    (a) $x = -1$
    (b) $x = 2$
47. $\frac{x - 6}{y + 8}$
    (a) $x = 6, y = 0$
    (b) $x = 0, y = -8$
48. $\frac{x + 9}{x^2 + 3x}$
    (a) $x = 3$
    (b) $x = -3$
49. $\frac{5x - 25}{x^2 - 10x + 25}$
    (a) $x = 5$
    (b) $x = 11.25$
50. $-\frac{x + 2y}{3x^2 + 6xy}$
    (a) $x = 2, y = -2$
    (b) $x = -2.45, y = 3.25$

Simplify.

51. $\frac{c^2 + 2c - 15}{c^2 + 8c + 15}$
52. $\frac{3x^2 + 3x - 6}{2x^2 + 6x + 4}$
53. $\frac{-x^3 + x}{x^3 - 2x^2 + x}$
54. $\frac{4x^2 + 12x + 9}{9 - 4x^2}$
55. $\frac{2n^2 + 7n + 3}{3 - 2n - n^2}$
56. $\frac{3n^2 + 5n - 2}{n^2 + 4n + 4}$
57. $\frac{6n^2 + 13n + 6}{2n^2 + 7n + 6}$
58. $\frac{4x^2 - 12x + 9}{4x^3 - 9x}$
59. $\frac{(x^2 - 9)(x^2 - 1)}{x^2 + 2x - 3}$

60. $\dfrac{8x^3 + 4x^2 - 24x}{2x^3 + 2x^2 - 4x}$ 61. $\dfrac{-5x - 15}{x^3 + 27}$ 62. $\dfrac{6 - 4a}{8a^3 - 27}$

63. $\dfrac{x^2 + 2x + xy + 2y}{x^2 + 4x + 4}$ 64. $\dfrac{a^2 - b^2}{a^2 - 6b - ab + 6a}$ 65. $\dfrac{a^2 - 16b^2}{a^3 + 64b^3}$

66. $\dfrac{x^2 + 2x^3 + 2 + 4x}{4x^2 - 1}$ 67. $\dfrac{16 - x^4}{x^3 - 2x^2 + 4x - 8}$ 68. $\dfrac{x^3 + 3x^2 - 2x - 6}{x^4 + 5x^2 - 14}$

Reduce to lowest terms. Assume that n is any positive integer.

69. $\dfrac{a^n b^{2n+1}}{a^{2n} b^{2n}}$ 70. $\dfrac{a^{n+1} - a^n}{a - 1}$ *71. $\dfrac{a^{2n} - b^{2n}}{a^{2n} + 2a^n b^n + b^{2n}}$

4.2 MULTIPLICATION AND DIVISION OF RATIONAL EXPRESSIONS

To multiply two fractions in algebra we follow the same rules that are used in arithmetic. The product is found as the product of the numerators divided by the product of the denominators.

Although the properties of fractions that we are studying here are given in terms of rational expressions, they apply to other kinds of fractions as well. For example:

$$\frac{\sqrt{x}}{5x} \cdot \frac{3}{\sqrt{x - 1}} = \frac{(3)(\sqrt{x})}{(5x)(\sqrt{x - 1})}$$

MULTIPLICATION OF RATIONAL EXPRESSIONS

For all real numbers a, b, c, and d, with $b \neq 0$ and $d \neq 0$,

$$\frac{a}{b} \cdot \frac{c}{d} = \frac{ac}{bd}$$

Here is an arithmetic example of this procedure:

$$\frac{3}{4} \cdot \frac{5}{7} = \frac{3 \times 5}{4 \times 7} = \frac{15}{28}$$

Here is an algebraic example:

Remember that division by zero is undefined. Therefore, $x \neq 2$ and $x \neq -2$.

$$\frac{x - 1}{x + 2} \cdot \frac{x + 1}{x - 2} = \frac{(x - 1)(x + 1)}{(x + 2)(x - 2)} \quad \text{or} \quad \frac{x^2 - 1}{x^2 - 4}$$

EXAMPLE 1 Multiply: $\dfrac{3a}{5b} \cdot \dfrac{5ac}{6a^2}$

Solution

Method 1. First multiply, then simplify.

$$\frac{3a}{5b} \cdot \frac{5ac}{6a^2} = \frac{15a^2c}{30a^2b}$$

$$= \frac{15a^2 \cdot c}{15a^2 \cdot 2b} \quad \text{(The GCF of the numerator and denominator is } 15a^2\text{.)}$$

$$= \frac{c}{2b}$$

CAUTION
Do you see what is **wrong** *with this?*

$$3\left(\frac{x+1}{x-1}\right) = \frac{3x+3}{3x-3}$$

This is the **correct** *procedure for this product:*

$$3\left(\frac{x+1}{x-1}\right) = \frac{3}{1}\cdot\frac{x+1}{x-1} = \frac{3x+3}{x-1}$$

Method 2. First divide (*cancel*) all common factors in numerators and denominators, then multiply.

$$\frac{\not{3}\not{a}}{\not{5}b}\cdot\frac{\not{5}\not{a}c}{\underset{2}{\not{6}}\not{a}^2} = \frac{c}{2b}$$ ■

Compare the two methods used in Example 1. In Method 1 the fractions were first multiplied and then the product was simplified. In Method 2 the simplification was done first. This was accomplished by dividing all common factors from the numerators and denominators of the given fractions, and then multiplying. The next example demonstrates these two procedures for two rational expressions whose numerators and denominators are not monomials.

CAUTION
Explain what is **wrong** *with this "simplification"*

$$\frac{\overset{x}{\not{x}^2} - \overset{1}{\not{9}}}{\underset{1}{\not{x}} + \underset{1}{\not{9}}} = x - 1$$

EXAMPLE 2 Multiply: $\dfrac{x^2-9}{x+9}\cdot\dfrac{x^2+x-2}{x^2-4x+3}$

Solution

Method 1. Multiply, factor, and then simplify.

$$\begin{aligned}\frac{x^2-9}{x+9}\cdot\frac{x^2+x-2}{x^2-4x+3} &= \frac{(x^2-9)(x^2+x-2)}{(x+9)(x^2-4x+3)}\\ &= \frac{(x+3)\cancel{(x-3)}(x+2)\cancel{(x-1)}}{(x+9)\cancel{(x-3)}\cancel{(x-1)}}\\ &= \frac{(x+3)(x+2)}{x+9}\end{aligned}$$

Method 2 is usually less work and will be used in the examples that follow.

Method 2. Factor, divide (cancel) common factors from the numerators and denominators, and then multiply.

$$\begin{aligned}\frac{x^2-9}{x+9}\cdot\frac{x^2+x-2}{x^2-4x+3} &= \frac{(x+3)\cancel{(x-3)}}{x+9}\cdot\frac{(x+2)\cancel{(x-1)}}{\cancel{(x-3)}\cancel{(x-1)}}\\ &= \frac{(x+3)(x+2)}{x+9}\end{aligned}$$ ■

EXAMPLE 3 Multiply: $\dfrac{x-y}{x+y}\cdot\dfrac{x^2-y^2}{x^2+y^2}\cdot\dfrac{2xy}{x^2-2xy+y^2}$

Solution Begin by factoring wherever possible.

$$\begin{aligned}&\frac{x-y}{x+y}\cdot\frac{(x-y)(x+y)}{x^2+y^2}\cdot\frac{2xy}{(x-y)(x-y)}\\ &= \frac{\cancel{x-y}}{\cancel{x+y}}\cdot\frac{\cancel{(x-y)}\cancel{(x+y)}}{x^2+y^2}\cdot\frac{2xy}{\cancel{(x-y)}\cancel{(x-y)}}\\ &= \frac{2xy}{x^2+y^2}\end{aligned}$$ ■

(Divide the common factors from the numerators and denominators.)

In summary, to multiply two or more rational expressions

1. Factor wherever possible.
2. Divide common factors in the numerators and denominators.
3. Multiply numerators and multiply denominators to find the product.

TEST YOUR UNDERSTANDING Think Carefully

Multiply.

1. $\frac{x}{3y} \cdot \frac{y^2}{x}$ **2.** $\frac{3a}{5b} \cdot \frac{b}{6a^2}$ **3.** $\frac{x-1}{x+1} \cdot \frac{3}{1-x}$

4. $\frac{2x-4}{2} \cdot \frac{x+2}{x^2-4}$ **5.** $\frac{x-2}{3(x+1)} \cdot \frac{x+2}{x^2+2x}$ **6.** $\frac{a}{b} \cdot \frac{ab+b^2}{a^2-ab}$

7. $\frac{a-b}{b-a} \cdot \frac{x+2}{x^2+2x}$ **8.** $\frac{x+y}{x-y} \cdot \frac{(x-y)^2}{x^2-y^2}$ **9.** $\frac{(a^2b)^2}{(ab)^3} \cdot \frac{b^4}{(a^3b^2)^2}$

(Answers: Page 176)

As with multiplication, division of two fractions in algebra follows the same rule as in arithmetic. To divide one fraction by a second one, the *divisor,* multiply the first fraction by the **reciprocal** of the divisor. Since the reciprocal of a fraction $\frac{c}{d}$ is $\frac{d}{c}$, the procedure for dividing two fractions can be stated as follows.

This rule is frequently described as "invert the divisor and multiply."

DIVISION OF RATIONAL EXPRESSIONS

For all real numbers a, b, c, and d, with b, c, and d not 0,

$$\frac{a}{b} \div \frac{c}{d} = \frac{a}{b} \cdot \frac{d}{c} = \frac{ad}{bc}$$

An important special case of this rule is obtained when $d = 1$. Thus

$$\frac{a}{b} \div c = \frac{a}{b} \cdot \frac{1}{c} = \frac{a}{bc}$$

EXAMPLE 4 Divide: **(a)** $\frac{3x^2}{5y} \div \frac{6x}{y^3}$ **(b)** $\frac{-10xy^2}{3} \div 15x^2y$

CAUTION
Do not invert the wrong fraction. This is **incorrect:**
$\frac{3x^2}{5y} \div \frac{6x}{y^3} = \frac{5y}{3x^2} \cdot \frac{6x}{y^3}$
Invert the **divisor** *as shown in Example 4(a).*

Solution

(a) $\frac{3x^2}{5y} \div \frac{6x}{y^3} = \frac{3x^2}{5y} \cdot \frac{y^3}{6x}$ (Invert the divisor.)

$= \frac{\overset{x}{\cancel{3x^2}}}{\cancel{5y}} \cdot \frac{\overset{y^2}{\cancel{y^3}}}{\underset{2}{\cancel{6x}}}$ (Cancel the common factors.)

$= \frac{xy^2}{10}$ (Multiply.)

(b) $\dfrac{-10xy^2}{3} \div 15x^2y = \dfrac{-10xy^2}{3} \cdot \dfrac{1}{15x^2y}$ (Think of $15x^2y$ as $\dfrac{15x^2y}{1}$ and invert.)

$= \dfrac{\overset{2\ \ y}{\cancel{-10}x\cancel{y^2}}}{3} \cdot \dfrac{1}{\underset{3\ \ x}{\cancel{15}\cancel{x^2}y}}$ (Cancel the common factors.)

$= -\dfrac{2y}{9x}$ (Multiply.) ■

EXAMPLE 5 Divide: $\dfrac{x^2 - 3x + 2}{x^2 - 4} \div \dfrac{x^2 + 3x}{x + 3}$

Solution

$$\frac{x^2 - 3x + 2}{x^2 - 4} \div \frac{x^2 + 3x}{x + 3} = \frac{x^2 - 3x + 2}{x^2 - 4} \cdot \frac{x + 3}{x^2 + 3x}$$

$$= \frac{(x - 1)\cancel{(x - 2)}}{\cancel{(x - 2)}(x + 2)} \cdot \frac{\cancel{(x + 3)}}{x\cancel{(x + 3)}}$$

$$= \frac{x - 1}{x(x + 2)}$$ ■

The next example demonstrates how the rules for multiplication and division of fractions apply when more than two fractions are involved.

EXAMPLE 6 Perform the indicated operations and simplify:

$$\left(\frac{a^2 - 4ab + 4b^2}{3ab^2} \div \frac{a^2 - 2ab}{6a + 6b}\right) \cdot \frac{-9a^2b}{a^2 - ab - 2b^2}$$

Solution

Explain each step in this solution.

$$\left(\frac{a^2 - 4ab + 4b^2}{3ab^2} \div \frac{a^2 - 2ab}{6a + 6b}\right) \cdot \frac{-9a^2b}{a^2 - ab - 2b^2}$$

$$= \frac{(a - 2b)^2}{3ab^2} \cdot \frac{6(a + b)}{a(a - 2b)} \cdot \frac{-9a^2b}{(a - 2b)(a + b)}$$

$$= -\frac{18}{b}$$ ■

EXERCISES 4.2

Perform the indicated operations and simplify.

1. $\dfrac{3}{8} \cdot \dfrac{5}{7}$
2. $\dfrac{4}{9} \cdot \dfrac{3}{5}$
3. $\dfrac{40}{9} \cdot \dfrac{18}{5}$
4. $\dfrac{21}{40} \div \dfrac{14}{20}$
5. $\dfrac{8}{19} \div \dfrac{3}{5}$
6. $\dfrac{14}{15} \div \dfrac{7}{10}$
7. $\dfrac{3x}{y} \cdot \dfrac{x}{2y}$
8. $\dfrac{2x^2}{y} \cdot \dfrac{y^2}{x^3}$
9. $\dfrac{15}{8x} \div 5x$
10. $5x \div \dfrac{15}{8x}$
11. $\dfrac{5x}{4y} \div \dfrac{1}{2y}$
12. $\dfrac{2}{3b} \div \dfrac{-8}{3b}$

13. $\frac{2x}{3y} \cdot \frac{4xy}{3x}$

14. $\frac{3a^2b}{5bc^2} \cdot \frac{5c^2}{3a^2}$

15. $\frac{ax^2}{2a^2} \cdot \frac{3b^2x}{ax^2}$

16. $\frac{-3abc^2}{b^2c} \cdot \frac{2b^2c^2}{3a^2b}$

17. $\frac{x^2}{y} \div \frac{x}{y^2}$

18. $\frac{ax}{ay} \div \frac{bx}{by}$

19. $\frac{2ab^2}{3b} \div \frac{2b^2}{3a}$

20. $\frac{3x^2}{2y^2} \div \frac{3x^3}{y}$

21. $\frac{a^2bc}{b^2c} \div \frac{b^2c^2}{2a^2}$

22. $\frac{-2b^3c^3}{3bc} \div \frac{ab^2c^2}{3c^3}$

23. $\frac{2a}{3} \cdot \frac{3}{a^2} \cdot \frac{1}{a}$

24. $\frac{a^2}{b^2} \cdot \frac{b}{c^2} \cdot \frac{1}{a}$

25. $\frac{x-1}{3} \cdot \frac{x^2+1}{x^2-1}$

26. $\frac{x^2-2x}{x-2} \cdot \frac{x-1}{x^2+x}$

27. $\frac{2ab}{a^2-25} \cdot \frac{a+5}{4b}$

28. $\frac{3s^2-6s}{s-3} \div \frac{9s^2}{3-s}$

29. $\frac{x^2-x-6}{x^2-3x} \cdot \frac{x^3+x^2}{x+2}$

30. $\frac{x^2-6x+9}{x^2-x-6} \cdot \frac{x^2+3x}{x^2-9}$

31. $\frac{x-1}{x+2} \div \frac{x^2-x}{x^2+2x}$

32. $\frac{x^2+3x}{x^2+4x+3} \div \frac{x^2-2x}{x+1}$

33. $\frac{x-3}{x+2} \div \frac{x^2-9}{x+3}$

34. $\frac{x-2}{x+2} \div \frac{x-y}{y-x}$

35. $\frac{x-1}{x^2+x} \div \frac{x-1}{x^2+2x}$

36. $\frac{x^3-x}{x^2-1} \div \frac{x-2}{x^2-x-2}$

37. $\frac{3st^2-19st+6s}{(st^2)^3} \cdot \frac{s^2t^5}{9t^2-1}$

38. $\frac{x^3-3x^2+2x}{x^2-5x+6} \div (x-1)$

39. $\frac{n^2+3n+2}{n^2+2n+1} \cdot \frac{n^2-n-2}{n^2-n-6}$

40. $\frac{n^2+n-6}{2n^2-3n-2} \cdot \frac{2n^2+9n+4}{n^2+7n+12}$

41. $\frac{a^2+2ab+b^2}{a^2-b^2} \div \frac{a^2+3ab+2b^2}{a^2-3ab+2b^2}$

42. $\frac{3a^3-3ab}{a^2-4ab+3b^2} \div \frac{2a^2b-2b^2}{a^2-9b^2}$

43. $\frac{x^2+2x+1}{2x^2+7x+3} \cdot \frac{3x^2+7x-6}{3x^2+x-2}$

44. $\frac{x^3+x^2-12x}{x^2-3x} \cdot \frac{3x^2-10x+3}{3x^2+11x-4}$

45. $\frac{a^2+ab-2b^2}{a^2-4b^2} \div \frac{a^2+4ab+3b^2}{a^2+ab-6b^2}$

46. $\frac{a^3+a^2b}{2a^2+5ab+3b^2} \div \frac{a^2-ab}{2a^2+ab-3b^2}$

47. $\frac{n^2+n}{2n^2+7n-4} \cdot \frac{4n^2-4n+1}{2n^2-n-3} \cdot \frac{2n^2+5n-12}{2n^3-n^2}$

48. $\frac{n^3-8}{n+2} \cdot \frac{2n^2+8}{n^3-4n} \cdot \frac{n^3+2n^2}{n^3+2n^2+4n}$

49. $\left(\frac{x^2+2xy-3y^2}{2x^2+3xy+y^2} \cdot \frac{2x^2+xy-y^2}{x^2-3xy+2y^2}\right) \div \frac{2x^2+5xy-3y^2}{x^2-4y^2}$

50. $\frac{a^3-27}{a^2-9} \div \left(\frac{a^2+2ab+b^2}{a^3+b^3} \cdot \frac{a^3-a^2b+ab^2}{a^2+ab}\right)$

51. $\left(\frac{xy-5x+3y-15}{(2xy^2)^2} \cdot \frac{2x^2-6x}{y^3-5y^2-y+5}\right) \div \frac{9-x^2}{4xy^3-4xy^2}$

52. Show that $\left(\frac{a}{b} \div \frac{c}{d}\right) \div \frac{e}{f} \neq \frac{a}{b} \div \left(\frac{c}{d} \div \frac{e}{f}\right)$ by performing the indicated operations and simplifying each side.

Perform the indicated operation. Then evaluate both the original expression and the simplified form for the given value.

53. $\frac{x^2-3x}{x+3} \cdot \frac{x^2+x-6}{x^2-2x}$; $x = 2.75$

54. $\frac{n^2+4n+4}{n^2+3n} \div \frac{n^2-3n-10}{n^2-2n-15}$; $n = 0.25$

Written Assignment: Justify the rule "invert and multiply" by considering the quotient $\dfrac{\frac{a}{b}}{\frac{c}{d}}$ and multiplying numerator and denominator by $\dfrac{d}{c}$.

4.3 ADDITION AND SUBTRACTION OF RATIONAL EXPRESSIONS

Addition and subtraction with fractions is relatively easy when the fractions have the same denominators. In that case we combine the numerators and use the common denominator.

These are the same rules used in arithmetic. For example:

$$\frac{3}{7}+\frac{2}{7}=\frac{3+2}{7}=\frac{5}{7}$$

$$\frac{7}{9}-\frac{2}{9}=\frac{7-2}{9}=\frac{5}{9}$$

ADDITION AND SUBTRACTION WITH LIKE DENOMINATORS

For all real numbers a, b, and c, with $c \neq 0$:

$$\frac{a}{c}+\frac{b}{c}=\frac{a+b}{c} \qquad \frac{a}{c}-\frac{b}{c}=\frac{a-b}{c}$$

These rules can easily be justified by using the distributive property. For example, consider the rule for addition of two fractions with the same denominator:

Note that $\dfrac{a}{c}=\dfrac{a}{1}\cdot\dfrac{1}{c}=a\left(\dfrac{1}{c}\right)$.

$$\begin{aligned}\frac{a}{c}+\frac{b}{c}&=a\left(\frac{1}{c}\right)+b\left(\frac{1}{c}\right)\\&=(a+b)\left(\frac{1}{c}\right)\\&=\frac{a+b}{c}\end{aligned}$$

The examples that follow illustrate the application of these rules.

EXAMPLE 1 Simplify: $\dfrac{x+2y}{x}+\dfrac{2x-y}{x}$

Solution Add the numerators to get the numerator of the sum. Use the common denominator x as the denominator of the sum.

$$\begin{aligned}\frac{x+2y}{x}+\frac{2x-y}{x}&=\frac{(x+2y)+(2x-y)}{x}\\&=\frac{3x+y}{x}\end{aligned}$$

■

The word *combine* is sometimes used when fractions are to be added or subtracted. This is used in the next example, which illustrates how to combine more than two fractions that have the same denominator.

EXAMPLE 2 Combine: $\dfrac{3x}{x-3} - \dfrac{x+10}{x-3} + \dfrac{4}{x-3}$

Solution The same operations between the fractions are used to form the numerator of the answer.

same sign

same sign

$$\frac{3x}{x-3} - \frac{x+10}{x-3} + \frac{4}{x-3} = \frac{3x - (x+10) + 4}{x-3}$$

$$= \frac{3x - x - 10 + 4}{x-3}$$

$$= \frac{2x-6}{x-3} = \frac{2(x-3)}{x-3}$$

$$= 2$$

CAUTION
The parentheses used here are very helpful. A common error is to write the following **incorrect** *numerator:*
$3x - x + 10 + 4 = 2x + 14$

In summary, to add or subtract rational expressions with a common denominator:

1. Combine the numerators by addition or subtraction.
2. Use this result as the numerator of the sum or difference.
3. Use the common denominator as the denominator of the sum or difference.
4. Reduce, if possible, by searching for common factors.

TEST YOUR UNDERSTANDING
Think Carefully

Combine as indicated and simplify.

1. $\dfrac{7}{9} + \dfrac{11}{9}$ **2.** $\dfrac{3}{8} - \dfrac{9}{8}$ **3.** $\dfrac{-4}{25} + \dfrac{9}{25}$

4. $\dfrac{x}{6} + \dfrac{4x}{6}$ **5.** $\dfrac{13y}{10} - \dfrac{3y}{10}$ **6.** $\dfrac{5}{x-1} + \dfrac{x}{x-1}$

7. $\dfrac{x}{x^2-4} + \dfrac{2}{x^2-4}$ **8.** $\dfrac{a^2}{a-2} - \dfrac{4a-4}{a-2}$ **9.** $\dfrac{2x^2}{x+3} + \dfrac{6x}{x+3}$

10. $\dfrac{9}{5} + \dfrac{7}{5} - \dfrac{6}{5}$ **11.** $\dfrac{5x+3}{x+1} - \dfrac{3x+1}{x+1}$ **12.** $\dfrac{t^2}{t-1} + \dfrac{t}{t-1} - \dfrac{2}{t-1}$

(Answers: Page 176)

When the fractions to be combined have different denominators we first change the fractions into equivalent forms having the same denominators, and then combine as before.

Compare the solutions for these two problems:

Observe the use of the fundamental principle of fractions in the reverse way from which we have been using it. That is, $\frac{3}{4} = \frac{3 \cdot 3}{4 \cdot 3}$ is an application of $\frac{a}{b} = \frac{ac}{bc}$, which says that the numerator and denominator of a fraction may be **multiplied** *by the same nonzero number.*

$$\frac{3}{4} + \frac{2}{3} \qquad\qquad \frac{3}{4x} + \frac{2}{3x}$$

$$\frac{3}{4} + \frac{2}{3} = \frac{3 \cdot 3}{4 \cdot 3} + \frac{2 \cdot 4}{3 \cdot 4} \qquad \frac{3}{4x} + \frac{2}{3x} = \frac{3 \cdot 3}{4x \cdot 3} + \frac{2 \cdot 4}{3x \cdot 4}$$

$$= \frac{9}{12} + \frac{8}{12} \qquad\qquad = \frac{9}{12x} + \frac{8}{12x}$$

$$= \frac{17}{12} \qquad\qquad = \frac{17}{12x}$$

To add or subtract rational expressions with different denominators, the first step is to find the least common denominator (LCD). Review the two preceding illustrations:

Although any common denominator could be used, it is less work to use the LCD.

$$\frac{3}{4} + \frac{2}{3} \qquad \text{The LCD is 12.}$$

$$\frac{3}{4x} + \frac{2}{3x} \qquad \text{The LCD is } 12x.$$

The least common denominator for two or more rational expressions can be found in this way:

1. Factor each denominator.
2. Use each factor to the highest power to which it appears in any of the denominators, and form their product.

EXAMPLE 3 Find the LCD for three denominators whose factored forms are

$$(x + 1)^2(x - 5), \quad (x + 1)(x - 5)^3, \quad \text{and} \quad (x + 1)^2(x - 5)^2(2x + 3)$$

Solution

The highest power for the factor $(x + 1)$ is $(x + 1)^2$.
The highest power for the factor $(x - 5)$ is $(x - 5)^3$.
The factor $(2x + 3)$ appears only to the first power.
Thus the LCD is $(2x + 3)(x + 1)^2(x - 5)^3$ ■

EXAMPLE 4 Add: $\frac{5}{x + 2} + \frac{3}{x - 1}$

Solution The LCD of the denominators is $(x + 2)(x - 1)$. Now use the fundamental principle to obtain *equivalent fractions* having a common denominator equal to the LCD.

Two fractions are **equivalent** *if they have the same values for all allowable replacements of the variable.*

$$\frac{5}{x + 2} = \frac{5(x + 1)}{(x + 2)(x + 1)} \qquad \frac{3}{x - 1} = \frac{3(x + 2)}{(x - 1)(x + 2)}$$

This is the complete solution:

$$\frac{5}{x+2}+\frac{3}{x-1}=\frac{5(x-1)}{(x+2)(x-1)}+\frac{3(x+2)}{(x-1)(x+2)}$$
$$=\frac{5(x-1)+3(x+2)}{(x+2)(x-1)}$$
$$=\frac{5x-5+3x+6}{(x+2)(x-1)}$$
$$=\frac{8x+1}{(x+2)(x-1)}$$

■

EXAMPLE 5 Subtract: $\dfrac{2x}{x^2-1}-\dfrac{x}{x-1}$

Solution First rewrite each fraction so that the denominators are in factored form; then note that the LCD is the product $(x-1)(x+1)$.

$$\frac{2x}{x^2-1}-\frac{x}{x-1}=\frac{2x}{(x-1)(x+1)}-\frac{x}{x-1}$$
$$=\frac{2x}{(x-1)(x+1)}-\frac{x(x+1)}{(x-1)(x+1)}$$
$$=\frac{2x-x(x+1)}{(x-1)(x+1)}$$
$$=\frac{2x-x^2-x}{(x-1)(x+1)}$$
$$=\frac{x-x^2}{(x-1)(x+1)}$$
$$=\frac{x(1-x)}{(x-1)(x+1)}$$
$$=-\frac{x}{x+1}$$

Note that $\dfrac{1-x}{x-1}=\dfrac{-1(x-1)}{x-1}=-1$

■

Let us use the procedures of these examples to find the sum $\dfrac{a}{b}+\dfrac{c}{d}$.

$$\frac{a}{b}+\frac{c}{d}=\frac{a}{b}\cdot\frac{d}{d}+\frac{c}{d}\cdot\frac{b}{b}$$
$$=\frac{ad}{bd}+\frac{bc}{bd}$$
$$=\frac{ad+bc}{bd}$$

The same procedure applies to the difference $\dfrac{a}{b}-\dfrac{c}{d}$ and leads to the following general rule.

Note: bd may not be the least common denominator.

ADDITION AND SUBTRACTION WITH UNLIKE DENOMINATORS

For all real numbers a, b, c, and d with $b \neq 0$ and $d \neq 0$:

$$\frac{a}{b} + \frac{c}{d} = \frac{ad + bc}{bd} \qquad \frac{a}{b} - \frac{c}{d} = \frac{ad - bc}{bd}$$

These rules are particularly useful when bd is the LCD. Thus, the work in Example 4 would be shortened since we would *begin* with this step:

$$\frac{5}{x+2} + \frac{3}{x-1} = \frac{5(x-1) + 3(x+2)}{(x+2)(x-1)}$$

However, when bd is not the LCD, these rules can create additional work and complexity. Try this in Example 5 and compare to the given solution.

TEST YOUR UNDERSTANDING
Think Carefully

Each of the following equations is ***incorrect*** *and represents an error that is commonly made when combining fractions. Replace the right side of each statement with the correct sum or difference.*

1. $\frac{2}{3} + \frac{x}{5} = \frac{2+x}{3+5}$ **2.** $\frac{1}{a} + \frac{1}{b} = \frac{1}{a+b}$ **3.** $\frac{x}{2} - \frac{5}{4} = \frac{x-5}{-2}$

4. $2 - \frac{x}{y} = \frac{2-x}{y}$ **5.** $\frac{x}{y} + 3 = \frac{x+3}{y}$ **6.** $\frac{x-2}{x^2-4} = \frac{1}{x-2}$

(Answers: Page 176)

EXERCISES 4.3

Perform the indicated operation and simplify.

1. $\frac{3}{5} + \frac{4}{5}$
2. $\frac{8}{9} - \frac{4}{9}$
3. $\frac{1}{2} + \frac{6}{5}$
4. $\frac{11}{8} - \frac{2}{3}$
5. $\frac{1}{2} + \frac{1}{3} + \frac{1}{4}$
6. $\frac{1}{2} + \frac{1}{5} - \frac{1}{10}$
7. $\frac{x}{5} + \frac{3x}{5}$
8. $\frac{x}{3} + \frac{2x}{3}$
9. $\frac{2x}{a} + \frac{3x}{a}$
10. $\frac{3x}{2} - \frac{x}{2}$
11. $\frac{8x}{5} - \frac{3x}{5}$
12. $\frac{7x}{a} - \frac{3x}{a}$
13. $\frac{x+2}{4} + \frac{x+3}{4}$
14. $\frac{x+1}{3} + \frac{x+2}{3}$
15. $\frac{2a-1}{5} + \frac{3a-4}{5}$
16. $\frac{2x+1}{3} - \frac{x+3}{3}$
17. $\frac{3x+4}{2} - \frac{x+2}{2}$
18. $\frac{5x-1}{4} - \frac{x-3}{4}$
19. $\frac{a+2b}{a} + \frac{3a+b}{a}$
20. $\frac{a+2b}{3a} + \frac{2a+b}{3a}$
21. $\frac{3x-y}{2x} - \frac{x-y}{2x}$
22. $\frac{a-2b}{2} - \frac{3a+b}{3}$
23. $\frac{2x+y}{5} + \frac{x+y}{3}$
24. $\frac{a+2}{2a} + \frac{3a-1}{3a}$
25. $\frac{3}{2x} + \frac{2}{x} + \frac{1}{3x}$
26. $\frac{7}{5x} - \frac{2}{x} + \frac{1}{2x}$
27. $\frac{x+2}{3x} + \frac{2x-3}{3x}$
28. $\frac{3x-1}{2x} - \frac{2x-3}{2x}$
29. $\frac{3x}{a+b} + \frac{2x}{a+b}$
30. $\frac{x+1}{x+3} + \frac{x+5}{x+3}$

31. $\dfrac{x^2}{x-1} - \dfrac{1}{x-1}$

32. $\dfrac{3x+17}{x+5} - \dfrac{x+7}{x+5}$

33. $\dfrac{3y}{y+1} + \dfrac{2y}{y-1}$

34. $\dfrac{3}{y^2+2y} + \dfrac{2}{y+2}$

35. $\dfrac{2a}{a^2-1} - \dfrac{a}{a+1}$

36. $\dfrac{a}{a-b} - \dfrac{b}{a+b}$

37. $\dfrac{3}{ab^2} + \dfrac{4}{a^3b} - \dfrac{1}{a^2b^2}$

38. $\dfrac{1}{x} - \dfrac{2}{3x^2} - \dfrac{4}{5x^3}$

39. $\dfrac{5a^2}{5a-2} - \dfrac{6-13a}{5a-2}$

40. $\dfrac{x^3+2}{x-1} - \dfrac{x^2-x}{x-1} - \dfrac{3}{x-1}$

41. $\dfrac{2x^2}{x^2+x} + \dfrac{x}{x+1}$

42. $\dfrac{3}{2x^2+4x} + \dfrac{1}{x^2-2x}$

43. $\dfrac{5x}{x^2+5x+6} - \dfrac{3}{x+2}$

44. $\dfrac{x}{x+3} - \dfrac{x^2-3x}{x^2+x-6}$

45. $\dfrac{x}{x+2} + \dfrac{x^2+3x}{x^2+3x+2}$

46. $\dfrac{3x+3}{2x^2-x-1} + \dfrac{1}{2x+1}$

47. $x - \dfrac{16x+8}{2x+1}$

48. $\dfrac{4x^2+2x-2}{x+1} - (3x-1)$

49. $\dfrac{2}{x^2+x-2} - \dfrac{x-8}{x^2-x-6}$

50. $\dfrac{5}{x^2-4} - \dfrac{3-x}{4-x^2}$

51. $\dfrac{2a-1}{3a} + \dfrac{a^2-2a}{4a^2} + \dfrac{5}{6a}$

52. $\dfrac{2x}{x^2-9} + \dfrac{x}{x^2+6x+9} - \dfrac{3}{x+3}$

53. $\dfrac{2}{x^2+x-6} + \dfrac{3}{x^3+2x^2-3x}$

54. $\dfrac{4x}{x^3-8} - \dfrac{2}{x-2}$

55. $\dfrac{1}{a^2-4} + \dfrac{3}{a-2} - \dfrac{2}{a+2}$

56. $\dfrac{a}{a-b} - \dfrac{2ab}{a^2-b^2} + \dfrac{b}{a+b}$

57. $\dfrac{35}{x^2-25} - \dfrac{2}{x-5} - \dfrac{7x}{25-x^2}$

58. $\dfrac{3x}{3x+7} - \dfrac{2}{x-2} + \dfrac{19x+14}{3x^2+x-14}$

59. $\dfrac{x}{x-1} - \dfrac{y}{y+1} - \dfrac{y+1}{xy+x-y-1}$

60. $\dfrac{1}{h-2} - \dfrac{h}{h^2+1} - \dfrac{h+3}{h^3-2h^2+h-2}$

Use the rule $\dfrac{a}{b} + \dfrac{c}{d} = \dfrac{ad \pm bc}{bd}$ *to find the following sums. Then find each sum by finding the LCD and compare your results.*

61. $\dfrac{3x}{x^2-4} + \dfrac{x}{x-2}$

62. $\dfrac{x}{x+3} + \dfrac{2x}{x^2+x-6}$

Evaluate the expression in the specified exercise using (a) $x = 5.42$ *and (b)* $x = -3.09$. *Compare these results to those obtained when these x-values are substituted into the simplified form previously obtained.*

63. Exercise 41

64. Exercise 43

CHALLENGE
Think Creatively

What is wrong with this story?

A man left 17 horses to his three children. He left one-half to the oldest, one-third to the middle child, and one-ninth to the youngest. Since 17 is not divisible by 2, 3, or 9, the children borrowed a horse from a neighbor in order to have a total of 18 horses. Then the oldest child received $\frac{1}{2} \times 18 = 9$ horses, the middle child received $\frac{1}{3} \times 18 = 6$ horses, and the youngest child received $\frac{1}{9} \times 18 = 2$ horses. Since $9 + 6 + 2 = 17$, the number of horses left to the three children, it was possible to return the extra horse to the neighbor!

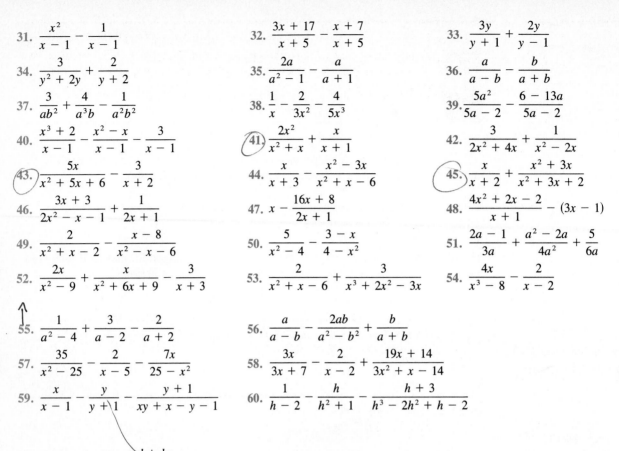

4.4 COMPLEX FRACTIONS

The fundamental properties of fractions can be used to simplify fractions whose numerators or denominators (or both) themselves contain fractions. Such fractions are sometimes called **complex fractions,** or **complex rational expressions.** We begin with a numerical illustration.

$$\frac{\frac{2}{3}}{\frac{5}{4}} = \frac{2}{3} \div \frac{5}{4} = \frac{2}{3} \cdot \frac{4}{5} = \frac{8}{15}$$

This simplification can also be done by first multiplying the numerator and denominator by 12, the LCD of the denominators of $\frac{2}{3}$ and $\frac{5}{4}$.

$$\frac{\frac{2}{3}}{\frac{5}{4}} = \frac{\left(\frac{2}{3}\right)12}{\left(\frac{5}{4}\right)12} = \frac{8}{15}$$

Similar procedures can be used when rational expressions are involved, as illustrated in the following examples.

EXAMPLE 1 Simplify:
$$\frac{\frac{a}{a+2}}{\frac{3a}{a+2}}$$

Solution Rewrite the complex fraction as a product, and multiply. That is, multiply the numerator of the complex fraction by the reciprocal of the denominator.

Verify this result mentally by multiplying the numerator and the denominator of the given fraction by a + 2.

$$\frac{\frac{a}{a+2}}{\frac{3a}{a+2}} = \frac{a}{a+2} \div \frac{3a}{a+2} = \frac{\not{a}}{\cancel{a+2}} \cdot \frac{\cancel{a+2}}{3\not{a}} = \frac{1}{3}$$ ■

EXAMPLE 2 Simplify:
$$\frac{x + \frac{1}{y}}{x - \frac{1}{y}}$$

Solution

Method 1. First combine the terms in the numerator and in the denominator.

$$\frac{x + \frac{1}{y}}{x - \frac{1}{y}} = \frac{\frac{xy+1}{y}}{\frac{xy-1}{y}}$$

Now convert to a multiplication problem and simplify.

$$\frac{\dfrac{xy+1}{y}}{\dfrac{xy-1}{y}} = \frac{xy+1}{y} \div \frac{xy-1}{y}$$

$$= \frac{xy+1}{y} \cdot \frac{y}{xy-1}$$

$$= \frac{xy+1}{xy-1}$$

Method 2. Multiply numerator and denominator by y and simplify.

Both methods shown are correct. In practice, Method 2 is usually more convenient to use.

$$\frac{x+\dfrac{1}{y}}{x-\dfrac{1}{y}} = \frac{\left(x+\dfrac{1}{y}\right)y}{\left(x-\dfrac{1}{y}\right)y}$$

$$= \frac{xy+1}{xy-1}$$ ■

EXAMPLE 3 Simplify: $\dfrac{\dfrac{1}{x^2}-\dfrac{1}{y^2}}{\dfrac{1}{x}-\dfrac{1}{y}}$

Solution Multiply both numerator and denominator by x^2y^2, the LCD of the denominators x^2, y^2, x, and y:

You should also solve Example 3 by first combining in the numerator and denominator, and then dividing.

$$\frac{\left(\dfrac{1}{x^2}-\dfrac{1}{y^2}\right)x^2y^2}{\left(\dfrac{1}{x}-\dfrac{1}{y}\right)x^2y^2} = \frac{y^2-x^2}{xy^2-x^2y}$$

$$= \frac{\cancel{(y-x)}(y+x)}{xy\cancel{(y-x)}}$$

$$= \frac{y+x}{xy}$$ ■

TEST YOUR UNDERSTANDING Think Carefully

Simplify.

1. $\dfrac{\dfrac{8}{15}}{\dfrac{2}{3}}$
2. $\dfrac{\dfrac{x^2}{3}}{\dfrac{x}{6}}$
3. $\dfrac{\dfrac{3x}{7}}{14}$
4. $\dfrac{24}{\dfrac{3}{x}}$
5. $\dfrac{\dfrac{a}{a-3}}{\dfrac{3a}{a-3}}$
6. $\dfrac{a-\dfrac{1}{b}}{a+\dfrac{1}{b}}$
7. $\dfrac{\dfrac{1}{a}-\dfrac{1}{3}}{a-3}$
8. $\dfrac{\dfrac{1}{a}+\dfrac{1}{b}}{\dfrac{1}{a^2}-\dfrac{1}{b^2}}$

(Answers: Page 176)

EXAMPLE 4 Simplify: $1 - \dfrac{1}{1 - \dfrac{1}{1 - \dfrac{1}{x}}}$

Solution Begin by simplifying the circled fraction.

$$1 - \frac{1}{1 - \dfrac{1}{1 - \dfrac{1}{x}}} = 1 - \frac{1}{1 - \dfrac{1 \cdot x}{\left(1 - \dfrac{1}{x}\right) \cdot x}}$$

$$= 1 - \frac{1}{1 - \dfrac{x}{x - 1}}$$

(Next simplify this circled fraction.)

$$= 1 - \frac{1 \cdot (x - 1)}{\left(1 - \dfrac{x}{x - 1}\right) \cdot (x - 1)}$$

$$= 1 - \frac{x - 1}{x - 1 - x}$$

$$= 1 - \frac{x - 1}{-1}$$

$$= 1 + x - 1$$

$$= x$$

The solution indicates that whatever value x is given in the original complex fraction, the simplified form will be that value. As a check, let $x = 5$ in the original fraction. Simplify the fraction and test that your answer will be 5. ■

EXAMPLE 5 Simplify: $\dfrac{\dfrac{1}{x + h} - \dfrac{1}{x}}{h}$

Solution First combine the terms in the numerator. Then invert and multiply.

Redo this problem by multiplying the numerator and the denominator by $x(x + h)$.

$$\frac{\dfrac{1}{x + h} - \dfrac{1}{x}}{h} = \frac{\dfrac{x - (x + h)}{x(x + h)}}{\dfrac{h}{1}}$$

$$= \frac{-h}{x(x + h)} \cdot \frac{1}{h}$$

$$= \frac{-1}{x(x + h)} = -\frac{1}{x(x + h)}$$ ■

Sometimes a fraction may contain variables with negative exponents that give rise to a complex fraction. This is illustrated in Example 6.

EXAMPLE 6 Simplify: $\dfrac{x + y^{-1}}{y + x^{-1}}$

Solution

$$\frac{x + y^{-1}}{y + x^{-1}} = \frac{x + \dfrac{1}{y}}{y + \dfrac{1}{x}}$$

$$= \frac{\left(x + \dfrac{1}{y}\right)xy}{\left(y + \dfrac{1}{x}\right)xy} \quad \text{(Multiply numerator and denominator by } xy.\text{)}$$

$$= \frac{x^2y + x}{xy^2 + y}$$

$$= \frac{x\cancel{(xy + 1)}}{y\cancel{(xy + 1)}} \quad \text{(Factor and cancel.)}$$

$$= \frac{x}{y}$$

Alternative Solution: Multiply numerator and denominator by xy:

$$\frac{x + y^{-1}}{y + x^{-1}} = \frac{(x + y^{-1})(xy)}{(y + x^{-1})(xy)}$$

$$= \frac{x^2y + x}{xy^2 + y}$$

$$= \frac{x\cancel{(xy + 1)}}{y\cancel{(xy + 1)}}$$

$$= \frac{x}{y}$$

■

EXERCISES 4.4

Simplify.

1. $\dfrac{\frac{1}{3}}{\frac{1}{3}}$

2. $\dfrac{\frac{3}{4}}{\frac{4}{3}}$

3. $\dfrac{\frac{8}{19}}{\frac{6}{15}}$

4. $\dfrac{\frac{7}{2}}{21}$

5. $\dfrac{18}{\frac{3}{5}}$

6. $\dfrac{-\frac{12}{13}}{\frac{8}{27}}$

7. $\dfrac{\frac{a}{b}}{c}$

8. $\dfrac{a}{\frac{b}{c}}$

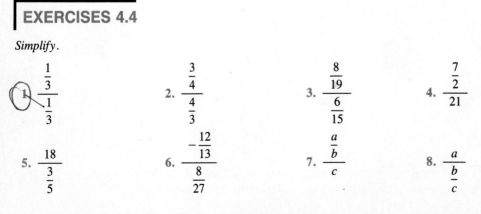

9. $\dfrac{\frac{a}{b}}{\frac{c}{d}}$ 10. $\dfrac{\frac{x}{2}}{\frac{x^2}{6}}$ 11. $\dfrac{\frac{2x}{5y}}{6y}$ 12. $\dfrac{-5a}{\frac{3b}{10a}}$

13. $\dfrac{\frac{x}{2y^2}}{\frac{3x^3}{y}}$ 14. $\dfrac{\frac{x}{x-1}}{\frac{2x}{x-1}}$ 15. $\dfrac{\frac{1}{x-3}}{\frac{2}{x^2-9}}$ 16. $\dfrac{\frac{a+b}{ab}}{\frac{a-b}{ab}}$

17. $\dfrac{2-\frac{1}{2}}{2+\frac{1}{2}}$ 18. $\dfrac{\frac{1}{2}+\frac{1}{3}}{6}$ 19. $\dfrac{\frac{1}{2}+\frac{3}{4}}{\frac{2}{3}-\frac{1}{2}}$ 20. $\dfrac{\frac{2}{5}-\frac{1}{5}}{\frac{4}{5}-\frac{3}{5}}$

21. $\dfrac{\frac{7}{10}-\frac{3}{10}}{\frac{11}{30}+\frac{7}{30}}$ 22. $\dfrac{1+\frac{1}{x}}{1+\frac{1}{y}}$ 23. $\dfrac{5-\frac{3}{7}}{\frac{9}{14}-1}$ 24. $\dfrac{x-\frac{2}{3}}{\frac{5}{6}-x}$

25. $\dfrac{x+\frac{3}{2}}{x-\frac{1}{2}}$ 26. $\dfrac{\frac{2x}{3}-2}{\frac{3x}{2}-3}$ 27. $\dfrac{\frac{1}{x}-\frac{1}{2}}{x-2}$ 28. $\dfrac{\frac{1}{3}-\frac{1}{9}}{\frac{1}{9}-\frac{1}{27}}$

29. $\dfrac{\frac{1}{x^2}-\frac{1}{9}}{x-3}$ 30. $\dfrac{\frac{1}{a^2}-\frac{1}{b^2}}{a+b}$ 31. $\dfrac{a-b}{\frac{1}{b^2}-\frac{1}{a^2}}$ 32. $\dfrac{\frac{2}{a}-\frac{2}{b}}{\frac{2}{b}-\frac{2}{a}}$

33. $\dfrac{\frac{1}{2+h}-\frac{1}{2}}{h}$ 34. $\dfrac{\frac{1}{x}+\frac{1}{y}}{\frac{1}{x^2}-\frac{1}{y^2}}$ 35. $\dfrac{\frac{x^2-25}{y}}{5x^2-x^3}$ 36. $\dfrac{\frac{1}{a+2}}{1+\frac{1}{a+2}}$

37. $\dfrac{\frac{a}{x-y}-\frac{a}{x+y}}{\frac{y}{x^2-y^2}}$ 38. $\dfrac{3-\frac{9}{a^2-1}}{\frac{3}{a-1}-\frac{1}{a+1}}$ 39. $\dfrac{x+\frac{6}{x-5}}{1-\frac{2}{x}}$ 40. $\dfrac{\frac{a+5}{a}-2}{\frac{a}{5}-\frac{5}{a}}$

41. $\dfrac{\frac{t}{t-3}+1}{\frac{2t-3}{t+3}}$

Simplify; combine fractions where possible.

42. $a^{-1}+b^{-1}$ 43. $a^{-2}-b^{-2}$ 44. $\dfrac{a^{-1}+b^{-2}}{a^{-2}+b^{-1}}$

45. $\dfrac{a^{-2}-b^{-1}}{a^{-1}-b^{-2}}$ 46. $1+\dfrac{1}{1+\dfrac{1}{1+\frac{1}{2}}}$ 47. $1-\dfrac{1}{1-\dfrac{1}{1-\frac{1}{2}}}$

48. $1-\dfrac{1}{1-\dfrac{1}{1-\dfrac{1}{1-\frac{1}{2}}}}$ 49. $1+\dfrac{1}{1+\dfrac{1}{1+\frac{1}{x}}}$ 50. $1-\dfrac{1}{1-\dfrac{1}{1-\dfrac{1}{1-\frac{1}{x}}}}$

51. There are three tests and a final exam given in a mathematics course. Let a, b, and c be the numerical grades of the tests and let d represent the exam grade.

(a) If the final grade is computed by allowing the average of the three tests and the exam to count the same, then show the final average is given by the expression $\dfrac{a + b + c + 3d}{6}$.

(b) Assume that the average of the three tests accounts for 60% of the final grade, and that the examination accounts for 40%. Show that the final average is given by the expression $\dfrac{a + b + c + 2d}{5}$.

52. Calculators frequently require that certain calculations be performed in a different manner so as to accommodate the machine. Show that in each case below the expression on the left can be computed by using the equivalent expression on the right.

(a) $\dfrac{A}{B} + \dfrac{C}{D} = \dfrac{\dfrac{A \cdot D}{B} + C}{D}$

(b) $A \cdot B + C \cdot D + E \cdot F = \left[\dfrac{\left(\dfrac{A \cdot B}{D} + C\right) \cdot D}{F} + E\right] \cdot F$

4.5 DIVISION OF POLYNOMIALS

The quotient of a polynomial by a monomial can be found in different ways. To illustrate this, consider the following division problem:

$$\frac{6x^2 + 9x^3 - 3x^5}{3x^2}$$

Usually the easiest way to find such a quotient is to divide each term in the polynomial by the monomial.

$$\frac{6x^2 + 9x^3 - 3x^5}{3x^2} = \frac{6x^2}{3x^2} + \frac{9x^3}{3x^2} - \frac{3x^5}{3x^2}$$
$$= 2 + 3x - x^3$$

As an alternative procedure, factor the numerator and simplify:

$$\frac{6x^2 + 9x^3 - 3x^5}{3x^2} = \frac{\cancel{3x^2}(2 + 3x - x^3)}{\cancel{3x^2}}$$
$$= 2 + 3x - x^3$$

In a division problem where N is divided by D to obtain Q, we write N/D = Q or N ÷ D = Q. N is called the **dividend,** *D is called the* **divisor,** *and Q is called the* **quotient.** *As a check, multiply Q times D to get N:*
$Q \cdot D = N$

To check the solution, multiply the *quotient* by the *divisor:*

$$3x^2(2 + 3x - x^3) = 6x^2 + 9x^3 - 3x^5$$

The result is the *dividend.*

EXAMPLE 1 Simplify: $\dfrac{6x^2y^2 - 4xy^2 + 8x^3y^4}{2xy^2}$

Solution Divide each term of the numerator by $2xy^2$.

$$\frac{6x^2y^2 - 4xy^2 + 8x^3y^4}{2xy^2} = \frac{6x^2y^2}{2xy^2} - \frac{4xy^2}{2xy^2} + \frac{8x^3y^4}{2xy^2}$$
$$= 3x - 2 + 4x^2y^2$$

Check this result by multiplying it by the divisor $2xy^2$. ■

TEST YOUR UNDERSTANDING Think Carefully

Perform each indicated division.

1. $(x^3 + 7x^2 - x) \div x$
2. $(8x^2 + 12x^4) \div 4x^2$
3. $(a^2b^3 + a^3b^2) \div ab^2$
4. $(a^6 + a^4 - a^2) \div a^2$
5. $\dfrac{6x^3y^2 - 9x^2y^2 + 12xy^2}{3xy}$
6. $\dfrac{15x^4 + 25x^6 - 10x^3}{5x^3}$
7. $\dfrac{-12a^2 + 8a^3 - 20a^4}{-4a^2}$
8. $\dfrac{a^2b^3 - a^3b^2 + a^4b^4}{a^2b^2}$
9. $\dfrac{12x^{12} - 8x^8 + 4x^6}{4x^4}$
10. $\dfrac{9y - 18y^2 + 27y^5}{-9y}$

(Answers: Page 176)

We can divide one polynomial by another quite easily if the divisor is a factor of the dividend. For example, here is one procedure for dividing $5x^2 - 22x + 21$ by $x - 3$:

$$\frac{5x^2 - 22x + 21}{x - 3} = \frac{(5x - 7)\cancel{(x - 3)}}{\cancel{(x - 3)}} = 5x - 7$$

However, a different procedure is needed when the divisor is not a factor of the dividend, or where the factorization of the dividend is not easily seen. For the preceding problem, the work can be completed in this form:

$$\begin{array}{r}
5x - 7 \\
x - 3\,\overline{)\,5x^2 - 22x + 21} \\
\underline{5x^2 - 15x} \\
-7x + 21 \\
\underline{-7x + 21} \\
0
\end{array}$$

The quotient is $5x - 7$ and the remainder is 0.

Rather than attempt to justify this procedure, we will compare it to the process used in long division in arithmetic. Compare the following solutions, step by step.

$$83\,\overline{)\,19505} \qquad\qquad x - 2\,\overline{)\,x^3 - x^2 - 5x + 6}$$

STEP 1. Divide 195 by 83. Place the result above the dividend.

STEP 1. Divide x^3 by x. Place the result above the dividend.

$$\begin{array}{r} 2 \\ 83\,\overline{)\,19505} \end{array} \qquad\qquad \begin{array}{r} x^2 \\ x - 2\,\overline{)\,x^3 - x^2 - 5x + 6} \end{array}$$

STEP 2. Multiply 83 by 2, and subtract.

$$\begin{array}{r} 2 \\ 83\overline{)19505} \\ \underline{166} \\ 29 \end{array}$$

STEP 2. Multiply $x - 2$ by x^2, and subtract:

$$\begin{array}{rl} & x^2 \\ x - 2 & \overline{)x^3 - x^2 - 5x + 6} \\ & \underline{x^3 - 2x^2} \\ & \phantom{x^3 - {}} x^2 \end{array}$$

STEP 3. Bring down the next digit, 0, and divide again: $290 \div 83$.

$$\begin{array}{r} 23 \\ 83\overline{)19505} \\ \underline{166} \\ 290 \end{array}$$

STEP 3. Bring down the next term, $-5x$, and divide again: $x^2 \div x = x$.

$$\begin{array}{rl} & x^2 + x \\ x - 2 & \overline{)x^3 - x^2 - 5x + 6} \\ & \underline{x^3 - 2x^2} \\ & \phantom{x^3 - {}} x^2 - 5x \end{array}$$

STEP 4. Repeat this process until you have used all digits. The completed division follows:

$$\begin{array}{r} 235 \\ 83\overline{)19505} \\ \underline{166} \\ 290 \\ \underline{249} \\ 415 \\ \underline{415} \end{array}$$

There is no remainder. To check, multiply 235 by 83. The result should be 19,505.

STEP 4. Repeat this process until you have used all terms. The completed division follows:

$$\begin{array}{rl} & x^2 + x - 3 \\ x - 2 & \overline{)x^3 - x^2 - 5x + 6} \\ & \underline{x^3 - 2x^2} \\ & \phantom{x^3 - {}} x^2 - 5x \\ & \phantom{x^3 - {}} \underline{x^2 - 2x} \\ & \phantom{x^3 - x^2 {}} -3x + 6 \\ & \phantom{x^3 - x^2 {}} \underline{-3x + 6} \end{array}$$

There is no remainder. To check, multiply $x^2 + x - 3$ by $x - 2$. The result should be $x^3 - x^2 - 5x + 6$.

EXAMPLE 2 Divide $5x + 3x^2 - 8$ by $x + 3$.

Solution First rewrite the dividend in *descending order*: $3x^2 + 5x - 8$

Note: When dividing polynomials it is important that the terms of the divisor and the dividend both be in descending order of powers of the variable. (Although not commonly done, they may also both be in ascending order.)

STEP 1. Divide: $3x^2 \div x = 3x$
Write $3x$ above the dividend.

STEP 2. Multiply: $(3x)(x + 3) = 3x^2 + 9x$

STEP 3. Subtract: $(3x^2 + 5x) - (3x^2 + 9x) = -4x$

STEP 4. Bring down the next term, -8.

$$\begin{array}{rl} & 3x \\ x + 3 & \overline{)3x^2 + 5x - 8} \\ & \underline{3x^2 + 9x} \\ & -4x - 8 \end{array}$$

Having completed these basic steps, we now repeat them, dividing x into $-4x$.

STEP 5. Divide: $-4x \div x = -4$
Write -4 above the dividend.

STEP 6. Multiply: $(-4)(x + 3) = -4x - 12$

STEP 7. Subtract: $(-4x - 8) - (-4x - 12) = 4$

$$\begin{array}{r} 3x - 4 \\ x + 3 \overline{\smash{)}\, 3x^2 + 5x - 8} \\ \underline{3x^2 + 9x\phantom{{}-8}} \\ -4x - 8 \\ \underline{-4x - 12} \\ 4 \end{array}$$

For a division problem in which there is a nonzero remainder R, we have
$\frac{N}{D} = Q + \frac{R}{D}$
Multiply by D to obtain
$N = QD + R$
Thus a check for such a problem consists of showing that $\boldsymbol{QD + R = N}$.

This process stops when the remainder has degree less than the degree of the divisor. Thus we stop because the degree of $4 = 4x^0$ is 0, which is less than the degree of $x + 3$, which is 1.

The quotient is $3x - 4$, and the remainder is 4. To check, the dividend should be the product of the quotient and the divisor, plus the remainder:

$$\begin{aligned}(x + 3)(3x - 4) + 4 &= 3x^2 + 5x - 12 + 4 \\ &= 3x^2 + 5x - 8\end{aligned}$$ ■

When writing a dividend in descending order, there may be some powers of the variable missing. When such is the case, we denote these by using a form of 0. For example, if there is no x^2 term, we write this as $0x^2$, as in Example 3.

EXAMPLE 3 Divide $x^3 + 5x - 7$ by $x - 1$.

Solution Write the dividend as $x^3 + 0x^2 + 5x - 7$. See if you can explain each step of the solution.

Check the result. Multiply the quotient $x^2 + x + 6$ by the divisor $x - 1$ and add the remainder -1 to the product. The result should be the dividend, $x^3 + 5x - 7$.

$$\begin{array}{r} x^2 + x + 6 \\ x - 1 \overline{\smash{)}\, x^3 + 0x^2 + 5x - 7} \\ \underline{x^3 - x^2\phantom{{}+5x-7}} \\ x^2 + 5x\phantom{{}-7} \\ \underline{x^2 - x\phantom{{}-7}} \\ 6x - 7 \\ \underline{6x - 6} \\ -1 \end{array}$$ ■

The final example demonstrates the division process when the divisor is a polynomial of degree more than 1.

EXAMPLE 4 Divide $2x^4 - 7x^3 + x - 4$ by $x^2 - 2x$.

Solution

$$\begin{array}{r} 2x^2 - 3x - 6 \\ x^2 - 2x \overline{\smash{)}\, 2x^4 - 7x^3 + 0x^2 + x - 4} \\ \underline{2x^4 - 4x^3\phantom{{}+0x^2+x-4}} \\ -3x^3 + 0x^2\phantom{{}+x-4} \\ \underline{-3x^3 + 6x^2}\phantom{{}+x-4} \\ -6x^2 + x\phantom{{}-4} \\ \underline{-6x^2 + 12x}\phantom{{}-4} \\ -11x - 4 \end{array}$$

The division process stops since the degree of the remainder is less than the degree of the divisor. The quotient is $2x^2 - 3x - 6$ and the remainder is $-11x - 4$. This result can also be written as

Check this solution by verifying that $QD + R = N$.

$$\frac{2x^4 - 7x^3 + x - 4}{x^2 - 2x} = 2x^2 - 3x - 6 + \frac{-11x - 4}{x^2 - 2x}$$ ■

EXERCISES 4.5

Divide.

1. $(21a^3 - 35a^2) \div 7a$
2. $(9b^3 + 12b^2 - 3b) \div 3b$
3. $\dfrac{45x^3 + 15x^2}{15x^2}$
4. $\dfrac{9y^2 + 12y^8 - 15y^6}{3y^2}$
5. $\dfrac{12x^3 + 8x^2 + 4x}{4x}$
6. $\dfrac{5a^2 - 10a^3 + 15a^4}{5a^2}$
7. $\dfrac{2a^5 + 4a^3 + 6a^2}{2a^2}$
8. $\dfrac{3x^5 - x^4 + 2x^3}{x^3}$
9. $\dfrac{8xy - 4x^2y^2 + 2x^3y^3}{2xy}$
10. $\dfrac{6a^4b^2 - 9a^2b^2 + 3ab}{-3ab}$
11. $\dfrac{x^2y^3 + x^4y^4}{x^2y^2}$
12. $\dfrac{4x^2y^2 - 6x^4y^4 + 8xy}{2xy}$
13. $\dfrac{a^2b^2 + ab^2 - a^2b^3}{ab^2}$
14. $\dfrac{-6a^3 + 9a^6 - 12a^9}{-3a^3}$
15. $\dfrac{6a^2x^2 - 8a^4x^6}{2a^2x^2}$
16. $\dfrac{-8a^3x^3 + 4ax^3 - 12a^2x^6}{-4ax^3}$

Divide and check your results.

17. $(x^2 + x - 6) \div (x - 2)$
18. $(x^2 + x - 6) \div (x + 3)$
19. $(2x^2 + x - 3) \div (x - 1)$
20. $(3x^2 + x - 10) \div (x + 2)$
21. $(5x^2 - 3x + 1) \div (x + 1)$
22. $(7x^2 + 5x - 2) \div (x - 1)$
23. $(12x^2 - 3x + 7) \div (x + 3)$
24. $(8x^2 - 7x + 4) \div (x - 2)$
25. $(x^3 - 2x^2 - 13x + 6) \div (x + 3)$
26. $(x^3 + 4x^2 + 3x - 2) \div (x + 2)$
27. $(x^3 - x^2 + 7) \div (x - 1)$
28. $(2x^3 + 5x - 3) \div (x + 2)$
29. $(4x^2 + 12x + 9) \div (2x + 3)$
30. $(2x^3 + 9x^2 - 3x - 1) \div (2x - 1)$
31. $(5x^2 - 7x + x^3 + 8) \div (x - 2)$
32. $(7x^2 - 2x + 1 - 3x^3 + x^4) \div (x + 2)$

Divide.

33. $\dfrac{9(x - y)^2 + 12(x - y)}{3(x - y)}$
34. $\dfrac{(3x + y)^3 + 2(3x + y)}{3x + y}$
35. $\dfrac{25(a^2 + ab + b^2) - 5(a + b)}{-5(a + b)}$
36. $\dfrac{8(4a^2 + 4ab + b^2) + 4(2a + b)}{2(b + 2a)}$
37. $\dfrac{3y^3 - 2y^2 + 3y - 2}{y^2 + 1}$
38. $\dfrac{6a^3 - 2a^2 - 15a + 5}{2a^2 - 5}$
39. $(a^3 - 27) \div (a - 3)$
40. $(a^3 + 27) \div (a + 3)$
41. $(x^4 - 16) \div (x - 2)$
42. $(x^4 - 16) \div (x + 2)$
43. $(a^3 - 27b^3) \div (a - 3b)$
44. $(a^3 + 27b^3) \div (a + 3b)$
45. $(x^5 - 32) \div (x - 2)$
46. $(x^5 - 32y^5) \div (x - 2y)$
47. $(4x^3 - 5x^2 + x - 7) \div (x^2 - 2x)$
48. $(8x^4 - 8x^2 + 6x + 6) \div (2x^2 - x)$
49. $(x^3 - x^2 - x + 10) \div (x^2 - 3x + 5)$
50. $(3x^3 + 4x^2 - 13x + 6) \div (x^2 + 2x - 3)$

51. (a) Find the remainder when the polynomial $p(x) = 2x^3 - 5x^2 + 2x - 3$ is divided by $x - 1$.
(b) Find $p(1)$, the value of the polynomial when $x = 1$.
(c) Compare your results in parts (a) and (b).

52. Repeat Exercise 51 for $p(x) = 5x^3 - 3x^2 + 4x - 1$ divided by $x - 2$ and for $p(2)$.

53. Repeat Exercise 51 for $p(x) = 3x^3 + x^2 - 5x + 2$ divided by $x + 3$ and for $p(-3)$.

Written Assignment: Assume a given polynomial $p(x)$ is divided by $x - c$. Use the results of Exercises 51 through 53 to formulate a conjecture regarding the division of a polynomial $p(x)$ by $x - c$.

4.6 EQUATIONS WITH RATIONAL EXPRESSIONS

The mathematics known by the ancient Egyptians was written by a scribe named Ahmes in about 1650 B.C. In 1927 an Englishman by the name of A. Henry Rhind uncovered a manuscript in Egypt and had a translation published that is now known as the *Rhind Papyrus*. In this manuscript we note that most fractions were expressed as the sum of *unit fractions*. Thus, for example, the fraction $\frac{3}{4}$ did not appear in the manuscript, but rather was presented as $\frac{1}{2} + \frac{1}{4}$.

The fraction $\frac{2}{7}$ did not appear, but was represented as the sum of $\frac{1}{4}$ and another fraction. To find that other fraction, using contemporary notation, we can write and solve the following **fractional equation:**

$$\frac{2}{7} = \frac{1}{4} + \frac{1}{x}$$

To solve for x, multiply both sides of the equation by $28x$, the LCD of the three fractions.

$$28x\left(\tfrac{2}{7}\right) = 28x\left(\tfrac{1}{4} + \tfrac{1}{x}\right) \qquad \text{(by the multiplication property of equality)}$$

$$8x = 7x + 28$$

$$x = 28$$

Check that this is correct: $\frac{2}{7} = \frac{1}{4} + \frac{1}{28}$

The properties of equality introduced in Chapter 3 can be applied to solve equations involving fractions. In particular, the multiplication property of equality is frequently used to clear the equations of fractions by multiplying through by the least common denominator (LCD). For example, consider this equation:

$$\frac{3x}{x+7} = \frac{8}{5}$$

The first step is to find the LCD of the two fractions. In this case, the LCD is $5(x + 7)$. Now use the multiplication property of equality and multiply each side of the equation by $5(x + 7)$.

Notice that multiplication by the LCD transforms the original equation into one that does not involve any fractions. Thereafter we are able to finish the solution using methods we have previously studied.

$$5(x+7)\cdot\frac{3x}{x+7} = 5(x+7)\cdot\frac{8}{5}$$

$$5\cdot 3x = (x+7)\cdot 8$$

$$15x = 8x + 56$$

$$7x = 56$$

$$x = 8$$

Check:

$$\frac{3x}{x+7} = \frac{8}{5}: \qquad \frac{3 \cdot 8}{8+7} = \frac{24}{15} = \frac{8}{5}$$

EXAMPLE 1 Solve for x: $\dfrac{x-1}{5} - \dfrac{2x-3}{3} = -2$

Solution The LCD is 15.

$$\begin{aligned} 15\left(\frac{x-1}{5}\right) - 15\left(\frac{2x-3}{3}\right) &= 15(-2) \\ 3(x-1) - 5(2x-3) &= -30 \\ 3x - 3 - 10x + 15 &= -30 \\ -7x + 12 &= -30 \\ -7x &= -42 \\ x &= 6 \end{aligned}$$

Check this solution by substitution in the original equation. ■

EXAMPLE 2 Solve for x: $\dfrac{5}{x-3} = \dfrac{3}{x+1}$

Solution The LCD is the product $(x-3)(x+1)$.

$$\begin{aligned} (x-3)(x+1)\left(\frac{5}{x-3}\right) &= (x-3)(x+1)\left(\frac{3}{x+1}\right) \\ 5(x+1) &= 3(x-3) \\ 5x + 5 &= 3x - 9 \\ 2x &= -14 \\ x &= -7 \end{aligned}$$

Check this result. ■

EXAMPLE 3 Solve for x: $\dfrac{6}{x} = 2 + \dfrac{3}{x+1}$

At times, multiplication by the LCD can give rise to a quadratic equation, as in Example 3. Try to explain each step in the solution.

Solution The LCD is $x(x+1)$.

$$\begin{aligned} x(x+1)\left(\frac{6}{x}\right) &= x(x+1)(2) + x(x+1)\left(\frac{3}{x+1}\right) \\ 6(x+1) &= 2x(x+1) + 3x \\ 6x + 6 &= 2x^2 + 2x + 3x \\ 0 &= 2x^2 - x - 6 \\ 0 &= (2x+3)(x-2) \end{aligned}$$

Thus

Recall that if $ab = 0$, then $a = 0$, or $b = 0$, or both $a = 0$ and $b = 0$.

$$2x + 3 = 0 \quad \text{or} \quad x - 2 = 0$$

$$x = -\tfrac{3}{2} \quad \text{or} \quad x = 2$$

Check that each of these values satisfies the original equation. The roots of the equation are $-\frac{3}{2}$ and 2. ■

It is especially important to check each solution of a fractional equation. The reason for this can best be explained through the use of another example.

EXAMPLE 4 Solve for x: $\dfrac{24}{x^2 - 16} - \dfrac{5}{x + 4} = \dfrac{3}{x - 4}$

$x^2 - 16$ can be factored as the difference of two squares: $x^2 - 16 = (x + 4)(x - 4)$

Solution Begin by multiplying each side by $(x + 4)(x - 4)$.

$$(x + 4)(x - 4) \cdot \frac{24}{x^2 - 16} - (x + 4)(x - 4) \cdot \frac{5}{x + 4} = (x + 4)(x - 4) \cdot \frac{3}{x - 4}$$

$$24 - (x - 4)5 = (x + 4)3$$

$$24 - 5x + 20 = 3x + 12$$

$$-8x = -32$$

$$x = 4$$

The logic used here began with the *assumption* that there was a value x for which the equation was true. This led to the value $x = 4$; that is, we argued that *if* there is a solution, then it must be 4. But if x is replaced by 4 in the given equation, we obtain:

In this example we could have noticed at the outset that $x - 4 \neq 0$, and thus $x \neq 4$. In other words, it is wise to notice such restrictions on the variable before starting the solution. These are the values of the variable that would cause division by zero.

$$\frac{24}{0} - \frac{5}{8} = \frac{3}{0}$$

Since division by 0 is meaningless, we have an impossible equation and therefore 4 cannot be a solution. We conclude that there is no replacement of x for which the equation is true. The solution is therefore the empty set. ■

TEST YOUR UNDERSTANDING
Think Carefully

Solve for x and check your results.

1. $\dfrac{x}{3} + \dfrac{x}{2} = 10$
2. $\dfrac{3}{x} + \dfrac{2}{x} = 10$
3. $\dfrac{x - 3}{8} = 4$
4. $\dfrac{8}{x - 3} = 4$
5. $\dfrac{x + 3}{x} = \dfrac{2}{3}$
6. $\dfrac{2}{x + 3} = \dfrac{3}{x + 3}$
7. $\dfrac{3x}{2} - \dfrac{2x}{3} = \dfrac{1}{4}$
8. $\dfrac{2}{3x} + \dfrac{3}{2x} = 4$
9. $\dfrac{3x - 1}{4} - \dfrac{x - 1}{2} = 1$
10. $\dfrac{1 - x}{2} - \dfrac{2x - 1}{3} = 2$
11. $\dfrac{3}{x - 3} + 2 = \dfrac{x}{x - 3}$
12. $\dfrac{2x}{x^2 - 4} + \dfrac{3}{x + 2} = \dfrac{3}{x - 2}$

(Answers: Page 176)

Numerous formulas involving fractions are found in mathematics as well as in areas such as science, business, and industry. When working with such formulas it is sometimes useful to solve for one of the variables in terms of the others. This is illustrated in the next example.

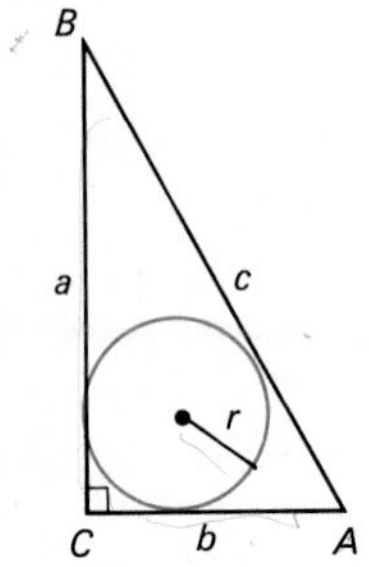

EXAMPLE 5 In the adjacent figure, the formula for the radius r of the inscribed circle of a right triangle in terms of the sides of the triangle is

$$r = \frac{ab}{a + b + c}$$

Solve for side b in terms of r, a, and c.

Solution Begin by clearing fractions:

$$r = \frac{ab}{a + b + c}$$

$$r(a + b + c) = ab$$

$$ra + rb + rc = ab$$

$$ra + rc = ab - rb \quad \text{(Bring all terms involving } b \text{ to one side.)}$$

$$r(a + c) = (a - r)b \quad \text{(Factor.)}$$

$$\frac{r(a + c)}{a - r} = b \quad \text{(Divide by } a - r.\text{)}$$

■

Check this answer by substituting for b in the given equation and simplifying the resulting complex fractions.

EXERCISES 4.6

Solve for x and check your results.

1. $\frac{x}{2} - \frac{x}{5} = 6$
2. $\frac{2}{x} - \frac{5}{x} = 6$
3. $\frac{x + 3}{5} = 2$
4. $\frac{5}{x + 3} = 2$
5. $\frac{x - 1}{2} = \frac{x + 2}{4}$
6. $\frac{2}{x - 1} = \frac{4}{x + 2}$
7. $\frac{5}{x} - \frac{3}{4x} = 1$
8. $\frac{5x}{2} - \frac{3x}{4} = 2$
9. $\frac{2x + 1}{5} - \frac{x - 2}{3} = 1$
10. $\frac{3x - 2}{4} - \frac{x - 1}{3} = \frac{1}{2}$
11. $\frac{6x - 19}{7} = x - 4$
12. $\frac{x}{3} + \frac{2x}{15} = \frac{49}{15}$
13. $\frac{8}{3x} - \frac{1}{3} = -1$
14. $\frac{2x}{x + 6} = 1$
15. $\frac{x + 4}{2x - 10} = \frac{8}{7}$
16. $\frac{10}{x} - \frac{1}{2} = \frac{15}{2x}$
17. $\frac{3x}{4} - \frac{3}{2} = \frac{x}{6} + \frac{4x}{3}$
18. $\frac{1}{3x - 1} + \frac{1}{3x + 1} = 0$
19. $\frac{x + 1}{x + 10} = \frac{1}{2x}$
20. $\frac{3}{x + 1} = \frac{9}{x^2 - 3x - 4}$
21. $\frac{x^2}{2} - \frac{3x}{2} + 1 = 0$
22. $\frac{x + 1}{x - 1} - \frac{2}{x(x - 1)} = \frac{4}{x}$
23. $\frac{5}{x^2 - 9} = \frac{3}{x + 3} - \frac{2}{x - 3}$
24. $\frac{3}{x^2 - 25} = \frac{5}{x - 5} + \frac{2}{x + 5}$
25. $\frac{1}{x^2 + 4} + \frac{1}{x^2 - 4} = \frac{18}{x^4 - 16}$
26. $\frac{3}{x^3 - 8} = \frac{1}{x - 2}$
27. $\frac{2x - 5}{x + 1} - \frac{3}{x^2 + x} = 0$

28. $\frac{2}{x} = -2x + 5$

29. $\frac{2}{x+6} - \frac{2}{x-6} = 0$

30. $\frac{2}{x-6} - \frac{3}{x+6} = 0$

31. $\frac{5}{2x+3} + \frac{2}{x-2} = \frac{4x-19}{2x^2-x-6}$

32. $\frac{1}{x-5} - \frac{4}{x^2-1} = \frac{16}{x^3-5x^2-x+5}$

33. $\frac{x}{a} = \frac{b}{c}$

34. $\frac{x}{a} + \frac{b}{c} = 0$

35. $\frac{a}{x} - \frac{b}{y} = 1$

36. $\frac{x}{a} + \frac{y}{b} = 1$

37. $\frac{a}{2x} - \frac{b}{x} = c$

38. $\frac{2x}{a} - \frac{x}{b} = c$

In Exercises 39 through 48, solve for the indicated variable.

39. $\frac{v^2}{K} = \frac{2g}{m}$, for m

40. $A = \frac{h}{2}(b + B)$, for B

41. $d = \frac{s-a}{n-1}$, for s

42. $S = \frac{n}{2}[2a + (n-1)d]$, for d

43. $\frac{1}{f} = \frac{1}{m} + \frac{1}{p}$, for m

44. $c = \frac{2ab}{a+b}$, for b

45. $V = \frac{1}{3}\pi h^2(3R - h)$, for R

46. $W = Rm\left(1 - \frac{R}{r}\right)$, for r

47. $S = 2(\ell w + hw + h\ell)$, for ℓ

48. $V = 100\left(\frac{Q_3 - Q_1}{Q_3 + Q_1}\right)$, for Q_1

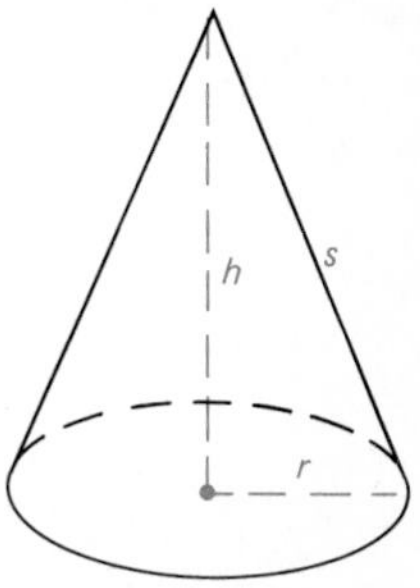

49. In the adjacent figure, the formula for the surface area S of a cone with slant height s and radius of the base r is $S = \pi r^2 + \pi rs$. Solve for s in terms of S, π, and r.

50. The formula for the volume V of the cone in the adjacent figure is $V = \frac{1}{3}\pi r^2 h$ where h is the height of the cone. Solve for h in terms of π and r.

CHALLENGE
Think Creatively

Consider these sums of unit fractions:

$$\frac{1}{9} + \frac{1}{18} = \frac{1}{6}$$

$$\frac{1}{12} + \frac{1}{24} = \frac{1}{8}$$

$$\frac{1}{15} + \frac{1}{30} = \frac{1}{10}$$

What do you notice about the two fractions being added? Try to write a formula for the sum of such unit fractions, and verify your result.

4.7 APPLICATIONS OF FRACTIONAL EQUATIONS

Diophantus was a Greek mathematician who lived in the third century A.D. An epitaph was written that provides an interesting puzzle about his final age:

> Diophantus passed one-sixth of his life in childhood, one-twelfth in youth, and one seventh-more as a bachelor. Five years after his marriage was born a son who died four years before his father at half his father's final age.

Let's try to write and solve and equation to find n, his final age:

Years		
$\frac{n}{6}$	in childhood	Combine these terms to form an equation:
$\frac{n}{12}$	in youth	
$\frac{n}{7}$	as a bachelor	$\frac{n}{6} + \frac{n}{12} + \frac{n}{7} + 5 + \frac{n}{2} + 4 = n$
5	after marriage	
$\frac{n}{2}$	son's life	
4	after son's death	

Multiply both sides of the equation by the LCD, 84; combine terms; and solve for n:

$$84\left(\frac{n}{6} + \frac{n}{12} + \frac{n}{7} + 5 + \frac{n}{2} + 4\right) = 84(n)$$

$$14n + 7n + 12n + 420 + 42n + 336 = 84n$$

$$75n + 756 = 84n$$

$$756 = 9n$$

$$84 = n$$

Diophantus lived to be 84 years old.

The general procedures for solving problems outlined in Section 3.2 apply to problems that involve fractions as well. You are advised to review that material at this time. Then study the following illustrative example.

EXAMPLE 1 The denominator of a certain fraction is 3 more than the numerator. If the numerator is decreased by 1 and the denominator is increased by 4, the value of the fraction is $\frac{1}{3}$. What is the original fraction?

Solution Read the problem several times to be sure of its meaning. Then refer to the first sentence only and introduce a variable:

Let x = the numerator

Then $x + 3$ = the denominator

The original fraction can be represented as $\frac{x}{x+3}$.

Now *decrease* the numerator by 1: $x - 1$

And *increase* the denominator by 4: $(x + 3) + 4 = x + 7$

Then the new fraction is represented by: $\frac{x-1}{x+7}$

Now set this fraction equal to $\frac{1}{3}$ and solve for x.

To solve the equation, multiply both sides by the LCD, $3(x + 7)$.

$$\frac{x-1}{x+7} = \frac{1}{3}$$

$$3(x+7)\left(\frac{x-1}{x+7}\right) = 3(x+7)\left(\frac{1}{3}\right)$$

$$3(x-1) = x + 7$$

$$3x - 3 = x + 7$$

$$2x = 10$$

$$x = 5$$

Don't stop now! We have solved for x, but were asked to find the original fraction, $\frac{x}{x+3}$. Thus the fraction is

$$\frac{5}{5+3} = \frac{5}{8}$$

Check this result by referring to the statement of the problem. ■

The equation that was derived in Example 1 is also called a **proportion.** A proportion is a statement that two ratios are equal, such as the following:

This is often read "a is to b as c is to d," and may be written in this form:

$$a:b = c:d$$

$$\frac{a}{b} = \frac{c}{d}$$

The proportion is in the form of a fractional equation and can be simplified by multiplying both sides by bd.

$$\frac{a}{b} = \frac{c}{d}$$

$$(bd)\frac{a}{b} = (bd)\frac{c}{d}$$

$$ad = bc$$

Sometimes we say that $ad = bc$ is obtained by ***cross-multiplying.*** *This description is used because ad and bc can be obtained by multiplying the numerators times the denominators along the lines in this cross.*

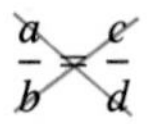

$ad = bc$

This relationship is used often and is called the **proportion property.**

PROPORTION PROPERTY

If $\frac{a}{b} = \frac{c}{d}$, then $ad = bc$.

Note that we could have used the proportion property when finding the solution for Example 1.

$$\text{If} \quad \frac{x-1}{x+7} = \frac{1}{3}, \quad \text{then} \quad 3(x-1) = 1(x+7).$$

EXAMPLE 2 How high is a tree that casts an 18-foot shadow at the same time that a 3-foot stick casts a 2-foot shadow?

Solution From the diagram we use similar triangles to write a proportion.

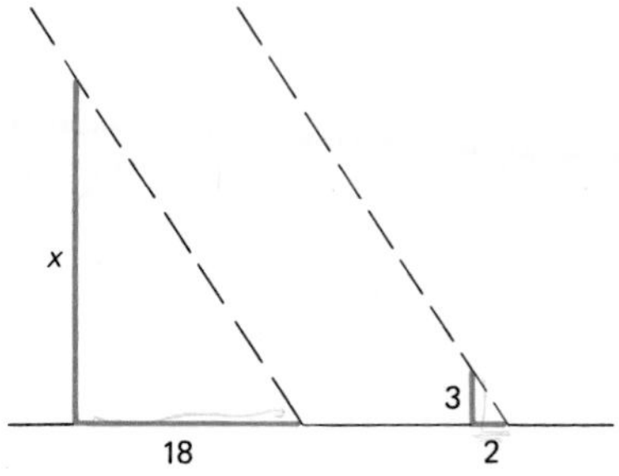

Let x represent the height of the tree. Then, since the triangles are similar,

Similar triangles have the same shape, but not necessarily the same size. If two triangles are similar, their corresponding angles are equal and their corresponding sides are proportional

$$\frac{x}{18} = \frac{3}{2}$$

That is, x is to 18 as 3 is to 2. By the proportion property:

$$2x = 54; \quad x = 27$$

The tree is 27 feet high. ■

EXAMPLE 3 A wildlife conservation team wants to determine the deer population in a state park. They capture a sample of 80 deer, tag each one, and then release them. After a sufficient amount of time has elapsed for the tagged deer to thoroughly mix with the others in the park, the team captures another sample of 100 deer, five of which had been previously tagged. Estimate the number of deer in the park.

Solution Let x be the total deer population. The 80 tagged deer were thoroughly mixed with the others. Thus, it is reasonable to expect the ratio of 80 tagged deer to the total deer population to be the same as the ratio of the tagged deer, captured a second time, to the total number in the second sample. Thus,

$$\frac{\text{total tagged deer}}{\text{total deer population}} = \frac{\text{tagged deer in second sample}}{\text{total deer in second sample}}$$

$$\frac{80}{x} = \frac{5}{100}$$

$$5x = 8000$$

$$x = 1600$$

There are approximately 1600 deer in the park. ■

TEST YOUR UNDERSTANDING
Think Carefully

Solve for x by using the proportion property.

1. $\frac{3}{x} = \frac{4}{5}$
2. $\frac{x}{5} = \frac{2}{3}$
3. $\frac{x+1}{3} = \frac{3}{2}$
4. $\frac{x-1}{2} = \frac{3}{4}$
5. $\frac{x-1}{x+1} = \frac{2}{3}$
6. $\frac{3}{x} = \frac{5}{x+2}$
7. $\frac{4}{x} = \frac{x}{16}$
8. $\frac{x}{2x-1} = \frac{1}{x}$
9. $\frac{x-1}{2} = \frac{3}{x}$
10. $\frac{2}{x^2+x+1} = \frac{x-1}{13}$
11. $\frac{x-2}{2x+1} = \frac{2x-1}{4x+2}$
12. $\frac{2x+10}{x+18} = \frac{x+3}{4-x}$

(Answers: Page 176)

EXAMPLE 4 For electric circuits, when two resistances R_1 and R_2 are wired *in parallel,* the single equivalent resistance R can be found by this relationship:

$$\frac{1}{R} = \frac{1}{R_1} + \frac{1}{R_2}$$

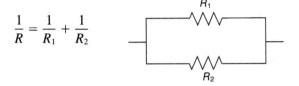

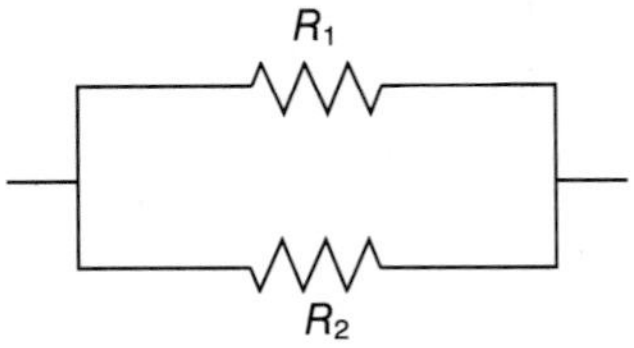

Find the single equivalent resistance for two resistances wired in parallel if these measure 2000 ohms and 3000 ohms, respectively.

Solution Let $R_1 = 2000$ and $R_2 = 3000$ and solve the following fractional equation for R:

$$\frac{1}{R} = \frac{1}{2000} + \frac{1}{3000}$$

$$6000 = 3R + 2R \qquad \text{(Multiply by the LCD} = 6000R.)$$

$$6000 = 5R$$

$$1200 = R$$

The single equivalent resistance is 1200 ohms. ■

Suppose that you are able to work at a steady rate and can paint a room in 6 hours. This means that in 1 hour you have painted $\frac{1}{6}$ of the room. In general, for a job that takes you n hours, the part of the job done in 1 hour is $\frac{1}{n}$. This idea is used to solve certain types of work problems that call for the solution of fractional equations.

EXAMPLE 5 Working alone Harry can mow a lawn in 3 hours. Elliot can complete the same job in 2 hours. How long will it take them working together, assuming they both start at the same time?

We assume for these work problems that there is a constant rate of work. In reality, such a constant rate of work would be difficult to achieve. But there is an average rate of work, and whether we assume a constant or average rate or work for n hours makes no difference. In either case, $\frac{1}{n}$ would be the part done in 1 hour.

Solution Working together should enable the boys to do the job in less time than it would take either of them to do the job alone. If Elliot can do the job alone in 2 hours, together the two boys should take less than this time.

To solve work problems of this type, we first consider the part of the job that can be done in 1 hour.

Let x = time (in hours) to do the job together

Then $\dfrac{1}{x}$ = portion of job done in 1 hour working together

Also: $\dfrac{1}{2}$ = portion of job done by Elliot in 1 hour

$\dfrac{1}{3}$ = portion of job done by Harry in 1 hour

Now we have that *each* of the expressions $\frac{1}{2} + \frac{1}{3}$ and $\frac{1}{x}$ represents the part of the job done in 1 hour. Therefore, these expressions are set equal to each other to form the required equation.

$$\frac{1}{2} + \frac{1}{3} = \frac{1}{x} \qquad \text{(Multiply by } 6x.)$$
$$3x + 2x = 6$$
$$5x = 6$$
$$x = 1\tfrac{1}{5}$$

Together they need $1\frac{1}{5}$ hours, or 1 hour and 12 minutes. ■

EXERCISES 4.7

Solve the proportion and check.

1. $\dfrac{x}{9} = \dfrac{5}{6}$
2. $\dfrac{2x}{15} = \dfrac{4}{-11}$
3. $\dfrac{x}{9} = \dfrac{1}{x}$
4. $\dfrac{x}{6} = \dfrac{-3x}{10}$
5. $\dfrac{2x+3}{2} = \dfrac{8}{2x-3}$
6. $\dfrac{10}{x} = \dfrac{x-3}{7}$
7. Solve for x assuming that the right triangles are similar.

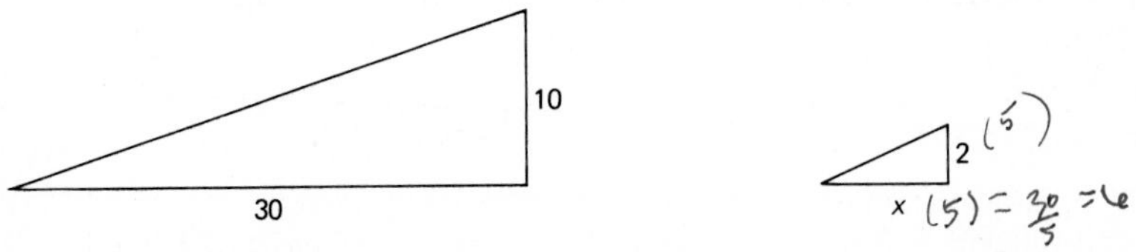

8. If x is to 8 as 3 is to 16, what is x?

9. The sum of a number and its reciprocal is $\frac{13}{6}$. Find all such numbers.

10. A positive number minus its reciprocal is $\frac{21}{10}$. Find the number.

11. What number must be subtracted from both the numerator and denominator of $\frac{35}{17}$ to obtain $\frac{5}{2}$?

12. The numerator of a fraction is 3 less than the denominator. If both numerator and denominator are increased by 1, the value of the fraction is $\frac{1}{2}$. Find the original fraction.

13. What number must be added to both the numerator and denominator of the fraction $\frac{3}{8}$ to give a fraction whose value is $\frac{1}{2}$?

14. A number diminished by its reciprocal is equal to $\frac{8}{3}$. Find all such numbers.

15. What number must be subtracted from both the numerator and denominator of the fraction $\frac{11}{23}$ to give a fraction whose value is $\frac{2}{5}$?

16. The denominator of a fraction is 3 more than the numerator. If 5 is added to the numerator and 4 is subtracted from the denominator, the value of the new fraction is 2. Find the original fraction.

17. One pipe can empty a tank in 3 hours. A second pipe takes 4 hours to complete the same job. How long will it take to empty the tank if both pipes are used?

18. Working together Amy and Julie can paint their room in 3 hours. If it takes Amy 5 hours to do the job alone, how long would it take Julie to paint the room working by herself?

19. Find two fractions whose sum is $\frac{2}{3}$ if the smaller fraction is $\frac{1}{2}$ the larger one. (*Hint:* Let x represent the larger fraction.)

20. Find two consecutive integers such that $\frac{2}{3}$ the smaller plus $\frac{3}{4}$ the larger is 22.

21. Find three consecutive integers such that $\frac{1}{4}$ the smallest, plus $\frac{1}{3}$ the second, plus $\frac{1}{2}$ the largest is 23.

22. A rope that is 20 feet long is cut into two pieces. The ratio of the smaller piece to the larger piece is $\frac{3}{5}$. Find the length of the shorter piece.

23. The shadow of a tree is 20 feet long at the same time when a 1-foot-high flower casts a 4-inch shadow. How high is the tree?

24. The area A of a triangle is given by $A = \dfrac{bh}{2}$, where b is the length of the base and h is the length of the altitude.
    (a) Solve for h in terms of A and b.
    (b) Find h when $A = 100.15$ and $b = 25.38$.

25. To find out how wide a certain river is, a pole 20 feet high is set straight up on one of the banks. Another pole 4 feet long is also set straight up, on the same side, some distance away from the embankment. The observer waits until the shadow of the 20-foot pole just reaches the other side of the river. At that time he measures the shadow of the 4-foot pole and finds it to be 34 feet. Use this information to determine the width of the river.

26. Estimate the total deer population in Example 3, page 160, if the number of deer captured each time is 100 and there were 2 deer that were tagged in the second sample.

27. A sample of 200 fish are caught, tagged, and returned to the lake from which they were caught. Some time later, when the 200 tagged fish had thoroughly mixed with the others, another 160 fish were caught including 4 that were tagged previously. Estimate the number of fish in the lake.

28. Estimate the total number of fish in the lake in Exercise 27 if the first catch contained 150 fish and the second catch had 120 fish, 3 of which were tagged previously.

29. If 58 is divided by a certain number, the quotient is 8 and the remainder is 2. Find the number. (*Hint:* For $N \div D$ we have $\dfrac{N}{D} = Q + \dfrac{R}{D}$, where Q is the quotient and R is the remainder.)

30. A student received grades of 72, 75, and 78 on three tests. What must her score on the next test be for her to have an average grade of 80 for all four tests?

31. The denominator of a certain fraction is 1 more than the numerator. If the numerator is increased by $2\frac{1}{2}$, the value of the new fraction will be equal to the reciprocal of the original fraction. Find the original fraction.

32. The sum of the reciprocals of two consecutive integers is $\frac{11}{30}$. What are the integers?

33. The sum of two numbers is 12, and the sum of their reciprocals is $\frac{3}{8}$. What are the two numbers?

34. Dan takes twice as long as George to complete a certain job. Working together they can complete the job in 6 hours. How long will it take Dan to complete the job by himself?

*35. Perry has a job stuffing envelopes that will take him 4 hours to complete. After working by himself for $1\frac{1}{2}$ hours, Kathy joins him and they complete the job in 1 hour. How long would it take Kathy to complete the entire job by herself?

*36. Prove:

$$\text{If} \quad \frac{a}{b} = \frac{c}{d}, \quad \text{then} \quad \frac{a}{a+b} = \frac{c}{c+d}$$

*37. Prove:

$$\text{If} \quad \frac{a}{b} = \frac{c}{d}, \quad \text{then} \quad \frac{a+b}{b} = \frac{c+d}{d}$$

38. What resistance, wired in parallel with a 20,000-ohm resistance, will give a single equivalent resistance of 4000 ohms?

39. When circuits are wired in parallel, an extended formula to find a single equivalent resistance is given by

$$\frac{1}{R} = \frac{1}{R_1} + \frac{1}{R_2} + \frac{1}{R_3} + \cdots + \frac{1}{R_n}$$

Find the single equivalent resistance if resistors of 200 ohms, 300 ohms, and 400 ohms are connected in parallel.

40. A formula used in optics relates the focal length of a lens, f, the distance of an object from the lens, p, and the distance from the lens to the image, q, in this way:

$$\frac{1}{f} = \frac{1}{p} + \frac{1}{q}$$

If the focal length of a lens is 15 centimeters and the distance of an object from the lens is 20 centimeters, find the distance of the image from the lens.

41. Find the focal length of a lens if the distance of an object from the lens is 30 centimeters and the distance of the image from the lens is 15 centimeters. (See Exercise 40.)

*42. A certain college gives 4 points per credit for a grade of A, 3 points per credit for B, 2 points per credit for a C, 1 point per credit for a D, and 0 for an F. A student is taking 15 credits for the semester. She expects A's in a 4 credit course and in a 3 credit course. She also expects a B in another 3 credit course and a D in a 2 credit course. What grade must she get in the fifth course in order to earn a 3.4 grade point average for the term? $\left(\textit{Hint: } \text{Grade point average} = \frac{\text{total points}}{\text{total credits}}.\right)$

*43. A bookstore has a stock of 30 paperback copies of *Algebra,* as well as 50 hardcover copies of the same book. They wish to increase their stock for the new semester. Based on past experience, they want their final numbers of paperback and hardcover copies to be in the ratio 4 to 3. However, the publisher stipulates that they will only sell the store 2 copies of the paperback edition for each copy of the hardcover edition ordered. Under these conditions, how many of each edition should the store order to achieve the 4 : 3 ratio?

EXPLORATIONS Think Critically

1. To solve an equation with fractions, we multiply through by the LCD. What happens, if instead, we just multiply by a common denominator? Thus in Example 4, page 155, is it possible to solve this equation by multiplying by the common denominator $(x^2 - 16)(x + 4)(x - 4)$? Explain your answer.
2. A sample of 100 deer is captured, tagged, and released. Later a second sample is captured and the number previously tagged counted. This is used as the basis for estimating the total number of deer in the park. Under what conditions will this be a good estimate? When might it not be a valid basis for estimating the total number in the park?
3. Explain the difference between the expressions $\dfrac{x^2 - 4}{x - 2}$ and $x + 2$.
4. Begin with a square sheet of paper and assume it has an area of one square unit. Cut the paper in half and place one half on the desk. Cut the remaining piece in half so that each part is $\frac{1}{4}$ of the original square unit, and place one of the pieces on the desk. Cut again to obtain pieces that are $\frac{1}{8}$ of a square unit, and place one of the pieces on the desk. Suppose you were able to continue in this way indefinitely. What would the pieces of paper on the desk represent? Explain your answer.
5. Suppose you travel from town A to town B at an average rate of speed of 45 miles per hour, and you return, averaging 30 mph. What is the average speed for the round trip? (*Hint:* Let s be the distance between the two towns so that the total distance travelled is $2s$.)
6. A car having a regular sized spare tire has travelled 40,000 miles. If the 5 tires were periodically rotated so that each of them travelled the same distance, what is the distance that each tire has travelled?

4.8 VARIATION

If a car is traveling at a constant rate of 40 miles per hour, then the distance d traveled in t hours is given by $d = 40t$. The change in the distance is "directly" affected by the change in the time; as t increases so does d. We say that *d is directly proportional to t*. This is because $d = 40t$ converts to the proportion $\dfrac{d}{t} = 40$. We also say that *d varies directly as t* and that 40 is the *constant of variation* or the *constant of proportionality*.

DIRECT VARIATION

y varies directly as x if $y = kx$ for some constant of variation k.

EXAMPLE 1

(a) Write the equation that expresses this direct variation: y varies directly as x, and y is 8 when x is 12.
(b) Find y for $x = 30$.

Solution

(a) Since y varies directly as x, we have $y = kx$ for some constant k. To find k, substitute the given values for the variables and solve.

$$8 = k(12)$$
$$\frac{2}{3} = k$$

Thus $y = \frac{2}{3}x$.

(b) For $x = 30$, $y = \frac{2}{3}(30) = 20$. ■

When y varies directly as x the equation $y = kx$ is equivalent to $\frac{y}{x} = k$ $(x \neq 0)$. Therefore, if each of the pairs x_1, y_1 and x_2, y_2 satisfies $y = kx$, then each of the ratios y_1 to x_1 and y_2 to x_2 is equal to k. Thus

$$\frac{y_1}{x_1} = \frac{y_2}{x_2}$$

This proportion can be used in some direct variation problems, if the constant of proportionality is not required. Thus part (b) of Example 1 can be answered without first writing the equation of variation. In particular, using $x_1 = 12$ with $y_1 = 8$, and $x_2 = 30$, we can solve for y_2 as follows:

$$\frac{y_2}{x_2} = \frac{y_1}{x_1}$$
$$\frac{y_2}{30} = \frac{8}{12}$$
$$y_2 = 30\left(\frac{8}{12}\right) = 20$$

Numerous examples of direct variation can be found in geometry. Here are some illustrations.

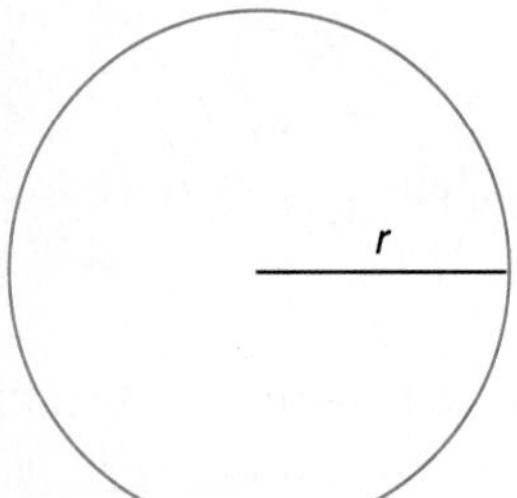

Circumference of a circle of radius r:
$C = 2\pi r$; C varies directly as the radius r;
2π is the constant of variation; $\frac{C}{r} = 2\pi$

Area of a circle of radius r:
$A = \pi r^2$; A varies directly as the square of r;
π is the constant of variation; $\frac{A}{r^2} = \pi$

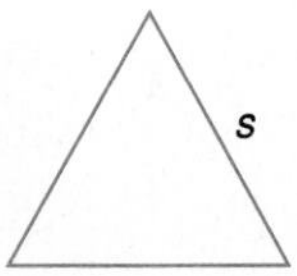

Area of an equilateral triangle of side s:
$A = \frac{\sqrt{3}}{4}s^2$; A varies directly as s^2;
$\frac{\sqrt{3}}{4}$ is the constant of proportionality; $\frac{A}{s^2} = \frac{\sqrt{3}}{4}$

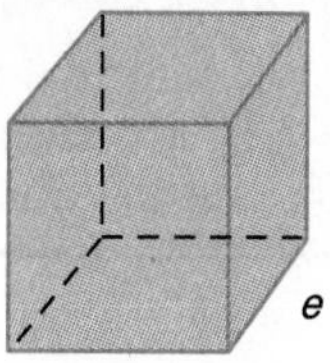

Volume of a cube of side e:

$V = e^3$; V varies directly as the cube of e (as e^3);

1 is the constant of variation; $\dfrac{V}{e^3} = 1$

The same principle applies to the force required to hold a spring when compressed x units within its natural length.

EXAMPLE 2 According to Hooke's Law, the force F required to hold a spring stretched x units beyond its natural length is directly proportional to x. If a force of 20 pounds is needed to hold a certain spring stretched 3 inches beyond its natural length, how far will 60 pounds of force hold the spring stretched beyond its natural length?

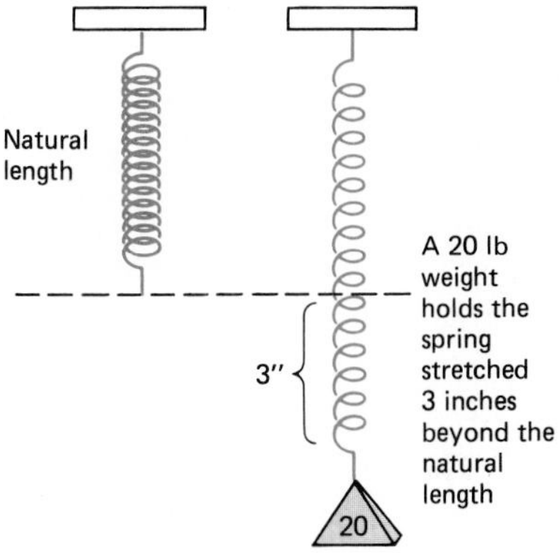

Note that the force F increases as x increases. This will always be the case in a direct variation situation provided that the constant of variation is positive.

Solution Since F varies directly as x, we have $F = kx$. Solve for k by substituting the known values for F and x.

$$20 = k(3)$$

$$\frac{20}{3} = k$$

Thus $F = \frac{20}{3}x$. Now let $F = 60$ and solve for x.

$$60 = \frac{20}{3}x$$

$$60\left(\frac{3}{20}\right) = x$$

$$9 = x$$

Thus 60 pounds is needed to hold the spring stretched 9 inches beyond its natural length. ■

When y varies directly as x, and the constant of variation is positive, the variables x and y increase or decrease together. There are other situations where one variable increases as the other decreases. We refer to this as *inverse variation*.

INVERSE VARIATION

y varies inversely as x if $y = \dfrac{k}{x}$ for some constant of variation k.

EXAMPLE 3 According to Boyle's law, the pressure P of a compressed gas is inversely proportional to the volume V. Suppose that there is a pressure of 25 pounds per square inch when the volume of gas is 400 cubic inches. Find the pressure when the gas is compressed to 200 cubic inches.

Solution Since P varies inversely as V, we have

$$P = \frac{k}{V}$$

Substitute the known values for P and V and solve for k.

$$25 = \frac{k}{400}$$

$$10{,}000 = k$$

Thus $P = \dfrac{10{,}000}{V}$, and when $V = 200$ we have

Note that the pressure increases as the volume decreases.

$$P = \frac{10{,}000}{200} = 50$$

The pressure is 50 pounds per square inch. ■

When y varies inversely as x the equation $y = \dfrac{k}{x}$ is equivalent to $xy = k$. Therefore, if each of the pairs x_1, y_1 and x_2, y_2 satisfies $y = \dfrac{k}{x}$, we get each of the products x_1y_1 and x_2y_2 equal to k. Thus

$$x_1y_1 = x_2y_2$$

This equation provides an alternative way to solve some inverse variation problems, if the equation of the variation is not required. Thus in Example 3 using $V_1 = 400$ with $P_1 = 25$ and $V_2 = 200$, we can solve for P_2 as follows:

$$V_2P_2 = V_1P_1$$

$$200P_2 = (400)(25)$$

$$P_2 = \frac{(400)(25)}{200} = 50$$

TEST YOUR UNDERSTANDING
Think Carefully

y varies directly as x.

1. $y = 5$ when $x = 4$. Find y for $x = 20$.
2. $y = 2$ when $x = 7$. Find y for $x = 21$.

y varies inversely as x.

3. $y = 3$ when $x = 5$. Find y for $x = 3$.
4. $y = 4$ when $x = 8$. Find y for $x = 2$.

(Answers: Page 176)

The variation of a variable may depend on more than one other variable. Here are some illustrations:

Observe that the word "jointly" is used to imply that z varies directly as the product of the two factors.

$z = kxy$	z varies *jointly* as x and y.
$z = kx^2y$	z varies *jointly* as x^2 and y.
$z = \dfrac{k}{xy}$	z varies *inversely* as x and y.
$z = \dfrac{kx}{y}$	z varies *directly* as x and *inversely* as y.
$w = \dfrac{kxy^3}{z}$	w varies *jointly* as x and y^3 and *inversely* as z.

EXAMPLE 4 Describe the variation given by these equations:

(a) $z = kx^2y^3$ **(b)** $z = \dfrac{kx^2}{y}$ **(c)** $V = \pi r^2 h$

Solution

(a) z varies joint as x^2 and y^3.
(b) z varies directly as x^2 and inversely as y.
(c) V varies jointly as r^2 and h. ■

EXAMPLE 5 Suppose that z varies directly as x and inversely as the square of y. If $z = \frac{1}{3}$ when $x = 4$ and $y = 6$, find z when $x = 12$ and $y = 4$.

Solution

$$z = \frac{kx}{y^2}$$

$$\frac{1}{3} = \frac{k(4)}{6^2}$$

$$\frac{1}{3} = \frac{k}{9}$$

$$3 = k$$

Thus $z = \dfrac{3x}{y^2}$. When $x = 12$ and $y = 4$ we have

$$z = \frac{3(12)}{4^2} = \frac{9}{4}$$ ■

EXERCISES 4.8

For Exercises 1 through 4, write the equation for the given variation and identify the constant of variation.

1. The perimeter P of a square varies directly as the side s.
2. The area of a circle varies directly as the square of the radius.
3. The area of a rectangle 5 centimeters wide varies directly as its length.
4. The volume of a rectangular-shaped box 10 centimeters high varies jointly as the length and width.

For Exercises 5 through 9, write the equation for the given variation using k as the constant of variation.

5. z varies jointly as x and y^3.
6. z varies inversely as x and y^3.
7. z varies directly as x and inversely as y^3.
8. w varies jointly as x and y and z.
9. w varies directly as x^2 and inversely as y and z.

In Exercises 10 through 14, find the constant of variation.

10. y varies directly as x; $y = 4$ when $x = \frac{2}{3}$.
11. s varies directly as t^2; $s = 50$ when $t = 10$.
12. y varies inversely as x; $y = 15$ when $x = \frac{1}{3}$.
13. u varies jointly as v and w; $u = 2$ when $v = 15$ and $w = \frac{2}{3}$.
14. z varies directly as x and inversely as the square of y; $z = \frac{7}{2}$ when $x = 14$ and $y = 6$.
15. a varies inversely as the square of b and $a = 10$, when $b = 5$. Find a when $b = 25$.
16. z varies jointly as x and y; $z = \frac{3}{2}$ when $x = \frac{5}{6}$ and $y = \frac{9}{20}$. Find z when $x = 2$ and $y = 7$.
17. s varies jointly as l and the square of w; $s = \frac{10}{3}$ when $l = 12$ and $w = \frac{5}{6}$. Find s when $l = 15$ and $w = \frac{9}{4}$.
18. The cost C of producing x number of articles varies directly as x. If it costs \$560 to produce 70 articles, what is C when $x = 400$?
19. If a ball rolls down an inclined plane, the distance traveled varies directly as the square of the time. If the ball rolls 12 feet in 2 seconds, how far will it roll in 3 seconds?
20. The volume V of a right circular cone varies jointly as the square of the radius r of the base, and the altitude h. If $V = 8\pi$ cubic centimeters when $r = 2$ centimeters and $h = 6$ centimeters, find the formula for the volume V.
21. A force of 2.4 pounds is needed to hold a spring stretched 1.8 inches beyond its natural length. Use Hooke's law to determine the force required to hold the spring stretched 3 inches beyond its natural length.
22. The force required to hold a metal spring compressed from its natural length is directly proportional to the change in the length of the spring. If 235 pounds is required to hold the spring compressed within its natural length of 18 inches to a length of 15 inches, how much force is required to hold it compressed to a length of 12 inches?
23. Fifty pounds per square inch is the pressure exerted by 150 cubic inches of a gas. Use Boyle's law to find the pressure if the gas is compressed to 100 cubic inches.
24. The gas in Exercise 23 expands to 500 cubic inches. What is the pressure?
25. If we neglect air resistance, the distance that an object will fall from a height near the surface of the earth is directly proportional to the square of the time it falls. If the object falls 256 feet in 4 seconds, how far will it fall in 7 seconds?
26. The volume of a right circular cylinder varies jointly as its height and the square of the radius of the base. The volume is 360π cubic centimeters when the height is 10 centimeters and the radius is 6 centimeters. Find V when $h = 18$ centimeters and $r = 5$ centimeters.
27. If the volume of a sphere varies directly as the cube of its radius and $V = 288\pi$ cubic inches when $r = 6$ inches, find V when $r = 2$ inches.
28. The resistance to the flow of electricity through a wire depends on the length and thickness of the wire. The resistance R is measured in *ohms* and varies directly as the length l and inversely as the square of the diameter d. If a wire 200 feet long with diameter 0.16 inch has a resistance of 64 ohms, how much resistance will there be if only 50 feet of wire is used?

29. A wire made of the same material as the wire in Exericse 28 is 100 feet long. Find R if $d = 0.4$ inch. Find R if $d = 0.04$ inch.

30. If z varies jointly as x and y, how does x vary with respect to y and z?

*31. A rectangular shaped beam is to be cut from a round log with a 2.5-foot diameter as shown in the adjacent figure. The strength s of the beam varies jointly as its height y and the square of its width x.

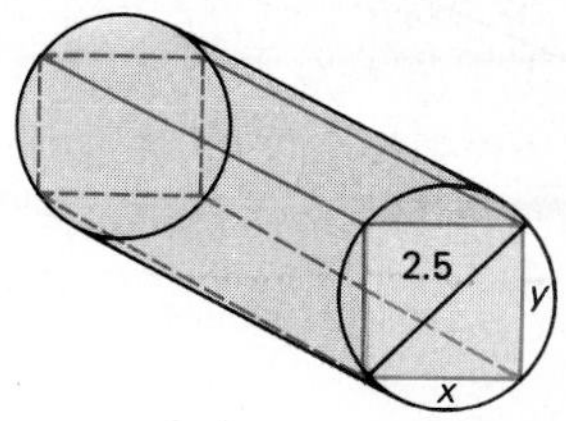

(a) Express s in terms of y.
(b) Express s in terms of x.
(c) Find s to two decimal places when $x = y$. Express the answer in terms of the constant of proportionality.

CHAPTER 4 SUMMARY

Review these key terms and concepts so that you are able to define or describe them. A clear understanding of these will be very helpful when reviewing the developments of this chapter.

Fundamental Principle of Fractions for real numbers a, b, and c, where $b \neq 0$ and $c \neq 0$:

$$\frac{ac}{bc} = \frac{a}{b}$$

To multiply rational expressions:

$$\frac{a}{b} \cdot \frac{c}{d} = \frac{ac}{bd} \quad \text{with } b \neq 0 \text{ and } d \neq 0$$

To divide, invert the divisor and multiply:

$$\frac{a}{b} \div \frac{c}{d} = \frac{a}{b} \cdot \frac{d}{c} = \frac{ad}{bc} \quad \text{with } b \neq 0,\ c \neq 0, \text{ and } d \neq 0$$

To add and subtract fractions with like denominators, add or subtract the numerators and place the result over the common denominator:

$$\frac{a}{c} + \frac{b}{c} = \frac{a+b}{c} \qquad \frac{a}{c} - \frac{b}{c} = \frac{a-b}{c} \qquad \text{with } c \neq 0$$

To add or subtract fractions with unlike denominators, find the least common denominator (LCD) of the denominators, write equivalent fractions, and combine.

To simplify a complex fraction, multiply numerator and denominator by the LCD of the denominator and simplify, as in this example:

$$\frac{\dfrac{1}{x} + \dfrac{1}{y}}{\dfrac{1}{x^2} - \dfrac{1}{y^2}} = \frac{\left(\dfrac{1}{x} + \dfrac{1}{y}\right)x^2y^2}{\left(\dfrac{1}{x^2} - \dfrac{1}{y^2}\right)x^2y^2} = \frac{xy^2 + x^2y}{y^2 - x^2} = \frac{xy(y + x)}{(y - x)(y + x)} = \frac{xy}{y - x}$$

Proportion Property: If $\frac{a}{b} = \frac{c}{d}$, then $ad = bc$.

Direct Variation: y varies *directly* as x if $y = kx$ for some *constant of variation* k.

Inverse Variation: y varies *inversely* as x if $y = \frac{k}{x}$ for some *constant of variation* k.

Joint Variation: z varies *jointly* as x and y if $z = kxy$ for some *constant of variation* k.

REVIEW EXERCISES

The solutions to the following exercises can be found within the text of Chapter 4. Try to answer each question before referring to the text.

Section 4.1

1. For what values is the following rational expression undefined? $\dfrac{x^2 + 3x}{x^2 - x - 2}$

Simplify.

2. $\dfrac{8x^2y^2}{12x^3y}$ 3. $\dfrac{x^2 + 5x + 6}{x^2 + 2x}$ 4. $\dfrac{5a - 3b}{3b - 5a}$ 5. $\dfrac{x^2 - 4}{2 - x}$ 6. $\dfrac{x^2 - 6x + 9}{3x^2 - x^3}$

Section 4.2

Perform the indicated operations.

7. $\dfrac{3a}{5b} \cdot \dfrac{5ac}{6a^2}$ 8. $\dfrac{x^2 - 9}{x + 9} \cdot \dfrac{x^2 + x - 2}{x^2 - 4x + 3}$ 9. $\dfrac{x - y}{x + y} \cdot \dfrac{x^2 - y^2}{x^2 + y^2} \cdot \dfrac{2xy}{x^2 - 2xy + y^2}$

10. $\dfrac{3x^2}{5y} \div \dfrac{6x}{y^3}$ 11. $\dfrac{-10xy^2}{3} \div 15x^2y$ 12. $\dfrac{x^2 - 3x + 2}{x^2 - 4} \div \dfrac{x^2 + 3x}{x + 3}$

13. $\left(\dfrac{a^2 - 4ab + 4b^2}{3ab^2} \div \dfrac{a^2 - 2ab}{6a + 6b}\right) \cdot \dfrac{-9a^2b}{a^2 - ab - 2b^2}$

Section 4.3

Perform the indicated operations.

14. $\dfrac{x + 2y}{x} + \dfrac{2x - y}{x}$ 15. $\dfrac{3x}{x - 3} - \dfrac{x + 10}{x - 3} - \dfrac{4}{x - 3}$

16. $\dfrac{5}{x + 2} + \dfrac{3}{x - 1}$ 17. $\dfrac{2x}{x^2 - 1} - \dfrac{x}{x - 1}$

18. Find the LCD for three denominators whose factored forms are

$$(x + 1)^2(x - 5), \quad (x + 1)(x - 5)^3, \quad \text{and} \quad (x + 1)^2(x - 5)^2(2x + 3)$$

Section 4.4

Simplify.

19. $\dfrac{\dfrac{a}{a + 2}}{\dfrac{3a}{a + 2}}$ 20. $\dfrac{x + \dfrac{1}{y}}{x - \dfrac{1}{y}}$ 21. $\dfrac{\dfrac{1}{x^2} - \dfrac{1}{y^2}}{\dfrac{1}{x} - \dfrac{1}{y}}$

22. $1 - \dfrac{1}{1 - \dfrac{1}{1 - \dfrac{1}{x}}}$ 23. $\dfrac{\dfrac{1}{x + h} - \dfrac{1}{x}}{h}$ 24. $\dfrac{x + y^{-1}}{y + x^{-1}}$

Section 4.5

Divide.

25. $\dfrac{6x^2y^2 - 4xy^2 + 8x^3y^4}{2xy^2}$ 26. $\dfrac{5x^2 - 22x + 21}{x - 3}$ 27. $(x^3 - x^2 - 5x + 6) \div (x - 2)$

28. $(5x + 3x^2 - 8) \div (x + 3)$ 29. $(x^3 + 5x - 7) \div (x - 1)$

30. $(2x^4 - 7x^3 + x - 4) \div (x^2 - 2x)$

Section 4.6

Solve for x.

31. $\frac{x-1}{5} - \frac{2x-3}{3} = -2$

32. $\frac{5}{x-3} = \frac{3}{x+1}$

33. $\frac{6}{x} = 2 + \frac{3}{x+1}$

34. $\frac{24}{x^2-16} - \frac{5}{x+4} = \frac{3}{x-4}$

35. Solve: $r = \frac{ab}{a+b+c}$ for b

Section 4.7

36. The denominator of a certain fraction is 3 more than the numerator. If the numerator is decreased by 1 and the denominator is increased by 4, the value of the fraction is $\frac{1}{3}$. What is the original fraction?
37. How high is a tree that casts an 18-foot shadow at the same time that a 3-foot stick casts a 2-foot shadow?
38. For electric circuits, when two resistances R_1 and R_2 are wired in parallel, the single equivalent resistance R can be found by: $\frac{1}{R} = \frac{1}{R_1} + \frac{1}{R_2}$. Find the single equivalent resistance for two resistances wired in parallel if these measure 2000 ohms and 3000 ohms, respectively.
39. Working alone Harry can mow a lawn in 3 hours. Elliot can complete the same job in 2 hours. How long will it take them working together?
40. A sample of 80 deer is tagged and released. Later, when the tagged deer were thoroughly mixed with the others, another sample of 100 was captured, of which five had been previously tagged. Estimate the number of deer in the park.

Section 4.8

41. **(a)** Write the equation that expresses the direct variation: y varies directly as x, and y is 8 when $x = 12$. **(b)** Find y for $x = 30$.
42. According to Hooke's law, the force F required to hold a spring stretched x units beyond its natural length is directly proportional to x. If a force of 20 pounds is needed to hold a certain spring stretched 3 inches beyond its natural length, how far will 60 pounds of force hold the spring stretched beyond its natural length?
43. According to Boyle's law, the pressure P of a compressed gas is inversely proportional to the volume V. Suppose that there is a pressure of 25 pounds per square inch when the volume of gas is 400 cubic inches. Find the pressure when the gas is compressed to 200 cubic inches.
44. Suppose z varies directly as x and inversely as the square of y. If $z = \frac{1}{3}$ when $x = 4$ and $y = 6$, find z when $x = 12$ and $y = 4$.

CHAPTER 4 TEST: STANDARD ANSWER

Use these questions to test your knowledge of the basic skills and concepts of Chapter 4. Then check your answers with those given at the back of the book.

Simplify.

1. $\frac{5a^2bc^5}{10ab^3c^2}$

2. $\frac{3x^2(x+2)}{2x(x+2)^3}$

3. $\frac{2x-x^2}{x^2-4}$

4. $\frac{x^2+4x+3}{2x^2+5x-3}$

Perform the indicated operations and reduce to lowest terms.

5. $\frac{2x-1}{x+2} + \frac{2-x}{x+2}$

6. $\frac{3x+5}{x+3} - \frac{x-1}{x+3}$

7. $\frac{2}{x(x-1)} - \frac{2}{x-1}$

8. $\dfrac{2x}{x+3} + \dfrac{x}{x-3}$

9. $\dfrac{x^2+4}{x^2-4} \cdot \dfrac{x-2}{x+2}$

10. $\dfrac{x^2+3x+2}{x^2+2x-3} \cdot \dfrac{x^2+x-6}{x^2-4}$

11. $\dfrac{x^3-2x^2}{x^3+3x^2+9x} \div \dfrac{x^3-4x^2+4x}{x^3-27}$

12. $\dfrac{8a^2b^4 - 4ab^2 + 6a^3b^6}{2ab^2}$

Simplify.

13. **(a)** $\dfrac{\frac{x}{3}}{\frac{x^2}{6}}$ **(b)** $\dfrac{\frac{1}{3} - \frac{1}{x}}{3-x}$

14. $\dfrac{x^{-1}+y^{-2}}{x^{-2}+y^{-1}}$

15. $\left(\dfrac{x^2-2xy+y^2}{3xy^2} \div \dfrac{x^2-xy}{x^2+y^2}\right) \cdot \dfrac{9x^2y}{x^2-y^2}$

Divide.

16. $(2x^3 - x^2 - 18x + 9) \div (x-3)$

17. $(x^4 + x^3 - 5x^2 + 2x + 5) \div (x^2 - x)$

Solve for x.

18. $\dfrac{x+3}{2} - \dfrac{x-2}{3} = 2$

19. $\dfrac{4}{x+2} - \dfrac{3}{4} = -\dfrac{x}{24}$

20. $\dfrac{6}{x^2-9} - \dfrac{2}{x+3} = \dfrac{1}{x-3}$

21. Use the proportion property to solve for x: $\dfrac{2x-1}{3x+2} = \dfrac{3}{8}$

22. The denominator of a certain fraction is 6 more than the numerator. If 3 is subtracted from the numerator and added to the denominator, the value of the fraction becomes $\frac{1}{4}$. Find the original fraction.

23. Working alone, Dave can wash his car in 45 minutes. If Ellen helps him, they can do the job together in 30 minutes. How long would it take Ellen to wash the car by herself?

24. Find two consecutive integers such that $\frac{2}{3}$ of the smaller one is 6 more than $\frac{2}{5}$ of the larger one.

25. z varies directly as x and inversely as y. If $z = \frac{2}{3}$ when $x = 2$ and $y = 15$, find z when $x = 4$ and $y = 10$.

CHAPTER 4 TEST: MULTIPLE CHOICE

1. Which is the correct simplification for $\dfrac{3x-2y}{2x-3y}$?

   (a) 1 **(b)** -1 **(c)** $-xy$ **(d)** $x - y$ **(e)** None of the preceding

2. Which of the following is *incorrect*?

   (a) $\dfrac{-x+2}{3} = -\dfrac{x-2}{3}$ **(b)** $\dfrac{-x+2}{3} = \dfrac{-(x-2)}{3}$ **(c)** $\dfrac{-x+2}{3} = -\dfrac{x+2}{3}$

   (d) $\dfrac{-x+2}{3} = \dfrac{x-2}{-3}$ **(e)** $-\dfrac{x+2}{3} = \dfrac{x+2}{-3}$

3. Which of the following is *incorrect*?

   (a) $\dfrac{a}{b} \div \dfrac{1}{c} = \dfrac{ac}{b}$ **(b)** $\dfrac{a}{b} \div c = \dfrac{a}{bc}$ **(c)** $a \div \dfrac{b}{c} = \dfrac{ac}{b}$ **(d)** $\dfrac{1}{a} \div \dfrac{b}{c} = \dfrac{ab}{c}$ **(e)** $ac \div b = \dfrac{ac}{b}$

4. Find the sum: $\dfrac{3x}{x-1} + \dfrac{2x}{1-x}$

   (a) $\dfrac{x}{x-1}$ **(b)** $\dfrac{x}{1-x}$ **(c)** $\dfrac{5x}{x^2-1}$ **(d)** $\dfrac{5x}{(x-1)^2}$ **(e)** None of the preceding

5. Simplify: $\dfrac{\dfrac{a}{b} - b}{\dfrac{a}{b} + b}$

(a) -1 **(b)** $\dfrac{a - b^2}{a + b^2}$ **(c)** $\dfrac{a - b}{a + b}$ **(d)** $\dfrac{a + b^2}{a - b^2}$ **(e)** None of the preceding

6. To solve the equation $\dfrac{5}{x - 1} = 3 + \dfrac{2}{x + 1}$, we mutiply by the LCD, which is:

(a) $x^2 - 1$ **(b)** $30(x - 1)(x + 1)$ **(c)** $3(x^2 - 1)$ **(d)** $x^2 + 1$ **(e)** None of the preceding

7. Each of 60 marbles, selected from a bag containing n marbles, is marked with an x and then returned into the bag. After the marbles are thoroughly mixed, 80 marbles are taken out of the bag at random. If 5 of the 80 marbles have an x, find the approximate value of n.

(a) 300 **(b)** 400 **(c)** 480 **(d)** 960 **(e)** 4800

8. Which of the following is equivalent to the product $5\left(\dfrac{x - 2}{x + 2}\right)$?

(a) $\dfrac{5x - 10}{x + 2}$ **(b)** $\dfrac{x - 2}{5x + 10}$ **(c)** $\dfrac{5x - 2}{5x + 2}$ **(d)** $\dfrac{5x - 10}{5x + 10}$ **(e)** None of the preceding

9. Which of the following is the simplification for $\dfrac{x^2 + 4}{x + 4}$?

(a) x **(b)** $x - 1$ **(c)** $x + 1$ **(d)** $x + 4$ **(e)** None of the preceding

10. Simplify: $\dfrac{\dfrac{1}{x} + \dfrac{1}{y}}{\dfrac{1}{x^2} - \dfrac{1}{y^2}}$

(a) $\dfrac{1}{x - y}$ **(b)** $\dfrac{y - x}{xy}$ **(c)** $\dfrac{xy}{y - x}$ **(d)** $\dfrac{xy}{x + y}$ **(e)** $\dfrac{1}{x + y}$

11. The equation $\dfrac{9x + 14}{2x^2 - x - 6} - \dfrac{1}{x - 2} = \dfrac{2}{2x + 3}$ has

(a) no solution **(b)** just one solution **(c)** exactly two solutions
(d) more than two solutions **(e)** None of the preceding

12. The denominator of a certain fraction is 5 less than the numerator, n. If the numerator is increased by 3 and the denominator is decreased by 2, the new fraction may be represented by:

(a) $\dfrac{n + 3}{n - 3}$ **(b)** $\dfrac{n + 3}{n - 7}$ **(c)** $\dfrac{n - 2}{n - 5}$ **(d)** $\dfrac{n + 5}{n - 2}$ **(e)** $\dfrac{n + 3}{n - 2}$

13. Working alone, Amy can complete a job in 5 hours. Julie can complete the same job in 4 hours. Which of the following equations can be used to find out how long it will take to complete the job together? (Use x for the time it takes them to complete the job working together.)

(a) $\dfrac{1}{4} + \dfrac{1}{x} = \dfrac{1}{5}$ **(b)** $\dfrac{1}{4} - \dfrac{1}{5} = \dfrac{1}{x}$ **(c)** $\dfrac{x}{4} - \dfrac{x}{5} = 1$ **(d)** $\dfrac{1}{4} + \dfrac{1}{5} = \dfrac{1}{x}$ **(e)** $\dfrac{1}{x} - \dfrac{1}{20} = \dfrac{4}{5}$

14. Suppose y varies directly as x^2 and inversely as z. Also, $y = 4$ when $x = 4$ and $z = 2$. Find y when $x = 2$ and $z = 4$.

(a) $\frac{1}{4}$ **(b)** 1 **(c)** 2 **(d)** 8 **(e)** None of the preceding

15. The volume V of a sphere varies directly as the cube of the radius, r. If $V = 36\pi$ cubic centimeters when $r = 3$ centimeters, what is the formula for the volume V?

(a) $V = \frac{3}{4}r^3$ **(b)** $V = \frac{4}{3}r^3$ **(c)** $V = \frac{4}{3}\pi r^3$ **(d)** $V = \frac{3}{4}\pi r^3$
(e) None of the preceding

ANSWERS TO THE TEST YOUR UNDERSTANDING EXERCISES

Page 129

1. $x = 0, x = 3$ 2. $x = -1, x = 5$ 3. $x = -2, x = 2$ 4. $\frac{3yz}{5}$ 5. $\frac{x^3}{2y}$
6. $-2xy$ 7. $\frac{2a^2b}{3}$ 8. $x + 2$ 9. $\frac{x}{x + 2}$ 10. $\frac{x + 3}{x - 2}$
11. $\frac{x + 5}{x - 2}$ 12. $\frac{1}{x^2 + 4}$ 13. $\frac{x + 2}{x}$ 14. $\frac{2x - 1}{2x + 3}$ 15. $\frac{x - 4}{2(3 + x)}$

Page 134

1. $\frac{y}{3}$ 2. $\frac{1}{10a}$ 3. $-\frac{3}{x + 1}$ 4. 1 5. $\frac{x - 2}{3x(x + 1)}$ 6. $\frac{a + b}{a - b}$ 7. $-\frac{1}{x}$
8. 1 9. $\frac{1}{a^5b}$

Page 138

1. 2 2. $-\frac{3}{4}$ 3. $\frac{1}{5}$ 4. $\frac{5x}{6}$ 5. y 6. $\frac{5 + x}{x - 1}$ 7. $\frac{1}{x - 2}$ 8. $a - 2$
9. $2x$ 10. 2 11. 2 12. $t + 2$

Page 141

1. $\frac{10 + 3x}{15}$ 2. $\frac{a + b}{ab}$ 3. $\frac{2x - 5}{4}$ 4. $\frac{2y - x}{y}$ 5. $\frac{x + 3y}{y}$ 6. $\frac{1}{x + 2}$

Page 144

1. $\frac{4}{5}$ 2. $2x$ 3. $\frac{3x}{98}$ 4. $8x$ 5. $\frac{1}{3}$ 6. $\frac{ab - 1}{ab + 1}$ 7. $-\frac{1}{3a}$ 8. $\frac{ab}{b - a}$

Page 149

1. $x^2 + 7x - 1$ 2. $2 + 3x^2$ 3. $ab + a^2$ 4. $a^4 + a^2 - 1$
5. $2x^2y - 3xy + 4y$ 6. $3x + 5x^3 - 2$ 7. $3 - 2a + 5a^2$
8. $b - a + a^2b^2$ 9. $3x^8 - 2x^4 + x^2$ 10. $-1 + 2y - 3y^4$

Page 155

1. 12 2. $\frac{1}{2}$ 3. 35 4. 5 5. -9 6. No solution
7. $\frac{3}{10}$ 8. $\frac{13}{24}$ 9. 3 10. -1 11. No solution 12. 6

Page 161

1. $3\frac{3}{4}$ 2. $3\frac{1}{3}$ 3. $3\frac{1}{2}$ 4. $2\frac{1}{2}$ 5. 5 6. 3
7. $-8, 8$ 8. 1 9. $-2, 3$ 10. 3 11. No solution 12. $-7, -\frac{2}{3}$

Page 168

1. 25 2. 6 3. 5 4. 16

CHAPTER 5 RATIONAL EXPONENTS AND RADICALS, COMPLEX NUMBERS, AND QUADRATIC EQUATIONS

5.1 RADICALS

You may recall our introductory work with radicals in Chapter 1. For example:

$$\sqrt{25} = 5 \qquad \text{since } 5^2 = 5 \cdot 5 = 25$$
$$\sqrt{81} = 9 \qquad \text{since } 9^2 = 9 \cdot 9 = 81$$

Now it is true also that $(-5)(-5) = 25$ and $(-9)(-9) = 81$, but we use the **radical sign** $\sqrt{}$ to denote the positive or **principal square root** of a number. We can denote a negative square root in this manner:

$$-\sqrt{25} = -5 \qquad -\sqrt{81} = -9$$

Note that the square root of a negative number is *not* a real number. For example, $\sqrt{-9}$ is not a real number since there is no real number b such that $b^2 = -9$. In general, we have the following definition:

Note that $\sqrt{a} = \sqrt[2]{a}$, is the principal square root of a and is usually referred to simply as the ***square root of a.***

DEFINITION OF $\sqrt{a}$ (the principal square root of a)
If $a > 0$, then $\sqrt{a}$ is the positive number b such that $b^2 = a$.

There are also principal cube roots, fourth roots, and so on. Here are two such illustrations:

$$\sqrt[3]{64} = 4 \text{ because } 4^3 = 64 \qquad \sqrt[5]{-32} = -2 \text{ because } (-2)^5 = -32$$

Even roots of negative numbers will be considered in Section 5.5.

However, an even root of a negative number does *not* produce a real number. For example, $\sqrt[4]{-16}$ is not a real number. Suppose we let $\sqrt[4]{-16} = b$, then $b^4 = -16$; but there is no real number whose fourth power is -16. In particular, $2^4 = 16$ and $(-2)^4 = 16$.

In general, the **principal *n*th root** of a number a is denoted by $\sqrt[n]{a}$ and is defined as follows.

*The principal nth root of a, $\sqrt[n]{a}$, is more simply referred to as the **nth root of a.** Since $n \geq 2$ this definition includes the definition of square root as a special case.*

DEFINITION OF $\sqrt[n]{a}$ (the principal nth root of a)

Let a be a real number and n a positive integer where $n \geq 2$.

1. $a > 0$ — $\sqrt[n]{a}$ is the positive number b such that $b^n = a$
2. $a < 0$, n odd — $\sqrt[n]{a}$ is the negative number b such that $b^n = a$
3. $a = 0$ — $\sqrt[n]{0} = 0$
4. $a < 0$, n even — $\sqrt[n]{a}$ is not a real number

The expression $\sqrt[n]{a}$ is called a **radical.** The number a in the **radical sign** $\sqrt[n]{\ }$ is called the **radicand,** and n is called the **index** or **root.** Note that when $\sqrt[n]{a} = b$, we have $b^n = a$, which may be rewritten as

$$(\sqrt[n]{a})^n = a$$

EXAMPLE 1 Evaluate those radicals that are real numbers and check. If an expression is not a real number, give a reason.

(a) $\sqrt[3]{-125}$ **(b)** $\sqrt{-9}$ **(c)** $\sqrt[4]{\frac{16}{81}}$ **(d)** $\sqrt{x^4}$

Solution

(a) $\sqrt[3]{-125} = -5$ *Check:* $(-5)^3 = -125$

(b) $\sqrt{-9}$ is not a real number since $-9 < 0$ and $n = 2$ is even.

(c) $\sqrt[4]{\frac{16}{81}} = \frac{2}{3}$ *Check:* $\left(\frac{2}{3}\right)^4 = \frac{2^4}{3^4} = \frac{16}{81}$

(d) $\sqrt{x^4} = x^2$ *Check:* $(x^2)^2 = x^4$ and $x^2 \geq 0$ ■

To learn how to multiply radicals we begin by investigating the specific products $\sqrt[3]{8} \cdot \sqrt[3]{-27}$. First evaluate each radical and then multiply:

$$\sqrt[3]{8} \cdot \sqrt[3]{-27} = (2)(-3) = -6$$

But -6 can also be obtained in this way:

$$\sqrt[3]{8(-27)} = \sqrt[3]{-216} = -6$$

Therefore,

$$\sqrt[3]{8} \cdot \sqrt[3]{-27} = \sqrt[3]{(8)(-27)}$$

This result leads to the following general rule. (A method of proving this rule is suggested in Exercise 93.)

> **MULTIPLICATION RULE FOR RADICALS**
>
> If $\sqrt[n]{a}$ and $\sqrt[n]{b}$ are real numbers, then
>
> $$\sqrt[n]{a} \cdot \sqrt[n]{b} = \sqrt[n]{ab}$$

The reason for requiring the variables to represent positive numbers is to avoid even roots of negative numbers.

EXAMPLE 2 Multiply, and simplify if possible. Assume that all variables represent positive numbers.

(a) $\sqrt{5} \cdot \sqrt{14}$ **(b)** $\sqrt{6x} \cdot \sqrt{7y}$ **(c)** $3\sqrt{50} \cdot 2\sqrt{\frac{1}{2}}$
(d) $\sqrt[3]{\frac{1}{24}} \cdot \sqrt[3]{-81}$ **(e)** $\sqrt[4]{8} \cdot \sqrt[4]{c}$ **(f)** $\sqrt[5]{16x} \cdot \sqrt[5]{-2x^4}$

Solution

In part (a), use a calculator to find $\sqrt{5}$ and $\sqrt{14}$ rounded to three decimal places. Then multiply and compare the result to $\sqrt{70}$, again to three decimal places.

(a) $\sqrt{5} \cdot \sqrt{14} = \sqrt{5 \cdot 14} = \sqrt{70}$
(b) $\sqrt{6x} \cdot \sqrt{7y} = \sqrt{6x \cdot 7y} = \sqrt{42xy}$
(c) $3\sqrt{50} \cdot 2\sqrt{\frac{1}{2}} = 3 \cdot 2 \cdot \sqrt{50} \cdot \sqrt{\frac{1}{2}} = 6\sqrt{50 \cdot \frac{1}{2}} = 6\sqrt{25} = 6 \cdot 5 = 30$
(d) $\sqrt[3]{\frac{1}{24}} \cdot \sqrt[3]{-81} = \sqrt[3]{(\frac{1}{24})(-81)} = \sqrt[3]{-\frac{27}{8}} = -\frac{3}{2}$
(e) $\sqrt[4]{8} \cdot \sqrt[4]{c} = \sqrt[4]{8c}$
(f) $\sqrt[5]{16x} \cdot \sqrt[5]{-2x^4} = \sqrt[5]{(16x)(-2x^4)} = \sqrt[5]{-32x^5} = -2x$ ■

Which requires less work to evaluate on a calculator: $\sqrt{24}$ or $2\sqrt{6}$?

The multiplication rule can be used to **simplify radicals.** To *simplify* a radical such as $\sqrt{24}$ means that we are to rewrite it so that the resulting radicand has no integer factors that are perfect squares, other than 1. Thus

$$\sqrt{24} = \sqrt{4 \cdot 6} = \sqrt{4} \cdot \sqrt{6} = 2\sqrt{6}$$

We say that $2\sqrt{6}$ is the *simplified form* of $\sqrt{24}$ since 6 has no factors that are perfect squares.

The solutions to parts (a) and (b) can be approximated by using a calculator. Thus, to three decimal places:

$$5\sqrt{2} = 5(1.414) = 7.070$$
$$-2\sqrt[3]{3} = -2(1.442) = -2.884$$

EXAMPLE 3 Simplify. Assume all variables represent positive numbers.
(a) $\sqrt{50}$ **(b)** $\sqrt[3]{-24}$ **(c)** $\sqrt[4]{64x^5}$ **(d)** $\sqrt{4x + 8}$

Solution

(a) $\sqrt{50} = \sqrt{25 \cdot 2} = \sqrt{25}\sqrt{2} = 5\sqrt{2}$
(b) Here we search for a perfect cube as a factor of the radicand.

$$\sqrt[3]{-24} = \sqrt[3]{(-8)3} = \sqrt[3]{-8} \cdot \sqrt[3]{3} = -2\sqrt[3]{3}$$

(c) Here we search for a perfect fourth power as a factor of the radicand.

$$\sqrt[4]{64x^5} = \sqrt[4]{16 \cdot 4 \cdot x^4 \cdot x} = \sqrt[4]{16x^4 \cdot 4x} = \sqrt[4]{16x^4} \cdot \sqrt[4]{4x} = 2x\sqrt[4]{4x}$$

(d) $\sqrt{4x + 8} = \sqrt{4(x + 2)} = \sqrt{4}\sqrt{x + 2} = 2\sqrt{x + 2}$ ■

TEST YOUR UNDERSTANDING
Think Carefully

Multiply, and simplify if possible. Assume that all variables represent positive numbers.

1. $\sqrt{3} \cdot \sqrt{7}$ **2.** $\sqrt{9} \cdot \sqrt{25}$ **3.** $\sqrt{4x} \cdot \sqrt{16x}$
4. $2\sqrt[3]{5} \cdot 3\sqrt[3]{4}$ **5.** $\sqrt[3]{8p} \cdot \sqrt[3]{27p^2}$ **6.** $\sqrt[3]{-8} \cdot \sqrt[3]{27}$
7. $4\sqrt{18} \cdot \sqrt{\frac{1}{2}}$ **8.** $\sqrt{\frac{1}{27}} \cdot \sqrt{3}$ **9.** $\sqrt[3]{\frac{1}{5}} \cdot \sqrt[3]{-40}$

Simplify. Assume that all variables represent positive numbers.

10. $\sqrt{40}$ **11.** $\sqrt{48}$ **12.** $\sqrt{75c^2}$
13. $\sqrt[3]{24}$ **14.** $\sqrt[3]{-16}$ **15.** $\sqrt[3]{128b^6}$

(Answers: Page 230)

Sometimes it is easier to simplify before multiplying. This is demonstrated in the next example.

EXAMPLE 4 Evaluate the product $\sqrt{28} \cdot \sqrt{63}$.

Solution If we multiply first, then

$$\sqrt{28} \cdot \sqrt{63} = \sqrt{1764}$$

Now it is not obvious that 1764 is a perfect square. However, you can verify that $1764 = 42^2$, and therefore $\sqrt{1764} = 42$. You can avoid the difficulty with the preceding method by first simplifying each radical:

$$\sqrt{28} \cdot \sqrt{63} = (2\sqrt{7})(3\sqrt{7}) = 6\sqrt{49} = 42$$ ■

To learn how to divide two radicals we begin by investigating a specific problem.

$$\frac{\sqrt{36}}{\sqrt{4}} = \frac{6}{2} = 3 \quad \text{and} \quad \sqrt{\frac{36}{4}} = \sqrt{9} = 3$$

Therefore, $\dfrac{\sqrt{36}}{\sqrt{4}} = \sqrt{\dfrac{36}{4}}$. In general, we have the following rule:

DIVISION RULE FOR RADICALS

If $\sqrt[n]{a}$ and $\sqrt[n]{b}$ are real numbers with $b \neq 0$, then

$$\frac{\sqrt[n]{a}}{\sqrt[n]{b}} = \sqrt[n]{\frac{a}{b}}$$

EXAMPLE 5 Divide and evaluate: **(a)** $\dfrac{\sqrt{20}}{\sqrt{5}}$ **(b)** $\dfrac{\sqrt[3]{81x^7}}{\sqrt[3]{-3x}}$

Solution

(a) $\dfrac{\sqrt{20}}{\sqrt{5}} = \sqrt{\dfrac{20}{5}} = \sqrt{4} = 2$

(b) $\dfrac{\sqrt[3]{81x^7}}{\sqrt[3]{-3x}} = \sqrt[3]{\dfrac{81x^7}{-3x}} = \sqrt[3]{-27x^6} = -3x^2$ ■

Here is another method for part (a):

$$\frac{\sqrt{20}}{\sqrt{5}} = \frac{\sqrt{4\cdot 5}}{\sqrt{5}} = \frac{2\sqrt{5}}{\sqrt{5}} = 2$$

The division rule can also be used to simplify radicals whose radicands are fractions. To simplify a radical such as $\sqrt[3]{\frac{16}{27}}$ means to rewrite it so that the result has no radical in the denominator, and the numerator is also in simplified radical form. Thus

$$\sqrt[3]{\frac{16}{27}} = \frac{\sqrt[3]{16}}{\sqrt[3]{27}} = \frac{\sqrt[3]{8\cdot 2}}{3} = \frac{\sqrt[3]{8}\cdot\sqrt[3]{2}}{3} = \frac{2\sqrt[3]{2}}{3}$$

You may find the following list of squares and cubes helpful when simplifying radicals.

Integer	Perfect squares	Perfect cubes
1	1	1
2	4	8
3	9	27
4	16	64
5	25	125
6	36	216
7	49	343
8	64	512
9	81	729
10	100	1000

EXAMPLE 6 Simplify: **(a)** $\sqrt{\dfrac{500}{49}}$ **(b)** $\sqrt[3]{-\dfrac{x}{125}}$

Solution

(a) $\sqrt{\dfrac{500}{49}} = \dfrac{\sqrt{500}}{\sqrt{49}} = \dfrac{\sqrt{100\cdot 5}}{7} = \dfrac{\sqrt{100}\sqrt{5}}{7} = \dfrac{10\sqrt{5}}{7}$

(b) $\sqrt[3]{-\dfrac{x}{125}} = \sqrt[3]{\dfrac{x}{-125}} = \dfrac{\sqrt[3]{x}}{\sqrt[3]{-125}} = \dfrac{\sqrt[3]{x}}{-5} = -\dfrac{\sqrt[3]{x}}{5}$ ■

At times a fraction can be simplified by a process known as **rationalizing the denominator.** This consists of eliminating a radical from the denominator of a fraction. For example, consider the fraction $\dfrac{4}{\sqrt{2}}$. To rationalize the denominator, multiply the numerator and denominator by $\sqrt{2}$.

$$\frac{4}{\sqrt{2}} = \frac{4\cdot\sqrt{2}}{\sqrt{2}\cdot\sqrt{2}} \quad \text{(Multiply numerator and denominator by } \sqrt{2}.)$$

$$= \frac{4\sqrt{2}}{2}$$

$$= 2\sqrt{2}$$

One reason for rationalizing denominators is to make computations easier. For example, suppose that we wish to evaluate $\dfrac{4}{\sqrt{2}}$ to three decimal places. It certainly is easier to multiply $\sqrt{2} = 1.414$ by 2 than to divide 4 by 1.414. Another reason for rationalizing denominators is to achieve a standard form for radical expressions in which they are more easily combined, as will be discussed in the next section.

If a calculator is to be used, $\frac{4}{\sqrt{2}}$ is just as easy to evaluate as $2\sqrt{2}$ and would therefore be an acceptable form.

EXAMPLE 7 Rationalize the denominator: **(a)** $\dfrac{6}{\sqrt{8}}$ **(b)** $\dfrac{5}{\sqrt[3]{2}}$

Solution

(a) Compare these two methods and note that the second one is less work.

$$\frac{6}{\sqrt{8}} = \frac{6\cdot\sqrt{8}}{\sqrt{8}\cdot\sqrt{8}} = \frac{6\sqrt{8}}{8} = \frac{3\sqrt{4\cdot 2}}{4} = \frac{3\cdot 2\sqrt{2}}{4} = \frac{3\sqrt{2}}{2}$$

$$\frac{6}{\sqrt{8}} = \frac{6\cdot\sqrt{2}}{\sqrt{8}\cdot\sqrt{2}} = \frac{6\sqrt{2}}{\sqrt{16}} = \frac{6\sqrt{2}}{4} = \frac{3\sqrt{2}}{2}$$

To help you find the radical multiplier when rationalizing such a denominator, keep in mind that ***after*** *you multiply in the denominator you want to have a radicand that is a perfect nth power.*

(b) Multiply numerator and denominator by $\sqrt[3]{4}$ in order to have a perfect cube in the denominator.

$$\frac{5}{\sqrt[3]{2}} = \frac{5\cdot\sqrt[3]{4}}{\sqrt[3]{2}\cdot\sqrt[3]{4}} = \frac{5\sqrt[3]{4}}{\sqrt[3]{8}} = \frac{5\sqrt[3]{4}}{2}$$ ■

CAUTION
Here is a common error you must avoid. Explain why the step
$\sqrt{(-4)(-4)} = \sqrt{-4}\sqrt{-4}$
is not permissible.

In applying the rules for radicals, care must be taken to avoid the type of error that results from the incorrect assumption of the existence of an nth root. For example, $\sqrt[2]{-4}$ is *not* a real number, but if this is not noticed, then *false* results such as the following occur:

$$4 = \sqrt{16} = \sqrt{(-4)(-4)} = \sqrt{-4}\cdot\sqrt{-4} = (\sqrt{-4})^2 = -4$$

EXERCISES 5.1

Evaluate those radicals that are real numbers and check. If an expression is not a real number, give a reason.

1. $\sqrt{81}$	2. $-\sqrt{81}$	3. $\sqrt[3]{27}$	4. $\sqrt[3]{-27}$
5. $-\sqrt[3]{-27}$	6. $\sqrt{\frac{4}{9}}$	7. $\sqrt{\frac{1}{25}}$	8. $\sqrt[3]{\frac{8}{125}}$
9. $\sqrt{-4}$	10. $\sqrt{-25}$	11. $\sqrt[3]{64x^3}$	12. $\sqrt[4]{16p^8}$
13. $\sqrt[4]{-81}$	14. $\sqrt[3]{-\dfrac{27}{8b^3}}$	15. $\sqrt[4]{0.0001}$	16. $\sqrt[6]{-64}$
17. $\sqrt[7]{-1}$	18. $\sqrt[8]{-1}$	19. $-\sqrt[4]{256}$	20. $\sqrt[4]{625}$
21. $\sqrt[3]{-0.008}$	22. $-\sqrt[5]{3125}$	23. $-\sqrt[5]{243}$	24. $\sqrt[6]{1{,}000{,}000}$

Simplify each radical. Also use Table I, or a calculator, to approximate to three decimal places.

25. $\sqrt{45}$	26. $-\sqrt{32}$	27. $\sqrt[3]{270}$	28. $\sqrt[3]{-875}$

Simplify. Assume that all variables represent positive numbers.

29. $\sqrt{56}$	30. $\sqrt[3]{128}$	31. $\sqrt{80x}$	32. $\sqrt{9x + 36}$
33. $\sqrt[3]{80x^4}$	34. $\sqrt[3]{-540}$	35. $\sqrt[3]{81x^6}$	36. $\sqrt[4]{64}$
37. $-\sqrt[4]{80}$	38. $\sqrt{605}$	39. $\sqrt{50y^3}$	40. $\sqrt[4]{625z^5}$
41. $\sqrt{\dfrac{128}{50}}$	42. $\sqrt[3]{\dfrac{250a^3}{b^3}}$	43. $\sqrt[3]{-\dfrac{2000}{125}}$	44. $\sqrt[4]{\dfrac{486}{625}}$

Perform the indicated operations and simplify. Assume that all variables represent positive numbers.

45. $\sqrt{6}\cdot\sqrt{3}$
46. $\sqrt{6}\cdot\sqrt{7}$
47. $\sqrt{5}\cdot\sqrt{10}$
48. $\sqrt{6x}\cdot\sqrt{12x}$
49. $\sqrt{3t}\cdot\sqrt{30t}$
50. $2\sqrt{10}\cdot 3\sqrt{20}$
51. $5\sqrt{17}\cdot\sqrt{17}$
52. $\sqrt{x}\cdot\sqrt{x^2+x}$
53. $3\sqrt{6x}\cdot\sqrt{30x^3}$
54. $5\sqrt[3]{2}\cdot 2\sqrt[3]{4}$
55. $\sqrt[3]{9}\cdot\sqrt[3]{-3}$
56. $\sqrt[3]{-4}\cdot\sqrt[3]{-12}$
57. $\dfrac{\sqrt{75}}{\sqrt{3}}$
58. $\dfrac{5\sqrt{5}}{\sqrt{80}}$
59. $\dfrac{\sqrt{300}}{5\sqrt{6}}$
60. $\dfrac{\sqrt{90}}{3\sqrt{2}}$
61. $\dfrac{\sqrt[3]{64}}{4\sqrt[3]{2}}$
62. $\dfrac{\sqrt{48x^3}}{\sqrt{6x}}$
63. $\dfrac{7\sqrt{400}}{2\sqrt{50}}$
64. $\dfrac{2\sqrt[3]{81}}{\sqrt[3]{-3}}$
65. $\dfrac{\sqrt[4]{0.0324}}{\sqrt[4]{4}}$
66. $\dfrac{\sqrt[5]{64b^{11}}}{\sqrt[5]{2b}}$
67. $\dfrac{\sqrt[3]{54s}}{\sqrt[3]{-2s^4}}$
68. $\dfrac{\sqrt[3]{-15}}{\sqrt[3]{-240}}$
69. $\sqrt{35}\cdot\sqrt{105}$
70. $\sqrt{77}\cdot\sqrt{99}$
71. $\sqrt[3]{40}\cdot\sqrt[3]{56}$
72. $\dfrac{\sqrt{11}}{\sqrt{2}}\cdot\dfrac{\sqrt{2}}{\sqrt{2}}$
73. $\dfrac{\sqrt{18}}{\sqrt{3}}\cdot\dfrac{\sqrt{2}}{\sqrt{3}}$
74. $\dfrac{\sqrt[3]{3}}{\sqrt[3]{2}}\cdot\dfrac{\sqrt[3]{9}}{\sqrt[3]{4}}$

Rationalize the denominator and simplify. Assume that all variables represent positive numbers.

75. $\dfrac{20}{\sqrt{5}}$
76. $\dfrac{24}{\sqrt{6}}$
77. $\dfrac{8x}{\sqrt{2}}$
78. $\dfrac{9y}{\sqrt{3}}$
79. $\dfrac{1}{\sqrt{18}}$
80. $\dfrac{1}{\sqrt{27}}$
81. $\dfrac{3}{\sqrt{12}}$
82. $\dfrac{6}{\sqrt{20}}$
83. $\dfrac{8}{\sqrt[3]{2}}$
84. $\dfrac{12}{\sqrt[3]{3}}$
85. $\dfrac{10}{\sqrt[4]{125}}$
86. $\dfrac{21}{\sqrt[5]{-27}}$
87. $\dfrac{9}{\sqrt{3+x}}$
88. $\dfrac{5}{\sqrt{x^3}}$
89. $\dfrac{\sqrt[3]{16}}{\sqrt[3]{2b^8}}$
90. $\dfrac{\sqrt[4]{ab}}{\sqrt[4]{a^9}}$

91. The diagonal d of a rectangle is given by the formula $d = \sqrt{\ell^2 + w^2}$, where ℓ is the length and w is the width.

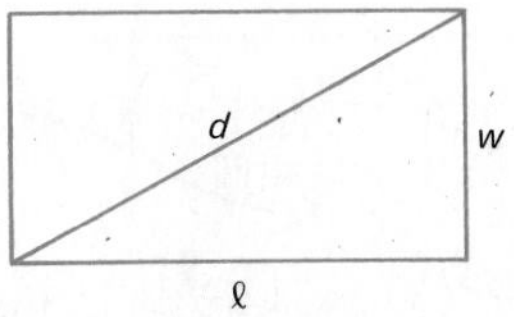

(a) Find d if $\ell = 20$ centimeters and $w = 15$ centimeters.

(b) Find d if $\ell = 16$ centimeters and $w = 10$ centimeters. Use a calculator and give the answer to two decimal places.

92. The formula $A = \sqrt{s(s-a)(s-b)(s-c)}$ is known as **Heron's formula.** It gives the area A of a triangle with sides of length a, b, c and semiperimeter $s = \frac{1}{2}(a+b+c)$. Show that for an equilateral triangle each of whose sides are of length a, Heron's formula gives $A = \dfrac{\sqrt{3}}{4}a^2$. (Heron of Alexandria, also known as Hero, was a prolific writer of mathematical and physical subjects who lived about 75 A.D.)

*93. Prove: $\sqrt[n]{a}\cdot\sqrt[n]{b} = \sqrt[n]{ab}$ assuming that all radicals are real numbers. (*Hint*: Let $\sqrt[n]{a} = x$ and $\sqrt[n]{b} = y$. Then $x^n = a$ and $y^n = b$.)

*94. Prove: $\dfrac{\sqrt[n]{a}}{\sqrt[n]{b}} = \sqrt[n]{\dfrac{a}{b}}$ assuming that all radicals are real numbers.

95. Verify each of the following and make a conjecture about a general result.

(a) $\sqrt[4]{\sqrt[2]{256}} = \sqrt[2]{\sqrt[4]{256}} = \sqrt[8]{256}$ **(b)** $\sqrt[3]{\sqrt[2]{729}} = \sqrt[2]{\sqrt[3]{729}} = \sqrt[6]{729}$

(c) $\sqrt[3]{\sqrt[3]{-512}} = \sqrt[9]{-512}$

*96. Prove that if each radical is a real number, then $\sqrt[m]{\sqrt[n]{a}} = \sqrt[mn]{a}$. (*Hint*: Let $y = \sqrt[mn]{a}$, then $y^{mn} = a$. Now use a rule for exponents.)

Written Assignment: What is the reason for rationalizing a denominator that contains a radical? Illustrate your explanation with a specific example.

CHALLENGE
Think Creatively

Consider the following *transformation formula:*

$$\frac{x}{y} \longrightarrow \frac{(x + y) + y}{x + y} = \frac{x + 2y}{x + y}$$

By this formula, the fraction $\frac{3}{4}$ is transformed to $\frac{11}{7}$ in the following way. Let $x = 3$, $y = 4$:

$$\frac{3}{4} \longrightarrow \frac{(3 + 4) + 4}{3 + 4} = \frac{11}{7}$$

Use the same formula to show that $\frac{11}{7}$ will transform into $\frac{25}{18}$. (Let $x = 11$ and $y = 7$.)

Repeat this process to obtain a sequence of five fractions. Then use a calculator to write each fraction as a decimal, correct to three decimal places.

Now begin again with several different fractions, such as $\frac{5}{7}$ and a fraction of your own choice, and follow the steps described above. See if you can discover an interesting pattern, and state your result.

5.2 COMBINING RADICALS

The rule for multiplication of radicals provides a way to find the product of any two radicals having the same index. For example,

$$\sqrt{4} \cdot \sqrt{9} = \sqrt{4 \cdot 9} = \sqrt{36} = 6$$

Recall: In order to be able to multiply two radicals, they must have the same index.

Does a similar pattern work for the addition of radicals? That is, is the sum of the square roots of two numbers equal to the square root of their sum? Does $\sqrt{4} + \sqrt{9} = \sqrt{4 + 9}$? This can easily be checked as follows:

$$\sqrt{4} + \sqrt{9} = 2 + 3 = 5$$

In general, $\sqrt{a} \pm \sqrt{b} \neq \sqrt{a \pm b}$.

But $\sqrt{4 + 9} = \sqrt{13}$. Therefore, $\sqrt{4} + \sqrt{9} \neq \sqrt{4 + 9}$.

In order to be able to add or subtract radicals, they must have the same index and the same radicand.

Consider, for example, this sum:

$$3\sqrt{5} + 4\sqrt{5}$$

Although we will usually perform the computation mentally, we can complete this example by letting x replace $\sqrt{5}$ and using the distributive property.

$$3x + 4x = (3 + 4)x = 7x$$
$$3\sqrt{5} + 4\sqrt{5} = (3 + 4)\sqrt{5} = 7\sqrt{5}$$

That is,

$$3\sqrt{5} + 4\sqrt{5} = 7\sqrt{5}$$

Sometimes radicals with the same index (or root) but with different radicands can be combined after the radicals are simplified.

A calculator can be used to check this solution. Thus, rounded to four decimal places, $\sqrt{50} - \sqrt{18} + \sqrt{45} = 9.5366$ *and* $2\sqrt{2} + 3\sqrt{5} = 9.5366$.

EXAMPLE 1 Combine: $\sqrt{50} - \sqrt{18} + \sqrt{45}$

Solution Although each radical has the same index, they do not have the same radicand. However, each can be simplified.

$$\sqrt{50} = \sqrt{25 \cdot 2} = 5\sqrt{2}$$
$$\sqrt{18} = \sqrt{9 \cdot 2} = 3\sqrt{2}$$
$$\sqrt{45} = \sqrt{9 \cdot 5} = 3\sqrt{5}$$

Thus

$$\begin{aligned}\sqrt{50} - \sqrt{18} + \sqrt{45} &= 5\sqrt{2} - 3\sqrt{2} + 3\sqrt{5} \\ &= 2\sqrt{2} + 3\sqrt{5}\end{aligned}$$

■

EXAMPLE 2 Combine: $\dfrac{6}{\sqrt{3}} + 2\sqrt{75} - \sqrt{3}$

Solution Rationalize the denominator in the first term:

$$\frac{6}{\sqrt{3}} = \frac{6 \cdot \sqrt{3}}{\sqrt{3} \cdot \sqrt{3}} = \frac{6\sqrt{3}}{3} = 2\sqrt{3}$$

Simplify the second term:

$$2\sqrt{75} = 2\sqrt{25 \cdot 3} = 10\sqrt{3}$$

Now combine:

$$\frac{6}{\sqrt{3}} + 2\sqrt{75} - \sqrt{3} = 2\sqrt{3} + 10\sqrt{3} - 1\sqrt{3} = 11\sqrt{3}$$

■

TEST YOUR UNDERSTANDING
Think Carefully

Combine, if possible.

1. $\sqrt{8} + \sqrt{32}$
2. $\sqrt{12} + \sqrt{48}$
3. $\sqrt{45} - \sqrt{20} + \sqrt{80}$
4. $2\sqrt{14} - \sqrt{22}$
5. $\sqrt[3]{16} + \sqrt[3]{54}$
6. $\sqrt{7} + \sqrt[3]{7} + \sqrt[4]{7}$
7. $\dfrac{8}{\sqrt{2}} + \sqrt{98}$
8. $\dfrac{9}{\sqrt{3}} + \sqrt{300}$
9. $3\sqrt{24} - \dfrac{6}{\sqrt{6}} - \sqrt{72}$
10. $3\sqrt{63} - \dfrac{14}{\sqrt{7}}$
11. $\dfrac{10}{\sqrt{10}} - \sqrt{10}$
12. $\sqrt{2} + \dfrac{1}{\sqrt{2}} + \dfrac{1}{\sqrt{8}}$

(Answers: Page 230)

*Do you think that $\sqrt{x^2} = x$? This is **not** always so. Study this explanation carefully; it is important for future work in mathematics.*

Sometimes it is necessary to simplify a radical that has an even index and contains a variable, such as $\sqrt{x^2}$. In this example, the usual reaction is to claim that $\sqrt{x^2} = x$. But suppose that x were negative, such as $x = -5$:

$$\text{If } \sqrt{x^2} = x, \quad \text{then} \quad \sqrt{(-5)^2} = -5$$
$$\text{But } \sqrt{(-5)^2} = \sqrt{25} = 5$$

It was stated earlier that the radical sign, $\sqrt{\ }$, means the *positive* square root of a number. Therefore, it is necessary that $\sqrt{(-5)^2}$ be equal to 5, and not -5. That is, we must have each of the following:

$$\sqrt{5^2} = 5 \qquad \sqrt{x^2} = x \text{ if } x \text{ is positive}$$
$$\sqrt{(-5)^2} = 5 \qquad \sqrt{x^2} = -x \text{ if } x \text{ is negative}$$

This leads to the following important result:

For all real numbers a, $\sqrt{a^2} = |a|$

This result can be extended as follows:

$$\sqrt[n]{a^n} = |a|, \qquad \text{if } n \text{ is even.}$$
$$\sqrt[n]{a^n} = a, \qquad \text{if } n \text{ is odd.}$$

EXAMPLE 3 Simplify:

(a) $\sqrt{(-10)^2}$ **(b)** $\sqrt{x^6}$ **(c)** $\sqrt[8]{(-\frac{1}{2})^8}$
(d) $\sqrt[7]{(-8)^7}$ **(e)** $\sqrt{75x^2}$ **(f)** $\sqrt[4]{(5x+1)^4}$

Solution

In part (b) we are using $a = x^3$ in the formula $\sqrt{a^2} = |a|$.

(a) $\sqrt{(-10)^2} = |-10| = 10$
(b) $\sqrt{x^6} = \sqrt{(x^3)^2} = |x^3|$
(c) $\sqrt[8]{(-\frac{1}{2})^8} = |-\frac{1}{2}| = \frac{1}{2}$
(d) $\sqrt[7]{(-8)^7} = -8$
(e) $\sqrt{75x^2} = \sqrt{25 \cdot 3} \cdot \sqrt{x^2} = 5\sqrt{3}|x|$
(f) $\sqrt[4]{(5x+1)^4} = |5x+1|$ ■

Sometimes the restrictions on a variable in a radicand are implied or assumed from the statement of the radical, without additional explanations. This is discussed in the next example.

EXAMPLE 4 Simplify: $\sqrt{48x^3}$

Solution

$$\begin{aligned}\sqrt{48x^3} &= \sqrt{16 \cdot 3 \cdot x^2 \cdot x} \\ &= \sqrt{16 \cdot x^2 \cdot 3x} \\ &= \sqrt{16} \cdot \sqrt{x^2} \cdot \sqrt{3x} \\ &= 4|x|\sqrt{3x}\end{aligned}$$

Note here that x cannot be negative if the given radical is to be a real number. If x were negative, then $48x^3$ would be negative, and the square root of a negative number is not a real number. Thus, *assuming* that the given radical is to be a real number, we have $x \geq 0$, and it is safe to delete the absolute value notation.

$$\sqrt{48x^3} = 4x\sqrt{3x}$$

■

When writing radicals whose radicands involve variables we automatically assume (unless otherwise stated) that the allowable values for the variables are those for which the radicand is a real number.

EXAMPLE 5 Find the allowable values of the variable in the given radicals.

(a) $\sqrt{x}$ **(b)** $\sqrt{x^4}$ **(c)** $\sqrt{x^5}$

(d) $\sqrt[3]{x}$ **(e)** $\sqrt[4]{x-2}$ **(f)** $\dfrac{1}{\sqrt{x+1}}$

Solution

(a) $x \geq 0$.
(b) All real numbers x.
(c) $x \geq 0$, since x^5 would be negative if $x < 0$.
(d) All real numbers x.
(e) $x \geq 2$, since $x - 2 \geq 0$.
(f) In order for the radicand to be nonnegative, $x + 1 \geq 0$ or $x \geq -1$. But $x = -1$ would produce a zero denominator. Therefore, the allowable values are $x > -1$. ■

EXAMPLE 6 Combine: $2\sqrt{8x^3} + 3x\sqrt{32x} - x\sqrt{18x}$

Note that the expression under the radicals would be negative for $x < 0$. Since the index is even, we must assume that $x \geq 0$ in order for these to have meaning.

Solution For this problem, $x \geq 0$. Thus we need not make use of absolute value notation.

$$2\sqrt{8x^3} = 2\sqrt{4 \cdot 2 \cdot x^2 \cdot x} = 4x\sqrt{2x}$$
$$3x\sqrt{32x} = 3x\sqrt{16 \cdot 2x} = 12x\sqrt{2x}$$
$$x\sqrt{18x} = x\sqrt{9 \cdot 2x} = 3x\sqrt{2x}$$

Since the radicands and the roots are the same, the distributive property can be used to combine.

$$\begin{aligned} 4x\sqrt{2x} + 12x\sqrt{2x} - 3x\sqrt{2x} &= (4x + 12x - 3x)\sqrt{2x} \\ &= 13x\sqrt{2x} \end{aligned}$$

■

The algebraic methods that were first studied for multiplying polynomials can be used for expressions involving radicals. For example, observe how the formula $(a - b)(a + b) = a^2 - b^2$ is applied for this product:

$$\underset{(a\ -\ \ b)\ \ (a\ +\ \ b)\ =\ \ a^2\ \ -\ \ \ b^2}{(\sqrt{x} - \sqrt{y})(\sqrt{x} + \sqrt{y}) = (\sqrt{x})^2 - (\sqrt{y})^2 = x - y}$$

EXAMPLE 7 Find the products and simplify:

(a) $\sqrt{5}(2\sqrt{5} - \sqrt{10})$ **(b)** $(2\sqrt[3]{8} + 7)(3\sqrt[3]{8} - 5)$

(c) $(\sqrt{11} - \sqrt{3})^2$ **(d)** $(\sqrt[3]{x^2} - 1)(\sqrt[3]{x^2} + 1)$

Solution

(a)
$$\begin{aligned} \sqrt{5}(2\sqrt{5} - \sqrt{10}) &= \sqrt{5}\cdot 2\sqrt{5} - \sqrt{5}\cdot\sqrt{10} \qquad \text{(Distributive property)} \\ &= 2\sqrt{25} - \sqrt{50} \\ &= 2\cdot 5 - 5\sqrt{2} = 10 - 5\sqrt{2} \end{aligned}$$

−35

$6\sqrt[3]{64}$

(b)
$$\begin{aligned} (2\sqrt[3]{8} + 7)(3\sqrt[3]{8} - 5) &= 6\sqrt[3]{64} - 10\sqrt[3]{8} + 21\sqrt[3]{8} - 35 \\ &= 48 + 11\sqrt[3]{8} - 35 \\ &= 13 + 11\sqrt[3]{8} = 13 + 22\sqrt{2} \end{aligned}$$

$21\sqrt[3]{8}$

$-10\sqrt[3]{8}$

(c) Use the form $(a - b)^2 = a^2 - 2ab + b^2$.

$$\begin{aligned} (\sqrt{11} - \sqrt{3})^2 &= (\sqrt{11})^2 - 2\sqrt{11}\sqrt{3} + (\sqrt{3})^2 \\ &= 11 - 2\sqrt{33} + 3 \\ &= 14 - 2\sqrt{33} \end{aligned}$$

(d)
$$\begin{aligned} (\sqrt[3]{x^2} - 1)(\sqrt[3]{x^2} + 1) &= (\sqrt[3]{x^2})^2 - 1^2 = \sqrt[3]{x^4} - 1 \\ &= \sqrt[3]{x^3\cdot x} - 1 = x\sqrt[3]{x} - 1 \end{aligned}$$

■

The formula $(a + b)(a - b) = a^2 - b^2$ can also be used to rationalize denominators that are sums or differences involving radicals. In this context each of the two factors $a + b$ and $a - b$ is called the **conjugate** of the other.

EXAMPLE 8 Rationalize the denominators.

(a) $\dfrac{5}{\sqrt{10} - 3}$ **(b)** $\dfrac{x}{\sqrt{x} + \sqrt{y}}$

Solution Multiply the numerator and denominator of each fraction by the conjugate of its denominator.

(a) The conjugate of the denominator is $\sqrt{10} + 3$.

$$\frac{5}{\sqrt{10} - 3} = \frac{5(\sqrt{10} + 3)}{(\sqrt{10} - 3)(\sqrt{10} + 3)} = \frac{5(\sqrt{10} + 3)}{10 - 9} = 5(\sqrt{10} + 3)$$

(b) The conjugate of the denominator is $\sqrt{x} - \sqrt{y}$.

$$\frac{x}{\sqrt{x} + \sqrt{y}} = \frac{x(\sqrt{x} - \sqrt{y})}{(\sqrt{x} + \sqrt{y})(\sqrt{x} - \sqrt{y})} = \frac{x(\sqrt{x} - \sqrt{y})}{x - y}$$ ■

CAUTION: Learn To Avoid These Mistakes

WRONG	RIGHT
$\sqrt{9 + 16} = \sqrt{9} + \sqrt{16}$	$\sqrt{9 + 16} = \sqrt{25}$
$\sqrt[3]{8} \cdot \sqrt[2]{8} = \sqrt[6]{64}$	$\sqrt[3]{8} \cdot \sqrt[2]{8} = 2 \cdot 2\sqrt{2} = 4\sqrt{2}$
$2\sqrt{x + 1} = \sqrt{2x + 1}$	$2\sqrt{x + 1} = \sqrt{4(x + 1)}$ $= \sqrt{4x + 4}$
$2 - \dfrac{1}{\sqrt{2}} = \dfrac{2 - 1}{\sqrt{2}}$	$2 - \dfrac{1}{\sqrt{2}} = 2 - \dfrac{\sqrt{2}}{2} = \dfrac{4 - \sqrt{2}}{2}$
$\sqrt{(x - 1)^2} = x - 1$	$\sqrt{(x - 1)^2} = \|x - 1\|$
$\sqrt{x^9} = x^3$	$\sqrt{x^9} = \sqrt{x^8 \cdot x} = x^4\sqrt{x}$

EXERCISES 5.2

Simplify.

1. $\sqrt{(-8)^2}$
2. $\sqrt[3]{(-17)^3}$
3. $\sqrt[4]{(-\frac{7}{2})^4}$
4. $\sqrt[3]{(-2)^9}$
5. $\sqrt[5]{(\frac{1}{2})^{10}}$
6. $\sqrt{17^6}$
7. $\sqrt[8]{x^8}$
8. $\sqrt{\dfrac{5x^2}{7y^2}}$
9. $\sqrt[6]{64s^6}$

Find the allowable values of the variables so that the radical expressions are real numbers.

10. $\sqrt{7x}$
11. $\dfrac{1}{\sqrt{b^2}}$
12. $\sqrt[3]{t - 3}$
13. $\sqrt[4]{1 - x}$
14. $\sqrt{-x}$
15. $\sqrt{\dfrac{x^3}{y^2}}$
16. $\sqrt{2p - 1}$
17. $\sqrt[4]{x^5}$
18. $\dfrac{1}{\sqrt[3]{x^2 - x}}$

Combine and simplify.

19. $\sqrt{2} + \sqrt{18}$
20. $\sqrt{48} - \sqrt{3}$
21. $\sqrt{25} + \sqrt{49}$
22. $\sqrt{64} - \sqrt{16}$
23. $\sqrt{12} - \sqrt{27}$
24. $\sqrt{32} + \sqrt{72}$
25. $\sqrt{18} + \sqrt{98}$
26. $\sqrt{48} + \sqrt{75}$
27. $2\sqrt{5} + 3\sqrt{125}$
28. $-5\sqrt{24} - 2\sqrt{54}$
29. $2\sqrt{200} - 5\sqrt{8}$
30. $3\sqrt{45} - 2\sqrt{20}$
31. $\sqrt[3]{128} + \sqrt[3]{16}$
32. $\sqrt[3]{24} + \sqrt[3]{81}$
33. $\sqrt{50} + \sqrt{32} - \sqrt{8}$
34. $\sqrt{12} - \sqrt{3} + \sqrt{108}$
35. $\dfrac{8}{\sqrt{2}} + 2\sqrt{50}$
36. $\dfrac{12}{\sqrt{3}} - \sqrt{12}$
37. $\dfrac{1}{\sqrt{3}} - \dfrac{2}{\sqrt{27}}$
38. $\sqrt{x} - \dfrac{1}{\sqrt{x}}$
39. $\dfrac{2}{\sqrt[3]{4}} + \sqrt[3]{2}$
40. $7\sqrt[3]{3} - \dfrac{6}{\sqrt[3]{9}}$
41. $\sqrt[3]{56x} + \sqrt[3]{7x}$
42. $\sqrt[3]{54x} + \sqrt[3]{250x}$
43. $3\sqrt{8x^2} - \sqrt{50x^2}$
44. $5\sqrt{75x^2} - 2\sqrt{12x^2}$
45. $\sqrt[4]{32} + \sqrt[4]{162}$
46. $\sqrt[5]{32} + \sqrt[5]{64}$
47. $\sqrt{24} + \sqrt{54} - \sqrt{18}$
48. $\sqrt{36} + \sqrt{28} + \sqrt{63}$
49. $3\sqrt{10} + 4\sqrt{90} - 5\sqrt{40}$
50. $3\sqrt{24} - \sqrt{54} + 2\sqrt{150}$
51. $\sqrt{18x} + \sqrt{50x} - \sqrt{2x}$
52. $10\sqrt{3x} - 2\sqrt{75x} + 3\sqrt{243x}$
53. $\dfrac{1}{\sqrt{2}} + 3\sqrt{72} - 2\sqrt{2}$
54. $\dfrac{2}{\sqrt{3}} + 10\sqrt{3} - 2\sqrt{12}$
55. $\sqrt{\dfrac{7}{3}} - \dfrac{21}{\sqrt{21}} + \dfrac{2\sqrt{21}}{3}$
56. $\dfrac{2}{\sqrt{2t}} - \dfrac{3\sqrt{2t}}{t} + \sqrt{\dfrac{2}{t}}$
57. $3\sqrt{9x^2} + 2\sqrt{16x^2} - \sqrt{25x^2}$
58. $\sqrt{2x^2} + 5\sqrt{32x^2} - 2\sqrt{98x^2}$
59. $\sqrt{x^2y} + \sqrt{8x^2y} + \sqrt{200x^2y}$
60. $\sqrt{72xy} + 2\sqrt{2xy} + \sqrt{128xy}$
61. $\sqrt{20a^3} + a\sqrt{5a} + \sqrt{80a^3}$
62. $\sqrt{12ab^3} + \sqrt{27ab^3} + 2b\sqrt{3ab}$
63. $\sqrt[3]{\dfrac{32}{x^2}} - \dfrac{2\sqrt[3]{x}}{\sqrt[3]{2x^3}}$
64. $\dfrac{\sqrt{72a^3}}{3b} - \dfrac{a\sqrt{50a}}{2b} + \dfrac{12a^2}{b\sqrt{2a}}$

Find the product and simplify.

65. $\sqrt{2}(\sqrt{6} - \sqrt{18})$
66. $5\sqrt{6}(\sqrt{24} + \sqrt{8} - 2\sqrt{18})$
67. $\sqrt{7}\left(\sqrt{14} - \dfrac{2}{\sqrt{14}}\right)$
68. $(\sqrt{7} + \sqrt{3})(\sqrt{7} - \sqrt{3})$
69. $(5\sqrt{3} + \sqrt{2})(\sqrt{3} - 3\sqrt{2})$
70. $(\sqrt{15} + \sqrt{3})^2$
71. $(2x - \sqrt{y})(3x - \sqrt{y})$
72. $(x - \sqrt[3]{4})(x + \sqrt[3]{4})$
73. $\left(\sqrt{x} - \dfrac{2}{\sqrt{x}}\right)^2$
74. $(\sqrt[3]{x} - 1)(\sqrt[3]{x^2} + \sqrt[3]{x} + 1)$

Rationalize the denominator.

75. $\dfrac{12}{\sqrt{5} - \sqrt{3}}$
76. $\dfrac{20}{3 - \sqrt{2}}$
77. $\dfrac{14}{\sqrt{2} - 3}$
78. $\dfrac{\sqrt{x}}{\sqrt{x} + \sqrt{y}}$
79. $\dfrac{\sqrt{x} + \sqrt{y}}{\sqrt{x} - \sqrt{y}}$
80. $\dfrac{1}{\sqrt{x} + 2}$
81. $\dfrac{3 - \sqrt{7}}{3 + \sqrt{7}}$
82. $\dfrac{\sqrt{3} - 3}{1 - \sqrt{3}}$
*83. $\dfrac{x + 2\sqrt{xy} + y}{\sqrt{x} + \sqrt{y}}$

*84. Rationalize the denominator: $\dfrac{1}{\sqrt[3]{2} - 1}$ (*Hint*: $a^3 - b^3 = (a - b)(a^2 + ab + b^2)$.)

In Exercises 85 through 87, rationalize the **numerator** *by multiplying the numerator and denominator by the conjugate of the numerator.*

85. $\dfrac{\sqrt{5}+3}{4}$ 86. $\dfrac{\sqrt{5}-\sqrt{3}}{2}$ 87. $\dfrac{\sqrt{7}+\sqrt{5}}{1-\sqrt{35}}$

88. Evaluate $x^2 + 2x - 1$ for $x = -1 + \sqrt{2}$.

89. Evaluate $x^2 - 6x + 8$ for $x = 3 - \sqrt{5}$.

90. Evaluate $x^2 - 3x + 1$ for $x = \dfrac{3+\sqrt{5}}{2}$.

91. Check the solution to Exercise 88 by using a calculator. First change the given x-value into decimal form, rounded to four decimal places. Then substitute this value into the given polynomial and compare the result with the value obtained in Exercise 88.

92. Apply the directions in Exercise 91 to Exercise 90.

Written Assignment: Give a convincing argument why $\sqrt{x^2} = |x|$ and not x.

5.3 RATIONAL EXPONENTS

We are now ready to make a further extension of the exponential concept to include fractional exponents. Once again our guideline will be to *preserve* the earlier rules for integral exponents. The first step is to consider exponents of the form $\frac{1}{n}$, where n is an integer, $n \geq 2$. That is, we wish to give meaning to the expression $b^{1/n}$. Now one of the rules for exponents that we developed earlier stated that $(b^m)^n = b^{mn}$. Therefore, if our rules for exponents are to work, we have

This is a good time to review the rules for exponents as stated on page 36.

$$(b^{1/n})^n = b^{(1/n)(n)} = b$$

Thus $b^{1/n}$ is the nth root of b (provided that such a root exists). This leads us to the following definition for $b^{1/n}$.

Since $\sqrt{-1}$ is not a real number, $(-1)^{1/2}$ is not defined. In general, $b^{1/n}$ is not defined within the set of real numbers when $b < 0$ and n is even.

DEFINITION OF $b^{1/n}$

For a real number b and a positive integer $n (n \geq 2)$,

$$b^{1/n} = \sqrt[n]{b}$$

provided that $\sqrt[n]{b}$ exists.

EXAMPLE 1
(a) Write $5^{1/2}$ in radical form. **(b)** Write $\sqrt{15}$ with a fractional exponent.

Solution
(a) $5^{1/2} = \sqrt{5}$ **(b)** $\sqrt{15} = 15^{1/2}$ ■

EXAMPLE 2 Evaluate, if possible:
(a) $9^{1/2}$ **(b)** $8^{1/3}$ **(c)** $(-8)^{1/3}$ **(d)** $(-16)^{1/4}$

Solution
(a) $9^{1/2} = \sqrt{9} = 3$ **(b)** $8^{1/3} = \sqrt[3]{8} = 2$
(c) $(-8)^{1/3} = \sqrt[3]{-8} = -2$ **(d)** $\sqrt[4]{-16}$ is not a real number since $b = -16 < 0$ and $n = 4$ is even. ■

Now that $b^{1/n}$ has been defined, we are able to define $b^{m/n}$, where $\frac{m}{n}$ is any rational number. Once again we want the earlier rules of exponents to apply. Observe, for example, the two ways that $8^{2/3}$ can be evaluated on the *assumption* that Rule 3 applies.

See Rule 3 for exponents on page 34.

$$8^{2/3} = 8^{(1/3)\cdot 2} = (8^{1/3})^2 = (\sqrt[3]{8})^2 = 2^2 = 4$$

$$\uparrow$$

Rule 3

$$\downarrow$$

$$8^{2/3} = 8^{2\cdot(1/3)} = (8^2)^{1/3} = \sqrt[3]{8^2} = \sqrt[3]{64} = 4$$

This leads to the following definition.

Note that a rational number can always be expressed with a positive denominator: for example,

$$\frac{2}{-3} = \frac{-2}{3}$$

DEFINITION OF $b^{m/n}$

Let $\frac{m}{n}$ be a rational number with $n \geq 2$. If b is a real number such that $\sqrt[n]{b}$ is defined, then

$$b^{m/n} = (\sqrt[n]{b})^m = \sqrt[n]{b^m}$$

Using only fractional exponents, we may also write

$$b^{m/n} = (b^{1/n})^m = (b^m)^{1/n}$$

EXAMPLE 3 Evaluate: $(-64)^{2/3}$

Solution Using $b^{m/n} = (\sqrt[n]{b})^m$

$$(-64)^{2/3} = (\sqrt[3]{-64})^2 = (-4)^2 = 16$$

Using $b^{m/n} = \sqrt[n]{b^m}$

$$(-64)^{2/3} = \sqrt[3]{(-64)^2} = \sqrt[3]{4096} = 16$$

Obviously, the first approach is less work. For most such problems *it is easier to take the nth root first and then raise to the mth power*, rather than the reverse. ■

Observe that the earlier definition $b^{-n} = \frac{1}{b^n}$ extends to the case $b^{-(m/n)}$ as follows:

$$b^{-(m/n)} = b^{-m/n} = (b^{1/n})^{-m} = \frac{1}{(b^{1/n})^m} = \frac{1}{b^{m/n}}$$

EXAMPLE 4 Evaluate: $8^{-2/3} + (-32)^{-2/5}$

Solution First rewrite each part using positive exponents. Then apply the definition and add.

Here is another way to evaluate $8^{-2/3}$.

$$8^{-2/3} = (8^{1/3})^{-2} = (\sqrt[3]{8})^{-2} = (2)^{-2} = \frac{1}{4}$$

Try this approach for $(-32)^{-2/5}$.

$$\begin{aligned} 8^{-2/3} + (-32)^{-2/5} &= \frac{1}{8^{2/3}} + \frac{1}{(-32)^{2/5}} \\ &= \frac{1}{(\sqrt[3]{8})^2} + \frac{1}{(\sqrt[5]{-32})^2} \\ &= \frac{1}{(2)^2} + \frac{1}{(-2)^2} \\ &= \frac{1}{4} + \frac{1}{4} = \frac{1}{2} \end{aligned}$$

■

TEST YOUR UNDERSTANDING
Think Carefully

Write each of the following as a radical.

1. $5^{1/2}$ **2.** $9^{1/3}$ **3.** $10^{1/4}$ **4.** $5^{2/3}$ **5.** $2^{3/4}$

Write each of the following using fractional exponents.

6. $\sqrt{7}$ **7.** $\sqrt[3]{-10}$ **8.** $\sqrt[4]{7}$ **9.** $\sqrt[3]{7^2}$ **10.** $(\sqrt[4]{5})^3$

Evaluate.

11. $25^{1/2}$ **12.** $64^{1/3}$ **13.** $(\frac{1}{36})^{1/2}$ **14.** $(-\frac{1}{125})^{1/3}$
15. $(\frac{9}{25})^{1/2}$ **16.** $49^{-1/2}$ **17.** $4^{3/2}$ **18.** $4^{-3/2}$
19. $(-8)^{2/3}$ **20.** $(-8)^{-2/3}$

(Answers: Page 230)

Observe that $(-8)^{2/6} = (-8)^{1/3} = \sqrt[3]{-8} = -2$. However, $(-8)^{2/6} \neq \sqrt[6]{(-8)^2} = \sqrt[6]{64} = 2$. Remember that to use the definition of $b^{m/n}$, $\sqrt[n]{b}$ *must* be a real number, and $\sqrt[6]{-8}$ is not a real number. This shows that we need to be careful when negative numbers are being used with rational exponents. In order to avoid such difficulties when using variables with fractional exponents, we will automatically assume that the variables represent positive numbers, unless it is indicated otherwise.

EXAMPLE 5 Simplify: **(a)** $(-3x^{2/5})(7x^{8/5})$ **(b)** $(x^2y^{-2})^{3/2}$

Solution

(a)

$$\begin{aligned} (-3x^{2/5})(7x^{8/5}) &= -3 \cdot 7 \cdot x^{2/5} \cdot x^{8/5} \\ &= -21x^{2/5+8/5} \\ &= -21x^{10/5} \\ &= -21x^2 \end{aligned}$$

(b) Apply the rules of exponents that state $(ab)^n = a^n b^n$ and $(b^m)^n = b^{mn}$.

$$(x^2y^{-2})^{3/2} = (x^2)^{3/2} \cdot (y^{-2})^{3/2} = x^3y^{-3}$$

Using only positive exponents the answer would be given as x^3/y^3, or $(x/y)^3$.

■

EXAMPLE 6 Simplify, and express the result with positive exponents only:

$$\frac{x^{2/3}y^{-2}z^2}{x^{1/2}y^{1/2}z^{-1}}$$

You might find it easier to rewrite the problem first, using positive exponents only.

Solution Use the rule for division with exponents, $b^m/b^n = b^{m-n}$.

$$\frac{x^{2/3}y^{-2}z^2}{x^{1/2}y^{1/2}z^{-1}} = x^{2/3-1/2}y^{-2-1/2}z^{2-(-1)}$$

$$= x^{1/6}y^{-5/2}z^3$$

Next rewrite the result using positive exponents, as required.

$$x^{1/6}y^{-5/2}z^3 = \frac{x^{1/6}z^3}{y^{5/2}}$$

Another method uses the preceding division rule as well as this variation:

$$\frac{b^m}{b^n} = \frac{1}{b^{n-m}}$$

Thus the solution may be obtained directly:

$$\frac{x^{2/3}y^{-2}z^2}{x^{1/2}y^{1/2}z^{-1}} = \frac{x^{(2/3)-(1/2)}z^{2-(-1)}}{y^{(1/2)-(-2)}} = \frac{x^{1/6}z^3}{y^{5/2}}$$ ■

EXAMPLE 7 Simplify: **(a)** $\dfrac{(s^{-3}t^2)^{-1/6}}{(s^4t^{-4})^{-1/2}}$ **(b)** $\left(\dfrac{-8a^3}{b^{-6}}\right)^{2/3}$

Solution

(a)

$$\frac{(s^{-3}t^2)^{-1/6}}{(s^4t^{-4})^{-1/2}} = \frac{(s^{-3})^{-1/6}(t^2)^{-1/6}}{(s^4)^{-1/2}(t^{-4})^{-1/2}}$$

$$= \frac{s^{1/2}t^{-1/3}}{s^{-2}t^2}$$

$$= \frac{s^{(1/2)-(-2)}}{t^{2-(-1/3)}}$$

$$= \frac{s^{5/2}}{t^{7/3}}$$

(b)

$$\left(\frac{-8a^3}{b^{-6}}\right)^{2/3} = \frac{(-8a^3)^{2/3}}{(b^{-6})^{2/3}}$$

$$= \frac{(-8)^{2/3}(a^3)^{2/3}}{(b^{-6})^{2/3}}$$

$$= \frac{\sqrt[3]{(-8)^2}a^2}{b^{-4}}$$

$$= 4a^2b^4 \qquad (\sqrt[3]{(-8)^2} = \sqrt[3]{64} = 4)$$ ■

The next two examples illustrate how algebraic procedures first used with polynomials apply to expressions with fractional exponents.

EXAMPLE 8 Multiply. Write the answers without negative exponents:
(a) $a^{-1/3}(a^2 - 4a^{1/3})$ **(b)** $(x^{1/2} + y^{-1/2})(x^{1/2} - y^{-1/2})$

Solution

(a) Use the distributive property.

$$\begin{aligned} a^{-1/3}(a^2 - 4a^{1/3}) &= a^{-1/3} \cdot a^2 - a^{-1/3} \cdot 4a^{1/3} \\ &= a^{-1/3+2} - 4a^{-1/3+1/3} \\ &= a^{5/3} - 4a^0 \\ &= a^{5/3} - 4 \end{aligned}$$

(b) Use the form $(a + b)(a - b) = a^2 - b^2$.

$$\begin{aligned} (x^{1/2} + y^{-1/2})(x^{1/2} - y^{-1/2}) &= (x^{1/2})^2 - (y^{-1/2})^2 \\ &= x^1 - y^{-1} \\ &= x - \frac{1}{y} \qquad \left(\text{or } \frac{xy - 1}{y}\right) \end{aligned}$$

■

EXAMPLE 9 Combine $5x^{-1/3} + x^{2/3}$ to obtain the form $\dfrac{5 + x}{\sqrt[3]{x}}$.

Solution First write without negative exponents and then combine.

$$\begin{aligned} 5x^{-1/3} + x^{2/3} &= \frac{5}{x^{1/3}} + x^{2/3} \\ &= \frac{5}{x^{1/3}} + \frac{x^{2/3} \cdot x^{1/3}}{x^{1/3}} \qquad (x^{1/3} \text{ is a common denominator.}) \\ &= \frac{5 + x}{x^{1/3}} \\ &= \frac{5 + x}{\sqrt[3]{x}} \end{aligned}$$

As an alternative solution, begin by factoring out the common factor $x^{-1/3}$:

$$\begin{aligned} 5x^{-1/3} + x^{2/3} &= x^{-1/3}(5 + x) \qquad (\text{Note: } x^{-1/3} \cdot x = x^{-(1/3)+1} = x^{2/3}.) \\ &= \frac{1}{\sqrt[3]{x}}(5 + x) \\ &= \frac{5 + x}{\sqrt[3]{x}} \end{aligned}$$

■

TEST YOUR UNDERSTANDING
Think Carefully

Each of the following simplifications is **incorrect** and represents an error commonly made. Find the correct simplification for each expression at the left.

1. $(25)^{1/2} = \pm 5$
2. $15^{3/4} = (\sqrt[3]{15})^4$
3. $(-2)^{-1/3} = 2^{1/3}$
4. $a^{-1/2} + b^{-1/2} = \dfrac{1}{\sqrt{a + b}}$
5. $(7^{1/3})(7^{1/2}) = 7^{1/6}$
6. $x^{-1/5} + x^{4/5} = x^{3/5}$
7. $(16^{-1/2})^2 = 16^{1/4}$
8. $(x^{1/2} + 3)^2 = x + 9$

(Answers: Page 230)

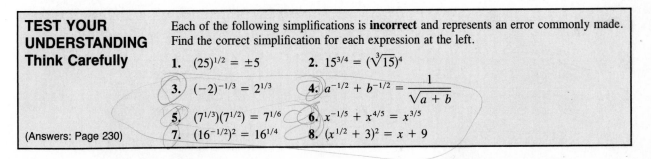

EXERCISES 5.3

Write in radical form.

1. $3^{1/2}$
2. $7^{1/3}$
3. $8^{1/2}$
4. $(-9)^{1/3}$
5. $15^{1/4}$
6. $6^{2/3}$
7. $5^{3/4}$
8. $2^{4/5}$
9. $7^{-2/3}$
10. $2^{-4/5}$
11. $\left(\frac{4}{5}\right)^{2/3}$
12. $\left(\frac{4}{5}\right)^{3/2}$

Write with fractional exponents.

13. $\sqrt{11}$
14. $\sqrt[3]{21}$
15. $\sqrt[4]{9}$
16. $\sqrt[3]{-10}$
17. $\sqrt[5]{-20}$
18. $\sqrt[3]{5^2}$
19. $(\sqrt[3]{-7})^2$
20. $(\sqrt[5]{2})^3$
21. $\dfrac{1}{\sqrt[3]{4^2}}$
22. $\dfrac{1}{\sqrt{17}}$
23. $\sqrt{\dfrac{1}{10}}$
24. $\sqrt[3]{-0.001}$

Evaluate, if possible.

25. $121^{1/2}$
26. $125^{1/3}$
27. $(-125)^{1/3}$
28. $81^{-1/2}$
29. $64^{-1/3}$
30. $64^{2/3}$
31. $64^{-2/3}$
32. $(-64)^{2/3}$
33. $(-64)^{-2/3}$
34. $(-125)^{3/2}$
35. $\left(\frac{8}{27}\right)^{2/3}$
36. $\left(\frac{8}{27}\right)^{-2/3}$
37. $\dfrac{9^{1/2}}{27^{1/3}}$
38. $\dfrac{9^{-1/2}}{27^{-1/3}}$
39. $\dfrac{9^{1/2}}{(-27)^{-1/3}}$
40. $\left(-\frac{1}{16}\right)^{1/4}$
41. $\left(\frac{1}{4}\right)^{3/2}\cdot\left(-\frac{1}{8}\right)^{2/3}$
42. $8^{2/5}\cdot 8^{8/5}$
43. $(-6)^{5/3}(-6)^{1/3}$
44. $\dfrac{12^{7/5}}{12^{2/5}}$
45. $\dfrac{3^{-1/2}}{3^{3/2}}$
46. $\dfrac{10^{2/3}\cdot 10^{8/3}}{10^{1/3}}$

Simplify, and express all answers with positive exponents. Assume that all variables represent positive numbers.

47. $\left(\dfrac{a^{-1}b^{1/2}}{c^3}\right)^{-2}$
48. $\left(\dfrac{a^2b^{-3}}{c^{-1/2}}\right)^2$
49. $(8a^3b^{-9})^{2/3}$
50. $(27a^{-3}b^9)^{-2/3}$
51. $(a^{-4}b^{-8})^{3/4}$
52. $(a^{2/3}b^{-1/2})(a^{1/3}b^{-1/2})$
53. $(a^{-1/2}b^{1/3})(a^{1/2}b^{-1/3})$
54. $(a^{3/4}b^{1/3})(a^{-1/4}b^{-2/3})$
55. $\dfrac{a^{-1/2}b^{2/3}}{a^{3/2}b^{4/3}}$
56. $\dfrac{(64a^6)^{2/3}}{(b^{-9})^{2/3}}$
57. $\left(\dfrac{a^3}{-8b^{-6}}\right)^{-2/3}$
58. $\left(\dfrac{49a^{-4}}{81b^6}\right)^{-1/2}$

Simplify, and express all answers with positive exponents. (Assume that n is a positive integer and that x and y are positive.)

59. $\dfrac{(x^n)^{1/2}}{(x^{n-2})^{1/2}}$
60. $\left(\dfrac{x^n}{x^{n-2}}\right)^{-1/2}$
61. $\left(\dfrac{x^{3n+1}y^n}{x^{3n+4}y^{4n}}\right)^{1/3}$

Simplify.

62. $\left(\frac{1}{8}+\frac{1}{27}\right)^{1/3}$
63. $\left(\frac{1}{8}\right)^{1/3}+\left(\frac{1}{27}\right)^{1/3}$
64. $\left(\frac{16}{81}\right)^{3/4}+\left(\frac{256}{625}\right)^{1/4}$
65. $\left(\frac{8}{27}\right)^{2/3}+\left(-\frac{32}{243}\right)^{2/5}$
66. $\left(-\frac{125}{8}\right)^{1/3}-\left(\frac{1}{64}\right)^{1/3}$
67. $\left(-\frac{125}{8}-\frac{1}{64}\right)^{1/3}$

Multiply and write all answers using only positive exponents. Assume that all variables represent positive numbers.

68. $x^{1/2}(x - 3x^{1/2})$
69. $2x^{-2/3}(5x + 2x^{2/3})$
70. $x^{4/3}(x^{-2/3} + x^{-4/3} + x^{-2})$
71. $(2a^{1/2} - 1)(2a^{1/2} + 1)$
72. $(s^{1/3} + t^{3/2})(s^{1/3} - t^{3/2})$
73. $(b^{1/2} + b^{-1/2})^2$
74. $(x^{1/2} - y^{1/2})^2$
75. $(3y^{-1/4} + 2)(y^{-1/4} - 5)$
76. $(x^{1/3} - 1)(x^{2/3} + x^{1/3} + 1)$
77. $(1 + y^{2/3})(1 - y^{2/3} + y^{4/3})$

Combine the expression at the left to obtain the expression at the right. Assume that all variables represent positive numbers.

78. $x^{-1/2} + x^{1/2}; \dfrac{1 + x}{\sqrt{x}}$
79. $y^{-1/2} - y^{3/2}; \dfrac{1 - y^2}{\sqrt{y}}$
80. $x^{-2/3} + x^{1/3}; \dfrac{1 + x}{\sqrt[3]{x^2}}$
81. $y^{-3/2} + y^{1/2}; \dfrac{1 + y^2}{\sqrt{y^3}}$
82. $a^{1/2} + \dfrac{1}{2}a^{-1/2}(a - 8); \dfrac{3a - 8}{2\sqrt{a}}$
83. $b^{2/3} + \dfrac{2}{3}b^{-1/3}(b - 10); \dfrac{5b - 20}{3\sqrt[3]{b}}$

Rewrite each fraction using only positive exponents and so that the denominator has only positive integer exponents. Assume that all variables represent positive numbers.

Example:
$$\frac{1}{x^{1/2} - 1} = \frac{1(x^{1/2} + 1)}{(x^{1/2} - 1)(x^{1/2} + 1)} = \frac{x^{1/2} + 1}{x - 1}$$

84. $\dfrac{2}{y^{1/2} + 3}$
85. $\dfrac{1}{a^{3/2} - 1}$
86. $\dfrac{x^{1/2} - 1}{x^{1/2} + 1}$
*87. $\dfrac{a^{1/2} + a^{-1/2}}{a^{1/2} - a^{-1/2}}$

88. Prove that if $\sqrt[n]{b}$ is a real number, then $\sqrt[n]{b^m}$ is also a real number, where m, n are integers, $n \geq 2$. (*Hint:* Consider the two cases: (1) $b > 0$; (2) $b < 0$ and n odd.)
89. Why is it wrong to write $(-16)^{2/4} = \sqrt[4]{(-16)^2}$?

5.4 EQUATIONS WITH RADICALS OR RATIONAL EXPONENTS

An equation in which a variable occurs in a radicand, such as $\sqrt{x + 2} + 2 = 5$, is called a **radical equation.** To solve this equation we first isolate the radical on one side and then square both sides to eliminate the radical.

$$\sqrt{x + 2} + 2 = 5$$
$$\sqrt{x + 2} = 3$$
$$(\sqrt{x + 2})^2 = 3^2 \qquad \text{(Square both sides.)}$$
$$x + 2 = 9$$
$$x = 7$$

You can check this result by substituting 7 for x in the original equation.

In general, when solving radical equations, we will be making use of the following principle:

If $a = b$, then $a^n = b^n$

This statement says that every solution of $a = b$ will also be a solution of $a^n = b^n$.

Sometimes the method of raising both sides of an equation to the same power can produce a new equation that has more solutions than the original equation. For example, the only solution of the simple equation

$$x = -3$$

is the number -3. However, squaring both sides produces

$$\begin{aligned} x^2 &= (-3)^2 \\ &= 9 \end{aligned}$$

which has the two solutions -3 and 3. The extra solution 3 is called an **extraneous solution** since it does not satisfy the original equation.

The following example shows how an extraneous solution may be introduced by squaring both sides of a radical equation.

EXAMPLE 1 Solve for x: $\sqrt{x+4} + 2 = x$

Solution

Note: If the radical in $\sqrt{x+4} + 2 = x$ is not first isolated, it is still possible to solve the equation, but the work will be more involved. Try it.

$$\begin{aligned} \sqrt{x+4} + 2 &= x \\ \sqrt{x+4} &= x - 2 \\ (\sqrt{x+4})^2 &= (x-2)^2 && \text{(Square both sides.)} \\ x + 4 &= x^2 - 4x + 4 \\ 0 &= x^2 - 5x \\ 0 &= x(x-5) \\ x = 0 &\text{ or } x = 5 && \text{(by the zero-product property)} \end{aligned}$$

Check: Since raising both sides of an equation to the same power may introduce extraneous solutions, all solutions must be checked in the original equation.

Let $x = 0$	Let $x = 5$:
$\sqrt{0+4} + 2 = 0$	$\sqrt{5+4} + 2 = 5$
$2 + 2 = 0$ No	$3 + 2 = 5$ Yes

We conclude that the only solution for the given equation is 5. ■

Radical equations may contain more than one radical. For such cases it is usually best to transform the equation first into one with as few radicals on each side as possible. Consider, for example, the following:

If you try to square each side of this equation, the left side will cause some difficulty. Try it!

$$\begin{aligned} \sqrt{x+5} + \sqrt{x} &= 5 \\ (\sqrt{x+5} + \sqrt{x}) - \sqrt{x} &= 5 - \sqrt{x} \\ \sqrt{x+5} &= 5 - \sqrt{x} \end{aligned}$$

Next square each side. Note that $5 - \sqrt{x}$ is a binomial, and is squared as follows:

$$\text{Since } (a - b)^2 = a^2 - 2ab + b^2$$
$$\text{then } (5 - \sqrt{x})^2 = 25 - 10\sqrt{x} + x$$

CAUTION
Find the algebraic mistake that has been made in the following.

$$\sqrt{x + 5} = 5 - \sqrt{x}$$
$$x + 5 = 25 - x$$
$$2x = 20$$
$$x = 10$$

Here is the complete solution:

$$\sqrt{x + 5} + \sqrt{x} = 5$$
$$\sqrt{x + 5} = 5 - \sqrt{x}$$
$$x + 5 = 25 - 10\sqrt{x} + x \qquad \text{(Square both sides.)}$$
$$10\sqrt{x} = 20$$
$$\sqrt{x} = 2$$
$$x = 4 \qquad \text{(Square both sides.)}$$

Check this solution by substituting $x = 4$ in the original equation.

EXAMPLE 2 Solve for x: $\sqrt{x - 7} - \sqrt{x} = 1$

Solution Supply the reasons for each step in this solution.

$$\sqrt{x - 7} - \sqrt{x} = 1$$
$$\sqrt{x - 7} = 1 + \sqrt{x}$$
$$(\sqrt{x - 7})^2 = (1 + \sqrt{x})^2$$
$$x - 7 = 1 + 2\sqrt{x} + x$$
$$-8 = 2\sqrt{x}$$
$$-4 = \sqrt{x}$$
$$16 = x$$

Check: $\sqrt{16 - 7} - \sqrt{16} = \sqrt{9} - \sqrt{16} = 3 - 4 = -1 \neq 1$. Therefore, this equation has no solution, which could have been observed at an earlier stage as well. For example, $-4 = \sqrt{x}$ has no solution. (Why not?) ■

Since our work produced only one possible solution, which did *not* check, we conclude there is no real number that satisfies the given equation. The solution set is the empty set.

TEST YOUR UNDERSTANDING
Think Carefully

Solve for x.

1. $\sqrt{x + 1} = 3$
2. $\sqrt{x} - 2 = 3$
3. $\sqrt{x + 2} = \sqrt{2x - 5}$
4. $\sqrt{x^2 + 9} = -5$
5. $\dfrac{1}{\sqrt{x}} = 3$
6. $\dfrac{4}{\sqrt{x} - 1} = 2$
7. $\sqrt{x^2 - 5} = 2$
8. $\sqrt{x} + \sqrt{x - 5} = 5$

(Answers: Page 230)

All the examples thus far have involved square roots. However, the method of raising both sides of an equation to the same power can also be applied to radical equations involving cube roots or fourth roots, and so on.

EXAMPLE 3 Solve: $\sqrt[3]{x + 3} = 2$

Solution In this case, cube each side of the equation.

$$(\sqrt[3]{x + 3})^3 = (2)^3$$

$$x + 3 = 8 \qquad [(\sqrt[3]{a})^3 = a]$$

$$x = 5$$

Thus if $\sqrt[3]{x + 3} = 2$, then $x = 5$.

Check: $\sqrt[3]{5 + 3} = \sqrt[3]{8} = 2$ ■

The equation in Example 3, as well as any other radical equation, can also be written using fractional exponents. Thus, for Example 3 we have the equation $(x + 3)^{1/3} = 2$, and to solve it we use the same steps as before.

$$(x + 3)^{1/3} = 2$$

$$[(x + 3)^{1/3}]^3 = 2^3 \qquad \text{(Cube both sides.)}$$

$$x + 3 = 8$$

$$x = 5$$

Check:

$$(5 + 3)^{1/3} = 8^{1/3} = 2$$

Note:

$$\frac{1}{2}x^{-1/2}(x - 27) = \frac{1}{2} \cdot \frac{1}{x^{1/2}}(x - 27) = \frac{x - 27}{2x^{1/2}}$$

EXAMPLE 4 Solve: $x^{1/2} + \frac{1}{2}x^{-1/2}(x - 27) = 0$

Solution First rewrite using only positive exponents.

$$x^{1/2} + \frac{x - 27}{2x^{1/2}} = 0$$

Multiply by $2x^{1/2}$ to clear fractions.

$$2x^{1/2} \cdot x^{1/2} + 2x^{1/2} \cdot \frac{x - 27}{2x^{1/2}} = 2x^{1/2} \cdot 0$$

$$2x + x - 27 = 0$$

$$3x = 27$$

$$x = 9$$

Check:

$$9^{1/2} + \frac{1}{2}(9^{-1/2})(9 - 27) = 3 + \frac{1}{2 \cdot 9^{1/2}}(-18)$$
$$= 3 + \frac{1}{6}(-18)$$
$$= 3 - 3$$
$$= 0$$

■

The final example illustrates a word problem that is solved by using a radical equation.

EXAMPLE 5 If 2 is subtracted from 3 times a number, the square root of the difference is 5. What is the number?

This problem can also be solved by using the equation $(3x - 2)^{1/2} = 5$.

Solution Let x = the number. Then $3x - 2$ represents 2 less than 3 times the number. The square root of this expression is equal to 5:

$$\sqrt{3x - 2} = 5$$
$$3x - 2 = 25$$
$$3x = 27$$
$$x = 9$$

The number is 9; check this result by referring back to the statement of the original problem. ■

EXERCISES 5.4

Solve for x and check.

1. $\sqrt{x - 1} = 4$
2. $\sqrt{2x + 1} = 3$
3. $\sqrt{3x - 2} = 5$
4. $\sqrt{4x + 9} = 7$
5. $\dfrac{12}{\sqrt{x}} = 4$
6. $\dfrac{8}{\sqrt{x + 2}} = 4$
7. $\sqrt{x - 1} = \sqrt{2x - 11}$
8. $\sqrt{3x + 1} = \sqrt{2x + 6}$
9. $\sqrt{x^2 - 9} = 4$
10. $\sqrt{x^2 + 2} = 3$
11. $\sqrt{x^2 + 5} = -3$
12. $\sqrt{x + 1} + 5 = x$
13. $2\sqrt{x} = 3\sqrt{2}$
14. $3\sqrt{x} = 2\sqrt{3}$
15. $3x = \sqrt{3 - 5x - 3x^2}$
16. $\dfrac{1}{\sqrt{2x - 1}} = \dfrac{3}{\sqrt{5 - 3x}}$
17. $2x - 5\sqrt{x} - 3 = 0$
18. $x = 8 - 2\sqrt{x}$
19. $\sqrt{x} + \sqrt{x - 7} = 7$
20. $\sqrt{x} + \sqrt{x - 5} = 5$
21. $\sqrt{x - 1} + \sqrt{3x - 2} = 3$
22. $\sqrt{10 - x} - \sqrt{x + 3} = 1$

23. $\sqrt{4x+1} + \sqrt{x+7} = 6$

24. $\sqrt{x+4} + \sqrt{3x+1} = 7$

25. $\sqrt[3]{2x+7} = 3$

26. $\sqrt[3]{3x+4} = 4$

27. $\sqrt{\dfrac{3x-1}{4}} = 2$

28. $\sqrt[3]{\dfrac{5x+4}{2}} = 3$

29. $\sqrt[4]{1-3x} = \dfrac{1}{2}$

30. $2 + \sqrt[3]{7x-4} = 0$

31. $\sqrt{x^2-6x} = x - \sqrt{2x}$

32. $\dfrac{5x^{2/3}}{3} - \dfrac{20}{3x^{1/3}} = 0$

33. $(5x-6)^{1/5} + \dfrac{x}{(5x-6)^{4/5}} = 0$

34. $\dfrac{3}{2}x^{1/2} - \dfrac{3}{2}x^{-1/2} = 0$

35. $x^{-3/2} - \dfrac{1}{9}x^{-1/2} = 0$

36. $x^{1/2} + \dfrac{1}{2}x^{-1/2}(x-9) = 0$

37. $x^{2/3} + \dfrac{2}{3}x^{-1/3}(x-10) = 0$

38. $\dfrac{1}{2}x^2(x+5)^{-1/2} + 2x(x+5)^{1/2} = 0$

*39. $\dfrac{(2x+2)^{3/2}}{(x+9)^2} - \dfrac{(2x+2)^{1/2}}{x+9} = 0$

*40. $4x^{2/3} - 12x^{1/3} + 9 = 0$ (*Hint*: Let $u = x^{1/3}$.)

41. The square root of the sum of 7 and twice a number is 5. What is the number?

42. The square root of the sum of 25 and the square of a positive number is 13. Find the number.

43. The cube root of the difference of 6 less than twice a number is 2. Find the number.

44. The cube root of the sum of 2 and the square of a number is 3. Find the number.

45. The radius of a sphere whose surface area is A is given by the formula $r = \frac{1}{2}\sqrt{\frac{A}{\pi}}$. Use this formula to solve for A in terms of π and r. Then find A when $r = 2$ centimeters.

46. The radius of a right circular cylinder with height h and volume V is given by this formula: $r = \sqrt{\frac{V}{\pi h}}$. Use this formula to solve for V in terms of r and h. Then find V when $r = 2$ centimeters and $h = 3$ centimeters.

47. The slant height s of a right circular cone is given by the formula $s = \sqrt{r^2 + h^2}$, where r is the radius of the base and h is the altitude. Solve for h in terms of r and s, and then find h when $s = 17.23$ cm and $r = 8.96$ cm, rounded to two decimal places.

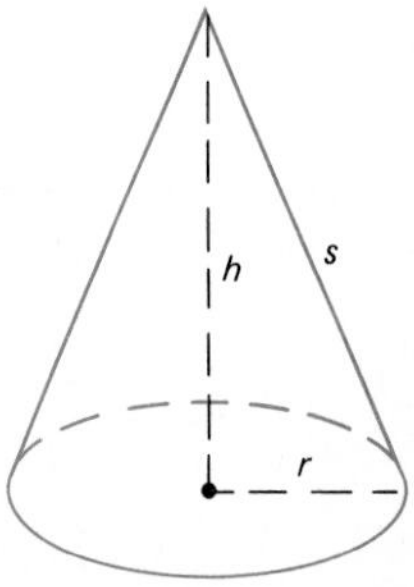

48. Solve $r = \dfrac{-b + \sqrt{b^2 - 4ac}}{2a}$ for c.

Written Assignment: Explain what is meant by an extraneous solution to an equation and how such a solution can be generated.

5.5 COMPLEX NUMBERS

In the definition of a radical, care was taken to avoid the even root of a negative number, such as $\sqrt{-4}$. This was necessary because there is no real number x whose square is -4. Consequently, there can be no real number that satisfies the equation $x^2 + 4 = 0$. Suppose, for the moment, we could solve $x^2 + 4 = 0$ using our algebraic methods. Then we might write the following:

$$\begin{aligned} x^2 + 4 &= 0 \\ x^2 &= -4 \\ x &= \pm\sqrt{-4} \\ &= \pm\sqrt{4(-1)} \\ &= \pm\sqrt{4}\sqrt{-1} \\ &= \pm 2\sqrt{-1} \end{aligned}$$

We could now claim that $2\sqrt{-1}$ is a solution of $x^2 + 4 = 0$. But it is certainly not a *real number solution*. What we now do is to formally introduce $\sqrt{-1}$ as a new kind of number; it will be the unit for a new set of numbers, the *imaginary numbers*. The symbol i is used to stand for this number and is defined as follows.

DEFINITION OF i

$$i = \sqrt{-1} \quad \text{and} \quad i^2 = -1$$

Using i, we can define the square root of a negative number.

For $n > 0$: $\sqrt{-n} = \sqrt{-1} \cdot \sqrt{n} = i\sqrt{n}$

EXAMPLE 1 Simplify: **(a)** $\sqrt{-16} + \sqrt{-25}$ **(b)** $\sqrt{-16} \cdot \sqrt{-25}$

In this example, $4i$ and $5i$ are combined by using the usual rules of algebra. You will see later that such procedures apply for this new kind of number.

Solution

(a)

$$\sqrt{-16} = \sqrt{-1} \cdot \sqrt{16} = i \cdot 4 = 4i$$

$$\sqrt{-25} = \sqrt{-1} \cdot \sqrt{25} = i \cdot 5 = 5i$$

Thus

$$\sqrt{-16} + \sqrt{-25} = 4i + 5i = 9i$$

(b)

$$\sqrt{-16} \cdot \sqrt{-25} = (4i)(5i) = 20i^2 = 20(-1) = -20$$

■

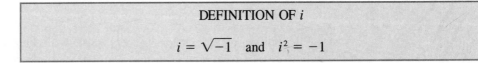

TEST YOUR UNDERSTANDING Think Carefully

Express as the product of a real number and i.

1. $\sqrt{-9}$ **2.** $\sqrt{-49}$ **3.** $\sqrt{-5}$ **4.** $-2\sqrt{-1}$ **5.** $\sqrt{-\frac{4}{9}}$

Simplify.

6. $\sqrt{-36} \cdot \sqrt{-121}$ **7.** $\sqrt{9} \cdot \sqrt{-49}$

8. $\sqrt{-50} + \sqrt{-32} - \sqrt{-8}$ **9.** $3\sqrt{-20} + 2\sqrt{-45}$

(Answers: Page 230)

An indicated product of a real number times the imaginary unit i, such as $7i$ or $\sqrt{2}i$, is called a **pure imaginary number.** The sum of a real number and a pure imaginary number is called a **complex number.**

A complex number has the form $a + bi$, where a and b are real and $i = \sqrt{-1}$.

We say that the real number a is the **real part** of $a + bi$ and the real number b is called the **imaginary part** of $a + bi$. In general, two complex numbers are equal only when both their real parts and their imaginary parts are equal. Thus

$$a + bi = c + di \quad \text{if and only if} \quad a = c \quad \text{and} \quad b = d$$

The collection of complex numbers contains all the real numbers, since any real number a can also be written as $a = a + 0i$. Similarly, if b is real, $bi = 0 + bi$, so that the complex numbers also contain the pure imaginaries.

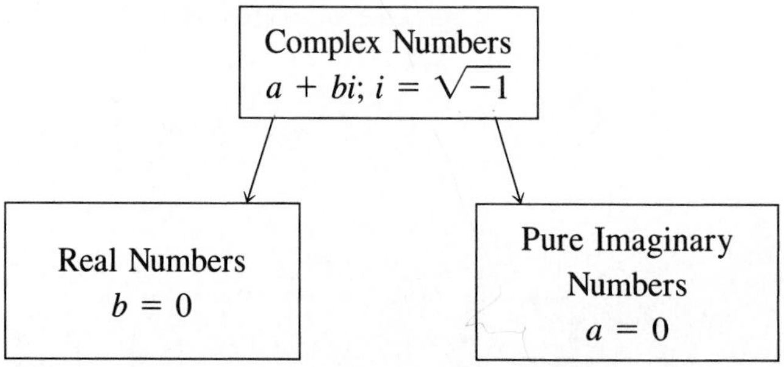

Various sets of numbers are described in Section 1.1.

EXAMPLE 2 Name the sets of numbers to which each of the following belongs:

(a) $\sqrt{\frac{4}{9}}$ **(b)** $\sqrt{-10}$ **(c)** $-4 + \sqrt{-12}$

Solution

(a) $\sqrt{\frac{4}{9}} = \frac{2}{3} = \frac{2}{3} + 0i$ is complex, real, and rational.

(b) $\sqrt{-10} = i\sqrt{10}$ is complex and pure imaginary.

(c) $-4 + \sqrt{-12} = -4 + \text{i}\sqrt{12} = -4 + 2\text{i}\sqrt{3}$ is complex. ■

How should complex numbers be added, subtracted, multiplied, or divided? In answering this question, it must be kept in mind that the real numbers are included in the collection of complex numbers. Thus the definitions we construct for the complex numbers must preserve the established operations for the reals.

We add and subtract complex numbers by combining their real and their imaginary parts separately, according to these definitions.

SUM AND DIFFERENCE OF COMPLEX NUMBERS

$$(a + bi) + (c + di) = (a + c) + (b + d)i$$

$$(a + bi) - (c + di) = (a - c) + (b - d)i$$

Actually, these procedures are quite similar to those used for combining polynomials. For example, compare these two sums:

$$(2 + 3x) + (5 + 7x) = (2 + 5) + (3 + 7)x = 7 + 10x$$

$$(2 + 3i) + (5 + 7i) = (2 + 5) + (3 + 7)i = 7 + 10i$$

Similarly, compare these subtraction problems:

$$(8 + 5x) - (3 + 2x) = (8 - 3) + (5 - 2)x = 5 + 3x$$

$$(8 + 5i) - (3 + 2i) = (8 - 3) + (5 - 2)i = 5 + 3i$$

To decide how to multiply two complex numbers we go back to the set of real numbers and extend the multiplication process. Recall the procedure for multiplying two binomials:

$$(a + b)(c + d) = ac + bc + ad + bd$$

Now compare these two products:

$$(3 + 2x)(5 + 3x) = 15 + 10x + 9x + 6x^2$$

$$(3 + 2i)(5 + 3i) = 15 + 10i + 9i + 6i^2$$

This last expression can be simplified by noting that $10i + 9i = 19i$ and $6i^2 = -6$. The result is $9 + 19i$.

In general, we can develop a rule for multiplication of two complex numbers by finding the product of $a + bi$ and $c + di$ as follows:

$$\begin{aligned}(a + bi)(c + di) &= ac + adi + bci + bdi^2\\ &= ac + (ad + bc)i + bd(-1)\\ &= (ac - bd) + (ad + bc)i\end{aligned}$$

Note: In practice, it is easier to find the product by using the procedure for multiplying binomials rather than by memorizing the formal definition.

PRODUCT OF COMPLEX NUMBERS

$$(a + bi)(c + di) = (ac - bd) + (ad + bc)i$$

EXAMPLE 3 Multiply: $(5 - 2i)(3 + 4i)$

Solution

$$\begin{aligned}(5 - 2i)(3 + 4i) &= 15 - 6i + 20i - 8i^2 \\ &= 15 - 6i + 20i + 8 \\ &= 23 + 14i\end{aligned}$$

■

Now consider the quotient of two complex numbers such as

$$\frac{2 + 3i}{3 + i}$$

The term "imaginary" is said to have been first applied to these numbers precisely because they seem mysteriously to vanish under certain multiplications.

Our objective is to express this quotient in the form $a + bi$. To do so, we use a method similar to rationalizing the denominator. Note what happens when $3 + i$ is multiplied by its **conjugate,** $3 - i$:

$$(3 + i)(3 - i) = 9 + 3i - 3i - i^2 = 9 - i^2 = 9 + 1 = 10$$

In general, $\boldsymbol{a + bi}$ and $\boldsymbol{a - bi}$ are conjugates of each other, and their product is a real number.

$$(a + bi)(a - bi) = a^2 - b^2i^2 = a^2 + b^2$$

We are now ready to complete the division problem.

$$\begin{aligned}\frac{2 + 3i}{3 + i} &= \frac{2 + 3i}{3 + i} \cdot \frac{3 - i}{3 - i} \\ &= \frac{6 + 9i - 2i - 3i^2}{9 - i^2} \\ &= \frac{6 + 9i - 2i + 3}{9 + 1} \\ &= \frac{9 + 7i}{10} \\ &= \frac{9}{10} + \frac{7}{10}i \longleftarrow \quad \text{(This is the quotient in the form } a + bi.)\end{aligned}$$

In general, multiplying the numerator and denominator of $\frac{a + bi}{c + di}$ by the conjugate of $c + di$ leads to the following definition for division (see Exercise 65).

Rather than memorize this definition, simply find quotients as in the preceding illustration.

QUOTIENT OF COMPLEX NUMBERS

$$\frac{a + bi}{c + di} = \frac{ac + bd}{c^2 + d^2} + \frac{bc - ad}{c^2 + d^2}i; \qquad c + di \neq 0$$

Although we will not go into the details here, it can be shown that some of the basic rules for real numbers apply for the complex numbers. For example, the commutative, associative, and distributive laws hold, whereas the rules of order do not apply.

It is also true that the rules for integer exponents apply for complex numbers. For example, $(2 - 3i)^0 = 1$ and $(2 - 3i)^{-1} = \dfrac{1}{2 - 3i}$. In particular, the integral powers of i are easily evaluated. Following are the first four powers of i.

$$i = \sqrt{-1}$$
$$i^2 = -1$$
$$i^3 = i^2 \cdot i = -1 \cdot i = -i$$
$$i^4 = i^2 \cdot i^2 = (-1)(-1) = 1$$

After the first four powers of i a repeating pattern exists, as can be seen from the next four powers of i:

The cycle for consecutive powers of i is: $i, -1, -i, 1$.

$$i^5 = i^4 \cdot i = 1 \cdot i = i$$
$$i^6 = i^4 \cdot i^2 = 1 \cdot i^2 = -1$$
$$i^7 = i^4 \cdot i^3 = 1 \cdot i^3 = -i$$
$$i^8 = i^4 \cdot i^4 = 1 \cdot i^4 = 1$$

To simplify i^n when $n > 4$, begin by finding the largest multiple of 4 in the integer n as in the next example.

EXAMPLE 4 Simplify: **(a)** i^{22} **(b)** i^{39}

Since $20 = 5(4)$, 5 is the largest multiple of 4 in 22. Since $36 = 9(4)$, 9 is the largest multiple of 4 in 39.

Solution

(a) $i^{22} = i^{20} \cdot i^2 = (i^4)^5 \cdot i^2 = 1^5 \cdot i^2 = i^2 = -1$

(b) $i^{39} = i^{36} \cdot i^3 = (i^4)^9 \cdot i^3 = 1^9 \cdot i^3 = i^3 = -i$ ■

The following example illustrates how to operate with a negative integral power of i.

EXAMPLE 5 Express $2i^{-3}$ as the indicated product of a real number and i.

Solution First note that $2i^{-3} = \dfrac{2}{i^3}$. Next multiply numerator and denominator by i to obtain a real number in the denominator.

$$2i^{-3} = \frac{2}{i^3} \cdot \frac{i}{i} = \frac{2i}{i^4} = \frac{2i}{1} = 2i$$

Alternative Solution

$$2i^{-3} = \frac{2}{i^3} = \frac{2}{-i}$$

$$= \frac{2}{-i} \cdot \frac{i}{i} \qquad (i \text{ is the conjugate of } -i.)$$

$$= \frac{2i}{-i^2} = \frac{2i}{-(-1)}$$

$$= 2i$$

■

EXERCISES 5.5

Classify each statement as true or false.

1. Every real number is a complex number.
2. Every complex number is a real number.
3. Every irrational number is a complex number.
4. Every integer can be written in the form $a + bi$.
5. Every complex number may be expressed as an irrational number.
6. Every negative integer may be written as a pure imaginary number.

Express each of the following numbers in the form $a + bi$.

7. $5 + \sqrt{-4}$
8. $7 - \sqrt{-7}$
9. -5
10. $\sqrt{25}$

Express in the form bi.

11. $\sqrt{-16}$
12. $\sqrt{-81}$
13. $\sqrt{-144}$
14. $-\sqrt{-9}$
15. $\sqrt{-\frac{9}{16}}$
16. $\sqrt{-3}$
17. $-\sqrt{-5}$
18. $-\sqrt{-8}$

Simplify.

19. $\sqrt{-9}\cdot\sqrt{-81}$
20. $\sqrt{4}\cdot\sqrt{-25}$
21. $\sqrt{-3}\cdot\sqrt{-2}$
22. $(2i)(3i)$
23. $(-3i^2)(5i)$
24. $(i^2)(i^2)$
25. $\sqrt{-9} + \sqrt{-81}$
26. $\sqrt{-12} + \sqrt{-75}$
27. $\sqrt{-8} + \sqrt{-18}$
28. $2\sqrt{-72} - 3\sqrt{-32}$
29. $\sqrt{-9} - \sqrt{-3}$
30. $3\sqrt{-80} - 2\sqrt{-20}$

Complete the indicated operations. Express all answers in the form $a + bi$.

31. $(7 + 5i) + (3 + 2i)$
32. $(8 + 7i) + (9 - i)$
33. $(8 + 2i) - (3 + 5i)$
34. $(7 + 2i) - (4 - 3i)$
35. $(7 + \sqrt{-16}) + (3 - \sqrt{-4})$
36. $(8 + \sqrt{-49}) - (2 - \sqrt{-25})$
37. $2i(3 + 5i)$
38. $3i(5i - 2)$
39. $(3 + 2i)(2 + 3i)$
40. $(\sqrt{5} + 3i)(\sqrt{5} - 3i)$
41. $(5 - 2i)(3 + 4i)$
42. $(\sqrt{3} + 2i)^2$
43. $\dfrac{3 + 5i}{i}$
44. $\dfrac{5 - i}{i}$
45. $\dfrac{5 + 3i}{2 + i}$
46. $\dfrac{7 - 2i}{2 - i}$
47. $\dfrac{3 - i}{3 + i}$
48. $\dfrac{8 + 3i}{3 - 2i}$
49. $(5 + 4i) + 2(2 - 3i) - i(1 - 5i)$
50. $2i(3 - 4i)(3 - 6i) - 7i$
51. $\dfrac{(2 + i)^2(3 - i)}{2 + 3i}$
52. $\dfrac{1 - 2i}{3 + 4i} - \dfrac{2i - 3}{4 - 2i}$

Express in the form bi.

53. $3i^3$ 54. $-5i^5$ 55. $2i^7$ 56. $3i^{-3}$ 57. $-4i^{18}$ 58. i^{-32}

Simplify and express each answer in the form a + bi.

59. $(3 + 2i)^{-1}$ 60. $(3 + 2i)^{-2}$

61. **(a)** Show that $x = \sqrt{2}i$ is a solution of $x^2 + 2 = 0$.
    (b) Find another solution.

62. The set of complex numbers satisfies the associative property for addition. Verify this by evaluating this expression in two different ways.

$$(3 + 5i) + (2 + 3i) + (7 + 4i)$$

63. Repeat Exercise 62 for multiplication, using this expression.

$$(3 + i)(3 - i)(4 + 3i)$$

64. Find the value of $x^2 + 3x + 5$ when $x = \dfrac{-3 + \sqrt{11}i}{2}$.

*65. Write $\dfrac{a + bi}{c + di}$ in the form $x + yi$. (*Hint:* Multiply the numerator and denominator by the conjugate of $c + di$.)

66. One of the basic rules for operating with radicals is that $\sqrt{ab} = \sqrt{a} \cdot \sqrt{b}$, where a and b are nonnegative real numbers. Prove that this rule does not work when both a and b are negative by showing that $\sqrt{(-4)(-9)} \neq \sqrt{-4} \cdot \sqrt{-9}$.

Use complex numbers to factor each polynomial.
Example:

$$x^2 + 9 = x^2 - (-9) = x^2 - (3i)^2$$
$$= (x + 3i)(x - 3i)$$

67. $x^2 + 1$ 68. $9x^2 + 4$ 69. $3x^2 + 75$ *70. $x^2 + 4ix - 3$

5.6 SOLVING QUADRATIC EQUATIONS BY COMPLETING THE SQUARE

In Section 3.6 we first learned how to solve quadratic equations by factoring. For example, after rewriting $x^2 + 2x - 8 = 0$ as $(x + 4)(x - 2) = 0$, we get $x = -4$ or $x = 2$. However, since many quadratic equations are not easily solved by this method, more general methods are needed. Such a method will now be developed. It uses the process of taking square roots according to this property:

The factoring method is not practical for solving an equation such as $x^2 + 2x - 9 = 0$ because $x^2 + 2x - 9$ is not factorable using integer coefficients.

SQUARE-ROOT PROPERTY FOR EQUATIONS

If $n^2 = k$, then $n = \sqrt{k}$ or $n = -\sqrt{k}$.

You can see that this property makes sense by observing that the values of $\sqrt{k}$ and $-\sqrt{k}$ satisfy the given equation $n^2 = k$. Thus $(\sqrt{k})^2 = k$ and $(-\sqrt{k})^2 = k$.

EXAMPLE 1 Use the square-root property for equations to solve for x.

(a) $x^2 = 25$ **(b)** $x^2 = 8$ **(c)** $(x + 2)^2 = 3$

Solution

(a)

$$x^2 = 25$$
$$x = \sqrt{25} \quad \text{or} \quad x = -\sqrt{25}$$
$$x = 5 \quad \text{or} \quad x = -5$$

(b)

$$x^2 = 8$$
$$x = \sqrt{8} \quad \text{or} \quad x = -\sqrt{8}$$
$$x = 2\sqrt{2} \quad \text{or} \quad x = -2\sqrt{2}$$

(c)

$$(x + 2)^2 = 3$$
$$x + 2 = \sqrt{3} \quad \text{or} \quad x + 2 = -\sqrt{3}$$
$$x = -2 + \sqrt{3} \quad \text{or} \quad x = -2 - \sqrt{3}$$

■

Now that complex numbers have been introduced, we are able to solve equations that involve the square root of a negative number. This is demonstrated in the following example.

EXAMPLE 2 Solve for x: $(x - 2)^2 = -9$

Solution

$$(x - 2)^2 = -9$$
$$x - 2 = \pm\sqrt{-9}$$
$$x - 2 = \pm 3i$$
$$x = 2 \pm 3i$$

The solutions are $2 + 3i$ and $2 - 3i$. ■

Here is a check of these values:

$$[(2 \pm 3i) - 2]^2 = (\pm 3i)^2$$
$$= 9i^2$$
$$= -9$$

Let us look again at the equation given in Example 1(c). This can be rewritten as follows:

$$(x + 2)^2 = 3$$
$$x^2 + 4x + 4 = 3$$
$$x^2 + 4x + 1 = 0$$

Note that $x^2 + 4x + 4$ is a perfect square trinomial.

Now suppose that we had *started* with $x^2 + 4x + 1 = 0$. By reversing the preceding steps we can see how to solve this equation. First rewrite the equation with the variable terms on one side and the constant on the other side.

$$x^2 + 4x + 1 = 0$$
$$x^2 + 4x = -1$$

Next, add 4 to each side to obtain a perfect square trinomial at the left.

$$x^2 + 4x + 4 = -1 + 4$$
$$(x + 2)^2 = 3$$

Finally, proceed as in Example 1(c) to solve for x.

The technique that we have just used is called **completing the square.** Study these illustrations of perfect squares that have been completed. Note that in each case the coefficient of the x^2-term is 1.

The process of completing the square makes use of one of these two identities:

$$(x + h)^2 = x^2 + 2hx + h^2$$
$$(x - h)^2 = x^2 - 2hx + h^2$$

In the trinomials, h^2 is the square of one-half the coefficient of the x-term (without regard to sign). That is, the third term $= [\frac{1}{2}(2h)]^2 = h^2$.

$$x^2 + 8x + \underline{\ ?\ } \longrightarrow x^2 + 8x + \underline{16} = (x + 4)^2$$
$$\left[\frac{1}{2}(8)\right]^2 = 4^2 = 16$$

$$x^2 - 3x + \underline{\ ?\ } \longrightarrow x^2 - 3x + \underline{\frac{9}{4}} = \left(x - \frac{3}{2}\right)^2$$
$$\left[\frac{1}{2}(-3)\right]^2 = \left(-\frac{3}{2}\right)^2 = \frac{9}{4}$$

$$x^2 + \frac{b}{a}x + \underline{\ ?\ } \longrightarrow x^2 + \frac{b}{a}x + \underline{\frac{b^2}{4a^2}} = \left(x + \frac{b}{2a}\right)^2$$
$$\left[\frac{1}{2}\left(\frac{b}{a}\right)\right]^2 = \left(\frac{b}{2a}\right)^2 = \frac{b^2}{4a^2}$$

Note that the roots of a quadratic equation will often be irrational numbers. The check for such numbers can be involved because of the radicals. However, it is good practice to check some of these answers not only to verify them, but also to strengthen your skills in working with radicals.

EXAMPLE 3 Solve for x: $x^2 + 5x - 3 = 0$

Solution First rewrite and complete the square

$$x^2 + 5x = 3$$
$$x^2 + 5x + \frac{25}{4} = 3 + \frac{25}{4}$$
$$\left(x + \frac{5}{2}\right)^2 = \frac{37}{4}$$

$$x + \frac{5}{2} = \frac{\sqrt{37}}{2} \quad \text{or} \quad x + \frac{5}{2} = -\frac{\sqrt{37}}{2} \qquad \text{(using the square-root property)}$$

$$x = -\frac{5}{2} + \frac{\sqrt{37}}{2} \quad \text{or} \quad x = -\frac{5}{2} - \frac{\sqrt{37}}{2}$$

Rational approximations can be computed for these solutions using $\sqrt{37} = 6.08$ from the square root table, or a calculator. Thus, to two decimal places

$$\frac{-5 + \sqrt{37}}{2} = 0.54$$
$$\frac{-5 - \sqrt{37}}{2} = -5.54$$

These roots may also be rewritten in this form:

$$x = \frac{-5 + \sqrt{37}}{2} \quad \text{or} \quad x = \frac{-5 - \sqrt{37}}{2}$$

Check: For $x = \dfrac{-5 + \sqrt{37}}{2}$.

$$\begin{aligned} x^2 + 5x - 3 &= \left(\frac{-5 + \sqrt{37}}{2}\right)^2 + 5\left(\frac{-5 + \sqrt{37}}{2}\right) - 3 \\ &= \frac{25 - 10\sqrt{37} + 37}{4} + \frac{-25 + 5\sqrt{37}}{2} - 3 \\ &= \frac{25 - 10\sqrt{37} + 37 + (-50 + 10\sqrt{37})}{4} - 3 \\ &= \frac{12}{4} - 3 = 0 \end{aligned}$$

Complete the check for $x = \dfrac{-5 - \sqrt{37}}{2}$.

■

TEST YOUR UNDERSTANDING Think Carefully

Supply the missing term to form a perfect square trinomial.

1. $x^2 + 4x +$ ____ **2.** $x^2 + 10x +$ ____ **3.** $x^2 - 6x +$ ____

4. $x^2 - 12x +$ ____ **5.** $x^2 + 3x +$ ____ **6.** $x^2 - 5x +$ ____

Solve for x by using the square-root property for equations.

7. $x^2 = 81$ **8.** $x^2 = -75$ **9.** $(x + 4)^2 = -10$

10. $x^2 + 4x + 4 = 9$ **11.** $x^2 + 4x + 5 = 3$ **12.** $x^2 - 6x + 3 = 0$

(Answers Page 231)

The method of completing the square can be adjusted to situations where the coefficient of x^2 is other than 1. Consider, for example, this equation:

$$4x^2 - 12x + 5 = 0$$

In Step 1, we can also multiply through by the reciprocal of the coefficient of x^2. In this example, multiply by $\frac{1}{4}$, which is the same as dividing through by 4.

STEP 1. Divide through by the coefficient of x^2. $\quad x^2 - 3x + \dfrac{5}{4} = 0$

STEP 2. Write the constant term on the right side of the equation. $\quad x^2 - 3x = -\dfrac{5}{4}$

STEP 3. Add the square of one-half of the coefficient of x to each side. $\quad x^2 - 3x + \dfrac{9}{4} = -\dfrac{5}{4} + \dfrac{9}{4}$

$$\left(\frac{1}{2} \cdot 3\right)^2 = \frac{9}{4}$$

STEP 4. Write the expression at the left as the square of a binomial. $\quad \left(x - \dfrac{3}{2}\right)^2 = 1$

STEP 5. Use the square-root property for equations and solve for x.

$$x - \frac{3}{2} = 1 \quad \text{or} \quad x - \frac{3}{2} = -1$$

$$x = \frac{5}{2} \quad \text{or} \quad x = \frac{1}{2}$$

STEP 6. Check the solutions in the original equation.

$$4\left(\frac{5}{2}\right)^2 - 12\left(\frac{5}{2}\right) + 5 = 25 - 30 + 5 = 0$$

$$4\left(\frac{1}{2}\right)^2 - 12\left(\frac{1}{2}\right) + 5 = 1 - 6 + 5 = 0$$

EXERCISES 5.6

Solve for x by using the square-root property for equations and check.

1. $x^2 = 36$
2. $x^2 = 144$
3. $x^2 = 0$
4. $x^2 = 7$
5. $x^2 = 12$
6. $x^2 = 27$
7. $(x - 1)^2 = 4$
8. $(x + 5)^2 = 0$
9. $(x + 3)^2 = -5$
10. $(2x - 1)^2 = -9$
11. $(2x + 1)^2 = 6$
12. $(3x + 2)^2 = 18$

Supply the missing term to form a perfect square trinomial.

13. $x^2 + 6x +$ ____
14. $x^2 - 8x +$ ____
15. $x^2 + 8x +$ ____
16. $x^2 - 20x +$ ____
17. $x^2 + x +$ ____
18. $x^2 - 3x +$ ____

Solve for x by completing the square.

19. $x^2 + 2x - 3 = 0$
20. $x^2 - 4x - 5 = 0$
21. $x^2 + 6x - 7 = 0$
22. $x^2 - 6x - 16 = 0$
23. $x^2 - 4x + 3 = 0$
24. $x^2 - 8x + 15 = 0$
25. $x^2 - 4x + 6 = 0$
26. $x^2 + 6x - 40 = 0$
27. $x^2 + 2x - 2 = 0$
28. $x^2 - x - 12 = 0$
29. $x^2 - x - 3 = 0$
30. $x^2 + 3x + 3 = 0$
31. $2x^2 + 2x + 1 = 0$
32. $2x^2 - 3x + 2 = 0$
33. $\frac{3}{2}x^2 + \frac{1}{2}x - 1 = 0$
34. $4x^2 - 4x + 3 = 4$
35. $\frac{3}{4}x^2 = \frac{3}{2}x - \frac{1}{4}$
36. $20x - 11 = 8x^2$
37. $\frac{x^2}{2} - \frac{3x}{2} - \frac{1}{4} = 0$
38. $\frac{x^2 + 1}{10} = \frac{x}{2}$
39. $\frac{7}{x^2} - \frac{4}{x} - 1 = 0$
40. $\frac{6x}{x + 2} - \frac{2}{x - 1} = 0$
41. $(x + 2)(x + 4) = 18$
*42. $x^2 + 4\sqrt{3}x = -3$

5.7 THE QUADRATIC FORMULA

Keep in mind that when $ax^2 + bx + c$ is easy to factor, then the quickest way to solve

$$ax^2 + bx + c = 0$$

is by the factoring method.

The method of completing the square can always be used to solve equations of the form $ax^2 + bx + c = 0$. However, as you have seen, this method soon becomes tedious and repetitious. To avoid some of this labor we will now develop an efficient formula that can be used in all cases. This will be done by solving the general quadratic equation $ax^2 + bx + c = 0$ for x in terms of the constants a, b, and c.

We begin with the general quadratic equation:

$$ax^2 + bx + c = 0 \qquad a \neq 0$$

Add $-c$ to each side:

$$ax^2 + bx = -c$$

Divide each side by $a(a \neq 0)$:

$$x^2 + \frac{b}{a}x = -\frac{c}{a}$$

Add $\left[\frac{1}{2}\left(\frac{b}{a}\right)\right]^2 = \frac{b^2}{4a^2}$ to each side:

$$x^2 + \frac{b}{a}x + \frac{b^2}{4a^2} = \frac{b^2}{4a^2} - \frac{c}{a}$$

Factor on the left and combine on the right:

$$\left(x + \frac{b}{2a}\right)^2 = \frac{b^2 - 4ac}{4a^2}$$

Take the square root of each side and solve for x.

$$x + \frac{b}{2a} = \pm\sqrt{\frac{b^2 - 4ac}{4a^2}} \qquad \text{(If } n^2 = k\text{, then } n = \pm\sqrt{k}.\text{)}$$

$$x + \frac{b}{2a} = \pm\frac{\sqrt{b^2 - 4ac}}{2a}$$

$$x = -\frac{b}{2a} \pm \frac{\sqrt{b^2 - 4ac}}{2a}$$

Combine terms to obtain the **quadratic formula.**

QUADRATIC FORMULA

If $ax^2 + bx + c = 0, a \neq 0,$

then $x = \dfrac{-b \pm \sqrt{b^2 - 4ac}}{2a}$

The values $x = \dfrac{-b + \sqrt{b^2 - 4ac}}{2a}$ and $x = \dfrac{-b - \sqrt{b^2 - 4ac}}{2a}$ are the *roots* of the quadratic equation.

This formula now allows you to solve any quadratic equation in terms of the constants used, as shown in these examples.

EXAMPLE 1 Solve for x: $2x^2 - 5x + 1 = 0$

Solution

$$2x^2 - 5x + 1 = 0 \qquad x = \frac{-b \pm \sqrt{b^2 - 4ac}}{2a}$$

$$a = 2 \qquad = \frac{-(-5) \pm \sqrt{(-5)^2 - 4(2)(1)}}{2(2)}$$

$$b = -5 \qquad = \frac{5 \pm \sqrt{17}}{4}$$

$$c = 1$$

Thus

$$x = \frac{5 + \sqrt{17}}{4} \quad \text{or} \quad x = \frac{5 - \sqrt{17}}{4}$$

Approximations of these roots can be found by calculator:

$x =$ (5) [+] (17) [√] [=] [÷] (4) [=] $2.280776 = 2.28$ (rounded to two decimal places)

Replacing [+] by [−] in the preceding gives the approximation 0.22 for the other root. ■

CAUTION
If the equation is not first written in the form

$$ax^2 + bx + c = 0$$

it could lead to incorrect assignments for some of the constants a, b, and c.

EXAMPLE 2 Solve for x: $4x^2 = 12x - 9$

Solution First rewrite the equation in the form $4x^2 - 12x + 9 = 0$ Now use $a = 4$, $b = -12$, and $c = 9$ in the quadratic formula.

$$x = \frac{-(-12) \pm \sqrt{(-12)^2 - 4(4)(9)}}{2(4)}$$

$$= \frac{12 \pm \sqrt{0}}{8}$$

$$= \frac{3}{2}$$

Since $\frac{3}{2}$ is the only solution, it is called a **double root** of the equation. ■

In each of the preceding examples the radicand $b^2 - 4ac$ was nonnegative. However, the quadratic formula also applies if this radicand is negative. When this occurs the imaginary unit $i = \sqrt{-1}$ is involved and the roots are imaginary numbers, as in Example 3.

EXAMPLE 3 Solve for x: $x^2 + 3x + 5 = 0$

Solution Use the quadratic formula with $a = 1$, $b = 3$ and $c = 5$.

$$x = \frac{-3 \pm \sqrt{3^2 - 4(1)(5)}}{2(1)}$$

$$= \frac{-3 \pm \sqrt{-11}}{2}$$

$$= \frac{-3 \pm \sqrt{11}i}{2}$$

Recall that
$\sqrt{-11} = \sqrt{11}\sqrt{-1}$
$= \sqrt{11}i$

These imaginary roots may also be written separately as:

$$-\frac{3}{2} + \frac{\sqrt{11}}{2}i \qquad -\frac{3}{2} - \frac{\sqrt{11}}{2}i$$ ■

TEST YOUR UNDERSTANDING Think Carefully	*Use the quadratic formula to solve for x.*	
	1. $x^2 + 3x - 10 = 0$	**2.** $6x^2 + x = 1$
	3. $x^2 - 9 = 0$	**4.** $x^2 - 2x - 2 = 0$
	5. $x^2 + 6x + 6 = 0$	**6.** $x^2 + 6x + 12 = 0$
(Answers: Page 231)	**7.** $2x^2 = 2x - 5$	**8.** $2x^2 + x - 4 = 0$

The three examples of this section illustrated quadratic equations with two real roots, one root, and two imaginary roots. As demonstrated by these examples, it is the value of the radicand $b^2 - 4ac$ that determines the types of roots the equation has. For this reason $b^2 - 4ac$ is called the **discriminant.** It can be used to predict the kinds of roots the equation has without actually finding the roots. This depends on the value of the discriminant being positive, zero, or negative, as shown in this table in which a, b, and c are real numbers, $a \neq 0$.

Quadratic Equation $ax^2 + bx + c = 0$	Value of the Discriminant $b^2 - 4ac$	Types of Roots
$2x^2 - 5x + 1 = 0$	17 (positive)	Two real roots
$4x^2 - 12x + 9 = 0$	0 (zero)	One real root
$x^2 + 3x + 5 = 0$	−11 (negative)	Two imaginary roots

Note also that when the discriminant $b^2 - 4ac > 0$, then the two solutions of $ax^2 + bx + c = 0$ will be rational numbers if $b^2 - 4ac$ is a perfect square. In case $b^2 - 4ac$ is not a perfect square, then the roots are irrational. We summarize as follows:

Using the Discriminant $b^2 - 4ac$, Where a, b, c Are Real Numbers

1. If $b^2 - 4ac > 0$, then $ax^2 + bx + c = 0$ has two real solutions. If the discriminant is a perfect square, these roots will be rational numbers; if not, they will be irrational.
2. If $b^2 - 4ac = 0$, the solution for $ax^2 + bx + c = 0$ is only one number (a double root).
3. If $b^2 - 4ac < 0$, $ax^2 + bx + c = 0$ has two imaginary solutions.

EXAMPLE 4 Use the discriminant to describe the type of roots for each of these equations.

(a) $3x^2 - 5x - 6 = 0$
(b) $5x^2 - 11x - 12 = 0$
(c) $2x^2 + 2x + 3 = 0$
(d) $9x^2 - 12x + 4 = 0$

Solution

	Equation	Value of $b^2 - 4ac$	Types of Roots
(a)	$3x^2 - 5x - 6 = 0$	$(-5)^2 - 4(3)(-6) = 97$	Two irrational roots since $97 > 0$ but 97 is not a perfect square
(b)	$5x^2 - 11x - 12 = 0$	$(-11)^2 - 4(5)(-12) = 361$	Two rational roots since $361 = 19^2$
(c)	$2x^2 + 2x + 3 = 0$	$2^2 - 4(2)(3) = -20$	Two imaginary roots since $-20 < 0$
(d)	$9x^2 - 12x + 4 = 0$	$12^2 - 4(9)(4) = 0$	A double root

Check these results by solving the given quadratic equations.

$9x^2 - 12x + 4 = 0.$

$-12^2 - 4(9)(4)$

EXERCISES 5.7

Use the quadratic formula to solve for x. When the roots are irrational numbers give their radical forms and also use a calculator to find approximations to two decimal places.

1. $x^2 - 3x - 10 = 0$
2. $x^2 - 3x - 4 = 0$
3. $2x^2 + 3x - 2 = 0$
4. $2x^2 - 5x - 3 = 0$
5. $x^2 - 6x + 9 = 0$
6. $2x^2 - 7x + 6 = 0$
7. $4x^2 - 9 = 0$
8. $9x^2 - 4 = 0$
9. $2x^2 - 3 = 0$
10. $3x^2 + 75 = 0$
11. $x^2 - 6x + 6 = 0$
12. $x^2 + 4x + 1 = 0$
13. $x^2 - 2x = 4$
14. $x^2 - 10x = -23$
15. $x^2 - 2x + 2 = 0$
16. $x^2 + 3x + 4 = 0$
17. $-x^2 + 6x - 14 = 0$
18. $-x^2 + 4x - 7 = 0$
19. $3x + 1 = 2x^2$
20. $3x^2 = -x + 1$
21. $2 - 5x + 2x^2 = 0$
22. $2 - 4x - x^2 = 0$
23. $6x^2 + 2x = -3$
24. $5x^2 = 8x - 8$

Use the discriminant to describe the solutions as **(a)** *a single real number,* **(b)** *two rational numbers,* **(c)** *two irrational numbers, or* **(d)** *two imaginary numbers.*

25. $x^2 - 8x + 16 = 0$
26. $x^2 + 3x + 5 = 0$
27. $x^2 + 2x - 8 = 0$
28. $9x^2 - 6x + 1 = 0$
29. $2x^2 - x + 1 = 0$
30. $-x^2 + 2x + 15 = 0$
31. $x^2 + 3x - 1 = 0$
32. $2x^2 + x - 5 = 0$
33. $6x^2 + 7x = 4$
34. $4x^2 + x - 3 = 0$
35. $2x^2 + 5 = x$
36. $2x + 3 = -4x^2$

Find the values of b so that the equation will have a double root. (Hint: Let $b^2 - 4ac = 0$.)

37. $x^2 + bx + 9 = 0$
38. $x^2 - bx + 7 = 0$
39. $4x^2 - bx + 9 = 0$
40. $9x^2 + bx + 4 = 0$

Find the values of k so that the equation will have two distinct real roots. (Hint: Let $b^2 - 4ac > 0$.)

41. $-x^2 + 4x + k = 0$
42. $2x^2 - 3x + 2k = 0$
43. $kx^2 - x - 1 = 0$
44. $kx^2 + 3x - 2 = 0$

Find the values of t so that the equation will have two imaginary roots. (Hint: Let $b^2 - 4ac < 0$.)

45. $x^2 - 6x + t = 0$
46. $2x^2 + tx + 8 = 0$

47. Show that the sum of the two roots of $ax^2 + bx + c = 0$ is $-\frac{b}{a}$, and that their product is $\frac{c}{a}$.

In Exercises 48 through 53, use the results of Exercise 47 to find the sum and product of the roots of the given equation. Then verify your answers by solving for the roots and finding their sum and product.

48. $x^2 - 3x - 10 = 0$
49. $6x^2 + 5x - 4 = 0$
50. $x^2 = 25$
51. $3x^2 + 35 = 26x$
52. $x^2 + 2x + 5 = 0$
53. $2x^2 + 6x + 9 = 0$

Solve for x.

54. $x^2 - \frac{1}{4} = \frac{3}{4}x$
55. $\frac{4x}{2x - 1} + \frac{3}{2x + 1} = 0$
56. $20x^{-2} - 19x^{-1} = 6$
57. $(2x - 1)^2 - 3(x^2 + 2) = 0$
58. $x^2 - 2x = (2x + 3)(x - 1)$
59. $x^2 + 2\sqrt{3}x + 2 = 0$
60. $2x^2 - 4\sqrt{2}x - 1 = 0$
61. $x^2 - ix + 2 = 0$
62. $8ix^2 - 2x + i = 0$
63. $x^4 - 5x^2 + 4 = 0$ (Let $u = x^2$ and first solve $u^2 - 5u + 4 = 0$.)
64. $2x^4 - 13x^2 - 7 = 0$ (Let $u = x^2$.)
65. $(x^2 + 2x)^2 - 7(x^2 + 2x) - 8 = 0$ (Let $u = x^2 + 2x$.)
66. $\left(x - \frac{2}{x}\right)^2 - 3\left(x - \frac{2}{x}\right) + 2 = 0$ $\left(\textit{Hint}\text{: Let } u = x - \frac{2}{x}.\right)$
67. $\sqrt{6x + 1} = \frac{3}{2}x + 1$ (Hint: Square each side.)
68. $\sqrt{5x + 1} = \sqrt{2x + 3} + 1$ (Hint: Square twice.)
69. $x^3 + 3x^2 - 4x - 12 = 0$ (*Hint*: Factor by grouping.)

*70. $x^3 - 1 - 2x(x - 1) = 0$ (*Hint*: Factor by grouping.)

Written Assignment: Use examples of your own and describe how the discriminant can be used to tell the nature of the roots of a quadratic equation of the form $ax^2 + bx + c = 0$.

EXPLORATIONS Think Critically

1. Many students claim that $\sqrt{81} = \pm 9$. Explain why this is incorrect.
2. Note that $\frac{12}{\sqrt{150}} = \frac{2}{5}\sqrt{6}$. Which is the preferred form? Explain.
3. Use the result $\sqrt{a^2} = |a|$ and the appropriate rules for radicals to prove that $|xy| = |x| \cdot |y|$ and $\left|\frac{x}{y}\right| = \frac{|x|}{|y|}$.
4. By completing the square show how the expression $2x^2 - 12x - 5$ can be converted to the form $2(x - 3)^2 - 23$. Use this result to explain why the smallest possible value of $2x^2 - 12x - 5$ is -23.
5. The fractions in $\frac{3}{4}x^2 - \frac{5}{6}x - \frac{2}{3} = 0$ make it clumsy to solve for the exact roots using the quadratic formula. How can this work be simplified?
6. Find three distinct irrational numbers, a, b, and c, for which $ax^2 + bx + c = 0$ has two roots that are integers.

CHALLENGE
Think Creatively

The ancient Greeks and some modern artists felt that the rectangular shape most pleasing to the eye is one whose ratio of length to width is approximately 1.62. This ratio is known as the **golden ratio**. The Parthenon in Greece, as well as many famous paintings, makes use of these dimensions.

The following *continued fraction* gives rise to the golden ratio x. Solve for x using the hint below.

$$x = 1 + \cfrac{1}{1 + \cfrac{1}{1 + \cfrac{1}{1 + \cfrac{1}{1 + \cdots}}}}$$

Hint: See the circled part shown above equal to x and solve the resulting equation, rounded to three decimal places. Also find the reciprocal of your solution, $\frac{1}{x}$, and comment on the result.

5.8 APPLICATIONS OF QUADRATIC EQUATIONS

The algebraic techniques developed in this chapter, can be used to solve word problems that involve quadratic equations. The examples that follow show a variety of such problems.

EXAMPLE 1 The length of a rectangular piece of cardboard is 2 inches more than its width. As in the following figure, an open box is formed by cutting out 4-inch squares from each corner and folding up the sides. If the volume of the box is 672 cubic inches, find the dimensions of the original cardboard.

Solution Let x be the width of the cardboard. Then $x + 2$ is the length.

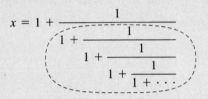

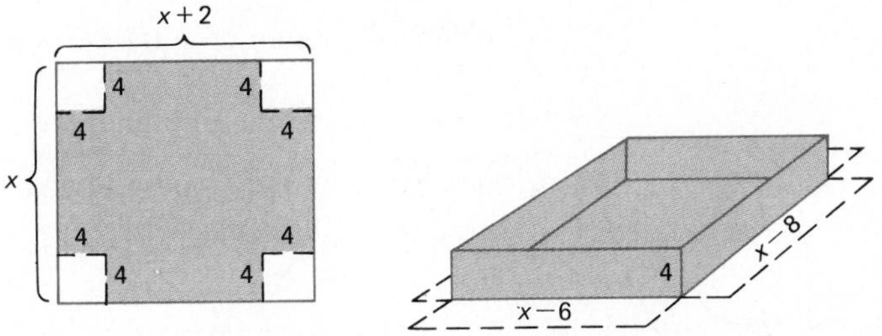

After the squares are cut off and the sides are folded up, the dimensions of the box are:

$$\text{length} = \ell = x + 2 - 8 = x - 6$$
$$\text{width} = w = x - 8$$
$$\text{height} = h = 4$$

Since the volume is to be 672 cubic inches, and $v = \ell wh$, we have

$$(x - 6)(x - 8)4 = 672$$
$$(x - 6)(x - 8) = 168$$
$$x^2 - 14x - 120 = 0$$
$$(x - 20)(x + 6) = 0$$
$$x = 20 \quad \text{or} \quad x = -6$$

Note that we reject the solution $x = -6$ since the dimensions must be positive numbers.

The dimensions of the original cardboard are: $w = 20$ inches, $\ell = 20 + 2 = 22$ inches. ■

The formula given here is really an approximation since it assumes that there is no air resistance.

EXAMPLE 2 When an object is thrown vertically upward from ground level with an initial velocity of v_0 feet per second, its height h in feet after t seconds is given by the formula $h = v_0t - 16t^2$. If $v_0 = 256$ feet per second, when will the object be 1008 feet above the ground?

Solution Since we are looking for the time t when $v_0 = 256$ and $h = 1008$, substitute these values into the given formula and solve for t.

$$h = v_0t - 16t^2$$
$$1008 = 256t - 16t^2$$
$$16t^2 - 256t + 1008 = 0$$
$$t^2 - 16t + 63 = 0 \qquad \text{(Divide by 16.)}$$
$$(t - 7)(t - 9) = 0$$
$$t = 7 \quad \text{or} \quad t = 9$$

The object will be at $h = 1008$ feet on the way up in 7 seconds, and again on the way down, 2 seconds later, when $t = 9$ seconds. ■

EXAMPLE 3 The sum of a number and its reciprocal is 5. Find the number.

Solution Let x represent the number, then $\frac{1}{x}$ is the reciprocal of the number. Then, since their sum is 5, we write and solve the following fractional equation:

$$x + \frac{1}{x} = 5$$

Multiply each side by x:

$$x\left(x + \frac{1}{x}\right) = (x)(5)$$

$$x^2 + 1 = 5x$$

$$x^2 - 5x + 1 = 0$$

Use the quadratic formula to solve for x:

$$x = \frac{5 \pm \sqrt{21}}{2}$$

There appear to be two possible answers, $\frac{5 + \sqrt{21}}{2}$ and $\frac{5 - \sqrt{21}}{2}$. Both solutions should be checked; here is a check for one of the possible answers. Let

$$x = \frac{5 + \sqrt{21}}{2}$$

Then

$$\begin{aligned} x + \frac{1}{x} &= \frac{5 + \sqrt{21}}{2} + \frac{2}{5 + \sqrt{21}} \\ &= \frac{(5 + \sqrt{21})(5 + \sqrt{21}) + 4}{2(5 + \sqrt{21})} \\ &= \frac{25 + 10\sqrt{21} + 21 + 4}{2(5 + \sqrt{21})} \\ &= \frac{50 + 10\sqrt{21}}{2(5 + \sqrt{21})} \\ &= \frac{10(5 + \sqrt{21})}{2(5 + \sqrt{21})} \\ &= 5 \end{aligned}$$

Check the other solution. ■

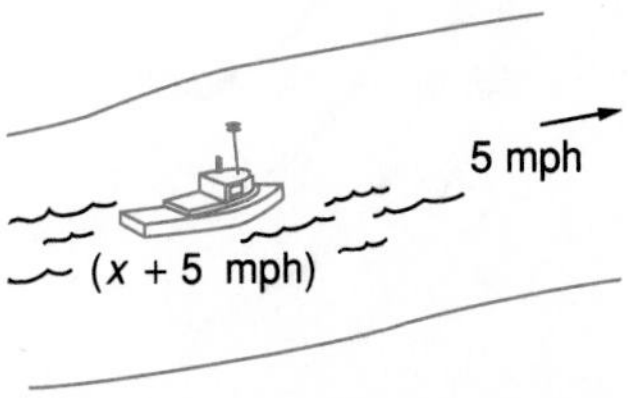

EXAMPLE 4 It takes a boat $1\frac{1}{2}$ hours to make a round trip traveling 10 miles with the current and returning 10 miles against the current. If the rate of the current is 5 mph, what is the rate of the boat in still water, and how long did it take to travel each way?

Solution Let x be the rate (in miles per hour) of the boat in still water. Then $x + 5$ is the rate going with the current, and $x - 5$ is the rate against the current. Expressions for the times involved are written in the table, which is based on the formula $d = rt$ (distance = rate × time).

	r	d	$t = \frac{d}{r}$
With the current	$x + 5$	10	$\frac{10}{x + 5}$
Against the current	$x - 5$	10	$\frac{10}{x - 5}$

Since the total time is $1\frac{1}{2}$ hours, we have

$$\underbrace{\frac{10}{x + 5}}_{\text{time with current}} + \underbrace{\frac{10}{x - 5}}_{\text{time against current}} = \underbrace{\frac{3}{2}}_{\text{total time}}$$

Multiply by $2(x + 5)(x - 5)$:

$$20(x - 5) + 20(x + 5) = 3(x^2 - 25)$$

$$3x^2 - 40x - 75 = 0$$

$$(x - 15)(3x + 5) = 0$$

We reject the solution $x = -\frac{5}{3}$ since the rate of the boat in still water must be positive.

$$x = 15 \quad \text{or} \quad x = -\frac{5}{3}$$

The boat's rate in still water is 15 mph.

The time it takes with the current is $\frac{10}{15 + 5} = \frac{1}{2}$ hour.

The time it takes against the current is $\frac{10}{15 - 5} = 1$ hour. ■

EXERCISES 5.8

Solve these word problems using quadratic equations. For those questions having irrational answers, give both the radical forms and their approximations to two decimal places.

1. Find two consecutive positive integers whose product is 210.
2. The sum of a number and its square is 56. Find the number. (There are two answers.)
3. The sum of a number and its reciprocal is 4. Find the number. (There are two answers.)
4. The sum of the squares of two consecutive positive integers is 113. Find the integers.
5. One positive integer is 3 greater than another. The sum of the squares of the two integers is 89. Find the integers.
6. Find two integers whose sum is 26 and whose product is 165. (*Hint:* Let the two numbers be represented by x and $26 - x$.)
7. How wide a border of uniform width should be added to a rectangle that is 8 feet by 12 feet in order to double the area?
8. The length of a rectangle is 3 centimeters greater than its width. The area is 70 square centimeters. Find the dimensions of the rectangle.

9. The area of a rectangle is 15 square centimeters and the perimeter is 16 centimeters. What are the dimensions of the rectangle? (*Hint:* If x represents the width, then $\frac{16 - 2x}{2} = 8 - x$ represents the length.)

10. The altitude of a triangle is 5 centimeters less than the base to which it is drawn. The area of the triangle is 21 square centimeters. Find the length of the base. (*Hint:* The area of a triangle is the product of $\frac{1}{2}$ the base times the altitude.)

11. When an object is dropped from rest, the distance s in feet is given by $s = 16t^2$, where t is the time in seconds. How long does it take for a stone dropped from the top of a 200-foot vertical cliff to reach the bottom?

12. After 20 congruent squares are punched out of a square metallic plate, the remaining part of the plate has an area of 320 square inches. If a side of each of the 20 squares is $\frac{1}{10}$ the side of the plate, what is the area of the original plate?

13. Find two consecutive even integers whose product is 80. (There are two possible sets of answers.)

14. The sum of two consecutive positive integers is subtracted from their product to obtain a difference of 71. What are the integers?

15. **(a)** A toy rocket is launched from ground level vertically upward with an initial velocity of 128 feet per second. When will the rocket be 240 feet above the ground? (See Example 2.)

    (b) How long will it take to return to the ground?

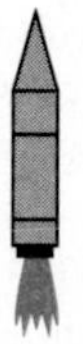

16. If the rocket in Exercise 15 is launched vertically upward from the roof of a building that is 100 feet high, then its height h in feet above the ground in t seconds is given by $h = 100 + 128t - 16t^2$. When will the rocket be 352 feet above the ground?

17. When will the rocket in Exercise 16 be 300 feet above the ground?

18. A backyard swimming pool is rectangular in shape, 10 meters wide and 18 meters long. It is surrounded by a walk of uniform width, whose area is 52 square meters. How wide is the walk?

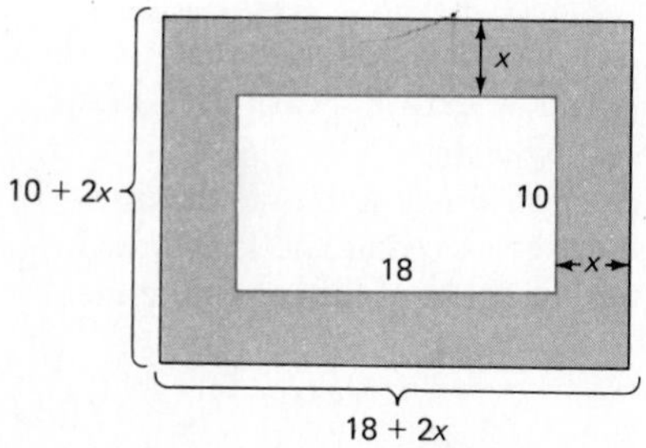

19. The sum S of the first n consecutive positive integers, 1, 2, 3, . . . , n, is given by the formula $S = \frac{1}{2}n(n + 1)$. Find n when $S = 120$ and check your answer by addition.

20. The sum S of the first n consecutive even positive integers, 2, 4, 6, . . . , $2n$, is given by the formula $S = n(n + 1)$. How many such consecutive positive even integers must be added to get a sum of 342?

21. The measures of the legs of a right triangle are consecutive odd integers. The hypotenuse is $\sqrt{130}$. Find the lengths of the legs. (*Hint:* Use the Pythagorean theorem.)

22. One positive integer is 3 greater than another. The difference of their reciprocals is $\frac{1}{6}$. Find the integers.

23. The sum of a number and twice the square of that number is 4. Find all such numbers.

24. If the length of one pair of opposite sides of a square is increased by 3 centimeters, and the length of the other pair is doubled, the area of the new figure will be 55 square centimeters greater than that of the original square. What is the length of the side of the square?

25. If the length of the sides of a square are increased by 2 centimeters, the newly formed square will have an area that is 36 square centimeters greater than the original one. Find the length of a side of the original square.

26. Wendy is 5 years older than Sharon. In 5 years the product of their ages will be $1\frac{1}{2}$ times as great as the product of their present ages. How old is Sharon now? (*Hint:* Let Sharon's age be represented by x and Wendy's age by $x + 5$. Then in 5 years their ages will be $x + 5$ and $x + 10$, respectively.

27. A boat travels downstream (with the current) for 36 miles and then makes the return trip upstream (against the current). The trip downstream took $\frac{3}{4}$ of an hour less than the trip upstream. If the rate of the current is 4 mph, find the rate of the boat in still water and the time for each part of the trip.

28. Two motorcycles each make the same 220-mile trip, but one of them travels 5 mph faster than the other. Find the rate of each motorcycle if the slower one takes 24 minutes longer to complete the trip than the faster one.

29. Two trains depart from the same station at the same time. One train travels due north at 60 mph and the other due east at 80 mph. When will they be 120 miles apart? (*Hint:* Use the Pythagorean theorem.)

30. A square piece of tin is to be used to form a box without a top by cutting off a 2-inch square from each corner, and then folding up the sides. The volume of the box will be 128 cubic inches. Find the length of a side of the original square.

31. José can complete a project in 4 hours less than it takes his younger brother. If it takes them $3\frac{3}{4}$ hours to complete the project working together, how long does it take each of them to do the project alone? (See Example 5, page 161.)

32. The sale price for a loaf of bread is 40¢ less than the regular price. If you can buy two more loaves for \$4.80 at the sale price than was possible at the regular price, what is the regular price?

*33. A rectangular solid has a square base and its height is 3 inches more than the length of the base. If 2 inches are added to each dimension, the volume is increased by 212 cubic inches. Find the volume of the original solid.

*34. A sporting goods store bought some baseball gloves from a manufacturer at a total cost of \$300. In one month's time the store sold all but 5 of the gloves at a \$10 profit per glove. If the total income on the sale of these gloves was \$440, what did the store pay per glove and how many did the store buy?

35. The following formula can be used to approximate the distance in feet, d, that it takes to stop a car after the brakes are applied for a car traveling on a dry road at the rate of r miles per hour: $d = 0.045r^2 + 1.1r$. To the nearest foot, how far will a car travel once the brakes are applied at (a) 40 mph, (b) 55 mph, and (c) 65 mph?

36. The police measured the skid marks made by a car that crashed into a tree. If the measurement gave a braking distance of 250 feet, was the driver exceeding the legal speed limit of 55 miles per hour? To the nearest mph, what was the speed of the car before the brakes were applied? (Use the formula in Exercise 35.)

CHAPTER 5 SUMMARY

Review these key terms and concepts so that you are able to define or describe them. A clear understanding of these will be very helpful when reviewing the developments of this chapter.

Definition of $\sqrt[n]{a}$:

If $a \geq 0$, then $\sqrt[n]{a}$ is the non-negative number b such that $b^n = a$.
If $a < 0$, then $\sqrt[n]{a}$ is the negative number b such that $b^n = a$, if n is odd. If n is even, then $\sqrt[n]{a}$ is not a real number.

In the following it is assumed that the radicals are real numbers.

Multiplication Rule for Radicals: $\sqrt[n]{a} \cdot \sqrt[n]{b} = \sqrt[n]{ab}$

Division Rule for Radicals: $\dfrac{\sqrt[n]{a}}{\sqrt[n]{b}} = \sqrt[n]{\dfrac{a}{b}}$

To add or subract two radicals, they must have the same index and the same radicand.

For all real numbers a: $\sqrt{a^2} = |a|$ $\quad\sqrt[n]{a^n} = |a|$, if n is even $\quad\sqrt[n]{a^n} = a$, if n is odd

Definition of $b^{1/n}$ $(n \geq 2)$: $b^{1/n} = \sqrt[n]{b}$

Definition of $b^{m/n}$ $(n \geq 2)$: $b^{m/n} = (\sqrt[n]{b})^m = \sqrt[n]{b^m}$

To solve radical equations, make use of this principle: If $a = b$, then $a^n = b^n$.

Definition of the imaginary number i: $i = \sqrt{-1}$ and $i^2 = -1 \quad (i^3 = -i,\ i^4 = 1)$

A complex number has the form $a + bi$, where a and b are real numbers and $i = \sqrt{-1}$.

Sum and Difference of Complex Numbers:

$$(a + bi) + (c + di) = (a + c) + (b + d)i$$

$$(a + bi) - (c + di) = (a - c) + (b - d)i$$

Product of Complex Numbers: $(a + bi)(c + di) = (ac - bd) + (ad + bc)i$

Quotient of Complex Numbers: $\dfrac{a + bi}{c + di} = \dfrac{ac + bd}{c^2 + d^2} + \dfrac{bc - ad}{c^2 + d^2}i$

Square-Root Property for Equations: If $n^2 = k$, then $n = \sqrt{k}$ or $n = -\sqrt{k}$

Quadratic Formula: If $ax^2 + bx + c = 0$, $a \neq 0$, then $x = \dfrac{-b + \sqrt{b^2 - 4ac}}{2a}$

Discriminant:

If $b^2 - 4ac > 0$, then $ax^2 + bx + c = 0$ has two real roots.
If $b^2 - 4ac = 0$, then $ax^2 + bx + c = 0$ has one solution called a double root.
If $b^2 - 4ac < 0$, then $ax^2 + bx + c = 0$ has two imaginary roots.

REVIEW EXERCISES

The solutions to the following exercises can be found within the text of Chapter 5. Try to answer each question before referring to the text.

Section 5.1

Evaluate.

1. $\sqrt[3]{-125}$ 2. $\sqrt[4]{\dfrac{16}{81}}$ 3. $\sqrt{x^4}$

Multiply and evaluate if possible.

4. $\sqrt{5} \cdot \sqrt{14}$ 5. $\sqrt{6x}\sqrt{7y}$ 6. $3\sqrt{50} \cdot 2\sqrt{\frac{1}{2}}$ 7. $\sqrt[3]{\dfrac{1}{24}} \cdot \sqrt[3]{-81}$

8. $\sqrt[4]{8} \cdot \sqrt[4]{c}$ 9. $\sqrt[5]{16x} \cdot \sqrt[5]{-2x^4}$ 10. $\sqrt{28} \cdot \sqrt{63}$

Simplify.

11. $\sqrt{50}$ 12. $\sqrt[3]{-24}$ 13. $\sqrt[4]{64x^5}$ 14. $\sqrt{4x + 8}$

Divide and evaluate.

15. $\dfrac{\sqrt{20}}{\sqrt{5}}$ 16. $\dfrac{\sqrt[3]{81x^7}}{\sqrt[3]{-3x}}$

Simplify.

17. $\sqrt{\dfrac{500}{49}}$ 18. $\sqrt[3]{-\dfrac{x}{125}}$

Rationalize the denominators.

19. $\dfrac{4}{\sqrt{2}}$ 20. $\dfrac{6}{\sqrt{8}}$ 21. $\dfrac{5}{\sqrt[3]{2}}$

Section 5.2

Combine.

22. $\sqrt{50} - \sqrt{18} + \sqrt{45}$ 23. $\dfrac{6}{\sqrt{3}} + 2\sqrt{75} - \sqrt{3}$

Simplify.

24. $\sqrt{(-10)^2}$ 25. $\sqrt{x^6}$ 26. $\sqrt[8]{(-\frac{1}{2})^8}$
27. $\sqrt[7]{(-8)^7}$ 28. $\sqrt{75x^2}$ 29. $\sqrt[4]{(5x + 1)^4}$

Find the allowable values of the variables.

30. $\sqrt{x}$ 31. $\sqrt{x^4}$ 32. $\sqrt{x^5}$ 33. $\sqrt[3]{x}$ 34. $\sqrt[4]{x - 2}$ 35. $\dfrac{1}{\sqrt{x + 1}}$

36. Combine: $2\sqrt{8x^3} + 3x\sqrt{32x} - x\sqrt{18x}$

Multiply.

37. $\sqrt{5}(2\sqrt{5} - \sqrt{10})$ 38. $(2\sqrt{8} + 7)(3\sqrt{8} - 5)$
39. $(\sqrt{11} - \sqrt{3})^2$ 40. $(\sqrt[3]{x^2} - 1)(\sqrt[3]{x^2} + 1)$

Rationalize the denominators.

41. $\dfrac{5}{\sqrt{10} - 3}$ 42. $\dfrac{x}{\sqrt{x} + \sqrt{y}}$

Section 5.3

Evaluate:

43. $9^{1/2}$ 44. $(-8)^{1/3}$ 45. $(-64)^{2/3}$ 46. $8^{-2/3} + (-32)^{-2/5}$

Simplify.

47. $(-3x^{2/5})(7x^{8/5})$ 48. $(x^2y^{-2})^{3/2}$ 49. $\dfrac{x^{2/3}y^{-2}z^2}{x^{1/2}y^{1/2}z^{-1}}$

50. $\dfrac{(s^{-3}t^2)^{-1/6}}{(s^4t^{-4})^{-1/2}}$ 51. $\left(\dfrac{-8a^3}{b^{-6}}\right)^{2/3}$

Multiply.

52. $a^{-1/3}(a^2 - 4a^{1/3})$ 53. $(x^{1/2} + y^{-1/2})(x^{1/2} - y^{-1/2})$

54. Combine: $5x^{-1/3} + x^{2/3}$

Section 5.4

Solve for x:

55. $\sqrt{x + 4} + 2 = x$ 56. $\sqrt{x + 5} + \sqrt{x} = 5$

57. $\sqrt{x - 7} - \sqrt{x} = 1$ 58. $\sqrt[3]{x + 3} = 2$

59. $x^{1/2} + \frac{1}{2}x^{-1/2}(x - 27) = 0$

60. If 2 is subtracted from 3 times a number, the square root of the difference is 5. What is the number?

Section 5.5

Simplify:

61. $\sqrt{-16} + \sqrt{-25}$ 62. $\sqrt{-16} \cdot \sqrt{-25}$

63. Name the sets of numbers to which each of the following belongs:
   (a) $\sqrt{\frac{4}{9}}$ **(b)** $\sqrt{-10}$ **(c)** $-4 + \sqrt{-12}$

Perform the indicated operations and express the answers in the form a + bi.

64. $(2 + 3i) + (5 + 7i)$ 65. $(8 + 5i) - (3 + 2i)$

66. $(5 - 2i)(3 + 4i)$ 67. $\dfrac{2 + 3i}{3 + i}$

Simplify.

68. i^{22} 69. i^{39} 70. $2i^{-3}$

Section 5.6

Use the square-root property for equations to solve for x.

71. $x^2 = 25$ 72. $x^2 = 8$ 73. $(x + 2)^2 = 3$ 74. $(x - 2)^2 = -9$

Add the appropriate number to complete the square and factor.

75. $x^2 + 8x +$ _____   76. $x^2 - 3x +$ _____

Solve for x by completing the square.

77. $x^2 + 5x - 3 = 0$ 78. $4x^2 - 12x + 5 = 0$

Section 5.7

79. State the quadratic formula.

Solve for x using the quadratic formula.

80. $2x^2 - 5x + 1 = 0$ 81. $4x^2 = 12x - 9$ 82. $x^2 + 3x + 5 = 0$

Find the discriminant and use it to describe the roots of each equation without solving for x.

83. $3x^2 - 5x - 6 = 0$ 84. $5x^2 - 11x - 12 = 0$ 85. $2x^2 + 2x + 3 = 0$

86. $9x^2 - 12x + 4 = 0$

Section 5.8

Solve each word problem using a quadratic equation.

87. The length of a rectangular piece of cardboard is 2 inches more than its width. A box open at the top is formed by cutting out 4-inch squares from each corner and folding up the sides. If the volume of the box is 672 cubic inches, find the dimensions of the original cardboard.
88. When an object is thrown vertically upward from ground level with an initial velocity of v_0 feet per second, its height h in feet after t seconds is given by the formula $h = v_0t - 16t^2$. If $v_0 = 256$ feet per second, when will the object be 1008 feet above the ground?
89. The sum of a number and its reciprocal is 5. Find the number.
90. It takes a boat $1\frac{1}{2}$ hours to make a round trip traveling 10 miles with the current and returning 10 miles against the current. If the rate of the current is 5 mph, what is the rate of the boat and how long did it take to travel each way?

CHAPTER 5 TEST: STANDARD ANSWER

Use these questions to test your knowledge of the basic skills and concepts of Chapter 5. Then check your answers with those given at the back of the book.

1. Evaluate: **(a)** $(-32)^{3/5}$ **(b)** $27^{-2/3}$

Simplify. Express the answers using positive exponents.

2. $(a^{1/2}b^{-1/2})(a^{-1/3}b^{1/3})$
3. $\dfrac{a^{2/3}b^{-1/2}}{a^{-1/2}b^{1/4}}$

Perform the indicated operations and simplify.

4. **(a)** $\sqrt{8} \cdot \sqrt{6}$ **(b)** $\dfrac{\sqrt{360}}{2\sqrt{2}}$ **(c)** $\dfrac{\sqrt[3]{-243x^8}}{\sqrt[3]{3x^2}}$
5. $\sqrt{50} + 3\sqrt{18} - 2\sqrt{8}$
6. $\dfrac{12}{\sqrt{3}} + 2\sqrt{3}$
7. $2\sqrt{x^2} + \sqrt{9x^2}$
8. $\sqrt{8x^2y^3} + \sqrt{18x^2y^3} - y\sqrt{2x^2y}$
9. $(3\sqrt{7} - 5)(2\sqrt{7} + 3)$

Solve for x.

10. $\sqrt{x} - 3 = 5$
11. $\sqrt{x + 7} - \sqrt{x} = 1$
12. The cube root of the sum of 7 and 4 times a number is 3. Find the number.

Simplify.

13. **(a)** $2\sqrt{-27} + 3\sqrt{-12}$ **(b)** $\sqrt{-9} \cdot \sqrt{-8}$

Complete the indicated operation, expressing your answer in the form $a + bi$.

14. $(3 + i)(2 - 3i)$
15. $\dfrac{3 + i}{2 - i}$

Solve for x by completing the square.

16. $x^2 - 8x + 12 = 0$
17. $x^2 + 3x - 2 = 0$

Use the quadratic formula to solve for x.

18. $x^2 - 3x - 6 = 0$
19. $3x^2 + 2x - 2 = 0$

20. $2x^2 + 3x = 5$ 21. $-x^2 + 4x - 8 = 0$

Use the discriminant to describe the solutions as a single real number, two rational numbers, two irrational numbers, or two imaginary numbers.

22. **(a)** $x^2 - 7x + 2 = 0$ **(b)** $2x^2 + x + 3 = 0$
23. Find the values of k so that the equation $2x^2 + 4x + k = 0$ will have two imaginary roots.
24. The sum of the squares of two consecutive odd integers is 290. What are the integers?
25. An object is thrown straight upward from ground level with an initial velocity of 160 feet per second. The formula $s = 160t - 16t^2$ gives its height in feet, s, after t seconds. When will the object be 256 feet above the ground?

CHAPTER 5 TEST: MULTIPLE CHOICE

1. Which of the following are true?

I. $a^{-1/2} + b^{-1/2} = \dfrac{1}{\sqrt{a}} + \dfrac{1}{\sqrt{b}}$ **II.** $(-x)^{-1/3} = x^{1/3}$ **III.** $x^{3/4} = (\sqrt[3]{x})^4$

(a) Only I **(b)** Only II **(c)** Only III **(d)** I, II, and III **(e)** None of the preceding

2. Which of the following are correct?

I. $x\sqrt{x^2 + 1} = \sqrt{x^3 + x}$ **II.** $x^{-1/5} + x^{3/5} = x^{-1/5}(1 + x^{4/5})$ **III.** $(x^2 - 3)^{1/3}(x^2 - 3)^{2/3} = (x^2 - 3)^{2/9}$

(a) Only I **(b)** Only II **(c)** Only III **(d)** Only I and II **(e)** None of the preceding

3. Evaluate: $(-64)^{2/3}$

(a) 16 **(b)** -16 **(c)** $-\dfrac{1}{16}$ **(d)** 512 **(e)** None of the preceding

4. Rationalize the denominator: $\dfrac{8}{\sqrt[3]{2}}$.

(a) $4\sqrt[3]{2}$ **(b)** $4\sqrt[3]{4}$ **(c)** $\sqrt[3]{4}$ **(d)** $8\sqrt[3]{2}$ **(e)** $2\sqrt[3]{2}$

5. Which of these statements are correct?

I. $\sqrt{(x + 1)^2} = x + 1$ **II.** $\sqrt{8 + 40} = \sqrt{8} + \sqrt{40}$ **III.** $\sqrt[3]{x - y} = \sqrt[3]{x} - \sqrt[3]{y}$

(a) Only I **(b)** Only I and II **(c)** Only II and III **(d)** I, II, and III **(e)** None of the preceding

6. Rationalize the denominator: $\dfrac{8}{\sqrt{5} - 1}$

(a) $8(\sqrt{5} + 1)$ **(b)** $2(\sqrt{5} - 1)$ **(c)** $2(\sqrt{5} + 1)$ **(d)** $2\sqrt{5} + 1$ **(e)** None of the preceding

7. Which of the following are true?

I. $2\sqrt{x + 1} = \sqrt{2x + 1}$ **II.** $\sqrt{(x - 1)^2} = |x - 1|$ **III.** $(x + y)^{1/3} = x^{1/3} + y^{1/3}$

(a) Only I **(b)** Only II **(c)** Only III **(d)** I, II, and III **(e)** None of the preceding

8. Multiply: $\sqrt[3]{\dfrac{1}{81}} \cdot \sqrt[3]{-24}$

(a) $\dfrac{2}{3}$ **(b)** $-\dfrac{2\sqrt[3]{3}}{9}$ **(c)** $-\dfrac{3}{2}$ **(d)** $-\dfrac{2}{3}$ **(e)** $\sqrt[6]{-\dfrac{8}{27}}$

9. Combine and simplify: $2\sqrt{28} - \dfrac{7}{\sqrt{7}} + 3\sqrt{63}$

(a) $4 + 9\sqrt{7}$ **(b)** $4\sqrt{7} + 9$ **(c)** $6\sqrt{7}$ **(d)** $12\sqrt{7}$ **(e)** None of the preceding

10. Write the quotient of $\dfrac{2 + i}{3 - i}$ in the form $a + bi$.

(a) $\dfrac{2}{3} - i$ **(b)** $\dfrac{1}{2} + \dfrac{1}{2}i$ **(c)** $\dfrac{1}{2} + 5i$ **(d)** $1 + i$ **(e)** $\dfrac{5}{8} + \dfrac{5}{8}i$

11. The equation $\sqrt{x^2 - 9x} + \sqrt{3x} = x$ has
(a) no solution (b) one solution (c) two solutions (d) three solutions (e) None of the preceding

12. The roots of $2x^2 + 4x - 5 = 0$ are:
(a) $\frac{-2 \pm \sqrt{11}}{2}$ (b) $\frac{-2 \pm \sqrt{14}}{2}$ (c) $-2 \pm \sqrt{14}$ (d) $\frac{-2 \pm \sqrt{6}i}{2}$ (e) $\frac{2 \pm \sqrt{14}}{2}$

13. Which of the following equations has two imaginary solutions?
I. $x^2 + 6x = -10$ **II.** $x^2 + 2x - 4 = 0$ **III.** $2x^2 = x + 1$
(a) Only I (b) Only II (c) Only III (d) I, II, and III (e) None of the preceding

14. For what values of $k\,(k \neq 0)$ will the equation $kx^2 - 4x + 2 = 0$ have two distinct real roots?
(a) $k > 2$ (b) $k = 2$ (c) $k < 2$ (d) $k > -2$ (e) None of the preceding

15. When a border of uniform width is put around a 6 foot by 10 foot rectangle, the area is enlarged by 80 square feet. Which equation can be used to find the uniform width x?
(a) $x(10 + 2x) + x(6 + 2x) = 80$ (b) $x^2 + 16x - 20 = 0$ (c) $x^2 + 16x - 80 = 0$
(d) $x^2 + 8x - 20 = 0$ (e) $(10 + 2x)(6 + 2x) = 20$

ANSWERS TO THE TEST YOUR UNDERSTANDING EXERCISES

Page 180

1. $\sqrt{21}$ 2. 15 3. $8x$ 4. $6\sqrt[3]{20}$ 5. $6p$ 6. -6
7. 12 8. $\frac{1}{3}$ 9. -2 10. $2\sqrt{10}$ 11. $4\sqrt{3}$ 12. $5c\sqrt{3}$
13. $2\sqrt[3]{3}$ 14. $-2\sqrt[3]{2}$ 15. $4b^2\sqrt[3]{2}$

Page 186

1. $6\sqrt{2}$ 2. $6\sqrt{3}$ 3. $5\sqrt{5}$ 4. Not possible
5. $5\sqrt[3]{2}$ 6. Not possible 7. $11\sqrt{2}$ 8. $13\sqrt{3}$
9. $5\sqrt{6} - 6\sqrt{2}$ 10. $7\sqrt{7}$ 11. 0 12. $\frac{7\sqrt{2}}{4}$

Page 193

1. $\sqrt{5}$ 2. $\sqrt[3]{9}$ 3. $\sqrt[4]{10}$ 4. $\sqrt[3]{25}$ 5. $\sqrt[4]{8}$
6. $7^{1/2}$ 7. $(-10)^{1/3}$ 8. $7^{1/4}$ 9. $7^{2/3}$ 10. $5^{3/4}$
11. 5 12. 4 13. $\frac{1}{6}$ 14. $-\frac{1}{5}$ 15. $\frac{3}{5}$
16. $\frac{1}{7}$ 17. 8 18. $\frac{1}{8}$ 19. 4 20. $\frac{1}{4}$

Page 195

1. 5 2. $\sqrt[4]{15^3} = (\sqrt[4]{15})^3$ 3. $\frac{1}{\sqrt[3]{-2}}$ 4. $\frac{1}{\sqrt{a}} + \frac{1}{\sqrt{b}}$
5. $7^{5/6}$ 6. $\frac{1 + x}{\sqrt[5]{x}}$ 7. $16^{-1} = \frac{1}{16}$ 8. $x + 6x^{1/2} + 9$

Page 199

1. $x = 8$ 2. $x = 25$ 3. $x = 7$ 4. No solution
5. $x = \frac{1}{9}$ 6. $x = 5$ 7. $x = -3$ or $x = 3$ 8. $x = 9$

Page 204

1. $3i$ 2. $7i$ 3. $\sqrt{5}i$ 4. $-2i$ 5. $\frac{2}{3}i$
6. -66 7. $21i$ 8. $7\sqrt{2}i$ 9. $12\sqrt{5}i$

Page 212

1. 4
2. 25
3. 9
4. 36
5. $\frac{9}{4}$
6. $\frac{25}{4}$
7. $x = \pm 9$
8. $x = \pm 5\sqrt{3}i$
9. $x = -4 \pm \sqrt{10}i$
10. $x = -5$ or $x = 1$
11. $x = -2 \pm \sqrt{2}$
12. $x = 3 \pm \sqrt{6}$

Page 216

1. $x = -5$ or $x = 2$
2. $x = -\frac{1}{2}$ or $x = \frac{1}{3}$
3. $x = -3$ or $x = 3$
4. $x = 1 \pm \sqrt{3}$
5. $x = -3 \pm \sqrt{3}$
6. $x = -3 \pm \sqrt{3}i$
7. $x = \dfrac{1 \pm 3i}{2}$
8. $x = \dfrac{-1 \pm \sqrt{33}}{4}$

CHAPTER 6 GRAPHING LINEAR EQUATIONS AND INEQUALITIES: THE FUNCTION CONCEPT

6.1 THE CARTESIAN COORDINATE SYSTEM

Drawing the pictures (*graphs*) for linear equations will help in understanding the relationships between the variables within the equations and will be of importance throughout much of the remaining chapters of this text. We begin by reviewing the structure of a **rectangular coordinate system.**

In a plane take any two lines that intersect at right angles and call their point of intersection the **origin.** Let each of these lines be a number line with the origin corresponding to zero for each line. Unless otherwise specified, the unit length is the same on both lines. On the horizontal line the positive direction is taken to be to the right of the origin; the negative direction is to the left. On the vertical line the positive direction is taken to be above the origin; the negative direction is below. Each of these two lines will be referred to as an **axis** of the system (plural: **axes**).

The union of algebra and geometry, credited to French mathematician René Descartes (1596–1650), led to the development of analytic geometry. In his honor, we often refer to the rectangular coordinate system as the ***Cartesian coordinate system,*** *or simply the* ***Cartesian plane.***

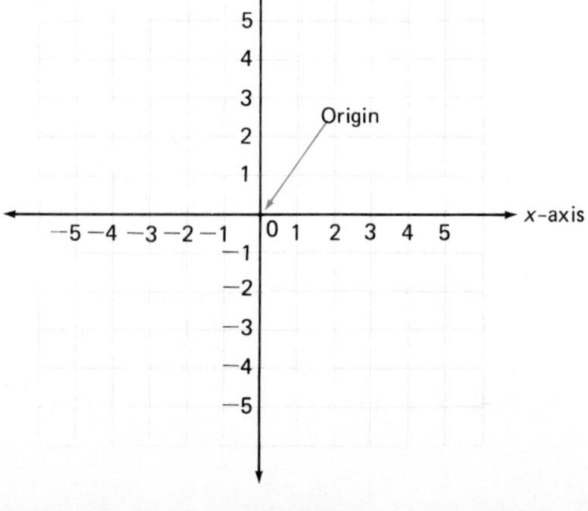

The horizontal line is usually called the **x-axis,** and the vertical line the **y-axis.** The axes divide the plane into four regions called **quadrants.** The quadrants are numbered in a counterclockwise direction as shown in the following figure.

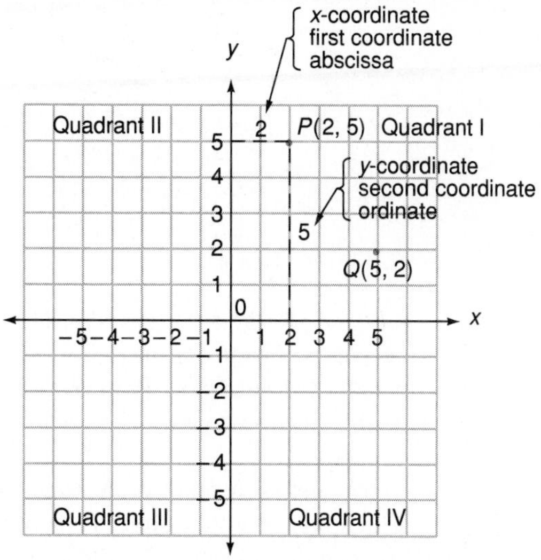

Note that the ordered pair (2, 5) is not the same as the pair (5, 2). Each gives the coordinates of a different point on the plane.

The points in the plane (denoted by the capital letters) are matched with pairs of numbers, referred to as the **coordinates** of these points. For example, starting at the origin, P can be reached by moving 2 units to the right, parallel to the x-axis; then 5 units up, parallel to the y-axis. Thus the first coordinate, 2, of P is called the **x-coordinate** (another name is **abscissa**) and the second coordinate, 5, is the **y-coordinate** (also called **ordinate**). We say that the *ordered pair of numbers* (2, 5) are the coordinates of P.

All points in the first quadrant are to the right and above the origin, and therefore have positive coordinates. Any point in quadrant II is to the left and above the origin and therefore has a negative x-coordinate and a positive y-coordinate. In quadrant III both coordinates are negative, and in the fourth quadrant they are positive and negative, respectively.

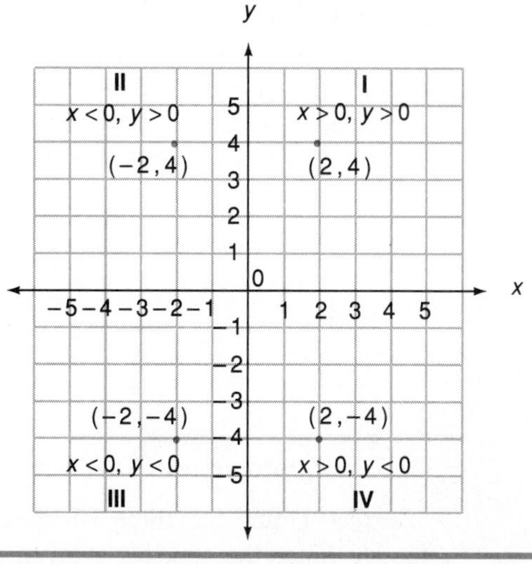

EXAMPLE 1 Find the coordinates of the given points. Also state the quadrant or axis in which each point is located.

Usually, the points on the axes are labeled with just a single number. In that case it is understood that the missing coordinate is 0. Thus in Example 1, B has coordinates (0, 2) and the coordinates of point F are (3, 0). The coordinates of the origin are (0, 0).

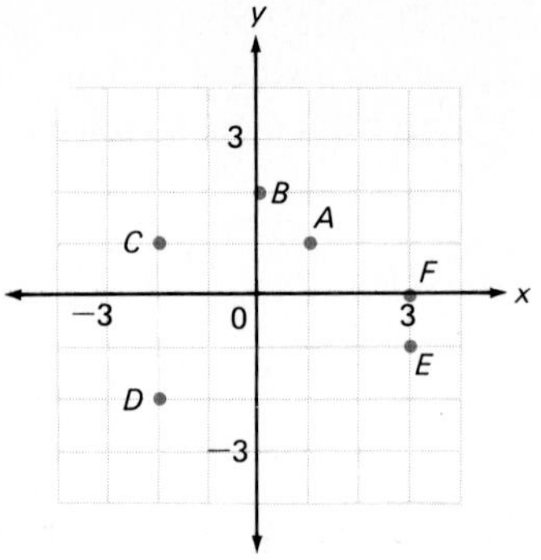

Solution Point A is located at (1, 1), in the first quadrant. Point B is at (0, 2), on the y-axis. Point C is at $(-2, 1)$, in the second quadrant. Point D is at $(-2, -2)$, in the third quadrant. Point E is at $(3, -1)$, in the fourth quadrant. Point F is at (3, 0), on the x-axis. ■

In locating a point on a graph it is customary to move in the x-direction first, and then in the y-direction. Thus to locate the point $(-2, 3)$ in Example 2 we begin at the origin and move 2 units to the left, and then 3 units up.

EXAMPLE 2 Graph the set of ordered pairs:

Note: $(-\sqrt{2}, 0)$ and (π, π) are graphed by using the approximations $-\sqrt{2} = -1.4$ and $\pi = 3.1$.

$$\left\{(4, 1),\ (0, 2),\ \left(-\frac{5}{2}, -3\right),\ \left(\frac{3}{2}, -3\right),\ (-2, 3),\ (-\sqrt{2}, 0),\ (\pi, \pi)\right\}$$

Solution

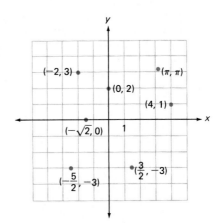

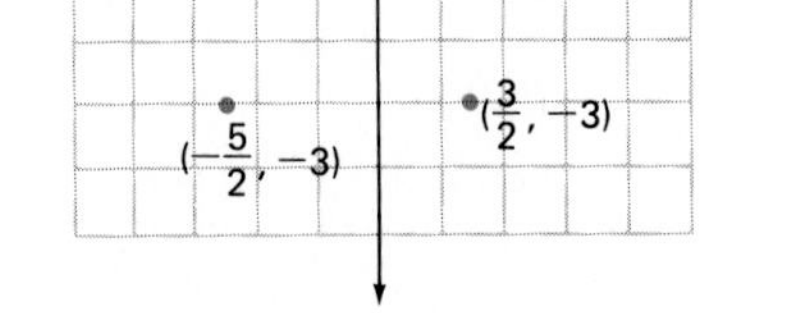

■

EXAMPLE 3 The base of a parallelogram has endpoints (0, 0) and (10, 0), and (2, 5) is a third vertex. If the fourth vertex is in quadrant I, find its coordinates.

Solution Graph the three vertices (0, 0), (10, 0), (2, 5) and connect them as shown. Since the opposite sides of a parallelogram are parallel and have the same length, and since the fourth vertex is in quadrant I, we start at (2, 5) and count 10 units horizontally to the right to locate the fourth vertex at the point (12, 5).

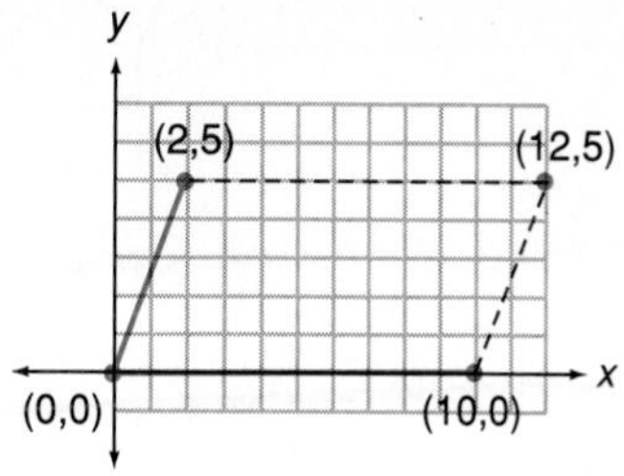

■

Our matching of points with pairs of numbers can be described as a *one-to-one correspondence* between the points in the plane and the set of all ordered pairs of real numbers. This means that for each point in the plane there can be found just one pair of real numbers (x, y); and for each pair of real numbers there is just one point.

A geometrical point is not the same as an ordered pair of real numbers. However, due to the one-to-one correspondence we may, without confusion, equate a point P with its coordinates (x, y). The expression "$P(x, y)$" means that point P has coordinates (x, y).

This one-to-one correspondence will eventually allow us to study geometric properties algebraically, and algebraic properties geometrically. The joint study of algebra and geometry through the use of the Cartesian (or rectangular) coordinate system is called **analytic geometry.** In this and later chapters you will see the beginning of this subject.

EXERCISES 6.1

Find the coordinates of the given points in Exercises 1 through 4.

1.

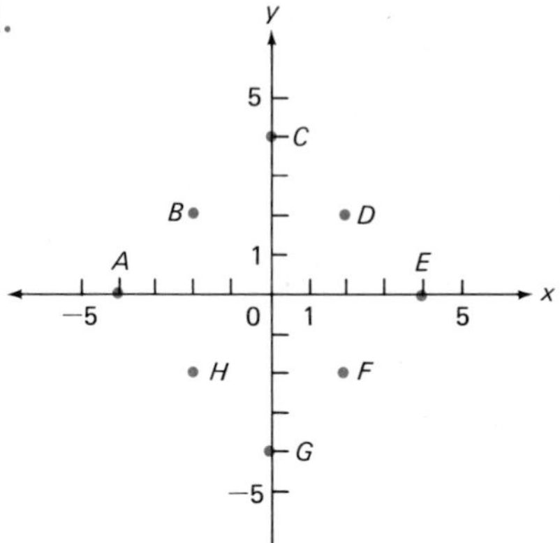

2.

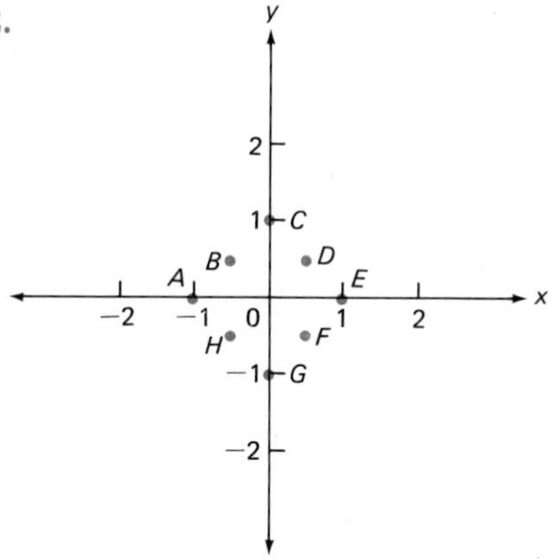

3.

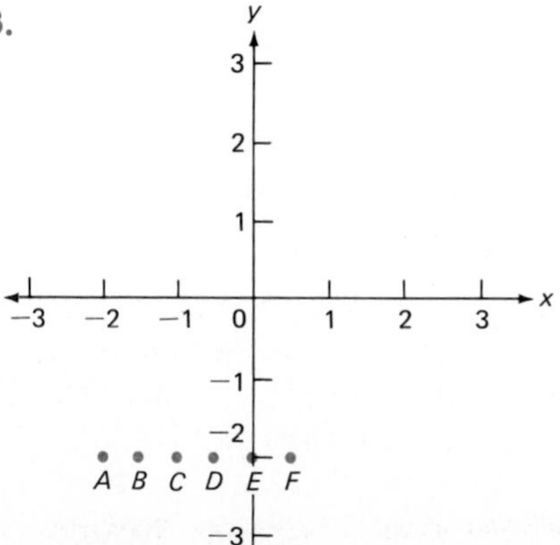

4.

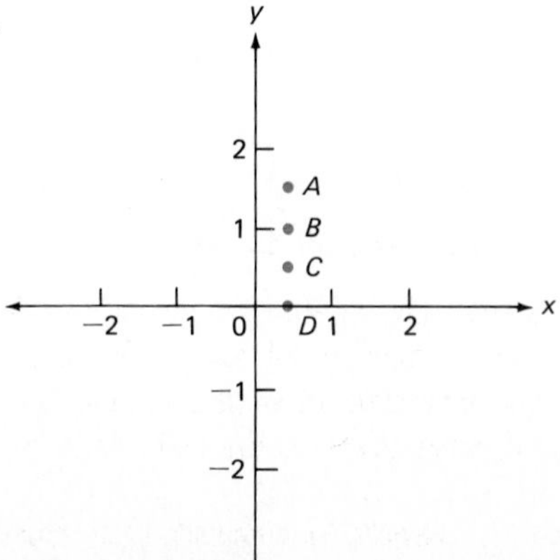

Graph each set of ordered pairs in Exercises 5 through 10.

5. $\{(0, 1), (1, 0), (0, -1), (-1, 0)\}$
6. $\{(2, 2), (2, -2), (-2, 2), (-2, -2)\}$
7. $\{(-5, \frac{1}{2}), (-5, 1), (-5, \frac{3}{2})\}$
8. $\{(\frac{1}{2}, -5), (\frac{3}{2}, -5), (\frac{5}{2}, -5)\}$
9. $\{(-2, -7), (0, -3), (\frac{1}{2}, -2), (1, -1), (2, 1)\}$
10. $\{(-2, 4), (0, 3), (1, \frac{5}{2}), (2, 2), (6, 0), (10, -2)\}$

Graph the points that satisfy each equation for the given values of x. For example, in Exercise 11 the point $(-2, 4)$ *satisfies the equation* $y = x^2$ *because when* $x = -2$, $y = (-2)^2 = 4$.

11. $y = x^2; x = -2, -1, -\frac{1}{2}, 0, \frac{1}{2}, 1, 2$
12. $y = |x|; x = -3, -2, -1, 0, 1, 2, 3$
13. $y = \dfrac{1}{x}; x = \frac{1}{2}, 1, \frac{3}{2}, 2, \frac{5}{2}, 3$
14. $y = \sqrt{x}; x = 0, \frac{1}{4}, 1, \frac{9}{4}, 4, 9$
15. $y = \pm\sqrt{25 - x^2}; x = -5, -4, -3, 0, 3, 4, 5$
16. $y = |1 - |x||;\ x = -2, -\frac{3}{2}, -1, -\frac{1}{2}, 0, \frac{1}{2}, 1, \frac{3}{2}, 2$

17. The following points satisfy an equation similar to the one given in Exercise 11. Find the equation. $(-2, -4), (-1, -1), (-\frac{1}{2}, -\frac{1}{4}), (0, 0), (\frac{1}{2}, -\frac{1}{4}), (1, -1), (2, -4)$
18. The following points satisfy an equation similar to the one given in Exercise 13. Find the equation. $(\frac{1}{2}, 4), (1, 2), (\frac{3}{2}, \frac{4}{3}), (2, 1), (\frac{5}{2}, \frac{4}{5}), (3, \frac{2}{3})$
19. If the coordinates of A and B are $(2, 3)$ and $(6, 7)$, respectively, and if ABC is a right triangle as shown in the following figure, what are the coordinates of C?

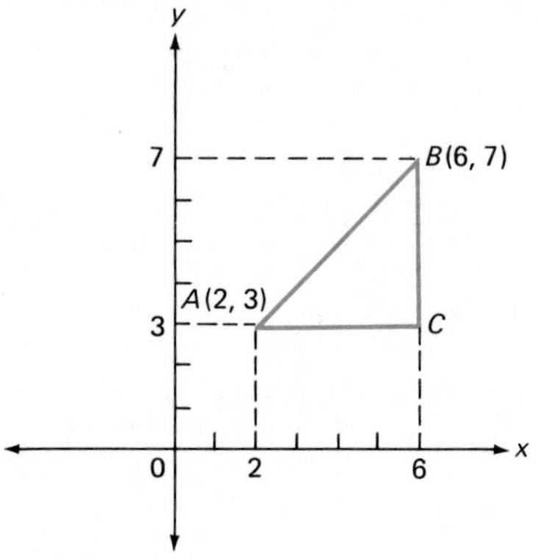

20. Find the coordinates of point C in right triangle ABC.

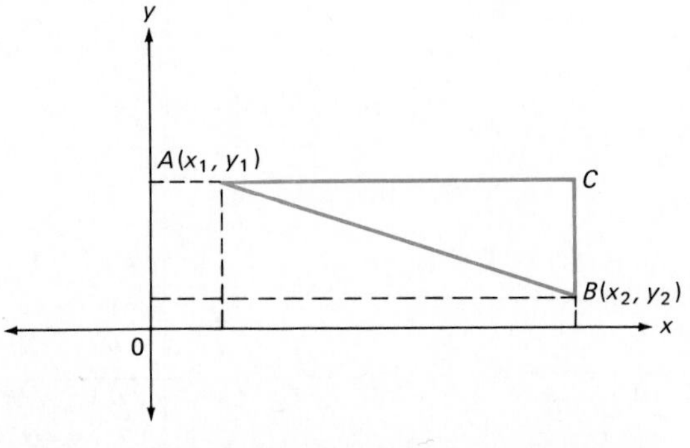

21. The base of a parallelogram has endpoints $(0, 0)$ and $(8, 0)$, and $(3, 4)$ is a third vertex. Find the coordinates of the fourth vertex if:

(a) It lies in the first quadrant.

(b) It lies in the second quadrant.

22. The center of a square is at the origin. Find the coordinates of the four vertices if:

(a) The four sides are parallel to the coordinate axes and their measures are 12 units.

(b) The four vertices are on the coordinate axes and each of the diagonals is 6 units long.

6.2 GRAPHING LINEAR EQUATIONS

The equality $y = x + 2$ is an equation in two variables. When a specific value for x is substituted into this equation, we get a corresponding y-value. For example, substituting 3 for x gives $y = 3 + 2 = 5$. We therefore say that the ordered pair (3, 5) *satisfies* the equation $y = x + 2$.

In the following **table of values** there are six more ordered pairs that satisfy this equation. Note that the ordered pairs in the table have been written without the usual parentheses. These ordered pairs have also been *plotted* in a rectangular coordinate system.

Note that a table of values can also be presented in horizontal form as well. (See Exercises 1–4 on page 241.)

x	$y = x + 2$
−3	−1
−2	0
−1	1
0	2
1	3
2	4

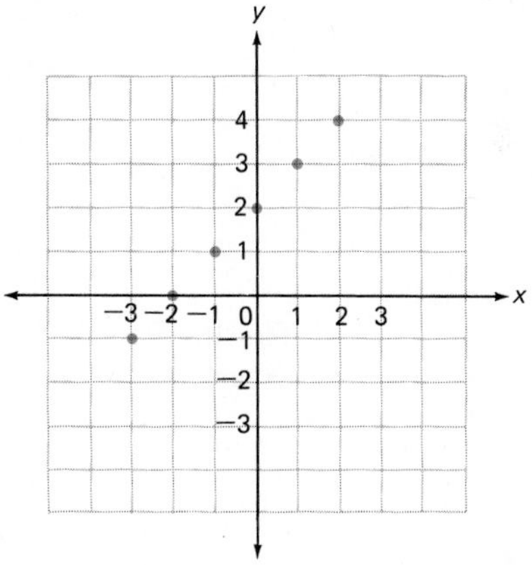

To graph an equation in the variables x and y means to locate all the points in a rectangular system whose coordinates satisfy the given equation.

There is an infinite number of ordered pairs that satisfy $y = x + 2$, and all are located on the same straight line. Since a line is endless, we draw a partial graph by joining the specific points previously located.

The arrowheads in the figure suggest that the line continues endlessly in both directions.

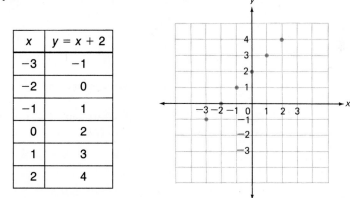

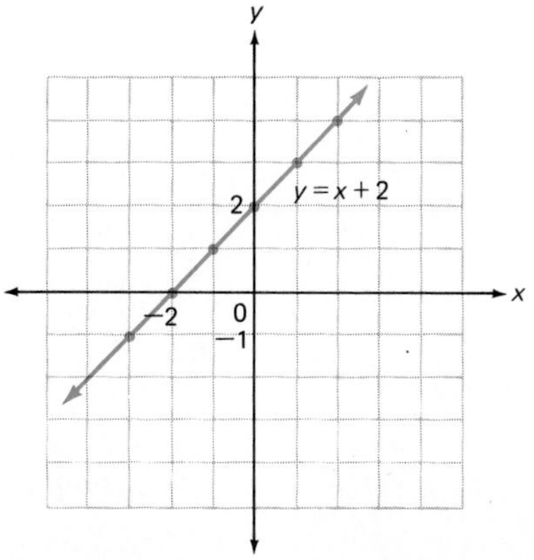

An equation of the form $Ax + By = C$ is said to be a **linear equation** in two variables. The graph of such an equation is a straight line. Inasmuch as two points determine a straight line, you really only need to locate two points in order to graph such an equation. It is usually wise, however, to locate at least a third point as a check of your work.

Two convenient points to use to graph a line are those where the graph crosses the axes. Thus the graph of $y = x + 2$ crosses the x-axis at the point $(-2, 0)$, where -2 is called the ***x*-intercept** of the graph. The graph crosses the y-axis at the point $(0, 2)$, where 2 is called the ***y*-intercept of the graph.**

EXAMPLE 1 Graph the linear equation $2x + y = 6$ by using the intercepts.

Solution To find the x-intercept, let $y = 0$.

$$2x + 0 = 6$$
$$2x = 6$$
$$x = 3$$
The x-intercept is 3. The line crosses the x-axis at the point $(3, 0)$.

To find the y-intercept, let $x = 0$.

$$2(0) + y = 6$$
$$y = 6$$
The y-intercept is 6. The line crosses the y-axis at the point $(0, 6)$.

Plot the points $(3, 0)$ and $(0, 6)$ and draw the straight line through them to determine the graph of $2x + y = 6$. ■

As a check, let $x = 1$ to locate a third point.

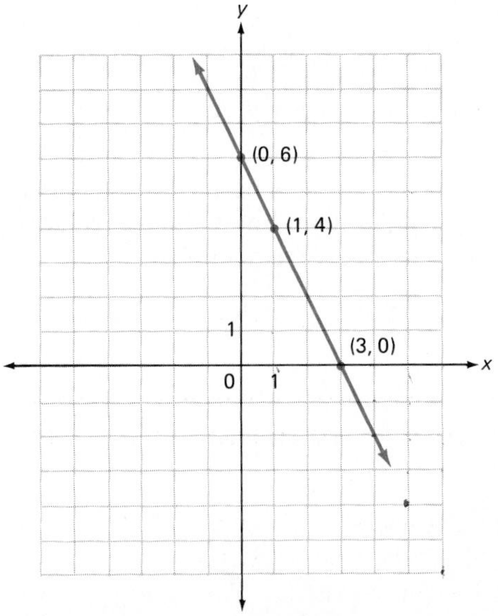

$$2(1) + y = 6$$
$$2 + y = 6$$
$$y = 4$$

TEST YOUR UNDERSTANDING
Think Carefully

(Answers: Page 284)

Name (a) the x-intercept and (b) the y-intercept for each of the following.

1. $x + y = 4$ **2.** $x - 3y = 6$ **3.** $3x + 4y = -12$
4. $x - \frac{1}{2}y = 3$ **5.** $\frac{1}{2}x + 2y = 2$ **6.** $3x - 2y = 5$
7. $y = 3x - 2$ **8.** $y = 5 - 3x$ **9.** $y + 3 = 3x$

Some lines have only one intercept. For example, the line $y = x$ passes through the origin. Both the x-intercept and y-intercept are 0.

x	y
−2	−2
0	0
2	2

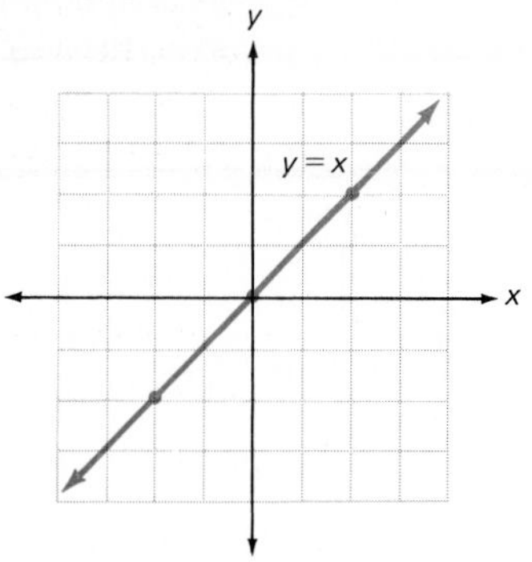

Lines that are parallel to the x and y axes also have only one intercept. Such lines have equations that are special cases of the general form $Ax + By = C$. For example:

Let $B = 0$ and $A \neq 0$:

$$Ax + By = C$$
$$Ax + 0y = C$$
$$Ax = C$$
$$\text{Thus} \qquad x = \frac{C}{A} \qquad \text{for } all \text{ values of } y$$

Let $A = 0$ and $B \neq 0$:

$$Ax + By = C$$
$$0x + By = C$$
$$By = C$$
$$\text{Thus} \qquad y = \frac{C}{B} \qquad \text{for } all \text{ values of } x$$

These special cases give rise to vertical and horizontal lines, as in the following example.

EXAMPLE 2 Graph on a plane: **(a)** $x = 2$ **(b)** $y = -1$

Solution

(a) In this case x is always 2; y can be any value. The graph is a vertical line parallel to the y-axis; the x-intercept is 2. There is no y-intercept.

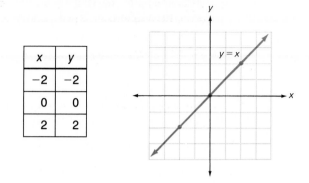

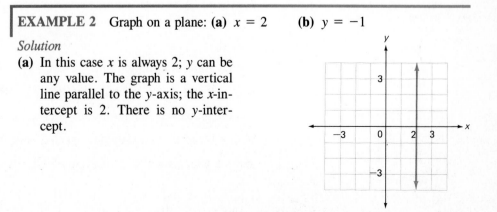

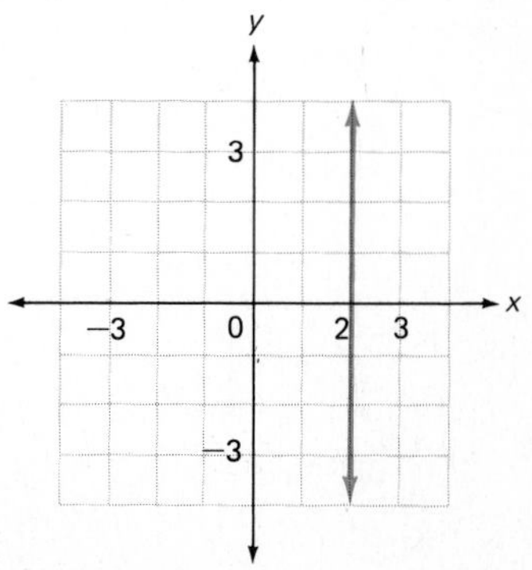

(b) Here the y-value is always -1; x can be any value. The graph is a horizontal line parallel to the x-axis; the y-intercept is -1. There is no x-intercept.

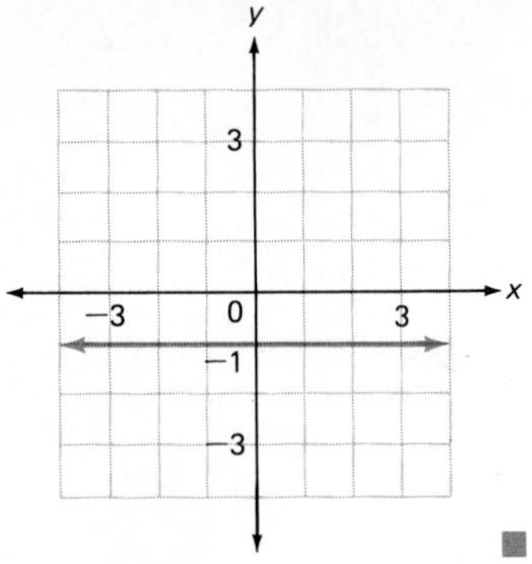

EXAMPLE 3 Write the equation of the line through the point (2, 3) that is parallel to **(a)** the x-axis and **(b)** the y-axis.

Solution

(a) All lines parallel to the x-axis are of the form $y = k$. That is, regardless of the x-coordinate, the line is always k units from the x-axis. Thus the equation here is $y = 3$.

(b) All lines parallel to the y-axis are of the form $x = h$. Regardless of the y-coordinate, the line is always h units from the y-axis. Thus the equation here is $x = 2$.

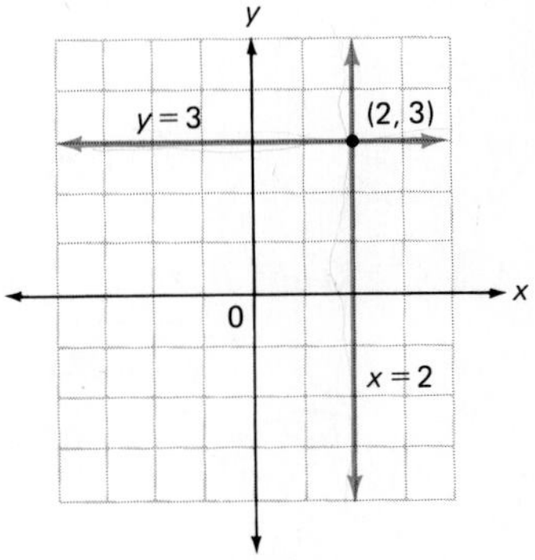

The results of this section can be summarized as follows:

These equations can also simply be written as $x = h$ and $y = k$, where h and k are constants.

> A general linear equation in two variables can be written as
>
> $$Ax + By = C$$
>
> and its graph is a straight line.
>
> If $B = 0$ and $A \neq 0$, then the equation is of the form $x = \frac{C}{A}$ and the graph is a straight line parallel to the y-axis.
>
> If $A = 0$ and $B \neq 0$, then the equation is of the form $y = \frac{C}{B}$ and the graph is a straight line parallel to the x-axis.

EXERCISES 6.2

Copy and complete each table of values. Then graph the line given by the equation.

1. $y = x - 2$

x	−3	−2	−1	0	1	2
y	−5	−4	−3	−2	−1	0

2. $y = -x + 1$

x	−3	−2	−1	0	2	3
y						

3. $y = 2x - 4$

x	−2	−1	0	1	2
y					

4. $y = -2x + 3$

x	−2	−1	0	1	2
y					

Write the equation for each graph.

5.

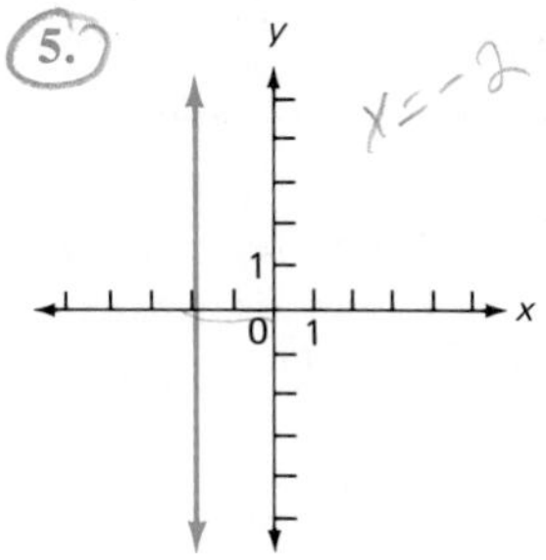

6.

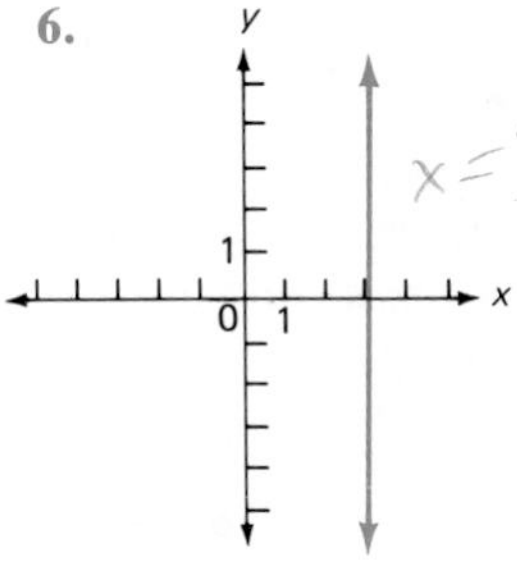

7.

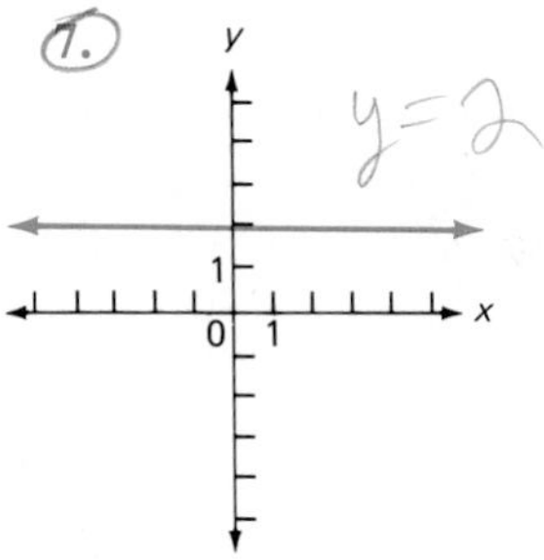

8.

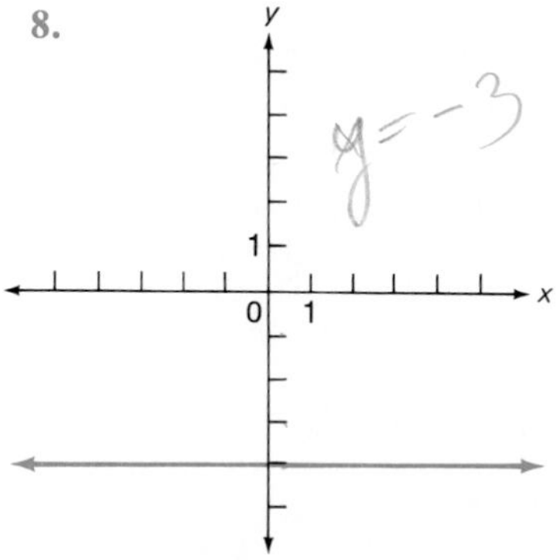

9.

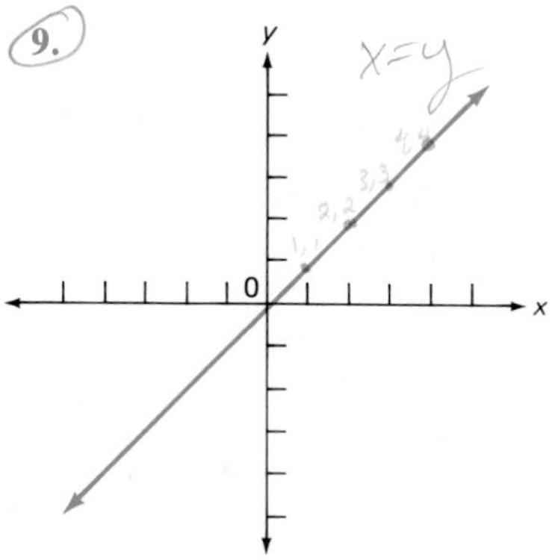

10.

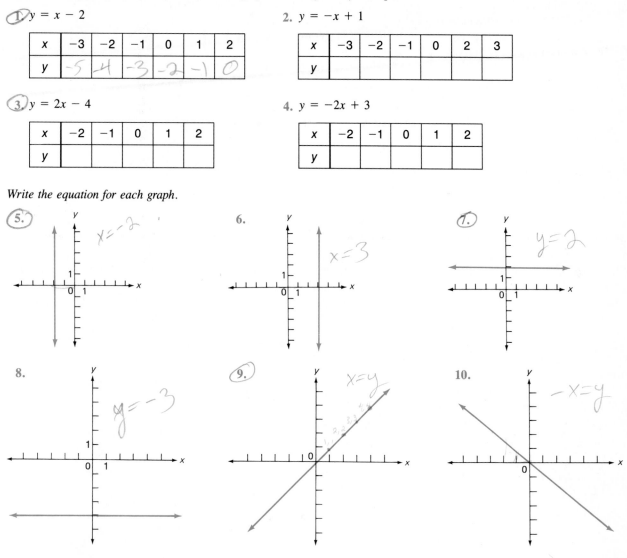

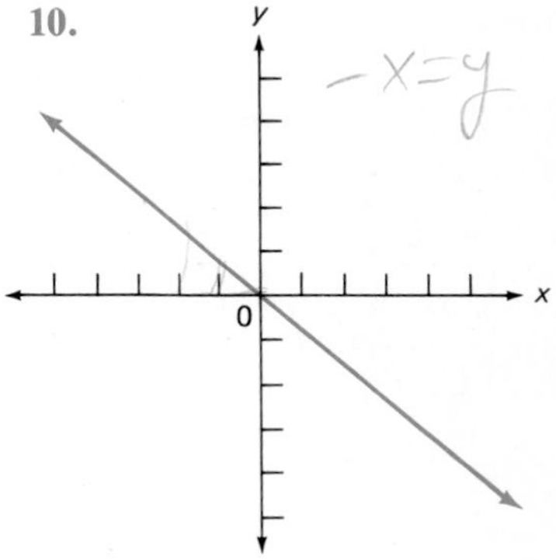

Name the x-and y-intercepts and use these to graph each of the following.

11. $x + 2y = 4$
12. $2x + y = 4$
13. $x - 2y = 4$
14. $2x - y = 4$
15. $2x - 3y = 6$
16. $3x + y = 6$
17. $y = -3x - 9$
18. $y + 2x = -5$
19. $y = 2x - 1$

20. Sketch the following on the same set of axes.
    (a) $y = x$
    (b) By adding 1 to each y-value (ordinate) in part (a), graph $y = x + 1$. In other words, *shift* each point of $y = x$ one unit up.
    (c) By subtracting 1 from each y-value in part (a), graph $y = x - 1$. That is, shift each point of $y = x$ one unit down.

21. Repeat Exercise 20 for **(a)** $y = -x$; **(b)** $y = -x + 1$; **(c)** $y = -x - 1$.

22. Sketch the following on the same set of axes.
   (a) $y = x$
   (b) By multiplying each y-value in part (a) by 2, graph $y = 2x$. In other words, *stretch* each y-value of $y = x$ to twice its size.
   (c) Graph $y = 2x + 3$ by shifting $y = 2x$ three units upward.

23. Sketch the following on the same set of axes.
   (a) $y = x$
   (b) By multiplying each y-value in part (a) by $\frac{1}{2}$, graph $y = \frac{1}{2}x$.
   (c) Graph $y = \frac{1}{2}x - 3$ by shifting $y = \frac{1}{2}x$ three units downward.

*24. Graph the equation $y = |x|$. (*Hint:* By definition, $|x| = x$ for $x \geq 0$ and $|x| = -x$ for $x < 0$.)

Written Assignment: Describe the graph of $Ax + By = C$ when **(a)** $A = 0$ and $B \neq 0$; **(b)** $A \neq 0$ and $B = 0$; **(c)** $A \neq 0$, $B \neq 0$, and $C = 0$; **(d)** $A = 0$, $B = 0$, and $C = 0$.

6.3 THE SLOPE OF A LINE

The following figure shows the graph of the linear equation $y = 2x - 4$, including the coordinates of four specific points. From the diagram you will see that the y-value increases 2 units each time that the x-value increases by 1 unit. The ratio of this change in y compared to the corresponding change in x is $\frac{2}{1} = 2$. Using the coordinates of the points (3, 2) and (2, 0), we have the following:

For convenience, the change y may be referred to as "Δy" (read "delta y"); the change in x is denoted as "Δx" (read "delta x").

$$\frac{\Delta y}{\Delta x} = \frac{2 - 0}{3 - 2} = \frac{2}{1} = 2$$

Show that the same ratio is obtained by using any other two points on the line, such as (1, −2) and (0, −4).

$$\frac{\text{change in } y\text{-values}}{\text{change in } x\text{-values}} = \frac{2 - 0}{3 - 2} = \frac{2}{1} = 2$$

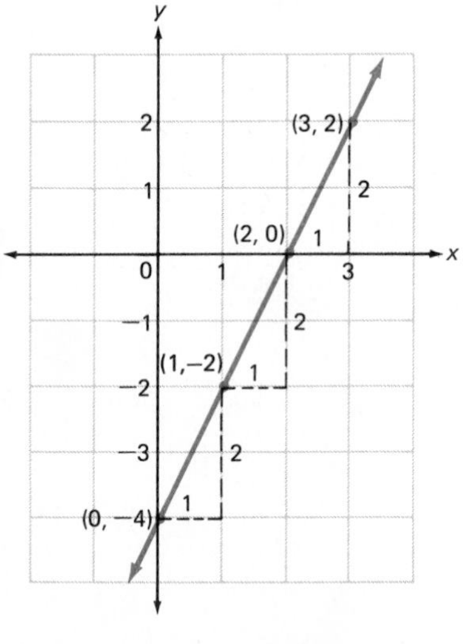

Let us find this ratio for the same line using (3, 2) and (1, − 2).

$$\frac{\Delta y}{\Delta x} = \frac{2 - (-2)}{3 - 1} = \frac{4}{2} = 2$$

Whether you evaluate this ratio using the points (3, 2) and (1, −2), or any other two points on the line, you will always find that the quotient is 2. We say that the **slope** of this line is 2, according to the following definition.

DEFINITION OF SLOPE

If two points (x_1, y_1) and (x_2, y_2) are on a line ℓ, then the slope m of line ℓ is defined by

$$m = \frac{y_2 - y_1}{x_2 - x_1}, \qquad x_2 \neq x_1$$

Notice that in the definition $x_2 - x_1$ cannot be zero; that is, $x_2 \neq x_1$. The only time that $x_2 = x_1$ is when the line is vertical.

In the following figure the coordinates of two points A and B have been labeled (x_1, y_1) and (x_2, y_2). The change in the y direction from A to B is given by the difference $y_2 - y_1$; the change in the x direction is $x_2 - x_1$. If a different pair of points is chosen, such as P and Q, then the ratio of these differences is still the same because the resulting triangles (ABC and PQR) are similar. That is, since corresponding sides of similar triangles are proportional, we have

This discussion shows that there can be only one slope for a given line.

$$m = \frac{y_2 - y_1}{x_2 - x_1} = \frac{AC}{CB} = \frac{PR}{RQ}$$

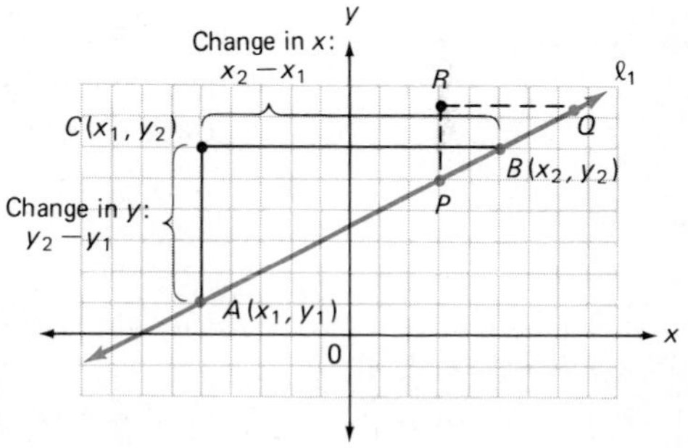

*Another descriptive language for slope is $m = \frac{\textbf{rise}}{\textbf{run}}$, where **rise** is the vertical change and **run** is the horizontal change.*

It may be helpful to think of the slope of a line in any of these ways:

$$m = \frac{y_2 - y_1}{x_2 - x_1} = \frac{\text{change in } y}{\text{change in } x} = \frac{\text{vertical change}}{\text{horizontal change}}$$

EXAMPLE 1 Find the slope of each line.

(a)

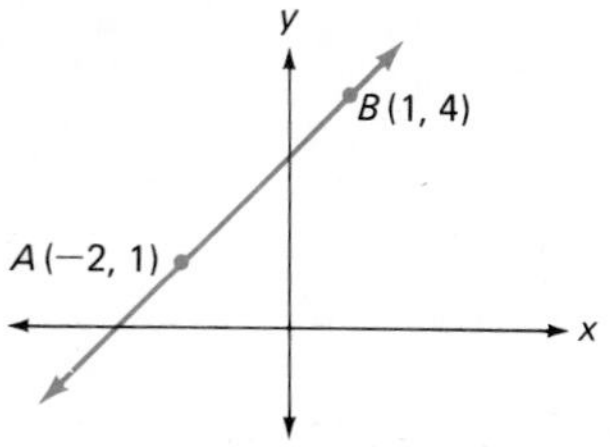

(b)

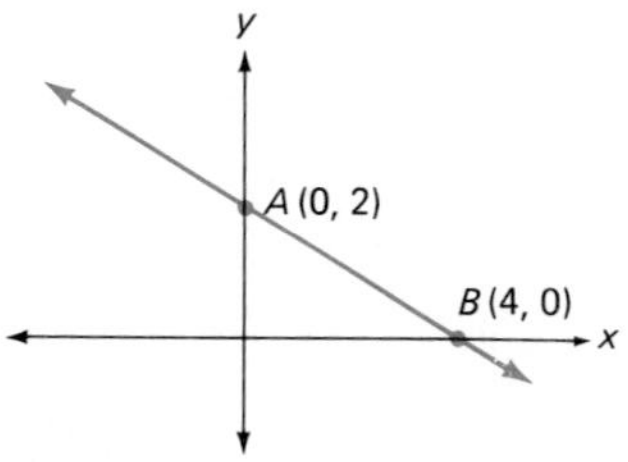

Note: It makes no difference which of the two points is called (x_1, y_1) or (x_2, y_2) since the ratio will be the same. Thus in Example 1(a), using (x_1, y_1) for B and (x_2, y_2) for A we have

$$m = \frac{1-4}{-2-1} = \frac{-3}{-3} = 1$$

Solution Using (x_1, y_1) for the coordinates of A and (x_2, y_2) for B, the slope formula gives:

(a) $$m = \frac{y_2 - y_1}{x_2 - x_1} = \frac{4 - 1}{1 - (-2)} = \frac{3}{3} = 1$$

(b) $$m = \frac{0 - 2}{4 - 0} = -\frac{1}{2}$$

■

CAUTION
Do not mix up the coordinates like this:

$$\frac{y_2 - y_1}{x_1 - x_2} = \frac{4 - (-6)}{1 - (-3)} = \frac{5}{2}$$

This is the ***negative*** *of the slope.*

Observe that the line in Example 1(a) is rising from left to right and its slope is positive. In Example 1(b) the line is falling from left to right and its slope is negative. In general, reading from left to right, rising lines have positive slope and falling lines have negative slope.

EXAMPLE 2 Find the slope of line ℓ through the points $(-3, 4)$ and $(1, -6)$.

Solution Use $(x_1, y_1) = (-3, 4)$; $(x_2, y_2) = (1, -6)$. Then

$$m = \frac{-6 - 4}{1 - (-3)} = -\frac{10}{4} = -\frac{5}{2}$$ ■

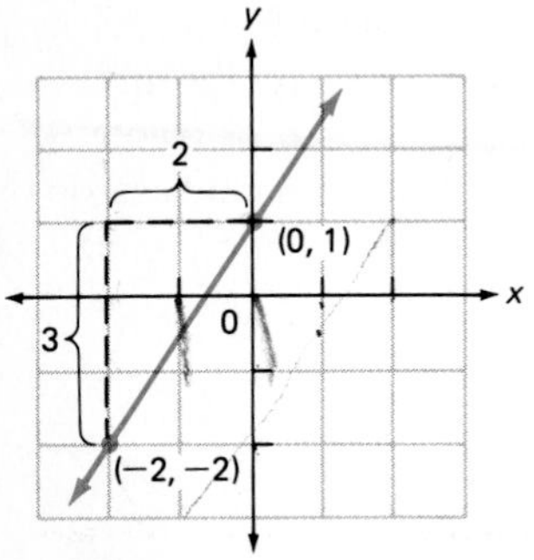

EXAMPLE 3 Graph the line with slope $\frac{3}{2}$ that passes through the point $(-2, -2)$.

Solution Think of $\frac{3}{2}$ as $\frac{\text{change in } y}{\text{change in } x}$. Now start at $(-2, -2)$ and move 3 units up and 2 units to the right. This locates the point $(0, 1)$. Draw the straight line through these two points. ■

TEST YOUR UNDERSTANDING
Think Carefully

Find the slopes of the lines determined by the given pairs of points:

1. $(1, 5)$; $(4, 6)$ **2.** $(3, -5)$; $(-3, 3)$

3. $(-2, -3)$; $(-1, 1)$ **4.** $(-1, 0)$; $(0, 1)$

Draw the line through the given point and with the given slope.

5. $(0, 0)$; $m = 2$ **6.** $(-3, 4)$; $m = -\frac{3}{2}$

(Answers: Page 284)

Next consider lines that are parallel to the axes, that is, horizontal and vertical lines.

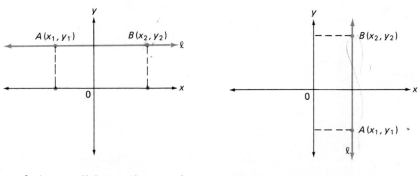

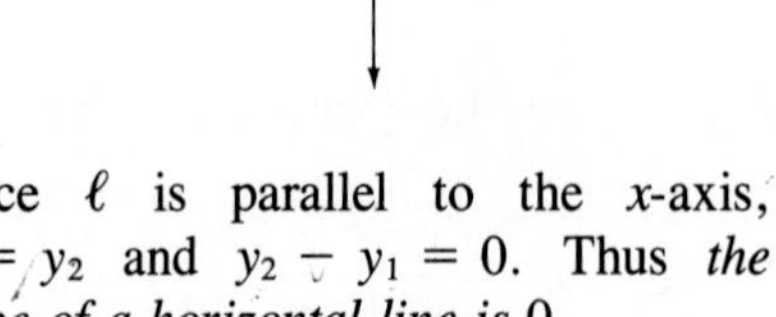

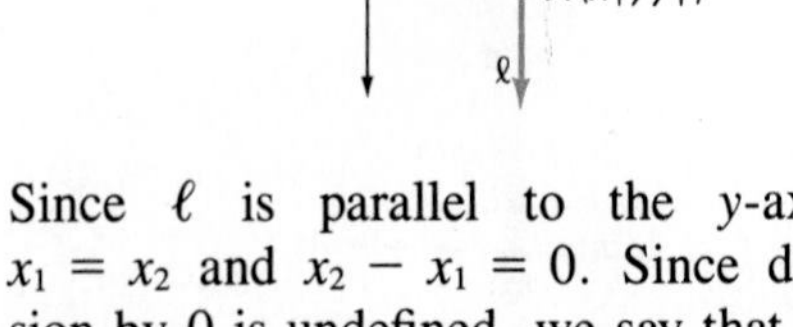

CAUTION
Do not confuse slope 0 for horizontal lines with undefined slope for vertical lines.

Since ℓ is parallel to the x-axis, $y_1 = y_2$ and $y_2 - y_1 = 0$. Thus *the slope of a horizontal line is* 0.

$$m = \frac{y_2 - y_1}{x_2 - x_1} = \frac{0}{x_2 - x_1} = 0$$

Since ℓ is parallel to the y-axis, $x_1 = x_2$ and $x_2 - x_1 = 0$. Since division by 0 is undefined, we say that *the slope of a vertical line is undefined;* that is, vertical lines do not have slopes.

Two nonvertical lines are parallel if and only if they have the same slope.

The slope property for perpendicular lines is not as obvious. The figure below suggests the following,

Two lines not parallel to the coordinate axes are perpendicular if and only if their slopes are negative reciprocals of one another.

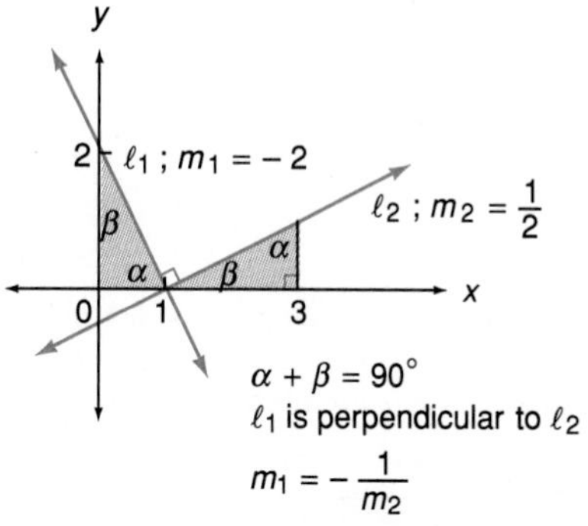

EXAMPLE 4 In the figure, line ℓ_1 has slope $\frac{2}{3}$ and is perpendicular to ℓ_2. If the lines intersect at $P\ (-1, 4)$, use the slope of ℓ_2 to find the coordinates of another point on ℓ_2.

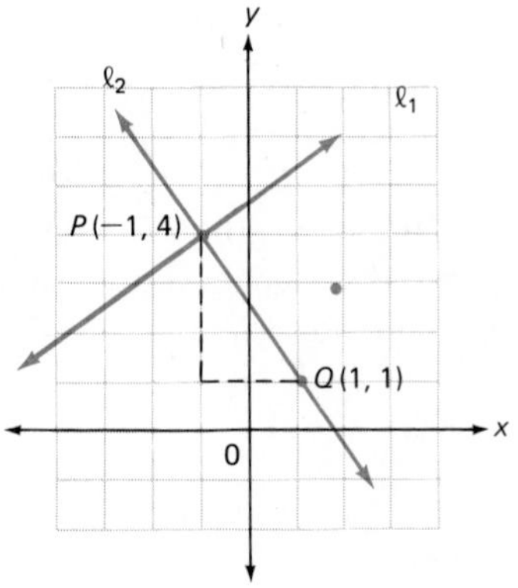

Solution Since the lines are perpendicular, the slope of ℓ_2 is $-\frac{3}{2}$. Now start at P and count 3 units downward and 2 units to the right to reach point (1, 1) on ℓ_2. Other solutions are possible. Can you locate a point on ℓ_2 that lies in the second quadrant? ■

In summary, the slope of a line is either positive, negative, zero, or undefined, as in the following figures.

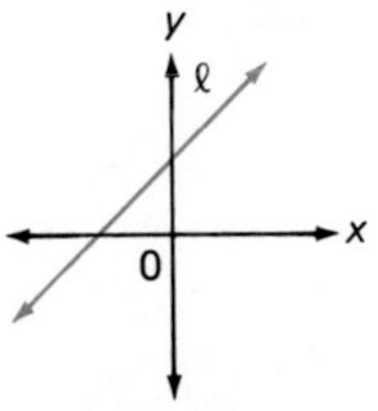

Line ℓ has a positive slope; ℓ is rising from left to right.

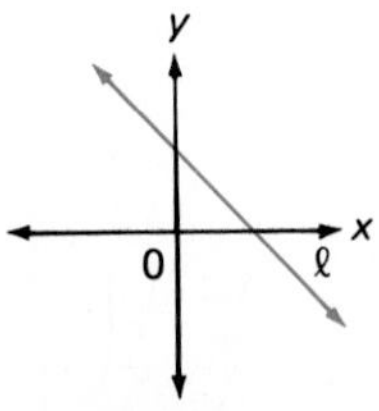

Line ℓ has a negative slope; ℓ is falling from left to right.

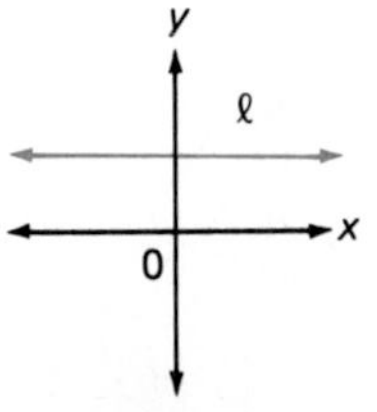

Line ℓ has a slope of 0; ℓ is horizontal.

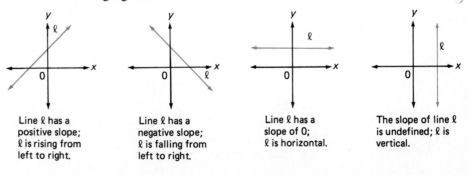

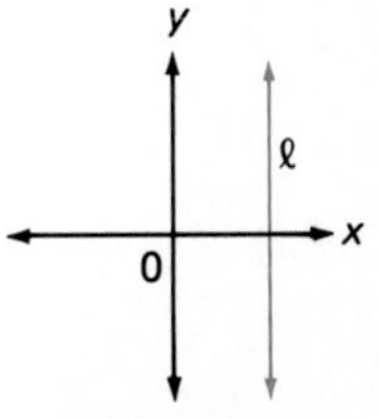

The slope of line ℓ is undefined; ℓ is vertical.

In business it is often noted that *demand* is related to the *selling price* of a commodity. That is, usually, the higher the price, the less the demand. Sometimes this relationship can be expressed as a linear equation having a *negative* slope. For example, suppose that for a certain calculator it is determined that the weekly demand, D, is related to the price x (in dollars) by this equation:

$$y = D = 500 - 20x \text{ where } 0 \leq x \leq 25$$

Note that the graph uses a different scale on each axis in order to accommodate the data shown.

From the graph we see that the demand is greatest when the price is zero dollars! The demand is zero when the price is $25.

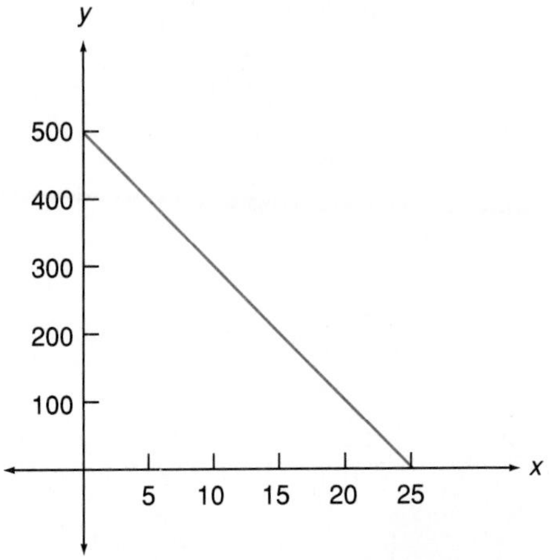

Now suppose that the weekly supply of calculators, S, is also dependent on the price, x, and is given by $y = S = 10x + 200$. When the graph of this supply equation is drawn in the same coordinate system as the demand equation, then the **equilibrium point** is the point where the two lines intersect. At this point, demand = supply for the same price x. To find this x-value, set $D = S$ and solve for the common value x.

$$500 - 20x = 10x + 200$$
$$300 = 30x$$
$$10 = x$$

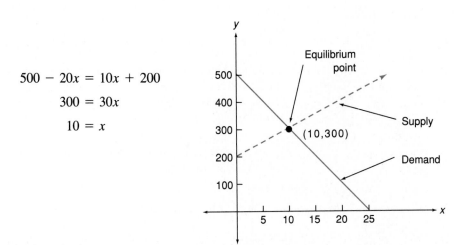

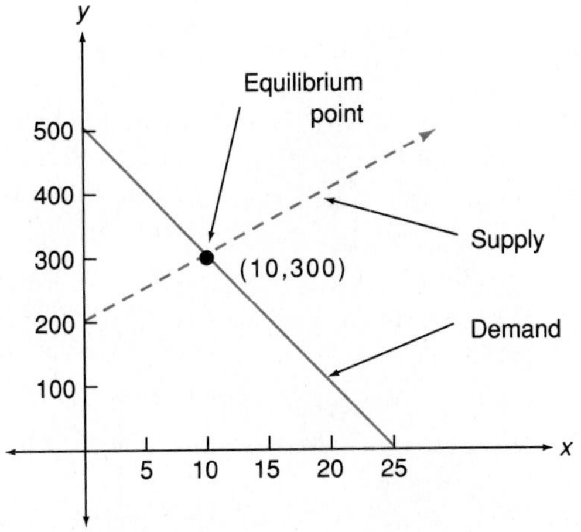

At $10 per calculator, 300 calculators will be supplied and sold per week. The result, $10, represents the price at which supply equals demand. The demand will be met, and no surplus will exist.

EXERCISES 6.3

1. Use the coordinates of each of the following pairs of points to find the slope of ℓ.
   (a) A, C
   (b) B, D
   (c) C, D
   (d) A, E
   (e) B, E
   (f) C, E

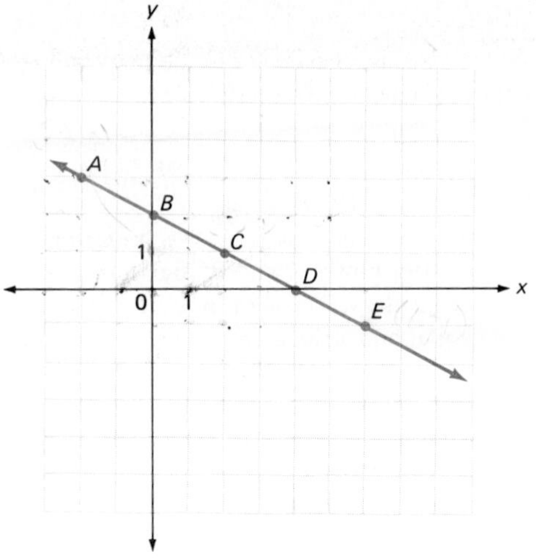

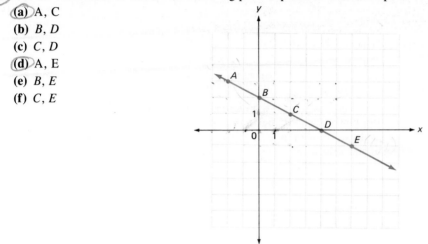

Find the slope of the line determined by the two given points.

2. (3, 2); (4, 5)
3. (1, 4); (3, 5)
4. (2, 0); (0, 2)
5. (−1, 2); (2, −1)
6. (0, −3); (2, −1)
7. (−2, 0); (0, −3)
8. (8, −5); (−3, 7)
9. $(0, \frac{1}{2})$; $(3, \frac{1}{2})$
10. $(-\frac{3}{2}, 5)$; $(2, -\frac{1}{2})$

In Exercises 11 through 20, draw the line through the given point having slope m.

11. (0, 0); $m = 2$
12. (0, 0); $m = -\frac{1}{2}$
13. (0, 2); $m = \frac{3}{4}$
14. $(-\frac{1}{2}, 0)$; $m = -1$
15. (−3, 4); $m = -\frac{1}{4}$
16. (3, −4); $m = 4$
17. $(-2, \frac{3}{2})$; $m = 0$
18. (1, 1); $m = 2$
19. (5, −3); slope undefined
20. $(-\frac{3}{4}, -\frac{1}{2})$; $m = 1$

21. In the same coordinate system draw five lines, each with slope −2, through points (−2, 0), (−1, 0), (0, 0), $(\frac{1}{2}, 0)$, and (1, 1), respectively.
22. Follow the instructions of Exercise 21 using $m = \frac{1}{3}$ and (2, 2), (−2, 2), (−2, −2), (2, −2), (0, 0).
23. In the same coordinate system draw the lines:
    (a) through (1, 0) with $m = -1$
    (b) through (0, 1) with $m = 1$
    (c) through (−1, 0) with $m = -1$
    (d) through (0, −1) with $m = 1$
24. Use the instructions of Exercise 23 for the following:
    (a) (1, 1); slope 0
    (b) $(0, \frac{1}{2})$; slope 0
    (c) (1, 1); slope $\frac{1}{2}$
    (d) $(2, \frac{1}{2})$; slope $\frac{1}{2}$
25. Why is the line determined by the points (6, −5) and (8, −8) parallel to the line through (−3, 12) and (1, 6)?
26. Verify that the points $A(1, 2)$, $B(4, -1)$, $C(2, -2)$, and $D(-1, 1)$ are the vertices of a parallelogram. Sketch the figure.
27. Tell whether or not the given pairs of lines are parallel or perpendicular. Explain your answer.
    (a) $3x - y = 6$
    $-6x + 2y = 5$
    (b) $2x + 5y = 10$
    $5x - 2y = 10$

28. Consider the four points $P(5, 11)$, $Q(-7, 16)$, $R(-12, 4)$, and $S(0, -1)$. Show that the four angles of the quadrilateral $PQRS$ are right angles. Also show that the diagonals are perpendicular.

29. Find t if the line through $(-1, 1)$ and $(3, 2)$ is parallel to the line through $(0, 6)$ and $(-8, t)$.

30. Find t if the line through $(-1, 1)$ and $(1, \frac{1}{2})$ is perpendicular to the line through $(1, \frac{1}{2})$ and $(7, t)$. Use the fact that two perpendicular lines have slopes that are negative reciprocals of one another.

31. Lines ℓ_1 and ℓ_2 are perpendicular and intersect at point $(-2, -6)$. ℓ_1 has slope $-\frac{2}{5}$. Use the slope of ℓ_2 to find the y-intercept of ℓ_2.

*32. **(a)** Prove that nonvertical parallel lines have equal slopes by considering two parallel lines ℓ_1, ℓ_2 as in the figure. On ℓ_1 select points A and B, and choose A' and B' on ℓ_2. Now form the appropriate right triangles ABC and $A'B'C'$ using points C and C' on the the x-axis. Prove they are similar and write a proportion to show the slopes of ℓ_1 and ℓ_2 are equal.

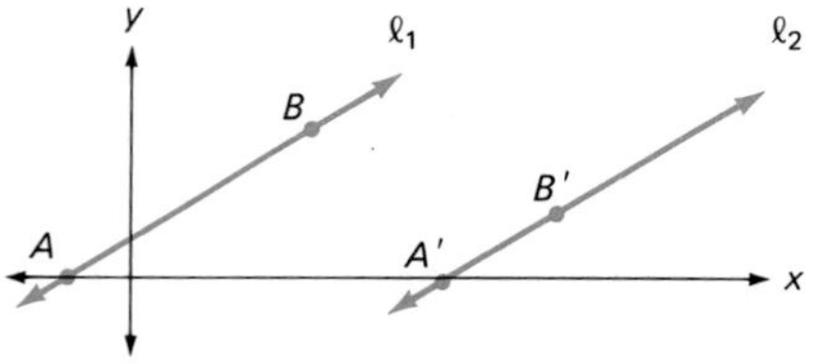

(b) Why is the converse of the fact given in part (a) also true?

33. The monthly demand, D, for a certain portable radio is related to the price, x, by the equation $D = 5000 - 25x$. The monthly supply is given by the formula $S = 10x + 275$. At what price will the supply equal the demand? How many radios will be sold at this price?

34. For a certain type of toy, the weekly demand, D, is related to the price, x, in dollars, by the equation $y = D = 1000 - 40x$. The weekly supply is given by the formula $y = S = 15x + 340$. Find the price at which the supply equals the demand. Graph both equations on the same set of axes and show the equilibrium point.

Written Assignment: Any horizontal line is perpendicular to any vertical line. Why were such lines excluded from the result, which states that lines are perpendicular if and only if their slopes are negative reciprocals?

CHALLENGE
Think Creatively

Use the following to *prove* that if two lines are perpendicular, then they have slopes that are negative reciprocals of one another.

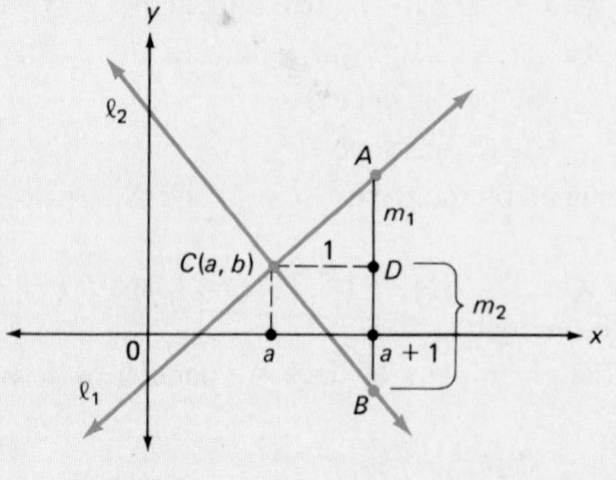

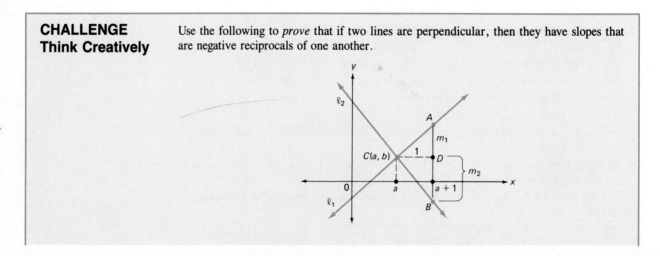

Let line ℓ_1 be perpendicular to line ℓ_2 at the point $C(a, b)$. Use m_1 for the slope of ℓ_1, and m_2 for ℓ_2. We want to show the following:

$$m_1 m_2 = -1 \quad \text{or} \quad m_1 = -\frac{1}{m_2}$$

Add 1 to the x-coordinate a of point C and draw the vertical line through $a + 1$ on the x-axis. This vertical line will meet ℓ_1 at some point A and ℓ_2 at some point B, forming right triangle ABC with right angle at C. Draw the perpendicular from C to AB meeting AB at D. Then CD has length 1.

(a) Using the right triangle CDA, show that $m_1 = DA$.

(b) Show that $m_2 = DB$. Is m_2 positive or negative?

(c) For right triangle ABC, CD is the mean proportional between segments BD and DA on the hypotenuse. Use this fact to conclude that $\frac{m_1}{1} = \frac{1}{-m_2}$, or $m_1 m_2 = -1$.

6.4 ALGEBRAIC FORMS OF A LINE

One algebraic form of a line is the general form $Ax + By = C$ that was introduced earlier. Other forms will be developed in this section.

Pictured below is a line with slope m and y-intercept b. To find the equation of ℓ we begin by considering any point $P(x, y)$ on ℓ other than $(0, b)$.

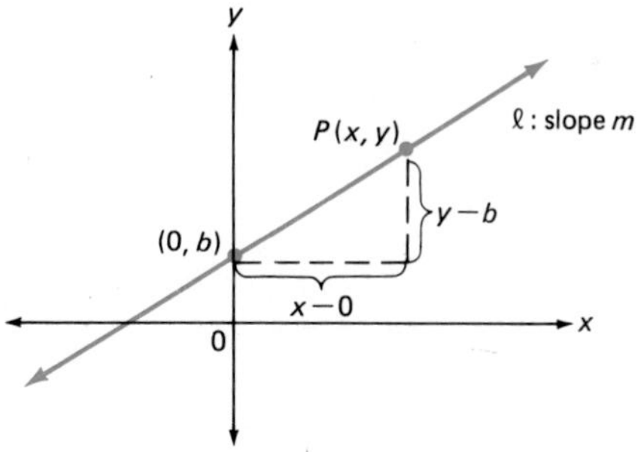

Since the slope of ℓ is given by any two of its points, we may use $(0, b)$ and (x, y) to write

$$m = \frac{y - b}{x - 0} = \frac{y - b}{x}$$

If both sides are multiplied by x, then

$$y - b = mx$$

or

$$\boldsymbol{y = mx + b} \longleftarrow \quad \text{(Note that the point } (0, b) \text{ also satisfies this final form.)}$$

This leads to the following **y-form of the equation of a line.**

A point (x, y) is on this line if and only if the coordinates satisfy this equation.

SLOPE-INTERCEPT FORM OF A LINE

$$y = mx + b$$

where m is the slope and b is the y-intercept.

EXAMPLE 1 Write, in slope-intercept form, the equation of the line with slope $\frac{2}{3}$ passing through the point $(0, -5)$. Show that the point $(3, 2)$ is *not* on this line.

Solution Since $m = \frac{2}{3}$ and $b = -5$, the slope-intercept form $y = mx + b$ gives

$$y = \frac{2}{3}x + (-5)$$

$$= \frac{2}{3}x - 5$$

To show that $(3, 2)$ is not on this line, verify that the coordinates do *not* satisfy the equation:

$$y = \frac{2}{3}x - 5$$

$$2 \stackrel{?}{=} \frac{2}{3}(3) - 5 \qquad \text{(Let } x = 3 \text{ and } y = 2.)$$

$$2 \neq -3$$

*In general, if for a point (x, y), $y \neq mx + b$, then the point (x, y) is **not** on the given line.*

EXAMPLE 2 A horizontal line has y-intercept 5. Write an equation for the coordinates of any point (x, y) on the line.

Solution The slope is 0. Then for (x, y) on the line, the slope-intercept form gives

$$y = 0(x) + 5 \quad \text{or} \quad y = 5$$

A special case of $y = mx + b$ is obtained when $m = 0$. Then

$$y = 0(x) + b \quad \textit{or} \quad y = b$$

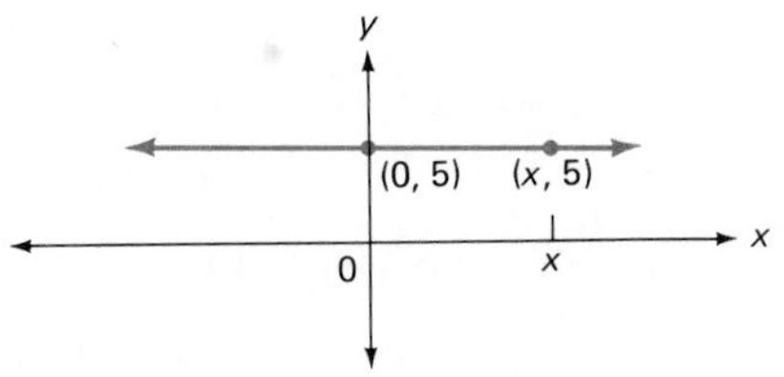

EXAMPLE 3 Use the slope-intercept form to sketch the line $y = -\frac{1}{2}x + \frac{3}{2}$.

Solution Using the slope-intercept form, we note that the slope is $-\frac{1}{2}$ and the y-intercept is $\frac{3}{2}$. Now locate point $(0, \frac{3}{2})$, move 2 units to the right and 1 unit down to reach $(2, \frac{1}{2})$. Connect $(0, \frac{3}{2})$ and $(2, \frac{1}{2})$ to get the graph of ℓ.

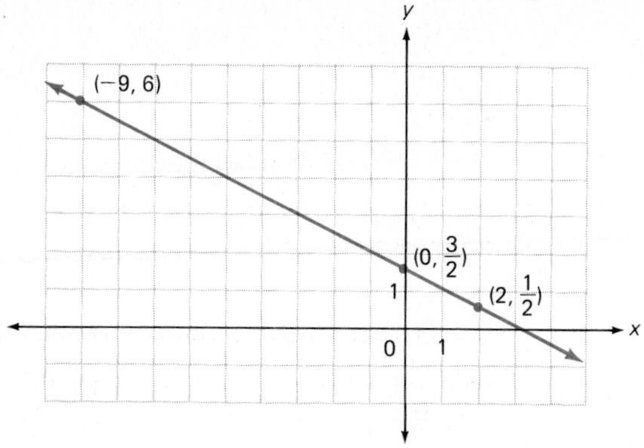

■

When the equation of a nonvertical line is not in slope-intercept form, we can find the slope and the y-intercept by first converting the equation into this form. Consider, for example, the equation $3x + 5y = 10$ and solve for y:

To write a linear equation in slope-intercept form, solve the equation for y.

$$3x + 5y = 10$$

$$\frac{3x}{5} + \frac{5y}{5} = \frac{10}{5}$$

$$\frac{3x}{5} + y = 2$$

$$y = -\frac{3}{5}x + 2$$

Thus the slope of the line is $-\frac{3}{5}$ and the y-intercept is 2.

In general we have the following:

In practice it is easier to solve Ax + By = C for y to find the slope and y-intercept than it is to memorize this result.

> The general equation $Ax + By = C$ $(B \neq 0)$ can be written in the form
>
> $$y = -\frac{A}{B}x + \frac{C}{B}$$
>
> and represents a line with slope $-\dfrac{A}{B}$ and y-intercept $\dfrac{C}{B}$.

TEST YOUR UNDERSTANDING
Think Carefully

Write each equation in slope-intercept form; give the slope and y-intercept.

1. $2x + y = 6$	**2.** $3x - y = 5$	**3.** $x + y = 2$
4. $x + 2y = 8$	**5.** $x - 2y = 4$	**6.** $2x - 4y = 8$
7. $3x - 2y = 5$	**8.** $3x = y + 5$	**9.** $2y = 7 - x$
10. $2y - 4 = 0$	**11.** $2x - 2y = \frac{1}{2}$	**12.** $3x - \frac{1}{2}y = 1$

(Answers: Page 284)

Now let ℓ be a line with slope m that passes through a specific point (x_1, y_1). We wish to determine the conditions on the coordinates of any point $P(x, y)$ that is on the line ℓ, as shown on the following page.

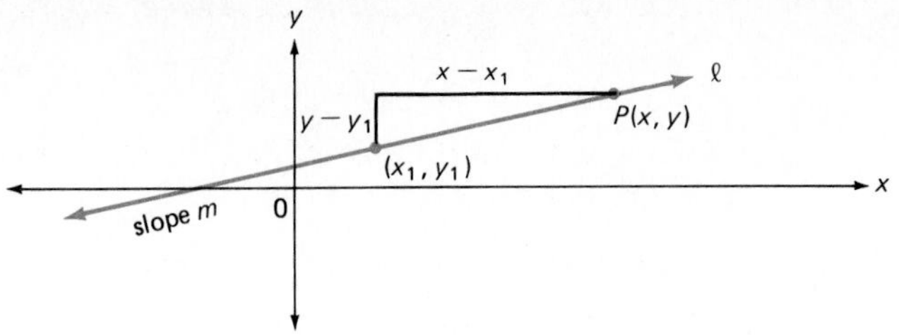

From the figure you can see that $P(x, y)$ will be on ℓ if and only if the ratio $\frac{y - y_1}{x - x_1}$ is the same as m. That is, P is on ℓ if and only if

$$m = \frac{y - y_1}{x - x_1}$$

Multiply both sides of this equation by $x - x_1$.

$$m(x - x_1) = y - y_1$$

This leads to another form for the equation of a straight line.

A point (x, y) is on this line if and only if the coordinates satisfy this equation.

POINT-SLOPE FORM OF A LINE

$$y - y_1 = m(x - x_1)$$

where m is the slope and (x_1, y_1) is a point on the line.

EXAMPLE 4 Write the point-slope form of the line ℓ with slope $m = 3$ that passes through the point $(-1, 1)$. Verify that $(-2, -2)$ is on the line.

Solution Since $m = 3$, and (x, y) on ℓ satisfies this equation where $x_1 = -1$ and $y_1 = 1$:

CAUTION
Pay attention to the minus signs on the coordinates when used in the point-slope form. Note the substitution of $x_1 = -1$ in this example.

$$y - 1 = 3[x - (-1)]$$
$$y - 1 = 3(x + 1)$$

To verify that $(-2, -2)$ is on the line, let $x = -2$:

$$y - 1 = 3(-2 + 1) = -3$$
$$y = -2$$

Thus $(-2, -2)$ is on the line. ■

The point-slope form can be used to write the equation of a line when the slope and any point of the line are given. When two points of the line are given, the slope can be found first and then the point-slope form used, as in the following example.

EXAMPLE 5 Find the equation of the line through the points (2, −3) and (3, −1). Write the equation:
(a) In point-slope form **(b)** In slope-intercept form
(c) In the form $Ax + By = C$.

Solution

(a) The slope m is $\dfrac{-1-(-3)}{3-2} = 2$. Use this slope and *either* point, such as (2, −3).

$$y - y_1 = m(x - x_1): \quad y - (-3) = 2(x - 2) \quad \text{point-slope form}$$

$$\text{or} \quad y + 3 = 2(x - 2)$$

(b) Use the solution for part (a) and solve for y.

$$y = mx + b: \quad y + 3 = 2(x - 2)$$
$$y + 3 = 2x - 4$$
$$y = 2x - 7 \quad \text{slope-intercept form}$$

(c) Rewrite the solution for part (b).

The form $Ax + By = C$ is not unique. For example, multiplying $2x - y = 7$ by ±3 gives the equivalent forms $6x - 3y = 21$ and $-6x + 3y = -21$.

$$Ax + By = C: \quad y = 2x - 7$$
$$-2x + y = -7 \quad \text{the form } Ax + By = C$$
$$\text{or} \quad 2x - y = 7$$

Note that all three forms are equivalent equations for the given line through (2, −3) and (3, −1). Show that $x = 3$ and $y = -1$ satisfies each form. ■

EXAMPLE 6 Write the equation of the line that is perpendicular to the line $5x - 2y = 2$ and that passes through the point (−2, −6).

Solution First find the slope of the given line by writing it in slope-intercept form.

$$5x - 2y = 2$$
$$-2y = -5x + 2$$
$$y = \tfrac{5}{2}x - 1 \quad \text{The slope is } \tfrac{5}{2}.$$

Recall that two perpendicular lines have slopes that are negative reciprocals of one another.

The perpendicular line has slope $-\dfrac{1}{\frac{5}{2}} = -\dfrac{2}{5}$. Since this line also goes through $P(-2, -6)$, the point-slope form gives

$$y + 6 = \frac{-2}{5}(x + 2)$$

$$y = -\frac{2}{5}x - \frac{34}{5}$$ ■

Following is a summary of the algebraic forms of a line that we have explored.

Slope-intercept form	Point-slope form	General linear equation
$y = mx + b$	$y - y_1 = m(x - x_1)$	$Ax + By = C$
Line with slope m and y-intercept b.	Line with slope m and through the point (x_1, y_1).	Line with slope $-\frac{A}{B}$ and y-intercept $\frac{C}{B}$, if $B \neq 0$.

CAUTION: Learn to Avoid These Mistakes

WRONG	RIGHT
The slope of the line through (2, 3) and (5, 7) is $m = \frac{7-3}{2-5} = -\frac{4}{3}$	The slope of the line through (2, 3) and (5, 7) is $m = \frac{7-3}{5-2} = \frac{4}{3}$
The slope of the line $2x - 3y = 7$ is $-\frac{3}{2}$.	The slope is $\frac{2}{3}$.
The line through $(-4, -3)$ with slope 2 is $y - 3 = 2(x - 4)$.	The equation is $y - (-3) = 2[x - (-4)]$ or $y + 3 = 2(x + 4)$.

slope y intercept
y= mx+b
5 3/5

EXERCISES 6.4

Write each equation in slope-intercept form; give the slope and the y-intercept.

1. $3x + y = 4$
2. $2x - y = 5$
3. $x + y = 5$
4. $4x + 2y = 10$
5. $6x - 3y = 1$
6. $5x - y = 8$
7. $3y - 5 = 0$
8. $2y = 4x + 7$
9. $x = y - 3$
10. $x + \frac{1}{2}y = 6$
11. $\frac{1}{2}x + y = 3$
12. $\frac{1}{4}x - \frac{1}{2}y = 1$

Write the equation of the line with the given slope m and y-intercept b.

13. $m = 2, b = 3$
14. $m = -2, b = 1$
15. $m = 1, b = 1$
16. $m = -1, b = 2$
17. $m = 0, b = 5$
18. $m = 0, b = -5$
19. $m = \frac{1}{2}, b = 3$
20. $m = -\frac{1}{2}, b = 2$
21. $m = \frac{1}{4}, b = -2$

Write the equation of the line passing through the given point (x, y) and with the given slope m. Use the point-slope form of a line.

22. $(3, 4); m = 2$
23. $(2, 3); m = 1$
24. $(1, 2); m = 3$
25. $(-2, 3); m = 4$
26. $(1, -3); m = -1$
27. $(-1, -2); m = \frac{1}{2}$
28. $(1, -2); m = 0$
29. $(-2, -2); m = -2$
30. $(4, -5); m = -\frac{1}{2}$

Write the equation of the line through the two given points in the form Ax + By = C.

x+y=0

31. (2, 3), (3, 2)
32. (−1, 2), (2, −1)
33. (3, 0), (0, −3)
34. (1, 1), (−1, −1)
35. (3, −4), (0, 0)
36. (−2, 3), (5, 0)
37. (−2, −2), (1, −3)
38. (4, −2), (−2, 4)
39. (−5, 5), (5, −5)

Write the equation of the line that is perpendicular to the given line and passes through the indicated point.

40. $y = -10x$; $(0, 0)$ **41.** $y = 3x - 1$; $(4, 7)$

42. $3x + 2y = 6$; $(6, 7)$ **43.** $y - 2x = 5$; $(-5, 1)$

44. Write the equation of the line parallel to $y = -3x - 6$ and with y-intercept 6.

45. Write the equation of the line parallel to $2x + 3y = 6$ that passes through the point $(1, -1)$.

46. Two lines, parallel to the coordinate axes, intersect at the point $(5, -7)$. What are their equations?

47. Find t such that $(2, 3)$, $(5, 8)$, and $(7, t)$ are collinear.

48. Four vertices of a parallelogram are located at $(-1, 1)$, $(0, 3)$, $(3, 3)$, and $(2, 1)$. Write the equations for the sides of the parallelogram.

49. In Exercise 48, write the equations of the two diagonals of the parallelogram.

50. The vertices of a rectangle are located at $(2, 2)$, $(6, 2)$, $(6, -3)$, and $(2, -3)$. What is the relationship between the slopes of the diagonals?

51. The vertices of a square are located at $(2, 2)$, $(5, 2)$, $(5, -1)$, and $(2, -1)$. What is the relationship between the slopes of the diagonals?

52. Any line having a nonzero slope that does not pass through the origin always has both an x- and a y-intercept. Let ℓ be such a line having equation $ax + by = c$.

(a) Why is $c \neq 0$?

(b) What are the x- and y-intercepts?

(c) Derive the equation $\frac{x}{p} + \frac{y}{q} = 1$, where p and q are the x- and y-intercepts, respectively. This is known as the *intercept form* of a line. (See part (d).)

(d) Use the intercept form to write the equation of the line passing through $(\frac{3}{2}, 0)$ and $(0, -5)$.

(e) Use the two points in part (d) to find the slope, write the slope-intercept form, and compare with the result in part (d).

***53.** Replace m in the point-slope form of a line through points (x_1, y_1) and (x_2, y_2) by $\frac{y_2 - y_1}{x_2 - x_1}$. Show that this gives the *two-point form* for the equation of a line:

$$\frac{y - y_1}{y_2 - y_1} = \frac{x - x_1}{x_2 - x_1}$$

54. Use the result of Exercise 53 to find the equation of the line through $(-2, 3)$ and $(5, -2)$. Write the equation in point-slope form.

Written Assignment: Select a point and a non-zero slope of your own choice. Then describe the various algebraic forms that can be used to identify the line through your given point and with your given slope.

6.5 THE DISTANCE AND MIDPOINT FORMULAS

One of the oldest and most important results in all of geometry is the **Pythagorean theorem.** This theorem, when adapted to the Cartesian plane, will allow us to find the length of any line segment. Consider a right triangle ABC, with right angle at C. The Pythagorean theorem states that the square of the hypotenuse is equal to the sum of the squares of the other two sides of the triangle. (See Exercise 23.)

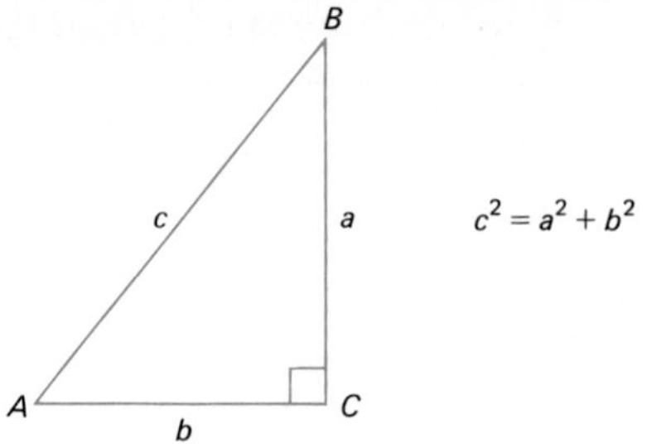

Now consider a line segment through the points $A(x_1, y_1)$ and $B(x_2, y_2)$ as in the figure below. Locate a third point C to complete the right triangle ABC.

Note that the diagram was set up so that $x_2 - x_1 > 0$ and $y_2 - y_1 > 0$. Other situations may have negative values, but it makes no difference because $(x_2 - x_1)^2 = (x_1 - x_2)^2$ and $(y_2 - y_1)^2 = (y_1 - y_2)^2$.

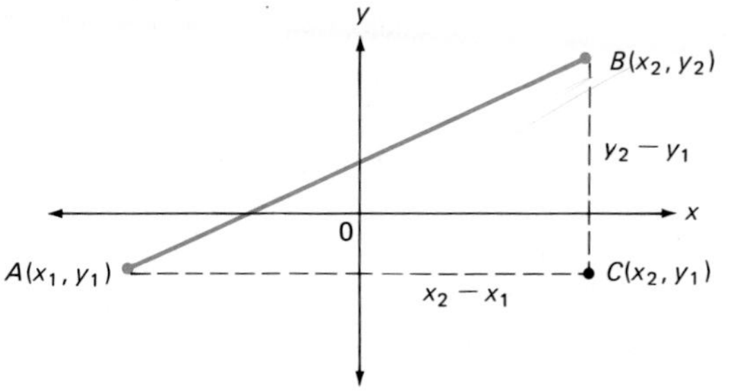

Since AB is the hypotenuse of the right triangle ABC, the Pythagorean theorem gives

$$(AB)^2 = (AC)^2 + (CB)^2$$

If A and B are on the same horizontal line, then $y_1 = y_2$ and $AB = \sqrt{(x_1 - x_2)^2} = |x_1 - x_2|$ (see page 186).

But $AC = x_2 - x_1$ and $CB = y_2 - y_1$. Thus

$$(AB)^2 = (x_2 - x_1)^2 + (y_2 - y_1)^2$$

Taking the positive square root produces the **distance formula.**

> DISTANCE FORMULA
>
> The distance d between any two points (x_1, y_1) and (x_2, y_2) is
>
> $$d = \sqrt{(x_2 - x_1)^2 + (y_2 - y_1)^2}$$

EXAMPLE 1 What is the distance between P and Q?

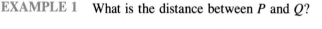

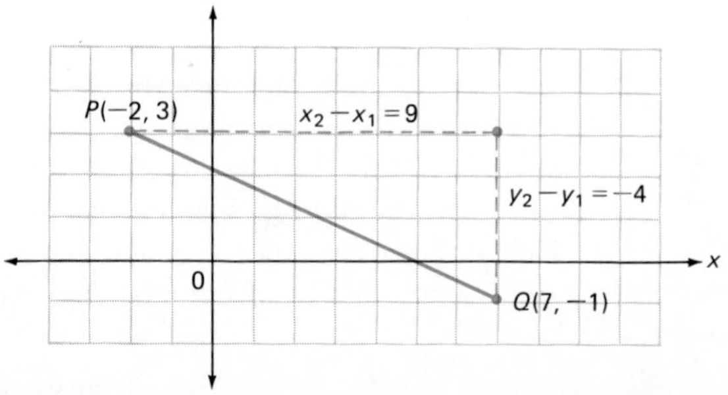

Solution Using $(-2, 3)$ and $(7, -1)$ for (x_1, y_1) and (x_2, y_2), respectively, the distance formula gives

$$\begin{aligned} PQ &= \sqrt{[7-(-2)]^2 + (-1-3)^2} \\ &= \sqrt{(9)^2 + (-4)^2} \\ &= \sqrt{97} \\ &= 9.85 \qquad \text{(approximately)} \end{aligned}$$

Use a calculator to obtain this approximation.

■

The **converse** of the Pythagorean theorem is also true. That is, if the sides of a triangle satisfy $a^2 + b^2 = c^2$, then the triangle is a right triangle with hypotenuse c. The next example uses this result.

EXAMPLE 2 Show that the points $P(3, -11)$, $Q(-6, -2)$, $R(8, 12)$ are the vertices of a right triangle.

Solution

$$\begin{aligned} (PQ)^2 &= (-6-3)^2 + (-2+11)^2 = 162 \\ (QR)^2 &= (-6-8)^2 + (-2-12)^2 = 392 \\ (PR)^2 &= (8-3)^2 + (12+11)^2 = 554 \end{aligned}$$

Thus PQR is a right triangle (with right angle at Q) by the Pythagorean theorem, since

$$(PR)^2 = 554 = 162 + 392 = (PQ)^2 + (QR)^2$$

■

TEST YOUR UNDERSTANDING
Think Carefully

(Answers: Page 284)

Find the distance between the given points.

1. (1, 1), (5, 4) **2.** (−3, 2), (5, 8)
3. (−2, 0), (3, −12) **4.** (0, 1), (1, 0)

The coordinate of the *midpoint of a line segment* on a number line is found by averaging the coordinates of the endpoints. This is shown for various segments AB in the following figure.

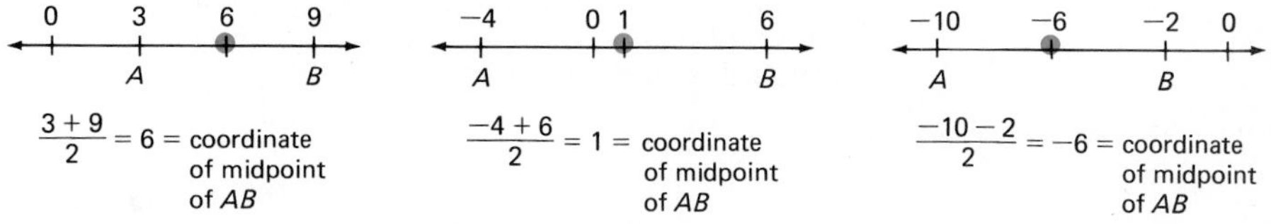

If x_1 and x_2 are used for the coordinates of the endpoints of a segment, then $\frac{x_1 + x_2}{2}$ is the coordinate of the midpoint. Now let PQ represent any segment in the

Cartesian plane and let M denote its midpoint, as illustrated below. Use (x', y') as the coordinates of M.

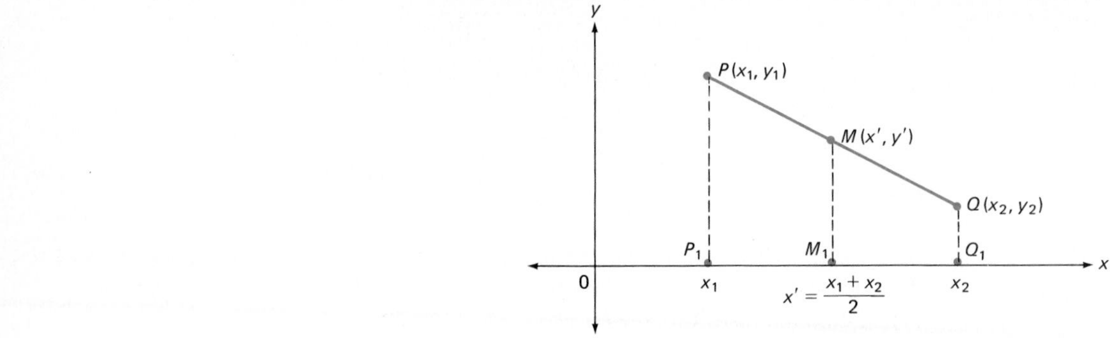

Draw vertical lines through P, M, and Q intersecting the x-axis at points P_1, M_1, and Q_1, respectively. Lines PP_1, MM_1, and QQ_1 are parallel. Lines PQ and P_1Q_1 are transversals cut by these three parallel lines. Parallel lines divide all transversals into proportional segments. Since M divides PQ in half, it follows that M_1 divides P_1Q_1 in half. Thus M_1 corresponds to $\frac{x_1 + x_2}{2}$ on the x-axis, which is also the first coordinate of M. Hence

$$x' = \frac{x_1 + x_2}{2}$$

A similar argument produces

$$y' = \frac{y_1 + y_2}{2}$$

In summary, we have the following **midpoint formula:**

MIDPOINT FORMULA

The midpoint of the line segment between (x_1, y_1) and (x_2, y_2) has coordinates

$$\left(\frac{x_1 + x_2}{2}, \frac{y_1 + y_2}{2}\right)$$

EXAMPLE 3 Find the midpoint of the segment between $(-2, -3)$ and $(6, 2)$.

Solution Use $(x_1, y_1) = (-2, -3)$, and $(x_2, y_2) = (6, 2)$. Then the midpoint has the following coordinates:

$$x' = \frac{-2 + 6}{2} = 2 \quad \text{and} \quad y' = \frac{-3 + 2}{2} = \frac{-1}{2} = -\frac{1}{2}$$ ■

EXAMPLE 4 Show that $A(2, 2)$, $B(5, 5)$, $C(8, 2)$, $D(5, -1)$ are the vertices of a square and verify that the diagonals of that square bisect each other.

Solution

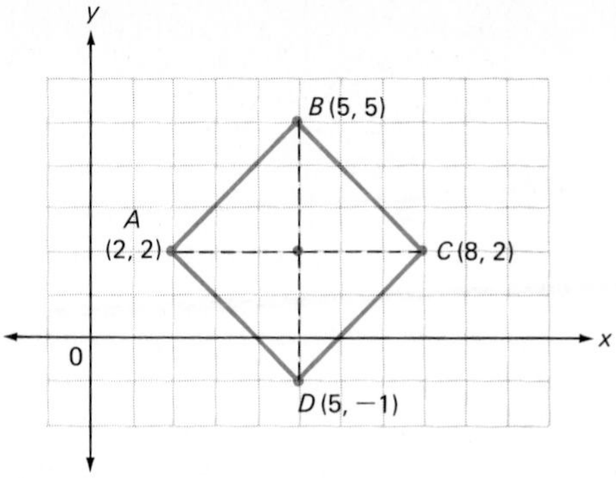

All sides have length $\sqrt{18}$. For example:

$$DC = \sqrt{(8-5)^2 + [2-(-1)]^2} = \sqrt{18}$$

Use the slope property of perpendicular lines as another way to show that the four angles are right angles.

The angles at A, B, C, D are right angles. This can be verified using the Pythagorean theorem. That is, ABC is a right triangle with right angle at B because

$$(AC)^2 = 6^2 = 18 + 18 = (AB)^2 + (BC)^2$$

The midpoints of diagonals AC and BD have coordinates as follows:

$$AC: \quad \left(\frac{8+2}{2}, \frac{2+2}{2}\right) = (5, 2)$$

$$BD: \quad \left(\frac{5+5}{2}, \frac{5+(-1)}{2}\right) = (5, 2)$$

Thus AC and BD intersect at (5, 2), their common midpoint. ■

EXERCISES 6.5

Find the length and midpoint of the line segments with the given endpoints.

1. (0, 0), (3, 4)
2. (−9, 3), (−6, 7)
3. (3, 8), (8, −4)
4. (−2, 0), (−7, −12)
5. $(-2, \frac{5}{4})$, $(10, \frac{5}{4})$
6. $(-2, \frac{5}{4})$, $(-2, 1)$
7. (−3, 5), (1, 7)
8. $(\frac{1}{2}, 4)$, $(\frac{9}{2}, -8)$
9. $(\sqrt{2}, 4\sqrt{2})$, $(3\sqrt{2}, -10\sqrt{2})$
10. (3, −5), (5, −3)
11. Verify that each of the following sets of three points determine a right triangle:
 (a) $P(1, 1)$, $Q(3, 2)$, $R(2, -1)$ **(b)** $P(5, -4)$, $Q(-2, 4)$, $R(-10, -3)$
12. Find the perimeter of the triangle determined by the points (−4, −5), (2, 3), and (−4, 3).
13. Show that $P(4, -3)$, $Q(5, 1)$, $R(2, 4)$, and $S(3, 8)$ are the vertices of a parallelogram by comparing lengths and slopes of the opposite sides.
14. What is the point of intersection of the diagonals of the parallelogram in Exercise 13? (*Hint*: The diagonals bisect each other.)
15. Show that the midpoints of the sides of the parallelogram in Exercise 13 form another parallelogram.

16. **(a)** Show that $A(4, -3)$, $B(8, 9)$, $C(16, 1)$ are the vertices of an isosceles triangle.
(b) Find the midpoint of the base of the triangle in part (a).
(c) How long is the altitude from the vertex to the base in the triangle in part (a)?

17. If $A(-1, -1)$, $B(2, -5)$, $C(3, 2)$ are three vertices of a square, find the fourth vertex D. (*Hint:* Slope CD = slope AB.)

18. **(a)** $Q(-6, -1)$ is the midpoint of a line segment with endpoints $P(-14, -4)$ and R. Find the coordinates of R.
(b) Find the coordinates of R if $P(-14, -4)$ is the midpoint of segment RQ and Q has coordinates $(-6, -1)$.

19. ABC is an isosceles triangle with $AB = 5 = AC$, and A has coordinates $(-1, \frac{1}{4})$. If base BC is on the line with equation $x = 3$, find the coordinates of B and C.

20. Triangle ABC is determined by the points $A(-2, -3)$, $B(1, 4)$, and $C(5, -1)$. Show that the line joining the midpoints of sides AB and BC is parallel to side AC and equal in length to one-half of AC.

21. Repeat Exercise 20 for triangle DEF located at $D(0, 0)$, $E(x_1, y_1)$, and $F(x_2, y_2)$ for the line joining the midpoints of sides DE and EF.

22. What are the coordinates of point C in parallelogram $ABCD$?

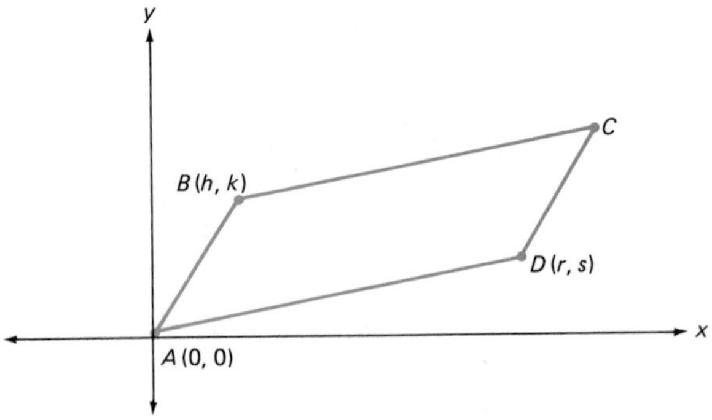

23. This exercise outlines a proof of the Pythagorean theorem.
(a) Begin with a right triangle as shown at the left. Construct a square as on the right, where each side has length $a + b$.

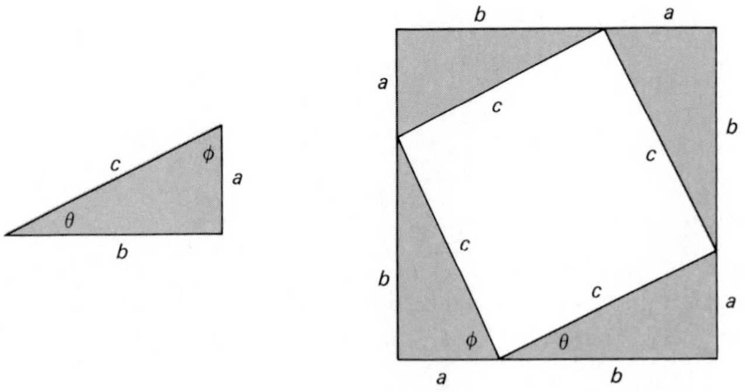

Why is the inner quadrilateral a square whose sides are of length c?
(b) Express the area of the outer square as the sum of the areas of the five parts.
(c) Since $(a + b)^2$ is also the area of the outer square, how can we conclude that $a^2 + b^2 = c^2$?

24. **(a)** In the following figure, ABC is a right triangle with $a = b$. Duplicate this figure on a

piece of cardboard, cut out the eight triangles forming the squares, and try to assemble them to form a square on side AB.

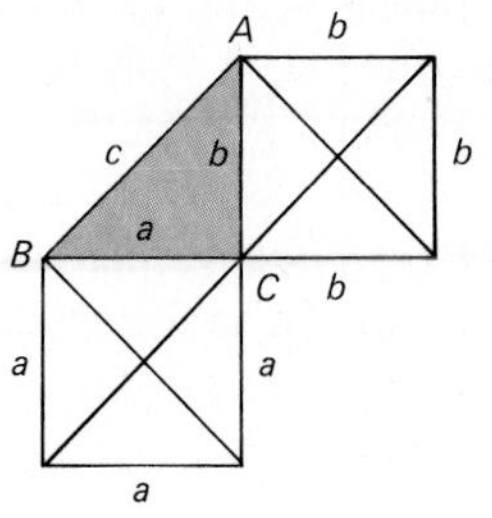

*(b) Follow the directions in part (a) for the figure below. In this figure, $a \neq b$ and AB is parallel to the line through points R, C, and V; also, PS and TW are parallel to AB. (*Hint:* Note that $RV = c$ and $PS + WT = c$.)

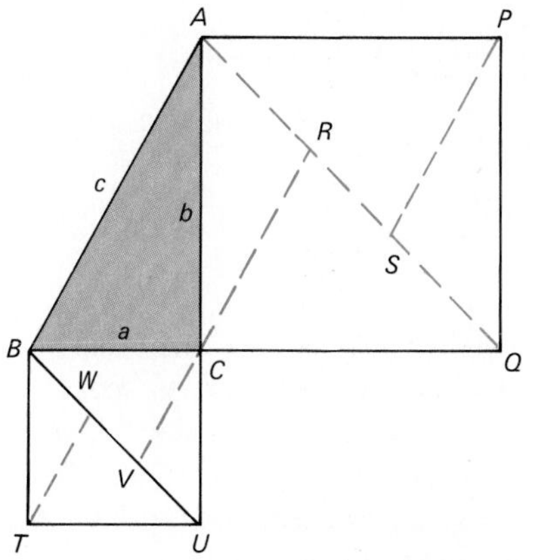

CHALLENGE
Think Creatively

A **median** in a triangle is a line segment drawn from a vertex to the midpoint of the opposite side. It can be proved that the three medians of a triangle meet at a single point, known as the *center of gravity* of that triangle. In a cardboard model of a triangle, the figure can be balanced by placing a pencil at that point.

For the specific triangle ABC below, show that the three medians meet at a point P and find the coordinates of that point.

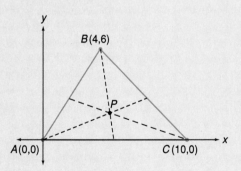

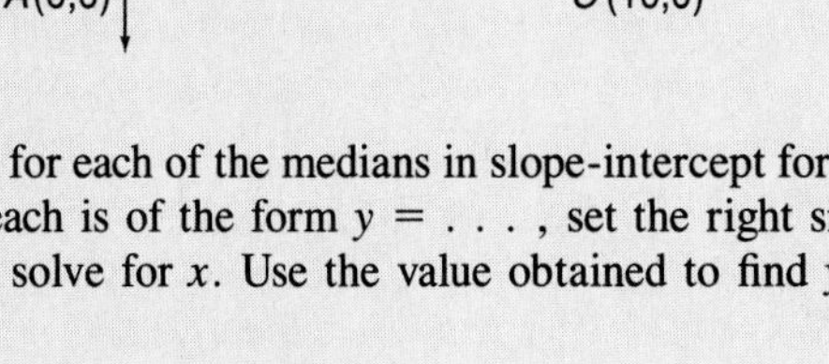

Hint: Write the equation for each of the medians in slope-intercept form. Select any two of these equations. Since each is of the form $y = \ldots$, set the right sides of the equations equal to each other and solve for x. Use the value obtained to find y. Then consider the third median.

6.6 LINEAR INEQUALITIES

A point on the number line divides the line into two parts known as half-lines. Similarly, in two dimensions, a line will divide the plane into two distinct **half-planes.** The half-planes determined by horizontal or vertical lines are easy to identify, as in the following figures.

Any point $P(x, y)$ *above* the horizontal line $y = 2$, shown as the shaded region, has a y-coordinate greater than 2; for such a point, $y > 2$.

Any point *to the right* of the vertical line $x = 3$, shown as the shaded region, has an x-coordinate greater than 3; for such a point, $x > 3$.

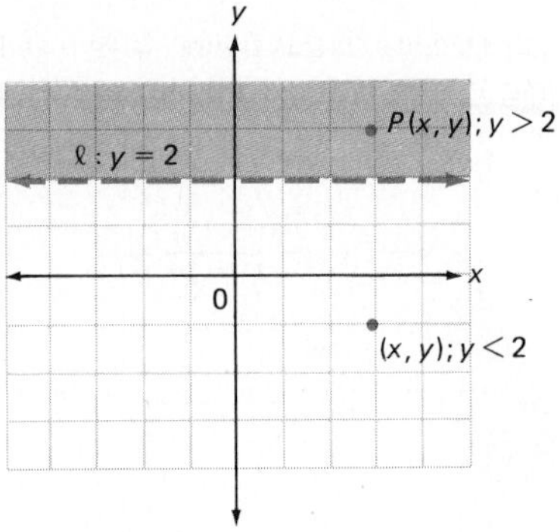

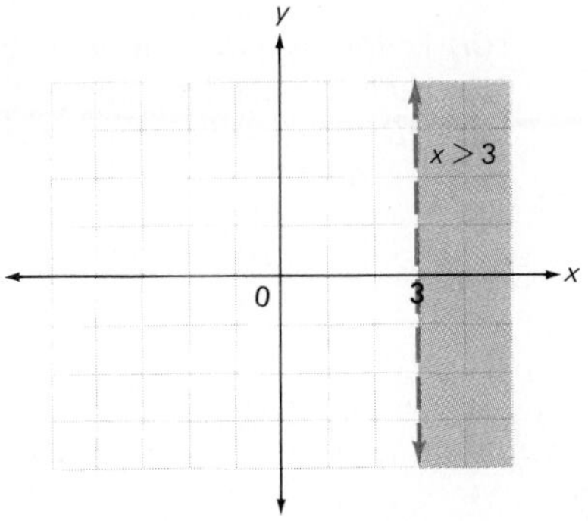

The half-plane *below* the line $y = 2$ consists of all points (x, y) such that $y < 2$.

The half-plane *to the left* of the line $x = 3$ consists of all points (x, y) such that $x < 3$.

For inequalities such as $y \geq 2$ or $x \geq 3$, we would use a solid rather than a dashed line in the drawing.

Next consider a line that is not parallel to one of the axes, such as $y = 2x - 1$. Again the line divides the plane into two half-planes. These two half-planes are the graphs for the statements of linear inequalities in two variables, $y < 2x - 1$ (*below the line*) and $y > 2x - 1$ (*above the line*). To show these graphs we use a dashed line and shade the appropriate half-plane as in the following figures.

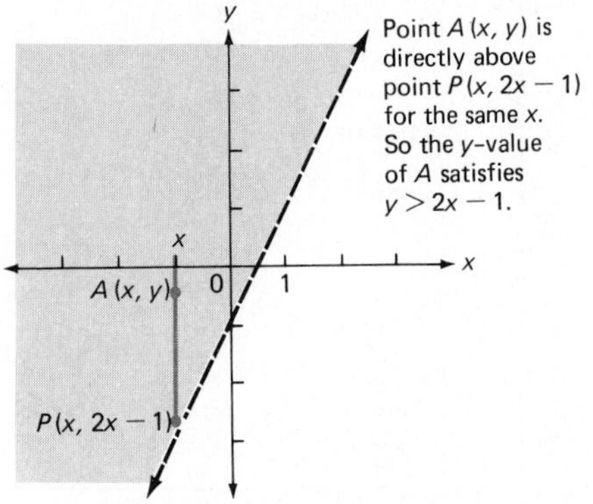

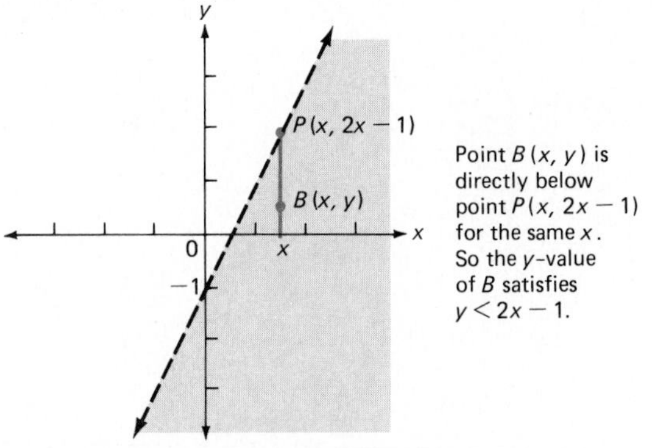

In general, the graph of $y > mx + b$ consists of the points *above* the line $y = mx + b$. The points *below* the line show the graph of $y < mx + b$.

In order to identify the region satisfying an inequality like $3x - y < -2$, it is useful first to graph the corresponding line $3x - y = -2$. Then there are a number of ways to proceed. Here are two methods:

1. Solve the given inequality for y; that is, $y > 3x + 2$. Therefore, the region is above the line ℓ.
2. After graphing the line, pick any convenient point, such as $(-2, 2)$, that is *not* on the line and substitute into the given inequality $3x - y < -2$. This gives the *true* statement $-8 < -2$. Therefore, $(-2, 2)$ must be on the correct side of ℓ, which may now be indicated by the shading in the following graph.

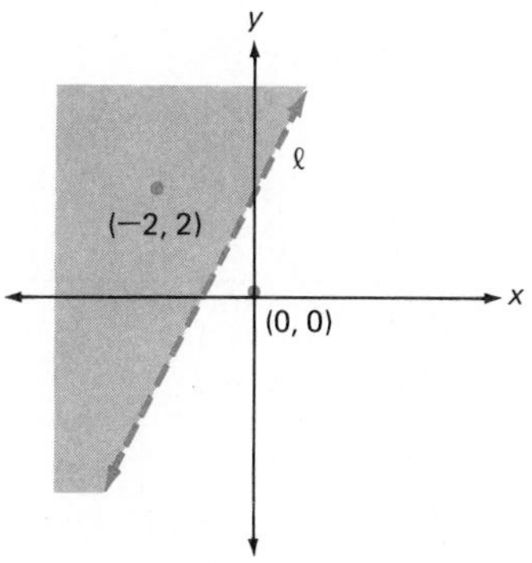

To show the graph of $3x - y \le -2$ draw the straight line as a solid rather than as a dashed line. The graph would consist of the same shaded half-plane together with the solid line.

As a check use a point on the other side, say, $(0, 0)$. Substituting into $3x - y < -2$ gives the *false* statement $0 < -2$. Consequently, $(0, 0)$ is on the wrong side, and you may safely conclude that $3x - y < -2$ works for the points on the other side.

EXAMPLE 1 Locate (graph) the region satisfying $2x + 3y < 1$.

Solution First sketch the graph of $2x + 3y = 1$ with a dashed line. Then use one of the following steps to obtain the required graph.

1. Solving for y produces $y < -\frac{2}{3}x + \frac{1}{3}$. Thus the required region satisfying the linear inequality is the half-plane below the line.

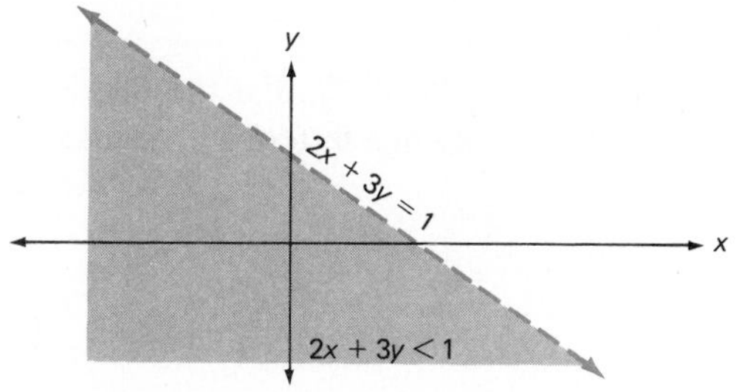

2. Substitute $(0, 0)$ into $2x + 3y < 1$ to get $0 < 1$. Hence $(0, 0)$ is on the correct side, namely, below the line $2x + 3y = 1$. ■

TEST YOUR UNDERSTANDING
Think Carefully

Match the shaded regions with the appropriate inequality. (Note: If a line is dashed, it is not part of the graph.)

1.

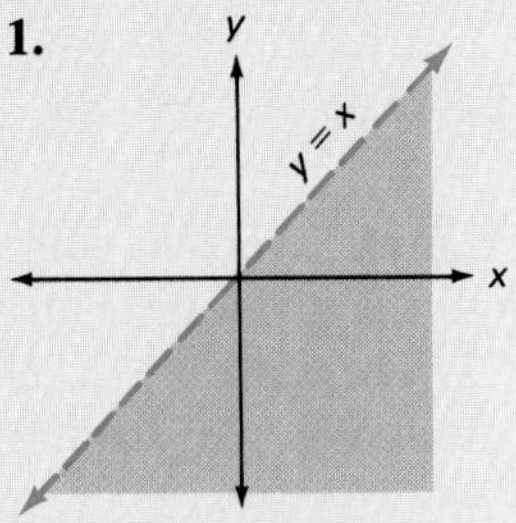

2.

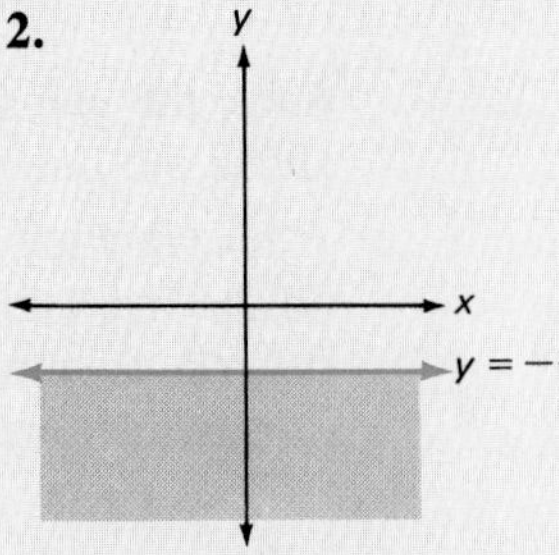

(a) $y > 4 - x$
(b) $y \leq -\frac{2}{3}$
(c) $y < -\frac{1}{10}x + \frac{1}{2}$
(d) $y < x$
(e) $y \geq 2x + 2$
(f) $x > -1$

3.

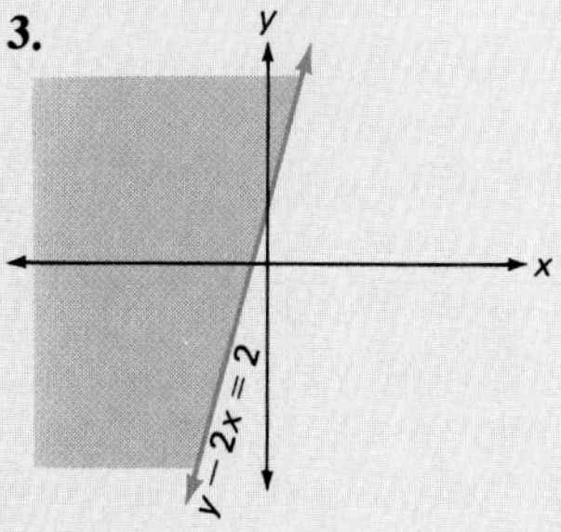

4.

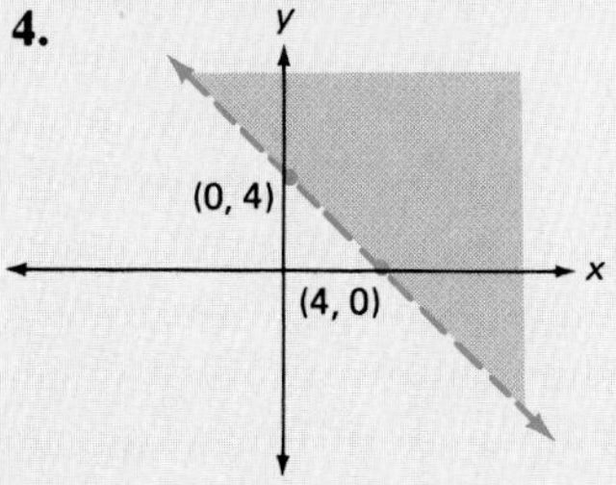

5.

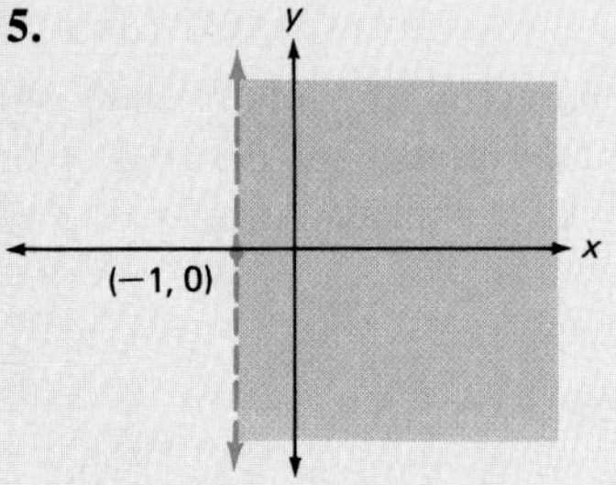

6.

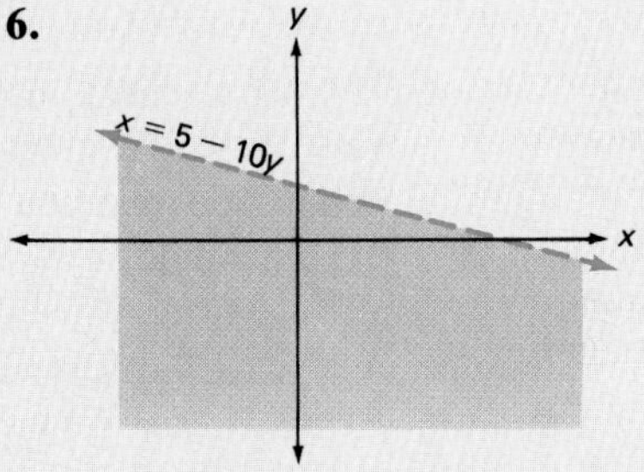

(Answers: Page 284)

Instead of graphing a given inequality, we can reverse the process by beginning with a line and then find the inequalities that identify the half-planes on either side of the line. This is illustrated in the next example.

EXAMPLE 2 For the line $2x - 4y = 5$ identify the half-planes above and below the line in terms of inequalities.

Solution

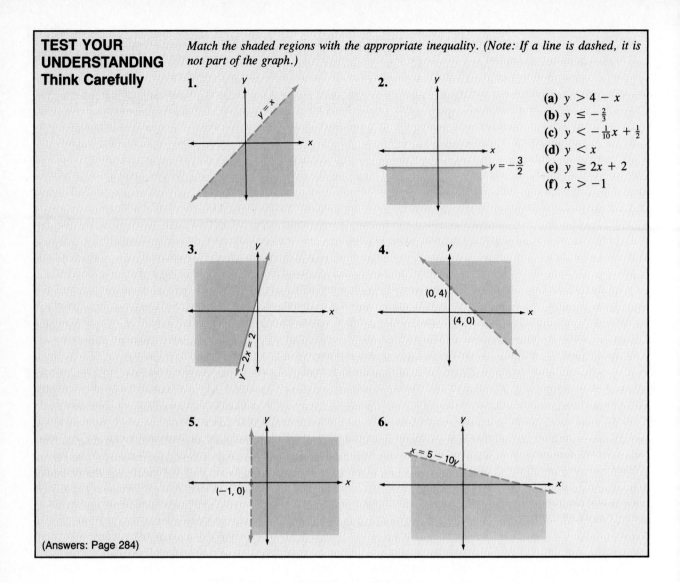

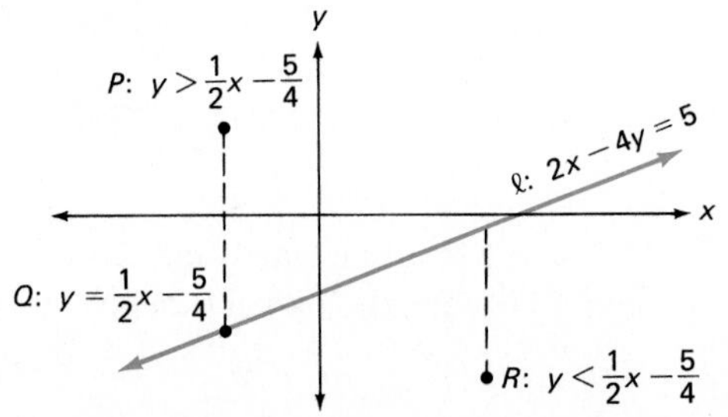

First solve for y:

$$y = \tfrac{1}{2}x - \tfrac{5}{4}$$

Then for all points $P(x, y)$ in the half-plane above ℓ,

$$y > \tfrac{1}{2}x - \tfrac{5}{4}$$

For all points $R(x, y)$ below ℓ,

$$y < \tfrac{1}{2}x - \tfrac{5}{4}$$

Note also that for all points $Q(x, y)$ on the line,

$$y = \tfrac{1}{2}x - \tfrac{5}{4}$$ ■

In an earlier chapter we learned how to graph inequalities such as $|x| < 2$ on a line (see page 110).

The solution is the interval $(-2, 2)$.

$\{x \mid |x| < 2\}$:

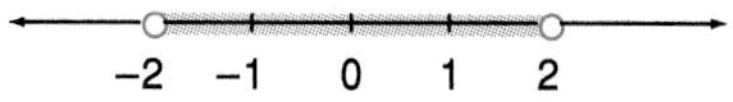

Now we will extend such graphs to a plane and consider the graph of

$$\{(x, y) \mid |x| < 2\} \qquad \text{(The set of all } (x, y) \text{ such that } |x| < 2.)$$

The ordered pair within the set-builder notation used here indicates that the graph is to be drawn on a plane rather than on a line.

Begin by drawing the two lines $x = -2$ and $x = 2$ with a dashed line, and then shade the region between the two lines.

The graph of $|x| \le 2$ on a plane would consist of the shaded region shown, but bounded by solid rather than dashed lines. What is the graph of $|x| > 2$?

$\{(x, y) \mid |x| < 2\}$:

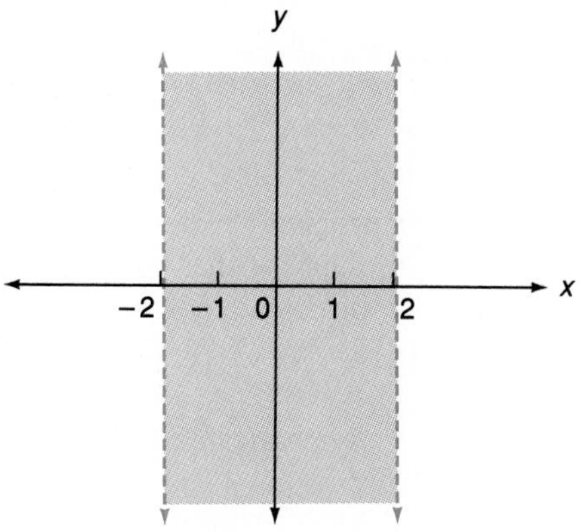

When asked to graph an inequality such as $|x| < 2$, you must know whether the graph is to be on a line or in the coordinate plane. This is often obvious from the context in which the inequality is being considered. Otherwise, it can be indicated either by a verbal statement or through the use of set-builder notation as just shown.

EXERCISES 6.6

Write the linear inequalities whose graph is the shaded region.

1.

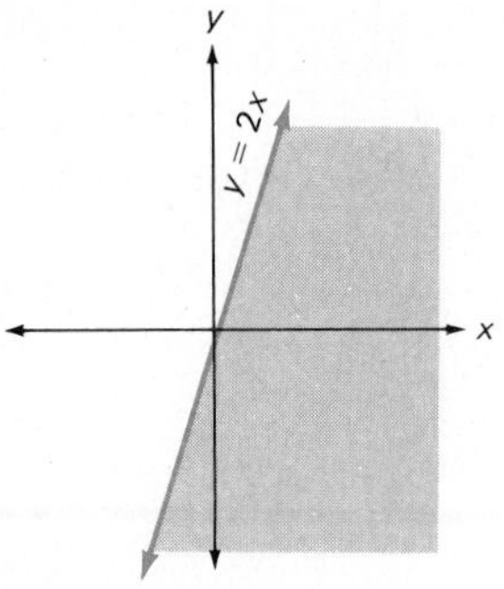

2.

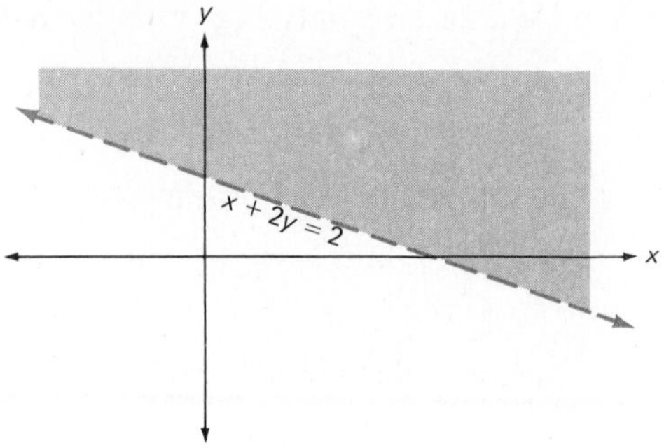

3.

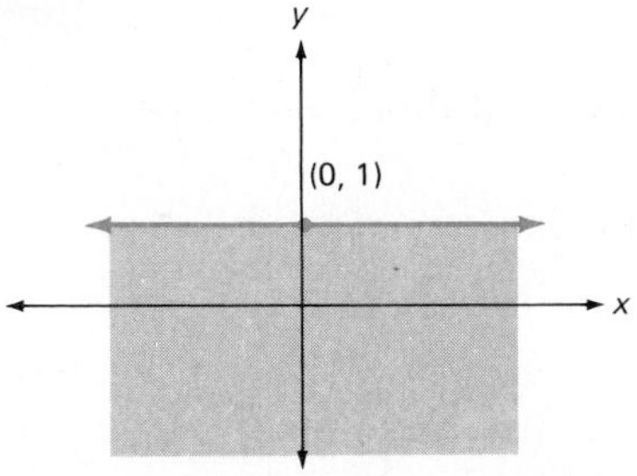

4.

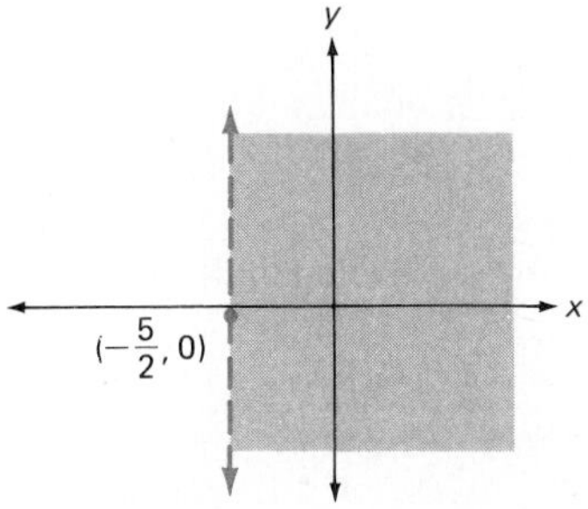

Graph the regions on a plane that satisfy the given linear inequalities.

5. $y > x + 2$
6. $y < x - 1$
7. $y \geq x - 2$
8. $y \leq x + 1$
9. $y > x$
10. $y < 2x + 4$
11. $y < -\frac{1}{2}x + 2$
12. $y \geq -\frac{1}{3}x - 2$
13. $2x + y > 6$
14. $3x - y < 6$
15. $x + y \leq 4$
16. $x - y \geq 2$
17. $2y + x - 4 < 0$
18. $x - 2y + 2 \leq 0$
19. $3y + x + 6 \geq 0$
20. $\frac{1}{2}x + y > 2$
21. $x - \frac{1}{2}y + 1 \geq 0$
22. $\frac{1}{2}x + \frac{1}{4}y < 1$
23. $|x| \leq 3$
24. $|x| \geq 3$
25. $|x| < 2$
26. $|y| > 1$
27. $|y| \leq 2$
28. $|y| \geq 2$
29. $|x - 2| \leq 1$
30. $|x + 2| > 1$
31. $|y - 1| \leq 2$

6.7 INTRODUCTION TO THE FUNCTION CONCEPT

Suppose that you are riding in a car that is averaging 40 miles per hour. Then the distance traveled is determined by the time traveled.

$$\text{distance} = \text{rate} \times \text{time}$$

Symbolically, this relationship can be expressed by the equation

$$s = 40t$$

where s is the distance traveled in time t (measured in hours). For $t = 2$ hours, the distance traveled is

$$s = 40(2) = 80 \text{ miles}$$

Similarly, for each specific value of $t \geq 0$ the equation produces *exactly one* value for s. This correspondence between the distance s and the time t is an example of a *functional relationship*. More specifically, we say that the equation $s = 40t$ defines s as a *function* of t.

In contrast, $y^2 = 12x$ does not define y as a function of x because ***more than one*** *y-value corresponds to some x-values. In particular, if $x = 3$, then $y^2 = 36$, which gives the two values $y = 6$ or $y = -6$.*

We say that $s = 40t$ defines s as a function of t because for *each* choice of t there corresponds *exactly one* value for s. We first choose a value of t. Then there is a corresponding value of s that depends on t; s is the **dependent variable** and t is the **independent variable** of the function defined by $s = 40t$.

Because the variable t represents time in the equation $s = 40t$, it is reasonable to say that $t \geq 0$. This set of allowable values for the independent variable is called the **domain** of the function. The set of corresponding values for the dependent variable is called the **range** of the function.

This is an important definition. Much of the work in mathematics deals with the study of functions.

DEFINITION OF FUNCTION

A **function** is a correspondence between two sets, the domain and the range, such that for each value in the domain there corresponds exactly one value in the range.

The specific letters used for the independent and dependent variables are of no consequence. Usually, we will use x for the independent variable and y for the dependent variable. Thus the equation $y = 40x$ can be used to define the same function as $s = 40t$. However, letters that are suggestive, such as t for time, can prove to be helpful.

Most of the algebraic expressions encountered earlier in this text can be used to define functions. As demonstrated in the examples that follow, the domain of such a function is determined by the expression being used.

EXAMPLE 1 Explain why the following equation defines y as a function of x and find the domain: $y = \dfrac{1}{x - 1}$.

Solution For each allowable x the expression $\dfrac{1}{x - 1}$ produces just one y-value. Therefore, the given equation defines a function. To find the domain, note that $x = 1$ produces division by zero and so must be excluded. Thus the domain consists of all real numbers except 1; that is, all $x \neq 1$. ■

For the functions that will be studied in this text we will assume that both the domain and the range must consist of real numbers; imaginary numbers are not to be used. This is illustrated in the next example.

EXAMPLE 2 Find the domain of the function given by $y = \sqrt{1 - x}$.

Solution Since the square roots of negative numbers are imaginary numbers, we need to avoid values for which $1 - x$ is negative. Thus the domain consists of those values x so that $1 - x \geq 0$, or $1 \geq x$. The domain is $\{x \mid x \leq 1\}$. ■

TEST YOUR UNDERSTANDING Think Carefully

Decide whether the given equation defines y to be a function of x. For each function, find the domain.

1. $y = -3x$ **2.** $y = x$ **3.** $y = 2x \pm 3$

4. $y = (x + 2)^2$ **5.** $y = \dfrac{1}{(x + 2)^2}$ **6.** $y = \dfrac{1}{x^2 + 2}$

7. $y = \sqrt{2x - 1}$ **8.** $y = \pm x$ **9.** $y^2 = x + 1$

(Answers: Page 284)

The function concept can be demonstrated geometrically, as in Example 3.

EXAMPLE 3 Graph the linear function given by $y = -\frac{1}{2}x + 5$ and identify the domain and range. Also display the domain value 6 and its corresponding range value.

Remember that the domain of the function is the set of numbers that the independent variable x represents, and the range of the function consists of all corresponding y-values.

Solution The graph of the function given by $y = -\frac{1}{2}x + 5$ is just the graph of this linear equation. Both the domain and the range of the function is the collection of all corresponding y-values. In each case this is the set of all real numbers.

For the domain value $x = 6$, the range value is $y = -\frac{1}{2}(6) + 5 = 2$, as shown.

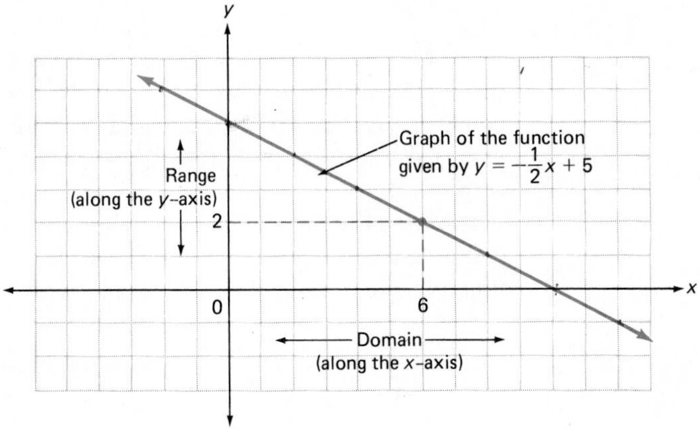

In Example 3 the domain for $y = -\frac{1}{2}x + 5$ consisted of all real numbers. The next example uses the same equation but with a different domain specified.

EXAMPLE 4 Let the function $y = -\frac{1}{2}x + 5$ have domain $\{x \mid 2 \le x \le 12\}$. Graph the function, find the range, and indicate both the domain and range on the graph.

Solution To graph this function first draw the line $y = -\frac{1}{2}x + 5$. Then use only those points between $x = 2$ and $x = 12$, inclusive.

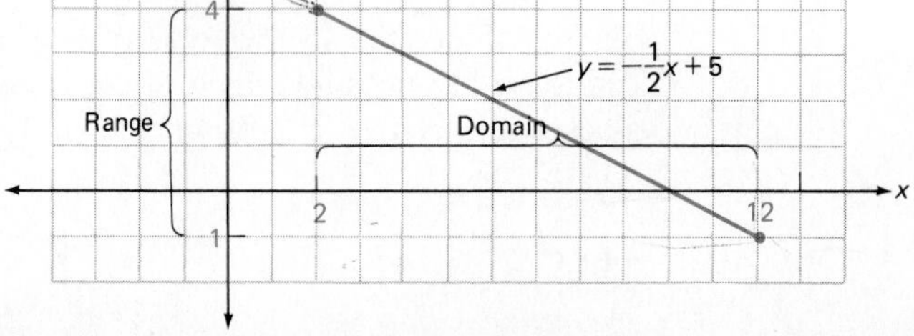

The preceding graph shows that the range consists of all values y such that $-1 \le y \le 4$; range $= \{y \mid -1 \le y \le 4\}$. ■

You have now seen how the equation $y = -\frac{1}{2}x + 5$ was used to define two different functions. The first had as its domain the set of all real numbers, and the second had domain $\{x \mid 2 \le x \le 12\}$. The next example shows how a function can be defined by using more than one equation.

EXAMPLE 5 Graph the function defined as follows:

$$y = \begin{cases} x + 2 & \text{for } -1 \le x \le 2 \\ -x + 3 & \text{for } 2 < x \le 5 \end{cases}$$

Solution First draw the part of the line $y = x + 2$ beginning with $x = -1$ and ending with $x = 2$. Then draw the part of the line $y = -x + 3$ for $2 < x \le 5$.

*The open dot at (2, 1) means that this point is **not** part of the graph. (Why not?) The closed dot at (2, 4) indicates that the point **is** part of the graph.*

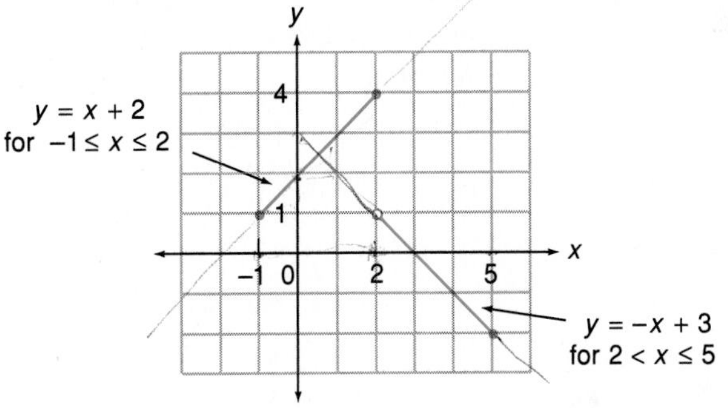

■

The absolute-value concept can also be used to define functions. For example, $y = |x|$ defines y to be a function of x because each real number has exactly one absolute value. The graph of this function is given in Example 6.

See Section 1.3 to review the meaning of absolute value.

EXAMPLE 6 Graph the absolute-value function $y = |x|$.

Solution Using the definition of absolute value, stated on page 14, $y = |x|$ can be written in terms of two equations as follows.

$$y = |x| = \begin{cases} x & \text{if } x \ge 0 \\ -x & \text{if } x < 0 \end{cases}$$

The graph of $y = |x|$ consists of two perpendicular rays intersecting at the origin as shown in the figure at the right.

To graph this function, first draw the line $y = -x$ and eliminate all those points on it for which x is positive. Then draw the line $y = x$ and eliminate the part for which x is negative. Now join these two parts to get the graph of $y = |x|$:

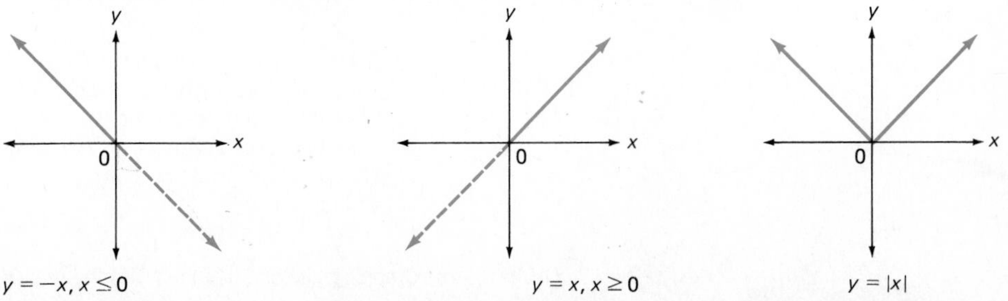

Note that the graph is symmetric about the y-axis. (If the paper were folded along the y-axis, the two parts would coincide.) This symmetry can be observed by noting that the y-values for x and $-x$ are the same. That is,

$$|x| = |-x| \qquad \text{for all } x$$

The domain is the set of all real numbers and the range consists of all $y \geq 0$. ■

As this course progresses, you will learn how to draw the graphs of a variety of equations. At times, it will not be obvious whether or not the given equation defines a function of x. However, if the graph is available, a simple geometric test can be used to make this determination.

Vertical Line Test for Functions

1. Consider the vertical line through any x-value.
2. If for each x the vertical line intersects the curve exactly once, then it is the graph of a function. Otherwise, if there is an x for which the vertical line intersects the curve more than once, then the curve is not the graph of a function.

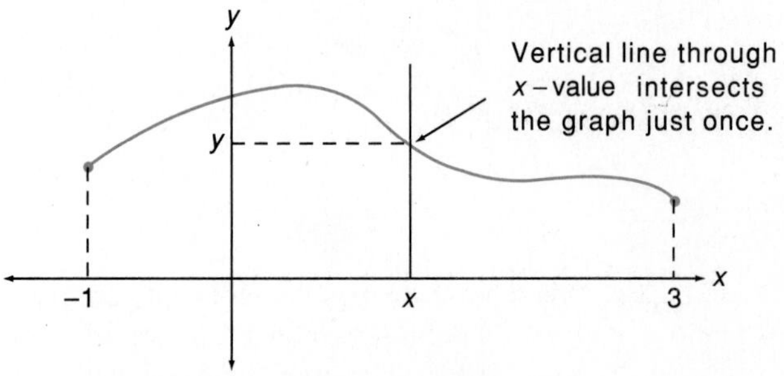

The vertical line through any point x, for $-1 \leq x \leq 3$, intersects the graph just once producing exactly one y-value. This is the graph of a function of x.

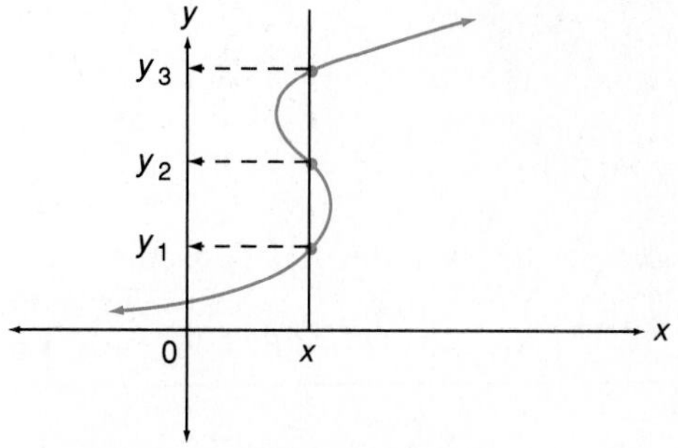

A vertical line through one of the x-values intersects the graph more than once, producing more than one y-value. This is *not* the graph of a function of x.

EXERCISES 6.7

Decide whether the given equation defines y to be a function of x. For each function, find the domain.

1. $y = x^3$
2. $y = \sqrt[3]{x}$
3. $y = \dfrac{1}{\sqrt{x}}$
4. $y = |2x|$
5. $y^2 = 2x$
6. $y = x \pm 3$
7. $y = \dfrac{1}{x + 1}$
8. $y = \dfrac{x - 2}{x^2 + 1}$
9. $y = \dfrac{1}{1 \pm x}$
10. $y = \sqrt{x - 4}$
11. $y = \dfrac{1}{\sqrt{x - 4}}$
12. $y = \dfrac{1}{\sqrt[3]{x^2 - 4}}$

In Exercises 13 through 18 graph the function defined by $y = 3x - 7$ for the given domain.

13. All real numbers
14. All $x \geq 0$
15. $\{x \mid -1 \leq x \leq 3\}$
16. $\{x \mid 4 \leq x \leq 5\}$
17. $\{-1, 0, 1, 2, 3\}$
18. $\{x \mid -7 \leq x \leq \frac{7}{3}\}$
19. If the domain of $y = -\frac{4}{3}x + \frac{2}{3}$ is $\{x \mid x \geq \frac{1}{2}\}$, find the range.
20. If y is a function of x where $y = -\frac{4}{3}x + \frac{2}{3}$ for the domain $\{x \mid -\frac{1}{4} \leq x \leq \frac{1}{2}\}$, find the range.
21. The line segment in the adjacent figure is the graph of a function.
    (a) What is the domain?
    (b) What is the range?
    (c) Find the equation that shows how y depends on x.
22. Explain why the equation $x = |y|$ does not define y to be a function of x.
23. Explain why the equation $y^2 = x - 2$ does not define y to be a function of x.

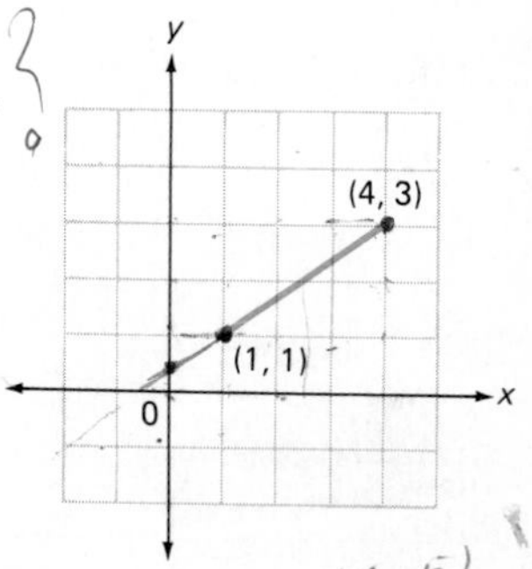

In each of Exercises 24 through 31 use the vertical line test to decide if the given graph can be the graph of a function having the indicated set A as the domain.

24.

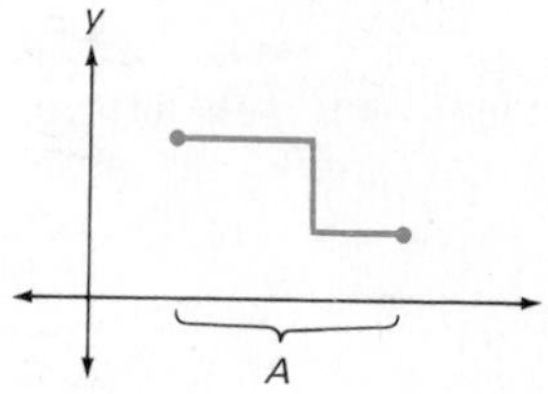

25.

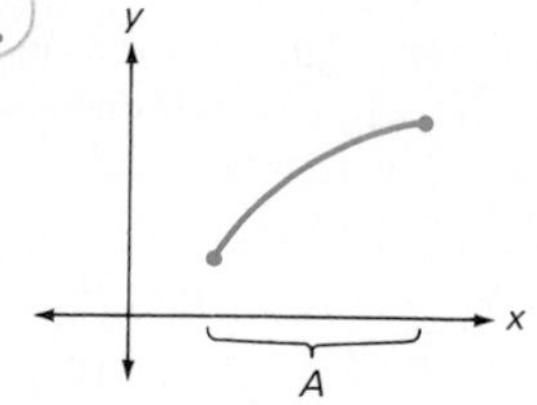

26.

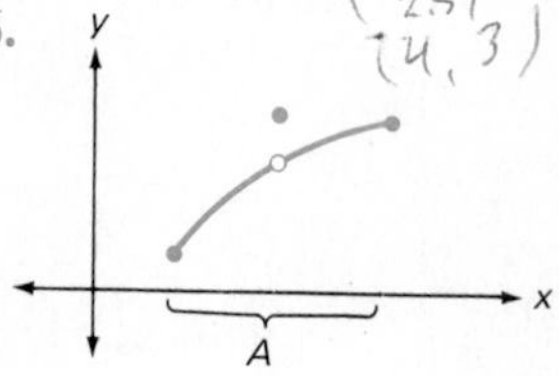

27.

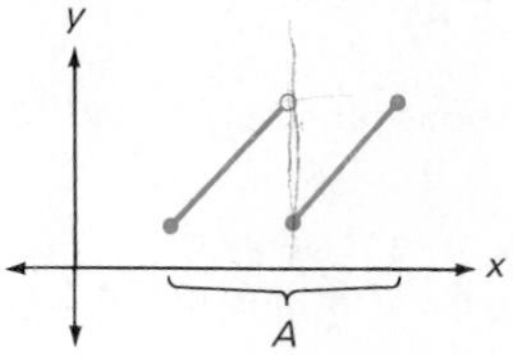

28.

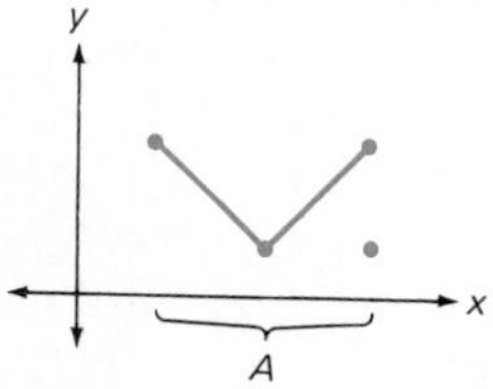

29.

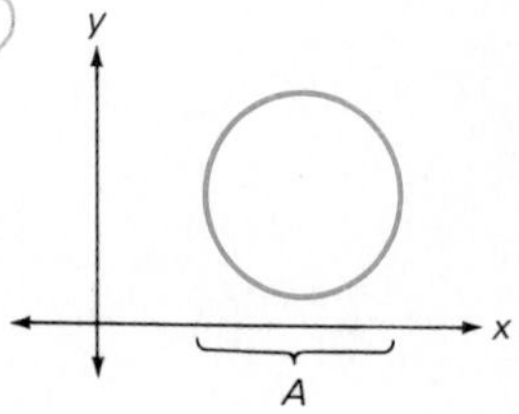

30.

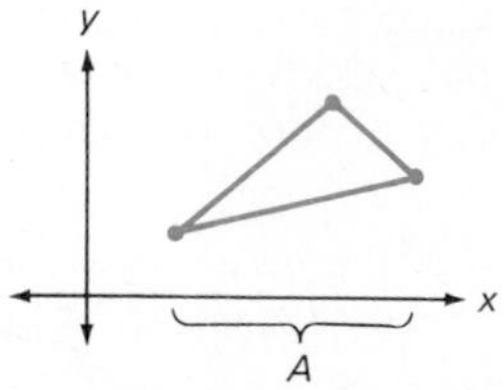

31.

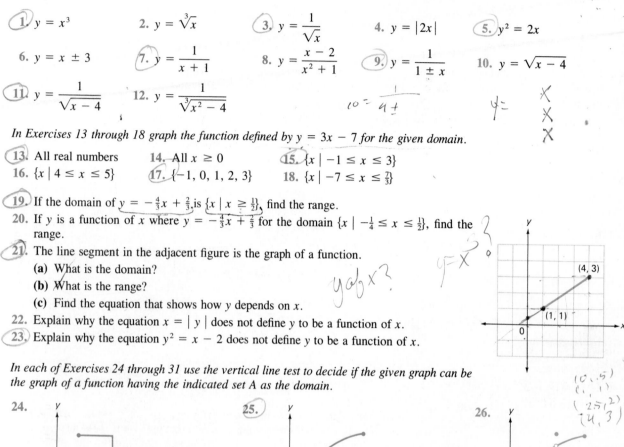

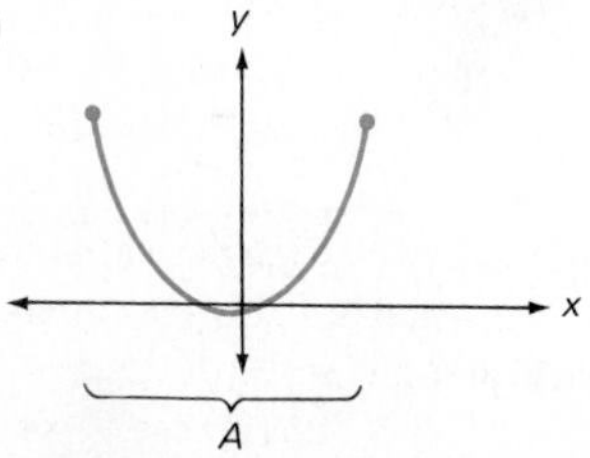

Graph each function and state the domain and range.

32. $y = \begin{cases} 3x & \text{for } -1 \le x \le 1 \\ -x & \text{for } 1 < x \end{cases}$

33. $y = \begin{cases} -2x + 3 & \text{for } x < 2 \\ x + 1 & \text{for } x > 2 \end{cases}$

34. $y = \begin{cases} x & \text{for } -2 < x \le 0 \\ 2x & \text{for } 0 < x \le 2 \\ -x + 3 & \text{for } 2 < x \le 3 \end{cases}$

35. $y = \begin{cases} x - 1 & \text{for } -3 < x \le -1 \\ -x + 3 & \text{for } -1 < x < 1 \\ 2x - 3 & \text{for } 1 \le x < 3 \end{cases}$

36. The sides of the parallelogram with vertices $(-1, 1)$, $(0, 3)$, $(2, 1)$, $(3, 3)$ are the graphs of four different functions. In each case find the equation that defines the function and state the domain.

37. Graph the function given by $y = |x|$ for the domain $\{x \mid -2 \le x \le 2\}$.

38. Graph the function given by $y = \frac{|x|}{x}$. What is the domain? (*Hint:* consider the two cases $x > 0$ and $x < 0$.)

CHALLENGE
Think Creatively

If $x = 2.3$, then the largest or greatest integer less than or equal to x is 2. That is, $2 \le 2.3$ and 2 is the largest integer for which this is true. The symbol used for this greatest integer of 2.3 is [2.3]. In general,

$$[x] = \text{greatest integer } \le x$$

In particular, $[3\frac{1}{2}] = 3$, $[0.75] = 0$, and $[-2.3] = -3$. Draw the graph of the **greatest integer function** $y = [x]$ for $-3 \le x \le 3$.

EXPLORATIONS
Think Critically

1. We often say that an equation such as $y = x^2 + 2x$ is a function. Explain why technically this is not correct.
2. Functions can be defined using tables. In each of the following tables, assume that each y-value in the second row corresponds to the x-value directly above it in the first row. Decide whether or not each table defined y to be a function of x. Justify your answer.

(a)

x	−1	0	2	3	10
y	3	5	−4	3	−10

(b)

x	4	1	5	2	1
y	−2	6	8	4	3

3. How can you prove, algebraically, that a specific point $P(x, y)$ is *not* located on a line with a given equation?
4. Consider the line defined by $y = mx + b$. Under what conditions on m and b will the line pass through quadrants 1, 3, and 4?
5. Consider the graphs of the equations $y = ax + c$ and $y = bx + d$. Under what conditions are the x-intercepts the same? Under what conditions are the two lines parallel but not vertical?
6. A procedure for graphing $y = |x|$ is shown on page 269. Another way that this can be done begins by drawing the line $y = x$. Then, since $y = |x|$ must be non-negative, take all points on the graph of $y = x$ that are below the x-axis (those having negative y-values) and reflect these through the x-axis. Use this procedure to draw the graph of $y = |2x - 5|$.
7. Consider any isosceles triangle whose vertex is either on the x-axis or on the y-axis. If the base of the triangle is on the other axis, what can you say about the slopes of the two sides of equal measure?

Written Assignment: Explain, in your own words, the meaning of a function as well as the domain and range of a function. Illustrate each with a specific function.

6.8 THE FUNCTION NOTATION

A useful way to refer to a function is to name it by using a specific letter. For example, the function given by $y = 5x^3 - 2x^2 + 9$ may be referred to as f and we write

$$f(x) = 5x^3 - 2x^2 + 9$$

The symbol $f(2)$ represents the range value of the function f corresponding to the domain value 2.

The symbol $f(x)$ is used to mean *the value of the function f at x*, and is usually read as "f of x" or "f at x." For example, $f(2)$ is obtained by substituting 2 for x in the expression $f(x)$. Thus, using

$$f(x) = 5x^3 - 2x^2 + 9$$

CAUTION
Note that $f(x)$ does not mean the product of f times x. When the letter f is written by itself it stands for the name of the function; it is not a number.

we have

$$f(2) = 5(2)^3 - 2(2)^2 + 9 = 40 - 8 + 9 = 41$$

In brief,

$$f(2) = 41 \qquad \text{(This is read as "}f\text{ of 2 is 41.")}$$

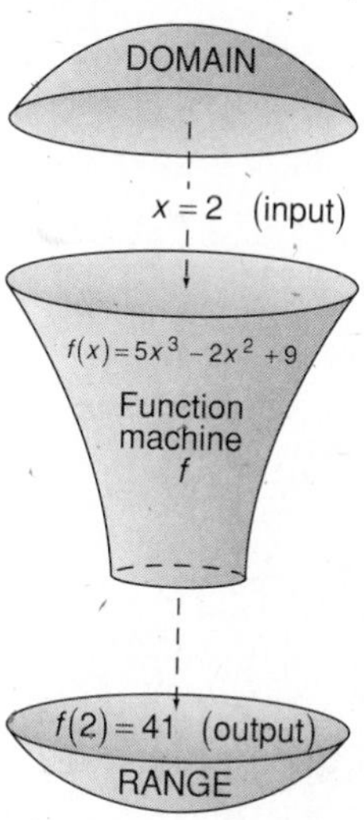

Similarly using domain values 0 and $-\frac{1}{2}$,

$$f(0) = 5(0)^3 - 2(0)^2 + 9 = 9$$

$$f(-\tfrac{1}{2}) = 5(-\tfrac{1}{2})^3 - 2(-\tfrac{1}{2})^2 + 9 = \frac{63}{8}$$

A function f may be thought of as a machine, as demonstrated in the adjoining diagram. The domain value x can be regarded as the raw material or "input" that goes into the machine f. The machine then processes x and produces the finished product or "output" $f(x)$, the range value corresponding to x.

Other letters than f may be used to name functions. For example, we may use g for the preceding function and write

$$g(x) = 5x^3 - 2x^2 + 9$$

EXAMPLE 1 Let h be the function given by $h(x) = x^2 + 2x$ and find:

(a) $h(0)$ **(b)** $h(1)$ **(c)** $h(2)$
(d) $h(3)$, and compare to $h(1) + h(2)$ **(e)** $3h(2)$ **(f)** $h(6)$, and compare to $3h(2)$

Solution

(a) $h(0) = 0^2 + 2(0) = 0$
(b) $h(1) = 1^2 + 2(1) = 3$
(c) $h(2) = 2^2 + 2 \cdot 2 = 8$
(d) $h(3) = 3^2 + 2 \cdot 3 = 15$; $h(1) + h(2) = 3 + 8 = 11$; $h(1) + h(2) \neq h(3)$
(e) $3h(2) = 3 \cdot 8 = 24$
(f) $h(6) = 6^2 + 2 \cdot 6 = 48$; $3h(2) = 24$; $3h(2) \neq h(6)$ ■

CAUTION
*Since $h(1) + h(2) \neq h(3)$ it would be **incorrect** to write $h(1 + 2) = h(1) + h(2)$. In general, the distributive property does not apply to the function notation.*

At times the variable x in $f(x)$ may be replaced by other variable forms. For example, suppose $f(x) = x^2 + 2x$ and we want to find $f(x - 2)$. To do this, all that needs to be done is to replace the x, wherever it appears, by $x - 2$. Thus using

$$f(x) = x^2 + 2x$$

we have

$$\begin{aligned} f(x-2) = (x-2)^2 + 2(x-2) &= (x^2 - 4x + 4) + (2x - 4) \\ &= x^2 - 2x \end{aligned}$$

EXAMPLE 2 For the function g defined by $g(x) = \frac{1}{x}$, find:

(a) $3g(x)$ **(b)** $g(3x)$ **(c)** $3 + g(x)$ **(d)** $g(3) + g(x)$

(e) $g(3 + x)$ **(f)** $g\left(\frac{1}{x}\right)$

Solution **(a)** $3g(x) = 3 \cdot \frac{1}{x} = \frac{3}{x}$ **(b)** $g(3x) = \frac{1}{3x}$

(c) $3 + g(x) = 3 + \frac{1}{x}$ **(d)** $g(3) + g(x) = \frac{1}{3} + \frac{1}{x}$

(e) $g(3 + x) = \frac{1}{3 + x}$ **(f)** $g\left(\frac{1}{x}\right) = \frac{1}{\frac{1}{x}} = x$ ■

TEST YOUR UNDERSTANDING
Think Carefully

Let f be the function given by $f(x) = -3x + 2$, *and find:*

1. $f(-3)$ **2.** $f(0)$ **3.** $f(\frac{2}{3})$ **4.** $f(3)$

For $h(x) = \frac{x}{x - 2}$, *find each of the following if it exists.*

5. $3h(10)$ **6.** $h(2)$ **7.** $h(\frac{1}{2}) + h(\frac{1}{3})$ **8.** $h(\frac{1}{2} + \frac{1}{3})$

Let $f(x) = x^2 - 3x$ *and find each of the following.*

9. $f(2x)$ **10.** $2f(x)$ **11.** $f(x - 3)$ **12.** $f(3 - x)$

13. $f\left(\frac{1}{x}\right)$ **14.** $\frac{1}{f(x)}$ **15.** $f(x^2)$ **16.** $[f(x)]^2$

(Answers: Page 284)

Let us see how functions can be combined. For example, let f and g be the functions given by

$$f(x) = x^3 - 7x^2 + x + 1 \quad \text{and} \quad g(x) = 8x^2 - 2x - 2$$

To add these functions we take each x in the common domain (the real numbers), find the corresponding range values $f(x)$ and $g(x)$, and add these two values to get $f(x) + g(x)$. For example, with $x = 2$, $f(2) = -17$ and $g(2) = 26$. Then

$$f(2) + g(2) = -17 + 26 = 9$$

and for *any x:*

$$f(x) + g(x) = (x^3 - 7x^2 + x + 1) + (8x^2 - 2x - 2)$$
$$= x^3 + x^2 - x - 1$$

It is this equation that defines the new function called the sum of f and g. It is the function that takes any number x and associates with it the following range value: $x^3 + x^2 - x - 1$.

This new function could be given a name, like s. Then s is the sum of functions f and g and is given by

Show that $s(2) = 9$.

$$s(x) = f(x) + g(x) = x^3 + x^2 - x - 1$$

Subtraction, multiplication, and division of functions are defined similarly. These operations are illustrated in the following examples.

EXAMPLE 3 Let $f(x) = x^3 - 7x^2 + x + 1$ and $g(x) = 8x^2 - 2x - 2$. Find $d(x)$, the difference of $f(x)$ minus $g(x)$, and find $d(1)$.

Solution For each real number x,

$$d(x) = f(x) - g(x)$$
$$= (x^3 - 7x^2 + x + 1) - (8x^2 - 2x - 2)$$
$$= x^3 - 15x^2 + 3x + 3$$

and

$$d(1) = 1 - 15 + 3 + 3 = -8$$

■

EXAMPLE 4 Let m be the function formed by multiplying functions f and g of Example 3. Find $m(x)$ for each x. Also find $f(1)$ and $g(1)$, and then verify that $m(1) = f(1) \cdot g(1)$.

Solution For each real number x,

$$m(x) = f(x) \cdot g(x)$$
$$= (x^3 - 7x^2 + x + 1)(8x^2 - 2x - 2)$$
$$= 8x^5 - 58x^4 + 20x^3 + 20x^2 - 4x - 2$$

Then

$$m(1) = 8 - 58 + 20 + 20 - 4 - 2$$
$$= -16$$

Also,

$$f(1) = -4; \qquad g(1) = 4$$

and

$$m(1) = -16 = (-4)(4) = f(1) \cdot g(1)$$

■

Note that the domain of q consists of all x common to the domains of f and g except those x for which $g(x) = 0$.

EXAMPLE 5 Let $q(x)$ be the quotient of $f(x) = 3x + 1$ divided by $g(x) = x^2 - 4$ and evaluate $q(0)$. Which values of x are not in the domain of q?

Solution

$$q(x) = \frac{f(x)}{g(x)} = \frac{3x + 1}{x^2 - 4}$$

Then $$q(0) = \frac{3(0) + 1}{0^2 - 4} = -\frac{1}{4}$$

When a number, such as 2, is not in the domain of function q we also say that q(2) is undefined, or that q(2) does not exist.

Since division by zero is undefined and the denominator $g(x) = x^2 - 4 = 0$ when $x = \pm 2$, the domain of q does not contain the values ± 2. ■

EXERCISES 6.8

In Exercises 1 through 10, find **(a)** $f(-1)$; **(b)** $f(0)$; **(c)** $f(\frac{1}{2})$; **(d)** $f(1)$; *and* **(e)** $f(2)$, *provided they exist.*

1. $f(x) = 2x - 1$
2. $f(x) = -5x + 6$
3. $f(x) = x^2 - 5x + 6$
4. $f(x) = x^3 - 1$
5. $f(x) = -3x^3 + \frac{1}{2}x^2 - 4x$
6. $f(x) = (x - 1)^2$
7. $f(x) = \dfrac{1}{x^2 - 1}$
8. $f(x) = \dfrac{x - 3}{2x - 1}$
9. $f(x) = \sqrt{4x + 8}$
10. $f(x) = \sqrt[3]{(2x - 1)^2}$

Use a calculator to evaluate the function values f(2.61), f(−0.97), and f(48) for each of the indicated functions. Round off to four decimal places where appropriate.

11. $f(x) = -3x^3 + \frac{1}{2}x^2 + 6$
12. $f(x) = \dfrac{1}{x^2 - 1}$
13. $f(x) = \sqrt{4x + 8}$
14. $f(x) = \sqrt[3]{(2x - 1)^2}$

For each of the following statements, use $f(x) = -x^2 + 3$. Classify each statement as true or false. If it is false, correct the right side to obtain a correct equation.

15. $f(3) = -6$
16. $3f(2) = -33$
17. $f(2)f(3) = -33$
18. $f(3) + f(-2) = 2$
19. $f(3) - f(2) = -5$
20. $f(2) - f(3) = 11$
21. $f(x) - f(4) = -(x - 4)^2 + 3$
22. $f(x) - f(4) = x^2 + 19$
23. $f(4 + h) = -h^2 - 8h - 13$
24. $f(4 + h) = -h^2 - 10$

25. Let $l(x) = 3x$.
    (a) Find $l(2) + l(3)$ and $l(5)$. **(b)** Find $4l(7)$ and $l(28)$. **(c)** Find $-2l(3) + 4l(-2)$ and $l(-14)$.
26. Let $h(x) = 2x^2 + x - 1$.
    (a) Find $h(1)$. **(b)** Find $2h(1)$. **(c)** Find $[h(2)]^2$. **(d)** Find $3h(2) - 2h(3)$.
27. For $f(x) = -3x + 9$ find:
    (a) $f(-x)$ **(b)** $f(x - 5)$ **(c)** $f(5 - x)$ **(d)** $f(\frac{1}{3}x - 3)$
28. For $g(x) = 5x + 2$ find:
    (a) $g(7x)$ **(b)** $g\left(\dfrac{x}{5}\right)$ **(c)** $g(x - 1)$ **(d)** $g(\frac{1}{5}x - \frac{2}{5})$
29. For $h(x) = x^2 - 3x + 1$ find:
    (a) $h(4x)$ **(b)** $h(-x)$ **(c)** $h(x + 4)$ **(d)** $h(x^2)$
30. For $f(t) = (t - 2)^3$ find:
    (a) $f(t + 2)$ **(b)** $f(t - 2)$ **(c)** $f(\sqrt[3]{t} + 2)$

31. For $g(u) = \frac{2}{u}$ find:

(a) $g\left(\frac{1}{u}\right)$ (b) $\frac{1}{g(u)}$ (c) $g\left(\frac{2}{u}\right)$

32. For $h(s) = \frac{1}{s^2 - 1}$ find:

(a) $h(a)$ (b) $h(-s)$ (c) $h(2s - 1)$

33. Let $f(x) = 2x - 3$ and $g(x) = 3x + 2$.
(a) Find $f(1)$, $g(1)$, and $f(1) + g(1)$.
(b) Letting $s(x) = f(x) + g(x)$, find $s(x)$.
(c) Use the result in part (b) to find $s(1)$.

34. Use f and g as in Exercise 33.
(a) Find $f(2)$, $g(2)$, and $f(2) - g(2)$.
(b) Letting $d(x) = f(x) - g(x)$, find $d(x)$.
(c) Use the result in part (b) to find $d(2)$.

35. Use f and g as in Exercise 33.
(a) Find $f(1)$, $g(1)$, and $f(1) \cdot g(1)$.
(b) Letting $p(x) = f(x) \cdot g(x)$, find $p(x)$.
(c) Use the result of part (b) to find $p(1)$.

36. Use f and g as in Exercise 33.

(a) Find $\frac{f(2)}{g(2)}$. (b) Find the quotient $q(x) = \frac{f(x)}{g(x)}$ and compute $q(2)$.

In Exercises 37 through 40 find:

(a) $f(x) + g(x)$ (b) $f(x) - g(x)$ (c) $g(x) - f(x)$ (d) $f(x) \cdot g(x)$ (e) $\frac{f(x)}{g(x)}$ (f) $\frac{g(x)}{f(x)}$

37. $f(x) = x - 5$; $g(x) = \frac{3}{x + 5}$

38. $f(x) = x^2 + 7$; $g(x) = \frac{1}{\sqrt[3]{x + 1}}$

39. $f(x) = \frac{x - 2}{x}$; $g(x) = \frac{x}{x^2 + 2x + 4}$

40. $f(x) = x^2$; $g(x) = x^{1/3}$

41. Let $f(x) = 5x + 1$ and $g(x) = \frac{1}{5}x - \frac{1}{5}$.
(a) Find $f(2)$ and $g(11)$.
(b) Find $g(5)$ and $f\left(\frac{4}{5}\right)$.
(c) Let a be a real number and find $f(a)$.
(d) Use the $f(a)$ found in part (c) and now find $g[f(a)]$.
(e) Find $g(2)$ and then find $f[g(2)]$.

42. Let $p(x) = 2x^3 - 10x + 6$ and $q(x) = x^2 - x + 5$.
(a) Find $p(-1) + q(-1)$.
(b) Find $s(x) = p(x) + q(x)$, and compute $s(-1)$.

43. Use p and q as in Exercise 42.
(a) Find $p(2) - q(2)$.
(b) Find $d(x) = p(x) - q(x)$, and compute $d(2)$.

44. Use p and q as in Exercise 42.
(a) Find $p(10) \cdot q(10)$.
(b) Find $f(x) = p(x) \cdot q(x)$, and compute $f(10)$.

45. Let $f(x) = 2x^3 + 8x^2 - 6$. If $f(x) = 2g(x)$, then what is $g(x)$?

46. Let $f(x) = x^2$.

(a) Find $f(x) - f(3)$.

(b) Show that $\dfrac{f(x) - f(3)}{x - 3} = x + 3$.

47. Let $f(x) = x^2 - 1$.

(a) Let h be a real number and find $f(2 + h)$.

(b) Show that $\dfrac{f(2 + h) - f(2)}{h} = 4 + h$.

48. Let $f(x) = 4x^2 - 12x + 9$. Find $g(x)$ so that $f(x) = [g(x)]^2$.

*49. In general, $f(a + b) = f(a) + f(b)$ is not true. However, for some special functions the preceding equality is true. Show that it is true for a linear function of the form $f(x) = mx$.

*50. In general, $f(cx) = cf(x)$ is not true. However, for some special functions, the preceding equality is true. Show that it is true for a linear function of the form $f(x) = mx$.

CHAPTER 6 SUMMARY

Review these key terms and concepts so that you are able to define or describe them. A clear understanding of these will be very helpful when reviewing the developments of this chapter.

Points on the coordinate plane are identified by ordered pairs of numbers, (x, y). The x-coordinate is called the **abscissa** and the y-coordinate is called the **ordinate.**

The **general form of a linear equation** is given by $Ax + By = C$, and its graph is a straight line. The x and y values of the points where the line crosses the axes are the ***x*** and **y-intercepts.**

The **slope** of a line through two points (x_1, y_1) and (x_2, y_2) is given as

$$m = \frac{y_2 - y_1}{x_2 - x_1} \qquad (x_2 \neq x_1)$$

Two parallel lines have the same slope; two perpendicular lines have slopes that are negative reciprocals of one another. The slope of a horizontal line is 0, and the slope of a vertical line is undefined.

Slope-intercept form: $y = mx + b$ where m is the slope and b is the y-intercept

Point-slope form: $y - y_1 = m(x - x_1)$ where m is the slope and (x_1, y_1) is a point on the line

The **distance formula** gives the distance, d, between two points (x_1, y_1) and (x_2, y_2):

$$d = \sqrt{(x_2 - x_1)^2 + (y_2 - y_1)^2}$$

The **midpoint formula** gives the coordinates of the midpoint of the line segment between two points (x_1, y_1) and (x_2, y_2):

$$\left(\frac{x_1 + x_2}{2}, \frac{y_1 + y_2}{2}\right)$$

A **function** is a correspondence between two sets, the **domain** and the **range,** such that for each value in the domain there corresponds exactly one value in the range. The symbol $\boldsymbol{f(x)}$ is used to mean the range value of the function at x. For example:

$$\text{If} \quad f(x) = 3x^2 - 5x + 7, \quad \text{then} \quad f(2) = 3(2)^2 - 5(2) + 7 = 1$$

REVIEW EXERCISES

The solutions to the following exercises can be found within the text of Chapter 6. Try to answer each question before referring to the text.

Section 6.1

1. Find the coordinates of the given points. Also state the quadrant in which each point is located.

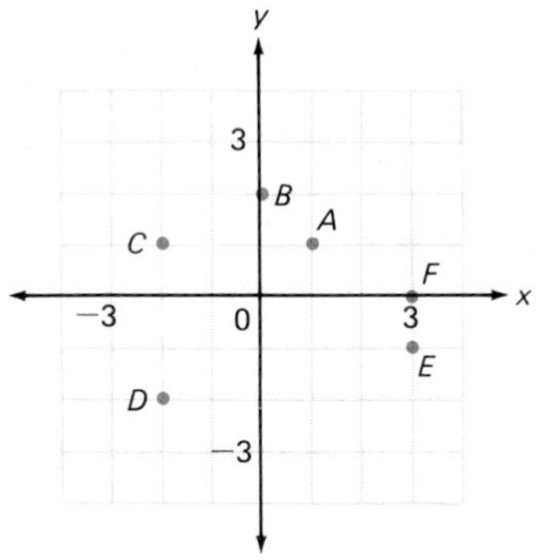

2. Graph the set of ordered pairs:
$\{(4, 1), (0, 2), (-\frac{5}{2}, -3), (\frac{3}{2}, -3), (-2, 3), (-\sqrt{2}, 0), (\pi, \pi)\}$
3. The base of a parallelogram has endpoints (0, 0) and (10, 0), and (2, 5) is a third vertex. If the fourth vertex is in quadrant I, find its coordinates.

Section 6.2

4. Graph the linear equation $2x + y = 6$ by using the intercepts.
5. Graph on a plane: **(a)** $x = 2$ **(b)** $y = -1$
6. Write the equation of the line through the point (2, 3) that is parallel to **(a)** the x-axis and **(b)** the y-axis.

Section 6.3

7. Find the slope of the line $y = 2x - 4$.
8. Find the slope of line ℓ determined by the points $(-3, 4)$ and $(1, -6)$.
9. Graph the line with slope $\frac{3}{2}$ that passes through the point $(-2, -2)$.
10. What is the slope of a horizontal line? Of a vertical line?
11. Line ℓ_1 has slope $\frac{2}{3}$ and is perpendicular to ℓ_2. If the lines intersect at $P(-1, 4)$, use the slope of ℓ_2 to find the coordinates of another point on ℓ_2.

Section 6.4

12. Find the equation of the line with slope $\frac{2}{3}$ passing through the point $(0, -5)$.
13. A horizontal line has y-intercept 5. Write an equation for the coordinates of any point (x, y) on the line.
14. Use the slope-intercept form to sketch the line $y = -\frac{1}{2}x + \frac{3}{2}$.
15. Solve the equation $3x + 5y = 10$ for y and identify the slope and the y-intercept.
16. Find the equation of the line through the points $(2, -3)$ and $(3, -1)$. Write the equation **(a)** in point-slope form, **(b)** in slope-intercept form, and **(c)** in the standard form $Ax + By = C$.
17. Write the equation of the line that is perpendicular to the line $5x - 2y = 2$ and that passes through the point $(-2, -6)$.

Section 6.5

18. Find the distance between $(-2, 3)$ and $(7, -1)$.
19. Show that the points $P(3, -11)$, $Q(-6, -2)$, and $R(8, 12)$ are the vertices of a right triangle.
20. Find the midpoint of the segment between $(-2, -3)$ and $(6, 2)$.
21. Show that $A(2, 2)$, $B(5, 5)$, $C(8, 2)$, and $D(5, -1)$ are the vertices of a square and verify that the diagonals of that square bisect each other.

Section 6.6

Graph the regions satisfying the given inequalities.

22. $y > 2$
23. $x > 3$
24. $3x - y < -2$
25. $2x + 3y < 1$
26. $|x| < 2$

Section 6.7

27. Explain why the following equation defines y as a function of x and find the domain: $y = \dfrac{1}{x - 1}$.
28. Find the domain of the function given by $y = \sqrt{1 - x}$.
29. Graph the function given by $y = -\frac{1}{2}x + 5$ and identify the domain and range. Also display the domain value 6 and its corresponding range value.
30. Let the function $y = -\frac{1}{2}x + 5$ have domain $\{x \mid 2 \le x \le 12\}$. Graph the function, find the range, and indicate both the domain and range on the graph.
31. Graph the function defined as follows:

$$y = \begin{cases} x + 2 & \text{for } -1 \le x \le 2 \\ -x + 3 & \text{for } 2 < x \le 5 \end{cases}$$

32. Graph the absolute-value function $y = |x|$.

Section 6.8

33. Let $h(x) = x^2 + 2x$. Find:
    (a) $h(0)$ **(b)** $h(1)$ **(c)** $h(2)$ **(d)** $h(3)$ **(e)** $3h(2)$ **(f)** $h(6)$
34. For $g(x) = \dfrac{1}{x}$, find:
    (a) $3g(x)$ **(b)** $g(3x)$ **(c)** $3 + g(x)$ **(d)** $g(3) + g(x)$ **(e)** $g(3 + x)$
    (f) $g\left(\dfrac{1}{x}\right)$
35. Let $f(x) = x^3 - 7x^2 + x + 1$ and $g(x) = 8x^2 - 2x - 2$. Find $s(x) = f(x) + g(x)$.
36. For the functions defined in Exercise 35, find $d(x) = f(x) - g(x)$ and find $d(1)$.
37. For the functions defined in Exercise 35, find $m(x) = f(x) \cdot g(x)$ and verify that $m(1) = f(1) \cdot g(1)$.
38. Let $q(x)$ be the quotient of $f(x) = 3x + 1$ divided by $g(x) = x^2 - 4$ and evaluate $q(0)$. Which values of x are not in the domain of q?

CHAPTER 6 TEST: STANDARD ANSWER

Use these questions to test your knowledge of the basic skills and concepts of Chapter 6. Then check your answers with those given at the back of the book.

1. Classify each statement as true or false.
   (a) The product of the slopes of two perpendicular lines not parallel to the axes is -1.
   (b) The graph of $x > 2y + 4$ is the half-plane above the line $x - 2y = 4$.
   (c) The domain of the function given by $y = \sqrt{x^2}$ is the set of real numbers.
   (d) A horizontal line cannot be the graph of a function.
   (e) A line whose y-intercept is 5 and whose slope is a positive number will have a negative x-intercept.
2. Write the equation of the line through the point $(-3, 5)$ with slope $-\frac{2}{5}$.
3. Use the x- and y-intercepts to graph the line $5x - 10y = 20$.
4. Find the slope of the line $2x + 7y = 1$.
5. Graph on the same set of axes: **(a)** $x = -3$ **(b)** $y = 4$
6. Write an equation of the line through the origin that is parallel to the line $3x + 4y = -2$.
7. Write an equation of the line through the point $(-2, 3)$ that is perpendicular to the line $2x - 3y = 6$.
8. Write the point-slope form of the line with slope -3 through the point $(2, -5)$.

Use the figure below for Exercises 9 through 15.

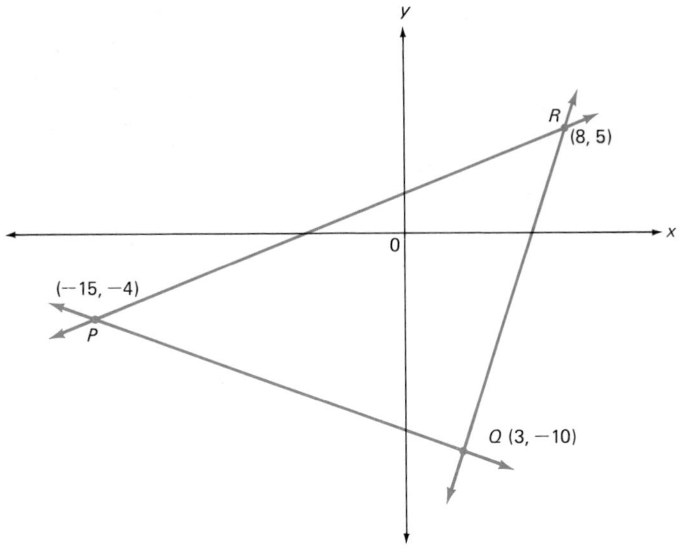

9. Write an equation for the line through P and Q.
10. Write an equation for the line through Q and R.
11. Find the coordinates of the midpoint of each segment: **(a)** PQ **(b)** RP
12. Evaluate: **(a)** $(PQ)^2$ **(b)** $(QR)^2$ **(c)** $(RP)^2$
13. Use the Pythagorean theorem to show that angle PQR is a right angle.
14. Make use of slopes to show that angle PQR is a right angle.
15. Line segment PR is the graph of a function. Find the domain and the range.
16. Find the equation of the line through $(-1, 2)$ and $(3, -2)$ in **(a)** point-slope form, **(b)** slope-intercept form, and **(c)** the form $Ax + By = C$.

17. A parallelogram $ABCD$ has three vertices located at $A(1, 1)$, $B(2, 3)$, and $C(5, 3)$. Find the coordinates of D and write the equation of line CD. Assume D is in quadrant I.

18. A parallelogram $PQRS$ has vertices located at $P(1, -1)$, $Q(3, 3)$, $R(7, 3)$, and $S(5, -1)$. Show that the line segments joining the midpoints of the sides of $PQRS$ form another parallelogram.

Graph each inequality.

19. $y - 3x < -2$ 20. $\{(x, y) \mid |x| > 5\}$

21. Explain why the equation $y^2 = x + 1$ does not define y to be a function of x.

22. Find the domain for each function: **(a)** $y = \sqrt{x - 3}$ **(b)** $y = \dfrac{1}{x^2 - 2}$

23. Graph the function defined as follows, and state the domain and range.

$$y = \begin{cases} -x + 2 & \text{for } -1 \le x < 1 \\ 2x + 1 & \text{for } 1 \le x < 2 \end{cases}$$

24. For the function g defined by $g(x) = \dfrac{1}{x^2}$, find **(a)** $g\left(\dfrac{1}{x}\right)$ **(b)** $g(x + 2)$

25. Let $f(x) = x^4 - x^3 + 2x^2 - 4x$ and $g(x) = x^2 + 3x - 5$.
Find **(a)** $f(x) - g(x)$ **(b)** $f(1) - g(3)$

CHAPTER 6 TEST: MULTIPLE CHOICE

1. Which of the following pairs of numbers are the x-intercept and y-intercept, respectively, for the equation $2y - 3x = 6$?
(a) 2; −3 **(b)** −3; 2 **(c)** −2; 3 **(d)** 3; −2 **(e)** None of the preceding

2. What is the equation of the straight line parallel to the x-axis that passes through the point $(-1, 4)$?
(a) $y = 4$ **(b)** $x = -1$ **(c)** $x + y = 3$ **(d)** $x = 4$ **(e)** $y = -1$

3. Find the slope of the line determined by the points $(-4, 2)$ and $(-2, -3)$.
(a) $-\frac{2}{5}$ **(b)** $\frac{2}{5}$ **(c)** $\frac{5}{2}$ **(d)** $-\frac{5}{2}$ **(e)** None of the preceding

4. What is the equation of the line that passes through the points $(0, -1)$ and $(-2, 1)$?
(a) $y = x - 1$ **(b)** $y = -x$ **(c)** $y = x + 1$ **(d)** $y = -x - 1$ **(e)** None of the preceding

5. Which of the following is the equation of a line with intercepts at $(0, -4)$ and $(2, 0)$?
(a) $2x + 4y = 8$ **(b)** $4x + 2y = 8$ **(c)** $4x + 2y = -8$
(d) $2x - 4y = 8$ **(e)** None of the preceding

6. Which of the following statements are correct?
   I. The slope of a horizontal line is undefined.
   II. The slope of a vertical line is 0.
   III. The slope of two perpendicular lines (not parallel to the coordinate axes) are negative reciprocals of one another.

   (a) Only I **(b)** Only II **(c)** Only III **(d)** Only I and II **(e)** Only I and III

7. Which of the following is the equation of a line perpendicular to $2x - 3y = 6$?
(a) $2x + 3y = 6$ **(b)** $3x + 2y = 6$ **(c)** $3x - 2y = 6$ **(d)** $3y - 2x = 6$ **(e)** None of the preceding

8. What is the slope-intercept form of the line through $(2, -3)$ and $(-1, 6)$?
(a) $y = 3x + 3$ **(b)** $y = -3x + 3$ **(c)** $y = -\frac{1}{3}x + \frac{11}{3}$ **(d)** $y = -3x - 3$ **(e)** $y = -\frac{1}{3}y - \frac{7}{3}$

9. Find the equation of the line parallel to $x + 2y = 4$ that passes through the point $(-2, 1)$.
(a) $y = -\frac{1}{2}x + 4$ **(b)** $y = \frac{1}{2}x - 4$ **(c)** $y = 2x - 4$ **(d)** $y = -2x - 4$ **(e)** None of the preceding

10. Find the distance between $(2, -3)$ and $(-3, -2)$.
(a) $\sqrt{2}$ **(b)** $\sqrt{24}$ **(c)** $\sqrt{26}$ **(d)** $\sqrt{50}$ **(e)** None of the preceding

11. What are the coordinates of the midpoint of the segment between $(-3, 4)$ and $(7, -2)$?
(a) $(4, 6)$ **(b)** $(5, 3)$ **(c)** $(1, 2)$ **(d)** $(4, -3)$ **(e)** None of the preceding

12. Which of the following inequalities is shown by the graph at the right?
(a) $y \geq x + 1$
(b) $y > x + 1$
(c) $y \leq x + 1$
(d) $y < x + 1$
(e) $y \geq x - 1$

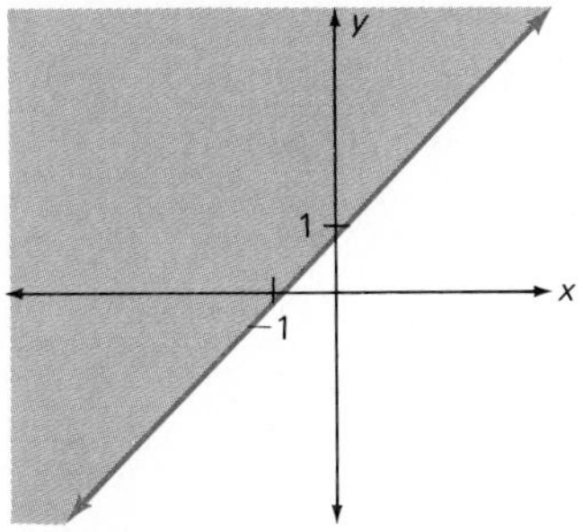

13. Which of the following statements is correct for the inequality $2x - 3y > 6$?
I. The graph consists of all points below the line $y = \frac{2}{3}x - 2$.
II. The graph consists of all points above the line $2x - 3y = 6$.
III. The graph consists of all points on or below the line $3y - 2x = -6$.
(a) Only I **(b)** Only II **(c)** Only III **(d)** Only I and III **(e)** None of the preceding

14. Which of the following are correct for the function defined by $f(x) = 2x^2 - 3$?
I. $f\left(\frac{1}{3}\right) = \frac{1}{f(3)}$
II. $f(-1) = f(1)$
III. $[f(3)]^2 = f(9)$
(a) Only I **(b)** Only II **(c)** Only III **(d)** Only II and III **(e)** I, II, and III

15. Consider the function defined below and find $f(-3) + f(1)$.

$$f(x) = \begin{cases} x - 2 & \text{if } -3 \leq x \leq 1 \\ -x + 1 & \text{if } 1 < x \end{cases}$$

(a) -6 **(b)** 5 **(c)** 4 **(d)** -5 **(e)** None of the preceding

ANSWERS TO THE TEST YOUR UNDERSTANDING EXERCISES

Page 238

1. (a) 4; (b) 4
2. (a) 6; (b) -2
3. (a) -4; (b) -3
4. (a) 3; (b) -6
5. (a) 4; (b) 1
6. (a) $\frac{5}{3}$; (b) $-\frac{5}{2}$
7. (a) $\frac{2}{3}$; (b) -2
8. (a) $\frac{5}{3}$; (b) 5
9. (a) 1; (b) -3

Page 244

1. $\dfrac{6-5}{4-1} = \dfrac{1}{3}$
2. $\dfrac{3-(-5)}{-3-3} = -\dfrac{8}{6} = -\dfrac{4}{3}$
3. $\dfrac{1-(-3)}{-1-(-2)} = \dfrac{4}{1} = 4$
4. $\dfrac{1-0}{0-(-1)} = \dfrac{1}{1} = 1$
5.

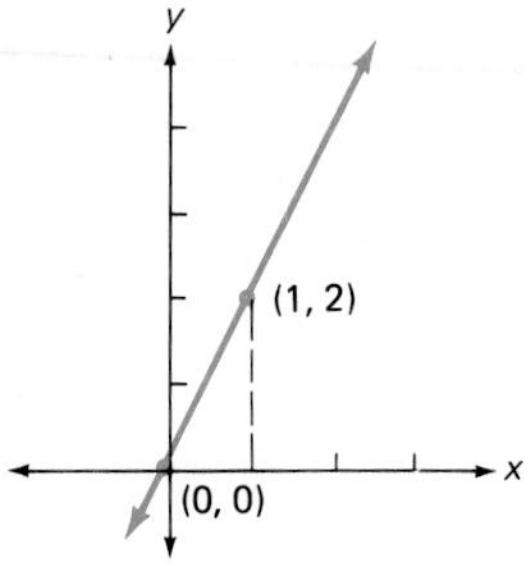

6.

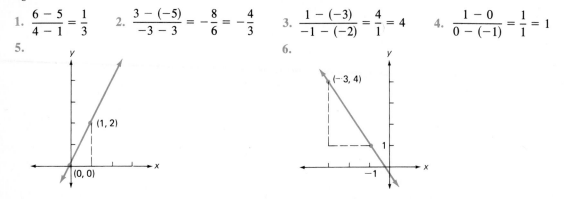

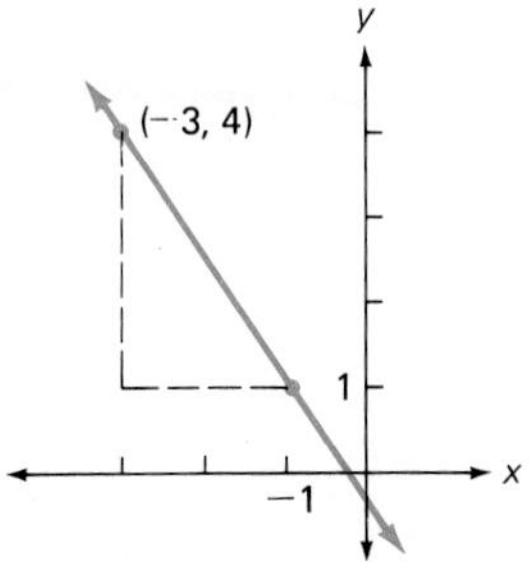

Page 251

1. $y = -2x + 6; -2; 6$
2. $y = 3x - 5; 3; -5$
3. $y = -x + 2; -1, 2$
4. $y = -\frac{1}{2}x + 4; -\frac{1}{2}; 4$
5. $y = \frac{1}{2}x - 2; \frac{1}{2}; -2$
6. $y = \frac{1}{2}x - 2; \frac{1}{2}; -2$
7. $y = \frac{3}{2}x - \frac{5}{2}; \frac{3}{2}; -\frac{5}{2}$
8. $y = 3x - 5; 3; -5$
9. $y = -\frac{1}{2}x + \frac{7}{2}; -\frac{1}{2}; \frac{7}{2}$
10. $y = 0x + 2; 0; 2$
11. $y = x - \frac{1}{4}; 1; -\frac{1}{4}$
12. $y = 6x - 2; 6; -2$

Page 257

1. $\sqrt{(1-5)^2 + (1-4)^2} = \sqrt{16+9} = 5$
2. $\sqrt{(-3-5)^2 + (2-8)^2} = \sqrt{64+36} = 10$
3. $\sqrt{(-2-3)^2 + (0-(-12))^2} = \sqrt{25+144} = 13$
4. $\sqrt{(0-1)^2 + (1-0)^2} = \sqrt{2} = 1.41$ (approximately)

Page 264

1. (d)
2. (b)
3. (e)
4. (a)
5. (f)
6. (c)

Page 268

1. Function, all reals
2. Function, all reals
3. Not a function
4. Function, all reals
5. Function, all $x \neq -2$
6. Function, all reals
7. Function, all $x \geq \frac{1}{2}$
8. Not a function
9. Not a function

Page 274

1. 11
2. 2
3. 0
4. -7
5. $\frac{15}{4}$
6. Does not exist
7. $-\frac{8}{15}$
8. $-\frac{5}{7}$
9. $4x^2 - 6x$
10. $2x^2 - 6x$
11. $x^2 - 9x + 18$
12. $x^2 - 3x$
13. $\dfrac{1-3x}{x^2}$
14. $\dfrac{1}{x^2 - 3x}$
15. $x^4 - 3x^2$
16. $x^4 - 6x^3 + 9x^2$

CUMULATIVE REVIEW QUESTIONS: CHAPTERS 4–6

1. Simplify: **(a)** $\dfrac{3x-2}{2-3x}$ **(b)** $\dfrac{2x^2+3x-2}{2x^2+5x+2}$

Perform the indicated operations and reduce to lowest terms.

2. $\dfrac{3x-2}{2x-3}-\dfrac{4-x}{2x-3}$

3. $\dfrac{4}{x^2-4}+\dfrac{2}{2-x}$

4. $\dfrac{a-1}{4a}+\dfrac{a+2}{3a^2}-\dfrac{a+1}{6a}$

5. $\dfrac{x}{x-y}-\dfrac{2xy}{x^2-y^2}-\dfrac{y}{x+y}$

6. $\dfrac{9a^2b^4-6a^4b^6+3a^2b^2}{3a^2b^2}$

7. $\dfrac{a-b}{a+b}\cdot\dfrac{a^2+b^2}{a^2-b^2}\cdot\dfrac{a^2+2ab+b^2}{a^2-2ab+b^2}$

8. $\dfrac{x^2-6x+9}{8x^3-10x^2+3x}\div\dfrac{x^2-2x-3}{2x^3+x^2-x}$

9. $\left(\dfrac{x^2-9}{9x^2-4}\cdot\dfrac{9x^2+12x+4}{2x^2+7x+3}\right)\div\dfrac{3-x}{6x^2-x-2}$

Simplify. (Express answers with positive exponents only.)

10. $\dfrac{\frac{1}{x}-\frac{1}{4}}{x-4}$

11. $\dfrac{x^{-3}+y^{-2}}{x^{-2}+y^{-3}}$

Solve for x.

12. $\dfrac{2}{x^2-4}+\dfrac{1}{x-2}=\dfrac{7}{x+2}$

13. $\dfrac{5x-3}{3x+1}=\dfrac{11}{8}$

14. Find two consecutive integers such that one-third of the smaller one is seven less than three-fourths of the larger one.

15. Working alone, Jennifer can complete a job in 5 hours. Juanita can complete the same job in 4 hours. How long will it take them working together?

16. z varies directly as x and inversely as y. If $z=\frac{3}{4}$ when $x=3$ and $y=15$, find z when $x=4$ and $y=10$.

17. The volume of a right circular cone varies jointly as its height and the square of the radius of the base. The volume is 32π cubic centimeters when $h=6$ cm and $r=4$ cm. Find the volume when $h=9$ cm and $r=5$ cm.

Evaluate.

18. **(a)** $(-27)^{-2/3}$ **(b)** $(16)^{3/4}$

Simplify. (Express answers with positive exponents only.)

19. $(a^{1/3}b^{-1/3})(a^{-1/2}b^{1/2})$

20. $\dfrac{a^{3/4}b^{-1/3}}{a^{-1/4}b^{-1/2}}$

Perform the indicated operations and simplify.

21. **(a)** $\sqrt{6}\cdot\sqrt{8}$ **(b)** $\dfrac{\sqrt{81}}{3\sqrt{3}}$

22. $\sqrt{45}+2\sqrt{20}-\dfrac{10}{\sqrt{5}}$

23. $\sqrt{12a^2b^3}+2\sqrt{3a^2b^3}-b\sqrt{27a^2b}$

Simplify.

24. $3\sqrt{-8}+5\sqrt{-18}$

25. $\sqrt{-25}\cdot\sqrt{-12}$

Complete the indicated operations and express your answer in the form $a + bi$.

26. $(5 - i)(3 + 2i)$ 27. $\dfrac{2 + i}{3 - i}$

Solve for x.

28. $2x^2 - 5x - 3 = 0$ 29. $9x^2 - 30x + 25 = 0$
30. $x^2 - 4x = 8$ 31. $2x^2 - 6x + 5 = 0$

Use the discriminant to describe the solutions for each equation.

32. $3x^2 - 5x + 1 = 0$ 33. $2x^2 + 4x + 5 = 0$

34. Find the values of k so that the equation $3x^2 - 2x + k = 0$ will have two real roots.
35. The sum of a number and its reciprocal is 3. Find the numbers.
36. The formula $s = 96t - 16t^2$ gives the height in feet, s, after t seconds of an object thrown straight upward from the ground with an initial velocity of 96 feet per second. When will the object be 128 feet above the ground?
37. Find the slope of the line $3x - 5y = 8$.

Write the equation for each of the following lines.

38. Through the points $(2, -3)$ and $(3, -4)$.
39. Through the point $(4, -2)$ with slope of $-\frac{3}{4}$.
40. Through the point $(3, -2)$ and perpendicular to the line $3x + 2y = 6$.
41. Through the origin and parallel to the line $2x + 4y + 6 = 0$.

Graph each inequality.

42. $2x + y < -4$ 43. $\{(x, y) \mid |x| < 3\}$

44. Find the domain: **(a)** $y = \sqrt{2 - x}$ **(b)** $y = \dfrac{1}{x^2 - 4}$
45. Given the points $(3, -5)$ and $(5, -3)$, find **(a)** the midpoint of the line segment between the points and **(b)** the distance between the points.
46. A quadrilateral is formed by taking the points $A(0, 0)$, $B(2, 3)$, $C(7, 3)$, and $D(5, 0)$ in order. Prove that the figure is a parallelogram by showing that sides AB and CD have the same slopes and are equal in length.
47. If $f(x) = \dfrac{1}{x^3}$, find **(a)** $f\left(\dfrac{1}{x}\right)$ and **(b)** $f(x + 1)$.
48. Let $f(x) = 2x^2 - 5x + 1$. Find $3f(2) + 2f(3)$.
49. Let $f(x) = x^2 - 3x + 2$ and $g(x) = x^3 - x^2 + 3$.
Find **(a)** $f(x) - g(x)$ and **(b)** $f(3) - g(2)$.
50. Graph the function defined as follows, and state the domain and the range.

$$f(x) = \begin{cases} x - 2 & \text{for } -2 \le x < 1 \\ 3 - x & \text{for } 1 \le x < 3 \end{cases}$$

CHAPTER 7

QUADRATIC FUNCTIONS AND THE CONIC SECTIONS, WITH APPLICATIONS

7.1 GRAPHING QUADRATIC FUNCTIONS

A function defined by a polynomial expression of degree 2 is referred to as a **quadratic function** in x. Thus the following are all examples of quadratic functions in x:

$$f(x) = -3x^2 + 4x + 1$$

$$g(x) = 7x^2 - 4$$

$$h(x) = x^2$$

If $a = 0$, then the resulting polynomial no longer represents a quadratic function; $f(x) = bx + c$ is a linear function.

The most general form of such a quadratic function is $f(x) = ax^2 + bx + c$ where a, b, and c represent constants, with $a \neq 0$.

The simplest quadratic function is given by $f(x) = x^2$. The graph of this quadratic function will serve as the basis for drawing the graph of any quadratic function $f(x) = ax^2 + bx + c$. We can save some labor by noting the *symmetry* that exists. For example, note the following:

$$f(-3) = f(3) = 9 \qquad f(-1) = f(1) = 1 \qquad f(-\tfrac{1}{2}) = f(\tfrac{1}{2}) = \tfrac{1}{4}$$

In general, for this function,

$$f(-x) = (-x)^2 = x^2 = f(x)$$

Note: When $f(-x) = f(x)$, the graph is said to be *symmetric* with respect to the y-axis.

Greater accuracy can be obtained by using more points. But since we can never locate an infinite number of points, we must admit that there is a certain amount of faith involved in connecting the points as we did.

The accompanying table of values gives several ordered pairs of numbers that are coordinates of points on the graph of $y = x^2$. When these points are located on a rectangular system and connected by a smooth curve, the graph of $y = f(x) = x^2$ is obtained. The curve is called a **parabola,** and every quadratic function of the form $y = ax^2 + bx + c$ has such a parabola as its graph. The domain of the function is the set of all real numbers. The range of the function depends on the constants a, b, and c. For the function $y = f(x) = x^2$, the range consists of all $y \geq 0$.

x	$y = x^2$
−3	9
−2	4
−1	1
0	0
1	1
2	4
3	9

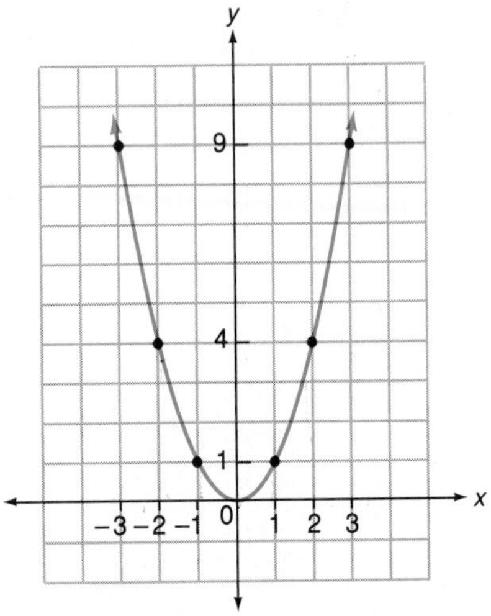

An important feature of such a parabola is that it is symmetric about a vertical line called its **axis of symmetry.** The graph of $y = x^2$ is symmetric with respect to the y-axis. This symmetry is due to the fact that $(-x)^2 = x^2$.

The parabola has a *turning point,* called the **vertex,** which is located at the intersection of the parabola with its axis of symmetry. For the preceding graph the coordinates of the vertex are (0, 0), and 0 is the *minimum value* of the function.

From the graph you can see that, reading from left to right, the curve is "falling" down to the origin and then is "rising." These features are technically described as f **decreasing** and f **increasing.**

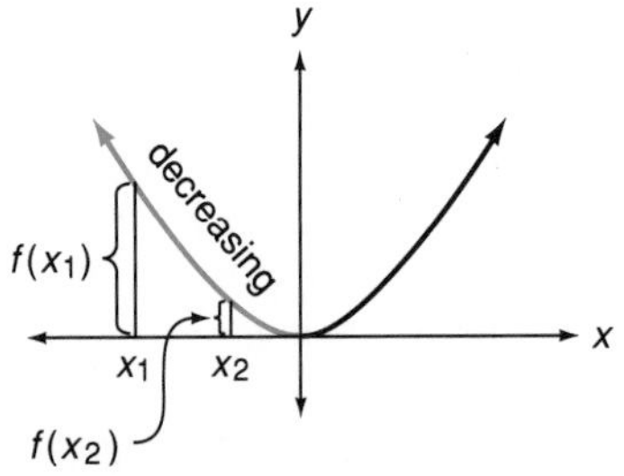

$f(x)$ is decreasing on $(-\infty, 0]$ because for *each* pair x_1, x_2 in this interval, if $x_1 < x_2$, then $f(x_1) > f(x_2)$.

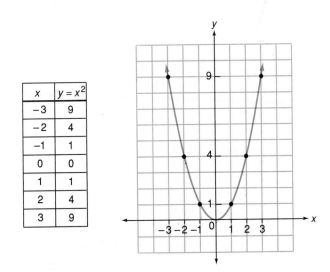

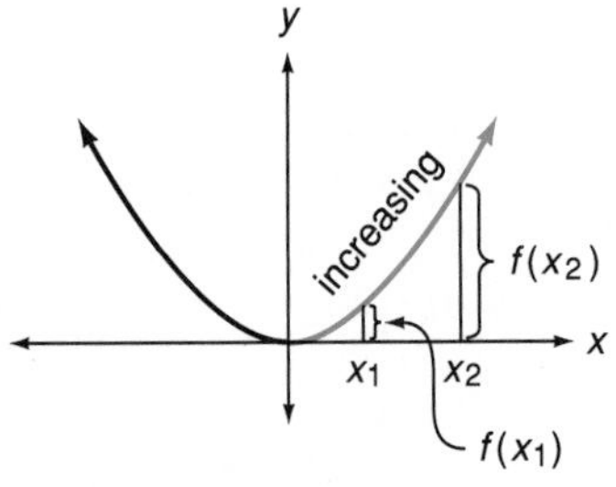

$f(x) = x^2$ is increasing on $[0, \infty)$ because for *each* pair x_1, x_2 in this interval, if $x_1 < x_2$ then $f(x_1) < f(x_2)$.

The graph of $y = x^2$ can be used as a guide to draw the graph of other quadratic functions. In the following illustrations, the graph of $y = x^2$ is shown as a dashed curve.

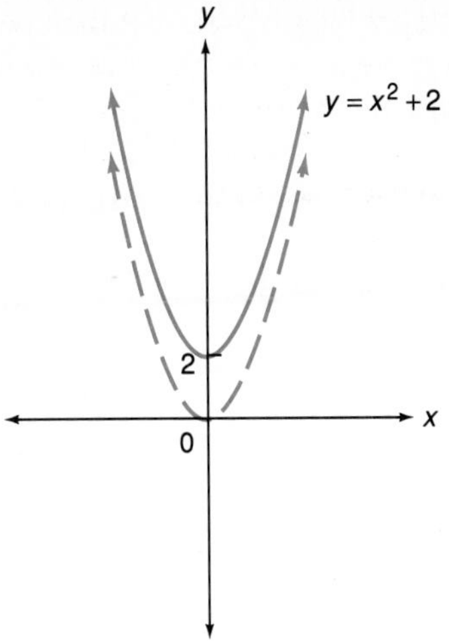

The graph of $y = x^2 + 2$ is congruent to that of $y = x^2$, but is shifted 2 units up.

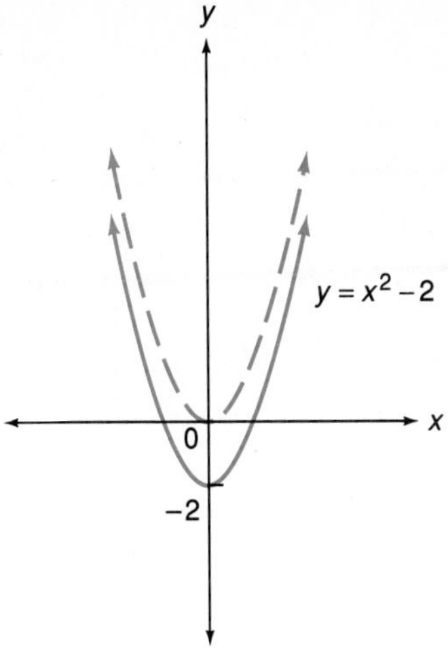

The graph of $y = x^2 - 2$ is congruent to that of $y = x^2$, but is shifted 2 units down.

Next consider the graph of $y = (x + 2)^2$. In this case first add 2 to x, and then square. For $x = -2$, $y = 0$ and the vertex of the curve is at $(-2, 0)$. The graph is congruent to that for $y = x^2$, but is shifted 2 units to the left. In a similar fashion, note that the graph of $y = (x - 2)^2$ is shifted 2 units to the right. Verify the entries in the table of values given beside the following graphs.

$y = (x + 2)^2$

x	−5	−4	−3	−2	−1	0	1
y	9	4	1	0	1	4	9

The axis of symmetry is the line $x = -2$

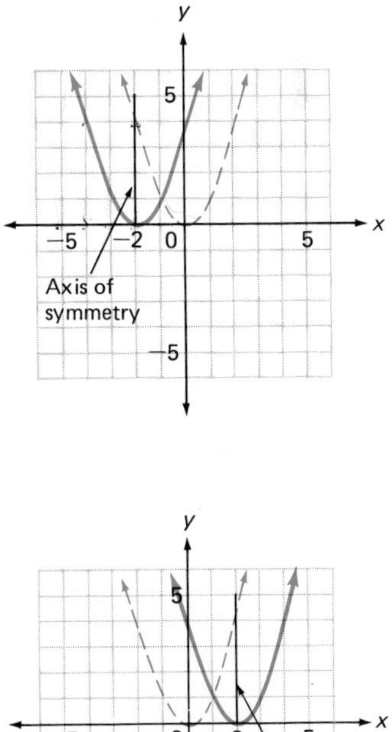

Both of these parabolas are congruent to the basic parabola $y = x^2$. Each may be graphed by translating *(shifting) the parabola $y = x^2$ by 2 units, to the right for $y = (x - 2)^2$ and to the left for $y = (x + 2)^2$.*

$y = (x - 2)^2$

x	−1	0	1	2	3	4	5
y	9	4	1	0	1	4	9

The axis of symmetry is the line $x = 2$

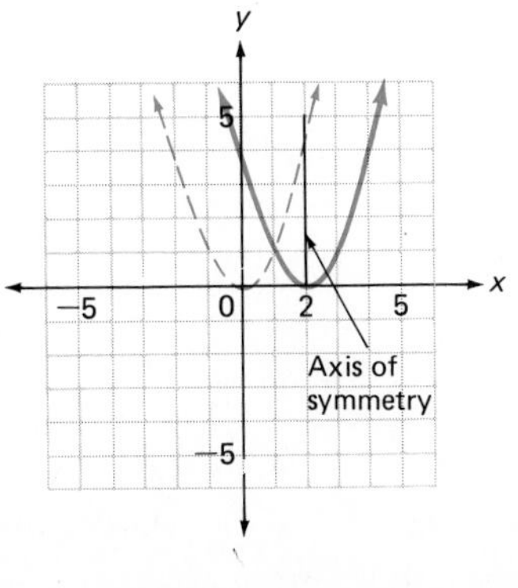

In each of the graphs drawn thus far, the coefficient of x^2 has been 1. If the coefficient is -1, it has the effect of reflecting the graph through the x-axis. The domain of the function is still the set of real numbers, but the range is the set of nonpositive real numbers. The vertex is at (0, 0) and 0 is the *maximum value* of the function. Again, the graph of $y = x^2$ is shown as a dashed curve. Since the graph of $y = x^2$ bends "upward," we say that the curve is **concave up.** Also, since $y = -x^2$ bends "downward," we say that the curve is **concave down.**

The graph of $y = -x^2$ may also be obtained by multiplying each of the ordinates of $y = x^2$ by -1. This step has the effect of "flipping" the parabola $y = x^2$ downward, a ***reflection*** *in the x-axis.*

$y = -x^2$

x	−3	−2	−1	0	1	2	3
y	−9	−4	−1	0	−1	−4	−9

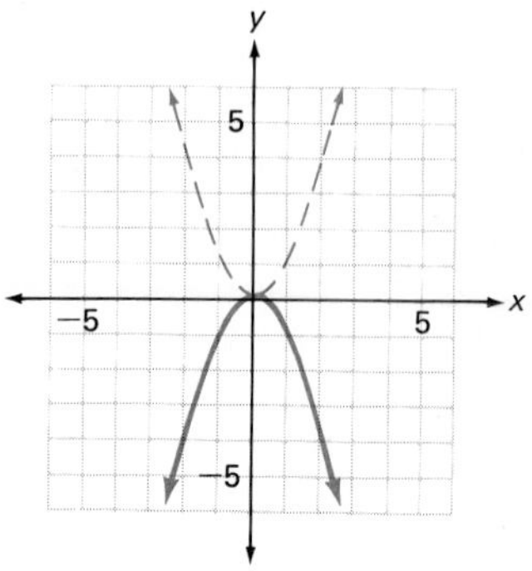

When the coefficient a in $y = ax^2$ is other than 1, then the graph of $y = ax^2$ can be obtained by multiplying each ordinate of $y = x^2$ by the number a.

EXAMPLE 1 Graph: **(a)** $y = \frac{1}{2}x^2$ **(b)** $y = 2x^2$

Note that the graph of $y = 2x^2$ is "steeper" than that of $y = x^2$; the graph of $y = \frac{1}{2}x^2$ is not as steep as that of $y = x^2$.

Solution

(a) $y = \frac{1}{2}x^2$

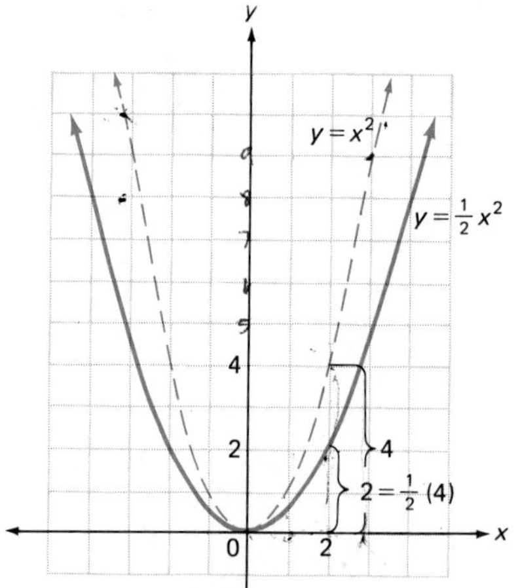

Each ordinate (y-value) is one-half that of the graph for $y = x^2$, shown as a dashed curve.

(b) $y = 2x^2$

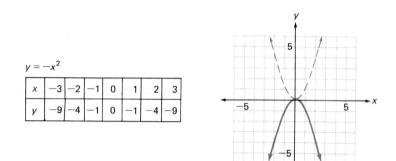

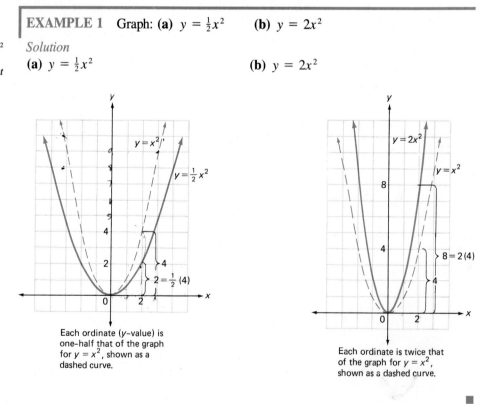

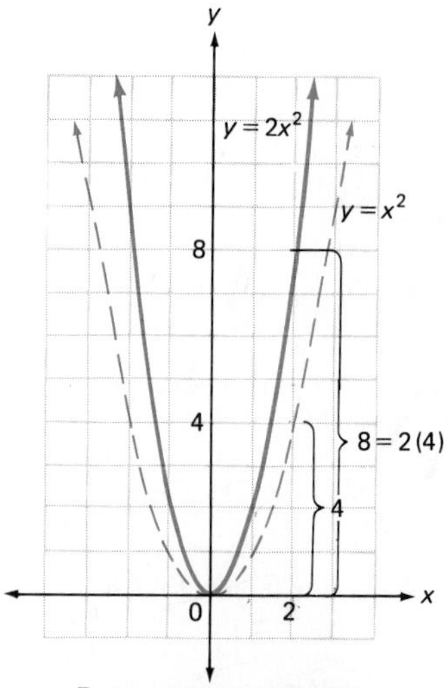

Each ordinate is twice that of the graph for $y = x^2$, shown as a dashed curve.

■

EXAMPLE 2 Graph: $y = f(x) = -x^2 + 2$. State where the function is increasing or decreasing. What is the concavity?

Solution Consider the graph of $y = -x^2$ and shift it up 2 units.

$y = -x^2 + 2$

x	-3	-2	-1	0	1	2	3
y	-7	-2	1	2	1	-2	-7

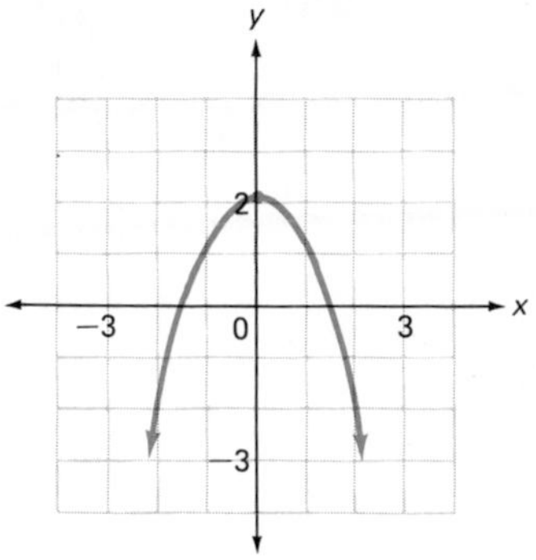

Observe that to say "f is increasing for all $x \leq 0$" means the same as saying "f is increasing on $(-\infty, 0]$." Similarly, we say that f is decreasing on $[0, \infty)$.

The function f is increasing for all $x \leq 0$, it is decreasing for all $x \geq 0$, and the curve is concave down. ■

TEST YOUR UNDERSTANDING
Think Carefully

Match each graph with one of the given quadratic equations.

(a) $y = x^2 - 1$
(b) $y = x^2 + 1$
(c) $y = (x - 1)^2$
(d) $y = (x + 1)^2$
(e) $y = -x^2 + 1$
(f) $y = -(x + 1)^2$

1.

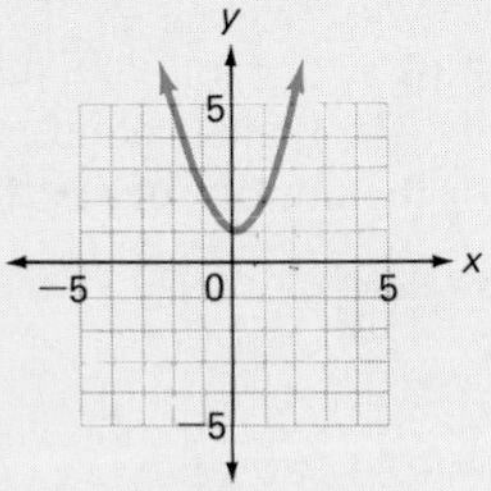

2.

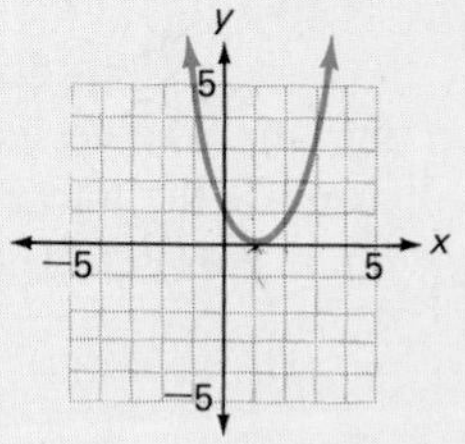

3.

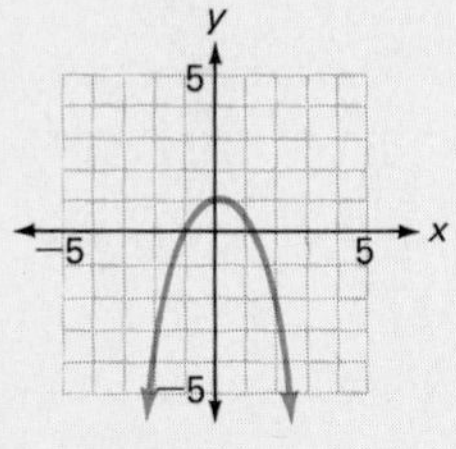

4.

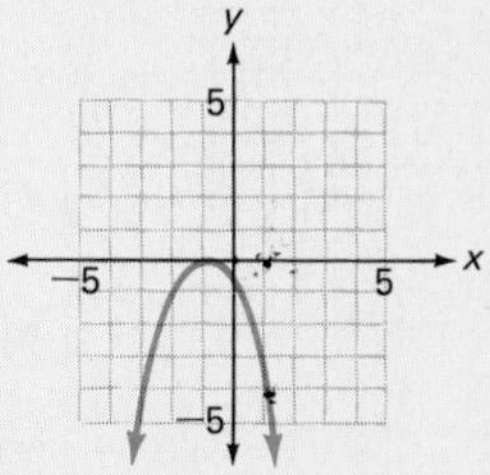

5.

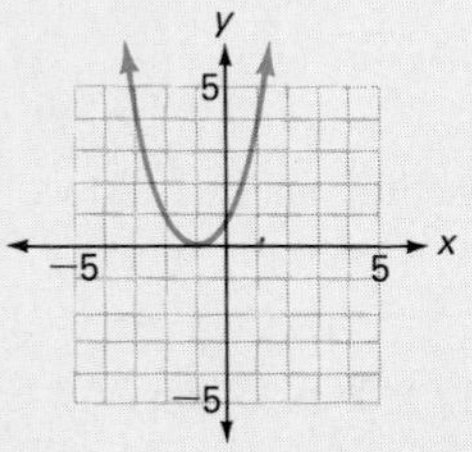

6.

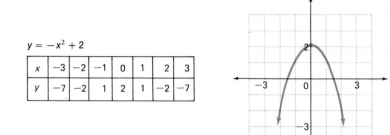

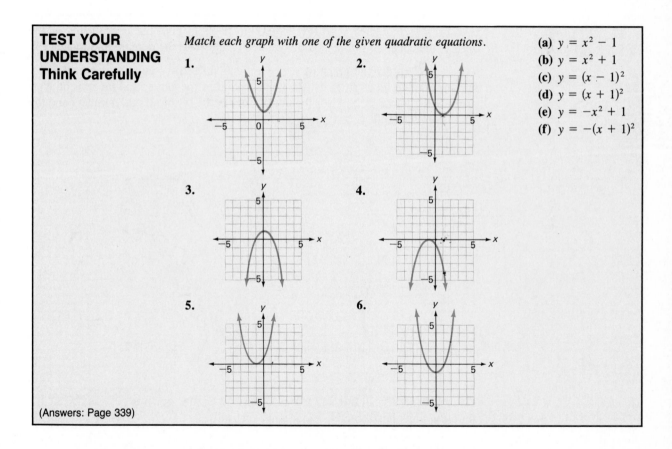

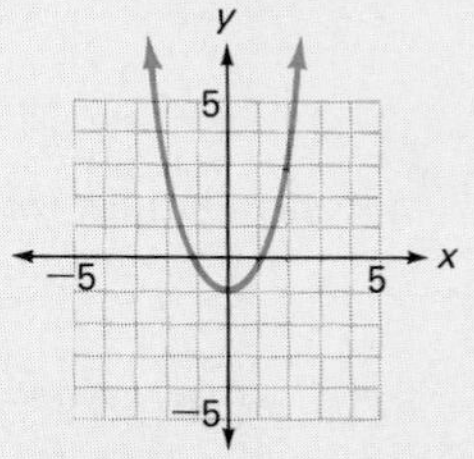

(Answers: Page 339)

Let us now put several ideas together and draw the graph of this function:

$$y = f(x) = (x + 2)^2 - 2$$

An effective way to do this is to begin with the graph of $y = x^2$, shift the graph 2 units to the left for $y = (x + 2)^2$, and then 2 units down for the graph of $f(x) = (x + 2)^2 - 2$, as in the following figures.

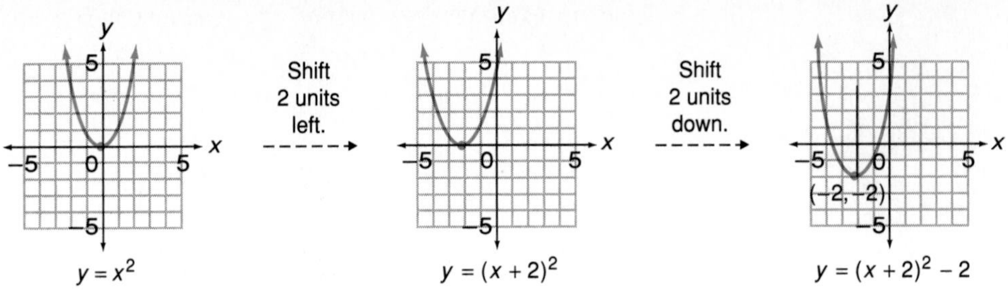

Note that the graph of $y = (x + 2)^2 - 2$ is congruent to the graph of $y = x^2$. The vertex of the curve is at $(-2, -2)$, and the axis of symmetry is the line $x = -2$. The minimum value of the function, -2, occurs at the vertex. Also observe that the domain consists of all numbers x, and the range consists of all numbers $y \geq -2$.

EXAMPLE 3 Graph $y = f(x) = -(x - 2)^2 + 1$, give the coordinates of the vertex, the equation of the axis of symmetry, and state the domain and range of f.

Solution Consider the graph of $y = -x^2$, and shift this 2 units to the right and 1 unit up. The vertex is at $(2, 1)$, the highest point of the curve, and the axis of symmetry is the line $x = 2$. Also, the domain of f is the set of all real numbers and the range consists of all $y \leq 1$.

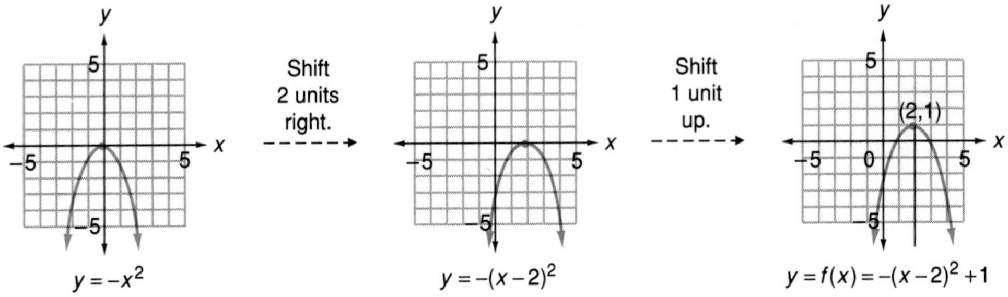

■

We may summarize our results thus far as follows:

All such parabolas may be described as being vertical since their axis of symmetry is vertical. Vertical parabolas open either upward or downward.

The graph of $y = a(x - h)^2 + k$ is congruent to the graph of $y = ax^2$, but is shifted h units horizontally, and k units vertically.

(a) The horizontal shift is to the right if $h > 0$, and to the left if $h < 0$.
(b) The vertical shift is upward if $k > 0$, and downward if $k < 0$.

The vertex is at (h, k) and the axis of symmetry is the line $x = h$.

(a) If $a < 0$, the parabola opens downward, and k is the maximum value.
(b) If $a > 0$, the parabola opens upward, and k is the minimum value.

EXAMPLE 4 Graph the parabola $y = f(x) = -2(x - 3)^2 + 4$.

Solution The graph will be a parabola congruent to $y = -2x^2$, with vertex at (3, 4) and with $x = 3$ as axis of symmetry. A brief table of values, together with the graph, is shown.

The function is increasing on $(-\infty, 3]$, decreasing on $[3, \infty)$, and the curve is concave down.

Vertex ↓ — Symmetric around $x = 3$ — Symmetric around $x = 3$

x	3	2	4	1	5
y	4	2	2	−4	−4

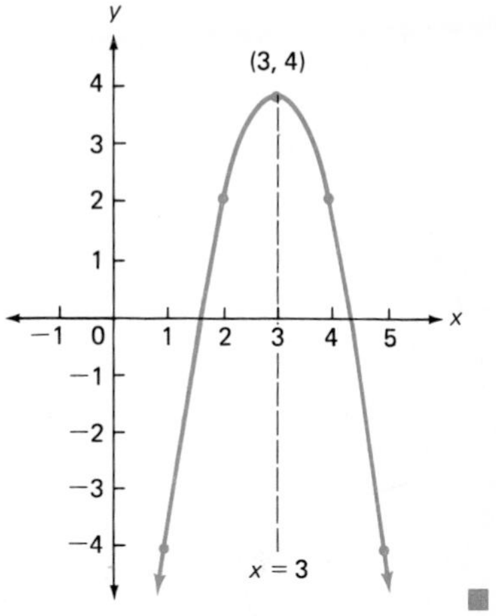

Interchanging the variables x and y in the equation $y = x^2$ produces $x = y^2$ whose graph is a horizontal parabola that opens to the right.

x	4	1	0	1	4
y	−2	−1	0	1	2

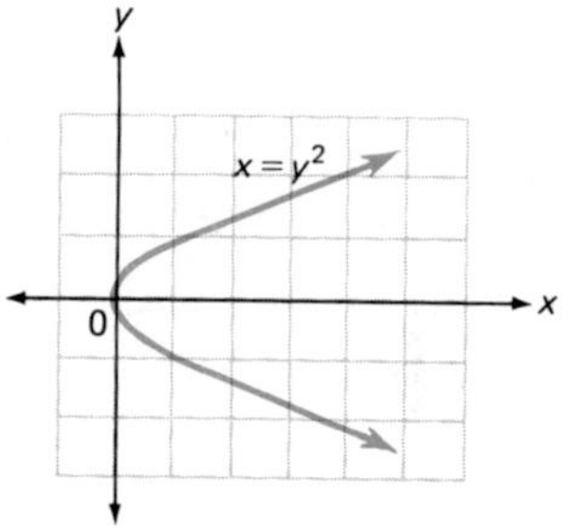

The graph of any quadratic equation of the form $x = ay^2 + by + c$ is a horizontal parabola that opens to the right when $a > 0$, or to the left when $a < 0$. These parabolas can be graphed using procedures similar to those used for the vertical parabolas. For example, the parabola $x = y^2 + 3$ can be obtained by shifting the basic horizontal parabola 3 units to the right. The axis of symmetry is the x-axis.

EXAMPLE 5 Graph $x = (y + 2)^2 - 4$ and identify the vertex and the axis of symmetry.

Solution Begin with the parabola $x = y^2$, shift 2 units downward and 4 units to the left. The vertex is $(-4, -2)$ and $y = -2$ is the axis of symmetry.

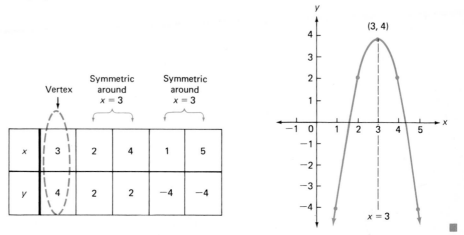

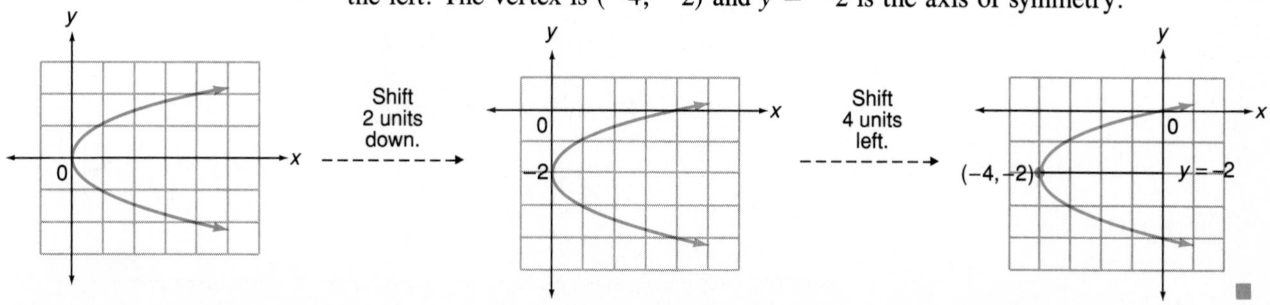

The vertical line test (page 270) applied to a horizontal parabola shows that such a parabola cannot be the graph of a function of x. Thus, the equations $x = y^2$ and $x = (y + 2)^2 - 4$ do not define functions of x. However, these equations, in fact *any* equation in the two variables x and y, are said to define a **relation.** Thus all functions are relations, but many relations are not functions.

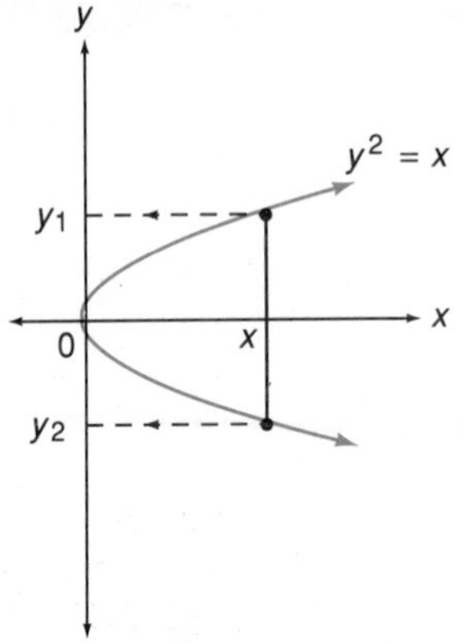

A vertical line through $x > 0$ intersects the graph more than once, producing more than one y-value. *This is not the graph of a function of* x.

The domain of a relation that is given by an equation in x and y consists of all allowable values of x, and the range is the set of all corresponding values of y. For example, from the preceding graph we see that the relation $x = (y + 2)^2 - 4$ has domain all $x \geq -4$ and range all real numbers y.

EXERCISES 7.1

Draw each set of graphs on the same axes.

1. **(a)** $y = x^2$ **(b)** $y = (x - 1)^2$ **(c)** $y = (x - 1)^2 + 3$
2. **(a)** $y = x^2$ **(b)** $y = (x + 1)^2$ **(c)** $y = (x + 1)^2 - 3$
3. **(a)** $y = -x^2$ **(b)** $y = -(x - 1)^2$ **(c)** $y = -(x - 1)^2 + 3$
4. **(a)** $y = -x^2$ **(b)** $y = -(x + 1)^2$ **(c)** $y = -(x + 1)^2 - 3$
5. **(a)** $y = x^2$ **(b)** $y = 2x^2$ **(c)** $y = 3x^2$
6. **(a)** $y = -x^2$ **(b)** $y = -\frac{1}{2}x^2$ **(c)** $y = -\frac{1}{2}x^2 + 1$

Draw the graph of each function.

7. $y = x^2 + 3$
8. $y = (x + 3)^2$
9. $y = -x^2 + 3$
10. $y = -(x + 3)^2$
11. $y = 3x^2$
12. $y = 3x^2 + 1$
13. $y = \frac{1}{4}x^2$
14. $y = \frac{1}{4}x^2 - 1$
15. $y = \frac{1}{4}x^2 + 1$
16. $y = -2x^2$
17. $y = -2x^2 + 2$
18. $y = -2x^2 - 2$

Graph each of the following functions. Where is the function increasing and decreasing? Describe the concavity.

19. $f(x) = (x - 1)^2 + 2$
20. $f(x) = (x + 1)^2 - 2$
21. $f(x) = -(x + 1)^2 + 2$
22. $f(x) = -(x + 1)^2 - 2$
23. $f(x) = 2(x - 3)^2 - 1$
24. $f(x) = 2(x + \frac{5}{4})^2 + \frac{5}{4}$

State **(a)** *the coordinates of the vertex,* **(b)** *the equation of the axis of symmetry,* **(c)** *the domain, and* **(d)** *the range for each of the following functions.*

25. $y = f(x) = (x - 3)^2 + 5$
26. $y = f(x) = (x + 3)^2 - 5$
27. $y = f(x) = -(x - 3)^2 + 5$
28. $y = f(x) = -(x + 3)^2 - 5$
29. $y = f(x) = 2(x + 1)^2 - 3$
30. $y = f(x) = \frac{1}{2}(x - 4)^2 + 1$
31. $y = f(x) = -2(x - 1)^2 + 2$
32. $y = f(x) = -\frac{1}{2}(x + 2)^2 - 3$
33. $y = f(x) = \frac{1}{4}(x + 2)^2 - 4$
34. $y = f(x) = 3(x - \frac{3}{4})^2 + \frac{4}{5}$

35. The graph of $y = ax^2$ passes through the point $(1, -2)$. Find a.

36. The graph of $y = ax^2 + c$ has its vertex at $(0, 4)$ and passes through the point $(3, -5)$. Find the values for a and c.

37. Find the value for k so that the graph of $y = (x - 2)^2 + k$ will pass through the point $(5, 12)$.

38. Find the value for h so that the graph of $y = (x - h)^2 + 5$ will pass through the point $(3, 6)$.

Graph each horizontal parabola.

39. $x = y^2 - 4$
40. $x = -y^2 + 3$
41. $x = (y - 3)^2$
42. $x = (y + 1)^2$
43. $x = 2y^2$
44. $x = -\frac{1}{2}y^2$

Draw each set of parabolas on the same axes.

45. **(a)** $x = y^2$ **(b)** $x = (y - 1)^2$ **(c)** $x = (y - 1)^2 - 5$

46. **(a)** $x = -y^2$ **(b)** $x = -(y + 3)^2$ **(c)** $x = -(y + 3)^2 + 4$

Write the equation of the parabola labeled P, which is obtained from the dashed curve by shifting it horizontally and vertically.

47.

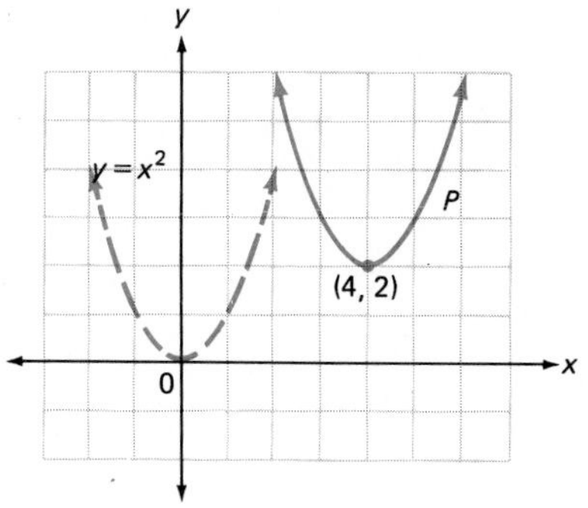

48.

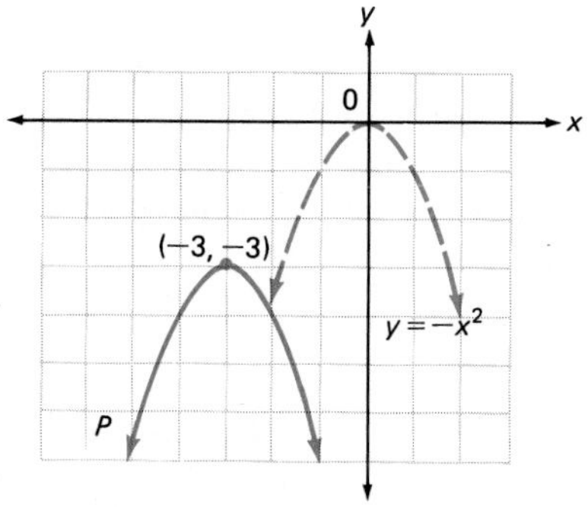

49.

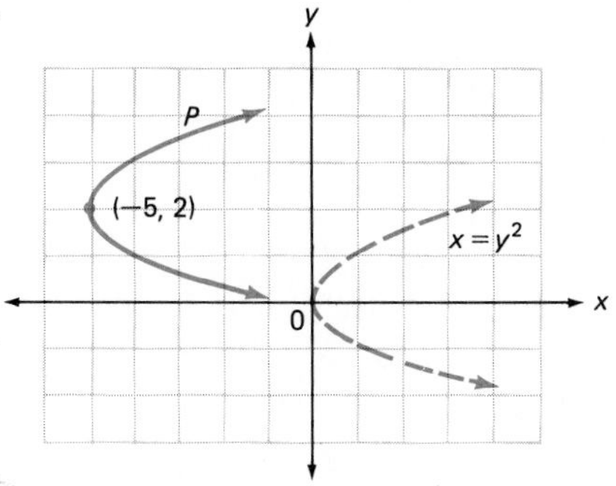

50.

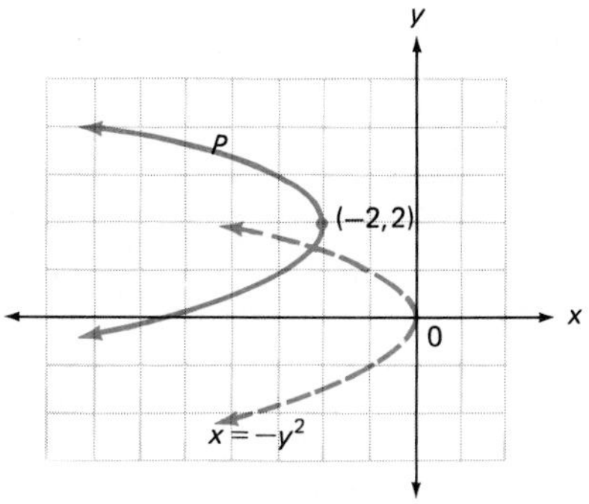

51.

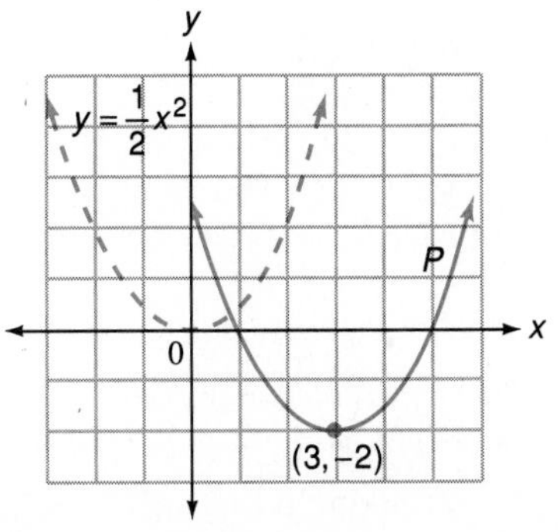

52.

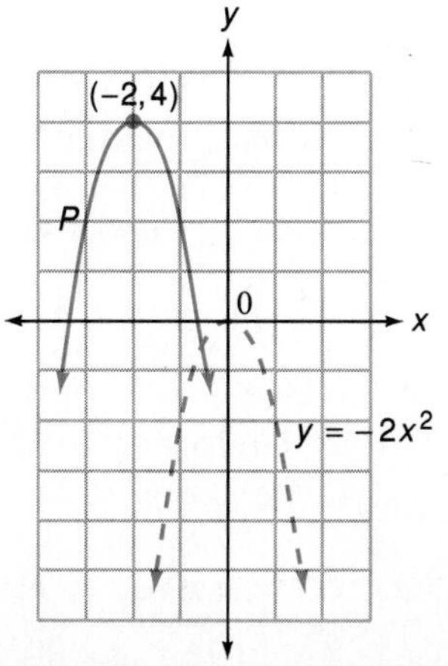

Written Assignments: If a curve is increasing, is it necessarily concave up? Explain your answer with a specific illustration.

Use a specific example of your own and describe in words how to draw the graph of $y = (x - h)^2 + k$ from the graph of $y = x^2$.

Explain, with specific examples, the distinction between a function and a relation.

7.2 APPLYING THE STANDARD FORM: $y = a(x - h)^2 + k$

How would you graph the quadratic equation $y = x^2 + 4x + 3$? At first glance it appears that we must go through the tedious process of plotting a sufficient number of points until the shape of the graph becomes clear. However, we can make use of our knowledge of completing the square to write the equation in this **standard form:**

$$y = a(x - h)^2 + k$$

To see how this is done, let us consider the quadratic function given by $y = x^2 + 4x + 3$. First rewrite the equation in this way:

$$y = (x^2 + 4x + \underline{?}) + 3$$

Note that if the question mark is replaced by 4, then we will have a perfect square within the parentheses. However, since this changes the given equation, we must also subtract 4. The completed work looks like this:

Observe that we have really added $4 - 4 = 0$ to the right side. This preserves the given equation without altering the form of the left side.

$$\begin{aligned} y &= x^2 + 4x + 3 \\ &= (x^2 + 4x + 4) + 3 - 4 \\ &= (x + 2)^2 - 1 \end{aligned}$$

From the form $y = (x + 2)^2 - 1$ you should recognize the graph to be a parabola with vertex at $(-2, -1)$, and $x = -2$ as axis of symmetry.

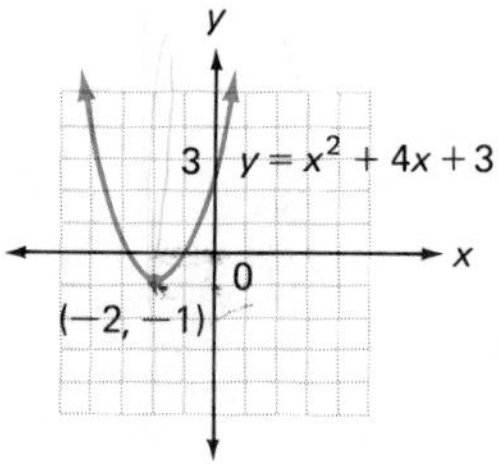

The x- and y-intercepts can also be useful in graphing parabolas. The y-intercept is found by letting $x = 0$. Thus $y = 3$ is the y-intercept of the preceding parabola. To find the x-intercepts we need to find where the curve intersects the x-axis. This occurs when $y = 0$. Thus we set $x^2 + 4x + 3 = 0$ and observe that the two roots $x = -1$ and $x = -3$ are also the x-intercepts of the parabola $y = x^2 + 4x + 3$.

EXAMPLE 1 Graph the function $y = x^2 - 4x + 4$ and find all its intercepts, the vertex, and the axis of symmetry. What are the domain and range?

Solution The standard form of this equation is $y = (x - 2)^2$. This is the equation of a parabola that opens upward with vertex at (2, 0), and with $x = 2$ as the axis of symmetry.

Note that the x-intercept 2 is also the double root of $x^2 - 4x + 4 = 0$.

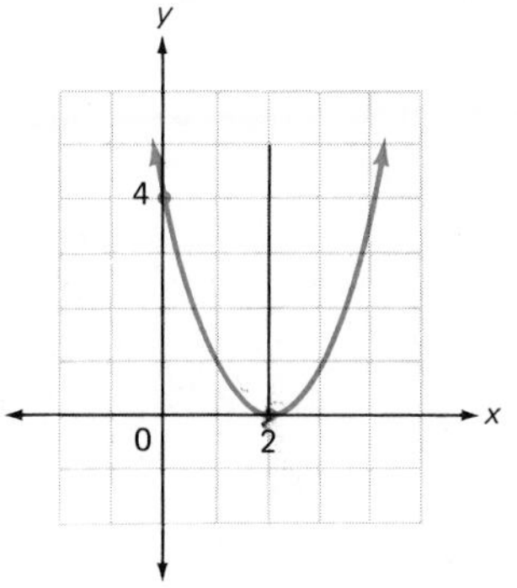

The x-axis is *tangent* to the graph at x = 2, the x-intercept.
When x = 0, y = 4, the y-intercept.
y
x
4
0
2

The domain is all real numbers x, and the range consists of all $y \geq 0$. ■

EXAMPLE 2 Graph the function $y = 2x^2 - 4x + 5$ and find the vertex, and the axis of symmetry. What is the range of the function?

Solution To convert the equation into standard form complete the square in the variable x. Since the coefficient of the x^2 term is not 1, begin by factoring its coefficient from the first two terms only.

$$y = 2x^2 - 4x + 5$$
$$= 2(x^2 - 2x + \underline{\quad\quad}) + 5$$

CAUTION
A common error here is to ignore the coefficient 2 in front of the parentheses and subtract 1 instead of 2(1), producing the false result
$$y = 2(x - 1)^2 + 4$$
To be safe, work backward and compare to the given equation.

Next add $[\frac{1}{2}(2)]^2 = 1$ inside the parentheses to obtain $x^2 - 2x + 1 = (x - 1)^2$. However, because of the factor 2 in front of the parentheses, we have really added $2(1) = 2$, and therefore 2 must also be subtracted.

$$y = 2(x^2 - 2x + 1) + 5 - 2$$
$$= 2(x - 1)^2 + 3$$

This shows the vertex to be (1, 3) and the axis of symmetry is $x = 1$. Since the parabola opens upward and the lowest point is the vertex (1, 3), the range of the function consists of all $y \geq 3$.

When finding the y-intercept it is usually simpler to substitute $x = 0$ into the form
$$y = ax^2 + bx + c$$
rather than in
$$y = a(x - h)^2 + k.$$

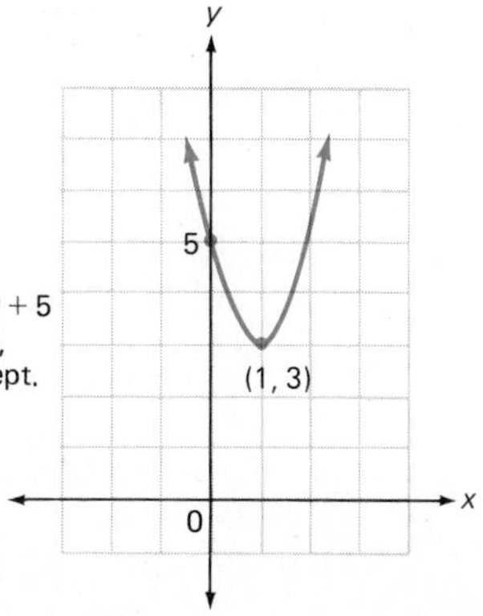

Let x = 0 in
y = 2x² − 4x + 5
to find y = 5,
the y-intercept.
y
x
5
(1, 3)
0

■

From the graph in Example 2 you can see that there are no x-intercepts. In general this occurs when the roots of $ax^2 + bx + c = 0$ are imaginary numbers. In this case we solve $2x^2 - 4x + 5 = 0$ by the quadratic formula with $a = 2$, $b = -4$, and $c = 5$.

Recall that

$$\sqrt{-24} = i\sqrt{24} = i\sqrt{4 \cdot 6} = 2i\sqrt{6}$$

See page 203.

$$x = \frac{-(-4) \pm \sqrt{(-4)^2 - 4(2)(5)}}{2(2)}$$

$$= \frac{4 \pm \sqrt{-24}}{4}$$

$$= \frac{4 \pm 2i\sqrt{6}}{4}$$

$$= 1 \pm \frac{\sqrt{6}}{2}i$$

The imaginary roots $1 + \frac{\sqrt{6}}{2}i$ and $1 - \frac{\sqrt{6}}{2}i$ indicate that there are no real roots and, consequently, no x-intercepts.

From our earlier work we know that the value of the discriminant $b^2 - 4ac$ *identifies the kinds of roots of* $ax^2 + bx + c = 0$ (see page 216). Also, we have just learned that when these roots are real numbers they are the x-intercepts of the parabola $y = ax^2 + bx + c$. Therefore, the discriminant can now be used to determine the number of x-intercepts of a parabola. The preceding examples illustrate the following three possibilities.

Similar observations can be made about the number of x-intercepts for parabolas that open downward.

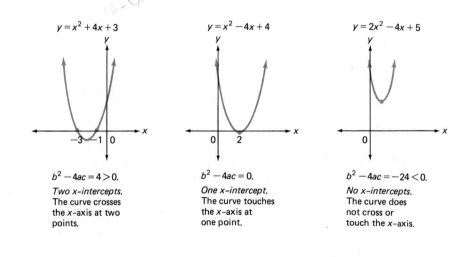

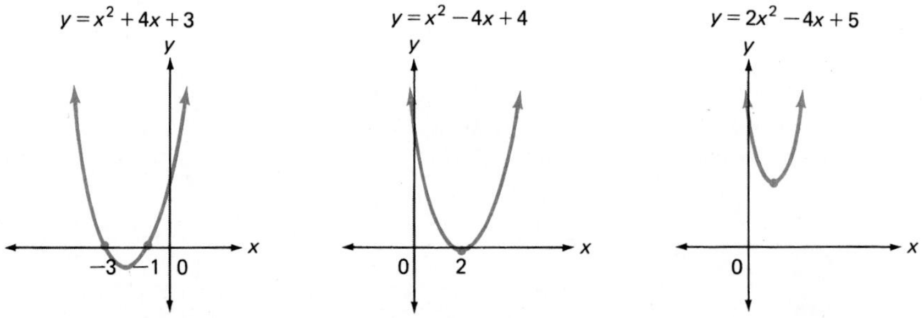

$b^2 - 4ac = 4 > 0$.
Two x-intercepts. The curve crosses the x-axis at two points.

$b^2 - 4ac = 0$.
One x-intercept. The curve touches the x-axis at one point.

$b^2 - 4ac = -24 < 0$.
No x-intercepts. The curve does not cross or touch the x-axis.

Using the Discriminant to Find the Number of x-intercepts of the Parabola $y = ax^2 + bx + c$

Discriminant	Number of x-intercepts
1. $b^2 - 4ac > 0$	two
2. $b^2 - 4ac = 0$	one
3. $b^2 - 4ac < 0$	none

TEST YOUR UNDERSTANDING
Think Carefully

Write each of the following in standard form. Then describe the graph of each by giving (a) the coordinate of the vertex, (b) the equation of the axis of symmetry, (c) the y-intercept, and (d) the x-intercepts.

1. $y = x^2 + x - 6$ **2.** $y = x^2 - 3x - 4$ **3.** $y = x^2 + 6x + 9$
4. $y = x^2 + 4x + 2$ **5.** $y = 2x^2 - 5x - 3$ **6.** $y = 3x^2 + 6x + 4$

Find the value of the discriminant and use it to determine the number of x-intercepts of the following.

7. $y = 9x^2 - 12x + 4$ **8.** $y = -3x^2 + 2x - 7$ **9.** $y = 4x^2 + x - 1$

(Answers: Page 339)

The standard form $y = f(x) = a(x - h)^2 + k$ can be used to solve applied problems that involve the vertex (h, k) of the parabola in which the second coordinate k is either a maximum or minimum value of $f(x)$. The figures show the two situations that were discussed on page 292 and are summarized below.

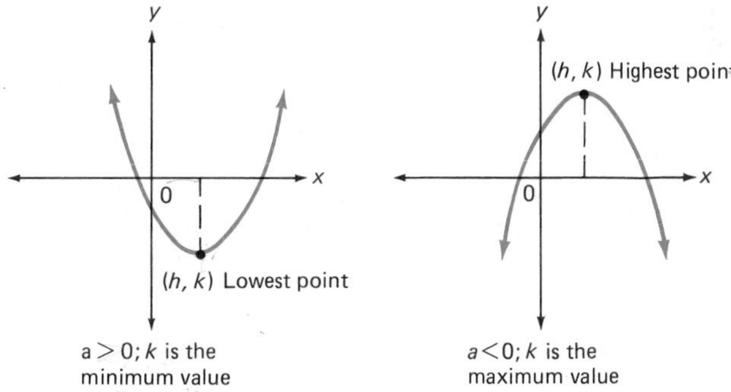

MINIMUM AND MAXIMUM VALUES OF QUADRATIC FUNCTIONS

$$y = f(x) = a(x - h)^2 + k \qquad (a \neq 0)$$

If $a > 0$, the graph is a parabola that opens upward and has a minimum value $f(h) = k$.

If $a < 0$, the graph is a parabola that opens downward and has a maximum value $f(h) = k$.

EXAMPLE 3 Find the maximum or minimum value of the function $f(x) = -2x^2 + 12x - 11$. At which x does $f(x)$ have this value?

Because of the factor -2 *we are really adding*
$-2(9) = -18$
by putting $+9$ *inside the parentheses. Therefore,* 18 *must be added outside:*
$-18 + 18 = 0$

Solution Convert to standard form:

$$\begin{aligned} y = f(x) &= -2x^2 + 12x - 11 \\ &= -2(x^2 - 6x) - 11 \\ &= -2(x^2 - 6x + 9) - 11 + 18 \\ &= -2(x - 3)^2 + 7 \end{aligned}$$

Since $a = -2 < 0$ the vertex $(h, k) = (3, 7)$ is the highest point, and the maximum value is $k = 7$ when $h = x = 3$; $f(3) = 7$. ■

For each x between 0 and 30, such a rectangle is possible. Here are a few.

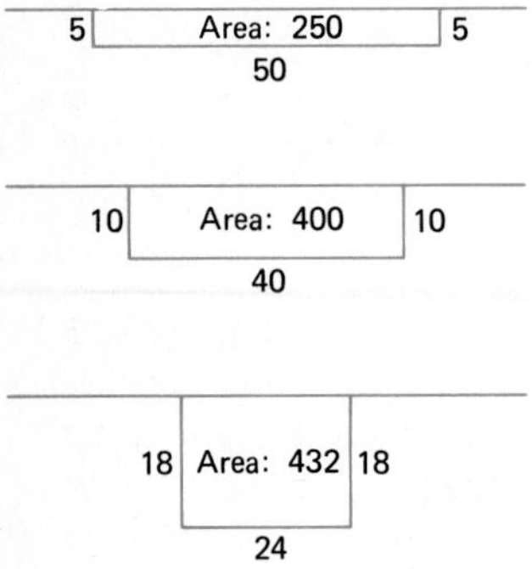

EXAMPLE 4 Suppose that 60 meters of fencing is available to enclose a rectangular garden, one side of which will be against the side of a house. What dimensions of the garden will guarantee a maximum area?

Solution From the sketch you can see that the 60 meters need only be used for three sides, two of which are of the same length x.

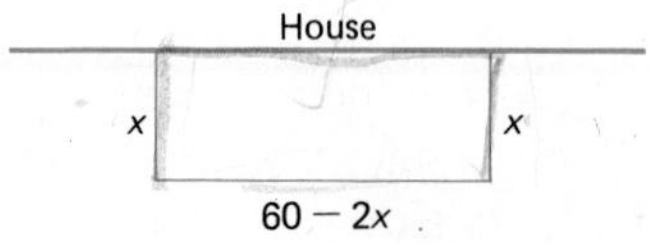

The remaining side has length $60 - 2x$, and the area A is given by

$$\begin{aligned} A(x) &= x(60 - 2x) \\ &= 60x - 2x^2 \end{aligned}$$

Example 4 shows how to select the rectangle of maximum area from an infinite collection of possibilities.

To "maximize" A, convert to the form $a(x - h)^2 + k$. Thus

$$\begin{aligned} A(x) &= -2(x^2 - 30x) \\ &= -2(x^2 - 30x + 225) + 450 \\ &= -2(x - 15)^2 + 450 \end{aligned}$$

Therefore, the maximum area of 450 square meters is obtained when the dimensions are $x = 15$ meters by $60 - 2x = 30$ meters. ■

EXAMPLE 5 A ball is thrown straight upward from ground level with an initial velocity of 32 feet per second. The formula $s = 32t - 16t^2$ gives its height in feet s, after t seconds. What is the maximum height reached by the ball?

Solution First complete the square in t.

$$\begin{aligned} s &= 32t - 16t^2 \\ &= -16t^2 + 32t \\ &= -16(t^2 - 2t) \\ &= -16(t^2 - 2t + 1) + 16 \\ &= -16(t - 1)^2 + 16 \end{aligned}$$

You should now recognize this as describing a parabola with vertex at (1, 16). Because the coefficient of t^2 is negative, the curve opens downward as shown on the next page. The maximum height, 16 feet, is reached after 1 second.

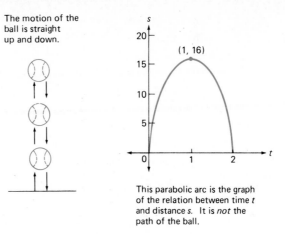

The motion of the ball is straight up and down.

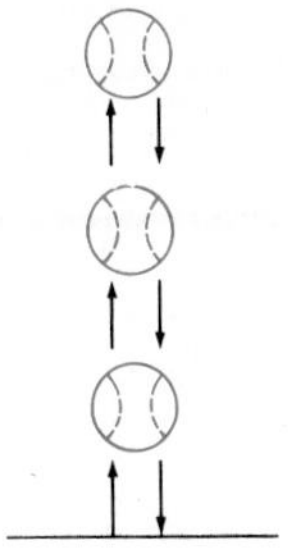

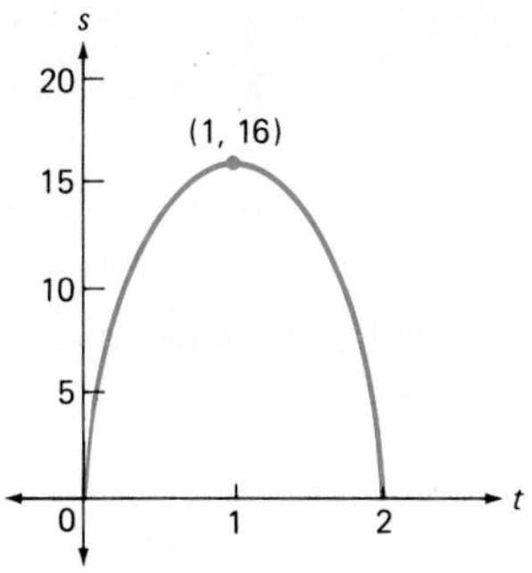

This parabolic arc is the graph of the relation between time t and distance s. It is *not* the path of the ball. ■

EXAMPLE 6 The marketing department of the TENRAQ Tennis Company found that, on the average, 600 tennis rackets will be sold monthly at the unit price of \$100. The department also observed that for each \$5 reduction in price, there will be an extra 50 rackets sold monthly. What price will bring the largest monthly income?

Solution Let x be the number of \$5 reductions in price for the racket. Then $5x$ is the total reduction and

$$100 - 5x = \text{reduced unit price}$$

For example, if $x = 2$ the reduction is \$10, the sales would be 700, and the income is 90(700) = 63,000.

Also, $50x$ is the increase in sales per month and

$$600 + 50x = \text{number of rackets sold monthly}$$

The monthly income, R, will be the unit price times the number of units sold. Then

$$\begin{aligned} R &= (\text{unit price})(\text{units sold}) \\ R(x) &= (100 - 5x)(600 + 50x) \\ &= 60{,}000 + 2000x - 250x^2 \\ &= -250(x^2 - 8x) + 60{,}000 \\ &= -250(x^2 - 8x + 16) + 60{,}000 + 4000 \\ &= -250(x - 4)^2 + 64{,}000 \qquad \text{(Change to the form } a(x - h)^2 + k.) \end{aligned}$$

Since $a = -250 < 0$, the maximum monthly income of \$64,000 is obtained when $x = 4$. The unit price should be set at $100 - 5(4)$ or \$80. ■

EXERCISES 7.2

Write in the standard form $y = a(x - h)^2 + k$.

1. $y = x^2 + 2x - 5$
2. $y = x^2 - 2x + 5$
3. $y = x^2 + 6x - 2$
4. $y = -x^2 - 6x + 2$
5. $y = x^2 - 4x + 1$
6. $y = x^2 + 4x - 1$

7. $y = x^2 - 3x + 4$
8. $y = -x^2 + 3x - 4$
9. $y = x^2 - 5x - 2$
10. $y = x^2 + 5x - 2$
11. $y = 2x^2 - 4x + 3$
12. $y = 2x^2 + 4x - 3$
13. $y = 3x^2 - 6x + 5$
14. $y = \frac{1}{2}x^2 + 12x - 4$
15. $y = -\frac{1}{3}x^2 - 6x + 5$

Write each of the following in standard form. Then describe the graph of each by giving **(a)** *the coordinates of the vertex,* **(b)** *the equation of the axis of symmetry,* **(c)** *the y-intercept, and* **(d)** *the x-intercepts.*

16. $y = x^2 - 6x + 5$
17. $y = x^2 + 2x - 1$
18. $y = x^2 - 4x + 7$
19. $y = -x^2 + 4x - 1$
20. $y = 2x^2 - 4x - 4$
21. $y = 3x^2 + 6x - 3$
22. $y = x^2 + 8x + 16$
23. $y = -x^2 + 2x - 1$
24. $y = x^2 + 2x - 2$
25. $y = 2x^2 + 2x + 1$
26. $y = -4x^2 - 4x - 1$
27. $y = x^2 + 3x + 1$

The graph for each of the following is a parabola. First use the discriminant to predict how many times, if any, the graph will cross the x-axis. Then give the **(a)** *coordinates of the vertex,* **(b)** *the y-intercept, and* **(c)** *the x-intercepts.*

28. $y = x^2 - 6x + 13$
29. $y = -x^2 + 2x + 3$
30. $y = 9x^2 - 6x + 1$
31. $y = 4x^2 + 6x + 9$
32. $y = -x^2 - 4x + 3$
33. $y = x^2 - 6x + 4$

Find the values of b so that the graph of the equation will be tangent to the x-axis (Hint: Let $b^2 - 4ac = 0$.)

34. $y = x^2 + bx + 25$
35. $y = x^2 - bx + 12$
36. $y = 16x^2 - bx + 9$
37. $y = 9x^2 + bx + \frac{1}{4}$

Find the maximum or minimum value of the quadratic function and state the x-value at which this occurs.

38. $f(x) = -x^2 + 10x - 18$
39. $f(x) = x^2 + 18x + 49$
40. $f(x) = 16x^2 - 64x + 100$
41. $f(x) = -\frac{1}{2}x^2 + 3x - 6$
42. $f(x) = 49 - 28x + 4x^2$
43. $f(x) = x(x - 10)$

44. A company's daily profit P, in dollars, is given by $P = -2x^2 + 120x - 800$, where x is the number of articles produced per day. Find x so that the daily profit is a maximum.
45. A manufacturer is in the business of producing small models of the Statue of Liberty. He finds that the daily cost in dollars, C, of manufacturing n statues is given by the formula $C = n^2 - 120n + 4200$. How many statues should be produced per day so that the cost will be a minimum? What is the minimal daily cost?
46. The sum of two numbers is 12. Find the two numbers if their product is to be a maximum. (*Hint:* Find the maximum point for $y = x(12 - x)$.)
47. The sum of two numbers is $\frac{1}{4}$. Find the numbers if their product is to be a maximum and also find this product.
48. The sum of two numbers is n. Find the two numbers such that their product will be a maximum.
49. The sum of a number and three times another number is 30. Find the two numbers so that their product is a maximum.
50. The difference of two numbers is 22. Find the numbers if their product is to be a minimum and also find this product.
51. A homeowner has 100 feet of wire and wishes to use it to enclose a rectangular garden. What should be the dimensions of the garden so as to enclose the largest possible area? (*Hint:* Use x and $50 - x$ as the dimensions. Then find the maximum for $y = x(50 - x)$).
52. Repeat Exercise 51, but this time assume that the side of the house is to be used as one boundary for the garden. Thus the wire is only needed for the other three sides.
53. A gardener has 300 feet of fencing to enclose three adjacent rectangular growing areas. If all three rectangles are to have the same dimensions, and if one side is to be against a

building as shown, what dimensions should be used so that the maximum growing area will be enclosed?

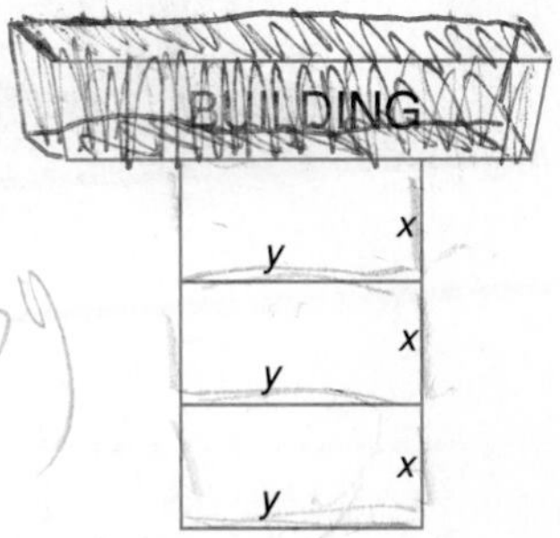

Hint: Write y in terms of x.

54. Find the lengths of the altitude and base of the triangle of maximum area if the sum of their lengths is 80 centimeters.
55. The sum of the lengths of the two perpendicular sides of a right triangle is 30 centimeters. What are their lengths if the square of the hypotenuse is a minimum?
56. Let s be the square of the distance from the origin to point $P(x, y)$ on the line through points (0, 4) and (2, 0). Find the coordinates of P such that s is a minimum value.

The formula $h = 128t - 16t^2$ gives the distance in feet above the ground, h, reached by an object in t seconds. Use this formula for Exercises 57 through 60.

57. What is the maximum height reached by the object?
58. How long does it take for the object to reach its maximum height?
59. How long does it take for the object to return to the ground?
60. In how many seconds will the object be at a height of 192 feet? (There are two possible answers.)

61. When a department store sold a certain style shirt for \$20, the average number of shirts sold per week was 100. The store observed that with each \$1 decrease in price, 10 more shirts were sold weekly. What unit price should be set for the shirts in order to realize the maximum weekly revenue? (*Hint:* If the price per shirt is reduced x dollars, then $100 + 10x$ is the number of shirts sold weekly.)
62. It is estimated that 14,000 people will attend a basketball game when the admission price is \$7.00. For each 25¢ added to the price, the attendance will decrease by 280. What admission price will produce the largest gate receipts? (*Hint:* If x quarters are added, the attendance will be $14,000 - 280x$.)
63. Suppose it is known that if 65 apple trees are planted in a certain size orchard, the average yield per tree would be 1500 apples per year. For each additional tree planted in the same orchard, the annual yield per tree drops by 20 apples. How many trees should be planted in order to produce the maximum crop of apples per year? (*Hint:* If n trees are added to the 65 trees, then the yield per tree is $1500 - 20n$.)
64. Convert $y = ax^2 + bx + c$, $a \neq 0$, into standard form and verify that the vertex has coordinates $\left(-\frac{b}{2a}, \frac{4ac - b^2}{4a}\right)$.

Use the result in Exercise 64 to find the coordinates of the vertex of each parabola and decide whether $\frac{4ac - b^2}{4a}$ is a maximum or minimum value.

65. $y = 2x^2 - 6x + 9$
66. $y = -3x^2 + 24x - 41$
67. $y = -\frac{1}{2}x^2 - \frac{1}{3}x + 1$
68. $y - x^2 + 5x = 0$

CHALLENGE
Think Creatively

Assume that the quadratic equation $ax^2 + bx + c = 0$ has a, b, c as integers. Explain why the discriminant $b^2 - 4ac$ is either a multiple of 4 or one more than a multiple of 4. Then decide if the numbers 215, 289, and 436 can be values of the discriminant. (*Hint:* Either b is even or b is odd. Use $b = 2n$ or $b = 2n + 1$ for b even or odd, respectively, where n is some integer in each case.)

7.3 QUADRATIC INEQUALITIES

The x-intercepts of a parabola can be used to solve a **quadratic inequality in one variable.** Consider, for example, the inequality $x^2 - x - 6 < 0$. To solve this inequality, first examine the graph of $y = x^2 - x - 6$. This is a parabola with x-intercepts at -2 and 3, as can be seen by writing the equation in factored form as $y = (x + 2)(x - 3)$.

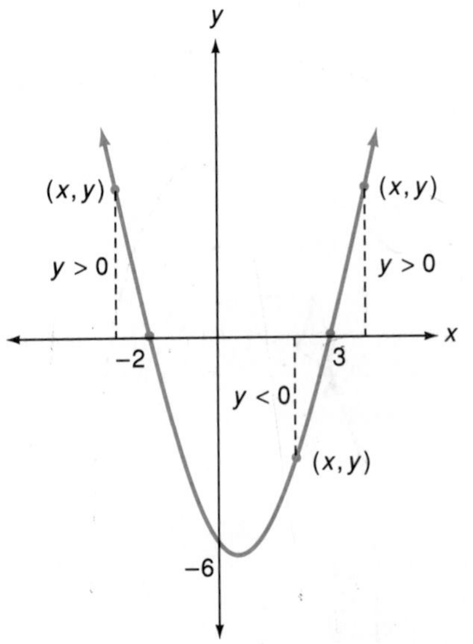

From the graph we note that the parabola $y = x^2 - x - 6$ is below the x-axis for $-2 < x < 3$. That is, $x^2 - x - 6 < 0$ for $-2 < x < 3$. Also, $x^2 - x - 6 > 0$ for $x < -2$ or $x > 3$.

Such quadratic inequalities can just as well be solved without actually drawing the parabola. This is done with the aid of the following properties, in which we assume that the related quadratic function $y = ax^2 + bx + c$ has two x-intercepts, say r_1 and r_2 with $r_1 < r_2$.

1. If $a > 0$, then the parabola opens upward and we have

$$y = ax^2 + bx + c < 0 \qquad \text{for } r_1 < x < r_2$$
$$y = ax^2 + bx + c > 0 \qquad \text{for } x < r_1 \text{ or } x > r_2$$

2. If $a < 0$, then the parabola opens downward and we have

$$y = ax^2 + bx + c < 0 \qquad \text{for } x < r_1 \text{ or } x > r_2$$
$$y = ax^2 + bx + c > 0 \qquad \text{for } r_1 < x < r_2$$

Notice that when the inequalities involve $\leq$ instead of $<$, the solution will include the numbers r_1 and r_2 (likewise when $\geq$ is used in place of $>$).

EXAMPLE 1 Solve for x and graph the solution on a number line:

$$-x^2 - x + 2 \geq 0$$

Solution Let $y = -x^2 - x + 2$. Since $a = -1 < 0$, the parabola opens downward. Also,

$$\begin{aligned} y &= -x^2 - x + 2 \\ &= -(x^2 + x - 2) \\ &= -(x + 2)(x - 1) \end{aligned}$$

The equation $y = -(x + 2)(x - 1)$ implies that the x-intercepts are -2 and 1. By property 2 on page 304, we get $-x^2 - x + 2 \geq 0$ for $-2 \leq x \leq 1$.

The graph of this solution may be shown as follows:

■

EXAMPLE 2 Solve for x: $x^2 + 2x - 2 > 0$

Solution Use the quadratic formula for the equation

$$x^2 + 2x - 2 = 0$$

$$\begin{aligned} x &= \frac{-2 \pm \sqrt{(2)^2 - 4(1)(-2)}}{2} \\ &= \frac{-2 \pm \sqrt{12}}{2} \\ &= \frac{-2 \pm 2\sqrt{3}}{2} \\ &= -1 \pm \sqrt{3} \end{aligned}$$

Thus $x = -1 - \sqrt{3}$ and $x = -1 + \sqrt{3}$ are the x-intercepts of the parabola $y = x^2 + 2x - 2$. Also, since $a = 1 > 0$, the parabola opens upward, and by Property 1 on page 304 the solution is

$$x < -1 - \sqrt{3} \quad \text{or} \quad x > -1 + \sqrt{3}$$

■

TEST YOUR UNDERSTANDING Think Carefully

(Answers: Page 339)

Solve for x.

1. $x^2 - 4 < 0$
2. $x^2 - 4 > 0$
3. $9 - x^2 \leq 0$
4. $9 - x^2 \geq 0$
5. $x^2 - x < 0$
6. $2x - x^2 \leq 0$
7. $x^2 - 5x + 6 \geq 0$
8. $6 - x - x^2 > 0$

When a parabola has less than two x-intercepts, similar reasoning can be used to solve the related inequality.

EXAMPLE 3 Solve: $x^2 + 6x > -9$

Solution First convert to an equivalent inequality having 0 on one side.

*Note: **Equivalent inequalities** have the same solution set.*

$$x^2 + 6x > -9$$
$$x^2 + 6x + 9 > 0 \qquad \text{(adding 9 to each side)}$$

Since $x^2 + 6x + 9 = (x + 3)^2$, the parabola $y = x^2 + 6x + 9$ has only one x-intercept: it is tangent to the x-axis at $x = -3$. Therefore, since the parabola opens upward, the solution consists of all $x \neq -3$. ■

EXAMPLE 4 Solve: $x^2 - 4x + 9 < 0$

Solution Let $y = x^2 - 4x + 9$. Since $x^2 - 4x + 9$ is not factorable, use the discriminant.

$$\begin{aligned} b^2 - 4ac &= (-4)^2 - 4(1)(9) \\ &= -20 < 0 \end{aligned}$$

Therefore, there are no x-intercepts. Furthermore, since the parabola opens upward there are no values x for which $y < 0$. The solution is the empty set.

Alternative Solution First convert to the standard form $a(x - h)^2 + k$.

$$\begin{aligned} y &= x^2 - 4x + 9 \\ &= (x^2 - 4x + 4) + 9 - 4 \\ &= (x - 2)^2 + 5 \end{aligned}$$

Thus, since $a = 1 > 0$, 5 is the minimum y-value which implies that $y \geq 5$ for all x. Hence, there are no x for which $y = x^2 - 4x + 9 < 0$. ■

Inequalities involving rational expressions such as $\dfrac{x+2}{x-5} < 0$ can be solved by making use of the preceding work with quadratic inequalities. This can be done by first multiplying the given fraction by the *square* of the denominator:

$$\frac{x+2}{x-5} \cdot (x - 5)^2 = (x + 2)(x - 5)$$

Observe that $(x - 5)^2$ is positive for all allowable values of x in the given fraction; 5 is not an allowable value. Therefore, whenever $\dfrac{x+2}{x-5}$ is negative, so is the result $(x + 2)(x - 5)$ because the product of the negative number $\dfrac{x+2}{x-5}$ times the positive number $(x - 5)^2$ must be negative. Likewise, whenever $\frac{x+2}{x-5}$ is positive, so is $(x + 2)(x - 5)$. In other words, the two forms $\frac{x+2}{x-5}$ and $(x + 2)(x - 5)$ are negative or positive together; they have the same signs for common x-values.

EXAMPLE 5 Solve: $\frac{x+2}{x-5} < 0$

The solution of $\frac{x+2}{x-5} \leq 0$ consists of all x where $-2 \leq x < 5$ since $\frac{x+2}{x-5} = 0$ when $x = -2$.

Solution From the preceding discussion, the signs of $\frac{x+2}{x-5}$ are the same as the signs of $\frac{x+2}{x-5} \cdot (x-5)^2 = (x + 2)(x - 5)$. But from our previous work, we know that $(x + 2)(x - 5) = x^2 - 3x - 10$ is negative for $-2 < x < 5$. Thus the solution set for the given inequality is $\{x \mid -2 < x < 5\}$. ■

Note: It is possible to also solve the inequality in Example 5 by multiplying through by $x - 5$. This would lead to a much more involved process because two cases need to be considered; when $x - 5 < 0$ and when $x - 5 > 0$. (See the Multiplication Property of Order on page 99.)

The preceding quadratic inequalities in one variable were solved by using the x-intercepts of the related parabolas. Parabolas can also be used to graph quadratic inequalities in *two variables*.

A parabola such as $y = x^2$ separates the plane into two regions. The graph of the quadratic inequality in two variables $y > x^2$ consists of all points in the plane *above* the parabola $y = x^2$; the graph of $y < x^2$ consists of all points *below*.

For the inequalities $y \geq x^2$ or $y \leq x^2$, the parabola $y = x^2$ would become part of the graph, and this would be indicated by using a solid rather than the dashed curve.

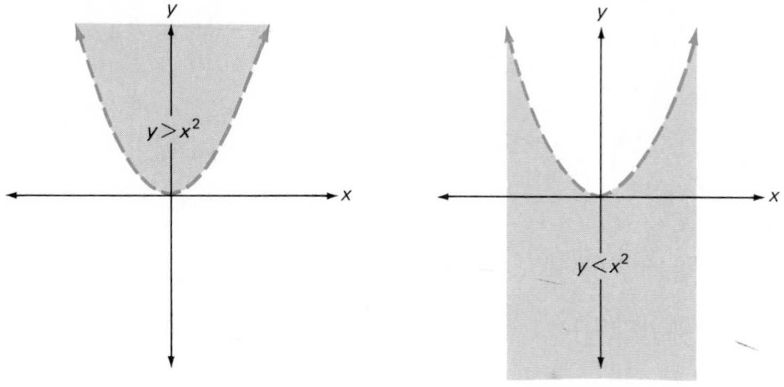

Note that a dashed curve is used to indicate that the points on the parabola are *not* part of the graph.

EXAMPLE 6 Graph the quadratic inequality $y \leq 1 + 4x - x^2$ by shading the appropriate region.

Here are the details for the computation shown:

$1 + 4x - x^2$
$= -(x^2 - 4x) + 1$
$= -(x^2 - 4x + 4) + 1 + 4$
$= -(x - 2)^2 + 5$

Solution First draw the related parabola,

$$y = 1 + 4x - x^2 = -(x - 2)^2 + 5$$

and then shade the region below the curve as follows.

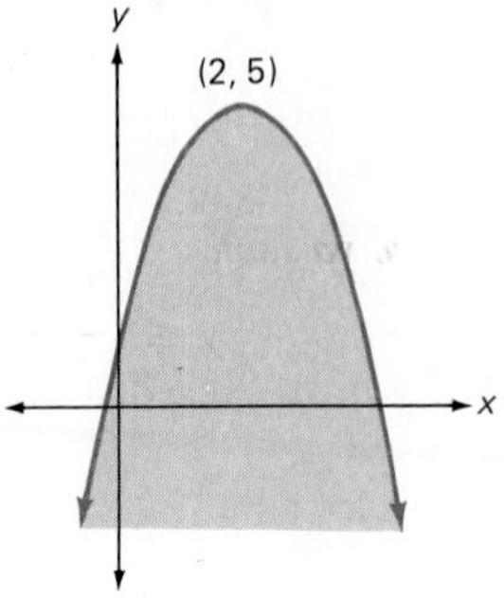

$y \leq 1 + 4x - x^2$

The graph consists of all points on or below the parabola. A solid curve is used to indicate that the points on the parabola are included.

EXERCISES 7.3

Solve for x.

1. $(x - 3)(x + 1) > 0$
2. $(x - 3)(x + 1) < 0$
3. $(x - 4)(x + 2) \leq 0$
4. $(x - 4)(x + 2) \geq 0$
5. $x^2 - 25 > 0$
6. $25 - x^2 > 0$
7. $16 - x^2 \leq 0$
8. $x^2 - 49 \leq 0$
9. $x^2 + 3x > 0$
10. $2x^2 - 6x < 0$
11. $7x - 14x^2 \leq 0$
12. $1 - 9x^2 \geq 0$
13. $4x^2 - 1 > 0$
14. $36x^2 - 25 < 0$
15. $8x^2 + 3x \leq 0$
16. $2x - 5x^2 \geq 0$
17. $x^2 + 3x - 10 > 0$
18. $x^2 + 3x - 10 < 0$
19. $x^2 + x - 12 < 0$
20. $x^2 + x - 12 > 0$
21. $15 + 2x - x^2 \leq 0$
22. $15 + 2x - x^2 \geq 0$
23. $2 - 3x - 2x^2 > 0$
24. $2 - 3x - 2x^2 < 0$
25. $2x^2 - 5x - 3 \geq 0$
26. $2x^2 - 5x - 3 \leq 0$
27. $x^2 - x - 30 > 0$
28. $x^2 + 5x - 14 < 0$
29. $3x^2 - 4x + 1 \leq 0$
30. $3x^2 + 4x + 1 \geq 0$
31. $6x^2 + x - 1 \geq 0$
32. $6x^2 - x - 1 \leq 0$
33. $x^2 + 4x > 2$
34. $x^2 + 2x > 4$
35. $x^2 + 3x < 1$
36. $x^2 + 5x < 2$
37. $2 - 2x^2 \leq x$
38. $2 - 3x^2 \leq -2x$
39. $x^2 + 4 \leq 2x$
40. $x^2 \leq -x - 2$
41. $6 < 5x + 6x^2$
42. $12 > 6x^2 + x$
43. $x^2 - 4x + 4 \geq 0$
44. $x^2 + 6x + 9 \leq 0$

*45. If $a > b$, solve $(x - a)(x - b) > 0$ for x.

*46. If $a > b$, solve $(x - a)(x - b) < 0$ for x.

Solve the rational inequalities.

47. $\dfrac{x - 3}{x - 2} < 0$
48. $\dfrac{x - 4}{x + 1} < 0$
49. $\dfrac{x}{x - 6} > 0$
50. $\dfrac{x + 3}{2x} > 0$
51. $\dfrac{2x + 1}{x - 1} \leq 0$
52. $\dfrac{3x - 4}{x + 5} \leq 0$
53. $\dfrac{1}{(x + 2)(x - 3)} > 0$
54. $\dfrac{5}{(2x - 1)x} \geq 0$
55. $\dfrac{4x + 5}{2x - 3} \geq 0$
56. $\dfrac{x^2}{x + 7} \geq 0$
57. $\dfrac{x - 2}{x^2(x - 6)} < 0$
58. $\dfrac{x^2(x + 6)}{x - 7} > 0$
59. $\dfrac{x}{x^3 - x} \geq 0$ (*Hint*: Factor the denominator and reduce.)

60. $\frac{x^3 + 8x^2}{x - 5} \leq 0$

61. $1 - \frac{3}{x} > 0$ (*Hint*: First combine.)

62. $2 + \frac{5}{x} < 0$

63. $\frac{3}{x} \leq -4$ (*Hint*: Add 4 to both sides and combine.)

64. $\frac{1}{2} > \frac{1}{2x}$

65. $\frac{x^3 + 9x^2}{x} < 0$

66. $\frac{2x}{x^2 - 4x + 4} \geq 0$

67. $\frac{3}{2x^2 + 3x - 5} > 0$

68. $\frac{x^2 - 5x}{x^2 - 25} \leq 0$

Graph each of the following quadratic inequalities in two variables.

69. $y \geq (x + 2)^2$
70. $y \leq (x + 2)^2$
71. $y \leq -x^2$
72. $y \leq -(x + 2)^2$
73. $y \geq x^2 - 4$
74. $y \geq -x^2 + 4$
75. $y < x^2 + 6x + 13$
76. $y \geq x^2 - 8x + 15$
77. $y > 2 + 2x - x^2$

Written Assignment: Describe, in words, the graph of the compound inequalities

(a) $y < x^2$ or $y > x^2$
(b) $y < -x^2$ or $y \geq x^2$
(c) $y < x^2$ and $y > x^2$
(d) $y \leq 1 - x^2$ and $y \geq x^2$

7.4 CONIC SECTIONS: THE CIRCLE

A **conic section** is a curve formed by the intersection of a plane with a double right-circular cone. These curves, also called **conics,** are known as the **circle, ellipse, parabola,** and **hyperbola.**

The figures indicate that the inclination of the plane in relation to the vertical axis of the cone determines the nature of the curve. These four curves have played a vital role in mathematics and its applications from the time of the ancient Greeks until the present day.

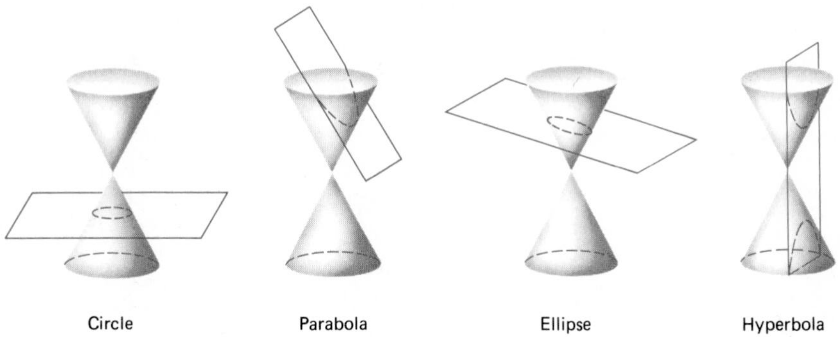

Each of these conic sections can also be defined in terms of sets of points in the plane. We begin with the study of circles.

> **DEFINITION OF A CIRCLE**
>
> A circle is the set of all points in the plane, each of which is at a fixed distance r from a given point called the center of the circle; r is the radius of the circle ($r > 0$).

The words "if and only if" here mean that if P is on the circle, then $OP = 5$ and if $OP = 5$ then P is on the circle.

Specifically, if the origin O is the center of a circle with radius 5, then the points on the circle are exactly 5 units from center O. Another way of saying this is that a point $P(x, y)$ is on the circle with center O and radius 5 if and only if $OP = 5$, as in the figure on the following page.

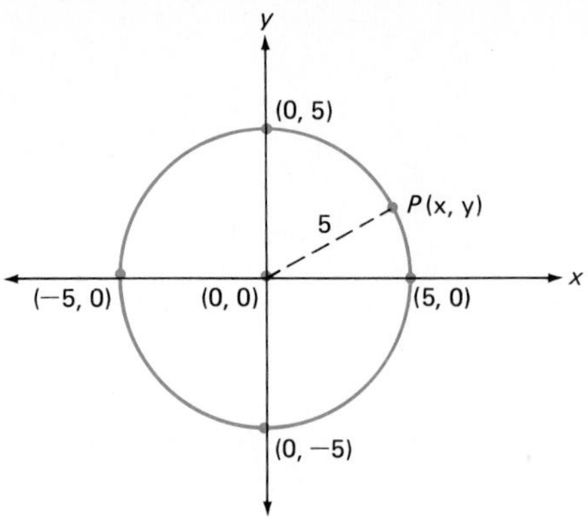

The distance formula is developed on page 256.

The relationship between x and y, where (x, y) is on this circle, can be found by using the distance formula as follows.

$$OP = \sqrt{(x - 0)^2 + (y - 0)^2}$$

or, since $OP = 5$,

$$\sqrt{x^2 + y^2} = 5$$

Since the variables in the equation of a circle are to the second power, the equation is also classified as being a quadratic equation.

Squaring the preceding equation produces

$$x^2 + y^2 = 25$$

In general, we conclude that $P(x, y)$ is on the circle with center at (0, 0) and radius r if and only if the coordinates of P satisfy the equation $r^2 = x^2 + y^2$.

STANDARD FORM FOR THE EQUATION OF A CIRCLE WITH CENTER AT THE ORIGIN AND RADIUS r

$$x^2 + y^2 = r^2$$

EXAMPLE 1 What is the equation of the circle with center O and radius 3?

Solution Using the general form, we get

$$x^2 + y^2 = 3^2 = 9$$

for each point (x, y) on the circle. ■

EXAMPLE 2 Draw the circle with equation $x^2 + y^2 = 2$.

Solution Write $x^2 + y^2 = (\sqrt{2})^2$ to fit the general form. Then the radius $r = \sqrt{2}$ and the equation is graphed as follows:

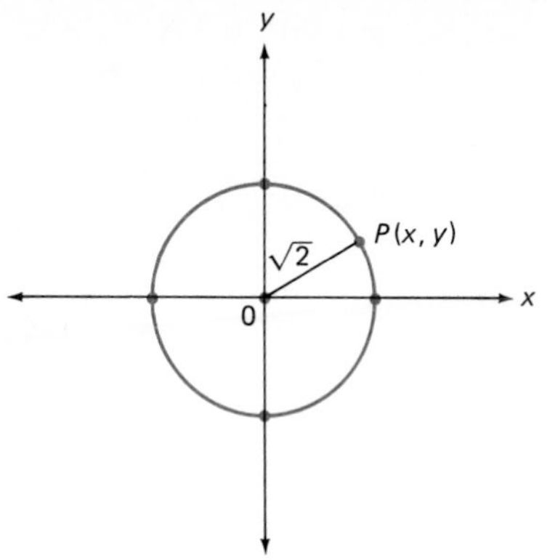

■

As you can see in the next figure, the first coordinate x of any point $P(x, y)$ on the circle with equation $x^2 + y^2 = 25$ must satisfy $-5 \leq x \leq 5$. Likewise, the second coordinate y satisfies $-5 \leq y \leq 5$.

The vertical line through $x = 3$ intersects the circle more than once. Therefore this (or any circle) is not the graph of a function. Consequently, $x^2 + y^2 = 25$ defines a relation that is not a function, with domain $-5 \leq x \leq 5$ and range $-5 \leq y \leq 5$.

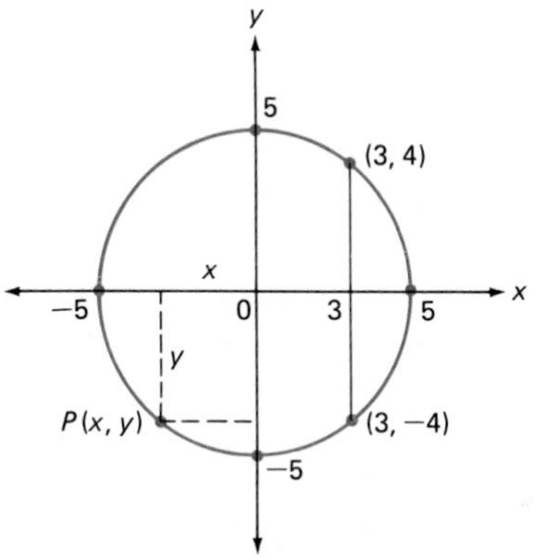

For $x = 3$ there are two points on the circle, as shown above. To find the second coordinates, substitute $x = 3$ into $x^2 + y^2 = 25$ to obtain

$$(3)^2 + y^2 = 25$$

This gives

$$y^2 = 25 - 9 = 16$$
$$y = 4 \quad \text{or} \quad y = -4$$

Therefore, (3, 4) and (3, −4) are points on the circle, as shown.

EXAMPLE 3 Find the two points on the circle $x^2 + y^2 = 25$ where $x = 2$.

Solution Substitute $x = 2$ into the given equation to get

$$4 + y^2 = 25 \quad \text{or} \quad y^2 = 21$$

Why are there no points (x, y) on this circle for $x = 6$?

Then $y = \pm\sqrt{21}$ and the points on the circle are $(2, \sqrt{21})$, $(2, -\sqrt{21})$. ■

A **tangent** *to a circle at a point P is the line through P and perpendicular to the radius at P*. Since perpendicular lines have slopes that are negative reciprocals of one another, we are able to write the equation of tangent lines to circles. For example, look at the tangent line to circle $x^2 + y^2 = 25$ at the point $P(3, 4)$.

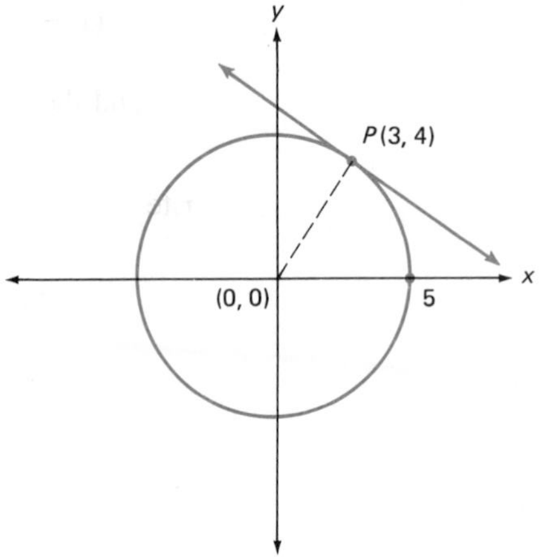

The algebraic forms of a line were developed in Section 6.4.

The radius OP has slope $\frac{4}{3}$. Then the tangent must have slope $-\frac{3}{4}$. Now use the point-slope form of a line to obtain the following as the equation of this tangent line:

$$y - 4 = -\tfrac{3}{4}(x - 3)$$

or

$$3x + 4y = 25$$

Now consider any circle of radius r, not necessarily one with the origin as center. Let the center C have coordinates (h, k). Then, using the distance formula, a point $P(x, y)$ is on this circle if and only if

$$CP = r = \sqrt{(x - h)^2 + (y - k)^2}$$

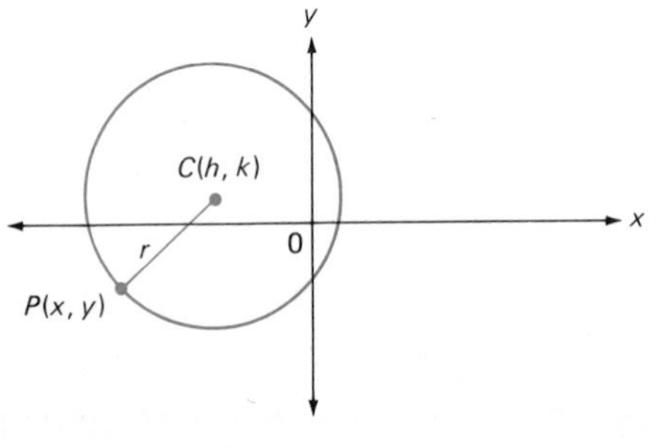

By squaring CP, we obtain the following:

STANDARD FORM FOR THE EQUATION OF A CIRCLE WITH CENTER AT (h, k) AND RADIUS r

$$(x - h)^2 + (y - k)^2 = r^2$$

EXAMPLE 4 Find the center and radius of the circle with this equation: $(x - 2)^2 + (y + 3)^2 = 4$.

Solution Using $y + 3 = y - (-3)$, rewrite the equation in this form:

$$(x - 2)^2 + [(y - (-3)]^2 = 2^2$$

Thus the radius $r = 2$ and the center is at $(2, -3)$. ■

EXAMPLE 5 Write the equation of the circle with center $(2, -5)$ and with radius 4.

Solution Use $h = 2$, $k = -5$, and $r = 4$ in the standard form. Thus

$$(x - h)^2 + (y - k)^2 = r^2$$
$$(x - 2)^2 + (y - (-5))^2 = 4^2$$
$$(x - 2)^2 + (y + 5)^2 = 16$$ ■

TEST YOUR UNDERSTANDING Think Carefully

Find the center and radius of each circle.

1. $x^2 + y^2 = 100$
2. $x^2 + y^2 = 10$
3. $(x - 1)^2 + (y + 1)^2 = 25$
4. $(x + \frac{1}{2})^2 + y^2 = 256$
5. $(x + 4)^2 + (y + 4)^2 = 50$
6. $2x^2 + 2(y - 5)^2 = 18$

Write the equation of the circle with the given center and radius in standard form.

7. Center at $(0, 4)$; $r = 5$
8. Center at $(1, -2)$; $r = \sqrt{3}$

(Answers: Page 339)

The equation in Example 4 can be written in another form.

$$(x - 2)^2 + (y + 3)^2 = 4$$
$$x^2 - 4x + 4 + y^2 + 6y + 9 = 4$$
$$x^2 - 4x + y^2 + 6y = -9$$

Note that the major reason for writing the equation of a circle in standard form is that this form enables us to identify the center and the radius of the circle. This information is sufficient to allow us to draw the circle.

This last equation no longer looks like the equation of a circle. Starting with such an equation we can convert it back into the standard form of a circle by completing the square in both variables, if necessary. For example, let us begin with

$$x^2 - 4x + y^2 + 6y = -9$$

Then complete the squares in x and y:

To complete the squares $(\frac{1}{2} \cdot 4)^2 = 4$ and $(\frac{1}{2} \cdot 6)^2 = 9$ are added to each side.

$$(x^2 - 4x + 4) + (y^2 + 6y + 9) = -9 + 4 + 9$$
$$(x - 2)^2 + (y + 3)^2 = 4$$

EXAMPLE 6 Convert $x^2 - 8x + y^2 - 9 = 0$ into the standard form of a circle and graph.

Solution First complete the square in the variable x.

$$x^2 - 8x + y^2 - 9 = 0$$
$$x^2 - 8x + y^2 = 9$$
$$(x^2 - 8x + 16) + y^2 = 9 + 16$$
$$(x - 4)^2 + y^2 = 25$$

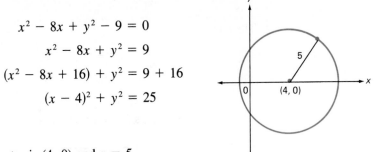

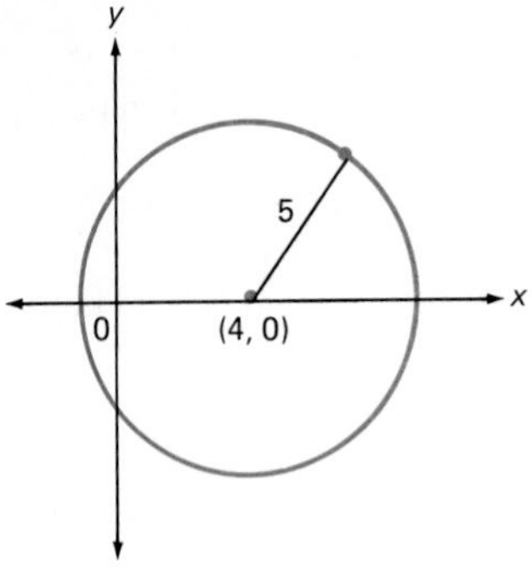

The center is (4, 0) and $r = 5$. ■

EXAMPLE 7 Find the center and radius of the circle whose equation is $9x^2 + 12x + 9y^2 = 77$.

Solution First divide by 9 so that the x^2 and y^2 terms each have a coefficient of 1.

$$x^2 + \tfrac{4}{3}x + y^2 = \tfrac{77}{9}$$

Complete the square in x; add $\frac{4}{9}$ to both sides of the equation.

$$(x^2 + \tfrac{4}{3}x + \tfrac{4}{9}) + y^2 = \tfrac{77}{9} + \tfrac{4}{9}$$
$$(x + \tfrac{2}{3})^2 + y^2 = 9$$

In standard form:

$$[x - (-\tfrac{2}{3})]^2 + (y - 0)^2 = 3^2$$

The center is at $(-\frac{2}{3}, 0)$ and $r = 3$. ■

The next example shows how to find the equation of a circle if only the two endpoints of one diameter are known.

EXAMPLE 8 Points $P(2, 5)$ and $Q(-4, -3)$ are the endpoints of a diameter of a circle. Find the center, radius, and equation of the circle.

See page 258 for the midpoint formula.

Solution The center is the midpoint of PQ whose coordinates (x', y') are

$$x' = \frac{2 + (-4)}{2} = -1, \qquad y' = \frac{5 + (-3)}{2} = 1$$

The center is located at $C(-1, 1)$. To find the radius, apply the distance formula to the points $C(-1, 1)$ and $P(2, 5)$.

$$r = \sqrt{(-1 - 2)^2 + (1 - 5)^2} = \sqrt{25} = 5$$

The equation of the circle is

$$(x + 1)^2 + (y - 1)^2 = 25$$ ■

CAUTION: Learn To Avoid These Mistakes	
WRONG	RIGHT
The circle $(x + 3)^2 + (y - 2)^2 = 7$ has center $(3, -2)$ and radius 7.	The circle has center $(-3, 2)$ and radius $\sqrt{7}$.
The equation of the circle with center $(-1, 0)$ and the radius 5 has equation $x^2 + (y + 1)^2 = 5$.	The circle has equation $(x + 1)^2 + y^2 = 25$.

EXERCISES 7.4

1. Graph these circles in the same coordinate system.
   (a) $x^2 + y^2 = 25$ **(b)** $x^2 + y^2 = 16$ **(c)** $x^2 + y^2 = 4$ **(d)** $x^2 + y^2 = 1$

Write the equation of the circle having center at the origin with the given radius r.

2. $r = 2$ 3. $r = \frac{3}{2}$ 4. $r = 10$

What is the radius of the circle described by each of the following equations?

5. $x^2 + y^2 = 36$ 6. $x^2 + y^2 = 121$ 7. $x^2 + y^2 = \frac{1}{4}$
8. $x^2 + y^2 = \frac{9}{16}$ 9. $x^2 + y^2 = 5$ 10. $x^2 + y^2 = 10$

One of the coordinates of a set of two points on the indicated circle is given in each of the following exercises. What are the coordinates of these points?

11. $x^2 + y^2 = 25; x = -4$ 12. $x^2 + y^2 = 25; y = 3$
13. $x^2 + y^2 = 25; x = 0$ 14. $x^2 + y^2 = 100; x = 8$
15. $x^2 + y^2 = 100; y = -6$ 16. $x^2 + y^2 = 36; x = 4$
17. $x^2 + y^2 = 36; y = 0$ 18. $x^2 + y^2 = 64; y = 7$

Write an equation of the line tangent to the given circle through point P.

19. $x^2 + y^2 = 80; P(-8, 4)$ 20. $x^2 + y^2 = 100; P(-6, 8)$
21. $x^2 + y^2 = 169; P(-5, -12)$ 22. $x^2 + y^2 = 9; P(-2, \sqrt{5})$

23. Draw the circle $x^2 + y^2 = 25$ and the tangent lines at the points $(3, 4)$, $(-3, 4)$, $(3, -4)$, and $(-3, -4)$. Write the equations of these tangent lines.
24. **(a)** Where are the two tangents to the circle $x^2 + y^2 = 4$ whose slopes equal 0? Write their equations.
    (b) Write the equations of the tangents to the circle $x^2 + y^2 = 4$ whose slope is undefined.
25. Graph these circles in the same coordinate system.
    (a) $(x - 3)^2 + (y - 3)^2 = 9$ **(b)** $(x + 3)^2 + (y - 3)^2 = 9$
    (c) $(x + 3)^2 + (y + 3)^2 = 9$ **(d)** $(x - 3)^2 + (y + 3)^2 = 9$

Write the equations of the circle with center C and radius r in standard form.

26. $C(-2, 3); r = 2$ 27. $C(-2, -3); r = 3$ 28. $C(2, -3); r = 9$
29. $C(2, 3); r = \sqrt{13}$ 30. $C(5, 0); r = 5$ 31. $C(0, -5); r = 5$
32. Explain why a circle cannot be the graph of a function.

Write the equations of each circle in the standard form $(x - h)^2 + (y - k)^2 = r^2$. Find the center and radius for each.

33. $x^2 - 4x + y^2 = 21$
34. $x^2 + y^2 + 8y = -12$
35. $x^2 - 2x + y^2 - 6y = -9$
36. $x^2 + 4x + y^2 + 10y + 20 = 0$
37. $x^2 - 4x + y^2 - 10y = -28$
38. $x^2 - 10x + y^2 - 14y = -25$
39. $x^2 - 8x + y^2 = -14$
40. $x^2 + y^2 + 2y = 7$
41. $x^2 - 20x + y^2 + 20y = -100$
42. $4x^2 - 4x + 4y^2 = 15$
43. $16x^2 + 24x + 16y^2 - 32y = 119$
44. $36x^2 - 48x + 36y^2 + 180y = -160$

Draw the given circle and the tangent line at the indicated point for each of the following. Write the equation of the tangent line.

45. $(x + 2)^2 + (y - 3)^2 = 73$; $(6, 0)$
46. $(x - 4)^2 + (y + 5)^2 = 45$; $(1, 1)$
47. $x^2 - 2x + y^2 - 2y = 8$; $(4, 2)$
48. $x^2 + 14x + y^2 + 18y = 39$; $(5, -4)$
49. Points $P(0, -3)$ and $Q(0, 11)$ are the endpoints of a diameter of a circle. Find the center, radius, and equation of the circle.
50. Points $P(3, -5)$ and $Q(-1, 3)$ are the endpoints of a diameter of a circle. Find the center, radius, and equation of the circle.
51. Write the equation of the tangent line to the circle $x^2 + y^2 = 80$ at the point in the first quadrant where $x = 4$.
52. Write the equation of the tangent line to the circle $x^2 + y^2 = 9$ at the point in the third quadrant where $y = -1$.
53. Write the equation of the tangent line to the circle $x^2 + 14x + y^2 + 18y = 39$ at the point in the second quadrant where $x = -2$.

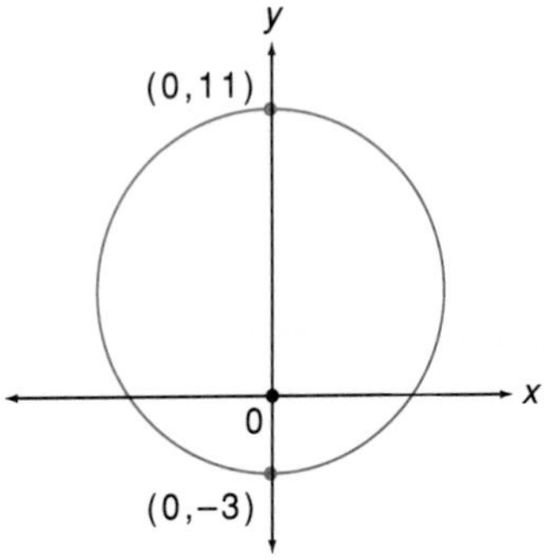

Written Assignment: Refer to a secondary geometry text, if necessary, and explain how to construct a line tangent to a given circle from a point outside the circle.

CHALLENGE Think Creatively

The points $A(-3, -3)$, $B(-1, 11)$, and $C(5, 13)$ are all on the same circle. What is the radius of this circle? (*Hint:* Find the center of the circle by making use of any two of the chords AB, BC, and AC.)

7.5 CONIC SECTIONS: THE ELLIPSE

The German astronomer Johannes Kepler, 1571–1630, discovered that the planets moved around the sun in elliptical orbits. This is also true for some comets and satellites that orbit the earth. The elliptical planetary orbits turn out to be nearly circular, so the earlier belief that these orbits were circular was not unreasonable. In fact, you will see that there are similarities in the definitions of ellipses and circles. (See Exercise 26(a) for another connection between ellipses and circles.)

DEFINITION OF AN ELLIPSE

An ellipse is the set of all points in a plane such that the sum of the distances from two fixed points (called the **foci**) is a constant.

The reason that the construction shown in the margin on the next page gives an ellipse can be seen from the figure. From any position of the pencil the sum of the

The definition suggests that an ellipse can be drawn as follows. Take a loop of string and place it around two thumbtacks. Use a pencil to pull the string taut to form a triangle and move the pencil around the loop.

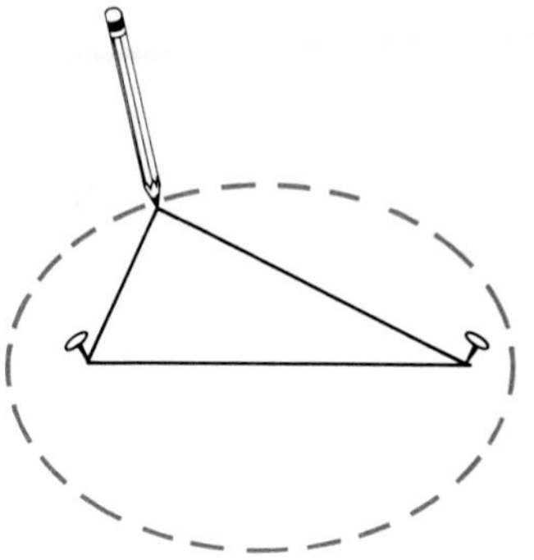

distances to the thumbtacks equals the length of the loop minus the distance between the thumbtacks. Since both the length of the loop and the distance between the thumbtacks are constants, their difference must also be a constant.

Now consider the ellipse whose foci are $F_1(-4, 0)$ and $F_2(4, 0)$, such that the sum of the distances to the foci from a point $P(x, y)$ on the ellipse is 10.

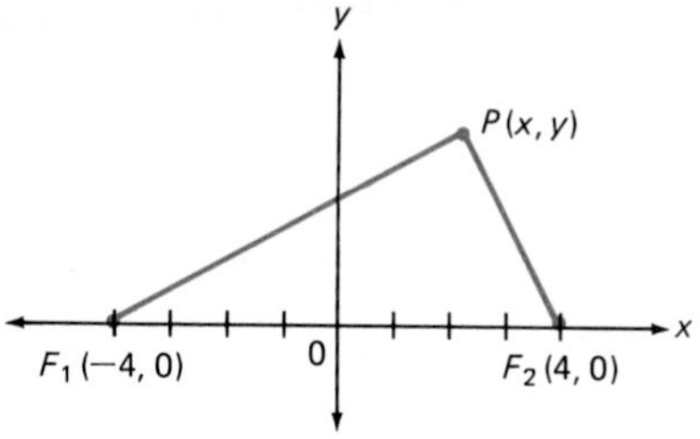

The distance formula gives

$$PF_1 = \sqrt{(x+4)^2 + y^2} \quad \text{and} \quad PF_2 = \sqrt{(x-4)^2 + y^2}$$

Since $PF_1 + PF_2 = 10$, we have

$$PF_1 = 10 - PF_2$$
$$\sqrt{(x+4)^2 + y^2} = 10 - \sqrt{(x-4)^2 + y^2}$$

In Exercise 27 you will be asked to complete the missing algebraic details in this derivation.

Square both sides and simplify to obtain

$$4x - 25 = -5\sqrt{(x-4)^2 + y^2}$$

Square again and simplify to get

$$9x^2 + 25y^2 = 225$$

Now divide by 225 to obtain these forms:

$$\frac{x^2}{25} + \frac{y^2}{9} = 1 \quad \text{or} \quad \frac{x^2}{5^2} + \frac{y^2}{3^2} = 1$$

Observe that the first coordinates of the foci, ± 4, are related to the denominators. That is, $(\pm 4)^2 = 4^2 = 5^2 - 3^2$.

When the preceding results are generalized, we obtain the following standard form of the equation of an ellipse whose foci are symmetric about the origin and on the x-axis.

STANDARD FORM FOR THE EQUATION OF AN ELLIPSE WITH FOCI AT $(-c, 0)$ AND $(c, 0)$

$$\frac{x^2}{a^2} + \frac{y^2}{b^2} = 1 \qquad \text{where } b^2 = a^2 - c^2$$

Note that the minor axis has length 2b, and 2b < 2a since

$$b = \sqrt{a^2 - c^2} < a$$

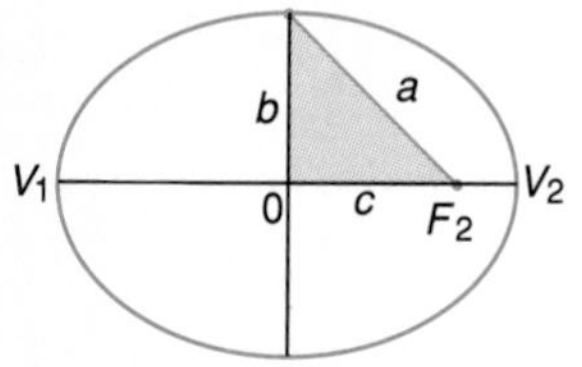

This figure shows the Pythagorean relationship of a, b, and c; namely $a^2 = b^2 + c^2$.

The geometric interpretations of a and b can be found from this equation. Letting $y = 0$ produces the x-intercepts, $x = \pm a$. The points $V_1(-a, 0)$ and $V_2(a, 0)$ are called the **vertices** of the ellipse. The **major axis** of the ellipse is the chord $V_1 V_2$, which has length $2a$. Letting $x = 0$ produces the y-intercepts, $y = \pm b$. The points $(0, -b)$ and $(0, b)$ are the endpoints of the **minor axis.** The intersection of the major and minor axes is the **center** of the ellipse; in this case the center is the origin.

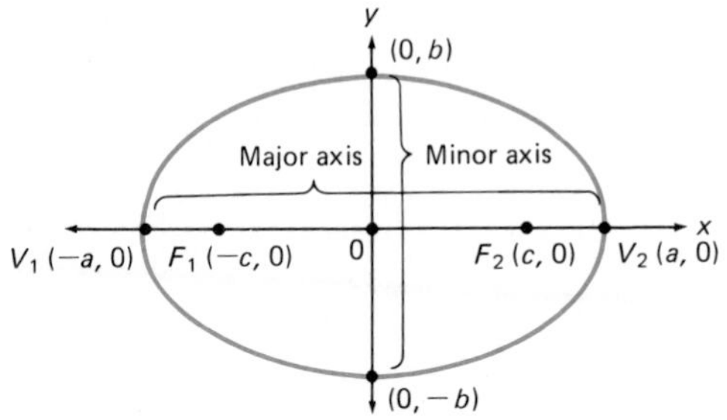

EXAMPLE 1 Sketch the graph of the ellipse $\dfrac{x^2}{25} + \dfrac{y^2}{9} = 1$.

Solution For $y = 0$, $x = \pm 5$. Therefore, the vertices are $(-5, 0)$ and $(5, 0)$, which are also the endpoints of the major axis. The endpoints of the minor axis are $(0, -3)$ and $(0, 3)$ since $y = \pm 3$ when $x = 0$. Locate these points, and several others, and draw the ellipse.

If $x = \pm 4$, then

$$\frac{(\pm 4)^2}{25} + \frac{y^2}{9} = 1$$

$$\frac{y^2}{9} = 1 - \frac{16}{25} = \frac{9}{25}$$

$$y^2 = \frac{81}{25}$$

$$y = \pm\frac{9}{5}$$

This gives the four points shown.

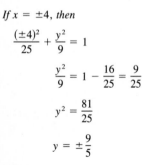

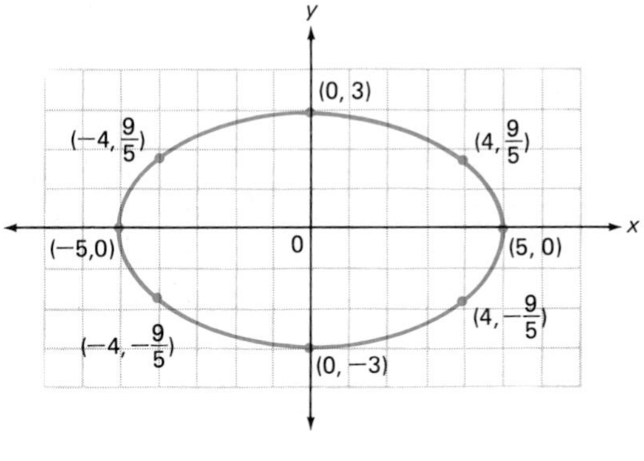

Observe that the additional four points, whose coordinates are computed in the margin, are useful in obtaining accuracy when drawing the ellipse. Also, note that only one of the four points, say $(4, \frac{9}{5})$ needs to be found, since the symmetry of the ellipse around the axes automatically gives the other three. ■

EXAMPLE 2 Write the equation of the ellipse in standard form having vertices $(\pm 10, 0)$ and foci $(\pm 6, 0)$.

Solution Since $a = 10$ and $c = 6$ we get

$$b^2 = a^2 - c^2 = 100 - 36 = 64.$$

Now substitute into the standard form.

$$\frac{x^2}{a^2} + \frac{y^2}{b^2} = 1$$

$$\frac{x^2}{100} + \frac{y^2}{64} = 1$$

■

When the foci of the ellipse are on the y-axis, a similar development produces this standard form:

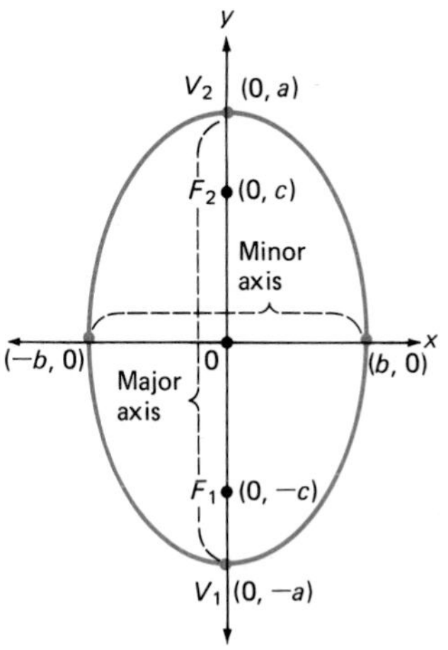

STANDARD FORM FOR THE EQUATION OF AN ELLIPSE WITH FOCI AT $(0, -c)$ AND $(0, c)$

$$\frac{x^2}{b^2} + \frac{y^2}{a^2} = 1 \qquad \text{where } b^2 = a^2 - c^2$$

The major axis is on the y-axis, and its endpoints are the vertices $(0, \pm a)$. The minor axis is on the x-axis and has endpoints $(\pm b, 0)$. Center: $(0, 0)$.

EXAMPLE 3 Change $25x^2 + 16y^2 = 400$ into standard form and graph.

Solution Divide both sides of $25x^2 + 16y^2 = 400$ by 400 to obtain

$$\frac{x^2}{16} + \frac{y^2}{25} = 1.$$

Since $a^2 = 25$, $a = 5$ and the major axis is on the y-axis with length $2a = 10$. Similarly, $b^2 = 16$ gives $b = 4$, and the minor axis has length $2b = 8$. Also, $c^2 = a^2 - b^2 = 25 - 16 = 9$, so that $c = 3$, which locates the foci at the points $(0, 3)$ and $(0, -3)$ as shown next.

When the equation of an ellipse is in standard form, the major axis is horizontal if the x^2-term has the largest denominator. It will be vertical if the y^2-term has the largest denominator. Another way to determine this is to let $x = 0$; then let $y = 0$ in the equation; and locate the endpoints of the two axes. The longer of the two is the major axis.

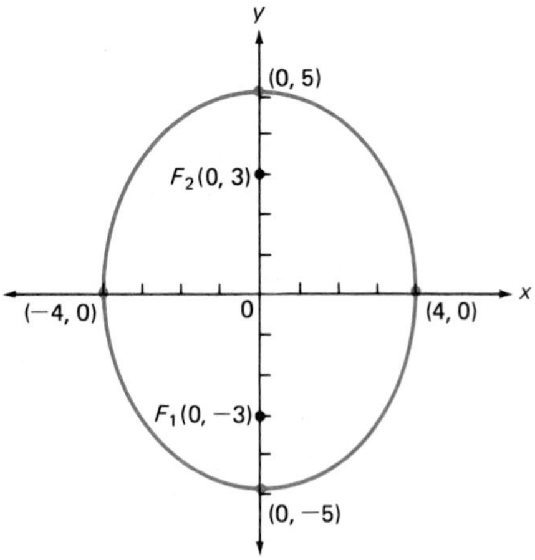

■

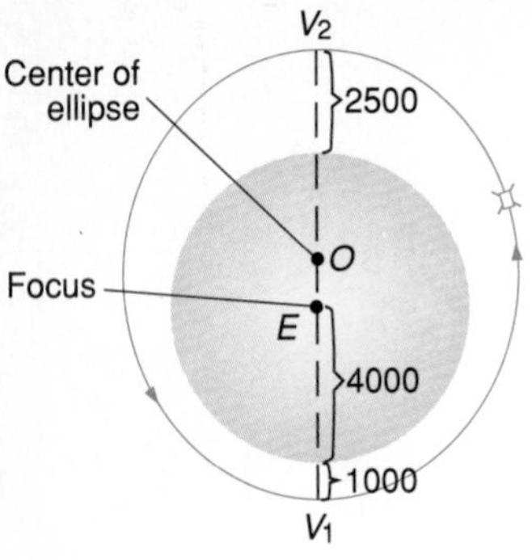

EXAMPLE 4 A satellite follows an elliptical orbit around the earth such that the center of the earth, E, is one of the foci. The figure indicates that the highest point that the satellite will be from the earth's surface is 2500 miles, and the lowest will be 1000 miles. Observe that these distances are measured along the major axis, which is assumed to be on the y-axis. Use 4000 miles as the radius of the earth and find the equation of the orbit.

Solution Since $2a = V_1V_2$ is the length of the major axis, we have

$$2a = 2500 + 8000 + 1000 = 11{,}500$$
$$a = 5750$$

Also,

$$\begin{aligned} c = OE &= OV_1 - EV_1 \\ &= a - 5000 \\ &= 5750 - 5000 = 750 \end{aligned}$$

Now find b^2.

$$\begin{aligned} b^2 &= a^2 - c^2 \\ &= 5750^2 - 750^2 \\ b &\approx 5701 \qquad \text{(by calculator)} \end{aligned}$$

Since the major axis is vertical, the equation of the orbit becomes

$$\frac{x^2}{5701^2} + \frac{y^2}{5750^2} = 1$$

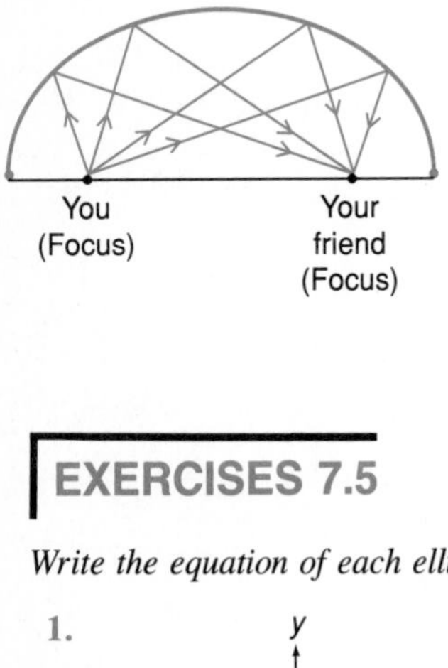

The ellipse has an interesting reflecting property. If a source of sound (or light) is positioned at one focus, the sound (or light) waves will reflect off the ellipse and pass through the other focus. Now suppose the ceiling of a large room is shaped like part of an **ellipsoid** (an ellipsoid is obtained by revolving an ellipse around its major axis) and you and a friend are each standing at a focus. If you whisper so that others in the room are unable to hear you, your friend will hear you because the sound waves will bounce off the ellipsoid and pass through the other focus. The capitol building in Washington, D.C. has such a "whispering gallery."

EXERCISES 7.5

Write the equation of each ellipse in standard form and find the coordinates of the foci.

1.

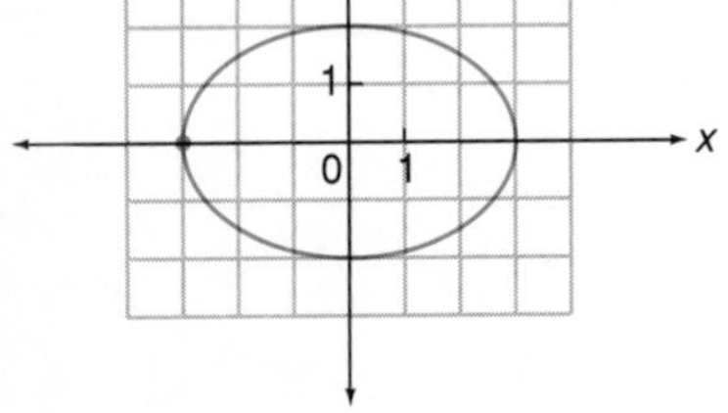

2.

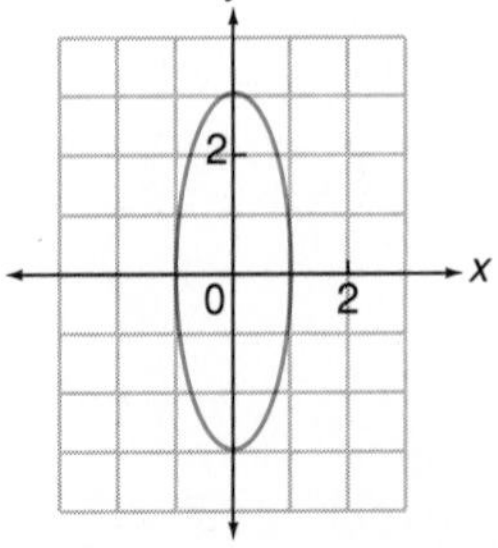

3.

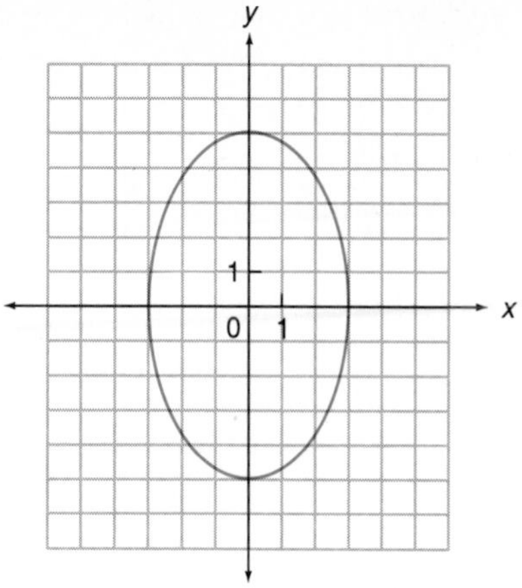

4.

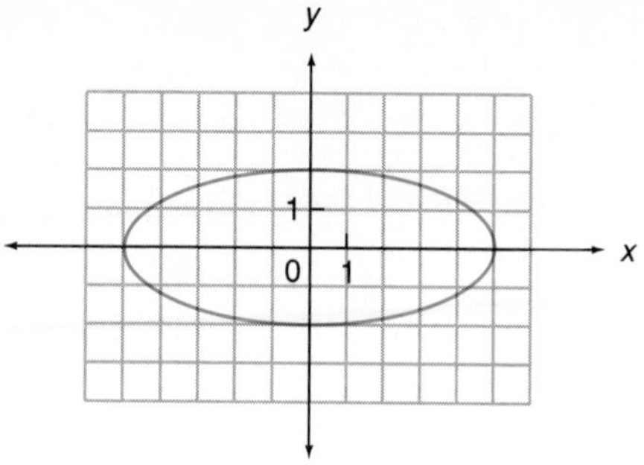

Graph each ellipse. State the coordinates of the center, vertices, and foci.

5. $\frac{x^2}{25} + \frac{y^2}{16} = 1$
6. $\frac{x^2}{16} + \frac{y^2}{25} = 1$
7. $\frac{x^2}{16} + \frac{y^2}{9} = 1$
8. $\frac{x^2}{9} + \frac{y^2}{16} = 1$
9. $\frac{x^2}{4} + \frac{y^2}{36} = 1$
10. $\frac{x^2}{36} + \frac{y^2}{4} = 1$
11. $9x^2 + y^2 = 9$
12. $x^2 + 9y^2 = 9$
13. $4x^2 + 9y^2 = 36$
14. $x^2 + 4y^2 = 4$
15. $25x^2 + 9y^2 = 225$
16. $16x^2 + 9y^2 = 144$

Write, in standard form, the equation of the ellipse having the given properties.

17. Endpoints of the major and minor axes are $(-8, 0)$, $(8, 0)$, and $(0, -4)$, $(0, 4)$.
18. Endpoints of the major and minor axes are $(0, -3)$, $(0, 3)$, and $(-1, 0)$, $(1, 0)$.
19. Center $(0, 0)$; horizontal major axis of length 10; minor axis of length 6.
20. Center $(0, 0)$; foci $(\pm 2, 0)$; vertices $(\pm 5, 0)$.
21. Vertices $(0, \pm 5)$; foci $(0, \pm 3)$
22. Foci $(0, \pm 4)$ and minor axis of length $4\sqrt{2}$.

23. In 1957, the Russians launched the first man-made satellite, Sputnik. Its orbit around the earth was elliptical with the center of the earth as one focus. The maximum height above the earth was about 580 miles and the minimum height was approximately 130 miles.
    (a) Assuming that the earth's radius is 4000 miles and that the major axis is along the y-axis, find the equation of Sputnick's orbit. (Leave the value b^2 in unsimplified form.)
    (b) Find the value of b to the nearest mile and rewrite the equation of the ellipse using this result.

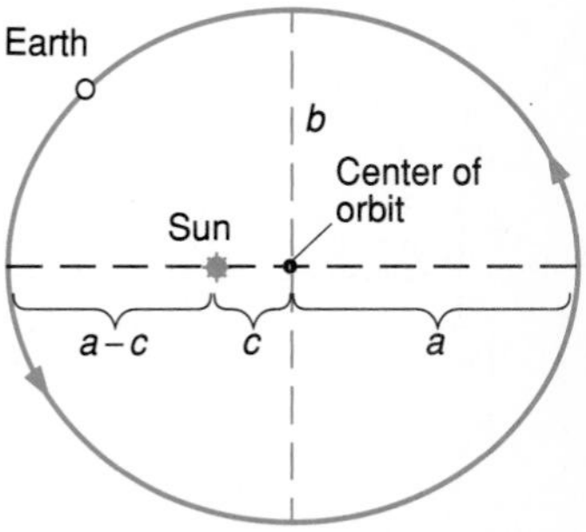

24. The orbit of the earth around the sun is elliptical with the sun being one of the foci. The earth's maximum distance from the sun is approximately 94.6 million miles, and the minimum distance is about 91.5 million miles.
    (a) Find a and b in millions of miles. (*Hint:* Use $a + c$ and $a - c$)
    (b) Compare a and b and comment on the comparison in view of Exercise 26(a).

25. The underside of a bridge over a two-lane roadway is in the shape of a semiellipse. The elliptical arch spans 60 feet (the length of the major axis) and the height at the center is 20 feet.
    (a) The outsides of the driving lanes are marked by lines that are 10 feet from the base of the bridge. What is the bridge clearance y above these lines? (Write the exact numerical expressions.)
    (b) Find the clearance rounded to the nearest tenth of a foot.

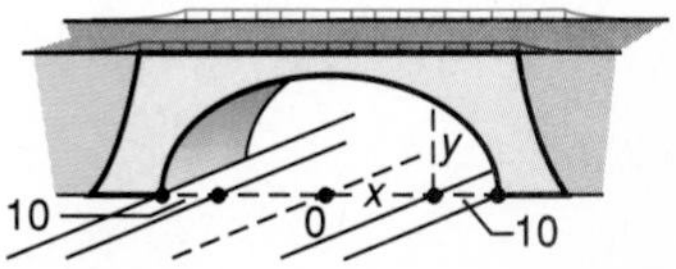

26. **(a)** The area of an ellipse with semimajor axis a and semiminor axis b is πab. Begin with an ellipse with such semiaxes, hold a fixed and allow the length b to get closer and closer to the length a. As b gets close to a, what curve do the ellipses seem to approach? What is the area of this figure?

(b) Find the areas of the ellipses given in Exercises 5 and 13.

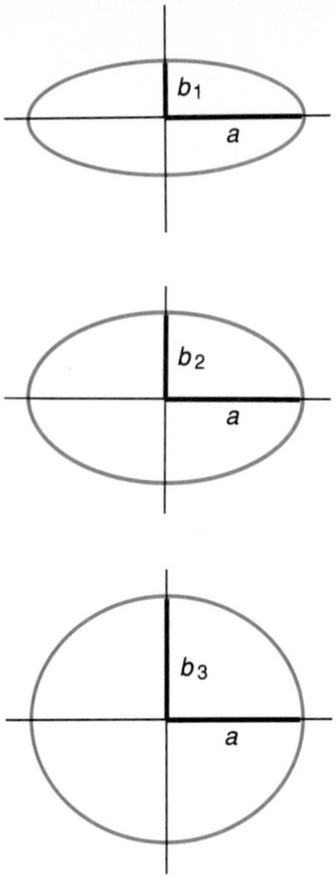

27. This exercise fills the algebraic details in the derivation of the standard form of the ellipse on page 317. Square both sides of

$$\sqrt{(x+4)^2+y^2} = 10 - \sqrt{(x-4)^2+y^2}$$

and obtain

$$4x - 25 = -5\sqrt{(x-4)^2+y^2}$$

Square again and simplify to get

$$9x^2 + 25y^2 = 225$$

When an ellipse with equation $\frac{x^2}{a^2}+\frac{y^2}{b^2}=1$ *is shifted h units horizontally and k units vertically the resulting ellipse is congruent to the original, has center* (h, k), *and has an equation of the form*

$$\frac{(x-h)^2}{a^2}+\frac{(y-k)^2}{b^2}=1$$

A similar observation can be made beginnning with an ellipse having equation $\frac{x^2}{b^2}+\frac{y^2}{a^2}=1$.

The values a, b and c are measured from the center (h, k) *just as they were previously measured from the center* $(0, 0)$. *Sketch the graph of each ellipse.*

28. $\frac{(x-3)^2}{64}+\frac{(y+3)^2}{100}=1$

29. $\frac{(x-2)^2}{25}+\frac{(y-5)^2}{16}=1$

30. $\frac{(x-1)^2}{9}+\frac{(y+2)^2}{4}=1$

31. $\frac{(x+2)^2}{4}+\frac{(y-3)^2}{9}=1$

Write, in standard form, the equation of the ellipse having the given properties.

32. Center $(-5, 0)$; foci $(-5, \pm 2)$; $b = 3$.

33. Center $(2, -3)$; vertical major axis of length 12; minor axis of length 8.

7.6 CONIC SECTIONS: THE HYPERBOLA

The hyperbola is another one of the conic sections (see page 309). Its definition is very similar to that of the ellipse. The definition of the ellipse uses the sum of the distances from the foci, whereas the definition of the hyperbola makes use of the difference.

DEFINITION OF A HYPERBOLA

A **hyperbola** is the set of all points in the plane such that the difference of the distances from two fixed points (called the **foci**) is a constant.

The two fixed points, F_1 and F_2, are called the **foci** of the hyperbola, and its **center** is the midpoint of the segment F_1F_2. It turns out that a hyperbola consists of two congruent branches which open in opposite directions.

Now consider a hyperbola whose foci on the x-axis are $F_1(-5, 0)$ and $F_2(5, 0)$ such that the difference of the distances from a point P on the hyperbola to the foci is 6.

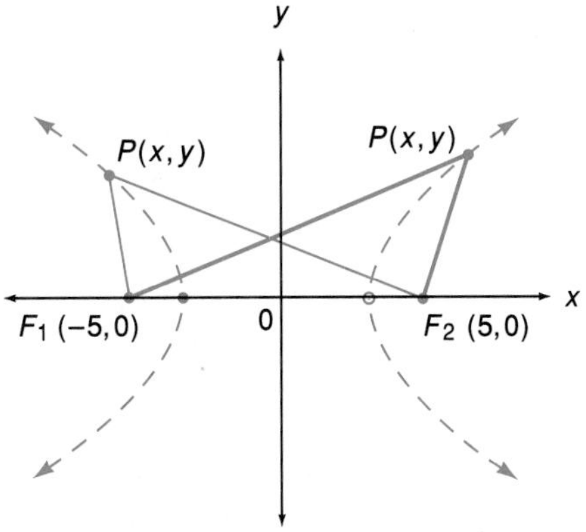

The same result is obtained for P on the left branch.

For $P(x, y)$ on the right branch, we have

$$PF_1 - PF_2 = 6 \quad \text{or} \quad PF_1 = 6 + PF_2$$

Using the distance formula, the preceding equation becomes

$$\sqrt{(x+5)^2 + y^2} = 6 + \sqrt{(x-5)^2 + y^2}$$

Then the same type of algebra used in deriving the equation of an ellipse (see page 317) produces these forms:

$$\frac{x^2}{9} - \frac{y^2}{16} = 1 \quad \text{or} \quad \frac{x^2}{3^2} - \frac{y^2}{4^2} = 1$$

Observe that the first coordinates of the foci, ± 5, are related to the preceding denominators. That is, $(\pm 5)^2 = 3^2 + 4^2$.

When this work is generalized, we obtain the following standard form of the equation of a hyperbola whose foci are symmetric about the origin on the x-axis.

The feature that distinguishes between the standard forms of hyperbolas and ellipses is the sign between the x^2 and y^2 terms. For ellipses the sign is +, and for hyperbolas the sign is −.

STANDARD FORM FOR THE EQUATION OF A HYPERBOLA WITH FOCI AT $(-c, 0)$ AND $(c, 0)$

$$\frac{x^2}{a^2} - \frac{y^2}{b^2} = 1 \qquad \text{where } b^2 = c^2 - a^2$$

The graph consists of two congruent branches which open in opposite directions. Letting $y = 0$, gives $x = \pm a$, and the points $V_1(-a, 0)$, $V_2(a, 0)$ are the vertices of the hyperbola. The segment V_1V_2 is called the **transverse axis** and its midpoint (the origin) is the center of the hyperbola, as shown on the next page.

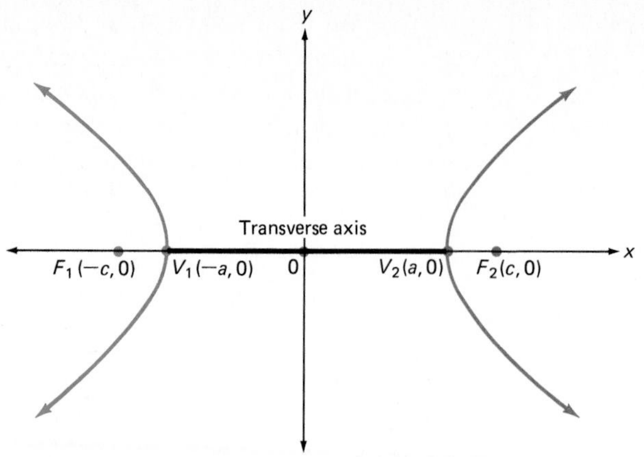

An efficient way to sketch a hyperbola is first to draw the rectangle that is $2a$ units wide and $2b$ units high, as shown in the following figure. Note that the center of the hyperbola is also the center of this rectangle. Draw the diagonals of the rectangle and extend them in both directions; these are the **asymptotes.** Now sketch the hyperbola by beginning at the vertices $(\pm a, 0)$ so that the two branches stay between, and approach the asymptotes whose equations are $y = \pm\frac{b}{a}x$.

Since $b^2 = c^2 - a^2$, or $a^2 + b^2 = c^2$, it follows that b can be used as a side of a right triangle having hypotenuse c. Thus if we construct a perpendicular at V_2 of length b, the resulting right triangle has hypotenuse c. We also see that this hypotenuse lies on the line $y = \frac{b}{a}x$.

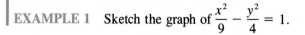

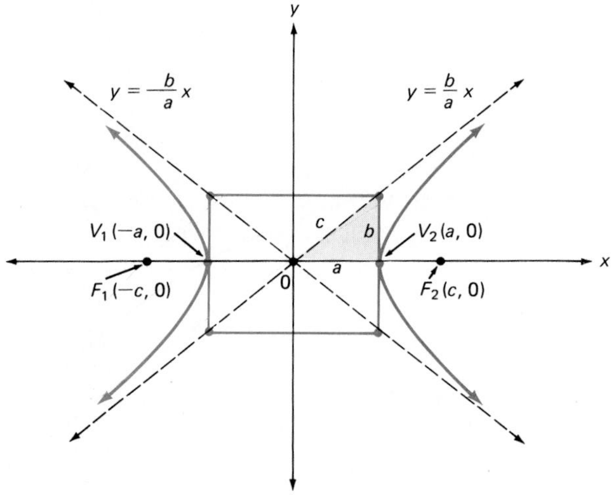

EXAMPLE 1 Sketch the graph of $\frac{x^2}{9} - \frac{y^2}{4} = 1$.

Solution Since $a^2 = 9$, the vertices are $(\pm 3, 0)$. Also,

$$b^2 = c^2 - a^2 \quad \text{or} \quad c^2 = a^2 + b^2 = 9 + 4 = 13$$

Therefore $(\pm\sqrt{13}, 0)$ are the foci. To sketch the hyperbola, first note that $2a = 6$, $2b = 4$ and draw the 6 by 4 rectangle with center $(0, 0)$. Draw the asymptotes by extending the diagonals and sketch the two branches, as follows.

The branches approach the asymptotes in the sense that as x gets larger (toward the right) the hyperbola gets closer to the asymptotes, but never touches them; similarly for x toward the left.

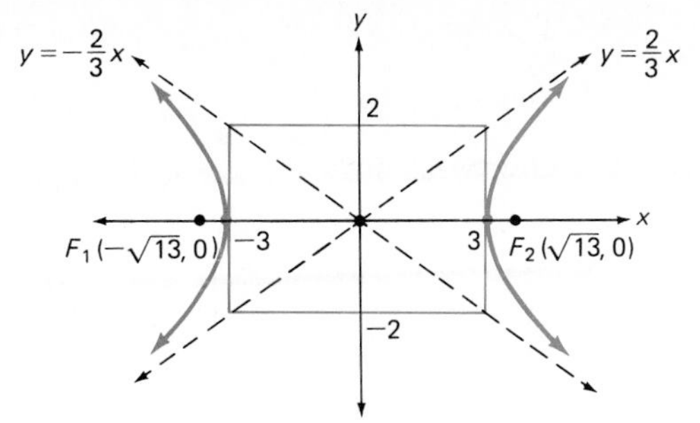

■

EXAMPLE 2 Write the equation of the hyperbola having vertices $(\pm 4, 0)$ and asymptotes $y = \pm\frac{3}{2}x$, and identify the foci.

Solution In general, the asymptotes are $y = \pm\dfrac{b}{a}x$. Therefore, $\dfrac{b}{a} = \dfrac{3}{2}$. And, since $a = 4$,

$$\frac{b}{4} = \frac{3}{2} \quad \text{or} \quad b = \frac{3}{2} \cdot 4 = 6$$

The equation is

$$\frac{x^2}{16} - \frac{y^2}{36} = 1 \quad \text{or} \quad \frac{x^2}{4^2} - \frac{y^2}{6^2} = 1$$

Also $c^2 = a^2 + b^2 = 16 + 36 = 52$, so that $c = \sqrt{52}$ and the foci are $(\pm 2\sqrt{13}, 0)$. ■

When the equation of a hyperbola is in standard form the branches open vertically if the minus sign precedes the x^2-term. They will open horizontally if the minus sign precedes the y^2-term.

EXAMPLE 3 Write in standard form and identify the foci and vertices: $16x^2 - 25y^2 = 400$.

Solution Divide through by 400 to place in standard form.

$$\frac{16x^2}{400} - \frac{25y^2}{400} = \frac{400}{400}$$

$$\frac{x^2}{25} - \frac{y^2}{16} = 1$$

Note that $a^2 = 25$ and $b^2 = 16$, so that $a = 5$ and $b = 4$. Then $c^2 = a^2 + b^2 = 25 + 16 = 41$ and $c = \pm\sqrt{41}$. The vertices of the hyperbola are located at $(-5, 0)$ and $(5, 0)$; the foci are at $(-\sqrt{41}, 0)$ and $(\sqrt{41}, 0)$. ■

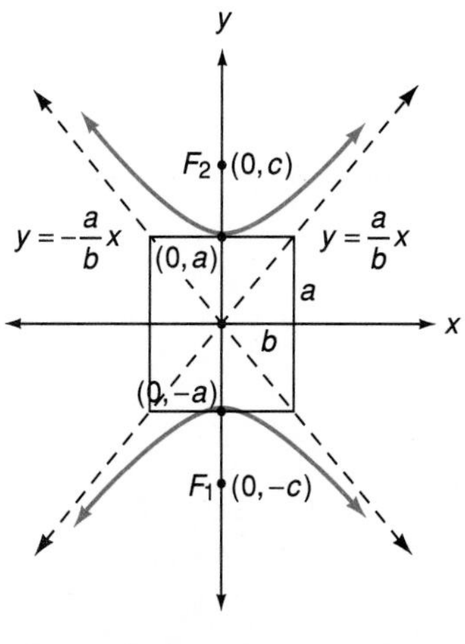

When the foci of a hyperbola are on the y-axis the equation has this standard form:

STANDARD FORM FOR THE EQUATION OF A HYPERBOLA WITH FOCI AT $(0, -c)$ AND $(0, c)$

$$\frac{y^2}{a^2} - \frac{x^2}{b^2} = 1 \qquad \text{where } b^2 = c^2 - a^2$$

The vertices are $(0, \pm a)$, the transverse axis is vertical and has length $2a$, and the asymptotes are the lines $y = \pm\frac{a}{b}x$. The branches of this hyperbola open upward and downward as in the figure in the margin on the preceding page.

EXAMPLE 4 Write the equation of the hyperbola with center (0, 0), foci $(0, \pm 5)$, and transverse axis of length 6. Sketch the graph.

Solution Since the foci are $(0, \pm 5)$ this is a hyperbola with a vertical transverse axis. The ends of the transverse axis are the vertices, and since $2a = 6$, $a = 3$ so that $(0, \pm 3)$ are the vertices. To find b use $c = 5$ and $a = 3$ in $b^2 = c^2 - a^2$; $b^2 = 25 - 9 = 16$, and $b = 4$. The equation is

$$\frac{y^2}{9} - \frac{x^2}{16} = 1$$

The asymptotes are $y = \pm\frac{a}{b}x = \pm\frac{3}{4}x$. ■

Additional points can be used to gain accuracy when drawing the hyperbola. For example, let $x = 5$ in the equation to obtain $y = 4.8$ to one decimal place. Then (5, 4.8) is on the hyperbola, and the symmetry through the axes also gives the points (−5, 4.8), (−5, −4.8), and (5, −4.8).

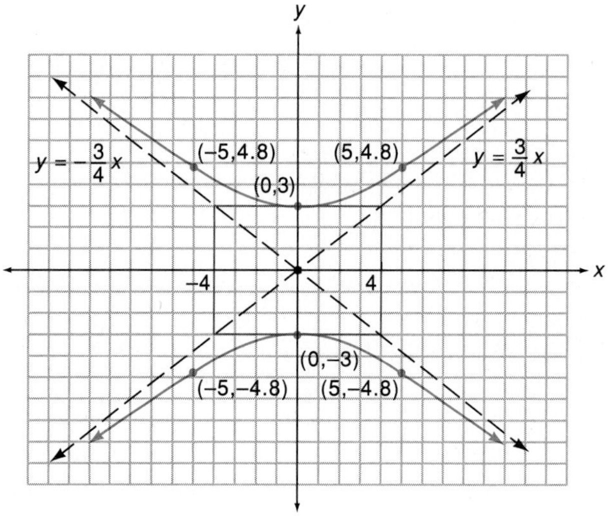

EXERCISES 7.6

Write the equation of each hyperbola in standard form. Also write the equations of the asymptotes and give the coordinates of the foci.

1.

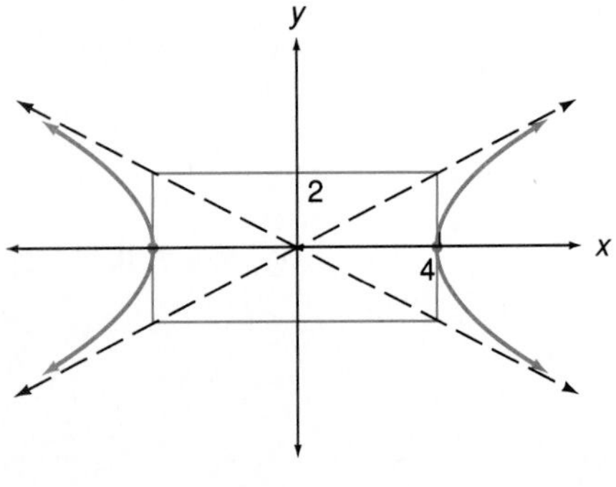

2.

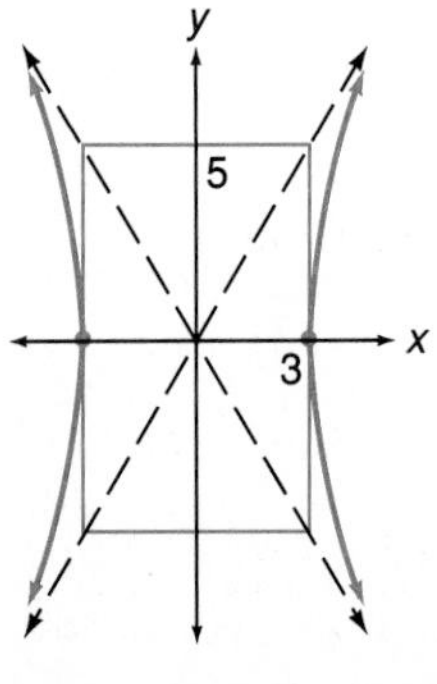

3.

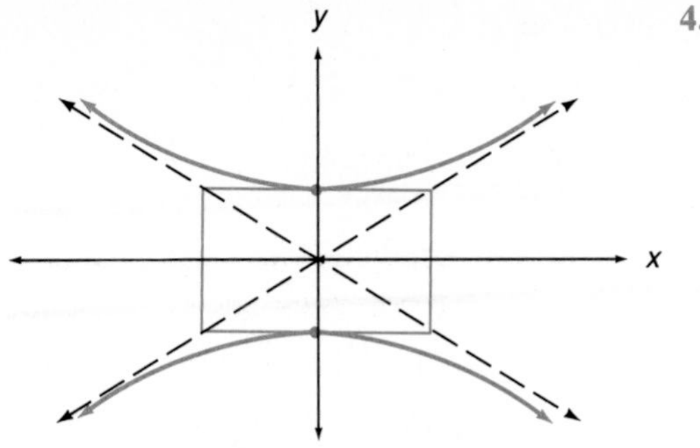

Dimensions of rectangle are 16 by 10.

4.

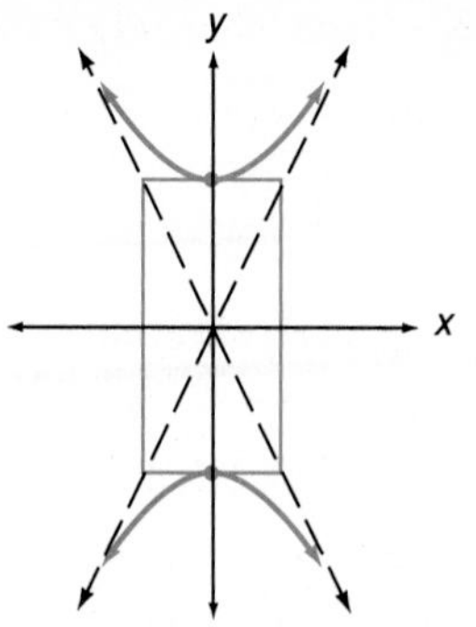

Dimensions of rectangle are 6 by 10.

Graph each hyperbola. State the coordinates of the center, vertices, and foci. Give the equations of the asymptotes.

5. $\frac{x^2}{25} - \frac{y^2}{9} = 1$
6. $\frac{x^2}{4} - \frac{y^2}{9} = 1$
7. $\frac{x^2}{16} - \frac{y^2}{25} = 1$
8. $\frac{y^2}{16} - \frac{x^2}{4} = 1$
9. $\frac{y^2}{36} - \frac{x^2}{9} = 1$
10. $\frac{y^2}{4} - \frac{x^2}{9} = 1$
11. $9x^2 - y^2 = 9$
12. $x^2 - 9y^2 = 9$
13. $4x^2 - 9y^2 = 36$
14. $9y^2 - 4x^2 = 36$
15. $25y^2 - 9x^2 = 225$
16. $9y^2 - 16x^2 = 144$

Write the equation of the hyperbola in standard form having the given properties.

17. Center (0, 0); foci (± 6, 0); vertices (± 4, 0).
18. Center (0, 0); foci (0, ± 4); vertices (0, ± 1).
19. Foci (0, ± 5) and transverse axis of length $6\sqrt{2}$.
20. Asymptotes $y = \pm\frac{4}{3}x$; vertices (0, ± 8).
21. Asymptotes $y = \pm\frac{1}{2}x$; vertices (± 4, 0).
*22. Asymptotes $y = \pm\frac{5}{12}x$; foci (± 13, 0)
*23. Asymptotes $y = \pm\frac{8}{15}x$; foci (0, ± 17)
24. Complete the derivation on page 323 by showing that the equation

$$\sqrt{(x + 5)^2 + y^2} = 6 + \sqrt{(x - 5)^2 + y^2}$$

produces $\frac{x^2}{9} - \frac{y^2}{16} = 1$. (*Hint:* Square both sides and simplify, then square again.)

When a hyperbola with equation $\frac{x^2}{a^2} - \frac{y^2}{b^2} = 1$ is shifted h units horizontally and k units vertically, the resulting hyperbola is congruent to the original, has center (h, k), and has an equation of the form

$$\frac{(x - h)^2}{a^2} - \frac{(y - k)^2}{b^2} = 1, \qquad b^2 = c^2 - a^2$$

A similar observation can be made beginning with the form $\frac{y^2}{a^2} - \frac{x^2}{b^2} = 1$. Sketch the graph of each hyperbola.

25. $\frac{(x - 2)^2}{9} - \frac{(y + 1)^2}{4} = 1$
26. $\frac{(x + 3)^2}{16} - \frac{(y - 2)^2}{9} = 1$
27. $\frac{(y - 4)^2}{36} - \frac{(x + 2)^2}{64} = 1$
28. $\frac{(x + 1)^2}{16} - \frac{(y + 5)^2}{9} = 1$

Write the equations, in standard form, of the hyperbola having the given properties.

29. Center $(-2, 3)$; vertical transverse axis of length 6; $c = 4$.
30. Center $(4, 4)$; vertex $(4, 7)$; $b = 2$.

EXPLORATIONS
Think Critically

1. What is the relationship between the graph of $x^2 + x - 6 = 0$ on a line and the graph of $y = x^2 + x - 6$ on a plane?
2. Explain how the inequality $\dfrac{x-2}{x-3} < 0$ can be solved by multiplying both sides by $x - 3$. *Hint:* Consider the two cases suggested by the multiplication property of order.
3. How can you determine the maximum or minimum value of a function $f(x) = ax^2 + bx + c$?
4. Suppose that a function is increasing on one side of a point P with coordinates (a, b) and decreasing on the other side of this point. What can be said about the point P? Explain.
5. Consider the graph of $y = (x + 2)^2 - 1$. How can you use this to determine the graph of $x = (y + 2)^2 - 1$?
6. Sketch the graph of the ellipse $\dfrac{x^2}{25} + \dfrac{y^2}{16} = 1$. Join the endpoints of the major axis to those of the minor axis. Find the equation of this new figure, said to be *analogous* to the ellipse.
7. The graphs of the following are said to be *conjugate hyperbolas:*

$$\frac{x^2}{25} - \frac{y^2}{16} = 1 \qquad \frac{x^2}{16} - \frac{y^2}{25} = 1$$

Graph both on the same set of axes and describe the relationship between them.

7.7 CONIC SECTIONS: THE PARABOLA

The parabola, like the other conic sections, can also be defined as a specific set of points in the plane, and it has numerous applications that are directly related to this definition.

Note that a considerable amount of work has already been done with parabolas in Sections 7.1 and 7.2.

DEFINITION OF A PARABOLA

A **parabola** is the set of all points in a plane equidistant from a given fixed line called the **directrix** and a given fixed point called the **focus.**

For each point P on the parabola in the following figure $PF = PQ$, where F is the focus and Q is the point on the directrix. The line through F and perpendicular to the directrix is called the **axis** of the parabola, and the point V, which is the intersection of the parabola with its axis, is called the **vertex.**

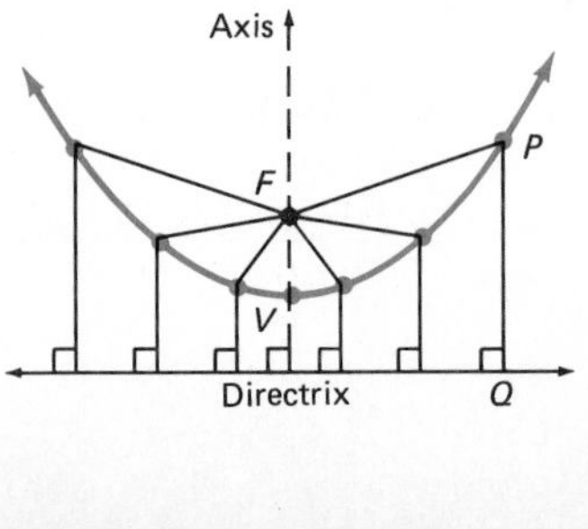

The parabolas we will consider will have either vertical or horizontal axes. We begin with parabolas whose axes are the y-axis.

In the figures that follow, let focus F have coordinates $(0, p)$, and let the directrix have equation $y = -p$ as indicated.

Observe that in each case the focus is within the parabola and the directrix is outside.

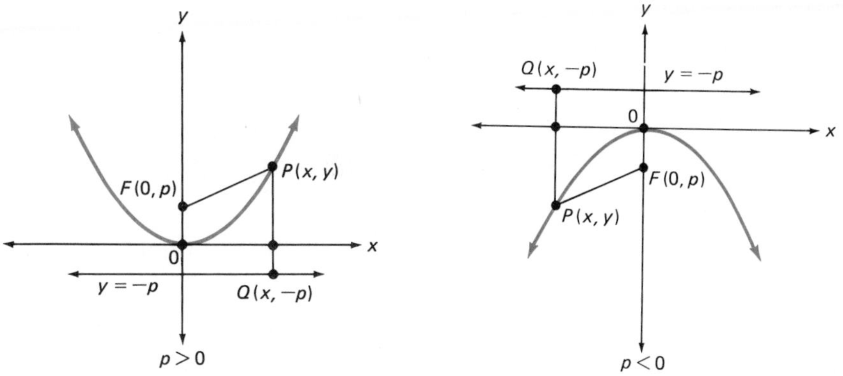

The origin, which is also the vertex, is on this parabola because it is the same distance from the focus and the directrix. In general, for any point $P(x, y)$ on the parabola the distance formula can be used to find PF and PQ. Then set these two distances equal to each other as indicated in the definition.

$$PF = \sqrt{(x-0)^2 + (y-p)^2} = \sqrt{x^2 + (y-p)^2}$$
$$PQ = \sqrt{(x-x)^2 + (y-(-p))^2} = \sqrt{(y+p)^2}$$
$$PF = PQ: \quad \sqrt{x^2 + (y-p)^2} = \sqrt{(y+p)^2} \quad \text{(Square each side.)}$$
$$x^2 + y^2 - 2py + p^2 = y^2 + 2py + p^2$$
$$x^2 = 4py$$

In summary, we have the following.

This equation can also be written as $y = \frac{1}{4p}x^2$. Letting $a = \frac{1}{4p}$ gives $y = ax^2$, the form we used in our earlier work with parabolas.

STANDARD FORM FOR THE EQUATION OF A PARABOLA WITH FOCUS $(0, p)$ AND DIRECTRIX $y = -p$

$$x^2 = 4py$$

The axis is the y-axis and the vertex is the origin.

This form for the equation of a parabola can be used to determine the coordinates of the focus and the equation of the directrix, as in Example 1.

EXAMPLE 1 Find the coordinates of the focus and the equation of the directrix for the parabola $x^2 = 4y$ and sketch the graph.

Solution Consider the general form $x^2 = 4py$, and let $4p = 4$. Thus $p = 1$ and we can locate the focus and directrix, as shown in the graph on the following page. The parabola has its focus at $(0, 1)$ and the equation of the directrix is $y = -p = -1$.

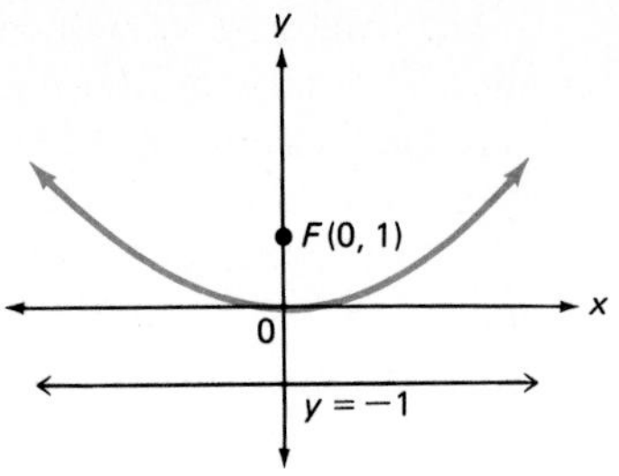

■

EXAMPLE 2 The focus of a parabola has coordinates $(0, -\frac{5}{2})$ and the vertex is at the origin. Find the equation of the parabola.

Solution Both the focus and vertex are on the y-axis. Therefore, the y-axis is the axis of the parabola. Since such parabolas have focus at $(0, p)$, we get $p = -\frac{5}{2}$. Then, using $x^2 = 4py$, the equation is

$$x^2 = 4(-\tfrac{5}{2})y = -10y$$

■

A parabola with vertex at the origin whose axis is the x-axis has an equation of the form $y^2 = 4px$. The derivation for this standard form is very similar to the derivation used to obtain the form $x^2 = 4py$.

STANDARD FORM FOR THE EQUATION OF A PARABOLA WITH FOCUS $(p, 0)$ AND DIRECTRIX $x = -p$

$$y^2 = 4px$$

The axis is the x-axis and the vertex is the origin.

EXAMPLE 3 The directrix of a parabola is $x = -2$ and the focus is $(2, 0)$. Find the equation and graph.

Solution Since the directrix is a vertical line, the axis of the parabola is horizontal. Also, since the focus $(2, 0)$ is on the x-axis, the parabola's axis is the x-axis. Such parabolas have focus $(p, 0)$ and therefore $p = 2$. Thus, using $y^2 = 4px$, the equation is

$$y^2 = 4(2)x = 8x$$

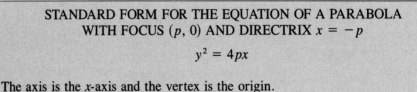

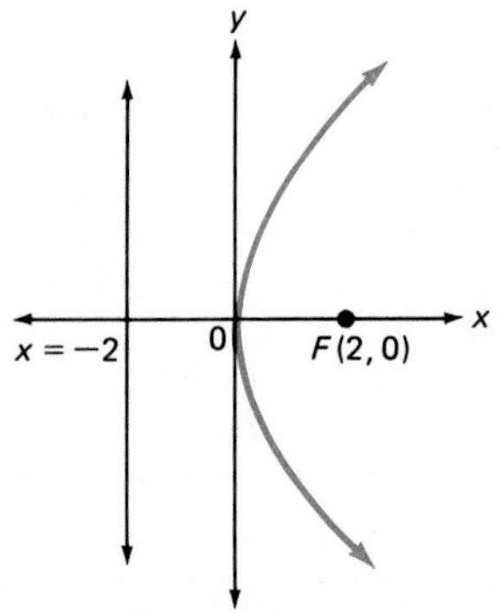

■

Parabolic reflectors are used in a wide variety of instruments such as reflecting telescopes, searchlights, microwave antennae, and solar energy devices. The surface of a parabolic reflector is obtained by revolving a parabola around its axis of symmetry.

The following figure at the left illustrates that when light rays, parallel to the axis, strike the surface they are reflected to the focus. Conversely, the figure at the right shows that when there is a light source at the focus of a parabolic reflector the light rays will reflect off the surface, forming a beam of light parallel to the axis.

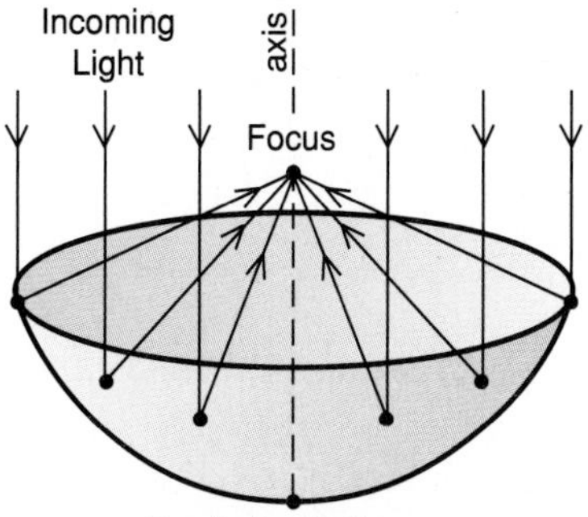

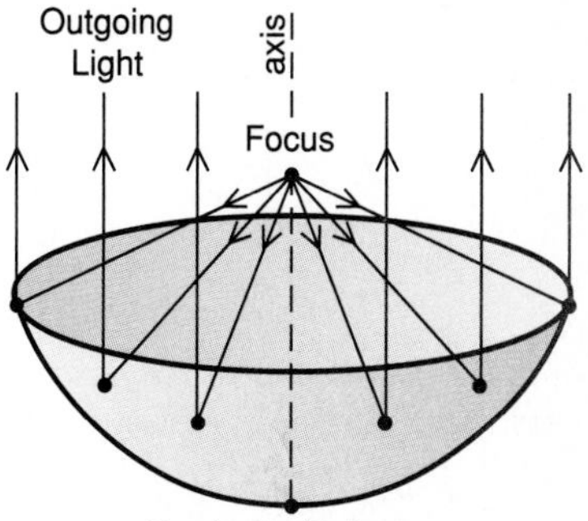

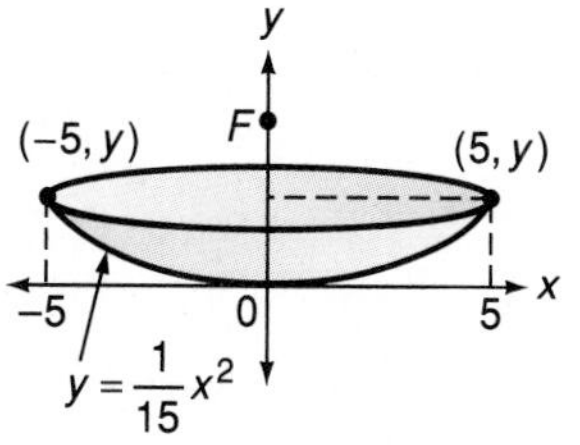

EXAMPLE 4 Suppose that the reflecting surface of a television antenna was formed by rotating the parabola $y = \frac{1}{15}x^2$ about its axis of symmetry for $-5 \le x \le 5$. Assuming that the measurements are made in feet, how far from the bottom or vertex of this "dish" antenna should the receiver be placed? How deep is this antenna?

Solution The receiver must be at the focus and the distance the focus is from the vertex is the value p in the standard form $x^2 = py$. Since $y = \frac{1}{15}x^2$, $x^2 = 15y$. Then,

$$4p = 15$$
$$p = \tfrac{15}{4} = 3.75$$

The receiver is 3.75 feet above the vertex along the axis.

The depth of the antenna is the y-value when $x = 5$. Thus, using 5 in $y = \frac{1}{15}x^2$, we get $y = \frac{25}{15} = \frac{5}{3}$. The antenna is $1\frac{2}{3}$ feet deep. ■

EXERCISES 7.7

Write the equation of each parabola. Give the equation of the directrix or the coordinates of the focus when they are not given.

1.

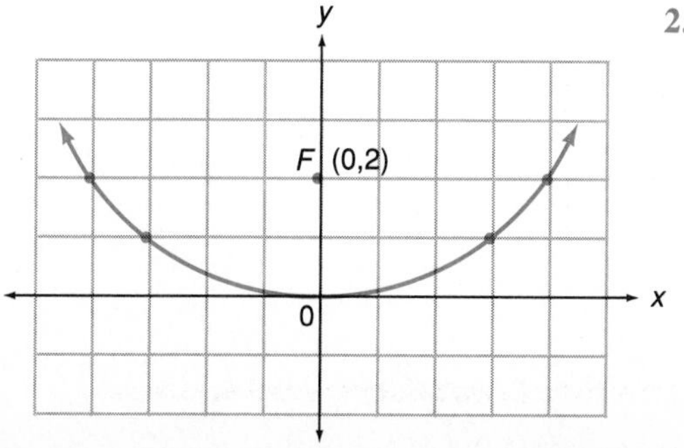

2.

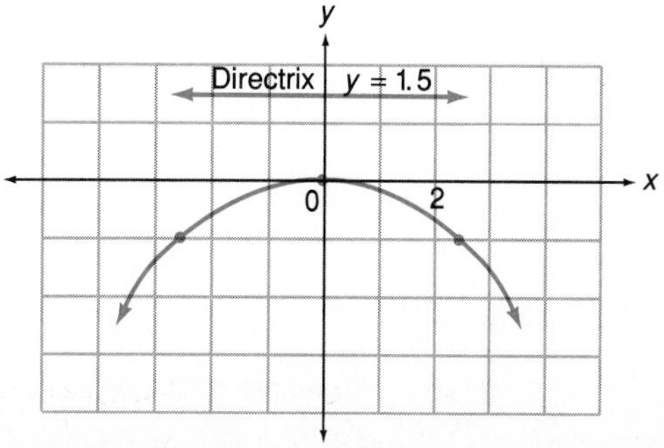

3.

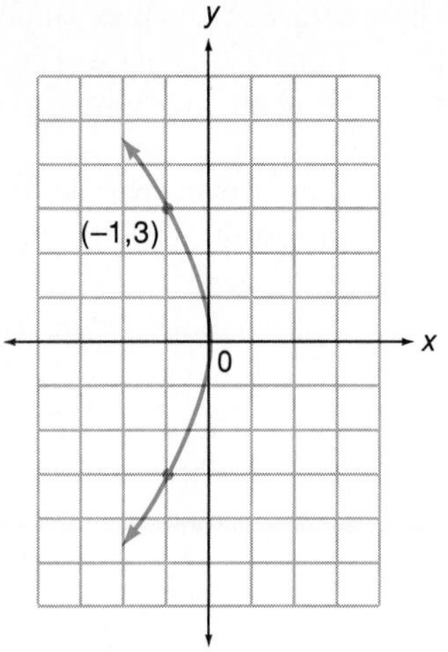

4.

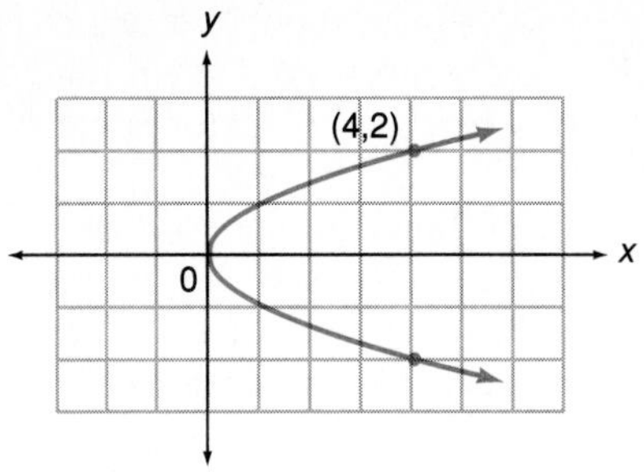

Find the coordinates of the focus and the equation of the directrix for each of the following parabolas.

5. $x^2 = \frac{1}{2}y$
6. $x^2 = -4y$
7. $x^2 = \frac{1}{4}y$
8. $y = \frac{1}{2}x^2$
9. $y = -4x^2$
10. $y = -\frac{1}{8}x^2$
11. $y^2 = 2x$
12. $y^2 = -2x$
13. $y^2 = -\frac{1}{3}x$
14. $x = -\frac{3}{2}y^2$
15. $-9x = 6y^2$
16. $\frac{1}{2}x = \frac{1}{8}y^2$

Write the equation of each parabola having the given properties and sketch the graph.

17. Focus $(0, -3)$; directrix $y = 3$.
18. Focus $(0, 3)$; directrix $y = -3$.
19. Directrix $y = -\frac{2}{3}$; vertex $(0, 0)$.
20. Vertex $(0, 0)$; vertical axis; $(2, -2)$ is on the parabola.
21. Focus $(-\frac{3}{4}, 0)$; directrix $x = \frac{3}{4}$.
22. Focus $(\frac{5}{8}, 0)$; vertex $(0, 0)$.
23. Vertex $(0, 0)$; horizontal axis; $(12, -6)$ is on the parabola.
24. Directrix $x = \frac{3}{2}$; vertex $(0, 0)$.

25. The reflecting surface of a radar antenna is generated by revolving the parabola $y = \frac{2}{9}x^2$ about its axis of symmetry for $-4 \le x \le 4$. Assuming the measurements are done in feet, how far from the bottom of the dish antenna is the receiver? What is the circumference of the antenna?
26. The reflecting surface of an antenna, as in Exercise 25, is generated by revolving a parabola of the form $x^2 = 4py$ about its axis of symmetry. If the antenna is 8 feet across the top (this is the length of a diameter) and $1\frac{1}{2}$ feet deep, where must the receiver be located?
27. The center cable of a suspension bridge forms a parabolic arc. The cable is suspended from the tops of the two support towers that are 800 feet apart. If the tops of the towers are 160 feet above the road, and if the cable just touches the road midway between the towers, find the height of the cable at a distance of 100 feet from a tower.

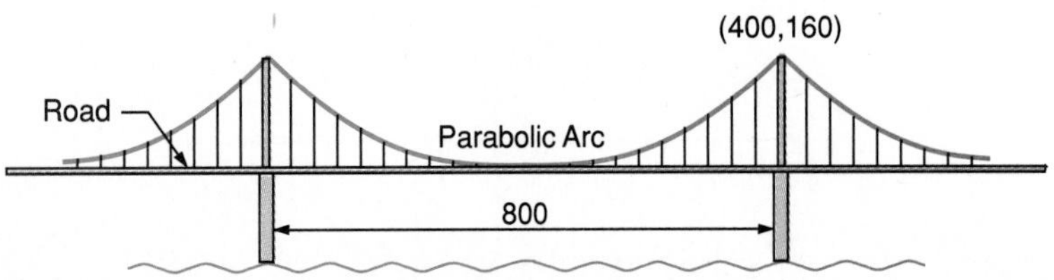

When a parabola with equation $y^2 = 4px$ is shifted h units horizontally and k units vertically, the resulting parabola is congruent to the original and has the following properties: equation, $(y - k)^2 = 4p(x - h)$; vertex, (h, k); axis of symmetry, $y = k$; focus, $(h + p, k)$; directrix, $x = h - p$. Identify the vertex, the axis of symmetry, the focus and directrix of each parabola.

28. $(y + 3)^2 = -2(x - 1)$ 29. $(y - 2)^2 = 4(x + 5)$

30. A parabola has directrix $x = -1$ and focus $(3, 3)$. Write the equation of the parabola and the axis of symmetry, and give the coordinates of the vertex.
31. A parabola has vertex $(-5, 2)$, focus at $(-3, 2)$, and directrix $x = -7$. Write its equation in the form $(y - k)^2 = 4p(x - h)$.
32. If the parabola $x^2 = 4py$ is shifted h units horizontally and k units vertically, what is the equation of the resulting parabola?

CHAPTER 7 SUMMARY

Review these key terms and concepts so that you are able to define or describe them. A clear understanding of these will be very helpful when reviewing the developments of this chapter.

Quadratic Functions

The graph of $y = f(x) = ax^2 + bx + c$, $a \neq 0$, is a parabola. Completing the square in x gives the standard form $y = a(x - h)^2 + k$.

Features

Axis of symmetry: $x = h$. **Vertex:** (h, k).

For $a > 0$, the parabola opens upward and $f(h) = k$ is the minimum value. For $a < 0$ it opens downward and $f(h) = k$ is the maximum value.

There are two x-intercepts, r_1 and r_2, if the discriminant $b^2 - 4ac > 0$; there is one if $b^2 - 4ac = 0$; and there are none if $b^2 - 4ac < 0$. For two x-intercepts, the following holds:

If $a > 0$, $ax^2 + bx + c < 0$ for $r_1 < x < r_2$, and $ax^2 + bx + c > 0$ for $x < r_1$ or $x > r_2$.
If $a < 0$, $ax^2 + bx + c < 0$ for $x < r_1$ or $x > r_2$, and $ax^2 + bx + c > 0$ for $r_1 < x < r_2$.

Multiplying the ordinates of $y = x^2$ by a, and translating h units horizontally and k units vertically, gives the graph of $y = a(x - h)^2 + k$ which is congruent to the graph of $y = ax^2$.

Circle

Conic Sections

A **circle** is the set of all points in the plane, each of which is at a fixed distance r from a given point called the **center** of the circle; r is the **radius** of the circle $(r > 0)$.

Circle with center $(0, 0)$ and radius r: $x^2 + y^2 = r^2$

Circle with center (h, k) and radius r: $(x - h)^2 + (y - k)^2 = r^2$

Ellipse

An **ellipse** is the set of all points in a plane such that the sum of the distances from two fixed points (called the **foci**) is a constant.

Ellipse with foci $(\pm c, 0)$ and center $(0, 0)$:

$$\frac{x^2}{a^2} + \frac{y^2}{b^2} = 1, \qquad b^2 = a^2 - c^2$$

Ellipse with foci $(0, \pm c)$ and center $(0, 0)$:

$$\frac{x^2}{b^2} + \frac{y^2}{a^2} = 1, \qquad b^2 = a^2 - c^2$$

Hyperbola

A **hyperbola** is the set of all points in the plane such that the difference of the distances from two fixed points (called the **foci**) is a constant.

Hyperbola with foci $(\pm c, 0)$, center $(0, 0)$, and asymptotes $y = \pm\frac{b}{a}x$:

$$\frac{x^2}{a^2} - \frac{y^2}{b^2} = 1, \qquad b^2 = c^2 - a^2$$

Hyperbola with foci $(0, \pm c)$, center $(0, 0)$, and asymptotes $y = \pm\frac{a}{b}x$:

$$\frac{y^2}{a^2} - \frac{x^2}{b^2} = 1, \qquad b^2 = c^2 - a^2$$

Parabola

A **parabola** is the set of all points in a plane equidistant from a given fixed line called the **directrix** and a given fixed point called the **focus.**

Parabola with focus $(0, p)$, directrix $y = -p$, center $(0, 0)$: $x^2 = 4py$

Parabola with focus $(p, 0)$, directrix $x = -p$, center $(0, 0)$: $y^2 = 4px$

REVIEW EXERCISES

The solution to the following exercises can be found within the text of Chapter 7. Try to answer each question before referring to the text.

Section 7.1

Graph each of the following.

1. $y = x^2$
2. $y = x^2 + 2$
3. $y = x^2 - 2$
4. $y = (x + 2)^2$
5. $y = (x - 2)^2$
6. $y = -x^2$

7. $y = \frac{1}{2}x^2$ 8. $y = 2x^2$ 9. $y = -x^2 + 2$
10. $y = f(x) = -(x - 2)^2 + 1$ 11. $y = f(x) = -2(x - 3)^2 + 4$ 12. $x = y^2$
13. $x = (y + 2)^2 - 4$

Section 7.2

14. Graph the function $y = x^2 + 4x + 3$; find all intercepts, the vertex, and the axis of symmetry.
15. Graph the function $y = x^2 - 4x + 4$; find all intercepts, the vertex, the axis of symmetry, and the domain and range.
16. Graph the function $y = 2x^2 - 4x + 5$; find the vertex, the axis of symmetry, and the range.
17. Find the maximum value of $f(x) = -2x^2 + 12x - 11$. At which x does $f(x)$ have this value?
18. Suppose that 60 meters of fencing is available to enclose a rectangular garden, one side of which will be against the side of a house. What dimensions of the garden will guarantee a maximum area?
19. A ball is thrown straight upward from ground level with an initial velocity of 32 feet per second. The formula $s = 32t - 16t^2$ gives its height in feet, s, after t seconds. What is the maximum height reached by the ball?
20. The marketing department of the TENRAQ Tennis Company found that, on the average, 600 tennis rackets will be sold monthly at the unit price of \$100. The department also observed that for each \$5 reduction in price, there will be an extra 50 rackets sold monthly. What price will bring the largest monthly income?

Section 7.3

21. Solve for x and graph the solution on a number line: $-x^2 - x + 2 \geq 0$

Solve each inequality.

22. $x^2 + 2x - 2 > 0$ 23. $x^2 + 6x > -9$ 24. $x^2 - 4x + 9 < 0$ 25. $\dfrac{x + 2}{x - 3} < 0$

26. Graph $y \leq 1 + 4x - x^2$ by shading the appropriate region.

Section 7.4

Write the equation of the circle in standard form with the given center and radius.

27. $C(0, 0)$; $r = 3$ 28. $C(2, -5)$; $r = 4$

Draw each circle.

29. $x^2 + y^2 = 2$ 30. $x^2 - 8x + y^2 - 9 = 0$

31. Find the two points on the circle $x^2 + y^2 = 25$ where $x = 2$.
32. Write the equation of the tangent to the circle in Exercise 31, at the point $(3, 4)$.
33. Find the center and radius of the circle $(x - 2)^2 + (y + 3)^2 = 4$.
34. Find the center and radius of the circle with equation $9x^2 + 12x + 9y^2 = 77$.
35. Points $P(2, 5)$ and $Q(-4, -3)$ are the endpoints of a diameter of a circle. Find the center, radius, and equation of the circle.

Section 7.5

36. Sketch the graph of the ellipse $\dfrac{x^2}{25} + \dfrac{y^2}{9} = 1$.

37. Write the equation of the ellipse in standard form having vertices $(\pm 10, 0)$ and foci $(\pm 6, 0)$.

38. Change $25x^2 + 16y^2 = 400$ into standard form and graph.

39. A satellite follows an elliptical orbit around the earth such that the center of the earth, E, is one of the foci. The highest point that the statellite will be from the earth's surface is 2500 miles, and the lowest will be 1000 miles. These distances are measured along the major axis, which is assumed to be on the y-axis. Use 4000 miles as the radius of the earth and find the equation of the orbit.

Section 7.6

40. Sketch the graph of $\frac{x^2}{9} - \frac{y^2}{4} = 1$.

41. Write the equation of the hyperbola having vertices $(\pm 4, 0)$ and asymptotes $y = \pm\frac{3}{2}x$, and identify the foci.

42. Write in standard form and identify the foci and the vertices of the hyperbola $16x^2 - 25y^2 = 400$.

43. Write the equation of the hyperbola with center $(0, 0)$, foci $(0, \pm 5)$, and transverse axis of length 6. Sketch the graph.

Section 7.7

44. Find the coordinates of the focus and the equation of the directrix for the parabola $x^2 = 4y$.

45. The focus of a parabola has coordinates $(0, -\frac{5}{2})$ and the vertex is at the origin. Find the equation of the parabola.

46. The directrix of a parabola is $x = -2$ and the focus is $(2, 0)$. Find the equation of the parabola and graph.

47. Suppose that the reflecting surface of a television antenna was formed by rotating the parabola $y = \frac{1}{15}x^2$ about its axis of symmetry for $-5 \le x \le 5$. Assuming that the measurements are made in feet, how far from the bottom or vertex of this "dish" antenna should the receiver be placed? How deep is this antenna?

CHAPTER 7 TEST: STANDARD ANSWER

Use these questions to test your knowledge of the basic skills and concepts of Chapter 7. Then check your answers with those given at the back of the book.

Draw the graph of each function.

1. $y = f(x) = (x - 2)^2 + 3$ 2. $y = f(x) = -(x + 1)^2 - 2$

3. Give **(a)** the coordinates of the vertex, **(b)** the equation of the axis of symmetry, **(c)** the domain, and **(d)** the range for the quadratic function $y = 2(x - 1)^2 - 5$.

4. State where the function $y = -\frac{1}{2}(x - 4)^2 + 6$ is increasing and where it is decreasing. Describe the concavity.

Write in the standard form $y = a(x - h)^2 + k$.

5. $y = x^2 + 4x - 9$ 6. $y = 2x^2 - 8x + 3$

Give the value of the discriminant and use this result to describe the x-intercepts. if any.

7. $y = x^2 + 3x + 1$ 8. $y = 6x^2 + 5x - 6$

9. Find the x-intercepts of $y = -x^2 + 4x + 7$.

10. A homeowner wishes to purchase a rectangular rug that will have a perimeter of 36 feet and at the same time cover the maximum possible area. What should be the dimensions of the rug to cover the largest possible area?

11. The formula $h = 64t - 16t^2$ gives the distance in feet above the ground, h, reached by an object in t seconds. What is the maximum height reached by the object?

Solve each inequality.

12. $16 + 6x - x^2 \leq 0$ 13. $4x^2 - 6x + 1 < 0$ 14. $\dfrac{x - 4}{2x + 1} > 0$

15. **(a)** Draw the circle $(x - 3)^2 + (y + 4)^2 = 25$.
(b) Write the equation of the tangent line to this circle at the point (6, 0).

16. Write the equation of the circle $x^2 + 10x + y^2 + 15 = 0$ in standard form and giv… center and radius.

17. Points $P(1, -5)$ and $Q(-3, 3)$ are the endpoints of a diameter of a circle. Fin… ter, radius, and equation of the circle.

Identify the conic section and sketch the graph. Find the foci, vertices, and asy… are any.

18. $100x^2 + 36y^2 = 3600$ 19. $9x^2 - 4y^2 = 36$

20. Write the equation of the ellipse in standard form with center at … axis of length 8, and minor axis of length 6.

21. Write the equation of the hyperbola in standard form w… $(\pm 8, 0)$, and vertices at $(\pm 6, 0)$.

22. Write the equation of the hyperbola with center (0, 0), … is of length 10.

23. Find the coordinates of the focus and the equa… arabola $y = -\frac{1}{8}x^2$.

24. A parabola has directrix $x = \frac{2}{3}$ and vertex … tion of the parabola and give the coordinates of the f…

25. A satellite follows an elliptical orbit aro… er of the earth is one of the foci. The highest point of t… rface is 3000 miles and the lowest is 1000 miles. Let th… ssume the earth's radius is 4000 miles to find the equ… unsimplified form.)

CHAPTER 7 TEST

1. The coordinates of … he axis of symmetry for the graph of $y = (x - 2)^2 - $…
(a) $(2, 5)$; $x = -2$ … 2   **(c)** $(2, -5)$; $x = 2$
(d) $(-2, 5)$; $x = -2$ … preceding

2. The range of the function g… $-2(x + 1)^2 - 3$ is
(a) all real numbers   **(b)** a… 2   **(c)** all $y \leq 3$ **(d)** all $y \geq -3$ **(e)** None of the preceding

3. Which of the following is the standard form for the equation $y = x^2 - 4x + 1$?
(a) $y = (x - 2)^2 - 3$ **(b)** $y = (x + 2)^2 - 3$ **(c)** $y = x(x - 4) + 1$
(d) $y = (x - 2)^2 + 1$ **(e)** None of the preceding

4. The graph of $y = 3(x + 1)^2 - 2$ is congruent to that for $y = 3x^2$ but is
(a) shifted 1 unit to the right and 2 units down
(b) shifted 1 unit to the left and 2 units down
(c) shifted 1 unit to the right and 2 units up
(d) shifted 1 unit to the left and 2 units up
(e) None of the preceding

5. The function $f(x) = x^2 - 8x + 10$ has
(a) the minimum value -6 when $x = 4$
(b) the minimum value 6 when $x = -4$
(c) the minimum value 26 when $x = 4$
(d) the minimum value 10 when $x = 8$
(e) None of the preceding

6. The parabola labeled P in the following figure is obtained by shifting the parabola $y = x^2$ both horizontally and vertically. The equation for P is:
(a) $y = (x + 5)^2 - 3$ **(b)** $y = (x - 5)^2 - 3$
(c) $y = (x + 3)^2 - 5$ **(d)** $y = (x - 3)^2 + 5$ **(e)** $y = -(x - 5)^2 - 3$

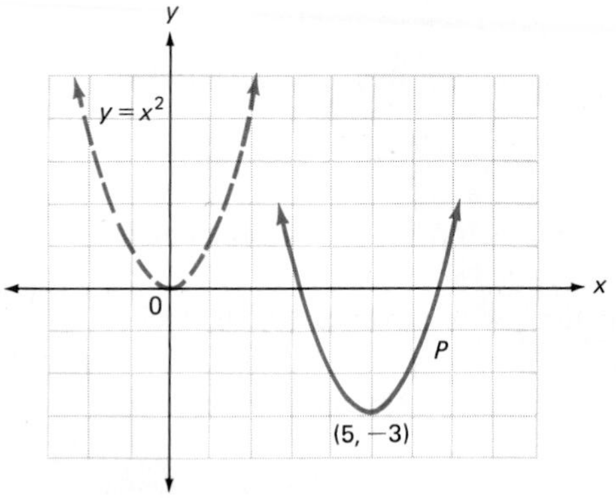

7. Which of the following is *true* for the parabola $y = -4x^2 + 20x - 25$?
(a) It opens downward and has two x-intercepts.
(b) It opens downward and has no x-intercepts.
(c) It opens downward and has one x-intercept.
(d) It opens to the left and has two y-intercepts.
(e) None of the preceding

8. The solution of the inequality $x^2 - 2x - 5 > 0$ is:
(a) $x < 1 - \sqrt{6}$ or $x > 1 + \sqrt{6}$ **(b)** $1 - \sqrt{6} < x < 1 + \sqrt{6}$
(c) $2 < x < 5$ **(d)** $x < 1$ or $x > 3$ **(e)** $x < 2$ or $x > 5$

9. If the sum of two numbers is 16 and the sum of their squares is a minimum, then this minimum value is:
(a) 36 **(b)** 72 **(c)** 128 **(d)** 144 **(e)** 256

10. The tangent line to the circle $x^2 + y^2 = 100$ at the point in the first quadrant where $x = 6$ is:
(a) $y - 8 = \frac{3}{4}(x - 6)$ **(b)** $y - 10 = -\frac{4}{3}(x - 6)$
(c) $y - 8 = -\frac{3}{4}(x - 6)$ **(d)** $4x + 3y = 48$ **(e)** $4x - 3y = 14$

11. The circle $x^2 + y^2 + 3y = -\frac{1}{4}$ has:
(a) Center $(0, \frac{3}{2})$ and radius $r = \frac{1}{2}$
(b) Center $(0, -3)$ and radius $r = 2$
(c) Center $(3, 0)$ and radius $r = \frac{1}{2}$
(d) Center $(0, -\frac{3}{2})$ and radius $r = \sqrt{2}$
(e) Center $(0, -3)$ and radius $r = \frac{1}{2}$

12. Which of the following is the equation of the ellipse with foci $(0, \pm 3)$ and major axis of length 10?

(a) $\frac{x^2}{9} + \frac{y^2}{100} = 1$ **(b)** $\frac{x^2}{100} + \frac{y^2}{9} = 1$ **(c)** $\frac{x^2}{16} + \frac{y^2}{25} = 1$ **(d)** $\frac{x^2}{25} + \frac{y^2}{9} = 1$ **(e)** None of the preceding

13. The ellipse with equation $x^2 + \frac{y^2}{4} = 1$ has foci given by:

(a) $(0, \pm\sqrt{5})$ **(b)** $(\pm 1, 0)$ **(c)** $(\pm 5, 0)$ **(d)** $(0, \pm 3)$ **(e)** $(0, \pm\sqrt{3})$

14. The graph of $\frac{x^2}{10} - \frac{y^2}{15} = 1$ is

(a) a hyperbola with asymptote $y = \pm\frac{3}{2}x$
(b) a hyperbola with foci $(\pm 5, 0)$
(c) an ellipse with a vertical major axis
(d) an ellipse with vertices $(0, \pm 5)$
(e) None of the preceding

15. The focus and directrix of the parabola $-2y^2 = x$ are

(a) $(-\frac{1}{8}, 0)$; $x = \frac{1}{8}$ **(b)** $(0, \frac{1}{8})$; $y = -\frac{1}{8}$ **(c)** $(-2, 0)$; $x = 2$ **(d)** $(0, 0)$; $x = -2$ **(e)** None of the preceding

ANSWERS TO THE TEST YOUR UNDERSTANDING EXERCISES

Page 291

1. **(b)** 2. **(c)** 3. **(e)** 4. **(f)** 5. **(d)** 6. **(a)**

Page 299

1. $y = (x + \frac{1}{2})^2 - \frac{25}{4}$; **(a)** $(-\frac{1}{2}, -\frac{25}{4})$, **(b)** $x = -\frac{1}{2}$, **(c)** -6, **(d)** $-3, 2$
2. $y = (x - \frac{3}{2})^2 - \frac{25}{4}$; **(a)** $(\frac{3}{2}, -\frac{25}{4})$, **(b)** $x = \frac{3}{2}$, **(c)** -4, **(d)** $-1, 4$
3. $y = (x + 3)^2$; **(a)** $(-3, 0)$, **(b)** $x = -3$, **(c)** 9, **(d)** -3
4. $y = (x + 2)^2 - 2$; **(a)** $(-2, -2)$, **(b)** $x = -2$, **(c)** 2, **(d)** $-2 - \sqrt{2}, -2 + \sqrt{2}$
5. $y = 2(x - \frac{5}{4})^2 - \frac{49}{8}$; **(a)** $(\frac{5}{4}, -\frac{49}{8})$, **(b)** $x = \frac{5}{4}$, **(c)** -3, **(d)** $-\frac{1}{2}, 3$
6. $y = 3(x + 1)^2 + 1$; **(a)** $(-1, 1)$, **(b)** $x = -1$, **(c)** 4 **(d)** None
7. 0; one **8.** -80; none **9.** 17; two

Page 305

1. $-2 < x < 2$ 2. $x < -2$ or $x > 2$ 3. $x \leq -3$ or $x \geq 3$ 4. $-3 \leq x \leq 3$
5. $0 < x < 1$ 6. $x \leq 0$ or $x \geq 2$ 7. $x \leq 2$ or $x \geq 3$ 8. $-3 < x < 2$

Page 313

1. $(0, 0)$; 10 2. $(0, 0)$; $\sqrt{10}$ 3. $(1, -1)$; 5 4. $(-\frac{1}{2}, 0)$; 16
5. $(-4, -4)$; $5\sqrt{2}$ 6. $(0, 5)$; 3 7. $x^2 + (y - 4)^2 = 25$ 8. $(x - 1)^2 + (y + 2)^2 = 3$

CHAPTER 8

GRAPHING FUNCTIONS: ROOTS OF POLYNOMIAL FUNCTIONS

8.1 GRAPHING POLYNOMIAL FUNCTIONS

The ***general polynomial function*** *of degree n can be written as*
$p(x) = a_nx^n + a_{n-1}x^{n-1} + \cdots + a_1x + a_0$
where the a_i are constants, $a_n \neq 0$.

The linear functions $y = f(x) = mx + b$, whose graphs are straight lines, and the quadratic functions $y = f(x) = ax^2 + bx + c$, whose graphs are parabolas, belong to the general category of **polynomial functions.** Now we will learn how to graph some additional fundamental polynomial functions by making use of symmetries, translations (shifting), reflections, and other procedures that were used in graphing the parabolas in Sections 7.1 and 7.2.

The concept of symmetry was used in Section 7.1 to graph parabolas. Recall that the graph of the function given by $y = f(x) = x^2$ is said to be *symmetric about the y-axis*. (The y-axis is the axis of symmetry.) Observe that points such as $(-a, a^2)$ and (a, a^2) are symmetric points about the axis of symmetry.

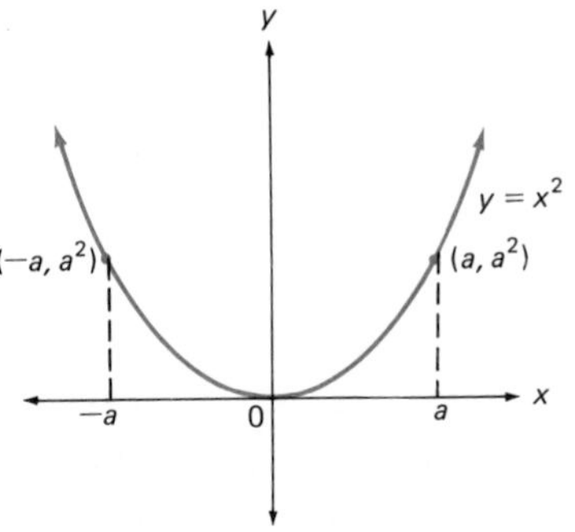

Now we turn our attention to a curve that is *symmetric with respect to a point*. As an illustration of a curve that has symmetry through a point we consider the fundamental *cubic function* given by $y = f(x) = x^3$. This function may also be referred to as the *cubing function* because for each domain value x, the corresponding range value is the cube of x.

A table of values is a very helpful aid for drawing the graph of a function. Several specific points are located and a smooth curve is drawn through them to show the graph of $y = x^3$.

$f(x) = x^3$ is increasing for all x. The curve is concave down on $(-\infty, 0)$ and concave up on $(0, \infty)$.

$y = f(x) = x^3$

x	y
-2	-8
-1	-1
$-\frac{1}{2}$	$-\frac{1}{8}$
0	0
$\frac{1}{2}$	$\frac{1}{8}$
1	1
2	8

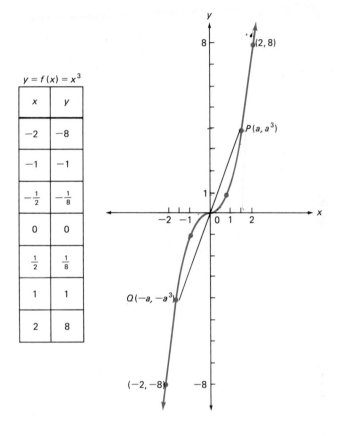

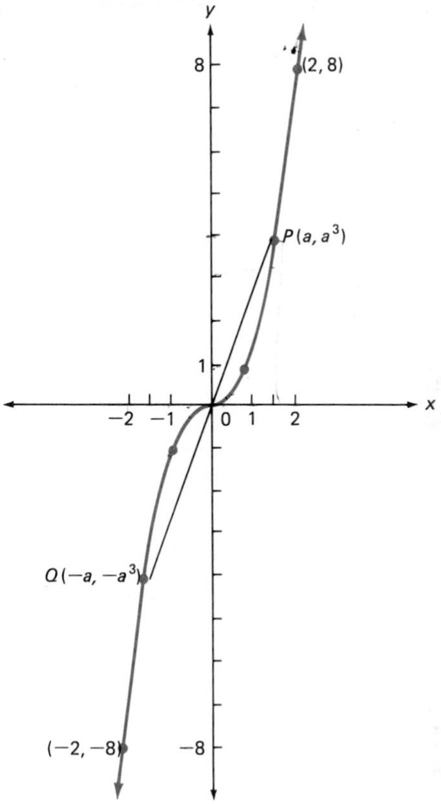

Domain: all real numbers x
Range: all real numbers y

The table and the graph reveal that the curve is symmetric through the origin. Geometrically this means that whenever a line through the origin intersects the curve at a point P, this line will also intersect the curve in another point Q (on the opposite side of the origin) so that the lengths of OP and OQ are equal. This means that both points (a, a^3) and $(-a, -a^3)$ are on the curve for each value $x = a$. These are said to be **symmetric points** through the origin. In particular, since (2, 8) is on the curve, then $(-2, -8)$ is also on the curve.

For the function $f(x) = x^3$, we have $f(-x) = (-x)^3 = -x^3 = -f(x)$. Thus $f(-x) = -f(x)$ and we have symmetry with respect to the origin.

In general, the graph of a function $y = f(x)$ is said to be *symmetric through the origin* if for all x in the domain of f, we have

$$f(-x) = -f(x)$$

Other techniques used for graphing quadratic functions in Chapter 7 can be used for these new functions as well. For example, the graph of $y = -x^3$ can be obtained by reflecting the graph of $y = x^3$ through the x-axis (or by multiplying the ordinates of $y = x^3$ by -1). This can also be seen by using a table of values as shown below.

$y = f(x) = -x^3$

x	y
−2	8
−1	1
$-\frac{1}{2}$	$\frac{1}{8}$
0	0
$\frac{1}{2}$	$-\frac{1}{8}$
1	−1
2	−8

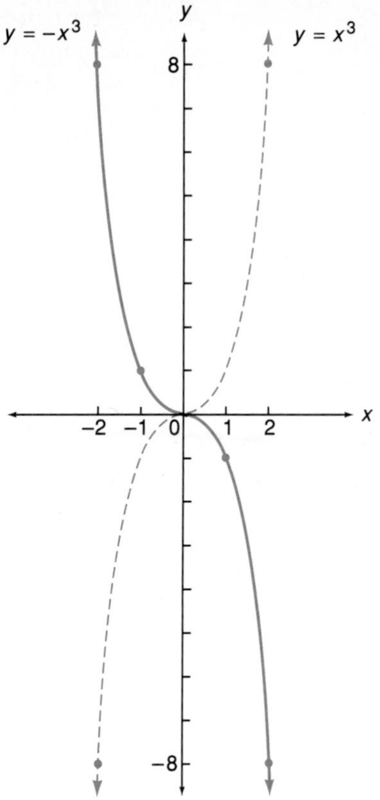

Translations (shifting), as done in Chapter 7, can be applied to the graph of $y = x^3$ as well as to the graphs of other functions.

EXAMPLE 1 Graph $y = g(x) = (x - 3)^3$.

Solution The graph of g is obtained by translating $y = x^3$ by 3 units to the right, as shown below.

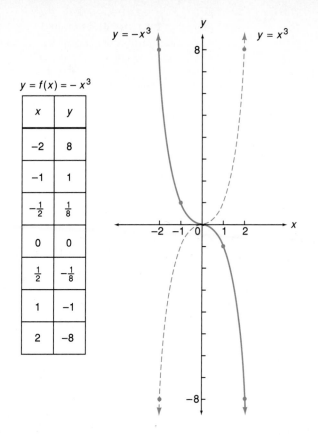

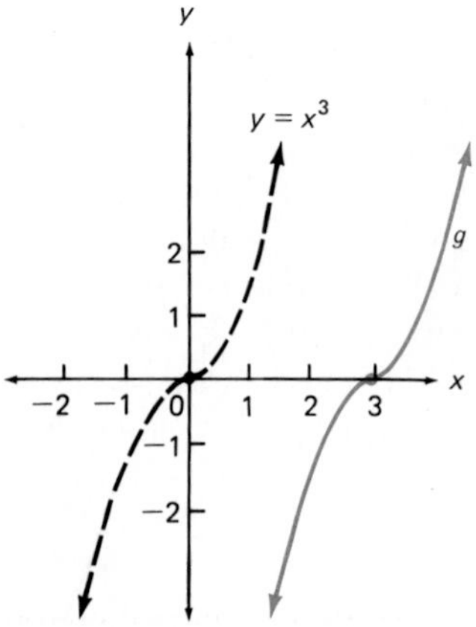

■

The following example shows how to graph the absolute value of a polynomial function.

The domain of f is all real numbers x, and the range is all $y \geq 0$.

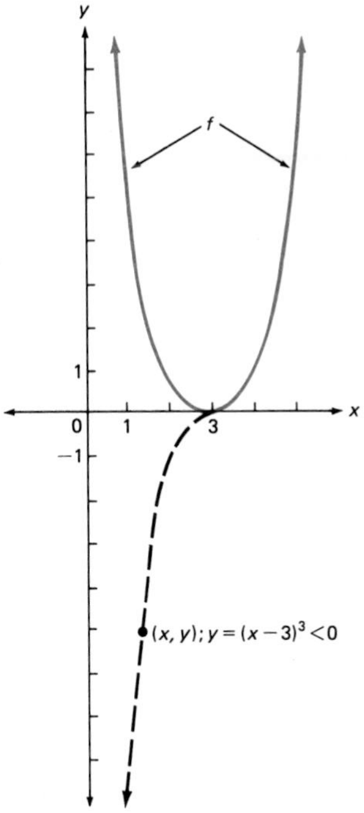

Explain how you can use the graph of f to draw the graph of $y = |x - 3|^3 + 2$.

Note that to shift a curve means the same as to translate a curve.

EXAMPLE 2 Graph $y = f(x) = |(x - 3)^3|$.

Solution First graph $y = (x - 3)^3$ as in Example 1. Then take the part of this curve that is below the x-axis $[(x - 3)^3 < 0]$ and reflect it through the x-axis as shown at the left. ■

It will be helpful to collect some of the general observations that are useful in graphing functions.

GRAPHING PROPERTIES AND PROCEDURES FOR FUNCTIONS

Symmetries

1. If $f(x) = f(-x)$, the curve is symmetric about the y-axis.
2. If $f(-x) = -f(x)$, the curve is symmetric through the origin.

Multiplying Ordinates and Reflections

3. The graph of $y = af(x)$ can be obtained by multiplying the ordinates of the curve $y = f(x)$ by the value a. The case $a = -1$ gives $y = -f(x)$, which is the reflection of $y = f(x)$ through the x-axis.

Absolute Value of a Function

4. The graph of $y = |f(x)|$ can be obtained from the graph of $y = f(x)$ by taking the part of $y = f(x)$ that is below the x-axis and reflecting it through the x-axis.

Horizontal Translations (h is positive)

5. The graph of $y = f(x - h)$ can be obtained by shifting $y = f(x)$ h units to the right.
6. The graph of $y = f(x + h)$ can be obtained by shifting $y = f(x)$ h units to the left.

Vertical Translations (k is positive)

7. The graph of $y = f(x) + k$ can be obtained by shifting $y = f(x)$ k units upward.
8. The graph of $y = f(x) - k$ can be obtained by shifting $y = f(x)$ k units downward.

EXAMPLE 3 In the figure on the following page the curve C_1 is obtained by shifting the curve C with equation $y = x^3$ horizontally, and C_2 is obtained by shifting C_1 vertically. What are the equations of C_1 and C_2?

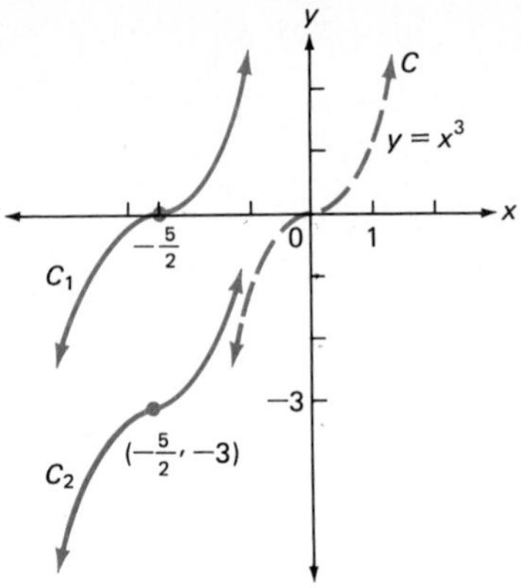

Solution

C_1: $y = (x + \frac{5}{2})^3$ ⟵ C is shifted $\frac{5}{2}$ units left.

C_2: $y = (x + \frac{5}{2})^3 - 3$ ⟵ C_1 is shifted 3 units down. ■

The next example gives the graph of the fundamental quartic (fourth degree) function $y = x^4$ and also demonstrates the procedure of multiplying ordinates.

EXAMPLE 4 Graph: **(a)** $y = f(x) = x^4$ **(b)** $y = 2x^4$

Solution

(a) Since $f(-x) = (-x)^4 = x^4 = f(x)$, the graph is symmetric about the y-axis. Use the table of values to locate the right half of the curve; the symmetry gives the rest, as shown next.

(b) Multiply each of the ordinates of $y = x^4$ by 2 to obtain the graph of $y = 2x^4$.

Observe that both functions are decreasing on $(-\infty, 0]$ and increasing on $[0, \infty)$. Also both curves are concave up for all x.

x	$y = x^4$	$y = 2x^4$
0	0	0
$\frac{1}{2}$	$\frac{1}{16}$	$\frac{1}{8}$
1	1	2
$\frac{3}{2}$	$\frac{81}{16}$	$\frac{81}{8}$
2	16	32

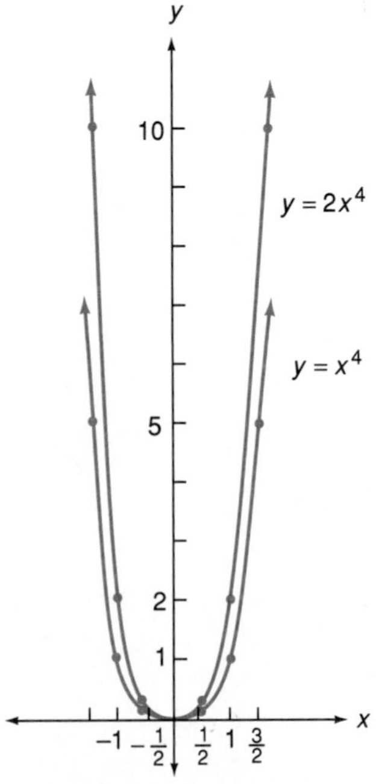

■

The final example demonstrates the use of several graphing procedures to find the graph of a function.

EXAMPLE 5 Graph $y = |x + 6| - 4$. Find the x- and y-intercepts.

See page 269 for another way to graph $y = |x|$.

Solution The graph of $y = |x|$ can be obtained by taking the part of the line $y = x$ that is below the x-axis and reflecting it through the x-axis. This results in the v-shaped graph for $y = |x|$, shown here using dashed lines. Shift this graph 6 units left, giving the graph of $y = |x + 6|$, shown as the solid black lines. Then shift 4 units down to obtain the graph of $y = |x + 6| - 4$ shown in red.

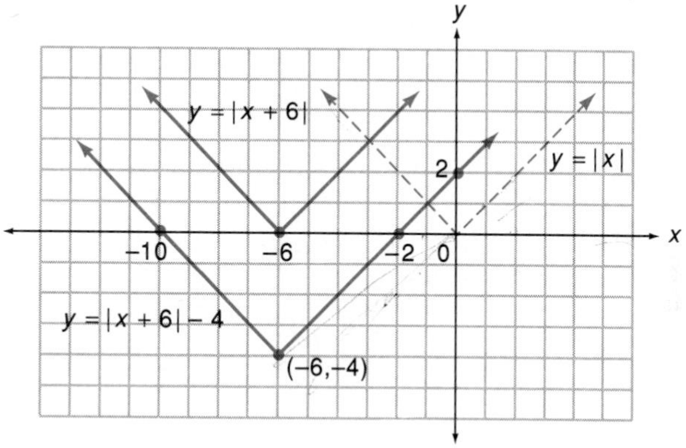

The given function is technically not a polynomial. However, it consists of two polynomial parts. Thus, if $x \geq -6$, $x + 6 \geq 0$ and $y = x + 2$, and if $x < -6$, $x + 6 < 0$ giving $y = -(x + 6) - 4 = -x - 10$.

The y-intercept is found by letting $x = 0$

$$y = |0 + 6| - 4 = 6 - 4 = 2$$

The x-intercept is found by letting $y = 0$

If $|a| = k$ then $a = \pm k$. See page 110.

$$0 = |x + 6| - 4$$

$$|x + 6| = 4$$

$$x + 6 = \pm 4$$

$$x = -10 \quad \text{or} \quad x = -2$$

The x-intercepts are -10 and -2; the y-intercept is 2. ■

EXERCISES 8.1

For Exercises 1–8 graph each set of curves in the same coordinate system. For each exercise use a dashed curve for the first equation and a solid curve for each of the others. For the last function given, state the domain and range, find where it is increasing or decreasing, and describe the concavity.

1. $y = x^2$, $y = (x - 3)^2$

2. $y = x^3$, $y = (x + 2)^3$

3. $y = x^3$, $y = -x^3$

4. $y = x^4$, $y = (x - 4)^4$

5. $f(x) = x^3$, $g(x) = \frac{1}{2}x^3$, $h(x) = \frac{1}{4}x^3$

6. $f(x) = x^3$, $g(x) = (x - 3)^3 - 3$, $h(x) = (x + 3)^3 + 3$

7. $f(x) = x^4$, $g(x) = (x - 1)^4 - 1$, $h(x) = (x - 2)^4 - 2$

8. $f(x) = x^4$, $g(x) = -\frac{1}{8}x^4$, $h(x) = -x^4 - 4$

9. Graph $y = |(x + 1)^3|$.

10. Graph $y = -|(x + 1)^3|$.

11. Graph $y = f(x) = |x|$, $y = g(x) = |x - 3|$, and $y = h(x) = |x - 3| + 2$, on the same axes. Find the x- and y-intercepts of f, g, and h, if any.

Graph each of the following by using translations and the method of multiplying ordinates.

12. $y = f(x) = (x + 1)^3 - 2$

13. $y = f(x) = (x - 1)^3 + 2$

14. $y = f(x) = 2(x - 3)^3 + 3$

15. $y = f(x) = 2(x + 3)^3 - 3$

Graph the following. What are the x- and y-intercepts?

16. $y = f(x) = -(x - 4)^4 + 16$

17. $y = f(x) = (x + 4)^4 - 16$

Find the equation of the curve C which is obtained from the dashed curve by a horizontal or vertical shift, or by a combination of the two.

18.

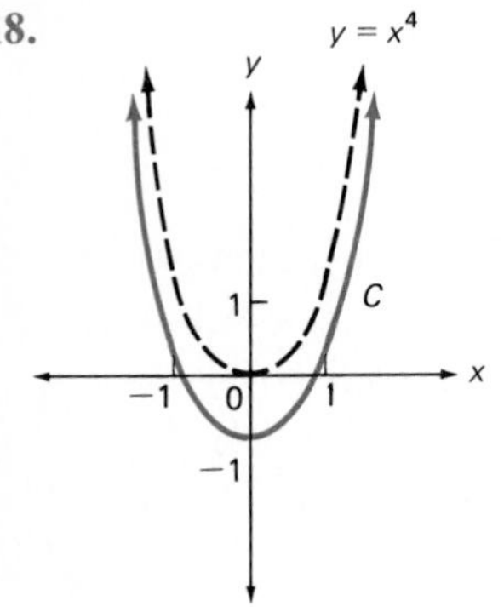

19.

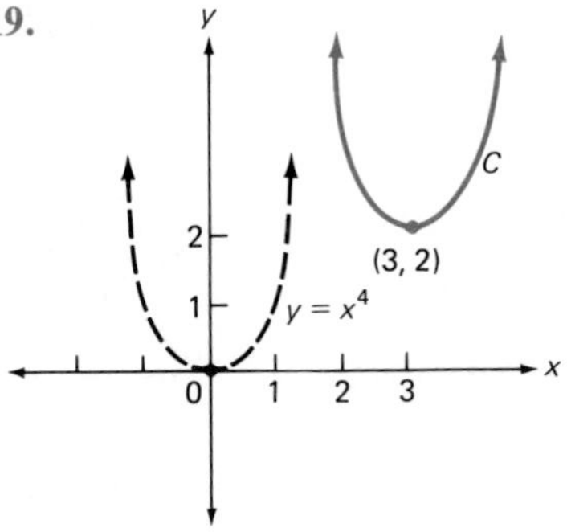

20.

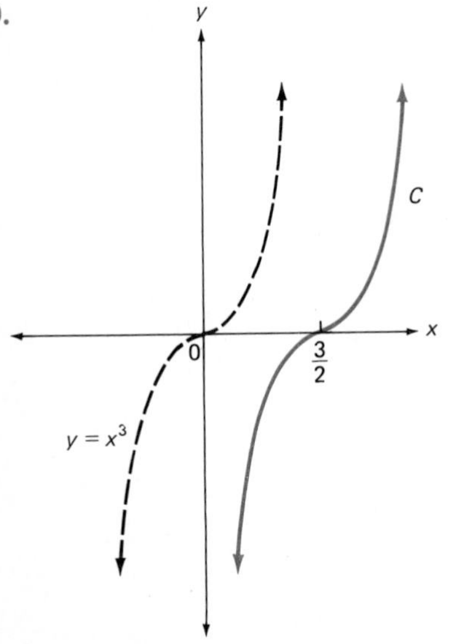

21.

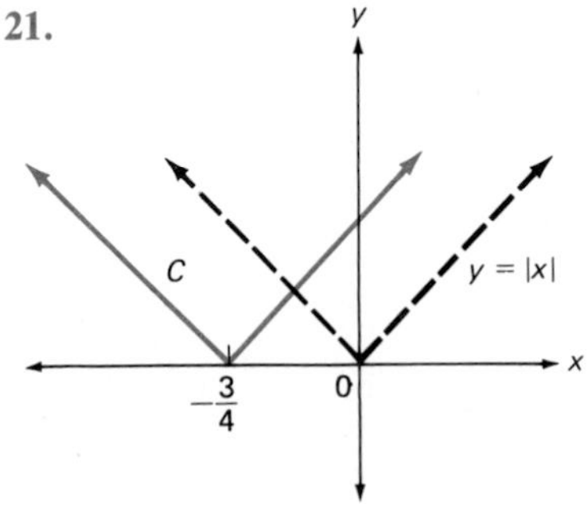

Find the equation of the curve C which is obtained from the dashed curve by using the indicated graphing procedures.

22. Translation and absolute value.

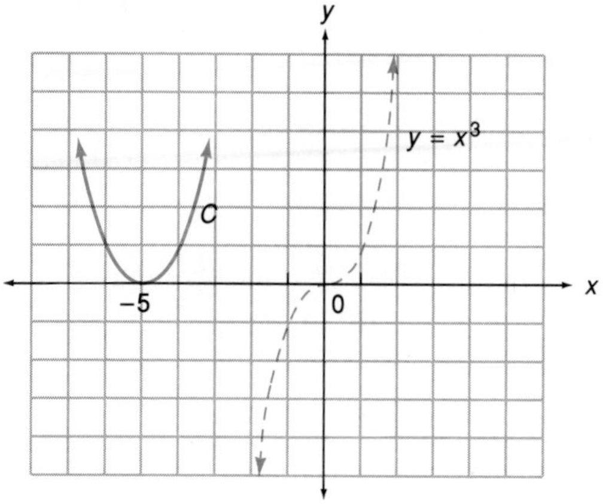

23. Translation *followed by* absolute value.

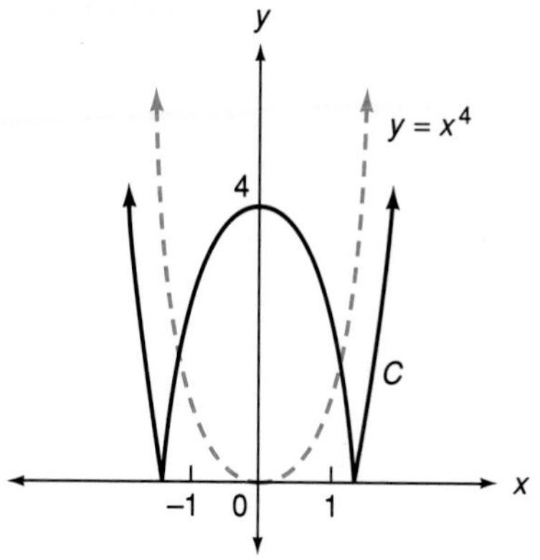

24. Reflection *followed* by translation.

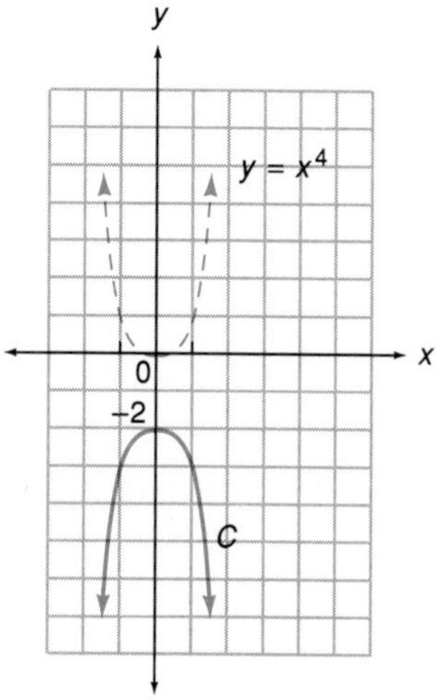

*25. Vertical translation *followed* by absolute value.

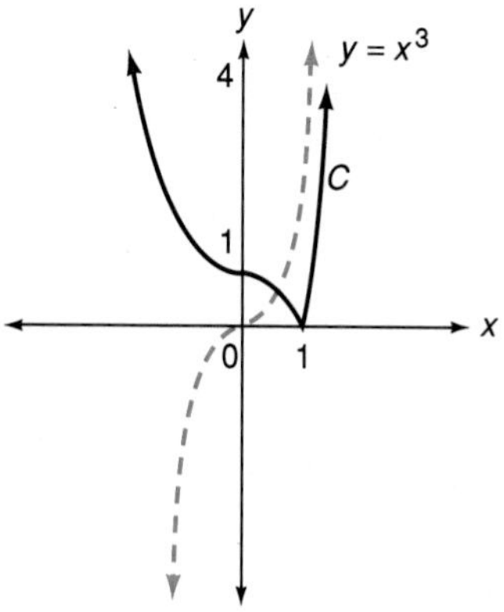

Graph each of the following. Show the x- and y-intercepts.

26. $y = |x^4 - 16|$ 27. $y = |x^3 - 1|$ 28. $y = |-1 - x^3|$

Graph each of the following. (Hint: Consider the expansion of $(a \pm b)^n$ *for appropriate values of n.)*

*29. $y = x^3 + 3x^2 + 3x + 1$ *30. $y = x^3 - 6x^2 + 12x - 8$

*31. $y = -x^3 + 3x^2 - 3x + 1$ *32. $y = x^4 - 4x^3 + 6x^2 - 4x + 1$

Evaluate and simplify the quotients.

33. $\dfrac{f(x) - f(3)}{x - 3}$ where $f(x) = x^3$

34. $\dfrac{f(-2 + h) - f(-2)}{h}$ where $f(x) = x^3$

35. $\dfrac{f(1 + h) - f(1)}{h}$ where $f(x) = x^4$

36. $\dfrac{f(x) - f(-1)}{x + 1}$ where $f(x) = x^4 + 1$

Written Assignment: Explain why the graph of a function of x cannot be symmetric about the x-axis.

8.2 GRAPHING RATIONAL FUNCTIONS

A *rational expression* is a ratio of polynomials and such expressions may be used to define rational functions. The function given by

$$y = f(x) = \frac{1}{x}$$

is a fundamental rational function whose domain and range each consists of all numbers except zero. The denominator x is a polynomial of degree 1, and the numerator $1 = 1x^0$ is a (constant) polynomial of degree zero.

To draw the graph of this function, first observe that it is symmetric through the origin because of the following:

$$f(-x) = \frac{1}{-x} = -\frac{1}{x} = -f(x)$$

Moreover, in $y = \frac{1}{x}$ both variables must have the same sign since $xy = 1$, a positive number. That is, x and y must both be positive or both be negative. Thus the graph will only appear in quadrants I and III. Next, we use a table of values to obtain points for the curve in the first quadrant. Finally, use the symmetry with respect to the origin to obtain the remaining portion of the graph in the third quadrant.

x	y
10	.1
100	.01
1000	.001
10,000	.0001
⋮	⋮
↓	↓

Getting very large — Getting close to 0

Horizontal asymptote $y = 0$

x	y
$\frac{1}{2}$	2
$\frac{1}{10}$	10
$\frac{1}{100}$	100
$\frac{1}{1000}$	1000
⋮	⋮
↓	↓

Getting close to 0 — Getting very large

Vertical asymptote $x = 0$

$y = f(x) = \frac{1}{x}$

x	y
$\frac{1}{5}$	5
$\frac{1}{2}$	2
1	1
2	$\frac{1}{2}$
3	$\frac{1}{3}$
4	$\frac{1}{4}$

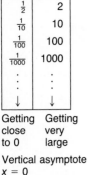

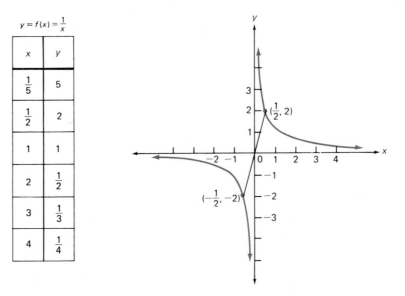

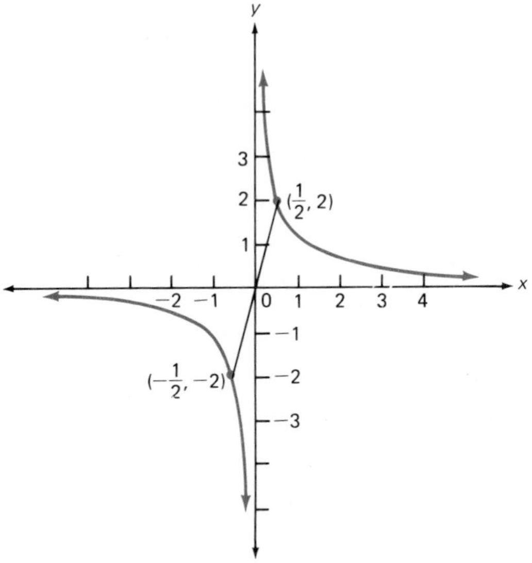

Domain: all $x \neq 0$
Range: all $y \neq 0$

Observe that the curve approaches the x-axis in quadrant I. That is, as the values for x become large, the values for y approach zero. Also, as the values for x approach zero in the first quadrant, the y-values become very large. A similar observation can be made about the curve in the third quadrant. We say that the axes are **asymptotes** for the curve; the curve is *asymptotic* to the axes.

EXAMPLE 1 Sketch the graph of $g(x) = \dfrac{1}{x-3}$. Find the asymptotes.

Solution Using $f(x) = \dfrac{1}{x}$, we have $f(x-3) = \dfrac{1}{x-3} = g(x)$. Therefore, the graph of g can be drawn by shifting the graph of $f(x) = \dfrac{1}{x}$ by 3 units to the right.

Describe how to obtain the graph of $y = h(x) = \dfrac{1}{x+3}$ using the graph of $f(x) = \dfrac{1}{x}$.

The x-axis and the vertical line $x = 3$ are the asymptotes for the graph. The domain is all $x \neq 3$, and the range is all $y \neq 0$.

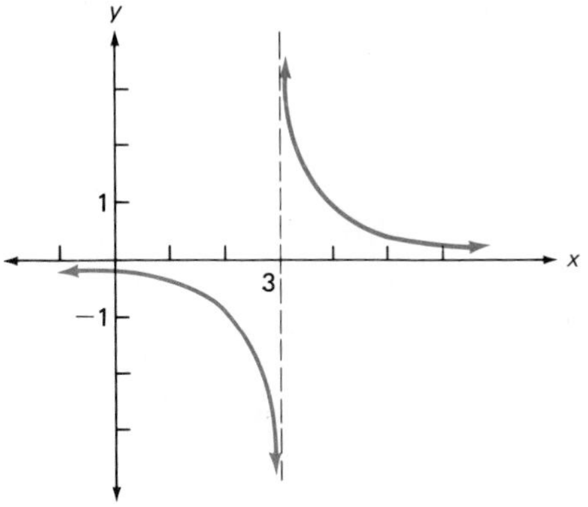

■

In symbols, we write "As $x \to \infty$, $y \to 0$." That is, as x becomes larger and larger in value, the values of y approach 0. Similarly, as $x \to 0$, $y \to \infty$.

$y = \dfrac{1}{x^2}$

x	y
$\frac{1}{2}$	4
$\frac{1}{10}$	100
$\frac{1}{100}$	10,000

As x is getting close to 0, y is getting very large.

Vertical asymptote $x = 0$

x	y
2	$\frac{1}{4}$
10	$\frac{1}{100}$
100	$\frac{1}{10,000}$

As x is getting large, y is getting close to 0.

Horizontal asymptote $y = 0$

EXAMPLE 2 Graph $y = \dfrac{1}{x^2}$ and find the asymptotes.

Solution First note that $x \neq 0$. For all other values of x we have $x^2 > 0$, so that the curve will appear in quadrants I and II only. In quadrant I, as the values for x become large, the values for $y = \dfrac{1}{x^2}$ get close to 0. Moreover, as x approaches zero, y becomes very large. Thus the axes are asymptotes to the curve. Note that the curve is symmetric about the y-axis. That is,

$$f(x) = \frac{1}{x^2} = \frac{1}{(-x)^2} = f(-x)$$

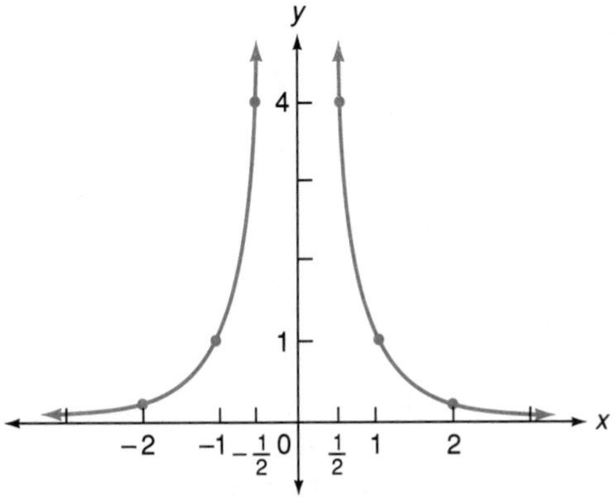

The domain is all $x \neq 0$, and the range is all $y > 0$. ■

EXAMPLE 3 Graph $y = \dfrac{1}{(x+2)^2} - 3$. What are the asymptotes? Find the domain and the range, describe the concavity of the curve, and state where it is increasing or decreasing.

Solution Shift the graph of $y = \dfrac{1}{x^2}$ in Example 2 by 2 units left and then 3 units down.

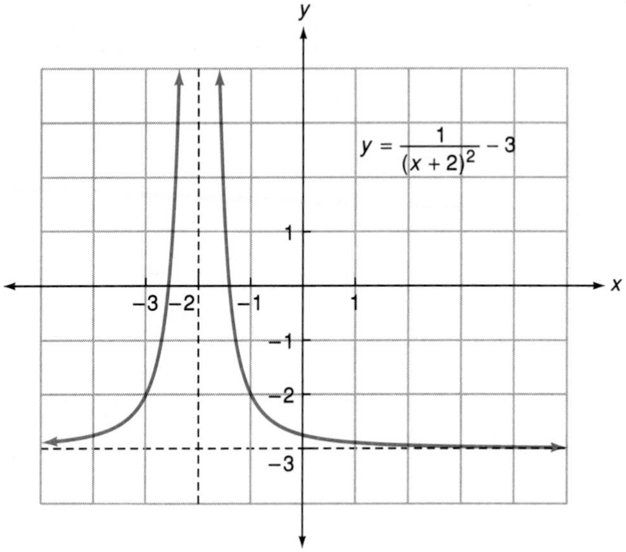

Find the x-intercepts by solving the equation

$$\frac{1}{(x+2)^2} - 3 = 0$$

Use a calculator to show that, to one decimal place, the solutions are

$x = 1.4$ *and* $x = -2.6$.

The curve is concave up and increasing on $(-\infty, -2)$, and concave up and decreasing on $(-2, \infty)$

Asymptotes: $x = -2$, $y = -3$ Domain: all $x \neq -2$
Range: all $y > -3$ ■

As $x \to -2$, $y \to \infty$;
as $x \to \infty$, $y \to -3$;
as $x \to -\infty$, $y \to -3$.

The discussion of finding the asymptotes of $y = \dfrac{1}{x}$ on page 348 and the similar work for the asymptotes of $y = \dfrac{1}{x^2}$ in Example 2 suggest the following procedure for finding horizontal asymptotes.

To find the horizontal asymptote, if there is one, allow x to become arbitrarily large or small. Then, if y gets closer and closer to a constant c, $y = c$ is the equation of the horizontal asymptote.

Note that "arbitrarily small" means negative values of x whose absolute values are getting very large—that is, values of x endlessly to the left.

The preceding examples also suggest that a vertical asymptote occurs when the denominator of the rational function is 0. Thus, in Example 1, for $g(x) = \dfrac{1}{x-3}$ the line $x = 3$ is a vertical asymptote. However, this applies only for rational functions written in simplest or reduced form. For example, the denominator in $f(x) = \dfrac{x+3}{x^2-9}$ is 0 when $x = \pm 3$, but

$$f(x) = \frac{x+3}{x^2-9} = \frac{x+3}{(x-3)(x+3)} = \frac{1}{x-3} \qquad (x \neq 3)$$

The point $(-3, -\frac{1}{6})$ is on the graph of $y = \dfrac{1}{x - 3}$, but not on the graph of the function $f(x)$.

shows that there is no vertical asymptote when x is -3. In fact, the graph of f is the same as the graph in Example 1, except that there would be an open circle at point $(-3, -\frac{1}{6})$ to indicate that this point is not on the graph of f. In general

A graph of a rational function has a vertical asymptote at each value a where the denominator is 0, and the numerator is not 0.

EXAMPLE 4 Graph $y = f(x) = \dfrac{x^2 + x - 6}{x - 2}$

Solution Factor the numerator and simplify the fraction.

$$f(x) = \frac{x^2 + x - 6}{x - 2} = \frac{(x + 3)(x - 2)}{x - 2}$$
$$= x + 3 \qquad (x \neq 2)$$

This indicates that the original rational function is the same as $y = x + 3$, except that $x \neq 2$ since division by 0 is not possible. Thus the graph of the function is the line $y = x + 3$, with an open circle at $(2, 5)$ to show that this point is *not* part of the graph.

Observe that the reduced form $y = x + 3$, $x \neq 2$, makes it easy to see that the graph of this function has no asymptotes. This was not obvious from the given form.

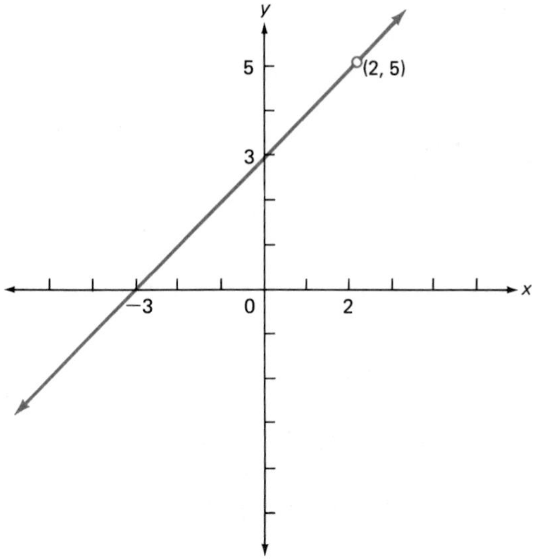

■

See the inside front cover for these and other useful curves.

The curves studied in this chapter, as well as the parabolas and straight lines discussed in earlier chapters, are very useful in the further study of mathematics. Having an almost instant recall of the graphs of the following functions will be helpful in future work. Not only should you know what these curves look like, but just as important, you should be able to obtain other curves from them by appropriate translations, reflections and multiplication of ordinates.

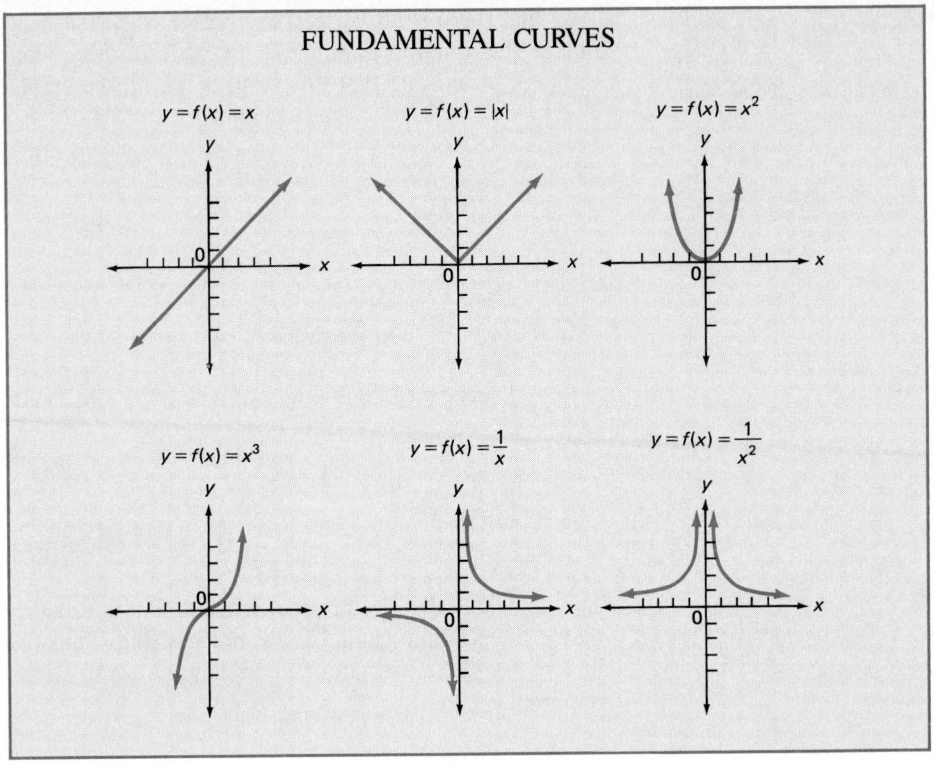

EXERCISES 8.2

Graph each set of curves in the same coordinate system. For each exercise use a dashed curve for the first equation and a solid curve for the other.

1. $y = \frac{1}{x}, y = \frac{2}{x}$
2. $y = \frac{1}{x}, y = -\frac{1}{x}$
3. $y = \frac{1}{x}, y = \frac{1}{x+2}$
4. $y = \frac{1}{x}, y = \frac{1}{x} + 5$
5. $y = \frac{1}{x}, y = \frac{1}{2x}$
6. $y = \frac{1}{x}, y = \frac{1}{x} - 5$
7. $y = \frac{1}{x^2}, y = \frac{1}{(x-3)^2}$
8. $y = \frac{1}{x^2}, y = -\frac{1}{x^2}$

Graph each of the following. Find all asymptotes, if any. State the domain and range, describe the concavity, and say where it is increasing or decreasing.

9. $y = -\frac{1}{x} + 2$
10. $y = -\frac{1}{x} - 2$
11. $y = \frac{1}{x+4} - 2$
12. $y = \frac{1}{x-4} + 2$
13. $y = -\frac{1}{x-2} + 1$
14. $y = -\frac{1}{x+1} - 2$
15. $y = \frac{1}{(x+1)^2} - 2$
16. $y = \frac{1}{(x-2)^2} + 3$
17. $y = \frac{1}{|x-2|}$

Graph each of the following. Find all asymptotes, if any.

18. $y = \frac{1}{x^3}$
19. $xy = 3$
20. $xy = -2$
21. $xy - y = 1$
22. $xy - 2x = 1$
23. $y = f(x) = \frac{x^2 - 9}{x - 3}$

24. $y = f(x) = \dfrac{x^2 - 9}{x + 3}$ 25. $y = f(x) = \dfrac{x^2 - x - 6}{x - 3}$ 26. $y = f(x) = \dfrac{1 + x - 2x^2}{x - 1}$

27. $y = f(x) = \dfrac{x + 1}{x^2 - 1}$ 28. $y = f(x) = \dfrac{x - 1}{x^2 + x - 2}$ 29. $y = f(x) = \dfrac{x^3 - 8}{x - 2}$

30. $y = f(x) = \dfrac{x^3 - 2x^2 - 3x + 6}{2 - x}$

31. Graph $y = \left|\dfrac{1}{x} - 1\right|$. 32. Graph $x = \dfrac{1}{y^2}$. Why is y not a function of x?

Find the quotients and simplify.

33. $\dfrac{f(x) - f(3)}{x - 3}$ where $f(x) = \dfrac{1}{x}$ 34. $\dfrac{f(1 + h) - f(1)}{h}$ where $f(x) = \dfrac{1}{x^2}$

Written Assignment: Use specific examples of your own to describe the conditions for the graph of a rational function to have vertical and horizontal asymptotes.

8.3 GRAPHING RADICAL FUNCTIONS

A radical expression in x, such as $\sqrt{x}$, may be used to define a function f, where $f(x) = \sqrt{x}$, the *square root function*. The domain of f consists of all real numbers $x \geq 0$ since the square root of a negative number is not a real number.

One way to obtain the graph of the fundamental radical function $y = \sqrt{x}$ begins by squaring both sides to obtain $y^2 = x$. Now recall that the graph of $x = y^2$ can be obtained from the graph of $y = x^2$ by interchanging variables; see page 293.

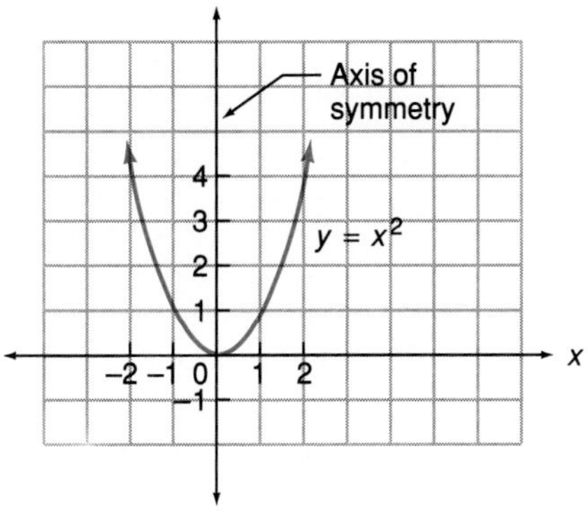

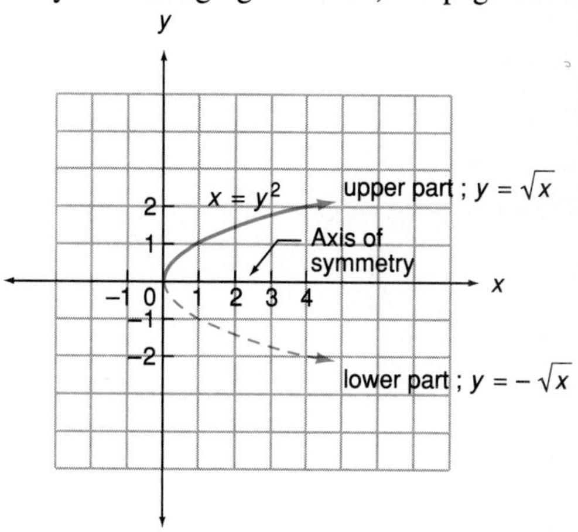

$x = y^2$ is not a function of x since for $x > 0$ there are two corresponding y-values.

Solving $y^2 = x$ for y gives $y = \pm\sqrt{x}$. Then, since $y = \sqrt{x} \geq 0$, the upper solid half of the sideways parabola is the graph of $y = \sqrt{x}$, and the lower dashed part is the graph of $y = -\sqrt{x}$.

A table of values may also be used to obtain the graph of $y = \sqrt{x}$. Verify that the specific points given in the following table satisfy the equation $y = \sqrt{x}$.

x	$y = \sqrt{x}$
0	0
$\frac{1}{4}$	$\frac{1}{2}$
1	1
$\frac{9}{4}$	$\frac{3}{2}$
3	$\sqrt{3}$ (1.7)
4	2
6	$\sqrt{6}$ (2.4)
9	3

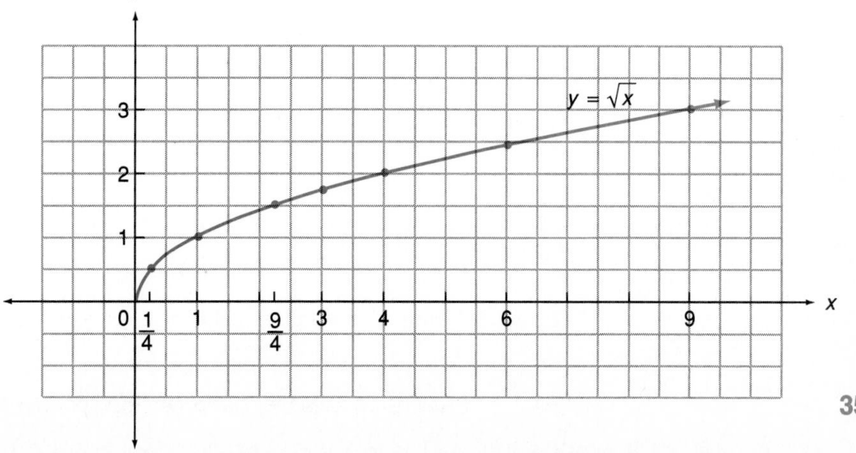

As x gets larger and larger $\sqrt{x}$ also becomes arbitrarily large suggesting the range to be $y \geq 0$.

In summary, the graph of $y = f(x) = \sqrt{x}$ is the upper half of the horizontal parabola $x = y^2$. The curve is increasing and concave down on $(0, \infty)$. The domain of f is all $x \geq 0$, and, as indicated by the graph, the range is all $y \geq 0$.

EXAMPLE 1 Find the domain and range of $y = g(x) = \sqrt{x - 2}$ and graph.

Note: *$x - 2 \geq 0$; thus $x \geq 2$.*

Solution Since the square root of a negative number is not a real number, the expression $x - 2$ must be nonnegative; therefore, the domain of g consists of all $x \geq 2$. The graph of g may be found by shifting the graph of $y = \sqrt{x}$ by 2 units to the right.

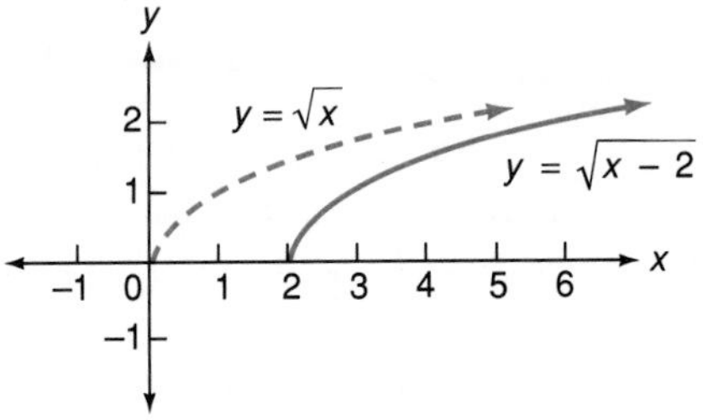

Since the range of $y = \sqrt{x}$ is all $y \geq 0$, and since the graph of g was obtained from $y = \sqrt{x}$ by a horizontal translation, the range of g is also all $y \geq 0$. ■

EXAMPLE 2 Find the domain and range of $y = f(x) = \sqrt{|x|}$ and graph.

Solution Since $|x| \geq 0$ for all x, the domain of f consists of all real numbers. To graph f, first note that

$$f(-x) = \sqrt{|-x|} = \sqrt{|x|} = f(x)$$

Therefore, the graph is symmetric about the y-axis. Thus we first find the graph for $x \geq 0$ and use symmetry to obtain the rest. For $x \geq 0$, we get $|x| = x$ and $f(x) = \sqrt{|x|} = \sqrt{x}$.

It is also helpful to locate a few specific points on the curve as an aid to graphing.

f is decreasing and concave down for $x < 0$, and it is increasing and concave down for $x > 0$.

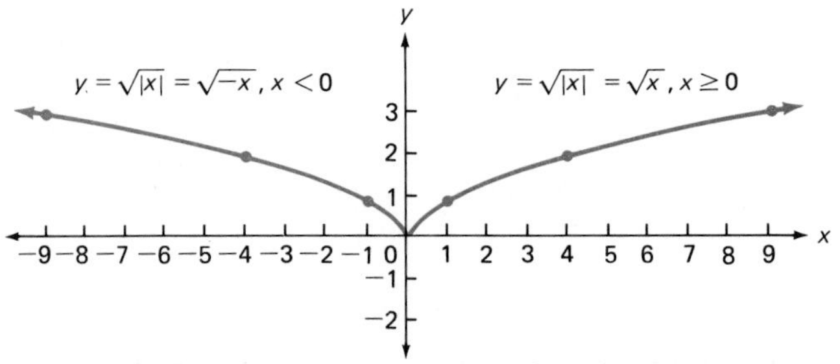

The range of f must be the same as for $y = \sqrt{x}$ and therefore the range consists of all $y \geq 0$. ■

EXAMPLE 3 Find the domain of $y = h(x) = x^{-1/2}$ and graph by using a table of values.

Solution Note that $h(x) = x^{-1/2} = \dfrac{1}{x^{1/2}} = \dfrac{1}{\sqrt{x}}$. Thus the domain consists of all

$x > 0$. Furthermore, $\dfrac{1}{\sqrt{x}} > 0$ for all x, so we know that the graph must be in the first quadrant only. Plot the points in the table and connect them with a smooth curve.

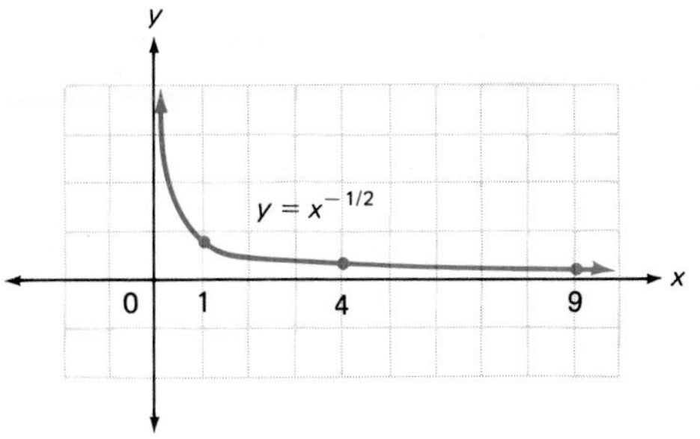

The equations of the two asymptotes are $x = 0$ and $y = 0$.

Observe that the closer x is to zero, the larger are the corresponding y-values. Also, as the values of x get larger, the corresponding y-values get closer to 0. These observations suggest that the coordinate axes are asymptotes to the curve $y = x^{-1/2}$. ■

EXAMPLE 4 Use the graph of $y = \dfrac{1}{\sqrt{x}}$ to graph $y = \dfrac{1}{\sqrt{x}} + 1$ and $y = \dfrac{1}{\sqrt{x+1}}$ and find the asymptotes.

Solution The graphing procedures developed earlier can be used here. Thus the graph of $y = \dfrac{1}{\sqrt{x}} + 1$ can be obtained from the graph in Example 3 by shifting up one unit. Then the asymptotes will be the y-axis and the line $y = 1$. The graph of $y = \dfrac{1}{\sqrt{x+1}}$ is found by shifting the graph for $y = \dfrac{1}{\sqrt{x}}$ to the left one unit. The equations of the asymptotes are $x = -1$ and $y = 0$.

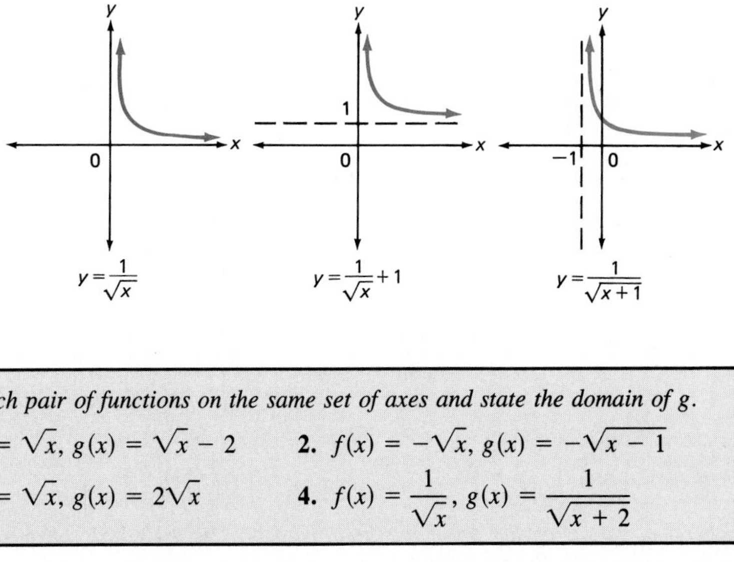

■

TEST YOUR UNDERSTANDING Think Carefully

(Answers: Page 377)

Graph each pair of functions on the same set of axes and state the domain of g.

1. $f(x) = \sqrt{x}$, $g(x) = \sqrt{x} - 2$

2. $f(x) = -\sqrt{x}$, $g(x) = -\sqrt{x-1}$

3. $f(x) = \sqrt{x}$, $g(x) = 2\sqrt{x}$

4. $f(x) = \dfrac{1}{\sqrt{x}}$, $g(x) = \dfrac{1}{\sqrt{x+2}}$

The graph of the **cube root function** $y = \sqrt[3]{x}$ can be found by a process similar to that used for graphing $y = \sqrt{x}$. First take $y = \sqrt[3]{x}$ and cube both sides to get

$y^3 = x$. Now recall the graph of $y = x^3$ and obtain the graph of $x = y^3$ by reversing the role of the variables.

y	$x = y^3$
-2	-8
-1	-1
0	0
1	1
2	8

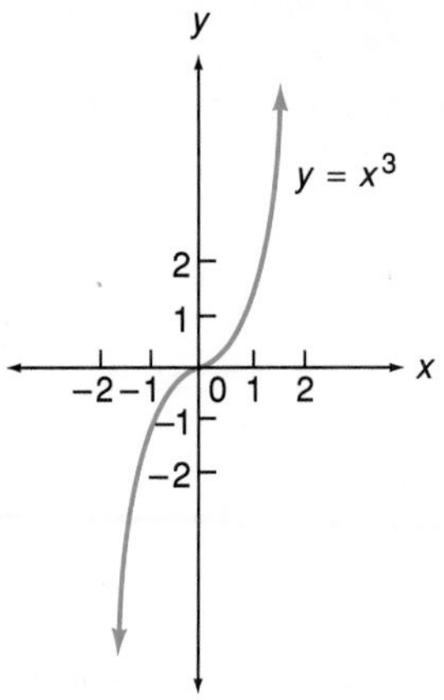

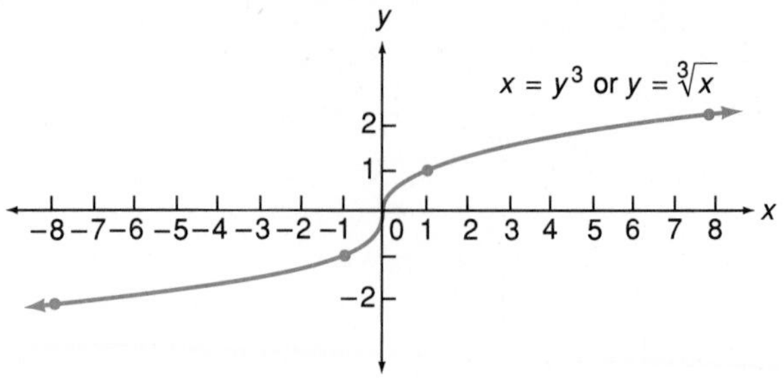

The next example illustrates a function that is defined by using both a polynomial and a radical expression.

EXAMPLE 5 Graph f defined on the domain $-2 \le x < 5$ as follows:

$$f(x) = \begin{cases} x^2 - 1 & \text{for } -2 \le x < 2 \\ \sqrt{x - 2} & \text{for } 2 \le x < 5 \end{cases}$$

Solution The first part of f is given by $f(x) = x^2 - 1$ for $-2 \le x < 2$. This is an arc of a parabola obtained by shifting the graph of $y = x^2$ downward one unit.

The second part of f is given by $f(x) = \sqrt{x - 2}$ for $2 \le x < 5$. This is an arc of the square root curve obtained by shifting the graph of $y = \sqrt{x}$ two units to the right.

When $x = 2$, the radical part of f is used; $f(2) = \sqrt{2 - 2} = 0$. So there is a solid dot at (2, 0) and an open dot at (2, 3). Also, there is an open dot for $x = 5$ since 5 is not a domain value of f.

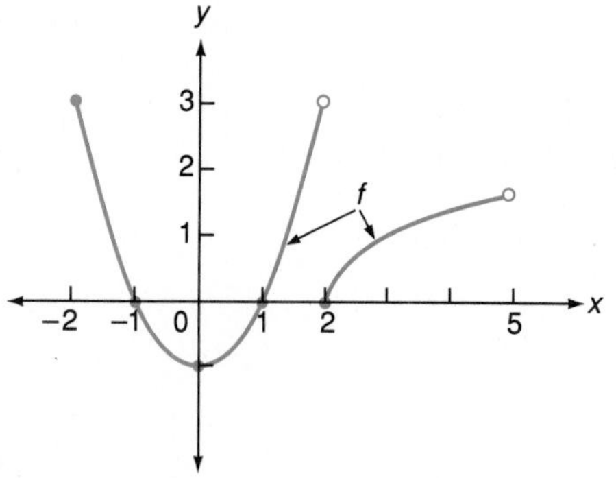

EXERCISES 8.3

Graph each set of curves on the same coordinate system. Use a dashed curve for the first equation and a solid curve for the second.

1. $f(x) = \sqrt{x}$, $g(x) = \sqrt{x - 1}$
2. $f(x) = -\sqrt{x}$, $g(x) = -2\sqrt{x}$
3. $f(x) = \sqrt{x}$, $g(x) = \sqrt{x + 1}$
4. $f(x) = -\sqrt{x}$, $g(x) = -\sqrt{x - 2}$

5. $f(x) = \sqrt[3]{x}$, $g(x) = \sqrt[3]{x+2}$
6. $f(x) = -\sqrt[3]{x}$, $g(x) = -\sqrt[3]{x-3}$
7. $f(x) = \sqrt[3]{x}$, $g(x) = \sqrt[3]{x-1}$
8. $f(x) = \sqrt[3]{x}$, $g(x) = -\sqrt[3]{x+1}$

Find the domain and range of f, sketch the graph, and give the equations of the asymptotic lines if there are any. State where f is increasing or decreasing and describe the concavity.

9. $f(x) = \sqrt{x+2}$
10. $f(x) = x^{1/2} + 2$
11. $f(x) = \sqrt{x-3} - 1$
12. $f(x) = -\sqrt{x} + 3$
13. $f(x) = \sqrt{-x}$
14. $f(x) = \sqrt{(x-2)^2}$
15. $f(x) = 2\sqrt[3]{x}$
16. $f(x) = |\sqrt[3]{x}|$
17. $f(x) = -x^{1/3}$
18. $f(x) = \sqrt[3]{-x}$
19. $f(x) = \dfrac{1}{\sqrt{x}} - 1$
20. $f(x) = \dfrac{1}{\sqrt{x-2}}$

21. **(a)** Explain why the graph of $f(x) = \dfrac{1}{\sqrt[3]{x}}$ is symmetric through the origin.
    (b) What is the domain of f?
    (c) Use a table of values to graph f.
    (d) What are the equations of the asymptotes?
22. Find the domain of $f(x) = \dfrac{1}{\sqrt[3]{x+1}}$, sketch the graph, and give the equations of the asymptotes.
23. Find the graph of the function $y = \sqrt[4]{x}$ by raising both sides of the equation to the fourth power and comparing to the graph of $y = x^4$.
24. Reflect the graph of $y = x^2$, for $x \geq 0$, through the line $y = x$. Obtain the equation of this new curve by interchanging variables in $y = x^2$ and solving for y.
25. Follow the instruction of Exercise 24 with $y = x^3$ for all values x.

In Exercises 26 and 27, the function f is defined by using more than one expression. Graph f on its given domain.

26. $f(x) = \begin{cases} \sqrt{x} & \text{for } 0 \leq x \leq 4 \\ 10 - \frac{1}{2}x^2 & \text{for } 4 < x < 6 \end{cases}$
27. $f(x) = \begin{cases} -2x - 1 & \text{for } -3 \leq x < 0 \\ \sqrt[3]{x-1} & \text{for } 0 \leq x \leq 2 \end{cases}$

Verify the equation involving the quotient for the given radical function.

28. $f(x) = \sqrt{x}$; $\dfrac{f(x) - f(25)}{x - 25} = \dfrac{1}{\sqrt{x} + 5}$ (Factor $x - 25$ as the difference of squares.)
29. $f(x) = \sqrt{x}$; $\dfrac{f(4+h) - f(4)}{h} = \dfrac{1}{\sqrt{4+h} + 2}$ (Rationalize the numerator.)
30. $f(x) = -\sqrt{x}$; $\dfrac{f(x) - f(9)}{x - 9} = -\dfrac{1}{\sqrt{x} + 3}$
31. A runner starts at point A, goes to point P that is x miles from B, and then runs to D. (Angles at B and C are right angles.)

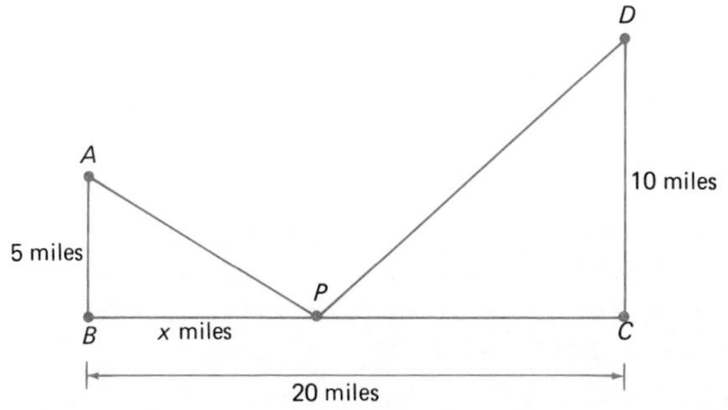

(a) Write the total distance d traveled as a function of x.

(b) The runner averages 12 miles per hour from A to P and 10 miles per hour from P to D. Write the time t for the trip as a function of x. (*Hint:* Use time = distance/rate.)

(c) Approximate, to the nearest tenth of a minute, the time for the trip when $x = 5$ miles.

32. **(a)** In the figure, AC is along the shoreline of a lake, and the distance from A to C is 12 miles. P represents the starting point of a swimmer who swims at 3 miles per hour along the hypotenuse PB. P is 5 miles from point A on the shore-line. After reaching B he walks at 6 miles per hour to C. Express the total time t of the trip as a function of x, where x is the distance from A to B (*Hint:* Use time = distance/rate and assume angle A is a right angle.)

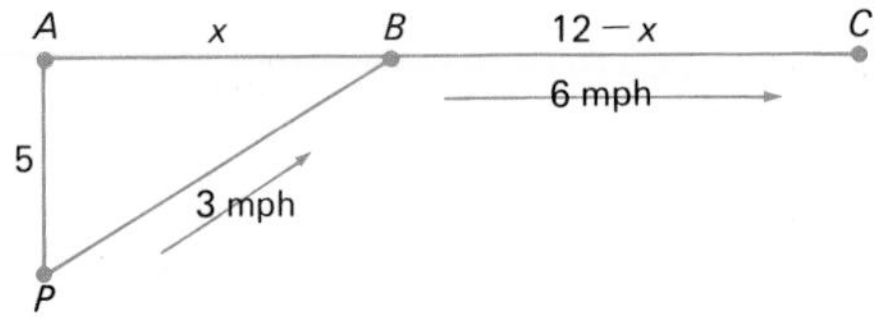

(b) Approximate, to the nearest tenth of a minute, the time for the trip when $x = 4$ miles.

Written Assignment: Explain the difference between the graphs of $y = \sqrt{x - 1}$ and $y^2 = x - 1$.

8.4 SYNTHETIC DIVISION

The x-intercepts for rational functions and some radical functions are also found by finding the roots of polynomial equations. This is discussed in Example 4 on page 368.

The graph of a linear function $y = mx + b$ has at most one x-intercept, and the graph of a quadratic function $y = ax^2 + bx + c$ has at most two x-intercepts. In general, the graph of a polynomial function of degree n has at most n x-intercepts.

Knowing the x-intercepts can be helpful in graphing and in solving applied problems. Finding the x-intercepts is a matter of setting $y = 0$. For example, the x-intercepts of $y = ax^2 + bx + c$ are the roots of $ax^2 + bx + c = 0$.

In this section, we will learn a special division technique called **synthetic division.** This method, in conjunction with the developments in the next two sections, will be helpful in finding the roots of polynomial equations of degree more than 2.

This special procedure is used for long-division problems with polynomials where the divisor is of the form $x \pm c$. To discover this procedure let us first examine the following long-division problem.

$$
\begin{array}{r|l}
 & x^2 + 3x - 5 \\
\hline
x - 2 & x^3 + x^2 - 11x + 12 \\
 & \underline{x^3 - 2x^2} \\
 & \quad + 3x^2 - 11x \\
 & \quad \underline{+ 3x^2 - 6x} \\
 & \qquad - 5x + 12 \\
 & \qquad \underline{- 5x + 10} \\
 & \qquad\qquad +2
\end{array}
$$

Now it should be clear that all of the work done involved the coefficients of the variables and the constants. Thus we could just as easily complete the division by omitting the variables, as long as we write the coefficients in the proper places. The division problem would then look like the following:

$$
\begin{array}{rrrrr}
 & 1 & +3 & -5 & \\ \cline{2-5}
1-2\,\big) & 1 & +1 & -11 & +12 \\
 & \textcircled{1} & -2 & & \\ \cline{2-3}
 & & +3 & -11 & \\
 & & +\textcircled{3} & -6 & \\ \cline{3-4}
 & & & -5 & +12 \\
 & & & -\textcircled{5} & +10 \\ \cline{4-5}
 & & & & +2
\end{array}
$$

Since the circled numerals are repetitions of those immediately above them, this process can be further shorted by deleting them. Moreover, since these circled numbers are the products of the numbers in the quotient by the 1 in the divisor, we may also eliminate this 1. Thus we have the following:

$$
\begin{array}{rrrrr}
 & 1 & +3 & -5 & \\ \cline{2-5}
-2\,\big) & 1 & +1 & -11 & +12 \\
 & & -2 & & \\ \cline{2-3}
 & & +3 & -11 & \\
 & & & -6 & \\ \cline{3-4}
 & & & -5 & +12 \\
 & & & & +10 \\ \cline{4-5}
 & & & & +2
\end{array}
$$

→ It is not necessary to bring down the −11 and 12.

$$
\begin{array}{rrrrr}
 & 1 & +3 & -5 & \\ \cline{2-5}
-2\,\big) & 1 & +1 & -11 & +12 \\
 & & -2 & & \\ \cline{2-3}
 & & +3 & & \\
 & & & -6 & \\ \cline{3-4}
 & & & -5 & \\
 & & & & +10 \\ \cline{4-5}
 & & & & +2
\end{array}
$$

→ Move the numerals upward.

$$
\begin{array}{rrrrr}
 & 1 & +3 & -5 & \\ \cline{2-5}
-2\,\big) & 1 & +1 & -11 & +12 \\
 & & -2 & -6 & +10 \\ \cline{2-5}
 & & +3 & -5 & +2
\end{array}
$$

When the top numeral 1 is brought down, then the last line contains the coefficients of the quotient and the remainder. So eliminate the line above the dividend.

$$
\begin{array}{r|rrrr}
-2 & 1 & +1 & -11 & +12 \\
 & & -2 & -6 & +10 \\ \hline
 & 1 & +3 & -5 & \multicolumn{1}{|r}{+2} \quad \text{remainder} \\
\text{quotient:} & \multicolumn{3}{r}{x^2 + 3x - 5} &
\end{array}
$$

We can further simplify this process by changing the sign of the divisor, making it +2 instead of −2. This change allows us to *add* throughout rather than subtract, as follows.

$$
\begin{array}{r|rrrr}
+2 & 1 & +1 & -11 & +12\} \quad \text{coefficients of dividend} \\
 & & +2 & +6 & -10 \\ \hline
\text{coefficients of quotient:} & 1 & +3 & -5 & \multicolumn{1}{|r}{+2} \quad \text{remainder} \\
\text{quotient:} & \multicolumn{3}{r}{x^2 + 3x - 5} &
\end{array}
$$

The long-division process has now been condensed to this short form. Doing a division problem by this short form is called **synthetic division,** as illustrated in the examples that follow.

EXAMPLE 1 Use synthetic division to find the quotient and remainder.

$$(2x^3 - 9x^2 + 10x - 7) \div (x - 3)$$

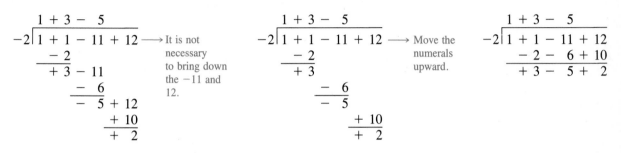

Solution Write the coefficients of the dividend in descending order. Change the sign of the divisor (change -3 to $+3$).

$$+3\,\big|\; 2 - 9 + 10 - 7$$

Now bring down the first term, 2, and multiply by $+3$.

$$\begin{array}{r|l} +3 & 2 - 9 + 10 - 7 \\ & \quad\;\, + 6 \\ \hline & 2 \end{array}$$

Add -9 and $+6$ to obtain the sum -3. Multiply this sum by $+3$ and repeat the process to the end. The completed example should look like this:

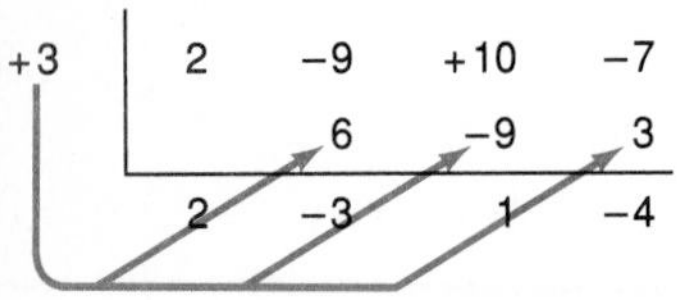

$$\begin{array}{r|l} +3 & 2 - 9 + 10 - 7 \\ & \quad\;\, + 6 - 9 + 3 \\ \hline \text{coefficients of quotient:} & 2 - 3 + 1 \,\big|\, -4 \quad \text{remainder} \end{array}$$

Since the original dividend began with x^3 (third degree). the quotient will begin with x^2 (second degree). Thus we read the last line as implying a quotient of $2x^2 - 3x + 1$ and a remainder of -4. Check this result by using the long-division process. ■

The synthetic division process has been developed for divisors of the form $x - c$. (Thus, in Example 1, $c = 3$.) A minor adjustment also permits divisors by polynomials of the form $x + c$. For example, a divisor of $x + 2$ may be written as $x - (-2)$; $c = -2$.

EXAMPLE 2 Divide $3x^3 - 2x^2 + 8$ by $x + 2$ using synthetic division.

Solution Write the dividend as $3x^3 - 2x^2 + 0x + 8$. Rewrite the divisor, $x + 2$, as $x - (-2)$ and use -2 as the divisor for the synthetic division.

When a power of x is missing in the dividend, insert a 0 in that position.

$$\begin{array}{r|l} -2 & 3 \quad - 2 \;+\; 0 \;+\; 8 \\ & \quad\;\; - 6 \;+ 16 - 32 \\ \hline & 3 \quad - 8 \;+ 16 \,\big|\, - 24 \quad \text{remainder} \end{array}$$

quotient $3x^2 - 8x + 16$

Check the result using $QD + R = N$ (see page 151.) ■

The final example illustrates synthetic division in which fractional coefficients are involved.

EXAMPLE 3 Use synthetic division to find the quotient and the remainder.

$$(-\tfrac{1}{3}x^4 + \tfrac{1}{6}x^2 - 7x - 4) \div (x + 3)$$

Solution Write $x + 3$ as $x - (-3)$. Since there is no x^3 term in the dividend, use $0x^3$.

Note that the quotient in a synthetic division problem is always a polynomial of degree one less than that of the dividend. This is so because the divisor has degree 1. The bottom line in the synthetic division process, except for the last entry on the right, gives the coefficients of the quotient: a polynomial in standard form.

$$\begin{array}{r|l} -3 & -\tfrac{1}{3} + 0 + \tfrac{1}{6} - 7 - 4 \\ & \qquad + 1 - 3 + \tfrac{17}{2} - \tfrac{9}{2} \\ \hline & -\tfrac{1}{3} + 1 - \tfrac{17}{6} + \tfrac{3}{2} \,\big|\, - \tfrac{17}{2} \quad \text{remainder} \end{array}$$

quotient: $-\frac{1}{3}x^3 + x^2 - \frac{17}{6}x + \frac{3}{2}$ remainder: $-\frac{17}{2}$

Check this result. ■

TEST YOUR UNDERSTANDING
Think Carefully

Use synthetic division, as indicated, to find each quotient and remainder.

1. $(x^3 - x^2 - 5x + 6) \div (x - 2)$
$+2 \,|\, 1 - 1 - 5 + 6$

2. $(2x^3 + 7x^2 + 5x - 2) \div (x + 2)$
$-2 \,|\, 2 + 7 + 5 - 2$

3. $(2x^3 - 3x^2 + 7) \div (x - 3)$
$+3 \,|\, 2 - 3 + 0 + 7$

4. $(3x^3 - 7x + 12) \div (x + 1)$
$-1 \,|\, 3 + 0 - 7 + 12$

(Answers: Page 377)

EXERCISES 8.4

Use synthetic division to find each quotient and remainder. Check each result.

1. $(x^3 - 2x^2 - 5x + 6) \div (x - 3)$
2. $(x^3 - x^2 - 5x + 2) \div (x + 2)$
3. $(2x^3 + x^2 - 3x + 7) \div (x + 1)$
4. $(3x^3 - 2x^2 + x - 1) \div (x - 1)$
5. $(x^3 + 5x^2 - 7x + 8) \div (x - 2)$
6. $(x^3 - 3x^2 + x - 5) \div (x + 3)$
7. $(x^4 - 3x^3 + 7x^2 - 2x + 1) \div (x + 2)$
8. $(x^4 + x^3 - 2x^2 + 3x - 1) \div (x - 2)$
9. $(2x^4 - 3x^2 + 4x - 2) \div (x - 1)$
10. $(3x^4 + x^3 - 2x + 3) \div (x + 1)$
11. $(x^3 - 27) \div (x - 3)$
12. $(x^3 - 27) \div (x + 3)$
13. $(x^3 + 27) \div (x + 3)$
14. $(x^3 + 27) \div (x - 3)$
15. $(x^4 - 16) \div (x - 2)$
16. $(x^4 - 16) \div (x + 2)$
17. $(x^4 + 16) \div (x + 2)$
18. $(x^4 + 16) \div (x - 2)$
19. $(x^4 - \frac{1}{2}x^3 + \frac{1}{3}x^2 - \frac{1}{4}x + \frac{1}{5}) \div (x - 1)$
20. $(4x^5 - x^3 + 5x^2 + \frac{3}{2}x - \frac{1}{2}) \div (x + \frac{1}{2})$

CHALLENGE
Think Creatively

Synthetic division can be applied to situations where the divisor has the form $ax + b$, $a \neq 0$. For example, to divide $2x^3 - 7x^2 + 8x + 6$ by $2x - 3$, first write the problem in fraction form and factor out $\frac{1}{2}$.

$$\frac{2x^3 - 7x^2 + 8x + 6}{2x - 3} = \frac{1}{2}\left(\frac{2x^3 - 7x^2 + 8x + 6}{x - \frac{3}{2}}\right)$$

Now use synthetic division for the fraction in parentheses, and then multiply by $\frac{1}{2}$ to obtain

$$x^2 - 2x + 1 + \frac{9}{2x - 3}$$

Use this procedure to do the following division problems.

1. $(6x^3 - 5x^2 - 3x + 4) \div (2x - 1)$
2. $(10x^3 - 3x^2 + 4x + 7) \div (2x + 1)$
3. $(6x^3 + 7x^2 + x + 8) \div (2x + 3)$
4. $(9x^4 + 6x^3 - 4x + 5) \div (3x - 1)$

8.5 THE REMAINDER AND FACTOR THEOREMS

When the polynomial $p(x) = 2x^3 - 9x^2 + 10x - 7$ is divided by $x - 3$, the quotient is the polynomial $q(x) = 2x^2 - 3x + 1$ and the remainder $r = -4$. (See Example 1, page 359.) As a check we see that

$$\underbrace{2x^3 - 9x^2 + 10x - 7}_{p(x)} = \underbrace{(2x^2 - 3x + 1)}_{q(x)} \cdot (x - 3) + \underbrace{(-4)}_{r}$$

Another form of this is
$\frac{p(x)}{x-c} = q(x) + \frac{r}{x-c}$.

In general, whenever a polynomial $p(x)$ is divided by $x - c$ we have

$$p(x) = q(x)(x - c) + r$$

where $q(x)$ is the quotient and r is the (constant) remainder. Since this equation holds for all x, we may let $x = c$ and obtain

$$\begin{aligned} p(c) &= q(c)(c - c) + r \\ &= q(c) \cdot 0 + r \\ &= r \end{aligned}$$

This result may be summarized as follows:

REMAINDER THEOREM

If a polynomial $p(x)$ is divided by $x - c$, the remainder is equal to $p(c)$.

EXAMPLE 1 Find the remainder when $p(x) = 3x^3 - 5x^2 + 7x + 5$ is divided by $x - 2$.

Solution By the remainder theorem, the answer is $p(2)$.

$$\begin{aligned} p(x) &= 3x^3 - 5x^2 + 7x + 5 \\ p(2) &= 3(2)^3 - 5(2)^2 + 7(2) + 5 \\ &= 23 \end{aligned}$$

Check this result by dividing $p(x)$ by $x - 2$.

■

EXAMPLE 2 Let $p(x) = x^3 - 2x^2 + 3x - 1$. Use synthetic division and the remainder theorem to find $p(3)$.

Solution According to the remainder theorem, $p(3)$ is equal to the remainder when $p(x)$ is divided by $x - 3$.

$$\begin{array}{r|rrrr} 3 & 1 & -2 & +3 & -1 \\ & & 3 & +3 & +18 \\ \hline & 1 & +1 & +6 & +17 \end{array} = \text{remainder} = p(3)$$

Check this result by substituting $x = 3$ in $p(x)$.

■

TEST YOUR UNDERSTANDING
Think Carefully

Use synthetic division to find the remainder r when $p(x)$ is divided by $x - c$. Verify that $r = p(c)$ by substituting $x = c$ into $p(x)$.

1. $p(x) = x^5 - 7x^4 + 4x^3 + 10x^2 - x - 5$; $x - 1$
2. $p(x) = x^4 + 11x^3 + 11x^2 + 11x + 10$; $x + 10$
3. $p(x) = x^4 + 11x^3 + 11x^2 + 11x + 10$; $x - 10$
4. $p(x) = 6x^3 - 40x^2 + 25$; $x - 6$

(Answers: Page 377)

Once again, we are going to consider the division of a polynomial $p(x)$ by a divisor of the form $x - c$. First recall that

$$p(x) = q(x)(x - c) + r$$

where $q(x)$ is the quotient and r is the (constant) remainder. Now suppose that $r = 0$. Then the remainder theorem gives $p(c) = r = 0$, and the preceding equation becomes

$$p(x) = q(x)(x - c)$$

If $u = vw$, then v and w are said to be factors of u.

It follows that $x - c$ is a *factor* of $p(x)$. Conversely, suppose that $x - c$ is a factor of $p(x)$. This means there is another polynomial, say $q(x)$, so that

$$p(x) = q(x)(x - c)$$

or

$$p(x) = q(x)(x - c) + 0$$

which tells us that when $p(x)$ is divided by $x - c$ the remainder is zero. These observations comprise the following result:

If $p(c) = 0$, then c is said to be a ***zero of the polynomial $p(x)$****, and it is also a root of $p(x) = 0$.*

FACTOR THEOREM

A polynomial $p(x)$ has a factor $x - c$ if and only if $p(c) = 0$.

EXAMPLE 3 Show that $x - 2$ is a factor of $p(x) = x^3 - 3x^2 + 7x - 10$.

Solution By the factor theorem we can state that $x - 2$ is a factor of $p(x)$ if $p(2) = 0$.

$$\begin{aligned} p(2) &= 2^3 - 3(2)^2 + 7(2) - 10 \\ &= 0 \end{aligned}$$

■

EXAMPLE 4 **(a)** Use the factor theorem to show that $x + 3$ is a factor of $p(x) = x^3 - x^2 - 8x + 12$. **(b)** Factor $p(x)$ completely.

Solution

(a) First write $x + 3 = x - (-3)$, so that $c = -3$. Then use synthetic division.

Note that $p(-3) = 0$. We use synthetic division here in order to be able to factor $p(x)$ as shown in part (b).

$$\begin{array}{r|rrrr} -3 & 1 & -1 & -8 & +12 \\ & & -3 & +12 & -12 \\ \hline & 1 & -4 & +4 & +0 \end{array}$$

Since $p(-3) = 0$, the factor theorem tells us that $x + 3$ is a factor of $p(x)$.

(b) Synthetic division has produced the quotient $x^2 - 4x + 4$. Therefore, since $x + 3$ is a factor of $p(x)$, we may write

$$x^3 - x^2 - 8x + 12 = (x^2 - 4x + 4)(x + 3)$$

To find the complete factored form, observe that $x^2 - 4x + 4 = (x - 2)^2$. Thus

$$x^3 - x^2 - 8x + 12 = (x - 2)^2(x + 3)$$

■

EXERCISES 8.5

Use synthetic division and the remainder theorem.

1. $f(x) = x^3 - x^2 + 3x - 2$; find $f(2)$.
2. $f(x) = 2x^3 + 3x^2 - x - 5$; find $f(-1)$.
3. $f(x) = x^4 - 3x^2 + x + 2$; find $f(3)$.
4. $f(x) = x^4 + 2x^3 - 3x - 1$; find $f(-2)$.
5. $f(x) = x^5 - x^3 + 2x^2 + x - 3$; find $f(1)$.
6. $f(x) = 3x^4 + 2x^3 - 3x^2 - x + 7$; find $f(-3)$.

Find the remainder for each division by substitution, using the remainder theorem. That is, in Exercise 7 (for example) let $f(x) = x^3 - 2x^2 + 3x - 5$ and find $f(2) = r$.

7. $(x^3 - 2x^2 + 3x - 5) \div (x - 2)$
8. $(x^3 - 2x^2 + 3x - 5) \div (x + 2)$
9. $(2x^3 + 3x^2 - 5x + 1) \div (x - 3)$
10. $(3x^4 - x^3 + 2x^2 - x + 1) \div (x + 3)$
11. $(4x^5 - x^3 - 3x^2 + 2) \div (x + 1)$
12. $(3x^5 - 2x^4 + x^3 - 7x + 1) \div (x - 1)$

Show that the given binomial $x - c$ is a factor of $p(x)$, and then factor $p(x)$ completely.

13. $p(x) = x^3 + 6x^2 + 11x + 6$; $x + 1$
14. $p(x) = x^3 - 6x^2 + 11x - 6$; $x - 1$
15. $p(x) = x^3 + 5x^2 - 2x - 24$; $x - 2$
16. $p(x) = -x^3 + 11x^2 - 23x - 35$; $x - 7$
17. $p(x) = -x^3 + 7x + 6$; $x + 2$
18. $p(x) = x^3 + 2x^2 - 13x + 10$; $x + 5$
19. $p(x) = 6x^3 - 25x^2 - 29x + 20$; $x - 5$
20. $p(x) = 12x^3 - 22x^2 - 100x - 16$; $x + 2$
21. $p(x) = x^4 + 4x^3 + 3x^2 - 4x - 4$; $x + 2$
22. $p(x) = x^4 - 8x^3 + 7x^2 + 72x - 144$; $x - 4$
23. $p(x) = x^6 + 6x^5 + 8x^4 - 6x^3 - 9x^2$; $x + 3$
24. $p(x) = 4x^5 - 40x^4 + 1000x^2 - 2500x$; $x - 5$

Use synthetic division to answer the following.

25. When $x^2 + 5x - 2$ is divided by $x + n$, the remainder is -8. Find all possible values of n and check by division.
26. Find d so that $x + 6$ is a factor of $x^4 + 4x^3 - 21x^2 + dx + 108$.
27. Find b so that $x - 2$ is a factor of $x^3 + bx^2 - 13x + 10$.
28. Find a so that $x - 10$ is a factor of $ax^3 - 25x^2 + 47x + 30$.
29. Find the complete factored form of $p(x) = x^5 + x^4 + 5x^2 - x - 6$ if $p(-2) = p(-1) = p(1) = 0$.
30. $p(x)$ is a fifth-degree polynomial in which 1 is the coefficient of x^5. Write the polynomial in standard form given that $p(0) = p(1) = p(2) = p(3) = p(4) = 0$.

Written Assignment: Use your own words and your own specific examples to explain both the remainder and the factor theorems.

EXPLORATIONS Think Critically

1. (a) Sketch the graph of a function having the following properties: $x = 1$ is a vertical asymptote, $y = -1$ is a horizontal asymptote, the domain is all $x \neq 1$, the range of all is all $y \neq -1$, the origin is the only x-intercept, f is increasing on both $(-\infty, 1)$ and $(1, \infty)$, and the graph is concave up on $(-\infty, 1)$ and concave down on $(1, \infty)$.
   (b) Sketch the graph of a function having the properties in part (a) above, except that there are no x-intercepts, the curve is decreasing and concave down on $(-\infty, 1)$, and the range is all $y < -1$.
2. Explain, with a specific example, why the following statement is *not* true:

   The graph of a rational function has a vertical asymptote at each value c where the denominator is 0.

3. Is it true that $\sqrt[3]{|x|} = |\sqrt[3]{x}|$ for all real numbers? Justify your answer.
4. Explain the difference between the graph of $f(x) = \dfrac{(x - a)(x - b)}{x - a}$ and that of $g(x) = x - b$.
5. The graph of a function is symmetric about the vertical line $x = c$ provided that $f(c + h) = f(c - h)$. Make a sketch that demonstrates this condition using $f(x) = (x - 3)^2$ where $c = 3$. Then use the condition to prove that $g(x) = 2(x - 5)^4 - 3$ is symmetric about the line $x = 5$.
6. Explain how the factor theorem can be used to decide that $x - y$ is a factor of $x^n - y^n$, where n is a positive integer. Use this procedure to decide the conditions on n so that $x + y$ is a factor of $x^n - y^n$. Now investigate when $x \pm y$ is a factor of $x^n + y^n$. Use synthetic division to find the quotient $(x^n - y^n) \div (x - y)$.

8.6 THE RATIONAL ROOT THEOREM

In order to see how the factor and remainder theorems may be applied to the solution of equations, let us first consider this polynomial equation:

$$(3x + 2)(5x - 4)(2x - 3) = 0$$

To find the roots, set each factor equal to zero.

$$3x + 2 = 0 \qquad 5x - 4 = 0 \qquad 2x - 3 = 0$$

$$x = -\frac{2}{3} \qquad x = \frac{4}{5} \qquad x = \frac{3}{2}$$

Now multiply the original three factors and keep careful note of the details of this multiplication. Your result should be

$$(3x + 2)(5x - 4)(2x - 3) = 30x^3 - 49x^2 - 10x + 24 = 0$$

which must have the same three rational roots.

As you analyze this multiplication it becomes clear that the constant 24 is the product of the three constants in the binomials, 2, −4, and −3. Also, the leading coefficient, 30, is the product of the three original coefficients of x in the binomials, namely 3, 5, and 2.

Furthermore, 3, 5, and 2 are also the denominators of the roots $-\frac{2}{3}$, $\frac{4}{5}$, and $\frac{3}{2}$. Therefore, the denominators of the rational roots are all factors of 30, and their numerators are all factors of 24.

These results are not accidental. It turns out that we have been discussing the following general result:

RATIONAL ROOT THEOREM

Let $f(x) = a_n x^n + a_{n-1}x^{n-1} + \cdots + a_1 x + a_0$ $(a_n \neq 0)$ be an nth-degree polynomial with integer coefficients. If $\dfrac{p}{q}$ is a rational root of $f(x) = 0$, where $\dfrac{p}{q}$ is in lowest terms, then p is a factor of a_0 and q is a factor of a_n.

Let us see how this theorem can be applied to find the rational roots of

$$f(x) = 4x^3 - 16x^2 + 11x + 10 = 0$$

Begin by listing all factors of the constant 10 and of the leading coefficient 4.

Possible numerators (factors of 10): $\pm 1, \pm 2, \pm 5, \pm 10$
Possible denominators (factors of 4): $\pm 1, \pm 2, \pm 4$
Possible rational roots (take each number in the first row and divide by each number in the second row):

Note that the theorem allows only these 16 numbers as ***possible*** *rational roots; no other rational numbers can be roots.*

$$\pm 1, \pm \tfrac{1}{2}, \pm \tfrac{1}{4}, \pm 2, \pm 5, \pm \tfrac{5}{2}, \pm \tfrac{5}{4}, \pm 10$$

To decide which (if any) of these are roots of $f(x) = 0$, we could substitute the values directly into $f(x)$. However, it is easier to use synthetic division because in most cases it leads to easier computations and also makes quotients available. Therefore, we proceed by using synthetic division with divisiors c, where c is a possible rational root.

If $f(c) = 0$, then c is a root; if $f(c) \neq 0$, then c is not a root.

Try $c = 1$:

$$\begin{array}{r|rrrr} 1 & 4 & -16 & +11 & +10 \\ & & +4 & -12 & -1 \\ \hline & 4 & -12 & -1 & \boxed{+9} \end{array}$$

Since $f(1) = 9 \neq 0$
1 is *not* a root.

Try $c = -1$:

$$\begin{array}{r|rrrr} -1 & 4 & -16 & +11 & +10 \\ & & -4 & +20 & -31 \\ \hline & 4 & -20 & +31 & \boxed{-21} \end{array}$$

Since $f(-1) = -21 \neq 0$
-1 is *not* a root.

Try $c = \frac{1}{2}$:

$$\begin{array}{r|rrrr} \frac{1}{2} & 4 & -16 & +11 & +10 \\ & & +2 & -7 & +2 \\ \hline & 4 & -14 & +4 & \boxed{+12} \end{array}$$

Since $f(\frac{1}{2}) = 12 \neq 0$
$\frac{1}{2}$ is *not* a root.

Try $c = -\frac{1}{2}$:

$$\begin{array}{r|rrrr} -\frac{1}{2} & 4 & -16 & +11 & +10 \\ & & -2 & +9 & -10 \\ \hline & 4 & -18 & +20 & \boxed{+0} \end{array}$$

Since $f(-\frac{1}{2}) = 0$,
$-\frac{1}{2}$ *is* a root.

By the factor theorem it follows that $x - (-\frac{1}{2}) = x + \frac{1}{2}$ is a factor of $f(x)$, and synthetic division gives the other factor, $4x^2 - 18x + 20$.

$$f(x) = (x + \tfrac{1}{2})(4x^2 - 18x + 20)$$

To find other roots of $f(x) = 0$ we could proceed by using the rational root theorem for $4x^2 - 18x + 20 = 0$. But this is unnecessary because the quadratic expression is factorable.

Contrary to the agreement about integer coefficients when factoring, on page 62, it is convenient here to use fractions in the factored form. However, we could just as well convert to the form $(2x + 1)(x - 2)(2x - 5)$ from which the same roots are obtained.

$$\begin{aligned} f(x) &= (x + \tfrac{1}{2})(4x^2 - 18x + 20) \\ &= (x + \tfrac{1}{2})(2)(2x^2 - 9x + 10) \\ &= 2(x + \tfrac{1}{2})(x - 2)(2x - 5) \end{aligned}$$

The solution of $f(x) = 0$ can now be found by setting each factor equal to zero. The roots are $x = -\frac{1}{2}$, $x = 2$, and $x = \frac{5}{2}$.

EXAMPLE 1 Factor $f(x) = x^3 + 6x^2 + 11x + 6$.

Whenever a polynomial has ±1 as the leading coefficient, then any rational root will be an integer that is a factor of the constant term of the polynomial.

Solution Since the leading coefficient is 1, whose only factors are ±1, the possible denominators of a rational root of $f(x) = 0$ can only be ±1. Hence the possible rational roots must all be factors of +6, namely ±1, ±2, ±3, and ±6. Use synthetic division to test these cases.

$$\begin{array}{r|rrrr} 1 & 1 & +\,6 & +\,11 & +\,6 \\ & & +\,1 & +\,7 & +\,18 \\ \hline & 1 & +\,7 & +\,18 & +\,24 = r \end{array}$$

Since $r = f(1) \neq 0$, $x - 1$ is *not* a factor of $f(x)$.

$$\begin{array}{r|rrrr} -1 & 1 & +\,6 & +\,11 & +\,6 \\ & & -\,1 & -\,5 & -\,6 \\ \hline & 1 & +\,5 & +\,6 & +\,0 = r \end{array}$$

Since $r = f(-1) = 0$, $x - (-1) = x + 1$ *is* a factor of $f(x)$.

$$x^3 + 6x^2 + 11x + 6 = (x + 1)(x^2 + 5x + 6)$$

Now factor the trinomial:

$$x^3 + 6x^2 + 11x + 6 = (x + 1)(x + 2)(x + 3)$$ ■

TEST YOUR UNDERSTANDING
Think Carefully

(Answers: Page 377)

For each $p(x)$, find **(a)** *the possible rational roots of $p(x) = 0$,* **(b)** *the factored form of $p(x)$, and* **(c)** *the roots of $p(x) = 0$.*

1. $p(x) = x^3 - 3x^2 - 10x + 24$

2. $p(x) = x^4 + 6x^3 + x^2 - 24x + 16$

3. $p(x) = 4x^3 + 20x^2 - 23x + 6$

4. $p(x) = 3x^4 - 13x^3 + 7x^2 - 13x + 4$

EXAMPLE 2 Find all real roots of $p(x) = 2x^5 + 7x^4 - 18x^2 - 8x + 8 = 0$.

Solution Begin by searching for rational roots. The possible rational roots are ±1, $\pm\frac{1}{2}$, ±2, ±4, and ±8. Testing these possibilities (left to right), the first root we find is $\frac{1}{2}$, as shown next.

$$\begin{array}{r|rrrrrr} \frac{1}{2} & 2 & +\,7 & +\,0 & -\,18 & -\,8 & +\,8 \\ & & +\,1 & +\,4 & +\,2 & -\,8 & -\,8 \\ \hline & 2 & +\,8 & +\,4 & -\,16 & -\,16 & +\,0 \end{array}$$

Therefore, $x - \frac{1}{2}$ is a factor of $p(x)$.

$$\begin{aligned} p(x) &= (x - \tfrac{1}{2})(2x^4 + 8x^3 + 4x^2 - 16x - 16) \\ &= 2(x - \tfrac{1}{2})(x^4 + 4x^3 + 2x^2 - 8x - 8) \end{aligned}$$

To find other roots of $p(x) = 0$ it now becomes necessary to solve the **depressed equation**

$$x^4 + 4x^3 + 2x^2 - 8x - 8 = 0$$

The possible rational roots for this equation are ± 1, ± 2, ± 4, and ± 8. However, values like ± 1 that were tried before, and produced nonzero remainders, need not be tried again. Why? We find that $x = -2$ is a root:

$$\begin{array}{r|l} -2 & 1 + 4 + 2 - 8 - 8 \\ & \quad - 2 - 4 + 4 + 8 \\ \hline & 1 + 2 - 2 - 4 \mid + 0 \end{array}$$

$$\begin{aligned} x^4 + 4x^3 + 2x^2 - 8x - 8 &= (x + 2)(x^3 + 2x^2 - 2x - 4) \\ &= (x + 2)[x^2(x + 2) - 2(x + 2)] \\ &= (x + 2)(x^2 - 2)(x + 2) \\ &= (x + 2)^2(x^2 - 2) \end{aligned}$$

Factoring by grouping was used here. Instead -2 could be tested again and produce the same result.

This gives

$$\begin{aligned} p(x) &= 2(x - \tfrac{1}{2})(x^4 + 4x^3 + 2x^2 - 8x - 8) \\ &= 2(x - \tfrac{1}{2})(x + 2)^2(x^2 - 2) \\ &= 2(x - \tfrac{1}{2})(x + 2)^2(x + \sqrt{2})(x - \sqrt{2}) \end{aligned}$$

This factored form of $p(x)$ can also be written using only integers as

$(2x - 1)(x + 2)^2(x^2 - 2)$

Setting each factor equal to zero produces the real roots of $p(x) = 0$ that consist of two rational and two irrational roots.

$$x = \tfrac{1}{2}, \qquad x = -2, \qquad x = -\sqrt{2}, \qquad x = \sqrt{2}$$ ■

These four roots of $p(x) = 0$ are also the x-intercepts of the polynomial function $p(x)$.

EXAMPLE 3 Solve for x: $q(x) = \frac{1}{3}x^5 + \frac{7}{6}x^4 - 3x^2 - \frac{4}{3}x + \frac{4}{3} = 0$

Solution To use the rational root theorem the polynomial must have integer coefficients. Therefore, multiply the given equation through by 6, the LCD, to obtain an equivalent equation having integer coefficients:

Equivalent equations have the same solutions.

$$6(\tfrac{1}{3}x^5 + \tfrac{7}{6}x^4 - 3x^2 - \tfrac{4}{3}x + \tfrac{4}{3}) = 6(0)$$
$$2x^5 + 7x^4 - 18x^2 - 8x + 8 = 0$$

Now the rational root theorem can be applied. However, since this is the same equation as in Example 2, we already know the solutions to be $\frac{1}{2}$, -2, $\pm\sqrt{2}$. ■

EXAMPLE 4 Find the x-intercepts for the graph of the rational function $f(x) = \dfrac{3x^4 - 7x^3 - 14x - 12}{x^2 - 16}$.

Solution A fraction can be 0 only when its numerator is 0. Thus $f(x) = 0$ precisely when $3x^4 - 7x^3 - 14x - 12 = 0$. Using the rational root theorem, the possible rational roots are

The x-intercepts of $g(x) = \sqrt{3x^4 - 7x^3 - 14x - 12}$ can be found using the same polynomial equation as in Example 4, because $g(x) = 0$ only when the radicand is 0.

$$\pm 1, \pm \tfrac{1}{3}, \pm 2, \pm \tfrac{2}{3}, \pm 3, \pm 4, \pm \tfrac{4}{3}, \pm 6, \pm 12$$

and we find that $-\frac{2}{3}$ and 3 are the only rational roots. Furthermore, the factor theorem gives

$$(x + \tfrac{2}{3})(x - 3)(3x^2 + 6) = 0$$
$$3(x + \tfrac{2}{3})(x - 3)(x^2 + 2) = 0$$

Then, since $x^2 + 2 = 0$ has no real roots, the x-intercepts are $-\frac{2}{3}$ and 3. ■

The roots of the equation $p(x) = 0$ in Example 2 are $\frac{1}{2}$, -2, $-\sqrt{2}$, and $\sqrt{2}$. Since the factored form of $p(x)$ contains the factor $(x + 2)^2$, we say that -2 is a *double root*. Therefore, counting the double root -2 as two roots, there are five roots for $p(x) = 0$, which is the same as the degree of $p(x)$. This is no coincidence. Such a result is true for any polynomial equation and is based upon the following result.

This was proved by Carl Friedrich Gauss in 1797 when he was 20 years old. He is recognized as one of the greatest mathematicians of all time.

FUNDAMENTAL THEOREM OF ALGEBRA

If $p(x)$ is a polynomial of degree $n \geq 1$, then $p(x) = 0$ has at least one real or complex root.

The proof of this theorem is beyond the level of this course, however, it leads to the next result that tells us that a polynomial equation of degree n cannot have more than n distinct roots.

THE N-ROOTS THEOREM

If $p(x)$ is a polynomial of degree $n \geq 1$, then $p(x) = 0$ has exactly n roots, provided that a root of multiplicity k is counted k times.

For example, the fifth degree equation in Example 2 has five roots, counting -2 twice because it is a double root.

Even though it can be very difficult to find all the roots of a polynomial equation, the preceding theorem does help us to the extent that we know that there can be no more than n roots when $p(x)$ has degree n. However, for illustrative purposes, we will consider polynomial equations whose roots can be found using the methods that have been studied.

EXAMPLE 5 Find all real and imaginary roots: $x^4 - 6x^2 - 8x + 24 = 0$.

Solution The possible rational roots are:

$$\pm 1, \pm 2, \pm 3, \pm 4, \pm 6, \pm 8, \pm 12, \pm 24$$

Trying these possibilities, left to right, shows that 2 is a root;

$$\begin{array}{r|rrrrr} 2 & 1 & +\,0 & -\,6 & -\,8 & +\,24 \\ & & +\,2 & +\,4 & -\,4 & -\,24 \\ \hline & 1 & +\,2 & -\,2 & -\,12 & +\,0 \end{array}$$

Now, by the factor theorem, we have

$$p(x) = (x - 2)(x^3 + 2x^2 - 2x - 12)$$

and any remaining rational root of the given equation must be a root of the *depressed equation*

$$x^3 + 2x^2 - 2x - 12 = 0$$

This work can be condensed as follows:

$$\begin{array}{r|rrrrr} 2 & 1 & +\,0 & -\,6 & -\,8 & +\,24 \\ & & 2 & +\,4 & -\,4 & -\,24 \\ \hline 2 & 1 & +\,2 & -\,2 & -\,12 & +\,0 \\ & & +\,2 & +\,8 & +\,12 & \\ \hline & 1 & +\,4 & +\,6 & +\,0 & \end{array}$$

Since ± 1 have already been eliminated, try 2 again.

$$\begin{array}{r|rrrr} 2 & 1 & +\,2 & -\,2 & -\,12 \\ & & +\,2 & +\,8 & +\,12 \\ \hline & 1 & +\,4 & +\,6 & +\,0 \end{array}$$

Then

$$x^3 + 2x^2 - 2x - 12 = (x - 2)(x^2 + 4x + 6)$$

and

$$\begin{aligned} p(x) &= (x - 2)(x - 2)(x^2 + 4x + 6) \\ &= (x - 2)^2(x^2 + 4x + 6) \end{aligned}$$

Using the quadratic formula, the roots of $x^2 + 4x + 6 = 0$ are found to be $-2 \pm \sqrt{2}i$. Thus $p(x) = 0$ has two imaginary roots and the double root 2; a total of four roots. Since the degree of $p(x)$ is 4, we know from the n-roots theorem that there can be no other roots; we have found them all. ■

EXERCISES 8.6

Find all real roots.

1. $x^3 + x^2 - 21x - 45 = 0$
2. $x^3 + 2x^2 - 29x + 42 = 0$
3. $3x^3 + 2x^2 - 75x - 50 = 0$
4. $x^3 + 3x^2 + 3x + 1 = 0$
5. $x^4 + 3x^3 + 3x^2 + x = 0$
6. $x^4 + 6x^3 + 7x^2 - 12x - 18 = 0$
7. $\frac{1}{4}x^4 + \frac{3}{2}x^3 + \frac{1}{2}x^2 - \frac{9}{2}x - \frac{15}{4} = 0$
8. $\frac{1}{3}x^3 - \frac{5}{2}x^2 + 4x + \frac{8}{3} = 0$

Find the x-intercepts for the graph of each function.

9. $p(x) = x^4 + 2x^3 - 7x^2 - 18x - 18$
10. $p(x) = x^4 - x^3 - 5x^2 - x - 6$
11. $p(x) = -x^5 + 5x^4 - 3x^3 - 15x^2 + 18x$
12. $p(x) = x^4 - 5x^3 + 3x^2 + 15x - 18$
13. $f(x) = \dfrac{2x^3 - 5x - 3}{x - 4}$
14. $f(x) = \dfrac{x^4 + 4x^3 - 7x^2 - 36x - 18}{2x - 5}$
15. $f(x) = \sqrt{3x^4 - 11x^3 - 3x^2 - 6x + 8}$
16. $f(x) = \sqrt[3]{6x^3 - 25x^2 + 21x + 10}$

Factor.

17. $-x^3 - 3x^2 + 24x + 80$
18. $x^3 - 3x^2 - 10x + 24$
19. $6x^4 + 9x^3 + 9x - 6$
20. $x^3 - 28x - 48$

Show that the given equation has no rational roots.

21. $x^4 - x^3 + 5x^2 - 2x + 16 = 0$
22. $2x^3 - 5x^2 - x + 8 = 0$

Find all real and imaginary roots.

23. $p(x) = x^4 - 3x^3 + 5x^2 - x - 10 = 0$
24. $p(x) = x^5 - 3x^4 - 3x^3 + 9x^2 - 10x + 30 = 0$
25. $p(x) = \frac{1}{10}x^3 - \frac{1}{6}x^2 + \frac{1}{15}x - \frac{4}{15} = 0$
26. $p(x) = 2x^4 - 5x^3 + x^2 + 4x - 4 = 0$
27. $p(x) = x^5 - 9x^4 + 31x^3 - 49x^2 + 36x - 10 = 0$
28. $p(x) = \frac{1}{5}x^6 + \frac{1}{2}x^5 + \frac{1}{10}x^4 + x^3 - \frac{2}{5}x^2 + \frac{1}{2}x - \frac{3}{10} = 0$

In Exercises 29 and 30 the factored form of the polynomial in the given equation is $(x \pm c)^n$. Find this form and graph the equation.

29. $y = x^4 - 8x^3 + 24x^2 - 32x + 16$
30. $y = x^4 + 12x^3 + 54x^2 + 108x + 81$

*31. Graph: $y = \frac{1}{8}x^3 - \frac{3}{4}x^2 + \frac{3}{2}x - 1$ (*Hint*: Convert to the form $y = a(x - c)^n$.)

Written Assignment: Use your own words and a specific example to describe the rational root theorem.

CHALLENGE Think Creatively Use the rational root theorem to prove that the real number $\sqrt{5}$ is an irrational number. (*Hint:* Consider the equation $x^2 - 5 = 0$.)

CHAPTER 8 SUMMARY

Review these key terms and concepts so that you are able to define or describe them. A clear understanding of these will be very helpful when reviewing the developments of this chapter.

Graphing Properties and Procedures for Functions

Symmetries

1. If $f(x) = f(-x)$, the curve is symmetric about the y-axis.
2. If $f(-x) = -f(x)$, the curve is symmetric through the origin.

Multiplying Ordinates and Reflections

3. The graph of $y = af(x)$ can be obtained by multiplying the ordinates of the curve $y = f(x)$ by the value a. The case $a = -1$ gives $y = -f(x)$, which is the reflection of $y = f(x)$ through the x-axis.

Absolute Value

4. The graph of $y = |f(x)|$ can be obtained from the graph of $y = f(x)$ by taking the part of $y = f(x)$ that is below the x-axis and reflecting it through the x-axis.

Horizontal Translations; $h > 0$

5. The graph of $y = f(x - h)$ can be obtained by shifting $y = f(x)$ h units to the right.
6. The graph of $y = f(x + h)$ can be obtained by shifting $y = f(x)$ h units to the left.

Vertical Translations; $k > 0$

7. The graph of $y = f(x) + k$ can be obtained by shifting $y = f(x)$ k units upward.
8. The graph of $y = f(x) - k$ can be obtained by shifting $y = f(x)$ k units downward.

A large variety of functions can be graphed by applying the preceding to fundamental polynomial, rational, or radical functions.

Fundamental Functions

Polynomial: $f(x) = x^2$, $f(x) = x^3$, $f(x) = x^4$

Rational: $f(x) = \dfrac{1}{x}$, $f(x) = \dfrac{1}{x^2}$

Radical: $f(x) = \sqrt{x}$, $f(x) = \sqrt[3]{x}$, $f(x) = \dfrac{1}{\sqrt{x}}$

Synthetic Division is a shortcut of the long-division process for dividing a polynomial by $x - c$. It gives the quotient and remainder as shown here for $(2x^3 - 9x^2 + 10x - 7) \div (x - 3)$.

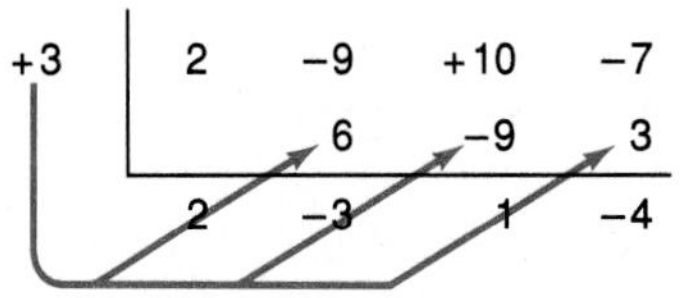

quotient: $2x^2 - 3x + 1$ remainder: -4

Remainder Theorem

If a polynomial $p(x)$ is divided by $x - c$, the remainder is equal to $p(c)$.

Factor Theorem

A polynomial $p(x)$ has a factor $x - c$ if and only if $p(c) = 0$.

Rational Root Theorem

Let $f(x) = a_nx^n + a_{n-1}x^{n-1} + \cdots + a_1x + a_0$ ($a_n \neq 0$) be an nth-degree polynomial with integer coefficients. If $\frac{p}{q}$ is a rational root of $f(x) = 0$, where $\frac{p}{q}$ is in lowest terms, then p is a factor of a_0 and q is a factor of a_n.

If the polynomial equation $p(x) = a_nx^n + a_{n-1}x^{n-1} + \cdots + a_1x + a_0 = 0$ has integer coefficients, the possible rational roots are found by dividing each factor of a_0 by each factor of a_n. Synthetic division and the remainder theorem can be used to test these possibilities to find the rational roots of $p(x) = 0$.

Fundamental Theorem of Algebra
If $p(x)$ is a polynomial of degree $n \geq 1$, then $p(x) = 0$ has at least one real or complex root.

The n-Roots Theorem
If $p(x)$ is a polynomial of degree $n \geq 1$, then $p(x) = 0$ has exactly n roots, provided that a root of multiplicity k is counted k times.

REVIEW EXERCISES

The solutions to the following exercises can be found within the text of Chapter 8. Try to answer each question before referring to the text.

Section 8.1

Graph each of the following.

1. $y = f(x) = x^3$
2. $y = f(x) = -x^3$
3. $y = f(x) = (x - 3)^3$
4. $y = f(x) = |(x - 3)^3|$
5. Graph $y = x^4$ and $y = 2x^4$ in the same coordinate system.
6. State where the function $f(x) = 2x^4$ is increasing and where it is decreasing. Describe the concavity.
7. Under what conditions is the graph of a function $y = f(x)$ said to be symmetric through the origin?
8. Under what conditions is the graph of $y = f(x)$ symmetric about the y-axis?
9. Explain how the graph of $y = f(x + h)$ can be obtained from the graph of $y = f(x)$ when $h > 0$, and also when $h < 0$.
10. Explain how the graph of $y = f(x) + k$ can be obtained from the graph of $y = f(x)$ when $k > 0$, and also when $k < 0$.
11. Graph $y = |x + 6| - 4$. Find the x- and y-intercepts.

Section 8.2

12. Draw the graph of $y = f(x) = \frac{1}{x}$. Describe the symmetry, find the equations of the asymptotes, and find the intervals on which the curve is concave down or concave up.
13. Sketch the graph of $g(x) = \frac{1}{x - 3}$. Find the asymptotes.
14. Graph $y = \frac{1}{x^2}$, describe the symmetry, and find the asymptotes.
15. Graph $x = \frac{1}{(x + 2)^2} - 3$. What are the asymptotes? Find the domain and range, describe the concavity of the curve, and state where it is increasing or decreasing.
16. Graph: $y = f(x) = \frac{x^2 + x - 6}{x - 2}$

Section 8.3

Graph each function and find the domain and range.

17. $y = \sqrt{x}$ 18. $y = \sqrt{x - 2}$ 19. $y = \sqrt{|x|}$

Graph each function. Find the domain and the equations of the asymptotes.

20. $y = x^{-1/2}$ 21. $y = \frac{1}{\sqrt{x}} + 1$ 22. $y = \frac{1}{\sqrt{x + 1}}$

23. Graph $y = x^3$ and $y = \sqrt[3]{x}$.
24. Graph: $f(x) = \begin{cases} x^2 - 1 & \text{if } -2 \le x < 2 \\ \sqrt{x - 2} & \text{if } 2 \le x < 5 \end{cases}$

Section 8.4

Use synthetic division to find each quotient.

25. $(x^3 + x^2 - 11x + 12) \div (x - 2)$ 26. $(2x^3 - 9x^2 + 10x - 7) \div (x - 3)$
27. $(-\frac{1}{3}x^4 + \frac{1}{6}x^2 - 7x - 4) \div (x + 3)$
28. State the rule for checking a division problem.

Section 8.5

29. State the remainder theorem. 30. State the factor theorem.
31. Find the remainder if $p(x) = 3x^3 - 5x^2 + 7x + 5$ is divided by $x - 2$.
32. Let $p(x) = x^3 - 2x^2 + 3x - 1$. Use synthetic division and the remainder theorem to find $p(3)$.
33. Show that $x - 2$ is a factor of $p(x) = x^3 - 3x^2 + 7x - 10$.
34. Use the factor theorem to show that $x + 3$ is a factor of $p(x) = x^3 - x^2 + 8x + 12$. Then factor $p(x)$ completely.

Section 8.6

35. State the rational root theorem.
36. What are the possible rational roots of $f(x) = 4x^3 - 16x^2 + 11x + 10 = 0$?
37. Factor $f(x) = x^3 + 6x^2 + 11x + 6$.
38. Find all real roots of $p(x) = 2x^5 + 7x^4 - 18x^2 - 8x + 8 = 0$.
39. Solve for x: $q(x) = \frac{1}{3}x^5 + \frac{7}{6}x^4 - 3x^2 - \frac{4}{3}x + \frac{4}{3} = 0$
40. Find the x-intercepts for the graph of the rational function
$$f(x) = \frac{3x^4 - 7x^3 - 14x - 12}{x^2 - 16}.$$
41. Find all real and imaginary roots: $p(x) = x^4 - 6x^2 - 8x + 24 = 0$.

CHAPTER 8 TEST: STANDARD ANSWER

Use these questions to test your knowledge of the basic skills and concepts of Chapter 8. Then check your answers with those given at the back of the book.

Graph each function and write the equation of the asymptotes if there are any.

1. $y = f(x) = x^3$ 2. $f(x) = (x + 2)^3 - \frac{3}{2}$ 3. $y = f(x) = -\frac{1}{x - 2}$
4. State the domain and range of f in Question 3. Where is f increasing or decreasing? Describe the concavity.
5. What is the domain and range of $y = x^4 - 16$? Find the x- and y-intercepts.

Graph each function, state its domain, and give the equations of the asymptotes if there are any.

6. $f(x) = \sqrt[3]{x + 2}$ 7. $g(x) = \dfrac{1}{\sqrt{x}} + 2$ 8. $f(x) = -\sqrt[3]{x - 2}$

9. What is the range of g in Question 7? Where is g increasing or decreasing? Describe the concavity.

Find the equations of the asymptotes for each curve.

10. $y = \dfrac{x - 2}{x^2 + x - 6}$ 11. $y = \dfrac{1}{x^2} + 3$

Explain how translations can be used to obtain the graph of g from the graph of f.

12. $f(x) = x^4,\ g(x) = (x - 5)^4 + 2$
13. $f(x) = \sqrt{x},\ g(x) = \sqrt{x + 4} - 3$

Explain how reflections and translations can be used to obtain the graph of g from the graph of f.

14. $f(x) = \sqrt[3]{x},\ g(x) = -\sqrt[3]{x} - 4$
15. $f(x) = x^3,\ g(x) = |x^3| - 1$
16. Explain how the graph of $g(x) = \frac{1}{2}x^4 - 3$ can be obtained from the graph of $f(x) = x^4$ by using translations and the method of multiplying ordinates.

Use synthetic division to find the quotient and remainder.

17. $(5x^3 - 12x^2 - 3x + 6) \div (x - 2)$
18. $(2x^5 + 5x^4 - x^2 - 21x + 7) \div (x + 3)$
19. $(x^5 + 32) \div (x + 2)$
20. **(a)** Let $p(x) = 27x^4 - 36x^3 + 18x^2 - 4x + 1$ and use the remainder theorem to evaluate $p(\frac{1}{3})$.
    (b) Use the result of part (a) and the factor theorem to determine whether or not $x - \frac{1}{3}$ is a factor of $p(x)$.
21. Show that $x + 4$ is a factor of $p(x) = 2x^3 + 3x^2 - 23x - 12$, and factor $p(x)$ completely.
22. Show that $x - 2$ is a factor of $p(x) = x^4 - 4x^3 + 7x^2 - 12x + 12$, and factor $p(x)$ completely.
23. Make use of the rational root theorem to factor $f(x) = x^4 + 5x^3 + 4x^2 - 3x + 9$.
24. Find the roots of $p(x) = 0$ for $p(x) = x^4 + 3x^3 - 3x^2 - 11x - 6$.
25. Find the x-intercepts for the graph of $f(x) = \dfrac{2x^3 + 15x^2 + 22x - 15}{x^2 + 7}$.

CHAPTER 8 TEST: MULTIPLE CHOICE

1. Which of the following are true?
   I. The graph of $f(x) = x^2$ is symmetric about the y-axis.
   II. The graph of $f(x) = x^3$ is symmetric through the origin.
   III. The graph of $f(x) = \dfrac{1}{x}$ is symmetric about the x-axis.
   (a) Only I **(b)** Only II **(c)** Only III **(d)** Only I and II **(e)** None of the preceding
2. The equation of the horizontal asymptote for $f(x) = \dfrac{1}{x - 3}$ is
   (a) $x = 0$ **(b)** $y = 0$ **(c)** $x = 3$ **(d)** $x = -3$ **(e)** None of the preceding

3. For $h > 0$, the graph of $y = f(x - h)$ can be obtained by shifting the graph of $y = f(x)$

(a) h units to the right **(b)** h units to the left **(c)** h units upward
(d) h units downward **(e)** None of the preceding

4. Which of the following are true for the graph of $y = -\frac{1}{x^2}$?

I. The horizontal asymptote is $x = 0$.
II. The vertical asymptote is $y = 0$.
III. The graph is symmetric about the y-axis.

(a) Only I **(b)** Only II **(c)** Only III **(d)** I, II, and III **(e)** None of the preceding

5. The equation of the vertical asymptote of $f(x) = \frac{x^2}{x^2 + x}$ is

(a) $y = -1$ **(b)** $y = 1$ **(c)** $x = 1$ **(d)** $x = 0$ **(e)** $x = -1$

6. The graph of $y = \frac{1}{(x - 1)^2} + 3$ is found by shifting the graph of $y = \frac{1}{x^2}$ by which of the following?

(a) One unit down and 3 units to the right
(b) One unit to the left and 3 units up
(c) One unit to the right and 3 units up
(d) One unit down and 3 units to the right
(e) None of the preceding

7. The graph of $y = \frac{1}{\sqrt{x} - 1}$ is found by shifting the graph of $y = \frac{1}{\sqrt{x}}$

(a) One unit downward **(b)** One unit upward **(c)** One unit to the left
(d) One unit to the right **(e)** None of the preceding

8. For which of the following functions is both the domain and the range the set of all real numbers greater than or equal to 0?

(a) $y = \sqrt{x}$ **(b)** $y = x^3$ **(c)** $y = |x|$ **(d)** $y = \sqrt[3]{x}$ **(e)** None of the preceding

9. What are the x-intercepts for the graph of $y = \sqrt{2x^2 - 8x - 24}$?

(a) 2, −6 **(b)** −4, 3 **(c)** −6, 4 **(d)** 0 **(e)** None of the preceding

10. The graph of which of the following can be obtained from $f(x) = \frac{1}{x}$ by a reflection and a horizontal translation?

(a) $y = \frac{1}{|x|} + 2$ **(b)** $y = \frac{1}{x + 2}$ **(c)** $y = \frac{2}{|x|}$ **(d)** $y = -\frac{1}{x - 2}$ **(e)** $y = -\frac{1}{x} + 2$

11. When the synthetic division at the right is completed, the remainder is

$$-2 \,|\, 1 + 3 - 5 + 7$$

(a) 0 **(b)** +7 **(c)** +21 **(d)** −7 **(e)** None of the preceding

12. When $p(x) = 3x^3 + 4x^2 - 8x - 14$ is divided by $x - 2$, the remainder is

(a) 38 **(b)** −14 **(c)** $p(-2)$ **(d)** $p(2)$ **(e)** None of the preceding

13. One of the factors of $p(x) = x^3 - 2x^2 - 5x + 6$ is

(a) $x + 1$ **(b)** $x - 6$ **(c)** $x - 3$ **(d)** $x - 2$ **(e)** None of the preceding

14. Which of the following is a set of possible rational roots of $f(x) = 2x^3 - 8x^2 + 7x - 10$?

(a) $\pm 1, \pm 2, \pm 5, \pm 10$ **(b)** $\pm 1, \pm\frac{1}{2}, \pm 2, \pm\frac{5}{2}, \pm 5, \pm 10$
(c) $\pm 2, \pm 10$ **(d)** $\pm 1, \pm 2, \pm 10$ **(e)** None of the preceding

15. The equation $x^4 - 6x^3 + 14x^2 - 16x + 8 = 0$ has 2 as a double root. The remaining roots are

(a) $1 \pm i$ **(b)** $1 \pm \sqrt{3}$ **(c)** $5 \pm 5i$ **(d)** ± 8 **(e)** None of the preceding

ANSWERS TO THE TEST YOUR UNDERSTANDING EXERCISES

Page 355

1. Domain of g; $x \geq 0$

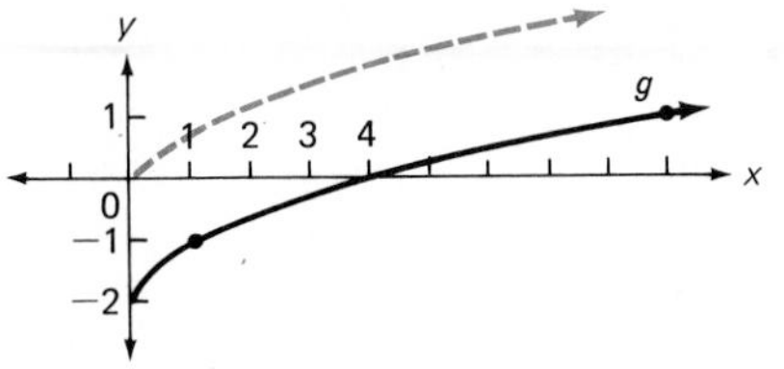

2. Domain of g: $x \geq 1$

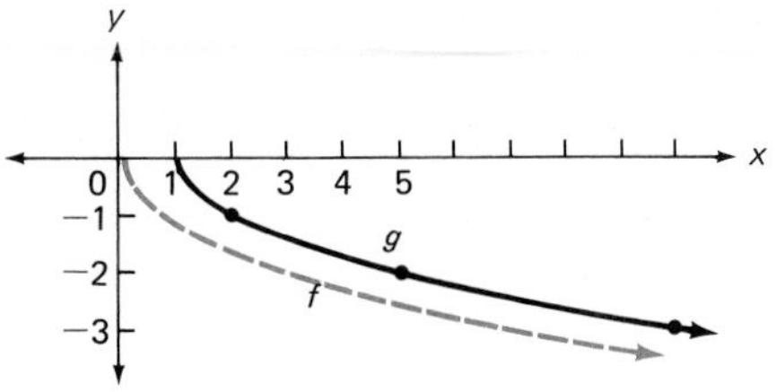

3. Domain of g: $x \geq 0$

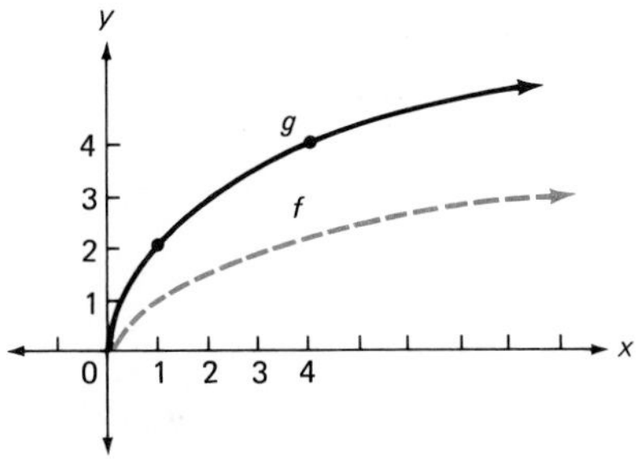

4. Domain of g: $x > -2$

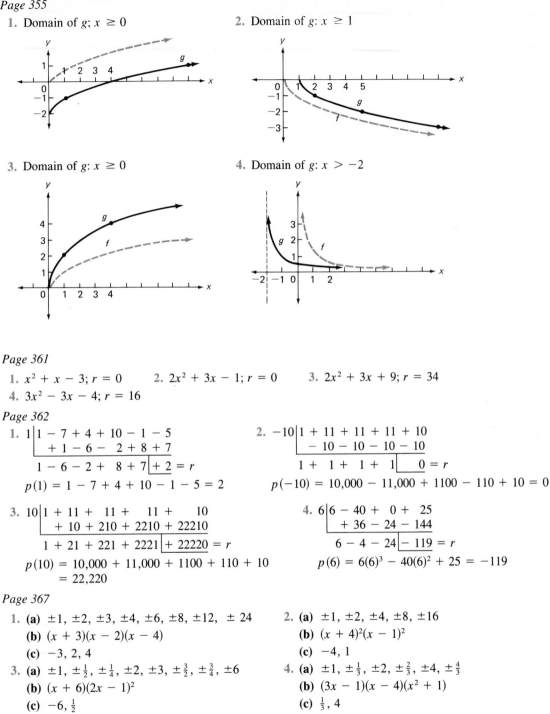

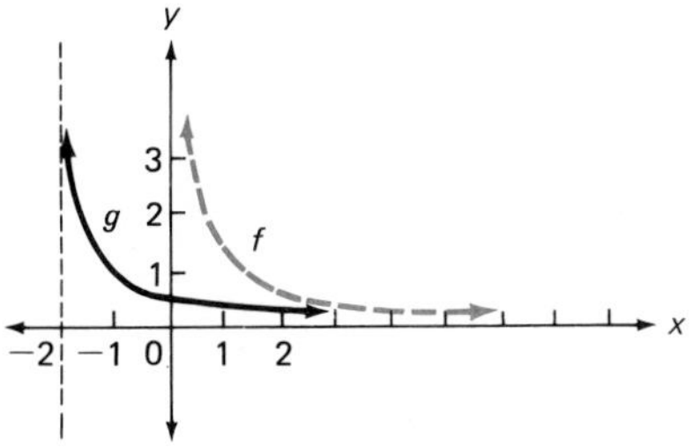

Page 361

1. $x^2 + x - 3$; $r = 0$
2. $2x^2 + 3x - 1$; $r = 0$
3. $2x^2 + 3x + 9$; $r = 34$
4. $3x^2 - 3x - 4$; $r = 16$

Page 362

1.
$$\begin{array}{r|l} 1 & 1 - 7 + 4 + 10 - 1 - 5 \\ & \quad + 1 - 6 - \; 2 + 8 + 7 \\ \hline & 1 - 6 - 2 + \; 8 + 7 \,|\, + 2 = r \end{array}$$
$p(1) = 1 - 7 + 4 + 10 - 1 - 5 = 2$

2.
$$\begin{array}{r|l} -10 & 1 + 11 + 11 + 11 + 10 \\ & \quad - 10 - 10 - 10 - 10 \\ \hline & 1 + \; 1 + \; 1 + \; 1 \,|\, \quad 0 = r \end{array}$$
$p(-10) = 10{,}000 - 11{,}000 + 1100 - 110 + 10 = 0$

3.
$$\begin{array}{r|l} 10 & 1 + 11 + \; 11 + \; 11 + \; 10 \\ & \quad + 10 + 210 + 2210 + 22210 \\ \hline & 1 + 21 + 221 + 2221 \,|\, + 22220 = r \end{array}$$
$p(10) = 10{,}000 + 11{,}000 + 1100 + 110 + 10$
$= 22{,}220$

4.
$$\begin{array}{r|l} 6 & 6 - 40 + \; 0 + \; 25 \\ & \quad + 36 - 24 - 144 \\ \hline & 6 - 4 - 24 \,|\, - 119 = r \end{array}$$
$p(6) = 6(6)^3 - 40(6)^2 + 25 = -119$

Page 367

1. **(a)** $\pm 1, \pm 2, \pm 3, \pm 4, \pm 6, \pm 8, \pm 12, \pm 24$
   (b) $(x + 3)(x - 2)(x - 4)$
   (c) $-3, 2, 4$
2. **(a)** $\pm 1, \pm 2, \pm 4, \pm 8, \pm 16$
   (b) $(x + 4)^2(x - 1)^2$
   (c) $-4, 1$
3. **(a)** $\pm 1, \pm \frac{1}{2}, \pm \frac{1}{4}, \pm 2, \pm 3, \pm \frac{3}{2}, \pm \frac{3}{4}, \pm 6$
   (b) $(x + 6)(2x - 1)^2$
   (c) $-6, \frac{1}{2}$
4. **(a)** $\pm 1, \pm \frac{1}{3}, \pm 2, \pm \frac{2}{3}, \pm 4, \pm \frac{4}{3}$
   (b) $(3x - 1)(x - 4)(x^2 + 1)$
   (c) $\frac{1}{3}, 4$

CHAPTER 9

EXPONENTIAL AND LOGARITHMIC FUNCTIONS

9.1 INVERSE FUNCTIONS

Two functions such as $f(x) = \sqrt{x}$ and $g(x) = \dfrac{1}{x-2}$ may be combined in a special way to form the *composition of f by g*. For example, take a specific value in the domain of g, say $x = 6$. Then the corresponding range value is $g(6) = \frac{1}{4}$. Take this range value and use it as a domain value for f to produce $f(\frac{1}{4}) = \sqrt{\frac{1}{4}} = \frac{1}{2}$. This work may be condensed in this way.

The output of g becomes the input of f: $g(6) = \frac{1}{4}$ comes out of g and goes into f.

$$f(g(6)) = f(\tfrac{1}{4}) = \sqrt{\tfrac{1}{4}} = \tfrac{1}{2}$$

Here is another illustration using $x = 10$.

Read this as "f of g of 10" $\longrightarrow f(g(10)) = f\left(\dfrac{1}{10-2}\right) = f\left(\dfrac{1}{8}\right) = \sqrt{\dfrac{1}{8}} = \dfrac{1}{2\sqrt{2}}$

The composition of functions does not always produce a real number. For instance, if $x = -3$, then $g(-3) = -\frac{1}{5}$; $f(-\frac{1}{5}) = \sqrt{-\frac{1}{5}}$ is not a real number. We therefore say that $f(g(-3))$ is undefined or that it does not exist.

The roles of f and g may be interchanged. Thus

Read this as "g of f of 10" $\longrightarrow g(f(10)) = g(\sqrt{10}) = \dfrac{1}{\sqrt{10}-2}$

TEST YOUR UNDERSTANDING
Think Carefully

For each pair of functions f and g, evaluate (if possible) each of the following:

(a) $g(f(1))$ **(b)** $f(g(1))$ **(c)** $f(g(0))$ **(d)** $g(f(-2))$

1. $f(x) = 3x - 1;\ g(x) = x^2 + 4$

2. $f(x) = \sqrt{x};\ g(x) = x^2$

3. $f(x) = \sqrt[3]{3x - 1};\ g(x) = 5x$

4. $f(x) = \dfrac{x + 2}{x - 1};\ g(x) = x^3$

(Answers: Page 424)

The preceding computations for specific values of x can be stated in terms of any allowable x. For instance, using $f(x) = \sqrt{x}$ and $g(x) = \dfrac{1}{x - 2}$ we have

$f(g(x))$ is read as "f of g of x."

$$f(g(x)) = f\left(\frac{1}{x - 2}\right) = \sqrt{\frac{1}{x - 2}} = \frac{1}{\sqrt{x - 2}}$$

This new correspondence between a domain value x and the range value $\dfrac{1}{\sqrt{x - 2}}$ is referred to as the **composite function of f by g**. This composite function is denoted by $f \circ g$ That is, for the given functions f and g, we form the composite of f by g, whose range values $(f \circ g)(x)$ are defined by

$$(f \circ g)(x) = f(g(x)) = \frac{1}{\sqrt{x - 2}}$$

Reversing the roles of the two functions gives the composite of g by f, where

$g(f(x))$ is read as "g of f of x."

$$(g \circ f)(x) = g(f(x)) = g(\sqrt{x}) = \frac{1}{\sqrt{x} - 2}$$

DEFINITION OF THE COMPOSITE OF TWO FUNCTIONS

For functions f and g the composite function g by f, denoted $g \circ f$, has range values defined by

$$(g \circ f)(x) = g(f(x))$$

and domain consisting of all x in the domain of f for which $f(x)$ is in the domain of g.

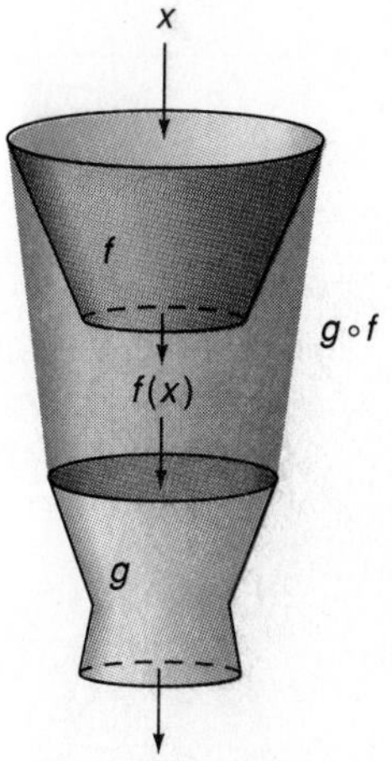

$g(f(x)) = (g \circ f)(x)$

It may help you to remember the construction of composites by looking at the following schematic diagram, as well as the figure in the margin.

$$x \xrightarrow{f} f(x) \xrightarrow{g} (f(x)) = (g \circ f)(x)$$

$$\underbrace{\hspace{6em}}_{g \circ f}$$

It is also helpful to view the composition $(g \circ f)(x) = g(f(x))$ as consisting of an "inner" function f and an "outer" function g.

EXAMPLE 1 Form the composites $g \circ f$ and $f \circ g$ where $f(x) = \dfrac{x}{x-1}$ and $g(x) = x^2$.

Solution

$$(g \circ f)(x) = g(f(x)) = g\left(\frac{x}{x-1}\right) = \left(\frac{x}{x-1}\right)^2$$

$$(f \circ g)(x) = f(g(x)) = f(x^2) = \frac{x^2}{x^2-1}$$ ■

Of particular interest to us are pairs of functions whose compositions have the special property demonstrated by the functions $f(x) = x^3$ and $g(x) = \sqrt[3]{x}$:

Function f cubes and function g "uncubes"; f and g are inverse functions.

$$(f \circ g)(x) = f(g(x)) = f(\sqrt[3]{x}) = (\sqrt[3]{x})^3 = x$$
$$(g \circ f)(x) = g(f(x)) = g(x^3) = \sqrt[3]{x^3} = x$$

In each case we obtained the same value x that we started with; whatever one of the functions does to a value x, the other function undoes. Whenever two functions act on each other in such a manner, we say that they are **inverse functions** or that either function is the inverse of the other.

DEFINITION OF INVERSE FUNCTIONS

Two functions f and g are said to be *inverse functions* if and only if:

1. For each x in the domain of g, $g(x)$ is in the domain of f and
$$(f \circ g)(x) = f(g(x)) = x$$
2. For each x in the domain of f, $f(x)$ is in the domain of g and
$$(g \circ f)(x) = g(f(x)) = x$$

f^{-1} is read as "the inverse of f" or as "f inverse."

The notation f^{-1} is also used to represent the inverse of f. Thus if $f(x) = x^3$, then $f^{-1}(x) = \sqrt[3]{x}$. Note that if $f(a) = b$, then $f^{-1}(b) = a$. For example, let $a = 2$.

$$f(x) = x^3: \qquad f(2) = 2^3 = 8$$
$$f^{-1}(x) = \sqrt[3]{x}: \qquad f^{-1}(8) = \sqrt[3]{8} = 2$$

CAUTION
In the expression $f^{-1}(x)$, -1 is not a negative exponent. Thus $f^{-1}(x) \neq \dfrac{1}{f(x)}$. The reciprocal of $f(x)$ can be correctly written in this way:
$$\frac{1}{f(x)} = [f(x)]^{-1}$$

Also, using this notation and the definition of inverse functions, we have

$$f(f^{-1}(x)) = x \quad \text{and} \quad f^{-1}(f(x)) = x$$

Thus

$$f(f^{-1}(8)) = 8 \quad \text{and} \quad f^{-1}(f(2)) = 2$$

If the paper were folded along the line $y = x$, the two curves would coincide.

If the variables in $y = x^3$ are interchanged, we obtain $x = y^3$, and solving for y gives $y = \sqrt[3]{x}$. Here are the graphs for these equations, together with the two graphs on the same axes. Because of this interchange of coordinates, the two curves are reflections of each other through the line $y = x$, as in part (c) of the figure. Another way to describe this relationship is to say that they are *mirror images* of one another through the "mirror line" $y = x$.

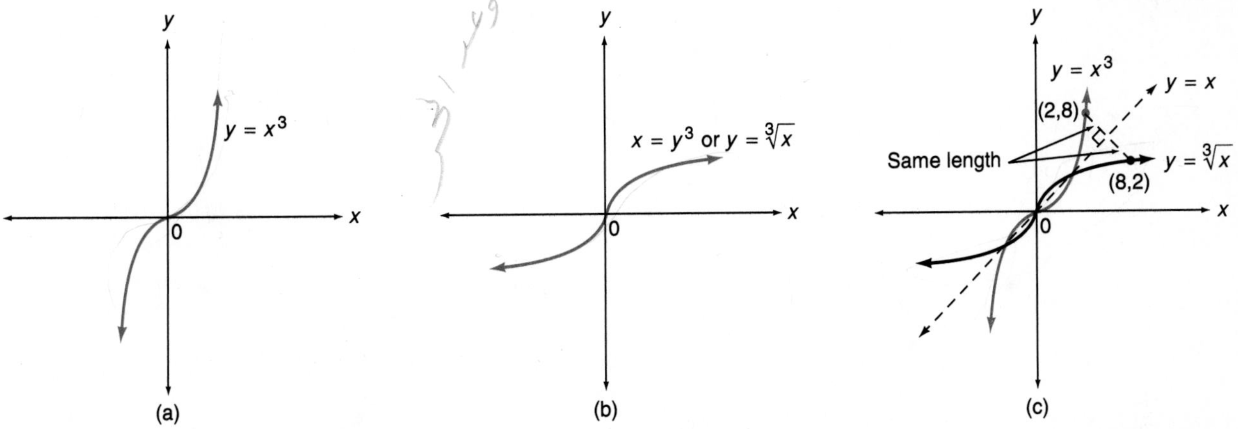

As demonstrated for the graphs of $y = x^3$ and $y = \sqrt[3]{x}$, the graphs of any pair of inverse functions are reflections of one another through the line $y = x$. Also, the method of interchanging variables to find the inverse, as demonstrated above with $y = x^3$ to find $y = \sqrt[3]{x}$, can be applied to find inverses of other functions. This is demonstrated in the next two examples.

To find the inverse g of f, begin with

$$y = f(x)$$

Then interchange variables

$$x = f(y)$$

and solve for y in terms of x, producing

$$y = g(x) = f^{-1}(x).$$

EXAMPLE 2 Find the inverse $g = f^{-1}$ of $y = f(x) = 2x + 3$. Then show that $(f \circ g)(x) = x$ and $(g \circ f)(x) = x$ and graph both functions on the same axes.

Solution Interchange variables in $y = 2x + 3$ and solve for y.

$$x = 2y + 3$$
$$2y = x - 3$$
$$y = f^{-1}(x) = \tfrac{1}{2}x - \tfrac{3}{2}$$

Using $y = g(x) = \frac{1}{2}x - \frac{3}{2}$, we have

$$\begin{aligned}(f \circ g)(x) = f(g(x)) &= f(\tfrac{1}{2}x - \tfrac{3}{2})\\ &= 2(\tfrac{1}{2}x - \tfrac{3}{2}) + 3\\ &= x - 3 + 3\\ &= x\end{aligned}$$

$$\begin{aligned}(g \circ f)(x) = g(f(x)) &= g(2x + 3)\\ &= \tfrac{1}{2}(2x + 3) - \tfrac{3}{2}\\ &= x + \tfrac{3}{2} - \tfrac{3}{2}\\ &= x\end{aligned}$$

The procedure used in Examples 2 and 3 to find the inverse has its limitations. Thus, if $y = f(x)$ is complicated, it may be algebraically difficult, or even impossible, to interchange variables and solve for the inverse. We will avoid such situations by limiting our work in this section to functions for which the procedure can be applied.

The graph of both functions are shown on the following page.

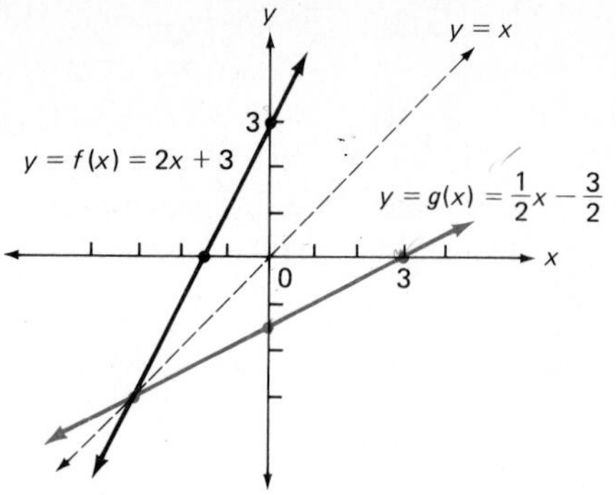

■

EXAMPLE 3 Find f^{-1} for the function $y = f(x) = \sqrt{x}$, graph both f and f^{-1} on the same axes, and verify that $(f \circ f^{-1})(x) = x$ and $(f^{-1} \circ f)(x) = x$.

Solution Interchange variables in $y = \sqrt{x}$ to get $x = \sqrt{y}$. At this point we see that x cannot be negative: $x \geq 0$. Solving for y by squaring produces $y = x^2$. Therefore the inverse function is $y = f^{-1}(x) = x^2$ with domain $x \geq 0$.

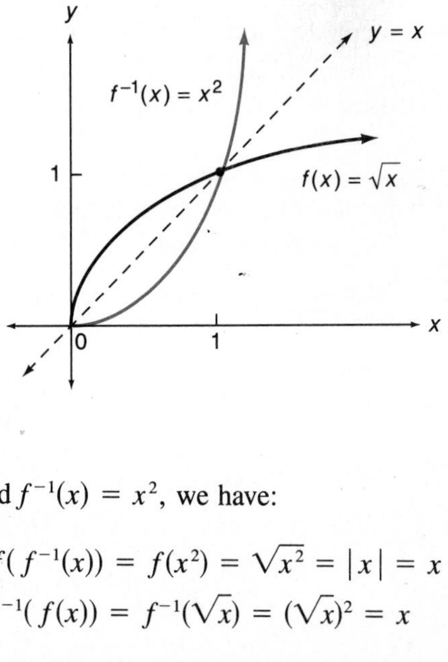

Using $f(x) = \sqrt{x}$ and $f^{-1}(x) = x^2$, we have:

$$(f \circ f^{-1})(x) = f(f^{-1}(x)) = f(x^2) = \sqrt{x^2} = |x| = x \qquad \text{(since } x \geq 0\text{)}$$

$$(f^{-1} \circ f)(x) = f^{-1}(f(x)) = f^{-1}(\sqrt{x}) = (\sqrt{x})^2 = x$$

■

Some functions do not have an inverse function; $y = x^2$, whose domain is *all* real numbers x, is such a function. You can see this by interchanging variables to obtain $x = y^2$. The graph is a horizontal parabola that cannot represent a function because, for $x > 0$, there are two y-values (see page 294). This occurs because for the given function $y = x^2$, there are two x-values for the same range value y as shown in the figure at the left on the next page. When to each range value y there corresponds only one domain value x, as for $y = x^3$, we say that the function is **one-to-one**. A *one-to-one function* has an inverse function; other kinds of functions do not.

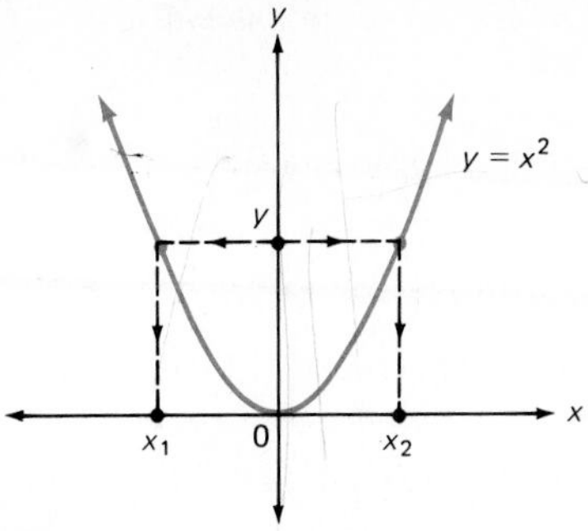

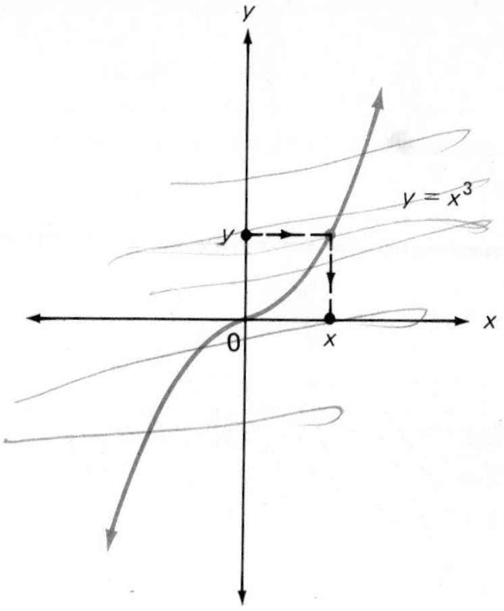

$y = x^2$ *is not* a one-to-one function. There are two domain values for a range value $y > 0$.

$y = x^3$ *is* a one-to-one function. There is exactly one domain value x for each range value y.

Once a graph of a function is known, there is a simple *horizontal line test* for determining the one-to-one property. Consider a horizontal line through each range value y, as in the preceding figures. If the line meets the curve exactly once, then we have a one-to-one function; otherwise, it is not one-to-one.

EXERCISES 9.1

In Exercises 1 and 2, let $f(x) = 2x - 3$ *and* $g(x) = 3x + 2$.

1. **(a)** Find $g(0)$ and $f(g(0))$. **(b)** Find $(f \circ g)(x)$. **(c)** Use the result in part (b) to evaluate $(f \circ g)(0)$.
2. **(a)** Find $f(0)$ and $g(f(0))$. **(b)** Find $(g \circ f)(x)$. **(c)** Use the result in part (b) to evaluate $(g \circ f)(0)$.

For each pair of functions, find the following:
(a) $f(g(6))$ **(b)** $f(g(8))$ **(c)** $(f \circ g)(x)$

3. $f(x) = x^2$, $g(x) = \sqrt{x}$
4. $f(x) = 5x - 1$, $g(x) = \dfrac{5}{1 + 3x}$
5. $f(x) = x^3 - 1$, $g(x) = \dfrac{1}{x}$
6. $f(x) = 3x - 1$, $g(x) = \frac{1}{3}x + \frac{1}{3}$
7. $f(x) = x^2 + 6x + 8$, $g(x) = \sqrt{x - 2}$
8. $f(x) = \sqrt[3]{x}$, $g(x) = x^2$

For each pair of functions, find $(f \circ g)(x)$ *and* $(g \circ f)(x)$.

9. $f(x) = -2x + 5$, $g(x) = 4x - 1$
10. $f(x) = |x|$, $g(x) = 3|x|$
11. $f(x) = 2x^2 - 1$, $g(x) = \dfrac{1}{2x}$
12. $f(x) = \sqrt{2x + 3}$, $g(x) = x^2 - 1$
13. $f(x) = x^2$, $g(x) = x - 1$
14. $f(x) = |x - 3|$, $g(x) = 2x + 3$
15. $f(x) = \dfrac{x}{x - 2}$, $g(x) = \dfrac{x + 3}{x}$
16. $f(x) = x^3 - 1$, $g(x) = \dfrac{1}{x^3 + 1}$
17. $f(x) = \sqrt{x + 1}$, $g(x) = x^4 - 1$
18. $f(x) = 2x^3 - 1$, $g(x) = \sqrt[3]{\dfrac{x + 1}{2}}$
19. $f(x) = \sqrt{x}$, $g(x) = 4$
20. $f(x) = \sqrt[3]{1 - x}$, $g(x) = 1 - x^3$

Use the horizontal line test to decide which of the following are one-to-one functions.

21. $y = x^2 - 2x + 1$ 22. $y = |x|$ 23. $y = \frac{1}{x}$

24. $y = 2x + 1$ 25. $y = \frac{1}{x^2}$ 26. $y = -x$

Show that f and g are inverse functions according to the criteria $(f \circ g)(x) = x$ and $(g \circ f)(x) = x$. Then graph both functions and the line $y = x$ on the same axes.

27. $f(x) = \frac{1}{3}x - 3$; $g(x) = 3x + 9$ 28. $f(x) = 2x - 6$; $g(x) = \frac{1}{2}x + 3$
29. $f(x) = (x + 1)^3$; $g(x) = \sqrt[3]{x} - 1$ 30. $f(x) = -(x + 2)^3$; $g(x) = -\sqrt[3]{x} - 2$

Find the inverse function g of the given function f.

31. $y = f(x) = (x - 5)^3$ 32. $y = f(x) = x^{1/3} - 3$ 33. $y = f(x) = \frac{2}{3}x - 1$
34. $y = f(x) = -4x + \frac{2}{5}$ 35. $y = f(x) = (x - 1)^5$ 36. $y = f(x) = -x^5$

Find $f^{-1}(x)$ and show that $f(f^{-1}(x)) = x$ and $f^{-1}(f(x)) = x$.

37. $f(x) = \frac{2}{x - 2}$ 38. $f(x) = -\frac{1}{x} - 1$ 39. $f(x) = \frac{3}{x + 2}$

40. $f(x) = \frac{x}{x + 1}$ 41. $f(x) = x^{-5}$ 42. $f(x) = \frac{1}{\sqrt[3]{x - 2}}$

Verify that the function is its own inverse by showing that $(f \circ f)(x) = x$.

43. $f(x) = \frac{1}{x}$ 44. $f(x) = \sqrt{4 - x^2}$; $0 \le x \le 2$ 45. $f(x) = \frac{x}{x - 1}$ 46. $f(x) = \frac{3x - 8}{x - 3}$

Find the inverse g of the given function f, and graph both in the same coordinate system.

47. $y = f(x) = (x + 1)^2$; $x \ge -1$ 48. $y = f(x) = x^2 - 4x + 4$; $x \ge 2$

49. $y = f(x) = \frac{1}{\sqrt{x}}$ 50. $y = f(x) = -\sqrt{x}$

*51. Aside from the linear function $y = x$, what other linear functions are their own inverses? (*Hint:* Inverse functions are reflections of one another through the line $y = x$.)

*52. Let $f(x) = x$. Find $(f \circ g)(x)$ and $(g \circ f)(x)$ for any function g.

The composition of two functions can be extended to the composition of three functions by defining $(f \circ g \circ h)(x) = f(g(h(x)))$. Use this definition in Exercises 53–56.

53. Let $f(x) = \frac{1}{x}$, $g(x) = 2x - 1$, and $h(x) = x^{1/3}$. Find the following:
 (a) $(f \circ g \circ h)(x)$ **(b)** $(g \circ f \circ h)(x)$ **(c)** $(h \circ f \circ g)(x)$

54. Let $f(x) = x + 2$, $g(x) = \sqrt{x}$, and $h(x) = x^3$. Find the following:
 (a) $(f \circ g \circ h)(x)$ **(b)** $(f \circ h \circ g)(x)$ **(c)** $(g \circ f \circ h)(x)$

55. Let $f(x) = x^2$, $g(x) = \frac{1}{x - 1}$, and $h(x) = 1 + \frac{1}{x}$. Find the following:
 (a) $(f \circ h \circ g)(x)$ **(b)** $(g \circ h \circ f)(x)$ **(c)** $(h \circ g \circ f)(x)$

56. Let $f(x) = \frac{1}{x}$. Find $(f \circ f)(x)$ and $(f \circ f \circ f)(x)$.

Written Assignment: Explain, with a specific example, what is meant by a one-to-one function. Also, give an example of a function that is *not* one-to-one and explain why it is not.

9.2 EXPONENTIAL FUNCTIONS

Perhaps you have heard that some populations grow exponentially. Many life forms, from rabbits to bacteria, reproduce at an exponential rate. Even if you don't know the technical meaning of this, you probably do know that it means that these populations are increasing very rapidly. **Exponential growth** is one of the subjects that will be studied in this chapter.

Note that the concept of exponential growth and decay was first introduced in Chapter 2. See page 37.

For illustrative purposes, imagine that a bacterial culture is growing at such a rate that after each hour the number of bacteria has doubled. Thus, if there were 10,000 bacteria present when the culture first started to grow, then after 1 hour it would have grown to 20,000, after 2 hours there would be 40,000 bacteria, and so on. Now consider these equalities:

$$20{,}000 = (10{,}000)(2^1)$$
$$40{,}000 = (10{,}000)(2^2)$$
$$80{,}000 = (10{,}000)(2^3)$$

At this point it becomes reasonable to say that the equation

$$y = f(x) = (10{,}000)2^x$$

gives the number y of bacteria present after x hours. This equation defines an *exponential function* with independent variable x and dependent variable y.

A function such as $f(x) = b^x$, with the variable as an exponent, is known as an **exponential function**. We shall study such functions, with the assumption that the *base number* $b > 0$. For example, let us consider the function $y = f(x) = 2^x$ and its graph. Note the following:

We use $b > 0$ in order to avoid even roots of negative numbers, such as $(-4)^{1/2} = \sqrt{-4}$.

1. The function is defined for all real values of x. When x is negative, we may apply the definition for negative exponents. Thus, for $x = -2$,

$$2^x = 2^{-2} = \frac{1}{2^2} = \frac{1}{4}$$

The domain of the function is the set of real numbers.

2. For all replacements of x, the function takes on a positive value. That is, 2^x can never represent a negative number, nor can 2^x be equal to 0. The range of the function is the set of positive real numbers.
3. Finally, a few specific ordered pairs of numbers can be located as an aid to graphing, as shown on the following page.

The function is increasing and the curve is concave up. The x-axis is a horizontal asymptote toward the left. All curves $y = b^x$ for $b > 1$ have this basic shape.

x	$y = 2^x$
-3	$\frac{1}{8}$
-2	$\frac{1}{4}$
-1	$\frac{1}{2}$
0	1
1	2
2	4
3	8

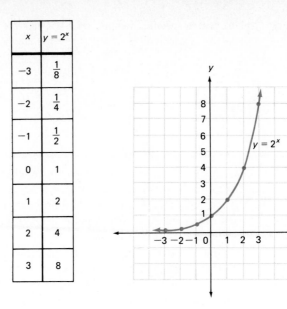

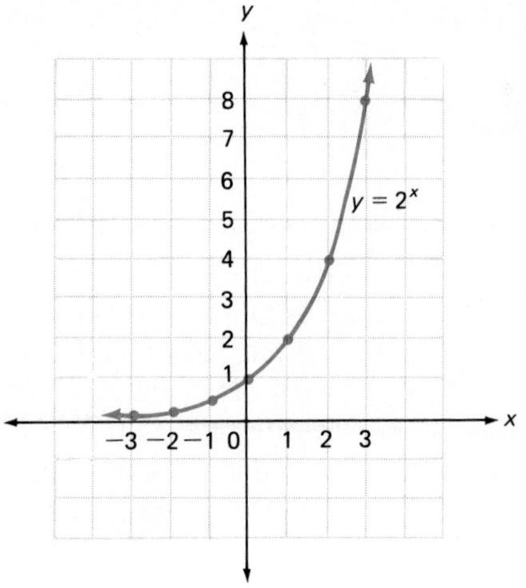

The value of $\sqrt{2}$ is given correct to two decimal places from Table 1.

If desired, the accuracy of this graph can be improved by using more points. For example, consider such rational values of x as $\frac{1}{2}$ or $\frac{3}{2}$:

$$2^{1/2} = \sqrt{2} \approx 1.4$$
$$2^{3/2} = (\sqrt{2})^3 \approx 2.8$$

Using irrational values for x, such as $\sqrt{2}$ or π, is another matter entirely. To give a precise meaning of such numbers is beyond the scope of this course. However, using a scientific calculator you can obtain approximations such as $(\sqrt{2}, 2^{\sqrt{2}}) \approx (1.41, 2.67)$ and $(\sqrt{5}, 2^{\sqrt{5}}) \approx (2.24, 4.71)$, and then verify that these points fit the curve.

When graphing $y = 2^x$ we made use of the property $2^{-n} = \dfrac{1}{2^n}$. This and other properties listed below have been studied previously for rational exponents r and s.

$$b^r b^s = b^{r+s} \qquad \frac{b^r}{b^s} = b^{r-s} \qquad (b^r)^s = b^{rs}$$

$$a^r b^r = (ab)^r \qquad b^0 = 1 \qquad b^{-r} = \frac{1}{b^r}$$

In more advanced work it can be shown that these rules hold for positive bases a and b and all real numbers r and s.

EXAMPLE 1 Graph the curve $y = 8^x$ on the interval $[-1, 1]$ by using a table of values.

Solution

x	y
-1	$\frac{1}{8}$
$-\frac{2}{3}$	$\frac{1}{4}$
$-\frac{1}{3}$	$\frac{1}{2}$
0	1
$\frac{1}{3}$	2
$\frac{2}{3}$	4
1	8

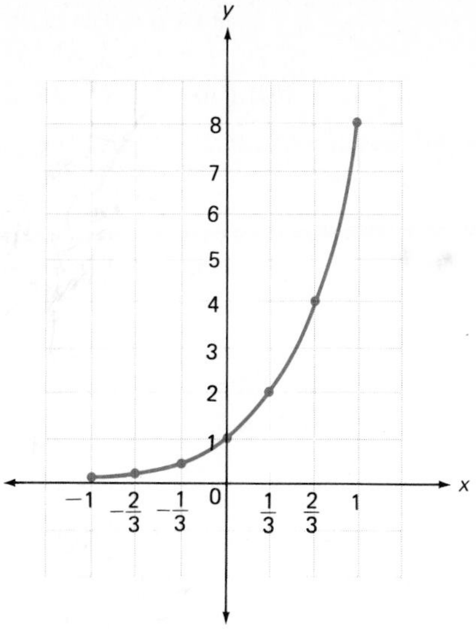

Recall that the decimal form of an irrational number is endless and nonrepeating.

The most important base number for exponential functions is the number denoted by the letter e. This is an irrational number approximately equal to 2.72. Here is the number e showing its first 15 decimal places:

$$e = 2.718281828459045\ldots$$

The larger n is taken, the closer $\left(1 + \frac{1}{n}\right)^n$ *gets to e.*
For example:
$(1 + \frac{1}{10})^{10} = 2.59374$
$(1 + \frac{1}{100})^{100} = 2.70481$
$(1 + \frac{1}{1000})^{1000} = 2.71692$
(See Exercise 38.)

Like the irrational number π, e also plays an important role in mathematics and its applications.

One way to approximate e is to evaluate the expression $\left(1 + \frac{1}{n}\right)^n$ for larger and larger values of n. Some of these calculations are shown in the margin. Since $2 < e < 3$, the graph of $y = e^x$ will be between the graphs of $y = 2^x$ and $y = 3^x$ as shown below.

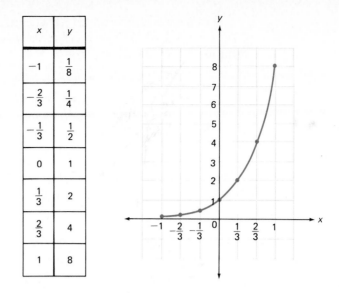

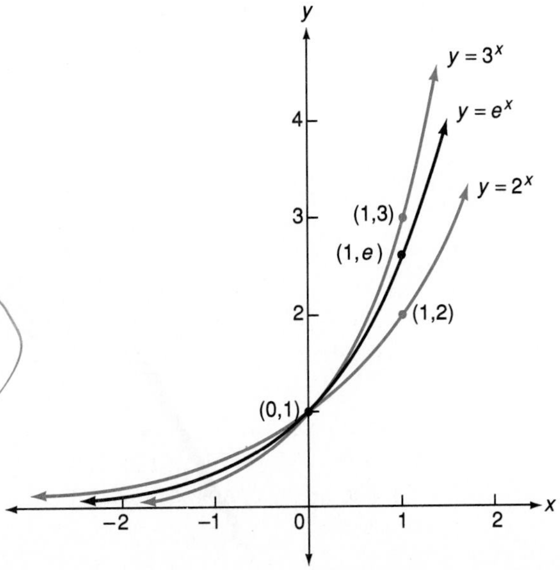

Thus far we have restricted our attention to exponential functions of the form $y = f(x) = b^x$, where $b > 1$. All of these have graphs that are of the same general shape as that for $y = 2^x$. For $b = 1$, $y = b^x = 1^x$ for all x. Since this is a constant function, $f(x) = 1$, we do not use the base $b = 1$ in the classification of exponential functions.

Now let us explore exponential functions $y = f(x) = b^x$ for which $0 < b < 1$. In particular, if $b = \frac{1}{2}$, we get $y = \left(\frac{1}{2}\right)^x = \frac{1}{2^x}$ or $y = 2^{-x}$.

All curves $y = b^x$ for $0 < b < 1$ have this same basic shape. The curve is concave up, the function is decreasing, and $y = 0$ is a horizontal asymptote toward the right.

x	$y = 2^{-x}$
−3	8
−2	4
−1	2
0	1
1	$\frac{1}{2}$
2	$\frac{1}{4}$
3	$\frac{1}{8}$

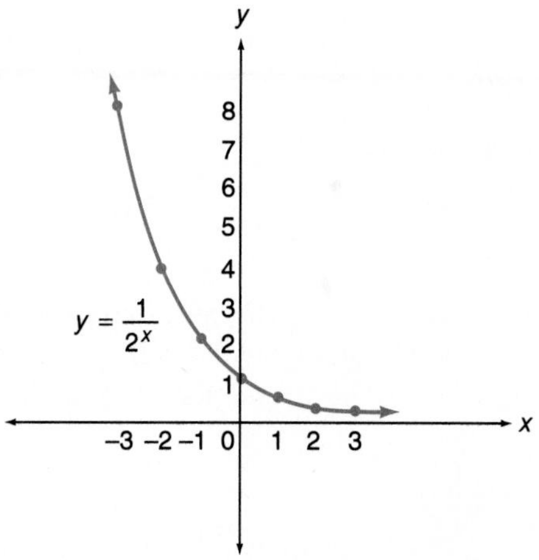

As shown below, the graph of $y = g(x) = \frac{1}{2^x}$ can also be found by comparing it to the graph of $y = f(x) = 2^x$. Since $g(x) = \frac{1}{2^x} = 2^{-x} = f(-x)$, the y-values for g are the same as the y-values for f *on the opposite side of the y-axis*. In other words, the graph of g is the *reflection* of the graph of f through the y-axis.

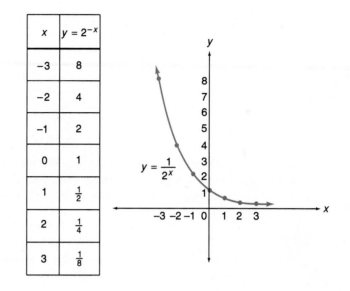

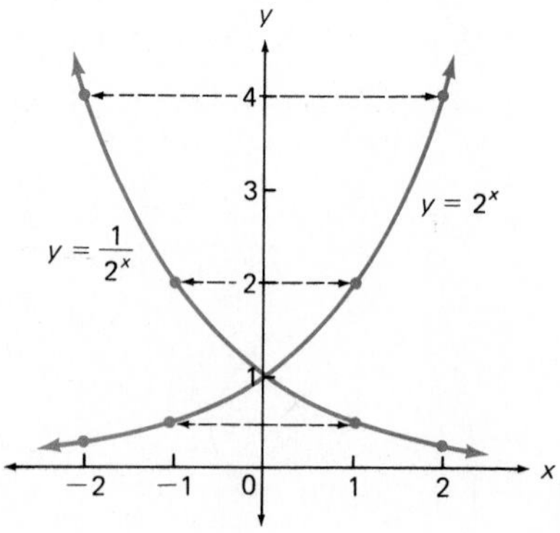

The next example demonstrates how the graphs of exponential functions can be translated.

EXAMPLE 2 Use the graph of $y = f(x) = 2^x$ to sketch the curves

$$y = g(x) = 2^{x-3} \quad \text{and} \quad y = h(x) = 2^x - 1$$

Solution Since $g(x) = f(x - 3)$, the graph of g can be obtained by shifting $y = 2^x$ by 3 units to the right. Moreover, since $h(x) = f(x) - 1$ the graph of h can be found by shifting $y = 2^x$ down 1 unit.

The graph of g is obtained by the ***translation*** *of the graph of f three units to the right. The graph of h is found by translating the graph of f one unit down.*

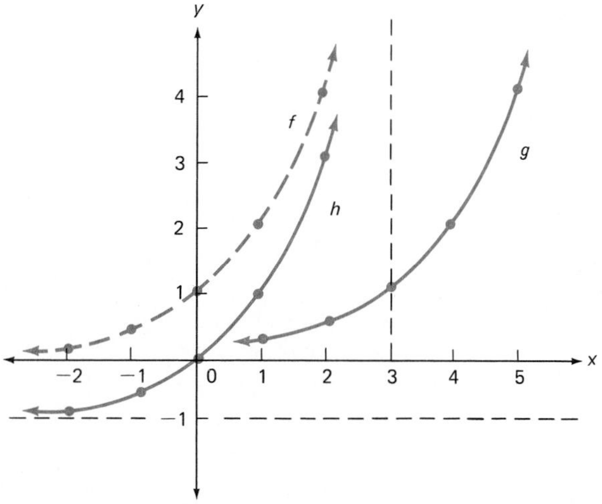

■

We have explored functions of the form $y = f(x) = b^x$ for specific values of b. In each case you should note that the graphs went through the point (0, 1) since $y = b^0 = 1$. Also, each such graph has the x-axis as a one-sided asymptote and there are no x-intercepts. These and other properties are summarized below.

PROPERTIES OF $y = f(x) = b^x$ for $b > 0$ and $b \neq 1$

1. The domain consists of all real numbers x.
2. The range consists of all positive numbers y.
3. The function is increasing (the curve is rising) when $b > 1$, and it is decreasing (the curve is falling) when $0 < b < 1$.
4. The curve is concave up for $b > 1$ and for $0 < b < 1$.
5. It is a one-to-one function.
6. The point (0, 1) is on the curve. There is no x-intercept.
7. The x-axis is a horizontal asymptote to the curve, toward the left for $b > 1$ and toward the right for $0 < b < 1$.
8. $b^{x_1}b^{x_2} = b^{x_1+x_2}$; $b^{x_1}/b^{x_2} = b^{x_1-x_2}$; $(b^{x_1})^{x_2} = b^{x_1x_2}$.

This form of the one-to-one property can sometimes be applied to the solutions of equations.

The one-to-one property of a function f may be stated in this way:

If $f(x_1) = f(x_2)$, then $x_1 = x_2$.

That is, since $f(x_1)$ and $f(x_2)$ represent the same range value there can only be one corresponding domain value; consequently, $x_1 = x_2$. Using $f(x) = b^x$, this statement means the following:

$$\text{If } b^{x_1} = b^{x_2}, \text{ then } x_1 = x_2.$$

CAUTION
a^{b^c} means $a^{(b^c)}$, and $(a^b)^c = a^{bc}$. Thus, in general $a^{b^c} = a^{(b^c)} \neq (a^b)^c$.

This property can be used to solve certain **exponential equations**, such as $5^{x^2} = 625$. First note that 625 can be written as 5^4.

$$5^{x^2} = 625$$
$$5^{x^2} = 5^4$$

By the one-to-one property applied to the function $f(t) = 5^t$, we may equate the exponents and solve for x.

$$x^2 = 4$$
$$x = \pm 2 \qquad (x = 2 \text{ or } x = -2)$$

To check these solutions, note that $5^{2^2} = 5^4 = 625$ and $5^{(-2)^2} = 5^4 = 625$.

The following examples illustrate further use of the one-to-one property to solve exponential equations.

EXAMPLE 3 Solve for x: $\dfrac{1}{3^{x-1}} = 81$

Solution Write 81 as 3^4 and $\dfrac{1}{3^{x-1}}$ as $3^{-(x-1)}$.

Note that the one-to-one property here is being applied to the function $f(t) = 3^t$.

$$3^{-(x-1)} = 3^4$$
$$-(x - 1) = 4 \qquad \text{(by the one-to-one property)}$$
$$-x + 1 = 4$$
$$-x = 3$$
$$x = -3$$

Check this result in the original equation. ■

EXAMPLE 4 Solve for x: $b^{x^2-x} = 1$

Solution Note that 1 can be written in the form b^0. Thus we have

$$b^{x^2-x} = b^0$$
$$x^2 - x = 0 \qquad (\text{If } b^{x_1} = b^{x_2}, \text{ then } x_1 = x_2.)$$
$$x(x - 1) = 0$$
$$x = 0 \quad \text{or} \quad x = 1$$

Check both of these results in the original equation. ■

TEST YOUR UNDERSTANDING
Think Carefully

Solve for x.

1. $2^{x-1} = 32$ **2.** $2^{x^2} = 16$ **3.** $8^{2x+1} = 64$

4. $\frac{1}{2^x} = 64$ **5.** $\frac{1}{5^{x+1}} = 125$ **6.** $\frac{1}{4^{x-2}} = 64$

7. $e^{3x+1} = 1$ **8.** $27^x = 9$ **9.** $125^x = 25$

10. $\left(\frac{1}{e^2}\right)^x = \frac{e}{e^{12x+26}}$ **11.** $\left(\frac{3}{5}\right)^x = \frac{27}{125}$ **12.** $\left(\frac{9}{25}\right)^x = \frac{5}{3}$

(Answers: Page 424)

EXERCISES 9.2

Graph the exponential function f by making use of a brief table of values. Then use this curve to sketch the graph of g. Indicate the horizontal asymptotes.

1. $f(x) = 2^x \quad g(x) = 2^{x+3}$
2. $f(x) = 3^x \quad g(x) = 3^x - 2$
3. $f(x) = 4^x \quad g(x) = -(4^x)$
4. $f(x) = 5^x \quad g(x) = (\frac{1}{5})^x$
5. $f(x) = (\frac{3}{2})^x \quad g(x) = (\frac{3}{2})^{-x}$
6. $f(x) = 8^x \quad g(x) = 8^{x-2} + 3$
7. $f(x) = 3^x \quad g(x) = 2(3^x)$
8. $f(x) = 3^x \quad g(x) = \frac{1}{2}(3^x)$

Sketch the curves on the same axes.

9. $y = e^x \quad y = e^{x-2}$
10. $y = e^x \quad y = 2e^x$
11. $y = e^x \quad y = e^x + 2$
12. $y = e^x \quad y = e^{-x}$
13. $y = (\frac{3}{2})^x \quad y = 2^x, \; y = (\frac{5}{2})^x$
14. $y = (\frac{1}{4})^x \quad y = (\frac{1}{3})^x \quad y = (\frac{1}{2})^x$

Use the one-to-one property of an appropriate exponential function to solve the indicated equation.

15. $2^x = 64$
16. $3^x = 81$
17. $2^{x^2} = 512$
18. $3^{x-1} = 27$
19. $5^{2x+1} = 125$
20. $2^{x^3} = 256$
21. $7^{x^2+x} = 49$
22. $b^{x^2+x} = 1$
23. $\frac{1}{2^x} = 32$
24. $\frac{1}{10^x} = 10{,}000$
25. $9^x = 3$
26. $64^x = 8$
27. $9^x = 27$
28. $64^x = 16$
29. $(\frac{1}{49})^x = 7$
30. $5^x = \frac{1}{125}$
31. $(\frac{27}{8})^x = \frac{9}{4}$
32. $(0.01)^x = 1000$
33. $e^{x^2} = e^x e^{3/4}$
34. $\frac{1}{e^{2x}} = e^{3x-1}$
35. Graph the functions $y = 2^x$ and $y = x^2$ in the same coordinate system for the interval $[0, 5]$. (Use a larger unit on the x-axis than on the y-axis.) What are the points of intersection?

*36. Solve for x: $(6^{2x})(4^x) = 1728$. ***37.** Solve for x: $(5^{2x+1})(7^{2x}) = 175$.

38. Use a calculator to complete the table. Round off the entries to four decimal places.

n	2	10	100	500	5000	10,000	100,000	1,000,000
$\left(1 + \frac{1}{n}\right)^n$								

Written Assignment: Explain why the function $f(x) = x^2$ is *not* an exponential function.

9.3 LOGARITHMIC FUNCTIONS

One of the properties of the exponential function $y = b^x$ is that it is a one-to-one function. This means that it has an inverse function whose graph can be obtained by reflecting the graph of $y = b^x$ through the line $y = x$. This has been done below for the case when $b > 1$ and also when $0 < b < 1$ in the following graph.

Recall from Section 9.1 that $f^{-1}(x)$ is a notation used to represent the inverse of a function f.

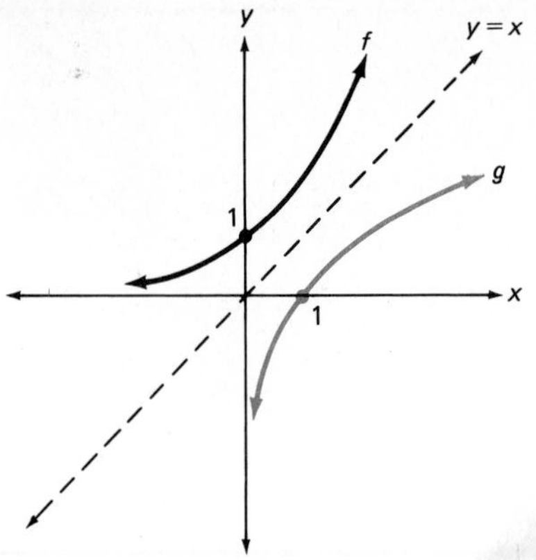

$y = f(x) = b^x \ (b > 1)$
$g = f^{-1}(x)$

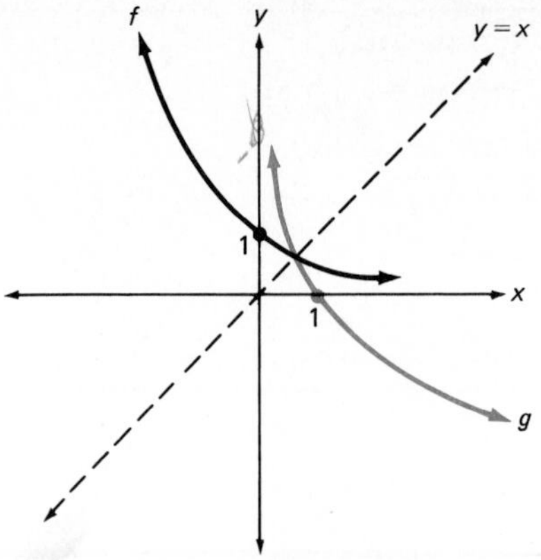

$y = f(x) = b^x \ (0 < b < 1)$
$g = f^{-1}(x)$

An equation for the inverse function g can be obtained by interchanging the roles of the variables as follows:

$$\text{Function, } f: \quad y = f(x) = b^x$$

$$\text{Inverse function, } g: \quad x = g(y) = b^y$$

Thus $x = b^y$ is an equation for g. Unfortunately, we have no way of solving $x = b^y$ to get y explicitly in terms of x. To overcome this difficulty we create some new terminology.

The equation $x = b^y$ tells us that *y is the exponent on b that produces x*. In situations like this the word **logarithm** is used in place of *exponent*. A logarithm, then, is an exponent. Now we may say that *y is the logarithm on b that produces x*. This description can be abbreviated to $y = \text{logarithm}_b\, x$ and abbreviating further we reach the final form

$$y = \log_b x$$

which is read "y equals log x to the base b" or "y equals log x base b."

It is important to realize that we are only defining (not proving) the equation $y = \log_b x$ to have the same meaning as $x = b^y$. In other words, these two forms are equivalent:

Exponential form:	$x = b^y$
Logarithmic form:	$y = \log_b x$

And since they are equivalent they define the same function g:

$$y = g(x) = \log_b x$$

Now we know that the exponential function $y = b^x$ and the **logarithmic function** $y = \log_b x$ are inverse functions. Consequently, we have the following:

$$f(g(x)) = f(\log_b x) = b^{\log_b x} = x \quad \text{and} \quad g(f(x)) = g(b^x) = \log_b(b^x) = x$$

Recall from Section 9.1 that for inverse functions, $f(g(x)) = x$, and $g(f(x)) = x$.

EXAMPLE 1 Write the equation of the inverse function g of $y = f(x) = 2^x$ and graph both on the same axes.

Solution The inverse g has equation $y = g(x) = \log_2 x$, and its graph can be obtained by reflecting $y = f(x) = 2^x$ through the line $y = x$.

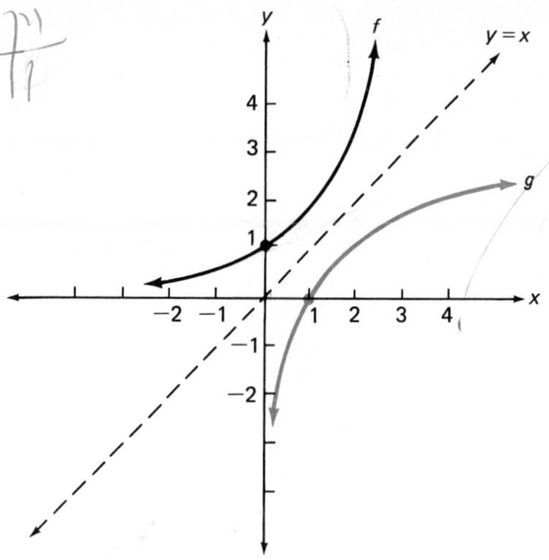

TEST YOUR UNDERSTANDING
Think Carefully

1. Find the equation of the inverse of $y = 3^x$ and graph both on the same axes.
2. Find the equation of the inverse of $y = (\frac{1}{3})^x$ and graph both on the same axes.

Let $y = f(x) = \log_5 x$. Describe how the graph of each of the following can be obtained from the graph of f.

3. $g(x) = \log_5(x + 2)$
4. $g(x) = 2 + \log_5 x$
5. $g(x) = -\log_5 x$
6. $g(x) = 2 \log_5 x$

(Answers: Page 424)

We found $y = \log_b x$ by interchanging the roles of the variables in $y = b^x$. As a consequence of this switching, the domains and ranges of the two functions are also interchanged. Thus

Domain of $y = \log_b x$ is the same as the range of $y = b^x$.

Range of $y = \log_b x$ is the same as the domain of $y = b^x$.

These results are incorporated into the following list of important properties.

PROPERTIES OF $y = f(x) = \log_b x$ for $b > 0$ and $b \neq 1$

1. The domain consists of all positive numbers x.
2. The range consists of all real numbers y.
3. The function increases (the curve is rising) for $b > 1$, and it decreases (the curve is falling) for $0 < b < 1$.
4. The curve is concave down for $b > 1$, and it is concave up for $0 < b < 1$.
5. It is one-to-one function; if $\log_b(x_1) = \log_b(x_2)$, then $x_1 = x_2$.
6. The point (1, 0) is on the graph. There is no y-intercept.
7. The y-axis is a vertical asymptote to the curve in the downward direction for $b > 1$ and in the upward direction for $0 < b < 1$.
8. $\log_b(b^x) = x$ and $b^{\log_b x} = x$.

EXAMPLE 2 Find the domain for $y = \log_2(x - 3)$.

Solution In $y = \log_2(x - 3)$ the quantity $x - 3$ plays the role that x does in $\log_2 x$. Thus $x - 3 > 0$ and the domain consists of all $x > 3$. ■

Since the base e is of special importance in mathematics, $\log_e x$ is given the special notation **ln *x***; that is, $\log_e x = \ln x$. Also, $y = \ln x$ is called the **natural logarithm function** and $\ln x$ is read as the "*natural log of x*."

When the results listed in Property 8 in the preceding table are written for the case $b = e$, we obtain

$f(x) = e^x$ and $g(x) = \ln x$ are inverse functions.

$$\ln e^x = x \quad \text{and} \quad e^{\ln x} = x$$

EXAMPLE 3 Solve for t: $e^{\ln(2t-1)} = 5$

Solution

$$\begin{aligned} e^{\ln(2t-1)} &= 5 \\ 2t - 1 &= 5 \qquad (e^{\ln x} = x) \\ 2t &= 6 \\ t &= 3 \end{aligned}$$

■

As noted earlier, the forms $\log_b x = y$ and $b^y = x$ are equivalent. The following table lists specific cases illustrating this equivalence.

CAUTION
*Do not confuse $x = b^y$ with its **inverse** $y = b^x$. These two forms are **not** equivalent.*

Logarithmic form $\log_b x = y$	Exponential form $b^y = x$
$\log_5 25 = 2$	$5^2 = 25$
$\log_{27} 9 = \frac{2}{3}$	$27^{2/3} = 9$
$\log_6 \frac{1}{36} = -2$	$6^{-2} = \frac{1}{36}$
$\ln \sqrt{e} = \frac{1}{2}$	$e^{1/2} = \sqrt{e}$

There are two special cases that apply to any base b:

$$\log_b 1 = 0 \qquad (\text{since } b^0 = 1)$$

$$\log_b b = 1 \qquad (\text{since } b^1 = b)$$

When one of x, y, or b in $\log_b x = y$ is unknown, it is often useful to convert to exponential form in order to find the unknown.

EXAMPLE 4 Evaluate: $\log_9 27$

Solution First write $\log_9 27 = y$ and convert to exponential form.

$$9^y = 27$$
$$(3^2)^y = 3^3 \quad \text{(Express each side as a power of 3.)}$$
$$3^{2y} = 3^3$$
$$2y = 3 \quad \text{(by the one-to-one property)}$$
$$y = \tfrac{3}{2}$$

Check: $\log_9 27 = \frac{3}{2}$ is correct since $9^{3/2} = 27$. ■

EXAMPLE 5 Solve for b: $\log_b 8 = \frac{3}{4}$

Solution Convert to exponential form.

$$b^{3/4} = 8$$

Raise both sides to the $\frac{4}{3}$ power.

$$(b^{3/4})^{4/3} = 8^{4/3}$$
$$b = (\sqrt[3]{8})^4 = 16$$

Check: $\log_{16} 8 = \frac{3}{4}$ since $16^{3/4} = (\sqrt[4]{16})^3 = 8$. ■

EXAMPLE 6 Solve for x: $\log_{49} x = -\frac{1}{2}$

Solution Convert to exponential form.

$$x = 49^{-1/2} = \frac{1}{\sqrt{49}} = \frac{1}{7} \quad \text{(Check this solution.)}$$ ■

EXAMPLE 7 Solve t: $e^{2t-1} = 5$

Solution Rewrite the exponential expression in logarithmic form.

$$e^{2t-1} = 5$$
$$2t - 1 = \ln 5 \quad (\log_e 5 = \ln 5 = 2t - 1)$$
$$2t = 1 + \ln 5$$
$$t = \tfrac{1}{2}(1 + \ln 5)$$

To three decimal places, $t = \frac{1}{2}(1 + \ln 5) = 1.305$

Check: $e^{2[1/2(1+\ln 5)]-1} = e^{1+\ln 5-1} = e^{\ln 5} = 5$ ■

EXERCISES 9.3

Sketch the graph of the function f. Reflect this curve through the line $y = x$ to obtain the graph of the inverse function g, and write the equation for g.

1. $y = f(x) = 4^x$ 2. $y = f(x) = 5^x$ 3. $y = f(x) = (\frac{1}{3})^x$ 4. $y = f(x) = (0.2)^x$

Describe how the graph of h can be obtained from the graph of g. Find the domain of h, and write the equation of the vertical asymptote.

5. $g(x) = \log_3 x;\ h(x) = \log_3(x + 2)$
6. $g(x) = \log_5 x;\ h(x) = \log_5(x - 1)$
7. $g(x) = \log_8 x;\ h(x) = 2 + \log_8 x$
8. $g(x) = \log_{10} x;\ h(x) = 2 \log_{10} x$

Sketch the graph of f and state its domain.

9. $f(x) = \log_{10} x$
10. $f(x) = -\log_{10} x$
11. $f(x) = |\log_{10} x|$
12. $f(x) = \log_{10}(-x)$
13. $f(x) = \log_{10}|x|$
14. $f(x) = \log_{1/10}(x + 1)$

Sketch each pair of functions on the same axes.

15. $y = \ln x;\ y = \frac{1}{2} \ln x$
16. $y = \ln x;\ y = \ln(x + 2)$
17. $y = \ln x;\ y = \ln(-x)$
18. $y = \ln x;\ y = \ln|x|$

Simplify.

19. $e^{\ln\sqrt{x}}$
20. $\ln(e^{3x})$
21. $e^{-2\ln x}$
22. $(e^{\ln x})^2$

Convert from the exponential to the logarithmic form.

23. $2^8 = 256$
24. $5^{-3} = \frac{1}{125}$
25. $(\frac{1}{3})^{-1} = 3$
26. $81^{3/4} = 27$
27. $17^0 = 1$
28. $(\frac{1}{49})^{-1/2} = 7$

Convert from the logarithmic form to the exponential form.

29. $\log_{10} 0.0001 = -4$
30. $\log_{64} 4 = \frac{1}{3}$
31. $\log_{\sqrt{2}} 2 = 2$
32. $\log_{13} 13 = 1$
33. $\log_{12} \frac{1}{1728} = -3$
34. $\log_{27/8} \frac{9}{4} = \frac{2}{3}$

Solve for the indicated quantity: y, x, or b.

35. $\log_2 16 = y$
36. $\log_{1/2} 32 = y$
37. $\log_{1/3} 27 = y$
38. $\log_7 x = -2$
39. $\log_{1/6} x = 3$
40. $\log_8 x = -\frac{2}{3}$
41. $\log_b 125 = 3$
42. $\log_b 8 = \frac{3}{2}$
43. $\log_b \frac{1}{8} = -\frac{3}{2}$
44. $\log_{100} 10 = y$
45. $\log_{27} 3 = y$
46. $\log_{1/16} x = \frac{1}{4}$
47. $\log_b \frac{16}{81} = 4$
48. $\log_8 x = -3$
49. $\log_b \frac{1}{27} = -\frac{3}{2}$
50. $\log_{\sqrt{3}} x = 2$
51. $\log_{\sqrt{8}} (\frac{1}{8}) = y$
52. $\log_b \frac{1}{128} = -7$
53. $\log_{0.001} 10 = y$
54. $\log_{0.2} 5 = y$

Solve for x.

55. $e^{-0.01x} = 27$
56. $e^{3x+5} = 100$
57. $e^{\ln(1-x)} = 2x$
58. $e^{\ln(6x^2-4)} = 5x$
59. $\ln e^{\sqrt{x+1}} = 3$
60. $\ln x = -2$

Evaluate each expression.

*61. $\log_2(\log_4 256)$
*62. $\log_{3/4}(\log_{1/27} \frac{1}{81})$

By interchanging the roles of the variables, find the inverse function g. Show that $(f \circ g)(x) = x$ and $(g \circ f)(x) = x$.

*63. $y = f(x) = 2^{x+1}$
*64. $y = f(x) = \log_3(x + 3)$

9.4 THE LAWS OF LOGARITHMS

From our knowledge of logarithms we have

$$\log_2 8 + \log_2 16 = 3 + 4$$
$$= 7 = \log_2 128 = \log_2 (8 \cdot 16)$$

$\log_2 8 = 3\ (2^3 = 8)$
$\log_2 16 = 4\ (2^4 = 16)$
$\log_2 128 = 7\ (2^7 = 128)$

Thus

$$\log_2 (8 \cdot 16) = \log_2 8 + \log_2 16$$

It turns out that this equation is a special case of the first law of logarithms.

Law 1 says that the log of a product is the sum of the logs of the factors. Can you give similar interpretations for Laws 2 and 3?

LAWS OF LOGARITHMS

If M and N are positive, $b > 0$, and $b \neq 1$, then

LAW 1. $\log_b MN = \log_b M + \log_b N$

LAW 2. $\log_b \dfrac{M}{N} = \log_b M - \log_b N$

LAW 3. $\log_b(N^k) = k \log_b N$

Since logarithms are exponents it is not surprising that these laws can be proved by using the appropriate rules of exponents. Following is a proof of Law 1; the proofs of Laws 2 and 3 are left as exercises.

Let

$$\log_b M = r \quad \text{and} \quad \log_b N = s$$

Recall: if $\log_b x = y$ then $b^y = x$.

Convert to exponential form:

$$M = b^r \quad \text{and} \quad N = b^s$$

Multiply the two equations:

$$MN = b^r b^s = b^{r+s}$$

Then convert to logarithmic form:

$$\log_b MN = r + s$$

Substitute for r and s to get the final result:

$$\log_b MN = \log_b M + \log_b N$$

EXAMPLE 1 For positive numbers A, B, and C, show that

$$\log_b = \frac{AB^2}{C} = \log_b A + 2 \log_b B - \log_b C$$

Solution

$$\begin{aligned} \log_b \frac{AB^2}{C} &= \log_b(AB^2) - \log_b C && \text{(Law 2)} \\ &= \log_b A + \log_b B^2 - \log_b C && \text{(Law 1)} \\ &= \log_b A + 2 \log_b B - \log_b C && \text{(Law 3)} \end{aligned}$$

■

Identify the laws of logarithms being used in Examples 2 and 3.

EXAMPLE 2 Express $\frac{1}{2}\log_b x - 3\log_b(x-1)$ as the logarithm of a single expression in x.

Solution

$$\begin{aligned}\tfrac{1}{2}\log_b x - 3\log_b(x-1) &= \log_b x^{1/2} - \log_b(x-1)^3\\ &= \log_b \frac{x^{1/2}}{(x-1)^3}\\ &= \log_b \frac{\sqrt{x}}{(x-1)^3}\end{aligned}$$ ■

EXAMPLE 3 Given: $\ln 2 = 0.6931$ and $\ln 3 = 1.0986$; find $\ln\sqrt{12}$.

Solution

Since $\ln 2 = \log_e 2$ we have $\log_e 2 = 0.6931$, so that $e^{0.6931} = 2$. Similarly, $e^{1.0986} = 3$ and $e^{1.2424} = \sqrt{12}$.

$$\begin{aligned}\ln\sqrt{12} &= \ln 12^{1/2} = \tfrac{1}{2}\ln 12\\ &= \tfrac{1}{2}\ln(3\cdot 4) = \tfrac{1}{2}[\ln 3 + \ln 4]\\ &= \tfrac{1}{2}[\ln 3 + \ln 2^2]\\ &= \tfrac{1}{2}[\ln 3 + 2\ln 2]\\ &= \tfrac{1}{2}\ln 3 + \ln 2\\ &= \tfrac{1}{2}(1.0986) + 0.6931\\ &= 1.2424\end{aligned}$$ ■

TEST YOUR UNDERSTANDING
Think Carefully

Convert the given logarithms into expressions involving $\log_b A$, $\log_b B$, and $\log_b C$.

1. $\log_b ABC$ **2.** $\log_b \frac{A}{BC}$ **3.** $\log_b \frac{(AB)^2}{C}$

4. $\ln AB^2C^3$ **5.** $\log_b \frac{A\sqrt{B}}{C}$ **6.** $\log_b \frac{\sqrt[3]{A}}{(BC)^3}$

Change each expression into the logarithm of a single expression in x.

7. $\log_b x + \log_b x + \log_b 3$ **8.** $2\log_b(x-1) + \frac{1}{2}\log_b x$

9. $\ln(2x-1) - 3\ln(x^2+1)$

10. $\log_b x - \log_b(x-1) - 2\log_b(x-2)$

Use the information given in Example 3 to find these logarithms.

11. $\ln 18$ **12.** $\ln\frac{16}{27}$

(Answers: Page 424)

Examples 4 to 6 illustrate how the laws of logarithms can be used to solve logarithmic equations.

EXAMPLE 4 Solve for x: $\log_8(x-6) + \log_8(x+6) = 2$

Solution First note that in $\log_8(x-6)$ we must have $x - 6 > 0$, or $x > 6$. Similary, $\log_8(x+6)$ calls for $x > -6$. Therefore, the only solutions, if there are any, must satisfy $x > 6$.

CAUTION
$\log_8(x^2 - 36) \neq \log_8 x^2 - \log_8 36$

$$\log_8(x - 6) + \log_8(x + 6) = 2$$
$$\log_8(x - 6)(x + 6) = 2 \qquad \text{(Law 1)}$$
$$\log_8(x^2 - 36) = 2$$
$$x^2 - 36 = 8^2 \qquad \text{(converting to exponential form)}$$
$$x^2 - 100 = 0$$
$$(x + 10)(x - 10) = 0$$
$$x = -10 \quad \text{or} \quad x = 10$$

The only possible solutions are -10 and 10. Our initial observation that $x > 6$ automatically eliminates -10. (If that initial observation had not been made, -10 could still have been eliminated by checking in the given equation.) The value $x = 10$ can be checked as follows:

$$\log_8(10 - 6) + \log_8(10 + 6) = \log_8 4 + \log_8 16$$
$$= \tfrac{2}{3} + \tfrac{4}{3} = 2$$

■

EXAMPLE 5 Solve for x: $\log_{10}(x^3 - 1) - \log_{10}(x^2 + x + 1) = 1$

Solution

See page 62 for the factorization of the difference of two cubes: $a^3 - b^3$.

$$\log_{10}(x^3 - 1) - \log_{10}(x^2 + x + 1) = 1$$
$$\log_{10}\frac{x^3 - 1}{x^2 + x + 1} = 1 \qquad \text{(Law 2)}$$
$$\log_{10}\frac{(x - 1)(x^2 + x + 1)}{x^2 + x + 1} = 1 \qquad \text{(by factoring)}$$
$$\log_{10}(x - 1) = 1$$
$$x - 1 = 10^1 \qquad \text{(Why?)}$$
$$x = 11$$

Check: $\log_{10}(11^3 - 1) - \log_{10}(11^2 + 11 + 1) = \log_{10} 1330 - \log_{10} 133$
$$= \log_{10}\tfrac{1330}{133}$$
$$= \log_{10} 10 = 1$$

■

EXAMPLE 6 Solve for x: $\ln 2x - \ln(x + 5) = 0$

Solution

$$\ln 2x - \ln(x + 5) = 0$$
$$\ln\frac{2x}{x + 5} = 0$$
$$\frac{2x}{x + 5} = e^0$$
$$\frac{2x}{x + 5} = 1$$
$$2x = x + 5$$
$$x = 5$$

Check: $\ln 2(5) - \ln(5 + 5) = \ln 10 - \ln 10 = 0$

■

Sometimes it is convenient to solve a logarithmic equation using the one-to-one property of logarithmic functions. This property (stated on page 393) says:

$$\text{If } \log_b M = \log_b N, \text{ then } M = N.$$

Here, for example, is the solution for the equation in Example 6 using this property.

$$\ln 2x - \ln(x+5) = 0$$
$$\ln 2x = \ln(x+5)$$
$$2x = x + 5 \qquad \text{(by the one-to-one property)}$$
$$x = 5$$

CAUTION: Learn To Avoid These Mistakes

WRONG	RIGHT
$\log_b A + \log_b B = \log_b(A+B)$	$\log_b A + \log_b B = \log_b AB$
$\log_b(x^2 - 4) = \log_b x^2 - \log_b 4$	$\log_b(x^2-4)$ $= \log_b(x+2)(x-2)$ $= \log_b(x+2) + \log_b(x-2)$
$(\log_b x)^2 = 2\log_b x$	$(\log_b x)^2 = (\log_b x)(\log_b x)$
$\log_b A - \log_b B = \dfrac{\log_b A}{\log_b B}$	$\log_b A - \log_b B = \log_b \dfrac{A}{B}$
If $2\log_b x = \log_b(3x+4)$, then $2x = 3x + 4$	If $2\log_b x = \log_b(3x+4)$, then $\log_b x^2 = \log_b(3x+4)$ and $x^2 = 3x + 4$.
$\log_b \dfrac{x}{2} = \dfrac{\log_b x}{2}$	$\log_b \dfrac{x}{2} = \log_b x - \log_b 2$
$\log_b(x^2+2) = 2\log_b(x+2)$	$\log_b(x^2+2)$ cannot be simplified further.

EXERCISES 9.4

Use the laws of logarithms (as much as possible) to convert the given logarithms into expressions involving sums, differences, and multiples of logarithms.

1. $\log_b \dfrac{3x}{x+1}$
2. $\log_b \dfrac{x^2}{x-1}$
3. $\log_b \dfrac{\sqrt{x^2-1}}{x}$
4. $\log_b \dfrac{1}{x}$
5. $\log_b \dfrac{1}{x^2}$
6. $\log_b \sqrt{\dfrac{x+1}{x-1}}$
7. $\ln \dfrac{(x-1)(x+3)^2}{\sqrt{x^2+2}}$
8. $\ln x\sqrt{x^2+1}$

Convert each expression into the logarithm of a single expression in x.

9. $\frac{1}{2}\ln x + \ln(x^2 + 5)$
10. $\ln 2 + \ln x - \ln(x - 1)$
11. $\log_b(x + 1) - \log_b(x + 2)$
12. $\log_b x + 2\log_b(x - 1)$
13. $\frac{1}{2}\log_b(x^2 - 1) - \frac{1}{2}\log_b(x^2 + 1)$
14. $\log_b(x + 2) - \log_b(x^2 - 4)$
15. $3\log_b x - \log_b 2 - \log_b(x + 5)$
16. $\frac{1}{3}\log_b(x - 1) + \log_b 3 - \frac{1}{3}\log_b(x + 1)$

Use the appropriate laws of logarithms to explain why each statement is correct.

17. $\log_b 27 + \log_b 3 = \log_b 243 - \log_b 3$
18. $\log_b 16 + \log_b 4 = \log_b 64$
19. $-2\log_b \frac{4}{9} = \log_b \frac{81}{16}$
20. $\frac{1}{2}\log_b 0.0001 = -\log_b 100$

Find the logarithms by using the laws of logarithms and the given information that $\log_b 2 = 0.3010$, $\log_b 3 = 0.4771$, *and* $\log_b 5 = 0.6990$. *Assume that all logs have base b.*

21. **(a)** $\log 4$ **(b)** $\log 8$ **(c)** $\log \frac{1}{2}$
22. **(a)** $\log \sqrt{2}$ **(b)** $\log 9$ **(c)** $\log 12$
23. **(a)** $\log 48$ **(b)** $\log \frac{2}{3}$ **(c)** $\log 125$
24. **(a)** $\log 50$ **(b)** $\log 10$ **(c)** $\log \frac{25}{6}$
25. **(a)** $\log \sqrt[3]{5}$ **(b)** $\log \sqrt{20^3}$ **(c)** $\log \sqrt{900}$
26. **(a)** $\log 0.2$ **(b)** $\log 0.25$ **(c)** $\log 2.4$

Solve for x and check.

27. $\log_{10} x + \log_{10} 5 = 2$
28. $\log_{10} x + \log_{10} 5 = 1$
29. $\log_{10} 5 - \log_{10} x = 2$
30. $\log_{10}(x + 21) + \log_{10} x = 2$
31. $\log_{12}(x - 5) + \log_{12}(x - 5) = 2$
32. $\log_3 x + \log_3(2x + 51) = 4$
33. $\log_{16} x + \log_{16}(x - 4) = \frac{5}{4}$
34. $\log_2(x^2) - \log_2(x - 2) = 3$
35. $\log_{10}(3 - x) - \log_{10}(12 - x) = -1$
36. $\log_{10}(3x^2 - 5x - 2) - \log_{10}(x - 2) = 1$
37. $\log_{1/7} x + \log_{1/7}(5x - 28) = -2$
38. $\log_{1/3} 12x^2 - \log_{1/3}(20x - 9) = -1$
39. $\log_{10}(x^3 - 1) - \log_{10}(x^2 + x + 1) = -2$
40. $2\log_{10}(x - 2) = 4$
41. $2\log_{25} x - \log_{25}(25 - 4x) = \frac{1}{2}$
42. $\log_3(8x^3 + 1) - \log_3(4x^2 - 2x + 1) = 2$
43. $\ln(x^2 - 4) - \ln(x + 2) = 0$
44. $\frac{1}{2}\ln(x + 4) = \ln(x + 2)$
45. $(e^{x+2} - 1)\ln(1 - 2x) = 0$
46. $\ln x = \frac{1}{2}\ln 4 + \frac{2}{3}\ln 8$

Explain how the graph of f can be obtained from the curve $y = \ln x$. *(Hint: First apply the appropriate laws of logarithms.)*

47. $f(x) = \ln ex$
48. $f(x) = \ln \dfrac{x}{e}$
49. $f(x) = \ln(x^2 - 1) - \ln(x + 1)$
50. $f(x) = \ln \dfrac{1}{x}$

*51. Solve for x: $\ln x = 2 + \ln(1 - x)$

52. Explain why $\log_b b = 1$.

*53. Prove Law 2. $\left(\textit{Hint}\text{: Follow the proof of Law 1 using } \dfrac{b^r}{b^s} = b^{r-s}.\right)$

*54. Prove Law 3. (*Hint*: Use $(b^r)^k = b^{rk}$.)

*55. Solve for x: $(x + 2)\log_b b^x = x$.

*56. Solve for x: $\log_{N^2} N = x$.

*57. Solve for x: $\log_x(2x)^{3x} = 4x$.

Written Assignment: State laws 2 and 3 of logarithms in your own words, that is, without symbols.

CHALLENGE
Think Creatively

1. Show that $\ln\left(\dfrac{\sqrt{x^2 + 9}}{9} - \dfrac{x}{9}\right) = -\ln(\sqrt{x^2 + 9} + x)$.
2. Solve for x: $\log_b(\log_b x) = 1$

9.5 EXPONENTIAL GROWTH AND DECAY

The number e to three decimal places is 2.718. This can be written as $e^1 = 2.718$. Other powers of e are given in Table II, rounded to various decimal places. Thus, from Table II,

$$e^{2.1} = 8.17 \quad \text{and} \quad e^{-2.1} = 0.122$$

In Table III, page 000, natural logarithms of x are given to three decimal places. For example, $\ln 2.1 = 0.742$.

A large variety of calculators are available. For this work, one of the keys [e^x] *or* [y^x] *and the key* [ln x] *are needed. (Sometimes the letter x is not included in the last key.)*

Such values can also be obtained using a calculator. If the calculator has an [e^x] key, then $e^{2.1}$ can be found using this calculator sequence of keystrokes:

(2.1) [e^x] = 8.166 (to 3 decimal places)
↑ enter ↑ press ↑ answer

For $e^{-2.1}$ use the sequence

(2.1) [±] [e^x] = 0.122 (to 3 decimal places)
↑ enter ↑ press (to change the sign) ↑ press ↑ answer

If the calculator does not have an [e^x] key, but it has the exponential key [y^x], use the following sequence.

The calculator sequences used here apply for many, but not all calculators. It may therefore be necessary to refer to your calculator instruction manual for the appropriate sequences.

(2.7183) [y^x] (2.1) [=] 8.166
↑ enter (e to 4 decimal places) ↑ press ↑ enter ↑ press ↑ answer

To find ln 2.1 on a calculator, use the sequence

(2.1) [ln x] = 0.742 (rounded to 3 decimal places)
↑ enter ↑ press ↑ answer

In the preceding calculator operations, we have circled a number to be entered and placed a rectangle around a key to be pressed. From now on, we will follow this convention without using the words "enter" and "press."

EXAMPLE 1 Solve the exponential equation $2^x = 35$ for x.

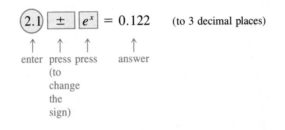

Solution (a) (using Table III)

$$2^x = 35$$
$$\ln 2^x = \ln 35 \qquad \text{(If } A = B\text{, then } \ln A = \ln B.)$$
$$x \ln 2 = \ln 35 \qquad \text{(Why?)}$$
$$x = \frac{\ln 35}{\ln 2}$$

Even though ln 35 is not given in the table, the first law of logarithms in conjunction with Table III can be used to find ln 35 as follows:

$$\ln 35 = \ln(3.5)(10) = \ln 3.5 + \ln 10$$
$$= 1.253 + 2.303 \qquad \text{(Table III)}$$
$$= 3.556$$

Note that the values found in the tables are approximations. For simplicity, however, we will use the equals sign (=).

Now we have

$$x = \frac{\ln 35}{\ln 2} = \frac{3.556}{0.693} = 5.13$$

Solution (b) (using a calculator)
From solution (a) we have

$$x = \frac{\ln 35}{\ln 2}$$

Now use this calculator sequence:

As a rough check note that 5.13 is a reasonable answer since $2^5 = 32$.

$x =$ (35) [ln x] [÷] (2) [ln x] [=] 5.13

Therefore $x = 5.13$ rounded to two decimal places. ■

As a variation of Example 1, consider the calculator solution of $2^{3x} = 35$. First isolate x as follows:

$$3x \ln 2 = \ln 35 \qquad \text{(Apply ln and use law 3 for logarithms.)}$$
$$x = \frac{\ln 35}{3 \ln 2}$$

Now use either of the following sequences. Note that the second sequence calls for the parentheses keys [(] and [)].

These two calculator solutions indicate that more than one correct procedure may be possible.

$x =$ (35) [ln x] [÷] (3) [÷] (2) [ln x] [=] 1.7098

or

$x =$ (35) [ln x] [÷] [(] (3) [×] (2) [ln x] [)] [=] 1.7098

Thus $x = 1.7098$ (to 4 decimal places).

TEST YOUR UNDERSTANDING Think Carefully

Solve each equation for x in terms of natural logarithms. Approximate the answer by using Table III or a calculator.

1. $4^x = 5$ **2.** $4^{-x} = 5$ **3.** $(\frac{1}{2})^x = 12$

4. $2^{3x} = 10$ **5.** $4^x = 15$ **6.** $63^{5x} = 4$

(Answers: Page 424)

At the beginning of Section 9.2 we developed the formula $y = (10{,}000)2^x$, which gives the number of bacteria present after x hours of growth; 10,000 is the initial number of bacteria. How long will it take for this bacterial culture to grow to 100,000? To answer this question we let $y = 100{,}000$ and solve for x.

$$
\begin{aligned}
(10{,}000)2^x &= 100{,}000 \\
2^x &= 10 \qquad \text{(Divide by 10,000.)} \\
x \ln 2 &= \ln 10 \\
x &= \frac{\ln 10}{\ln 2} \\
&= 3.32 \qquad \text{(by Table III or by calculator)}
\end{aligned}
$$

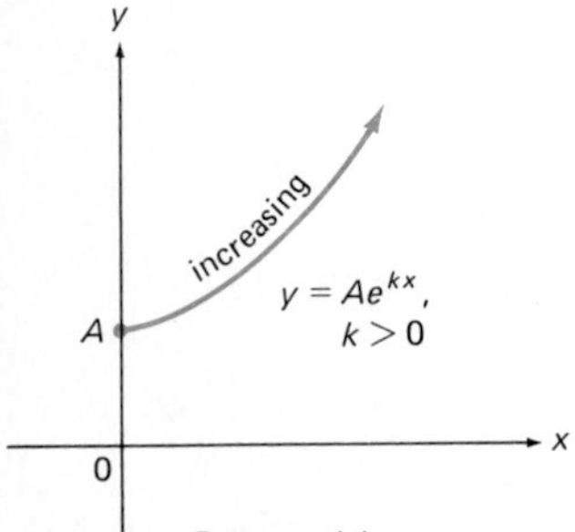

Exponential Growth

It will take about 3.3 hours.

In the preceding illustration the exponential and logarithmic functions were used to solve a problem of *exponential growth*. Many problems involving **exponential growth**, or **exponential decay** can be solved by using the general formula.

$$y = f(x) = Ae^{kx}$$

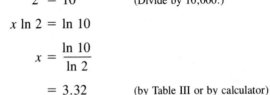

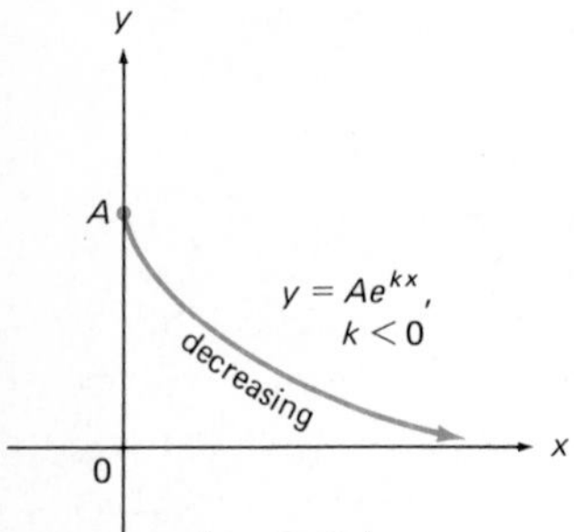

Exponential Decay

which shows how the amount of a substance y depends on the time x. Since $f(0) = A$, A represents the initial amount of the substance and k is a constant. In a given situation $k > 0$ signifies that y is growing (increasing) with time. For $k < 0$ the substance is decreasing. (Compare to the graphs of $y = e^x$ and $y = e^{-x}$.)

The preceding bacterial problem also fits this general form. This can be seen by substituting $2 = e^{\ln 2}$ into $y = (10{,}000)2^x$:

$$y = (10{,}000)2^x = (10{,}000)(e^{\ln 2})^x = 10{,}000e^{(\ln 2)x}$$

EXAMPLE 2 A radioactive substance is decaying (it is changing into another element) according to the formula $y = Ae^{-0.2x}$, where y is the amount of material remaining after x years.

(a) If the initial amount $A = 80$ grams, how much is left after 3 years?

(b) The **half-life** of a radioactive substance is the time it takes for half of it to decompose. Find the half-life of the substance in which $A = 80$ grams.

Solution

(a) Since $A = 80$, $y = 80e^{-0.2x}$. We need to solve for the amount y when $x = 3$.

$$
\begin{aligned}
y &= 80e^{-0.2x} \\
&= 80e^{-0.2(3)} \\
&= 80e^{-0.6} \xrightarrow{\text{Using Table II}} = 80(0.549) = 43.92
\end{aligned}
$$

The results using a calculator are usually more accurate than those obtained by using the tables. To one decimal place, however, the answers are the same in part (a).

Using a calculator with $\boxed{y^x}$:

$$y = (80)\ \boxed{\times}\ (2.7183)\ \boxed{y^x}\ (0.6)\ \boxed{\pm}\ \boxed{=}\ 43.90$$

There will be about 43.9 grams after 3 years.

(b) This question calls for the time x at which only half of the initial amount is left. Consequently, the half-life x is the solution to $40 = 80e^{-0.2x}$. Divide each side by 80:

$$\tfrac{1}{2} = e^{-0.2x}$$

Take the natural log of both sides, or change to logarithmic form, to obtain $-0.2x = \ln \frac{1}{2}$. Since $\ln \frac{1}{2} = \ln 1 - \ln 2 = -\ln 2$, we solve for x as follows.

$$-0.2x = -\ln 2$$

$$x = \frac{\ln 2}{0.2} \xrightarrow{\text{Using Table III}} = \frac{0.693}{0.2} = 3.465$$

Using a calculator with $\boxed{\ln x}$:

$$x = (2)\ \boxed{\ln x}\ \boxed{\div}\ (0.2)\ \boxed{=}\ 3.466$$

The half-life is approximately 3.47 years. ■

This process of finding the age of the remains is referred to as ***radioactive carbon dating.***

Carbon-14, also written as ^{14}C, is a radioactive isotope of carbon with a half-life of about 5750 years. By finding how much ^{14}C is contained in the remains of a formerly living organism, it becomes possible to determine what percentage this is of the original amount of ^{14}C at the time of death. Once this information is given, the formula $y = Ae^{kx}$ will enable us to date the age of the remains. The dating will be done after we solve for the constant k. Since the amount of ^{14}C after 5750 years will be $\frac{A}{2}$, we have the following:

Explain each step in this solution.

$$\frac{A}{2} = Ae^{5750k}$$

$$\frac{1}{2} = e^{5750k}$$

$$5750k = \ln \frac{1}{2}$$

$$k = \frac{\ln 0.5}{5750}$$

Substitute this value for k into $y = Ae^{kx}$ to obtain the following formula for the amount of carbon-14 remaining after x years:

$$y = Ae^{(\ln 0.5/5750)x}$$

EXAMPLE 3 An animal skeleton is found to contain one third of its original amount of ^{14}C. How old is the skeleton?

Solution Let x be the age of the skeleton. Then

$$\tfrac{1}{3}A = Ae^{(\ln 0.5/5750)x}$$

$$\tfrac{1}{3} = e^{(\ln 0.5/5750)x}$$

$$\left(\frac{\ln 0.5}{5750}\right)x = \ln\frac{1}{3} = -\ln 3$$

$$x = \frac{(5750)(-\ln 3)}{\ln 0.5}$$

$$= \frac{-5750 \ln 3}{\ln 0.5} \xrightarrow{\text{Using Table III}} = -\frac{5750(1.099)}{-0.693} = 9119$$

$\downarrow$

Using a calculator:

$x =$ (5750) [±] [×] (3) [ln x] [÷] (0.5) [ln x] [=] 9114

To the nearest five hundred years, the skeleton is about 9000 years old. ■

The formulas used in the evaluation of **compound interest** are also applications of exponential growth. The statement that an investment earns compound interest means that the interest earned over a fixed period of time is added to the initial investment, and then the new total earns interest for the next investment period, and so on. For example, suppose an investment of P dollars earns interest at a yearly rate of r percent, and the interest is compounded annually. Then, after the first year the total value is the sum of the initial investment P plus the interest Pr (r is being used as a decimal fraction). Thus, the total after one year is

$$P + Pr = P(1 + r)$$

After the second year the total amount is $P(1 + r)$ plus the interest this amount earns, which is $P(1 + r)r$. Then the total after two years is

Note that this expression is simplified by factoring out $P(1 + r)$.

$$P(1 + r) + P(1 + r)r = P(1 + r)(1 + r) = P(1 + r)^2$$

Similarly, after three years the total is

$$P(1 + r)^2 + P(1 + r)^2 r = P(1 + r)^2(1 + r) = P(1 + r)^3$$

and after t years the final amount A is given by

$$A = P(1 + r)^t$$

The interest periods for compound interest are usually less than one year. They could be quarterly (4 times per year), monthly, or daily, and so forth. For such cases the interest rate per period is the annual rate r divided by the number of interest periods per year. Thus, if the interest is compounded quarterly, the rate per period is

$r/4$. Now, following the reasoning used to obtain $A = P(1 + r)^t$, the final amount A, after one year (4 interest periods) is

$$A_1 = P\left(1 + \frac{r}{4}\right)^4$$

If there are n interest periods per year, the rate per period is r/n, and after one year we have

$$A_1 = P\left(1 + \frac{r}{n}\right)^n$$

Likewise, after t years the final amount A_t is given by

This result can be derived from the preceding result. See Exercise 46.

$$A_t = P\left(1 + \frac{r}{n}\right)^{nt}$$

EXAMPLE 4 A \$5000 investment earns interest at the annual rate of 8.4% compounded monthly. Answer the following:
(a) What is the investment worth after one year?
(b) What is it worth after 10 years?
(c) How much interest was earned in 10 years?

Solution

(a) Since the annual rate $r = 8.4\% = 0.084$, and the compounding is done monthly, the interest rate per month is $r/n = 0.084/12 = 0.007$. Substitute this, together with $P = 5000$ and $n = 12$ into $A = P\left(1 + \frac{r}{n}\right)^n$.

$$A = 5000(1 + 0.007)^{12} = 5000(1.007)^{12}$$
$$= \text{(5000)} \boxed{\times} \text{(1.007)} \boxed{y^x} \text{(12)} \boxed{=} 5436.55$$

To the nearest dollar, the amount on deposit after one year is \$5437.

(b) Use $A_t = P\left(1 + \frac{r}{n}\right)^{nt}$ where $P = 5000$, $\frac{r}{n} = 0.007$, $n = 12$, and $t = 10$.

As an illustration, let $r = 0.2$ and use a calculator to verify the following computations, rounded to 5 decimal places. They demonstrate that $\left(1 + \frac{0.2}{n}\right)^n$ approaches $e^{0.2}$ as n gets larger.

$$\left(1 + \frac{0.2}{10}\right)^{10} = 1.21899$$
$$\left(1 + \frac{0.2}{100}\right)^{100} = 1.22116$$
$$\left(1 + \frac{0.2}{1000}\right)^{1000} = 1.22138$$

Also, $e^{0.2} = 1.22140$

$$A = 5000(1.007)^{12(10)} = 5000(1.007)^{120}$$
$$= \text{(5000)} \boxed{\times} \text{(1.007)} \boxed{y^x} \text{(120)} \boxed{=} 11{,}547.99$$

The amount after 10 years is approximately \$11,548.

(c) After 10 years the interest earned is

$$11548 - 5000 = 6548 \text{ dollars}$$

■

The marginal note on page 387 points out that the values of $\left(1 + \frac{1}{n}\right)^n$ approach the number e as n gets larger and larger. It is also true that $\left(1 + \frac{r}{n}\right)^n$ approaches e^r

as n gets larger and larger. These observations, when made mathematically precise, lead to the following formula for **continuous compound interest**.

$$A = Pe^{rt}$$

where P is the initial investment, r is the annual rate of interest, and t is the number of years. Thus if \$1000 is invested at 10%, compounded continuously for 10 years, the amount on deposit will be

$$\begin{aligned} A &= 1000e^{(0.10)(10)} = 1000e^{1} \\ &= 1000(2.718) \\ &= 2718 \end{aligned}$$

After 10 years, despite continuous compounding, the amount on deposit (to the nearest dollar) will not grow beyond \$2718.

EXAMPLE 5 Suppose that \$1000 is invested at 10% interest, compounded continuously. How long will it take for this investment to double?

Solution We wish the final amount on deposit to be \$2000. Thus we have the following equation and need to solve for t:

$$\begin{aligned} 2000 &= 1000e^{(0.10)t} \\ 2 &= e^{(0.1)t} && \text{(Divide by 1000.)} \\ \ln 2 &= (0.1)t && \text{(Write in logarithmic form.)} \\ \frac{\ln 2}{0.1} &= t && \text{(Divide by 0.1.)} \\ 6.93 &= t && \text{(Use Table III or a calculator.)} \end{aligned}$$

It will take approximately 7 years for the investment to double in value. As a check, note from Table II that $e^{(0.1)(7)} = e^{0.7} = 2.01$, which is approximately equal to 2. ■

EXERCISES 9.5

Evaluate each of the following to three decimal places.

1. $\dfrac{\ln 6}{\ln 2}$
2. $\dfrac{\ln 10}{\ln 5}$
3. $\dfrac{\ln 8}{\ln 0.2}$
4. $\dfrac{\ln 0.8}{\ln 4}$
5. $\dfrac{\ln 15}{2 \ln 3}$
6. $\dfrac{\ln 25}{3 \ln 5}$
7. $\dfrac{\ln 100}{-4 \ln 10}$
8. $\dfrac{\ln 80}{-5 \ln 8}$

Estimate the value of y in $y = Ae^{kx}$ for the given values of A, k, and x.

9. $A = 100$, $k = 0.75$, $x = 4$
10. $A = 25$, $k = 0.5$, $x = 10$

11. $A = 1000$, $k = -1.8$, $x = 2$ 12. $A = 12.5$, $k = -0.04$, $x = 50$

Solve for k. Leave the answer in terms of natural logarithms.

13. $5000 = 50e^{2k}$ 14. $75 = 150e^{10k}$ 15. $\frac{A}{3} = Ae^{4k}$ 16. $\frac{A}{2} = Ae^{100k}$

17. A bacterial culture is growing according to the formula $y = 10{,}000e^{0.6x}$, where x is the time in days. Estimate the number of bacteria after 1 week.
18. Estimate the number of bacteria after the culture in Exercise 17 has grown for 12 hours.
19. How long will it take for the bacterial culture in Exercise 17 to triple in size?
20. How long will it take until the number of bacteria in Exercise 17 reaches 1,000,000?
21. A certain radioactive substance decays according to the exponential formula

$$S = S_0e^{-0.04t}$$

where S_0 is the initial amount of the substance and S is the amount of the substance left after t years. If there were 50 grams of the radioactive substance to begin with, how long will it take for half of it to decay?

22. Show that when the formula in Exercise 21 is solved for t, the result is

$$t = -25 \ln \frac{S}{S_0}$$

23. A radioactive substance is decaying according to the formula $y = Ae^{kx}$, where x is the time in years. The initial amount $A = 10$ grams, and 8 grams remain after 5 years.
 (a) Find k. Leave the answer in terms of natural logs.
 (b) Estimate the amount remaining after 10 years.
 (c) Find the half-life to the nearest tenth of a year.

24. The half-life of radium is approximately 1690 years. A laboratory has 50 milligrams of radium.
 (a) Use the half-life to solve for k in $y = Ae^{kx}$. Leave your answer in terms of natural logs.
 (b) To the nearest 10 years, how long does it take until there are 40 milligrams left?
25. Suppose that 5 grams of a radioactive substance decrease to 4 grams in 30 seconds. What is its half-life to the nearest tenth of a second?
26. How long does it take for two-thirds of the radioactive material in Exercise 25 to decay? Give your answer to the nearest tenth of a second.
27. When the population growth of a certain city was first studied, the population was 22,000. It was found that the population P grows with respect to time t (in years) by the exponential formula

$$P = (22{,}000)(10^{0.0163t})$$

How long will it take for the city to double its population?

28. How long will it take for the population of the city described in Exercise 27 to triple?
29. An Egyptian mummy is found to contain 60% of its ^{14}C. To the nearest 100 years, how old is the mummy? (*Hint*: If A is the original amount of ^{14}C, then $\frac{3}{5}A$ is the amount left.)
30. A skeleton contains one-hundredth of its original amount of ^{14}C. To the nearest 1000 years, how old is the skeleton?
31. Answer the question in Exercise 30 if one-millionth of its ^{14}C is left.

Use a calculator having an exponential key and a natural logarithm key to answer the following questions.

32. Suppose that a \$10,000 investment earns compound interest at the annual interest rate of 9%. If the time of deposit of the investment is one year ($t = 1$), find the value of the investment for each of the following types of compounding.
    (a) $n = 4$ (quarterly) **(b)** $n = 12$ (monthly) **(c)** $n = 52$ (weekly)
    (d) $n = 365$ (daily) **(e)** continuously.
33. Follow the instructions in Exercise 32 but change the time of deposit for the investment to 5 years.
34. Compute the interest earned for each case in Exercise 32.
35. Follow the instructions in Exercise 32 but change the time of deposit for the investment to 3.5 years.
36. Suppose that \$1500 is invested at the annual interest rate of 8%, compounded continuously. How much will be on deposit after 5 years? After 10 years?
37. Mrs. Kassner deposits \$5000 at the annual interest rate of 9% compounded continuously. How long will it take her investment to double in value? How long would it take if the interest rate were 12%?
38. How long would it take for a \$1000 investment to double if it earns 12% interest annually, compounded continuously? How long would it take to triple?
39. A \$1000 investment earns interest at the annual rate of r%, compounded continuously. If the investment doubles in 5 years, what is r?
40. How long does it take a \$4000 investment to double if it earns interest at the annual rate of 8%, compounded quarterly?
41. In Exercise 40, how long would it take if the compounding were done monthly?
42. An investment P earns 9% interest per year, compounded continuously. After 3 years, the value of the investment is \$5000. Find the initial amount P. (*Hint:* Solve $A = Pe^{rt}$ for P.)
43. Answer the question in Exercise 42 using 6 years as the time of deposit.
44. An investment P earns 8% interest per year, compounded quarterly. After one year the value of the investment is \$5000. Find the initial amount P. (*Hint:* Solve $A_1 = P(1 + \frac{r}{n})^n$ for P.)
45. How many dollars must be invested at the annual rate of 12%, compounded monthly, in order for the value of the investment to be \$20,000 after 5 years? (*Hint:* Solve $A_t = P(1 + \frac{r}{n})^{nt}$ for P.)
46. Explain how the result $A_t = P(1 + \frac{r}{n})^{nt}$ can be obtained from $A_1 = P(1 + \frac{r}{n})^n$. [*Hint:* A_2, the value of the investment after 2 years, is obtained when A_1 has earned interest for one year compounded n times: Thus $A_2 = A_1(1 + \frac{r}{n})^n$.]

Written Assignment: Contact at least two local banks and determine how much interest you earn on a certificate of deposit of \$5000 for five years. Find out what the annual interest rate is and how it is compounded. Use appropriate results developed in this section to compute the interest earned over the five-year period and compare your results with the information obtained from the bank.

CHALLENGE
Think Creatively

Solve for y: $x = -\frac{1}{k}(\ln A - \ln y)$. Use your result to describe the relationship between the variables x and y, assuming that x represents time, A is a positive constant, and k is a nonzero constant.

9.6 COMMON LOGARITHMS AND APPLICATIONS (OPTIONAL)

Logarithms were developed about 350 years ago. Since then they have been widely used to simplify involved numerical computations. Much of this work can now be done more efficiently with the aid of computers and calculators. However, logarithmic computations will help us to better understand the theory of logarithms, which plays an important role in many parts of mathematics (including calculus) and in its applications.

Scientific notation was developed in Section 2.2.

For scientific and technical work, numbers are often written in scientific notation and we will therefore be using logarithms to the base 10, called **common logarithms**.

Below is an excerpt of Appendix Table IV. It contains the common logarithms of three-digit numbers from 1.00 to 9.99. To find a logarithm, say $\log_{10} 3.47$, first find the entry 3.4 in the left-hand column under the heading x. Now in the row for 3.4 and in the column headed by the digit 7 you will find the entry .5403. This is the common logarithm of 3.47. We write

Note that the values found in the tables of logarithms are approximations. For the sake of simplicity, however, we will use the equals sign (=).

$$\log_{10} 3.47 = 0.5403 \qquad \text{(Recall that this means } 3.47 = 10^{0.5403}.)$$

By reversing this process we can begin with the $\log_{10} x = 0.5403$ and find x.

x	0	1	2	3	4	5	6	7	8	9
.	.	.	.	.	.	.	.	.	.	.
.	.	.	.	.	.	.	.	.	.	.
.	.	.	.	.	.	.	.	.	.	.
3.3	.5185	.5198	.5211	.5224	.5237	.5250	.5263	.5276	.5289	.5302
3.4	.5315	.5328	.5340	.5353	.5366	.5378	.5391	.5403	.5416	.5428
3.5	.5441	.5453	.5465	.5478	.5490	.5502	.5514	.5527	.5539	.5551
.	.	.	.	.	.	.	.	.	.	.
.	.	.	.	.	.	.	.	.	.	.
.	.	.	.	.	.	.	.	.	.	.

The common logarithms in Table IV are four-place decimals between 0 and 1. Except for the case $\log_{10} 1 = 0$, they are all approximations. The fact that they are between 0 and 1 will be taken up in the exercises.

Since common logarithms are always considered to be to the base 10, we can simplify the notation and drop the subscript 10 from the logarithmic statements. Thus we will write $\log N$ instead of $\log_{10} N$. Verify the following entries from Table IV:

$$\log 3.07 = 0.4871 \qquad \log 8.88 = 0.9484$$

$$\text{If } \log x = 0.7945, \text{ then } x = 6.23$$

To find $\log N$, where N may not be between 1 and 10, we first write N in scientific notation, $N = x(10^c)$. This form of N, in conjunction with Table IV, will allow us to find $\log N$. In general,

See page 412.

$$\begin{aligned} \log N &= \log (x10^c) \\ &= \log x + \log 10^c \qquad \text{(Law 1 for logs)} \\ &= \log x + c \qquad \text{(Why?)} \end{aligned}$$

The integer c is the **characteristic** of $\log N$, and the four-place decimal $\log x$ is its **mantissa**. Using $N = 62{,}300$, we have

Note the distinction:
log N ← common logarithm, base 10
ln N ← natural logarithm, base e

$$\begin{aligned} \log 62{,}300 &= \log 6.23(10^4) = \log 6.23 + \log 10^4 \\ &= \log 6.23 + 4 \\ &= 0.7945 + 4 \qquad \text{(Table IV)} \\ &= 4.7945 \end{aligned}$$

EXAMPLE 1 Find log 0.419.

Solution

$$\begin{aligned} \log 0.0419 &= \log 4.19(10^{-2}) \\ &= \log 4.19 + \log 10^{-2} \\ &= 0.6222 + (-2) \end{aligned}$$

■

Suppose that in Example 1 the mantissa 0.6222 and the negative characteristic are combined:

$$\begin{aligned} 0.6222 + (-2) &= -1.3778 = -(1 + 0.3778) \\ &= -1 + (-0.3778) \end{aligned}$$

Since Table IV does not have negative mantissas, like -0.3778, we avoid such combining and preserve the form of log 0.0419 so that its mantissa is positive. For computational purposes there are other useful forms of $0.6222 + (-2)$ in which the mantissa 0.6222 is preserved. Note that $-2 = 8 - 10$, $18 - 20$, and so forth. Thus

$$0.6222 + (-2) = 0.6222 + 8 - 10 = 8.6222 - 10 = 18.6222 - 20$$

Similarly,

$$\log 0.00569 = 7.7551 - 10 = 17.7551 - 20$$

$$\log 0.427 = 9.6304 - 10 = 29.6304 - 30$$

An efficient way to find N, if $\log N = 6.1239$, is to find the three-digit number x from Table IV corresponding to the mantissa 0.1239. Then multiply x by 10^6. Thus, since $\log 1.33 = 0.1239$, we have

$$N = 1.33(10^6) = 1{,}330{,}000$$

In the following explanation you can discover why this technique works.

$$\begin{aligned} \log N &= 6.1239 \\ &= 6 + 0.1239 \\ &= 6 + \log 1.33 \\ &= \log 10^6 + \log 1.33 \\ &= \log 10^6(1.33) \\ &= \log 1{,}330{,}000 \end{aligned}$$

Therefore, $\log N = \log 1{,}330{,}000$, and we conclude that $N = 1{,}330{,}000$.

TEST YOUR UNDERSTANDING Think Carefully

Find the common logarithm.

1. log 267 **2.** log 26.7 **3.** log 2.67
4. log 0.267 **5.** log 0.0267 **6.** log 42,000
7. log 0.000813 **8.** log 7990 **9.** log 0.00111

Find N.

10. $\log N = 2.8248$ **11.** $\log N = 0.8248$
12. $\log N = 9.8248 - 10$ **13.** $\log N = 0.8248 - 3$
14. $\log N = 7.7126$ **15.** $\log N = 18.9987 - 20$

(Answers: Page 424)

Note: Unless otherwise stated, log N will always mean $\log_{10} N$.

EXAMPLE 2 Estimate $P = (963)(0.00847)$ by using (common) logarithms.

Solution

$$\log P = \log(963)(0.00847)$$
$$= \log 963 + \log 0.00847 \qquad \text{(Law 1)}$$

Now use Table IV.

For easy reference:
Law 1.
$\log MN = \log M + \log N$
Law 2.
$\log \frac{M}{N} = \log M - \log N$
Law 3.
$\log N^k = k \log N$

$$\left.\begin{aligned} \log 963 &= 2.9836 \\ \log 0.00847 &= 7.9279 - 10 \end{aligned}\right\} \text{(Add.)}$$
$$\log P = 10.9115 - 10 = 0.9115$$
$$P = 8.16(10^0) = 8.16$$

Note: The mantissa 0.9115 is not in Table IV. In this case we use the closest entry, namely 0.9117, corresponding to $x = 8.16$. Such approximations are good enough for our purposes. ■

For a more accurate procedure, see Exercise 39. Exercise 38 shows how to find log x when $0 \le x < 1$ and x has more than three digits.

EXAMPLE 3 Use logarithms to estimate $Q = \dfrac{0.00439}{0.705}$.

Solution We find $\log Q = \log 0.00439 - \log 0.705$ (by Law 2). Now use the table.

(This form is used to avoid a negative mantissa when subtracting in the next step.)

$$\left.\begin{aligned} \log 0.00439 &= 7.6425 - 10 = 17.6425 - 20 \\ \log 0.705 &= 9.8482 - 10 = 9.8482 - 10 \end{aligned}\right\} \text{(Subtract.)}$$
$$\log Q = 7.7943 - 10$$
$$Q = 6.23(10^{-3})$$
$$= 0.00623$$ ■

EXAMPLE 4 Use logarithms to estimate $R = \sqrt[3]{0.0918}$.

Solution

$$\begin{aligned}
\log R &= \log(0.0918)^{1/3} \\
&= \tfrac{1}{3}\log 0.0918 && \text{(Law 3)} \\
&= \tfrac{1}{3}(8.9628 - 10) \\
&= \tfrac{1}{3}(28.9628 - 30) \longleftarrow && \text{(We avoid a fractional characteristic by changing to } 28.9628 - 30.) \\
&= 9.6543 - 10 \\
R &= 4.51(10^{-1}) = 0.451
\end{aligned}$$

■

EXAMPLE 5 To determine how much a paint dealer should charge for a gallon of paint, he needs to find out how much the paint cost him per gallon in the first place. The paint is stored in a cylindrical drum $2\frac{1}{2}$ feet in diameter and $3\frac{3}{4}$ feet high. If he paid \$400 for this quantity of paint, what did it cost him per gallon? (Use 1 cubic foot = 7.48 gallons.)

Solution The volume of the drum is the area of the base times the height. Thus there are

Volume of a cylinder: $V = \pi r^2 h$

$$\pi(1.25)^2(3.75)$$

cubic feet of paint in the drum. Then the number of gallons is

$$\pi(1.25)^2(3.75)(7.48)$$

Since the total cost was \$400, the cost per gallon is given by

$$C = \frac{400}{\pi(1.25)^2(3.75)(7.48)}$$

We use $\pi = 3.14$ to do the computation, using logarithms:

$$\log C = \log 400 - (\log 3.14 + 2\log 1.25 + \log 3.75 + \log 7.48)$$

$$\begin{array}{lrl}
 & & \log 400 = 2.6021 \\
 & \log 3.14 = 0.4969 & \\
\log 1.25 = 0.0969 \rightarrow & 2\log 1.25 = 0.1938 & \text{(Add.)} \qquad \text{(Subtract.)} \\
 & \log 3.75 = 0.5740 & \\
 & \underline{\log 7.48 = 0.8739} & \\
 & 2.1386 & \rightarrow \quad \underline{2.1386} \\
 & & \log C = 0.4635 \\
 & & C = 2.91 \times 10^0 \\
 & & \quad = 2.91
\end{array}$$

The paint cost the dealer approximately \$2.91 per gallon. ■

EXERCISES 9.6

Find the common logarithm.

1. log 457 2. log 45.7 3. log 0.457 4. log 0.783 5. log 72.9 6. log 8.56

Find N.

7. $\log N = 0.5705$ 8. $\log N = 0.8904$ 9. $\log N = 1.8331$
10. $\log N = 2.9523$ 11. $\log N = 9.1461 - 10$ 12. $\log N = 8.6972 - 10$

Estimate by using common logarithms.

13. (512)(84,000) 14. (906)(2330)(780) 15. $\dfrac{(927)(818)}{274}$ 16. $\dfrac{274}{(927)(818)}$

17. $\dfrac{(0.421)(81.7)}{(368)(750)}$ 18. $\dfrac{(579)(28.3)}{\sqrt{621}}$ 19. $\dfrac{(28.3)\sqrt{621}}{579}$ 20. $\left[\dfrac{28.3}{(579)(621)}\right]^2$

21. $\sqrt{\dfrac{28.3}{(579)(621)}}$ 22. $\dfrac{(0.0941)^3(0.83)}{(7.73)^2}$ 23. $\dfrac{\sqrt[3]{(186)^2}}{(600)^{1/4}}$ 24. $\dfrac{\sqrt[4]{600}}{(186)^{2/3}}$

Use common logarithms to solve the remaining problems.

25. After running out of gasoline, a motorist had her gas tank filled at a cost of \$16.93. What was the cost per gallon if the gas tank's capacity is 14 gallons?
26. Suppose that a spaceship takes 3 days, 8 hours, and 20 minutes to travel from the earth to the moon. If the distance traveled was one-quarter of a million miles, what was the average speed of the spaceship in miles per hour?
27. A spaceship, launched from the earth, will travel 432,000,000 miles on its trip to the planet Jupiter. If its average velocity is 21,700 miles per hour, how long will the trip take? Give the answer in years.

Exercises 28–30 make use of the compound interest formulas studied in Section 9.5.

28. When P dollars is invested in a bank that pays compound interest at the rate of r percent (expressed as a decimal) per year, the amount A after t years is given by the formula

$$A = P(1 + r)^t$$

   (a) Find A for $P = 2500$, $r = 0.09$ (9%), and for $t = 3$.
   (b) An investment of \$3750 earns compound interest at the rate of 11.2% per year. Find the amount A after 5 years.

29. The formula $P = \dfrac{A}{(1 + r)^t}$ gives the initial investment P in terms of the current amount of money A, the annual compound interest rate r, and the number of years t. How much money was invested at 12.8% if after 6 years there is now \$8440 in the bank?
30. If P dollars is invested at an interest rate r and the interest is compounded n times per year, the amount A after t years is given by

$$A = P\left(1 + \frac{r}{n}\right)^{nt}$$

   (a) Use this formula to compute A for $P = \$5000$ and $r = 0.08$ if the interest is compounded semiannually for 3 years.

(b) Find A in part (a) with interest compounded quarterly.

(c) Find A in part (a) with $n = 8$.

31. An oil tanker carries 253,000 barrels of crude oil. This oil will produce 1,830,000 gallons of a certain kind of fuel. How many gallons of fuel are produced by 1 gallon of crude oil? (1 barrel = 31.5 gallons.)

32. The dimensions of a rectangular-shaped container are 2.75 by 5.35 by 4.4 feet. How many gallons can this container hold? (Use 1 cubic foot = 7.48 gallons.) If this container is filled with water, how many pounds of water does it hold? (Use 1 cubic foot water = 62.4 pounds.)

33. The volume V of a sphere with radius r is given by $V = \frac{4}{3}\pi r^3$. Use $\pi = 3.14$ to find the volume of a sphere with a radius of 12 centimeters.

34. The surface area S of a sphere is given by $S = 4\pi r^2$. What is the surface area of the sphere in Exercise 33?

35. The period P of a simple pendulum is the time (in seconds) it takes to make one full swing. The period is given by $P = 2\pi\sqrt{\frac{l}{32}}$ where l is the length of the pendulum. Find the period of a pendulum with a length of $3\frac{3}{4}$ feet.

36. The area of a triangle A can be given in terms of its three sides. The formula is

$$A = \sqrt{s(s-a)(s-b)(s-c)}$$

in which a, b, and c are the lengths of the three sides and s is half the perimeter; $s = \frac{1}{2}(a + b + c)$.

(a) Use this formula to find the area of a triangle whose sides are 346, 330, and 104 centimeters.

(b) Use the Pythagorean theorem to show that this is a right triangle. (*Do not* use logarithms to do this.)

(c) Use the formula $A = \frac{1}{2}$(base)(height) and compare to the result in part (a).

***37.** Explain why the mantissas in Table IV are between 0 and 1. (*Hint:* Take $1 \le x < 10$ and now consider the common logarithms of 1, x, and 10.)

***38.** Here is a computation for finding log 6.477. Study this procedure carefully and then find the logarithms below in the same manner.

$$\begin{array}{ccc} & N & & \log N & \\ 0.010\left\{\begin{array}{l} 0.007\left\{\begin{array}{l}6.470\\6.477\end{array}\right.\\ \quad 6.480\end{array}\right. & \begin{array}{c}\longrightarrow\\\longrightarrow\\\longrightarrow\end{array} & \left.\begin{array}{l}\left.\begin{array}{l}0.8109\\ \underline{\quad ? \quad}\end{array}\right\}d\\ 0.8116\end{array}\right\}0.0007 \end{array}$$

$$\frac{0.007}{0.010} = \frac{d}{0.0007}$$

$$0.7 = \frac{d}{0.0007}$$

$$d = (0.7)(0.0007) = 0.00049$$

$$\log N = 0.8109 + 0.00049$$

$$= 0.8114 \quad \text{(rounded to four decimal places)}$$

(a) log 3.042 **(b)** log 7.849 **(c)** log 1.345 **(d)** log 5.444

(e) log 6.803 **(f)** log 2.711 **(g)** log 4.986 **(h)** log 9.008

The method used here is called **linear interpolation**. The rationale behind this method is suggested by the following figure.

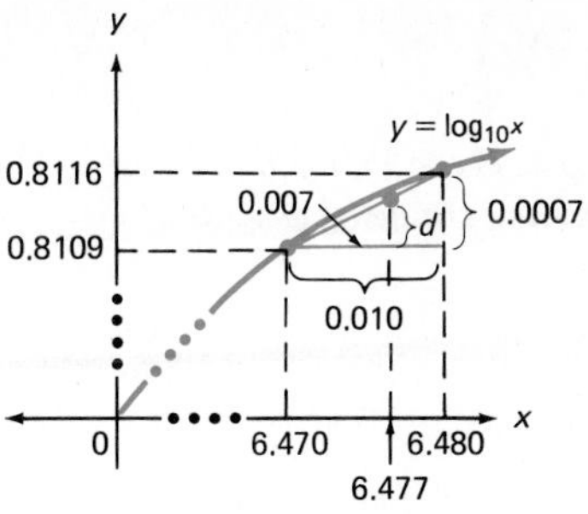

*39. The method in Exercise 38 can be adapted for finding the number when the given logarithm is not an exact table entry. Study the following procedure for finding N in $\log N = 0.7534$, and then find the numbers N below in the same manner.

$$
\begin{array}{cccc}
 & \log N & & N \\
0.0008\left\{\begin{array}{l} 0.0006\left\{\begin{array}{l}0.7528 \\ 0.7534\end{array}\right. \\ \quad\;\; 0.7536 \end{array}\right. & & \begin{array}{c}\longrightarrow \\ \longrightarrow \\ \longrightarrow\end{array} & \left.\begin{array}{c} \left.\begin{array}{c}5.660 \\ \underline{\;?\;}\end{array}\right\} d \\ 5.670 \end{array}\right\} 0.010
\end{array}
$$

$$\frac{0.0006}{0.0008} = \frac{d}{0.01}$$

$$0.75 = \frac{d}{0.01}$$

$$d = (0.01)(0.75) = 0.0075$$

$$N = 5.660 + 0.0075$$

$$= 5.668 \quad \text{(rounded off to three decimal places)}$$

(a) $\log N = 0.4510$ **(b)** $\log N = 0.9672$ **(c)** $\log N = 0.1391$
(d) $\log N = 0.7395$ **(e)** $\log N = 0.6527$ **(f)** $\log N = 0.8749$
(g) $\log N = 0.0092$ **(h)** $\log N = 0.9781$ **(i)** $\log N = 0.3547$

EXPLORATIONS Think Critically

1. Sketch the graphs of two functions f and g such that the domain of f is equal to the range of g, and the domain of g is equal to the range of f. Furthermore, for each function the domain and range are not the same set of numbers.

2. In order to make the graphing of $y = (\sqrt{5})^{2x}$ efficient what should first be done with the form of $(\sqrt{5})^{2x}$?

3. Consider the inequalities $2^x < 3^x$, $2^x > 3^x$, and the equation $2^x = 3^x$. Decide where each of them is true and where they are false.

4. For the function $f(x) = \log_2 x$ form the composition $(f \circ f \circ f)(x) = f(f(f(x)))$ and determine x for which $f(f(f(x))) = 2$.

5. Make use of your knowledge of logarithmic functions to determine the values of c and d, in $y = f(x) = \log_b(cx + d)$, so that the graph of f has the vertical asymptote $x = -\frac{1}{2}$, and x-intercept 0.

6. What is the half-life of a substance that decays exponentially according to $y = Ae^{kt}$, $k < 0$? Use this result to verify the work in Exercise 21, page 000. What is the doubling-time for a substance that grows exponentially according to $y = Ae^{kt}$, $k > 0$?

CHAPTER 9 SUMMARY

Review these key terms and concepts so that you are able to define or describe them. A clear understanding of these will be very helpful when reviewing the developments of this chapter.

Composition of Functions

For functions f and g the composite function g by f, denoted $g \circ f$, has range values defined by

$$(g \circ f)(x) = g(f(x))$$

and domain consisting of all x in the domain of f for which $f(x)$ is in the domain of g.

Inverse Functions

Two functions f and g are said to be inverse functions if and only if:

1. For each x in the domain of g, $g(x)$ is in the domain of f and

$$(f \circ g)(x) = f(g(x)) = x$$

2. For each x in the domain of f, $f(x)$ is in the domain of g and

$$(g \circ f)(x) = g(f(x)) = x$$

To find the inverse $g = f^{-1}$ of f interchange variables in $y = f(x)$ and, if possible, solve for y to obtain f^{-1}.

One-to-One Functions

If to each range value y there corresponds exactly one x-value, then the function is one-to-one and it has an inverse function.

PROPERTIES OF $y = f(x) = b^x$, for $b > 0$, $b \neq 1$

1. The domain consists of all real numbers x.
2. The range consists of all positive numbers y.
3. The function is increasing (the curve is rising) when $b > 1$, and it is decreasing (the curve is falling) when $0 < b < 1$.
4. The curve is concave up for $b > 1$ and for $0 < b < 1$.
5. It is a one-to-one function.
6. The point (0, 1) is on the curve. There is no x-intercept.
7. The x-axis is a horizontal asymptote to the curve, toward the left for $b > 1$ and toward the right for $0 < b < 1$.
8. $b^{x_1}b^{x_2} = b^{x_1+x_2}$; $b^{x_1}/b^{x_2} = b^{x_1-x_2}$; $(b^{x_1})^{x_2} = b^{x_1x_2}$.

$y = b^x$ and $y = \log_b x$ are *inverse functions*. When $b = e$, $\log_b x = \ln x$ is called the **natural logarithm**, and when $b = 10$, $\log_{10} x$ is called the **common logarithm**.

PROPERTIES OF $y = f(x) = \log_b x$ for $b > 0$, $b \neq 1$

1. The domain consists of all positive numbers x.
2. The range consists of all real numbers y.
3. The function increases (the curve is rising) for $b > 1$, and it decreases (the curve is falling) for $0 < b < 1$.
4. The curve is concave down for $b > 1$, and it is concave up for $0 < b < 1$.
5. It is a one-to-one function; if $\log_b(x_1) = \log_b(x_2)$, then $x_1 = x_2$.
6. The point $(1, 0)$ is on the graph. There is no y-intercept.
7. The y-axis is a vertical asymptote to the curve in the downward direction for $b > 1$ and in the upward direction for $0 < b < 1$.
8. $\log_b(b^x) = x$ and $b^{\log_b x} = x$.

The equations $\log_b x = y$ and $b^y = x$ are *equivalent*.

Exponential Growth or Decay

$y = Ae^{kx}$, $k > 0$, defines exponential growth, and $y = Ae^{kx}$, $k < 0$, defines exponential decay.

Compound Interest

$A_t = P\left(1 + \frac{r}{n}\right)^{nt}$ gives the amount A_t when the initial investment P earns compound interest for t years, at the annual rate r, compounded n times per year.

Continuous Compounding

$A = Pe^{rt}$ gives the amount A when the initial investment P earns compound interest for t years at the annual rate r, compounded continuously.

REVIEW EXERCISES

The solutions to the following exercises can be found within the text of Chapter 9. Try to answer each equation before referring to the text.

Section 9.1

1. Evaluate $g(f(10))$ and $f(g(10))$, where $g(x) = \frac{1}{x - 2}$ and $f(x) = \sqrt{x}$.
2. Find $(g \circ f)(x)$ for f and g in Exercise 1.
3. Find $(f \circ g)(x)$ for f and g in Exercise 1.
4. Form the composite $g \circ f$ where $f(x) = \frac{x}{x - 1}$ and $g(x) = x^2$.
5. Find $(f \circ g)(x)$ for f and g in Exercise 4.
6. State the definition of inverse functions.
7. Find the inverse g of $y = f(x) = 2x + 3$. Then show that $(f \circ g)(x) = x$ and $(g \circ f)(x) = x$, and graph both functions on the same axes.

8. Find the inverse f^{-1} of $y = f(x) = \sqrt{x}$. Graph both functions on the same axes, and show that $(f \circ f^{-1})(x) = x$ and $(f^{-1} \circ f)(x) = x$.

9. State the definition of a one-to-one function.

10. Describe the horizontal line test for one-to-one functions.

Section 9.2

11. List the important properties of the exponential function $f(x) = b^x$ for $b < 1$, and for $0 < b < 1$.

12. Use a table of values to sketch $y = 8^x$ on the interval $[-1, 1]$.

13. Sketch $y = 2^x$ and $y = (\frac{1}{2})^x$ on the same axes.

14. Explain how to obtain the graphs of $y = 2^{x-3}$ and $y = 2^x - 1$ from $y = 2^x$.

15. If f is a one-to-one function and $f(x_1) = f(x_2)$, then what can you say about x_1 and x_2?

16. Solve for x: $\dfrac{1}{3^{x-1}} = 81$

17. Solve for x: $b^{x^2-x} = 1$

Section 9.3

18. Which of the following statements are true?
    (a) If $0 < b < 1$, the function $f(x) = b^x$ decreases.
    (b) The point $(0, 1)$ is on the curve $y = \log_b x$.
    (c) $y = \log_b x$, for $b > 1$, increases and the curve is concave down.
    (d) The domain of $y = b^x$ is the same as the range of $y = \log_b x$.
    (e) The x-axis is an asymptote to $y = \log_b x$ and the y-axis is an asymptote to $y = b^x$.

19. Write the equation of the inverse of $y = 2^x$ and graph both on the same axes.

20. Find the domain for $y = \log_2(x - 3)$.

21. Change to logarithmic form: $27^{2/3} = 9$

22. Change to exponential form: $\log_6 \frac{1}{36} = -2$

23. Solve for b: $\log_b 8 = \frac{3}{4}$

24. Solve for x: $\log_{49} x = -\frac{1}{2}$

25. Find the inverse of $f(x) = e^x$.

26. Solve for t: $e^{\ln(2t-1)} = 5$

27. Solve for t: $e^{2t-1} = 5$

Section 9.4

28. Write the three laws of logarithms.

29. Express $\log_b \dfrac{AB^2}{C}$ in terms of $\log_b A$, $\log_b B$, and $\log_b C$.

30. Express $\frac{1}{2} \log_b x - 3 \log_b(x - 1)$ as the logarithm of a single expression in x.

31. Given that $\ln 2 = 0.6931$ and $\ln 3 = 1.0986$, find $\ln \sqrt{12}$.

32. Solve for x: $\log_8(x - 6) + \log_8(x + 6) = 2$.

33. Solve for x: $\log_{10}(x^3 - 1) - \log_{10}(x^2 + x + 1) = 1$

34. Solve for x: $\ln 2x - \ln(x + 5) = 0$.

Section 9.5

35. Evaluate the following using a calculator:
    (a) $e^{2.1}$ **(b)** $\ln 2.1$

36. Find an approximate value for x in $2^x = 35$.

37. Use a calculator to solve for x: $2^{3x} = 35$

38. Solve for x: $(10{,}000)2^x = 100{,}000$

39. A radioactive material is decreasing according to the formula $y = Ae^{-0.2x}$, where y is the amount of material remaining after x years. If the initial amount $A = 80$ grams, how much is left after 3 years?

40. Find the half-life of the radioactive substance in Exercise 39.

41. Solve for k in $\dfrac{A}{2} = Ae^{5750k}$. Leave your answer in terms of natural logs.

42. Use the formula $y = Ae^{(\ln 0.5/5750)x}$ to estimate the age of a skeleton that is found to contain one-third its original amount of carbon-14.

43. A \$5000 investment earns interest at the annual rate of 8.4% compounded monthly. What is the investment worth after one year? After 10 years? How much interest was earned in 10 years?

44. Suppose that \$1000 is invested at 10% interest, compounded continuously. How much will be on deposit in 10 years?

45. How long will it take for the investment in Exercise 44 to double?

Section 9.6

46. Find log 0.0419.

47. Find N if $\log N = 6.1239$.

48. Use common logarithms to estimate $P = (963)(0.00847)$.

49. Use common logarithms to estimate $Q = \dfrac{0.00439}{0.705}$.

50. Use common logarithms to estimate $R = \sqrt[3]{0.0918}$.

51. To determine how much a paint dealer should charge for a gallon of paint, he needs to find out how much the paint cost him per gallon in the first place. The paint is stored in a cylindrical drum $2\frac{1}{2}$ feet in diameter and $3\frac{3}{4}$ feet high. If he paid \$400 for this quantity of paint, what did it cost him per gallon? (Use 1 cubic foot = 7.48 gallons.)

CHAPTER 9 TEST: STANDARD ANSWER

Use these questions to test your knowledge of the basic skills and concepts of Chapter 9. Then test your answers with those given at the back of the book.

1. For $f(x) = \dfrac{1}{1 - x^2}$ and $g(x) = \sqrt{x}$ find the composites $(f \circ g) - (x)$ and $(g \circ f)(x)$.

2. Find the inverse $f^{-1}(x)$ of $f(x) = 3x - 2$. Graph both functions on the same axes.

3. Find the inverse g of $y = f(x) = \sqrt[3]{x} - 1$ and show that $(f \circ g)(x) = x$ and $(g \circ f)(x) = x$.

4. Match each curve with one of the given equations listed below.

(a) $y = b^x;\ b > 1$
(b) $y = b^x;\ 0 < b < 1$
(c) $y = \log_b x;\ b > 1$
(d) $y = \log_b x;\ 0 < b < 1$
(e) $y = \log_b(x + 1);\ b > 1$
(f) $y = b^x - 3;\ b > 1$

(i) (ii) (iii)

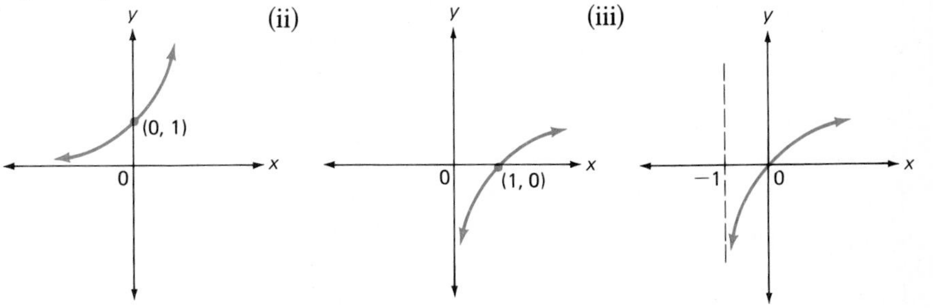

(iv) (v) (vi)

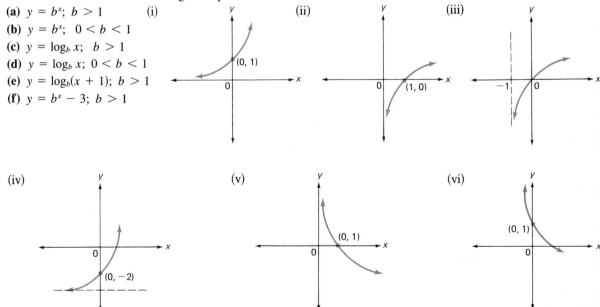

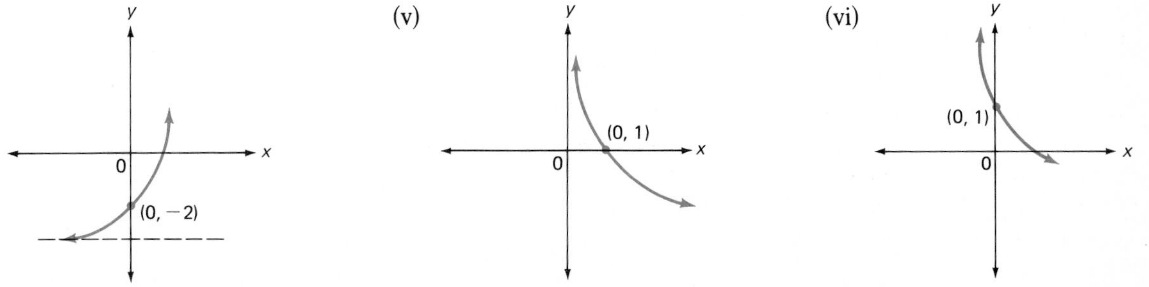

5. Write in exponential form:

(a) $\log_5 125 = 3$ (b) $\log_9 \frac{1}{81} = -2$

6. Write in logarithmic form:

(a) $16^{3/4} = 8$ (b) $7^{-2} = \frac{1}{49}$

7. (a) Solve for b: $\log_b \frac{27}{8} = -3$ (b) Evaluate: $\log_{10} 0.01$

8. Solve for x: (a) $81^x = 9$ (b) $e^{\ln x} = 9$

9. Solve for x: $2^{x^2} = 128$

10. Solve for x: $\frac{1}{2^{x+1}} = 64$

11. Solve for x in $9^{2x} = 321$ and express the answer in terms of natural logarithms.

Find the domain of each function and give the equation of the vertical or the horizontal asymptote.

12. $y = f(x) = 2^x - 4$

13. $y = f(x) = \log_3(x - 4)$

Use the laws of logarithms (as much as possible) to write the following as an expression involving sums, differences, and multiples of logarithms:

14. $\log_b x(x^2 + 1)^{10}$

15. $\ln \frac{x^3}{(x + 1)\sqrt{x^2 + 2}}$

Express as the logarithm of a single expression in x:

16. $\log_7 x + \log_7 2x + \log_7 5$

17. $\frac{1}{3} \log_b x - 2 \log_b(x + 2)$

Solve for x:

18. $\log_{10} x + \log_{10}(3x + 20) = 2$

19. $\log_{25} x^2 - \log_{25}(2x - 5) = \frac{1}{2}$

20. Sketch the graphs of $y = e^{-x}$ and its inverse on the same axes. Write an equation of the inverse in the form $y = g(x)$.

21. A radioactive substance decays according to the formula $y = Ae^{-0.04t}$, where t is the time in years. If the initial amount $A = 50$ grams, find the half-life. Leave the answer in terms of natural logs.

22. A \$2000 investment earns interest at the annual rate of 8%, compounded quarterly. Write the expression that equals the value of the investment after 6 years. (Leave the answer in exponential form.)

23. If \$6000 is invested at the annual interest rate of 7%, compounded continuously, what is the investment worth in 10 years?

24. If $\log_{10} 24 = 1.38$, $\log_{10} 84 = 1.92$, and $\log_{10} 29 = 1.46$, find $\log_{10} \frac{(24)^2}{84\sqrt{29}}$.

25. Solve for x: $\ln 2 - \ln(1 - x) = 1 - \ln(x + 1)$

CHAPTER 9 TEST: MULTIPLE CHOICE

1. Let $f(x) = \sqrt[3]{x}$ and $g(x) = \frac{1}{x^3 - 1}$. Then $g(f(x)) =$

(a) $\frac{1}{\sqrt[3]{x^3 - 1}}$ (b) $\frac{1}{\sqrt[3]{x} - 1}$ (c) $\sqrt[3]{x^3 - 1}$ (d) $\frac{1}{x - 1}$ (e) None of the preceding

2. Which of the following is the inverse of $y = f(x) = -3x + 2$?

(a) $y = -2x + 3$ (b) $y = -\frac{1}{3}x + \frac{2}{3}$ (c) $y = 2 - 3x$ (d) $y = \frac{1}{3}x + 2$ (e) $\frac{1}{3x - 2}$

3. Which of the following are true for the function $y = f(x) = b^x$; $b > 0$ and $b \neq 1$?

I. The domain consists of all real numbers x.

II. The range consists of all positive numbers y.

III. It is a one-to-one function.

(a) Only I **(b)** Only II **(c)** Only III **(d)** I, II, and III **(e)** None of the preceding

4. Solve for x: $b^{x^2-2x} = 1$

(a) $x = 0, x = 2$ **(b)** $x = 1, x = 3$ **(c)** $x = 1$ **(d)** $x = 0$ **(e)** None of the preceding

5. For which of the following is the point $(-1, \frac{1}{4})$ on the graph of the function?

(a) $y = f(x) = -\log_2(x - 1)$ **(b)** $y = f(x) = \log_2(x - 1)$

(c) $y = f(x) = 2^{x-1}$ **(d)** $y = f(x) = -2^{x-1}$ **(e)** $y = f(x) = \dfrac{1}{2^{x+1}}$

6. Which of the following is the domain for the function $y = f(x) = \log_2(x + 3)$?

(a) $x < 3$ **(b)** $x \geq 3$ **(c)** $x < -3$ **(d)** $x \geq -3$ **(e)** None of the preceding

7. Solve for x: $\log_3 x + \log_3(2x + 51) = 4$

(a) $\frac{3}{2}$ **(b)** -27 **(c)** $\frac{47}{3}$ **(d)** 10 **(e)** None of the preceding

8. Solve for b: $\log_b \frac{1}{64} = -\frac{3}{2}$.

(a) $\dfrac{1}{16}$ **(b)** 8 **(c)** 16 **(d)** 512 **(e)** $\dfrac{1}{512}$

9. Which of the following are true?

I. $(\log_b x)^2 = 2 \log_b x$

II. $\log_b A + \log_b B = \log_b(A + B)$

III. $\log_b A - \log_b B = \dfrac{\log_b A}{\log_b B}$

(a) Only I **(b)** Only II **(c)** Only III **(d)** I, II, and III **(e)** None of the preceding

10. Express as the logarithm of a single expression in x: $2 \ln(x - 1) - \ln(x^2 - 1)$

(a) $\dfrac{\ln(x - 1)}{\ln(x + 1)}$ **(b)** $\ln \dfrac{x - 1}{x + 1}$ **(c)** $\ln \dfrac{x + 1}{x - 1}$ **(d)** $\dfrac{\ln(x - 1)}{x + 1}$ **(e)** $2 \ln(x - x^2)$

11. Which of the following is true for the function $y = f(x) = \log_2 x$?

I. For $0 < x < 1$, $y < 0$.

II. The domain is the set of real numbers.

III. The range is the set of positive real numbers.

(a) I only **(b)** II only **(c)** I and II only **(d)** I, II, and III **(e)** All three are false.

12. Solve for x: $9^{x-1} = 4$

(a) $\dfrac{\ln 4}{\ln 9} + 1$ **(b)** $\dfrac{\ln 5}{\ln 9}$ **(c)** $\ln 4 - \ln 9 + 1$ **(d)** $\dfrac{\ln 9}{\ln 4} + 1$ **(e)** $\ln \dfrac{4}{9} - 1$

13. A radioactive substance decays according to the formula $y = Ae^{kx}$, where the initial amount $A = 40$ grams. If after 8 years 30 grams remain, then the amount y remaining after x years is given by

(a) $y = 30e^{(x \ln 0.75)/8}$ **(b)** $y = 40e^{(x \ln 0.75)/8}$ **(c)** $y = 40e^{(x \ln 1.3)/8}$

(d) $y = 40e^{(x \ln 0.75)/\ln 8}$ **(e)** None of the preceding

14. A \$3000 investment earns interest at the annual rate of 8.4% compounded monthly. After five years the total amount of the investment, A_5, is given by

(a) $A_5 = 3000(1.084)^{60}$ **(b)** $A_5 = 3000(1.084)^5$ **(c)** $A_5 = 3000(1.007)^{60}$

(d) $A_5 = 3000(1.007)^{12}$ **(e)** None of the preceding

15. If $Q = \dfrac{(\sqrt[5]{409})(0.0058)}{7.29}$, then $\log Q$ is which of the following?

(a) $\dfrac{1}{5}(\log 409 + \log 0.0058) - \log 7.29$ **(b)** $5 \log 409 + \log 0.0058 - \log 7.29$

(c) $\dfrac{\frac{1}{5} \log 409 + \log 0.0058}{\log 7.29}$ **(d)** $\dfrac{1}{5} \log 409 + \log 0.0058 - \log 7.29$ **(e)** $\dfrac{(\frac{1}{5} \log 409)(\log 0.0058)}{\log 7.29}$

ANSWERS TO THE TEST YOUR UNDERSTANDING EXERCISES

Page 379

1. (a) $g(f(1)) = g(2) = 8$
   (b) $f(g(1)) = f(5) = 14$
   (c) $f(g(0)) = f(4) = 11$
   (d) $g(f(-2)) = g(-7) = 53$
2. (a) $g(f(1)) = g(1) = 1$
   (b) $f(g(1)) = f(1) = 1$
   (c) $f(g(0)) = f(0) = 0$
   (d) $g(f(-2))$ is undefined.
3. (a) $5\sqrt[3]{2}$ (b) $\sqrt[3]{14}$ (c) -1 (d) $-5\sqrt[3]{7}$
4. (a) undefined (b) undefined (c) -2 (d) 0

Page 391

1. 6
2. 2 or -2
3. $\frac{1}{2}$
4. -6
5. -4
6. -1
7. $-\frac{1}{3}$
8. $\frac{2}{3}$
9. $\frac{2}{3}$
10. $-\frac{5}{2}$
11. 3
12. $-\frac{1}{2}$

Page 393

1.

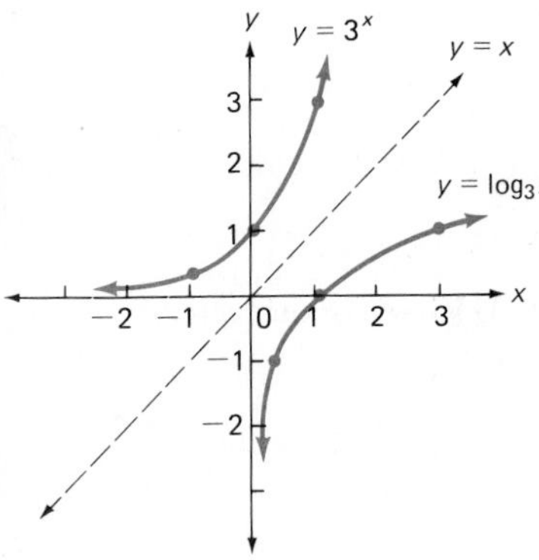

2.

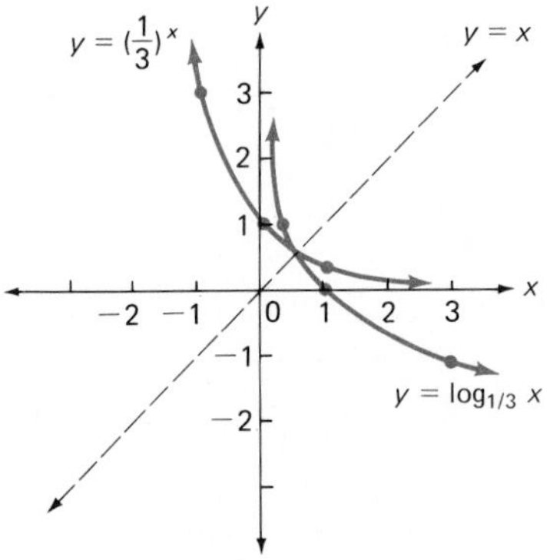

3. Shift 2 units left.
4. Shift 2 units up.
5. Reflect through x-axis.
6. Double the size of each ordinate.

Page 398

1. $\log_b A + \log_b B + \log_b C$
2. $\log_b A - \log_b B - \log_b C$
3. $2 \log_b A + 2 \log_b B - \log_b C$
4. $\ln A + 2 \ln B + 3 \ln C$
5. $\log_b A + \frac{1}{2} \log_b B - \log_b C$
6. $\frac{1}{3} \log_b A - 3 \log_b B - 3 \log_b C$
7. $\log_b 3x^2$
8. $\log_b[\sqrt{x}(x-1)^2]$
9. $\ln \dfrac{2x-1}{(x^2+1)^3}$
10. $\log_b \dfrac{x}{(x-1)(x-2)^2}$
11. 2.8903
12. -0.5234

Page 404

1. $\dfrac{\ln 5}{\ln 4} = \dfrac{1.609}{1.386} = 1.16$
2. $-\dfrac{\ln 5}{\ln 4} = -\dfrac{1.609}{1.386} = -1.16$
3. $\dfrac{\ln 12}{\ln 0.5} = \dfrac{2.485}{-0.693} = -3.59$
4. $\dfrac{\ln 10}{3 \ln 2} = \dfrac{2.303}{3(0.693)} = 1.11$
5. $\dfrac{\ln 15}{\ln 4} = \dfrac{2.708}{1.386} = 1.95$
6. $\dfrac{\ln 4}{5 \ln 67} = \dfrac{1.386}{5(4.205)} = 0.066$

Page 413

1. 2.4265
2. 1.4265
3. 0.4265
4. 9.4265 − 10
5. 8.4265 − 10
6. 4.6232
7. 6.9101 − 10
8. 3.9025
9. 7.0453 − 10
10. 668
11. 6.68
12. 0.668
13. 0.00668
14. 51,600,000
15. 0.0997

CUMULATIVE REVIEW QUESTIONS: CHAPTERS 7–9

1. Explain how the graph of the following can be obtained by applying translations or reflections to the graph of $y = x^2$. Also, find the x- and y-intercepts.
   (a) $y = (x - 3)^2 - 5$ **(b)** $y = -x^2 + \frac{1}{2}$
2. For the following, find (i) the coordinates of the vertex, (ii) the equation of the axis of symmetry, (iii) the domain and range, (iv) where f is increasing or decreasing, and (v) describe the concavity.
   (a) $f(x) = 2x^2 - 3$ **(b)** $f(x) = -(x + 1)^2 + 4$
3. Find the value(s) of h so that the point $(3, -5)$ is on the graph of $y = (x - h)^2 - 9$.
4. Find the discriminant of $y = -2x^2 + 5x - 4$ and determine how many times, if any, the graph intersects the x-axis.
5. Write $y = 2x^2 - 12x - 5$ in the standard form $y = a(x - h)^2 + k$, determine the maximum or minimum value of the function, and find the range.
6. The perimeter of a rectangle is 64 cm. What must be its dimensions so that the square of a diagonal is a minimum?

Solve each inequality.

7. $2x^2 + 9x - 5 < 0$
8. $\dfrac{x - 2}{3x} > 0$
9. Write the equation of the circle with center $(-7, 4)$ and radius 9 in standard form.
10. Find the center and radius of the circle given by $x^2 - 6x + y^2 + 10y = 2$.
11. Write an equation of the tangent line to the circle $x^2 + y^2 = 169$ at the point $P(12, -5)$.
12. Write in standard form the equation of the ellipse having vertices $(\pm 9, 0)$ and foci $(\pm 6, 0)$.
13. Find the foci and vertices of the ellipse $36x^2 + 4y^2 = 144$. What is the length of the minor axis?
14. Suppose we have a miniature model of a satellite orbiting a planet in an elliptical orbit so that the center P of the planet is one focus of the ellipse. Assume that the origin is the center of the ellipse and that the major axis is on the y-axis. The maximum and minimum distances, measured along the major axis, that the satellite is from point P are 25 units and 17 units, respectively. Find the equation of the orbit.
15. Write the equation of the hyperbola having foci $(\pm 8, 0)$ and vertices $(\pm 5, 0)$ in standard form.
16. Write the equation of the hyperbola with asymptotes $y = \pm\frac{2}{3}x$ and vertices $(\pm 3, 0)$ in standard form.
17. Find the foci and vertices of the hyperbola $16y^2 - 9x^2 = 144$, and give the equations of the asymptotes.
18. Find the coordinates of the focus and the equation of the directrix for the parabola $x^2 = 2y$.
19. Write the equation of the parabola with vertex $(0, 0)$ and directrix $x = \frac{5}{3}$. Find the coordinates of the focus.
20. Explain how the graphs of the following can be obtained by applying translations or reflections to the graph of $y = x^3$.
    (a) $y = (x + 2)^3 - 4$ **(b)** $y = |(x - 5)^3|$
21. Find where the function in Question 20(b) is increasing or decreasing, and find the x- and y-intercepts.
22. Find the equation of the curve obtained by shifting the graph of $y = \dfrac{1}{x}$ three units left and four units down. What are the equations of the asymptotes?

23. Find the domain and range of $y = \frac{1}{(x-2)^2} + 1$, describe the concavity, and state where the curve is increasing or decreasing.

24. Write the equation of the curve obtained by shifting the graph of $y = \sqrt{x}$ two units right and five units up.

25. Find the domain of each function, state where it is increasing or decreasing, and write the equations of the asymptotes.

(a) $y = \frac{1}{\sqrt{x+2}}$ **(b)** $y = \frac{1}{\sqrt{x}} + 2$ **(c)** $y = \sqrt[3]{x-1}$

26. Which of the following are symmetric around the y-axis, and which are symmetric through the origin?

(a) $y = x^2$ **(b)** $y = x^3$ **(c)** $y = x^4$ **(d)** $y = \frac{1}{x}$

(e) $y = \frac{1}{x^2}$ **(f)** $y = \sqrt{x}$ **(g)** $y = \sqrt[3]{x}$ **(h)** $y = -\frac{1}{x}$

(i) $y = |x|$ **(j)** $y = \sqrt{|x|}$ **(k)** $y = -2x^4$ **(l)** $y = \sqrt[3]{-x}$

27. Use synthetic division to find the quotient and remainder when $p(x) = 2x^5 - 3x^4 + 7x^2 - 12x + 6$ is divided by $x - 2$.

28. Describe two different methods of finding the remainder when $p(x) = 5x^3 + x^2 - 8x - 4$ is divided by $x + 3$.

29. Use the factor theorem to decide if $x - 1$ is a factor of $p(x) = -x^3 + 5x^2 - 10x + 7$.

30. Find d so that $x - 3$ is a factor of $p(x) = x^4 - 7x^3 + 9x^2 + dx + 15$.

31. Factor $x^3 + 3x^2 - 6x - 8$ completely.

32. Find all roots of $p(x) = 2x^5 - 9x^4 + 14x^3 - 13x^2 + 12x - 4 = 0$.

33. Find the x-intercepts for the graph of $f(x) = \frac{3x^3 - 7x^2 - 43x + 15}{x^2 + x + 1}$.

34. For $f(x) = \frac{2x}{x-5}$ and $g(x) = \frac{x+1}{x}$ find the following and simplify your results.

(a) $f(g(1))$ and $g(f(-1))$ **(b)** $(f \circ g)(x)$ **(c)** $(g \circ f)(x)$

35. Find the inverse f^{-1} for $y = f(x) = -\frac{2}{x} + 3$

36. Describe the graph of $y = 3^x$ by answering the following:

(a) What are the domain and range?
(b) Where is the function increasing or decreasing?
(c) Describe the concavity.
(d) Find the x- and y-intercepts, if any.
(e) Write the equations of the asymptotes, if any.

37. Find the equation of the curve obtained from $y = \frac{1}{2^x}$ by multiplying the ordinates by 3 and shifting the graph 3 units to the right.

38. Use the one-to-one property of exponential functions to solve $\frac{1}{3^{2x-x^2}} = 27$.

39. Find the inverse g of $f(x) = 5^x$ and state the domain of g.

40. **(a)** Solve for x: $e^{\ln(x^2-x)} = 6$.
(b) Simplify: $80e^{3\ln 1/2}$

41. **(a)** Solve for b: $\log_b 27 = \frac{3}{2}$
(b) Solve for y: $\log_8 4 = y$

42. Solve $e^{2x-1} = 25$ for x and state the solution in terms of natural logarithms.

43. Use the laws of logarithms, as much as possible, to convert $\log_b \frac{x^3\sqrt{x^2+1}}{5x-1}$ into an expression involving sums, differences, and multiples of logarithms in base b.

Solve each equation.

44. $\log_8(x-6) + \log_8(x+6) = 2$

45. $\log_5(3x^2 + x - 2) - \log_5(x+1) = 2$

46. Explain how the graph of $y = \ln \frac{1}{x}$ can be obtained from the graph of $y = \ln x$ by using translations or reflections.

47. **(a)** Solve for x in terms of natural logarithms: $3^{2x} = 12$.
    (b) Use a calculator to find x in part (a) to two decimal places.

48. A bacterial culture is growing according to the formula $y = 50{,}000e^{0.8x}$, where x is the time in hours. Use a calculator to find how long it will take for the culture to grow to 500,000 bacteria.

49. A \$5000 investment earns compound interest at the annual rate of 8%. Find the value of the investment (to the nearest dollar) after two years for each of the following kinds of compounding.
    (a) Weekly **(b)** Continuous

50. Evaluate $\frac{(0.24)^3(\sqrt{9.56})}{289}$ using common logarithms.

CHAPTER 10 SYSTEMS OF EQUATIONS AND INEQUALITIES

10.1 SYSTEMS OF LINEAR EQUATIONS IN TWO VARIABLES

Whenever there are two nonparallel lines in a plane, they must intersect in exactly one point. When the equations of two such lines are given, this point of intersection can be found by graphing the **system of equations**, as in the figure.

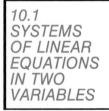

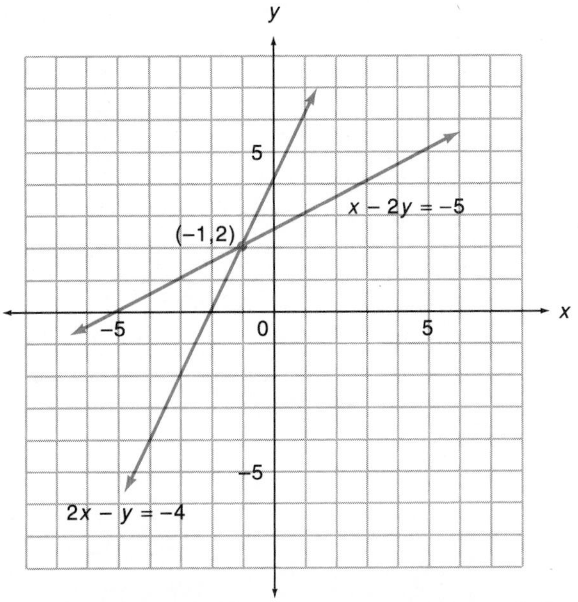

The coordinates of P in the figure *appear* to be $(-1, 2)$. Can we be certain of this? Remember that every line in the plane is the graph of a linear equation, and that a point is on the line if and only if its coordinates satisfy the equation. Therefore, since P is on both lines, its coordinates must satisfy *both* equations. We can check this solution by substituting $x = -1$ and $y = 2$ in each equation.

For ℓ_1:

$$\begin{aligned} 2x - y &= -4 \\ 2(-1) - (2) &= -4 \\ -2 - 2 &= -4 \\ -4 &= -4 \ \checkmark \end{aligned}$$

For ℓ_2:

$$\begin{aligned} x - 2y &= -5 \\ (-1) - 2(2) &= -5 \\ -1 - 4 &= -5 \\ -5 &= -5 \ \checkmark \end{aligned}$$

Thus $(-1, 2)$ satisfies both equations and we can conclude that the ordered pair $(-1, 2)$ is the solution of this system.

Solving a system of equations by graphing is laborious, if not impossible, especially if the point of intersection does not involve integers. An alternative approach is to use the **substitution method**. This consists of solving one of the equations for one of the variables, and then substituting this value into the second equation.

For the point of intersection of two lines, the x and y values are the same for both equations. This provides the rationale for the substitution.

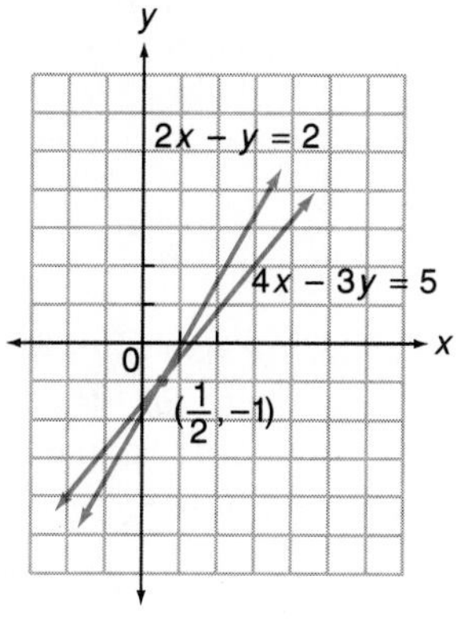

EXAMPLE 1 Solve the system.

$$\begin{aligned} 4x - 3y &= 5 \\ 2x - y &= 2 \end{aligned}$$

Solution Here it is convenient to use the second equation to solve for y.

$$y = 2x - 2$$

Now substitute this y-value into the first equation and solve for x.

$$\begin{aligned} 4x - 3(2x - 2) &= 5 \\ 4x - 6x + 6 &= 5 \\ -2x &= -1 \\ x &= \tfrac{1}{2} \end{aligned}$$

The corresponding y-value can be found by using $y = 2x - 2$.

$$y = 2x - 2 = 2(\tfrac{1}{2}) - 2 = -1$$

The solution is $(\frac{1}{2}, -1)$; that is, $x = \frac{1}{2}$ and $y = -1$.

As a check, substitute these values into the first equation.

Check:

$$\begin{aligned} 4x - 3y &= 5 \\ 4(\tfrac{1}{2}) - 3(-1) &= 2 + 3 = 5 \ \checkmark \end{aligned}$$

Show that $(\frac{1}{2}, -1)$ satisfies the second equation as well.

■

Although the substitution method will suffice to solve any system of linear equations, there is an another method that proves helpful in many cases. Consider, for example, this system:

$$2x + 3y = 16$$
$$5x - 3y = 19$$

Note that the term $3y$ is added in the first equation and subtracted in the second. Thus, if the two equations are added, the y-terms will be *eliminated*. This will lead to a single equation in one variable.

$$\begin{aligned} 2x + 3y &= 16 \\ 5x - 3y &= 19 \\ \hline 7x \quad &= 35 \quad \text{(By addition of equals)} \\ x \quad &= 5 \end{aligned}$$

We can then find the corresponding value for y by substituting into either of the two equations.

$$\begin{aligned} 2x + 3y &= 16 \\ 2(5) + 3y &= 16 \\ 10 + 3y &= 16 \\ 3y &= 6 \\ y &= 2 \end{aligned}$$

The solution of the system is the ordered pair (5, 2). This can be checked by substituting these values into the second equation.

$$\begin{aligned} 5x - 3y &= 19 \\ 5(5) - 3(2) = 25 - 6 &= 19 \end{aligned}$$

When the coefficients of one of the variables in two equations are opposites of one another, the system can be solved by addition. If the coefficients are the same, the system can be solved by subtraction. Otherwise, we need to search for a suitable multiplication that will accomplish this objective as illustrated by this system of equations:

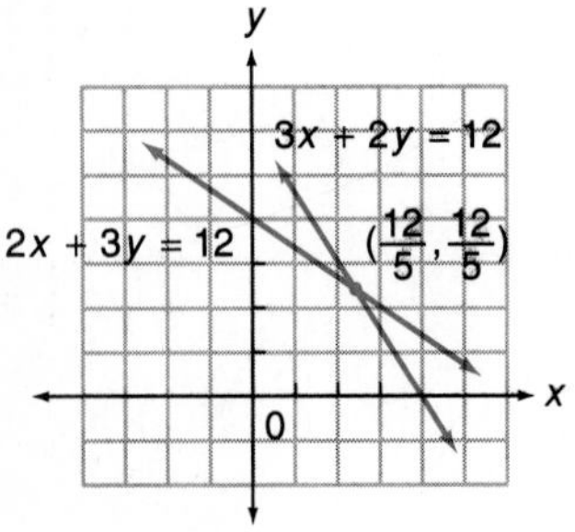

$$2x + 3y = 12$$
$$3x + 2y = 12$$

The idea here is to alter the equations so that the coefficients of one of the variables are either negatives of one another or equal to each other. We may, for example, multiply the first equation by 3 and multiply the second by -2. In the resulting system, we may add equals to equals to eliminate the variable x.

$$\begin{aligned} 6x + 9y &= 36 \\ -6x - 4y &= -24 \\ \hline 5y &= 12 \\ y &= \tfrac{12}{5} \end{aligned}$$

Substituting $y = \frac{12}{5}$ into one of the original equations gives $x = \frac{12}{5}$ so that the ordered pair $(\frac{12}{5}, \frac{12}{5})$ is the solution of the system.

If in the preceding solution the second equation is multiplied by 2 instead of -2, the system becomes

$$6x + 9y = 36$$
$$6x + 4y = 24$$

Now x can be eliminated by subtracting the equations.

The method just illustrated is called the **multiplication-addition** (or **multiplication-subtraction) method** for solving a system of linear equations. As in the next example, this method can be condensed into a compact form.

EXAMPLE 2 Solve the system by the multiplication-addition method.

$$\tfrac{1}{3}x - \tfrac{2}{5}y = 4$$
$$7x + 3y = 27$$

Solution

Multiply both sides of the first equation by 15, and both sides of the second equation by 2. Observe that $2(7x + 3y = 27)$ *is an abbreviation for* $2(7x + 3y) = 2(27)$.

$$15(\tfrac{1}{3}x - \tfrac{2}{5}y = 4) \Rightarrow 5x - 6y = 60$$
$$2(7x + 3y = 27) \Rightarrow \underline{14x + 6y = 54}$$
$$\text{Add:}\quad 19x = 114$$
$$x = 6$$

Substitute to solve for y.

$$7x + 3y = 27 \Rightarrow 7(6) + 3y = 27$$
$$3y = -15$$
$$y = -5$$

Check: $\frac{1}{3}(6) - \frac{2}{5}(-5) = 4$; $7(6) + 3(-5) = 27$.

The solution is $(6, -5)$. ■

TEST YOUR UNDERSTANDING Think Carefully

Use the substitution method to solve each linear system.

1. $y = 3x - 1$, $y = -5x + 7$

2. $y = 4x + 16$, $y = -\frac{2}{5}x + \frac{14}{5}$

3. $4x - 3y = 11$, $y = 6x - 13$

4. $2x + 2y = \frac{4}{5}$, $-7x + 2y = -1$

5. $x + 7y = 3$, $5x + 12y = -8$

6. $4x - 2y = 40$, $-3x + 3y = 45$

Use the multiplication-addition (or subtraction) method to solve each linear system.

7. $3x + 4y = 5$, $5x + 6y = 7$

8. $-8x + 5y = -19$, $4x + 2y = -4$

9. $\frac{1}{7}x + \frac{5}{2}y = 2$, $\frac{1}{2}x - 7y = -\frac{17}{4}$

(Answers Page 490)

When a linear system has a unique solution, as in the preceding examples, we say that the system is **consistent**. Graphically, this means that the lines intersect. There are two other possibilities as demonstrated next.

An **inconsistent** system:

$$39x - 91y = -28$$
$$6x - 14y = 7$$

Note that the slope of each line is the same; $m = \frac{3}{7}$.

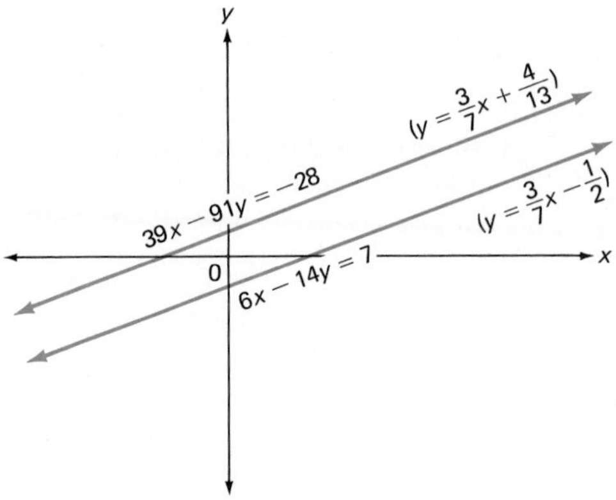

The two lines have the same slope but different y-intercepts. Therefore, the lines are parallel and there is no point of intersection. An inconsistent system has no solution.

Since there is no (x, y) common to both equations, the substitution step in this attempted solution leads to a false result.

An inconsistent system can also be identified algebraically by obtaining a false result when *trying* to solve such a system. Thus for the preceding system, if the second equation is solved for y, we obtain

$$y = \tfrac{3}{7}x - \tfrac{1}{2}$$

Now substitute into the first equation.

Arriving at a false result, as shown here, is not a wasted effort. As long as there are no computational errors, such a false conclusion tells us that the given system has no solution. That is, ***the system has been solved by learning that it has no solution****.*

$$39x - 91(\tfrac{3}{7}x - \tfrac{1}{2}) = -28$$
$$39x - 39x + \tfrac{91}{2} = -28$$
$$\tfrac{91}{2} = -28 \qquad \text{False}$$

This false result tells us that the system is inconsistent.

A **dependent** system:

$$y = -\tfrac{2}{3}x + 5$$
$$2x + 3y = 15$$

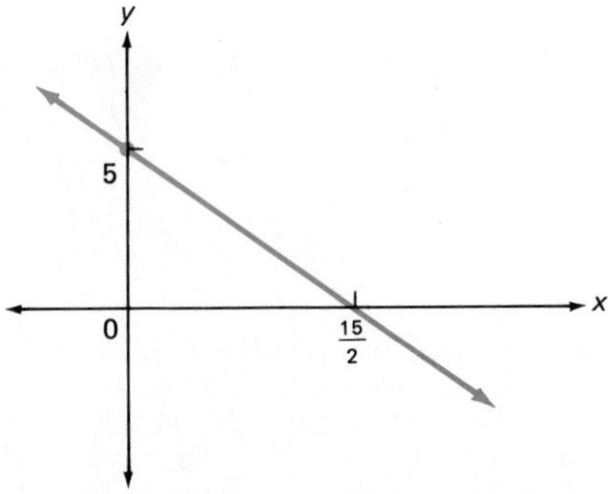

When the second equation is solved for y, we have:

$$2x + 3y = 15$$
$$3y = -2x + 15$$
$$y = -\tfrac{2}{3}x + 5$$

The graph is a single line. Each equation in the given system can be changed into the algebraic form of the other. In a dependent system, each pair (x, y) that satisfies one equation must also satisfy the other; the equations are equivalent. Therefore, a dependent system has an infinite number of solutions. Any attempt at "solving" such a system will result in some statement that is always true (an **identity**). For example, using the preceding system, we have

$$3(y = -\tfrac{2}{3}x + 5) \Rightarrow 3y = -2x + 15 \Rightarrow 2x + 3y = 15$$
$$2x + 3y = 15 \Rightarrow 2x + 3y = 15 \Rightarrow \underline{2x + 3y = 15}$$
$$\text{Subtract:} \quad 0 = 0 \quad \text{True}$$

This true result tells us that the system is dependent.

EXERCISES 10.1

Solve the given system by graphing the equations and locating the coordinates of the point of intersection.

1. $x - 2y = 3$, $x + y = 0$
2. $x - 2y = 3$, $y - x = -4$
3. $2x - y = 6$, $x + 2y = -2$

Solve each system by the substitution method.

4. $x - 2y = 3$, $x + y = 0$
5. $x - 2y = 3$, $y - x = -4$
6. $4x + 3y = 10$, $6x + 9y = 21$
7. $2x + y = -10$, $6x - 3y = 6$
8. $-3x + 6y = 0$, $4x + y = 9$
9. $x - y = 14$, $3x + y = 2$
10. $x - 2y = 7$, $x + 2y = 3$
11. $3x - y = 11$, $2x + y = 4$
12. $2x + 4y = -2$, $4x - 2y = -24$
13. $3x + 2y = -15$, $x - 3y = -16$
14. $5x - 2y = 9$, $10x - 5y = 10$
15. $-4x + 6y = 1$, $8x - 9y = 0$
16. $3y = 5 - x$, $7x = 11 - 9y$
17. $x + y = \frac{5}{2}$, $2x - 3y = 1$
18. $6x + \frac{5}{2}y = \frac{7}{2}$, $\frac{1}{2}x - 5y = \frac{4}{3}$

Solve each system by the multiplication-addition (or subtraction method.)

19. $y = x - 1$, $y = -x + 5$
20. $2x + 3y = -3$, $-x + 3y = 6$
21. $2x + 3y = 2$, $4x - 3y = 1$
22. $-3x + y = 16$, $2x - y = 10$
23. $2x - 3y = -2$, $2x + y = 14$
24. $2x + 4y = 24$, $-3x + 5y = -25$
25. $2x + 3y = 4$, $5x + 6y = 7$
26. $3y - 9x = 30$, $8x - 4y = 24$
27. $2x - 6y = -16$, $5x - 3y = 8$
28. $4x = 7y - 6$, $9y = -12x + 12$

    $x = y - 7$, $3y = 2x + 16$

30. $\frac{1}{2}x + 3y = 6$, $-x - 8y = 18$

Solve the given system in each of Exercises 31 through 42 using any method.

31. $2x + y = 6$
$3x - 4y = 12$

32. $-3x + 8y = 16$
$16x - 5y = 103$

33. $3a - 8b = -16$
$7a + 19b = -188$

34. $16m - 5n = 103$
$7m + 19n = -188$

35. $v = -4u + 1$
$u = \frac{1}{4}v - \frac{1}{4}$

36. $3s + t - 3 = 0$
$2s - 3t - 2 = 0$

37. $\frac{1}{4}r + \frac{1}{3}s = \frac{5}{12}$
$\frac{1}{2}r + s = 1$

38. $0.1x + 0.2y = 0.7$
$0.01x - 0.01y = 0.04$

39. $0.2x - 0.5y = 0.9$
$0.01x + 0.03y = -0.01$

40. $\dfrac{x-2}{5} + \dfrac{y+1}{10} = 1$
$\dfrac{x+2}{3} - \dfrac{y+3}{2} = 4$

41. $2(x - y - 1) = 1 - 2x$
$6(x - y) = 4 - 3(3y - x)$

42. $\frac{1}{3}x - \frac{2}{3}y = 1$
$\frac{3}{2}y - \frac{9}{2}x = -21$

Decide whether the given systems are consistent, inconsistent, or dependent.

43. $x + y = 2$
$x + y = 3$

44. $3x - y = 7$
$-9x + 3y = -21$

45. $2x - 3y = 8$
$-8x + 12y = 33$

46. $\frac{1}{2}x + 5y = -4$
$7x - 3y = 17$

47. $-6x + 3y = 9$
$10x + 5y = -1$

48. $20x + 36y = -27$
$\frac{5}{27}x + \frac{1}{3}y = -\frac{1}{4}$

49. $4x - 12y = 3$
$x + \frac{1}{3}y = 3$

50. $-7x + y = 2$
$28x - 4y = 2$

51. $2x + 5y = -20$
$x = -\frac{5}{2}y - 10$

52. $x - y = 3$
$-\frac{1}{3}x + \frac{1}{3}y = 1$

53. $x - 5y = 15$
$0.01x - 0.05y = 0.5$

54. $2x = 3$
$-4y = 8$

55. Points $(-8, -16)$, $(0, 10)$, and $(12, 14)$ are three vertices of a parallelogram. Find the coordinates of the fourth vertex if it is located in the third quadrant.

56. Find the point of intersection for the diagonals of the parallelogram given in Exercise 55.

57. A line with equation $ax + by = 3$ passes through $(6, 3)$ and $(-1, -1)$. Find a and b without finding the slope.

58. A line with equation $y = mx + b$ passes through the points $(-\frac{1}{3}, -6)$ and $(2, 1)$. Find m and b by substituting the coordinates into the equation and solving the resulting system.

For each nonlinear system let $u = \dfrac{1}{x}$ and $v = \dfrac{1}{y}$, solve the resulting linear system for u and v, and then solve for x and y.

59. $-\dfrac{6}{x} + \dfrac{8}{y} = -10$
$\dfrac{5}{x} - \dfrac{7}{y} = 6$

60. $\dfrac{1}{x} + \dfrac{3}{y} = 5$
$\dfrac{2}{x} + \dfrac{4}{y} = 6$

61. $\dfrac{6}{x} - \dfrac{1}{y} = -5$
$-\dfrac{3}{x} + \dfrac{2}{3y} = 4$

62. $\dfrac{4}{x} - \dfrac{8}{y} = 5$
$\dfrac{6}{x} + \dfrac{5}{y} = -1$

63. $\dfrac{4}{x} - \dfrac{5}{y} = 3$
$\dfrac{16}{x} + \dfrac{2}{y} = 3$

64. $-\dfrac{3}{x} - \dfrac{7}{y} = 8$
$\dfrac{1}{2x} + \dfrac{7}{10y} = 1$

Written Assignment: Explain, with examples, the distinction between a consistent and an inconsistent system of linear equations in two variables.

CHALLENGE
Think Creatively

Find the coordinates of the given triangle ABC and then answer the questions below.

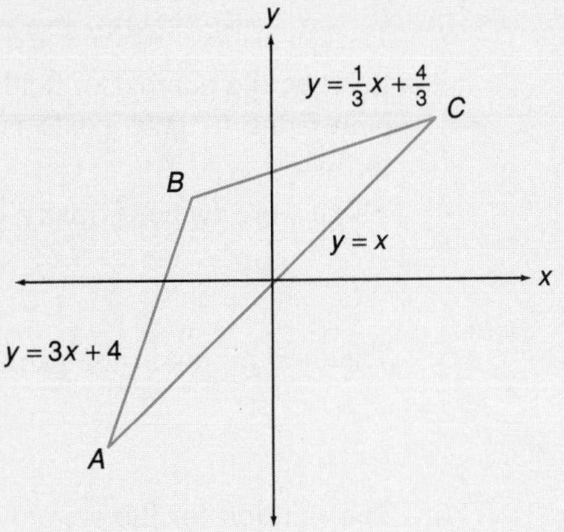

1. An altitude of a triangle is the perpendicular segment from a vertex to the opposite side. In triangle ABC the altitude from B to side AC is internal to the triangle and the other two altitudes are external. Find the equations determined by the altitudes and prove that they all pass through a common point P. (*Hint:* Find the point of intersection P of two altitudes and then show that the third one also goes through P.)
2. The line that is perpendicular to the side of a triangle at its midpoint is referred to as the perpendicular bisector of the side. Find the equations of the three perpendicular bisectors of triangle ABC and prove that they all pass through a common point Q.
3. A median of a triangle is the line segment connecting a vertex to the midpoint of the opposite side. Prove that the three medians all pass through a common point R.
4. Prove that points P, Q, and R are on the same line. (This line is known as the **Euler line** for triangle ABC. There is one such line for every triangle.)

10.2 APPLICATIONS OF LINEAR SYSTEMS

There are many word problems that can be solved by using systems of linear equations. In some cases only one equation in one variable need be used. However, it is worthwhile to learn how to use more than one variable in order to simplify the process of translating from the verbal to the mathematical form.

Solving most word problems consists of two major parts. The first job is to translate the prose into mathematical symbolism; appropriate algebraic methods must then be used to get the answer. The problems that we will now encounter can all be translated into systems of two linear equations. Since we have developed the techniques for solving these systems, the most challenging part of solving such word problems lies in finding correct translations.

The guidelines given in Section 3.2, where you learned how to solve word problems leading to one equation in one variable, can be adjusted and used here. The solution to Example 1 is based on these guidelines.

EXAMPLE 1 A field goal in basketball is worth 2 or 3 points and a free throw is worth 1 point. In a recent game the school basketball team scored 85 points. There were no 3-point field goals, but there were twice as many 2-point field goals as free throws. How many of each were there?

Solution

1. Read the problem once or twice.
2. The quantities involved are numbers of field goals and of free throws.
3. Let x be the number of field goals and y the number of free throws. Then the points due to field goals is $2x$, and y is the number of points due to free throws.
4. There were twice as many field goals as free throws, so

$$x = 2y$$

The total points in the game was 85; therefore,

$$2x + y = 85$$

5. The solution for the system

$$x = 2y$$
$$2x + y = 85$$

is (34, 17).

6. *Check:* 34 field goals (at 2 points each) gives 68 points; and 17 free throws (1 point each) produces a total of 68 + 17 = 85 points. Also, there are twice as many field goals as free throws. ■

Notice that the answer in Example 1 was checked by returning to the language of the original problem rather than to any of the equations. It is essential to do this for word problems because the equations themselves may be incorrect translations of the problem.

It is not essential to follow a prescribed set of general guidelines in solving word problems. With some experience you will probably develop your own methods. The following examples are intended to help you gain such experience.

EXAMPLE 2 The sum of two numbers is 101. Five times the smaller is one more than the larger. Find the numbers.

Solution Let x = smaller number and y = larger number. Then

$$\underbrace{x + y}_{\left(\substack{\text{sum of the}\\ \text{two numbers}}\right)} = 101$$

and

$$\underbrace{5x}_{\left(\substack{\text{five times}\\ \text{the smaller}}\right)} = \underbrace{y + 1}_{\left(\substack{\text{1 more than}\\ \text{the larger}}\right)}$$

Solving the system

$$x + y = 101$$
$$5x = y + 1$$

we find that $x = 17$ and $y = 84$. You can check these results by returning to the statement of the problem. ■

EXAMPLE 3 There is a two-digit number that equals four times the sum of its digits. Also, if 27 is added to the number, then its digits would be reversed. What is the number?

Solution The value of a two-digit number is obtained by multiplying its tens' digit by ten and adding the units' digit; for example, $53 = 5(10) + 3$.

Let t be the tens' digit and u the units' digit. Then the value of the number is $10t + u$. The sum of the digits is $t + u$, and, since the number is four times this sum,

$$10t + u = 4(t + u) \quad (1)$$

The value of the number obtained by reversing the digits is $10u + t$. Then, according to the second sentence of the stated problem,

$$(10t + u) + 27 = 10u + t \quad (2)$$

When equations (1) and (2) are simplified, we get the system

$$6t - 3u = 0$$
$$9t - 9u = -27$$

Check the result of Example 3 in the original statement of the problem.

Solving this system gives $t = 3$ and $u = 6$. The number is 36. ■

EXAMPLE 4 For her participation in a recent "walk-for-poverty," Ellen collected \$2 per mile for a total of \$52. She recorded that for a certain time she walked at the rate of 3 miles per hour (mph) and the rest at 4 mph. Afterward she mentioned that it was too bad that she did not have the energy to reverse the rates. For if she could have walked 4 mph for the same time that she actually walked 3 mph, and vice versa, she would have collected a total of \$60. How long did her walk take?

Solution The problem asks for the total time of the walk. This time is broken into two parts: the time she walked at 3 mph and the time at 4 mph.

Let x = time at 3 mph

and y = time at 4 mph

We want to find $x + y$.

Since **distance = rate × time**, $3x$ is the distance at the 3-mph rate and $4y$ is the distance at 4 mph; the total distance is the sum $3x + 4y$. This total must equal 26 because she earned \$52 at \$2 per mile. Thus

$$3x + 4y = 26$$

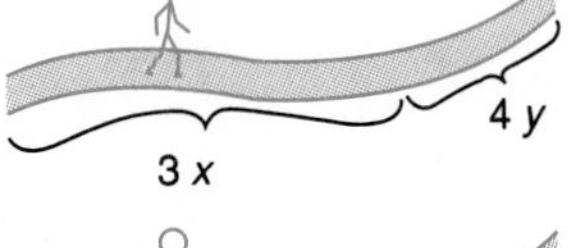

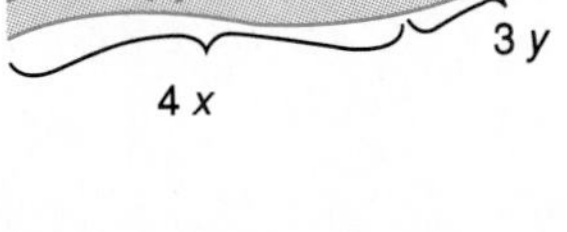

The \$60 she could have earned would have required walking 30 miles. And this 30 miles, she said, would have been possible by reversing the rates for the actual times that she did walk. This says that $4x + 3y = 30$. We now have the system

$$3x + 4y = 26$$
$$4x + 3y = 30$$

The common solution is (6, 2), and therefore the total time for the walk is 6 + 2, or 8 hours.

Check: She walked $3 \times 6 = 18$ miles at 3 mph, and $4 \times 2 = 8$ miles at 4 mph, for a total of 26 miles. At \$2 per mile, she collected \$52. Reversing the rates gives $(4 \times 6) + (3 \times 2) = 30$ miles, for which she would have earned \$60. ■

EXAMPLE 5 A grocer sells Brazilian coffee at \$5.00 per pound and Colombian coffee at \$8.50 per pound. How many pounds of each should he mix in order to have a blend of 50 pounds that he can sell at \$7.10 per pound?

Solution

Let x = the number of pounds of Brazilian coffee to be used.
Let y = the number of pounds of Colombian coffee to be used.

It is frequently helpful, although not necessary, to summarize the given information in a table such as the following.

	Number of Pounds	Cost per Pound	Total Cost
Brazilian	x	5.00	$5.00x$
Colombian	y	8.50	$8.50y$
Blend	50	7.10	50(7.10)

Use the information in the first and third columns to write this system of equations:

$$x + y = 50$$
$$5.00x + 8.50y = 355 \qquad 50(7.10) = 355$$

To check this solution show that 20 pounds of the \$5.00 coffee and 30 pounds of the \$8.50 coffee will give the same income as 50 pounds of a \$7.10 blend.

Show that the solution for the system is $x = 20$ and $y = 30$. Thus the grocer should mix 20 pounds of the Brazilian coffee with 30 pounds of the Colombian coffee. This will give him 50 pounds of a blend to sell at \$7.10 per pound. ■

The next example demonstrates how a company can find the **break-even point** for a product that they produce and sell. This is the point at which the companies' expenses equal their revenues. Beyond this point the company makes a profit, and below it the company has a loss.

EXAMPLE 6 It costs the Roller King Company \$8 to produce one skateboard. In addition, there is a \$200 daily fixed cost for building maintenance. **(a)** Find the total daily cost for producing x skateboards per day. **(b)** Find the total daily revenue if the company sells x skateboards per day for \$16 each. **(c)** Find the daily break-even point. That is, find the coordinates of the point at which the cost equals the revenue.

Solution

(a)

$$\text{Total daily cost} = \begin{pmatrix}\text{Cost per}\\ \text{item}\end{pmatrix}\begin{pmatrix}\text{Number}\\ \text{of items}\end{pmatrix} + \text{Fixed cost}$$

$$y = 8 \cdot x + 200$$

(b) Total daily revenue $=$ (Sale price per item)(Number of items)

$$y = 16 \cdot x$$

(c) The daily profit or loss can be observed from the graphs of the cost and revenue lines drawn in the same coordinate system.

For $x > 25$, the revenue is greater than the cost, resulting in a profit. For $x < 25$, the cost is more than the revenue, resulting in a loss.

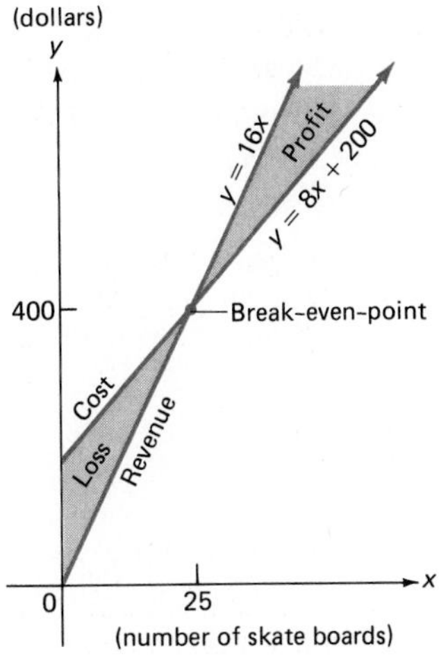

The point where the lines intersect is the break-even point. At this point the cost for producing x skateboards is the same as the revenue for selling x skateboards on a daily basis. To find the break-even point, we solve this system.

$$\text{Cost:} \quad y = 8x + 200$$

$$\text{Revenue:} \quad y = 16x$$

Some of these exercises can be solved by using just one equation with one variable. However, in most cases it is easier to use two equations and two variables.

Thus, $16x = 8x + 200$, which gives $x = 25$ and $y = 16(25) = 400$. The break-even point is (25, 400). When the company produces and sells 25 skateboards per day, their cost and revenue are \$400 each. ■

EXERCISES 10.2

1. The sum of two numbers is 56 and their difference is 30. Find the numbers.
2. The total points that a basketball team scored was 96. If there were two-and-a-half times as many field goals as free throws, how many of each were there? (Field goals count 2 points; free throws count 1 point. There were no 3-point field goals.)
3. The perimeter of a rectangle is 60 centimeters. If the length is three more than twice the width, find the dimensions.
4. Three times the larger of two numbers is 10 more than twice the smaller. Five times the smaller is 11 less than four times the larger. What are the numbers?
5. The sum of two numbers is 63. If three times the smaller is four more than twice the larger, what are the numbers?
6. The perimeter of an isosceles triangle is 35 inches. If five times the base is one less than one of the equal sides, how long are the sides?

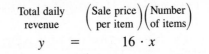

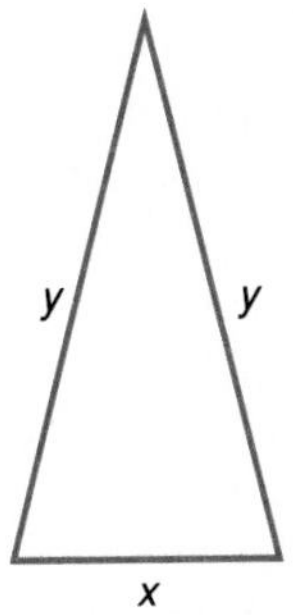

7. A shopper pays \$3.82 for $9\frac{1}{2}$ pounds of vegetables consisting of potatoes and string beans. If potatoes cost 35¢ per pound and string beans cost 68¢ per pound, how much of each was purchased?

8. The difference between two numbers is 32. If 40 added to the larger equals twice the smaller, what are the numbers?

9. During a round of golf a player only scored fours and fives per hole. If he played 18 holes and his total score was 80, how many holes did he play in four strokes and how many in five?

10. A football team scored 54 points. They scored touchdowns at 6 points each, some points-after-touchdown at 1 point each, and some field goals at 3 points each. If there were two-and-one-third as many touchdowns as field goals, and just as many points-after-touchdown as field goals, how many of each were there?

11. The tens' digit of a two-digit number is three less than the units digit. If two were added to the number, the result would be seven times the units' digit. What is the number?

12. A two-digit number is seven times the sum of its digits. If 36 is subtracted from the number, then its digits will be reversed. Find the number.

13. The tuition fee at a college plus the room and board is \$8400 per year. The tuition is \$1200 less then twice the room and board. How much is the tuition, and what does the room and board cost?

14. Harry's age is two more than five times Wendy's age. In 17 years Harry will be twice as old as Wendy. How old is each now?

15. A college student had a work-study scholarship that paid \$4.20 per hour. She also made \$2.00 per hour babysitting. Her income one week was \$80.00 for a total of $23\frac{1}{2}$ hours employment. How many hours did she spend on each of the two jobs?

16. Karin walked at a steady rate for a half-hour, and then rode a bicycle for another half-hour, also at a constant rate. The total distance traveled was 7 miles. The next day, going at the same rates, she covered 6 miles, only she walked for two-thirds of an hour and rode for a third of an hour. What was her speed walking and riding?

17. David walked 19 kilometers. The first part of the walk was at 5 kph (kilometers per hour) and the rest at 3 kph. He would have covered 2 kilometers less if he had reversed the rates; that is, if he had walked at 3 kph and at 5 kph for the same times that he actually walked at 5 kph and 3 kph, respectively. How long did it take him to walk the 19 kilometers?

18. The treasurer of the student body reported that the receipts for the last concert totaled \$916 and that 560 people attended. If students paid \$1.25 per ticket and nonstudents paid \$2.25 per ticket, how many students attended the concert?

19. An airplane, flying with a tail wind, takes 2 hours and 40 minutes to travel 1120 miles. The plane makes the return trip against the wind in 2 hours and 48 minutes. What is the wind velocity and what is the speed of the plane in still air? (Assume that both velocities are constant; add the velocities for the downwind trip, and subtract them for the return trip.)

20. The cost of 10 pounds of potatoes and 4 pounds of apples is \$6.16, while 4 pounds of potatoes and 8 pounds of apples cost \$6.88. What is the cost per pound of potatoes and of apples?

21. The units' digit of a two-digit number is one less than twice the tens' digit. The difference of the digits is 4. Find the number.

22. There are only nickels and quarters in a child's bank. There are 22 coins in the bank having a total value of \$3.90. How many nickels and how many quarters are there?

23. A swimmer going downstream takes 1 hour, 20 minutes to travel a certain distance. It takes the swimmer 4 hours to make the return trip against the current. If the river flows at the rate of $1\frac{1}{2}$ miles per hour, find the rate of the swimmer in still water and the distance traveled one way.

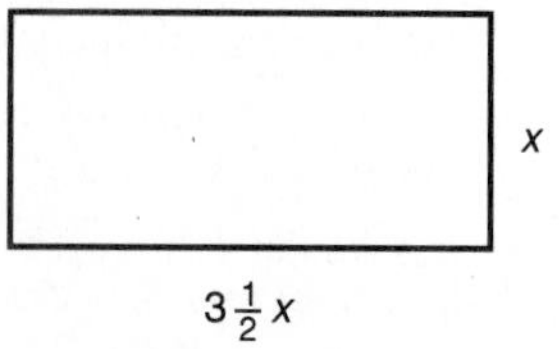

24. The perimeter of a rectangle is 72 inches. The length is three-and-one-half times as large as the width. Find the dimensions.

25. Twice the larger of two numbers plus the smaller is -25. The larger minus three times the smaller is 138. Find the numbers.

26. Ellen and Robert went to the store to buy some presents. They had a total of \$22.80 to spend and came home with \$6.20. If Ellen spent two-thirds of her money and Robert spent four-fifths of his money, how much did they each have to begin with?

27. A store paid \$299.50 for a recent mailing. Some of the letters cost 45¢ postage and the rest needed 25¢ postage. How many letters at each rate were mailed if the total number sent out was 910?

28. A wholesaler has two grades of oil that ordinarily sell for \$1.14 per quart and 94¢ per quart. He wants a blend of the two oils to sell at \$1.05 per quart. If he anticipates selling 400 quarts of the new blend, how much of each grade should he use? [One of the equations makes use of the fact that the total income will be 400(1.05), or \$420.]

29. The annual return on two investments totals \$464. One investment gives 8% interest and the other $7\frac{1}{2}$%. How much money is invested at each rate if the total investment is \$6000?

30. A salesperson said that it did not matter whether one pair of shoes was sold for \$31 or two pairs for \$49, because the profit was the same on each sale. How much does one pair of shoes cost the salesperson and what is the profit?

31. A student in a chemistry laboratory wants to form a 32-milliliter mixture of two solutions to contain 30% acid. Solution A contains 42% acid and solution B contains 18% acid. How many milliliters of each solution must be used? (*Hint:* Use the fact that the final mixture will have $0.30(32) = 9.6$ milliliters of acid.)

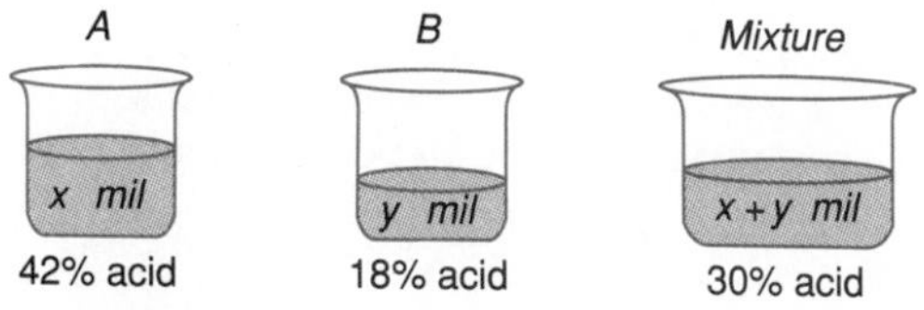

32. It costs \$16.25 for a 16-word phone-delivery telegram that is delivered within two hours. Another such telegram of 21 words costs \$19.00. These rates consist of a basic rate for the first 10 words and a fixed amount for each additional word. Find the basic rate and the cost for each additional word.

33. To go to work a commuter first averages 36 miles per hour driving her car to the train station, and then rides the train which averages 60 miles per hour. The entire trip takes 1 hour and 22 minutes. It costs the commuter 15¢ per mile to drive the car and 6¢ per mile to ride the train. If the total cost is \$5.22, find the distances traveled by car and by train.

34. A purchase of 6 dozen oranges and 10 pounds of peaches costs \$10.60. If the price per dozen oranges increases 10%, and if peaches increase 5% per pound, the same order would cost \$11.46. What is the initial cost of a pound of peaches and of a dozen oranges?

35. How many answers are there for the following puzzle?

> The difference between two numbers is 3. The larger number decreased by 1 is the same as $\frac{1}{3}$ the sum of the smaller plus twice the larger.

36. Suppose that someone asked you to find the two numbers in the following puzzle:

> The larger of two numbers is 16 more than twice the smaller. The difference between $\frac{1}{4}$ of the larger and $\frac{1}{2}$ the smaller is 2.

Why can you say that there is no answer possible for this puzzle?

*37. A bag containing a mixture of 6 oranges and 12 tangerines sold for \$1.44. A smaller bag containing 2 oranges and 4 tangerines sold for 47¢. An alert shopper asked the salesclerk if it was a better buy to purchase the larger bag. The clerk wasn't sure, but said that it really made no difference because the price of each package was based on the same unit price for each kind of fruit. Why was the clerk wrong?

*38. If Sue gave Sam one of her dollars, then Sam would have half as many dollars as Sue. If Sam gave Sue one of his dollars, then Sue would have five times as many dollars as Sam. How many dollars did each of them have?

39. Refer to Example 6, page 438, and determine the daily profit when 40 skateboards are produced and sold per day.

40. Refer to Example 6, page 438, and determine the loss when 20 skateboards are produced and sold per day.

41. Refer to Example 6, page 438, and determine how many skateboards must be made and sold per day to obtain a $520 profit.

42. Find the break-even point for each pair of cost and revenue equations. (Assume the equations represent dollars.)

	Cost:	Revenue:
(a)	$y = 15x + 450$	$y = 30x$
(b)	$y = 240x + 1600$	$y = 400x$

43. The Electro Calculator Company can produce a calculator at a cost of $12. Their daily fixed cost is $720, and they plan to sell each calculator for $20. Graph the cost and revenue equations in the same coordinate system and find the break-even point.

44. What is the profit or loss for 200 calculators? For 300 calculators? (Refer to Exercise 43.)

45. For how many calculators will there be a $200 daily profit? (Refer to Exercise 43.)

46. A bakery makes extra large chocolate chip cookies at a cost of $0.12 per cookie and sells them for $0.60 a piece. If the daily overhead expenses are $60, find the break-even-point.

10.3 SYSTEMS OF LINEAR EQUATIONS IN THREE VARIABLES WITH APPLICATIONS

There is an important feature that the various procedures for solving a system of two linear equations have in common. They all begin by *eliminating* one of the two variables. Thus you soon reach one equation in one unknown. This basic strategy of reducing the number of unknowns can be applied to "larger" linear systems. For instance, here is a system of three linear equations in three variables:

$$\begin{aligned} &(1) \qquad & 2x - 5y + z &= -10 \\ &(2) \qquad & x + 2y + 3z &= 26 \\ &(3) \qquad & -3x - 4y + 2z &= 5 \end{aligned}$$

To solve this system we may begin by eliminating the variable x from the first two equations.

$$\left.\begin{aligned} 2x - 5y + z &= -10 \\ -2(x + 2y + 3z &= 26) \end{aligned}\right\} \Rightarrow \quad \begin{aligned} 2x - 5y + z &= -10 \\ -2x - 4y - 6z &= -52 \\ \hline \text{Add:} \quad -9y - 5z &= -62 \end{aligned}$$

Another equation in y and z can be obtained from equations (2) and (3) by eliminating x:

$$\left.\begin{aligned} 3(x + 2y + 3z &= 26) \\ -3x - 4y + 2z &= 5 \end{aligned}\right\} \Rightarrow \quad \begin{aligned} 3x + 6y + 9z &= 78 \\ -3x - 4y + 2z &= 5 \\ \hline \text{Add:} \quad 2y + 11z &= 83 \end{aligned}$$

Now we have this system in two variables:

$$9y + 5z = 62$$
$$2y + 11z = 83$$

You can solve this system to find $y = 3$ and $z = 7$. Now substitute these values into an earlier equation, say (2).

$$x + 2y + 3z = 26$$
$$x + 2(3) + 3(7) = 26$$
$$x = -1$$

The remaining equations can be used for checking:

(1) $\quad 2(-1) - 5(3) + 7 = -10$

(3) $\quad -3(-1) - 4(3) + 2(7) = 5$

Three planes intersecting at a common point.

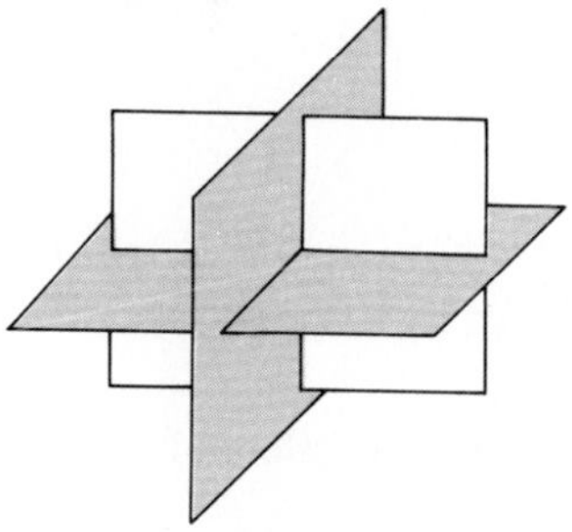

The solution is $x = -1$, $y = 3$, $z = 7$.

Geometrically, an equation with three variables of the form $ax + by + cz = d$ can be interpreted as the equation of a plane. In this example the three planes have a single point in common, the *ordered triple* $(-1, 3, 7)$.

EXAMPLE 1 Solve the system.

$$3x - 2y + z = 1$$
$$x - y - z = 2$$
$$6x - 4y + 2z = 3$$

Three planes having no common point of intersection.

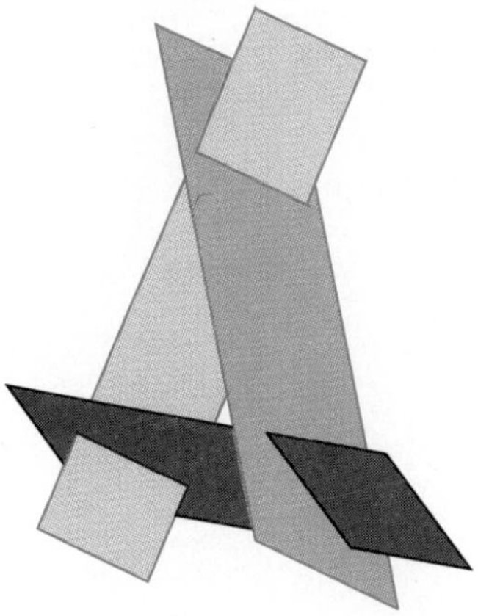

Solution Add the first two equations to eliminate z.

$$\begin{aligned} 3x - 2y + z &= 1 \\ x - y - z &= 2 \\ \hline 4x - 3y \quad &= 3 \end{aligned}$$

Now eliminate z from the second and third equations. To do so, multiply the second equation by 2 and add.

$$\begin{aligned} 2x - 2y - 2z &= 4 \\ 6x - 4y + 2z &= 3 \\ \hline 8x - 6y \quad &= 7 \end{aligned}$$

We now have the following system of two equations in two variables to solve:

$$4x - 3y = 3$$
$$8x - 6y = 7$$

Three planes, two of which are parallel. No common point of intersection.

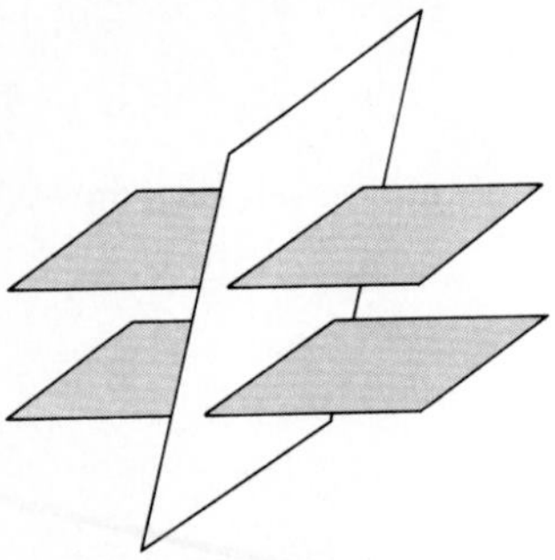

However, note what happens if you multiply the first of these equations by 2 and subtract:

$$\begin{array}{r} 8x - 6y = 6 \\ 8x - 6y = 7 \\ \hline 0 = -1 \end{array}$$

This conclusion, which is false, indicates that the original system of equations is inconsistent, and has no solution. Geometrically, in this example, two of the three planes represented by the given equations are parallel.

TEST YOUR UNDERSTANDING
Think Carefully

Solve each linear system.

1. $\begin{aligned} x + y + z &= 6 \\ 2x + y - z &= 1 \\ 3x - 2y + z &= 2 \end{aligned}$

2. $\begin{aligned} x + y - 2z &= 3 \\ 2x - y + z &= -3 \\ 3x + 3y - 6z &= -4 \end{aligned}$

3. $\begin{aligned} 2x + 3y + 4z &= 2 \\ x - 2y + 3z &= 3 \\ 3x + 5y + 5z &= 1 \end{aligned}$

4. $\begin{aligned} 2x + y + z &= 2 \\ 4x - 2y - z &= 4 \\ 3x + 3y + 2z &= \tfrac{1}{2} \end{aligned}$

(Answers Page 490)

The next example is an application that calls for the solution of a system of three equations in three variables.

Some word problems are best solved by using a system of three equations, as shown in Example 2.

EXAMPLE 2 A veterinarian wants to control the diet of an animal so that on a monthly basis the animal consumes (besides hay, grass, and water) 60 pounds of oats, 75 pounds of corn, and 55 pounds of soybeans. The veterinarian has three feeds available, each consisting of oats, corn, and soybeans as shown in the table. How many pounds of each feed should be used to obtain the desired mix?

	Oats	Corn	Soybeans
1 lb of feed *A*	6 oz	5 oz	5 oz
1 lb of feed *B*	6 oz	6 oz	4 oz
1 lb of feed *C*	4 oz	7 oz	5 oz

Solution

$$\begin{aligned} \text{Let } x &= \text{pounds of feed } A \\ y &= \text{pounds of feed } B \\ z &= \text{pounds of feed } C \end{aligned}$$

Then

$$\begin{aligned} 6x &= \text{ounces of oats in } x \text{ pounds of feed } A \\ 6y &= \text{ounces of oats in } y \text{ pounds of feed } B \\ 4z &= \text{ounces of oats in } z \text{ pounds of feed } C \end{aligned}$$

Since the total number of ounces of oats required is 60(16) = 960, we have

$$6x + 6y + 4z = 960$$

Similar analysis for the total ounces of corn and soybeans leads to this linear system:

$$\begin{aligned} 6x + 6y + 4z &= 960 \qquad \text{(oats)} \\ 5x + 6y + 7z &= 1200 \qquad \text{(corn)} \\ 5x + 4y + 5z &= 880 \qquad \text{(soybeans)} \end{aligned}$$

To solve this system we may begin by eliminating x from the second and third equations, by subtraction, and obtain

$$2y + 2z = 320$$

or

$$y + z = 160$$

Eliminating x from the first two equations leads to this system in y and z:

$$\begin{aligned} y + z &= 160 \\ 3y + 11z &= 1200 \end{aligned}$$

Solving for y and z, we have $y = 70$ and $z = 90$. Then using Equation 1, we get

$$\begin{aligned} 6x + 6(70) + 4(90) &= 960 \\ 6x &= 180 \\ x &= 30 \end{aligned}$$

Therefore, 30 pounds of feed A, 70 pounds of feed B, and 90 pounds of feed C should be combined to obtain the desired mix. ■

EXERCISES 10.3

Solve each linear system.

1. $\begin{aligned} x + 2y + 3z &= 5 \\ -4x + z &= 6 \\ 3x - y &= -3 \end{aligned}$

2. $\begin{aligned} x + y + z &= 2 \\ x - y + 3z &= 12 \\ 2x + 5y + 2z &= -2 \end{aligned}$

3. $\begin{aligned} x + 2y + 3z &= -4 \\ 4x + 5y + 6z &= -4 \\ 7x - 15y - 9z &= 4 \end{aligned}$

4. $\begin{aligned} -3x + 3y + z &= -10 \\ 4x + y + 5z &= 2 \\ x - 8y - 2z &= 12 \end{aligned}$

5. $\begin{aligned} x + 2y - z &= 3 \\ 2x - 3y + 3z &= \\ y - 2z &= 6 \end{aligned}$

6. $\begin{aligned} 2x + y &= 5 \\ -3x + 2z &= 7 \\ 3y - 8z &= 5 \end{aligned}$

7. $4x - 2y - z = 1$
$2x + y + 2z = 9$
$x - 3y - z = \frac{3}{2}$

8. $5x - y - 2z = 2$
$3y + 2z = 5$
$-5x + 4y + 4z = -1$

9. $x + y + z = 4$
$3x - 2y + z = 6$
$2x + 5y + 3z = 11$

10. $2x - 2y + z = -7$
$3x + y + 2z = -2$
$5x + 3y - 3z = -7$

11. $2x + 3y - z = 8$
$x + y + z = 5$
$-4x - 6y + 2z = 3$

12. $2x + 2y - 2z = 9$
$3x - y + 4z = 0$
$6x + 4y + 6z = 13$

13. $0.1x + 0.2y + 0.2z = 0.2$
$0.3x + 0.5y + 0.1z = -0.1$
$0.2x - 0.3y - 0.5z = 0.7$

14. $\frac{1}{2}x - \frac{1}{2}y - z = 11$
$\frac{3}{4}x + \frac{3}{2}y - \frac{1}{4}z = -2$
$\frac{3}{2}x + \frac{1}{3}y + 2z = 2$

Write a system of three equations for each problem, and solve.

15. Daniel has \$575 in one-dollar, five-dollar, and ten-dollar bills. Altogether he has 95 bills. The number of one-dollar bills plus the number of ten-dollar bills is five more than twice the number of five-dollar bills. How many of each type of bill does he have?

16. The sum of three numbers is 33. The largest number is one less than twice the smallest number. Three times the smallest number is one less than the sum of the other two numbers. Find the three numbers.

17. The sum of the angles of a triangle is 180°. The largest angle is equal to the sum of the other two angles. Twice the smallest angle is 10° less than the largest angle. Find the measures for each angle.

18. The treasurer of a club invested \$5000 of their savings into three different accounts, at annual yields of 8%, 9%, and 10%. The total interest earned for the year was \$460. The amount earned by the 10% account was \$20 more than that earned by the 9% account. How much was invested at each rate?

19. A grocer sells peanuts a \$2.80 per pound, pecans at \$4.50 per pound, and brazil nuts at \$5.40 per pound. He wants to make a mixture of 50 pounds of mixed nuts to sell at \$4.44 per pound. The mixture is to contain as many pounds of brazil nuts as the other two types combined. How many pounds of each type must he use in this mixture?

20. Answer the question in Example 2 on page 444 after making the following changes. The monthly consumption of oats, corn and soybeans is 45, 60, and 45 pounds, respectively. Also, the contents of the feeds are given in this table:

	Oats	Corn	Soybeans
1 lb of feed *A*	4 oz	6 oz	6 oz
1 lb of feed *B*	8 oz	4 oz	4 oz
1 lb of feed *C*	3 oz	8 oz	5 oz

21. A dietician wants to combine three foods so that the resulting mixture contains 900 units of vitamins, 750 units of minerals, and 350 units of fat. The units of vitamins, minerals, and fat contained in each gram of the three foods are shown in the table. How many grams of each food should be combined to obtain the required mixture?

	Vitamins	Minerals	Fat
1 gram of food *A*	35 units	15 units	10 units
1 gram of food *B*	10 units	20 units	10 units
1 gram of food *C*	20 units	15 units	5 units

22. The equation $ax + by + cz = -3$ has the solutions $(-1, 2, 4)$, $(2, -1, 2)$ and $(3, -6, -8)$ for (x, y, z). Find the equation by identifying a, b, c.

Written Assignment: What are the various ways that three planes can intersect?

10.4 SOLVING LINEAR SYSTEMS USING SECOND-ORDER DETERMINANTS

The general linear equation has the form $ax + by = c$. Since we are now dealing with two linear equations, it is appropriate to use subscripts to distinguish the constants in the first equation from those in the second. Let

$$a_1x + b_1y = c_1$$
$$a_2x + b_2y = c_2$$

represent any system of linear equations, and let us refer to the system by using the letter S.

Suppose that system S is a consistent system. Then S has a unique solution that can be found by the multiplication-addition method.

Multiply the first equation by b_2 and the second by $-b_1$.

$$a_1b_2x + b_1b_2y = c_1b_2$$
$$-a_2b_1x - b_1b_2y = -c_2b_1$$

Add to eliminate y.

$$a_1b_2x - a_2b_1x = c_1b_2 - c_2b_1$$

Factor.

$$(a_1b_2 - a_2b_1)x = c_1b_2 - c_2b_1$$

To solve for x it must be the case that $a_1b_2 - a_2b_1 \neq 0$. Then

$$x = \frac{c_1b_2 - c_2b_1}{a_1b_2 - a_2b_1}$$

Similarly, multiplying the first and second equations of S by $-a_2$ and a_1, respectively, will produce this solution for y.

$$y = \frac{a_1c_2 - a_2c_1}{a_1b_2 - a_2b_1}$$

We have found that if $a_1b_2 - a_2b_1 \neq 0$, then system S has the common solution given above. The situation $a_1b_2 - a_2b_1 = 0$ will be discussed in detail at the end of this section.

EXAMPLE 1 Use the general solution to solve the system

$$8x - 20y = 3$$
$$4x + 10y = \tfrac{3}{2}$$

Solution

$$a_1 = 8,\ b_1 = -20,\ c_1 = 3,\ a_2 = 4,\ b_2 = 10,\ c_2 = \tfrac{3}{2}.$$

$$x = \frac{(3)(10) - (\frac{3}{2})(-20)}{(8)(10) - (4)(-20)} = \frac{30 + 30}{80 + 80} = \frac{60}{160} = \frac{3}{8}$$

$$y = \frac{(8)(\frac{3}{2}) - (4)(3)}{(8)(10) - (4)(-20)} = \frac{12 - 12}{160} = \frac{0}{160} = 0$$ ■

The work in Example 1 amounts to nothing more than substituting into the general forms and simplifying. But keeping track of the position of each constant in the general form is tedious. We can simplify this process with the introduction of some special symbolism.

First notice that the denominator for both x and y is the same value $a_1b_2 - a_2b_1$. This value is given the name **determinant**, and is assigned the special symbolism as shown in the following definition.

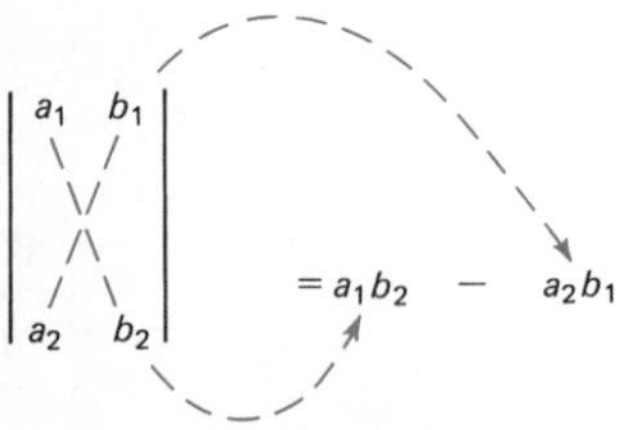

You can use this diagram to help remember the definition.

DEFINITION OF A SECOND-ORDER DETERMINANT

$$\begin{vmatrix} a_1 & b_1 \\ a_2 & b_2 \end{vmatrix} = a_1b_2 - a_2b_1$$

Here is an illustration of this definition.

$$\begin{vmatrix} 8 & -20 \\ 4 & 10 \end{vmatrix} = (8)(10) - (4)(-20) = 160$$

The arrangement of the four entries is important because with different arrangements the same four numbers may give different determinants. For example, compare these two determinants:

$$\begin{vmatrix} -20 & 8 \\ 10 & 4 \end{vmatrix} = (-20)(4) - (10)(8) = -160$$

$$\begin{vmatrix} 10 & -20 \\ 8 & 4 \end{vmatrix} = (10)(4) - (8)(-20) = 200$$

TEST YOUR UNDERSTANDING
Think Carefully

Evaluate each of the determinants.

1. $\begin{vmatrix} 1 & 2 \\ 3 & 4 \end{vmatrix}$ **2.** $\begin{vmatrix} 1 & 3 \\ 2 & 4 \end{vmatrix}$ **3.** $\begin{vmatrix} 2 & 1 \\ 4 & 3 \end{vmatrix}$ **4.** $\begin{vmatrix} 1 & 3 \\ 4 & 2 \end{vmatrix}$

5. $\begin{vmatrix} 2 & 3 \\ 1 & 4 \end{vmatrix}$ **6.** $\begin{vmatrix} 4 & 3 \\ 1 & 2 \end{vmatrix}$ **7.** $\begin{vmatrix} 10 & -5 \\ 2 & 1 \end{vmatrix}$ **8.** $\begin{vmatrix} \frac{1}{2} & 6 \\ 0 & 4 \end{vmatrix}$

9. $\begin{vmatrix} -8 & -4 \\ -7 & -3 \end{vmatrix}$ **10.** $\begin{vmatrix} -1 & 0 \\ 0 & -1 \end{vmatrix}$ **11.** $\begin{vmatrix} 0 & 2 \\ 2 & 0 \end{vmatrix}$ **12.** $\begin{vmatrix} \frac{1}{3} & -\frac{1}{4} \\ 8 & -6 \end{vmatrix}$

(Answers: Page 490)

The numerator in the general form for x is $c_1b_2 - c_2b_1$. Using our new symbolism, this number is the *second-order determinant:*

The general forms for x and y are given on page 447.

$$\begin{vmatrix} c_1 & b_1 \\ c_2 & b_2 \end{vmatrix} = c_1b_2 - c_2b_1$$

Similarly, the numerator for y is

$$\begin{vmatrix} a_1 & c_1 \\ a_2 & c_2 \end{vmatrix} = a_1c_2 - a_2c_1$$

In summary, we have the following, known as ***Cramer's rule***, for solving such a system of linear equations.

CRAMER'S RULE

The system of equations S

$$a_1x + b_1y = c_1$$
$$a_2x + b_2y = c_2$$

has the unique solution

$$x = \frac{\begin{vmatrix} c_1 & b_1 \\ c_2 & b_2 \end{vmatrix}}{\begin{vmatrix} a_1 & b_1 \\ a_2 & b_2 \end{vmatrix}} \quad \text{and} \quad y = \frac{\begin{vmatrix} a_1 & c_1 \\ a_2 & c_2 \end{vmatrix}}{\begin{vmatrix} a_1 & b_1 \\ a_2 & b_2 \end{vmatrix}}$$

provided that $a_1b_2 - a_2b_1 \neq 0$.

Another way of stating that $a_1b_2 - a_2b_1 \neq 0$ is to say that system S is consistent.

Cramer's rule becomes easier to apply after making a few observations. First, the determinant

$$\begin{vmatrix} a_1 & b_1 \\ a_2 & b_2 \end{vmatrix}$$

is the denominator for each fraction. The first column of the determinant consists of the coefficients of the x-terms in S, and the second column contains the coefficients of the y-terms. To write the fraction for x, first record the denominator. Then, to get the determinant in the numerator, simply remove the first column (the coefficients of x) and replace it with constants c_1 and c_2, as shown in the following figure. Similarly, the replacement of the second column by c_1, c_2 gives the numerator in the fraction for y.

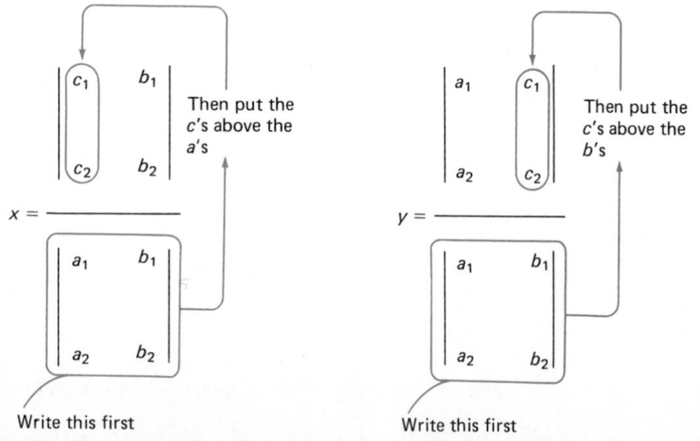

EXAMPLE 2 Solve the following system using Cramer's Rule.

$$5x - 9y = 7$$
$$-8x + 10y = 2$$

Solution

$$x = \frac{\begin{vmatrix} 7 & -9 \\ 2 & 10 \end{vmatrix}}{\begin{vmatrix} 5 & -9 \\ -8 & 10 \end{vmatrix}} = \frac{70 - (-18)}{50 - 72} = \frac{88}{-22} = -4$$

$$y = \frac{\begin{vmatrix} 5 & 7 \\ -8 & 2 \end{vmatrix}}{\begin{vmatrix} 5 & -9 \\ -8 & 10 \end{vmatrix}} = \frac{10 - (-56)}{-22} = -3$$

■

EXAMPLE 3 Use determinants to solve the given system.

$$3x = 2y + 22$$
$$2(x + y) = x - 2y - 2$$

CAUTION
It is important to have both equations in the general form
$$ax + by = c$$
since Cramer's rule is based on this form.

Solution First write the system in standard form.

$$3x - 2y = 22$$
$$x + 4y = -2$$

$$x = \frac{\begin{vmatrix} 22 & -2 \\ -2 & 4 \end{vmatrix}}{\begin{vmatrix} 3 & -2 \\ 1 & 4 \end{vmatrix}} = \frac{84}{14} = 6 \qquad y = \frac{\begin{vmatrix} 3 & 22 \\ 1 & -2 \end{vmatrix}}{14} = \frac{-28}{14} = -2$$

■

The general system

$$a_1x + b_1y = c_1$$
$$a_2x + b_2y = c_2$$

Recall that a system of two linear equations in two variables is dependent when the two equations are equivalent. Inconsistent means that there are no (common) solutions.

is either dependent or inconsistent when the determinant of the coefficients is zero, that is, when

$$\begin{vmatrix} a_1 & b_1 \\ a_2 & b_2 \end{vmatrix} = 0$$

For example, consider these two systems:

$$\textbf{(a)} \quad \begin{aligned} 2x - 3y &= 5 \\ -10x + 15y &= 8 \end{aligned} \qquad \textbf{(b)} \quad \begin{aligned} 2x - 3y &= 8 \\ -10x + 15y &= -40 \end{aligned}$$

In each case we have the following:

Verify these conclusions by solving systems (a) and (b) by the multiplication-addition method, (see Section 10.1).

$$\begin{vmatrix} a_1 & b_1 \\ a_2 & b_2 \end{vmatrix} = \begin{vmatrix} 2 & -3 \\ -10 & 15 \end{vmatrix} = 0$$

System (a) turns out to be inconsistent and (b) is dependent.

EXERCISES 10.4

Evaluate each determinant.

1. $\begin{vmatrix} 5 & -1 \\ -3 & 4 \end{vmatrix}$ 2. $\begin{vmatrix} 1 & 2 \\ 3 & 4 \end{vmatrix}$ 3. $\begin{vmatrix} 17 & -3 \\ 20 & 2 \end{vmatrix}$ 4. $\begin{vmatrix} -7 & 9 \\ -5 & 5 \end{vmatrix}$

5. $\begin{vmatrix} 10 & 5 \\ 6 & -3 \end{vmatrix}$ 6. $\begin{vmatrix} 6 & 11 \\ 0 & -9 \end{vmatrix}$ 7. $\begin{vmatrix} 16 & 0 \\ -9 & 0 \end{vmatrix}$ 8. $\begin{vmatrix} a & b \\ 3a & 3b \end{vmatrix}$

Solve each system using Cramer's rule.

9. $3x + 9y = 15$, $6x + 12y = 18$
10. $x - y = 7$, $-2x + 5y = -8$
11. $-4x + 10y = 8$, $11x - 9y = 15$
12. $7x + 4y = 5$, $-x + 2y = -2$
13. $5x + 2y = 3$, $2x + 3y = -1$
14. $3x + 3y = 6$, $4x - 2y = -1$
15. $\frac{1}{3}x + \frac{3}{8}y = 13$, $x - \frac{9}{4}y = -42$
16. $x - 3y = 7$, $-\frac{1}{2}x + \frac{1}{4}y = 1$
17. $3x + y = 20$, $y = x$
18. $3x + 2y = 5$, $2x + 3y = 0$
19. $\frac{1}{2}x - \frac{2}{7}y = -\frac{1}{2}$, $-\frac{1}{3}x - \frac{1}{2}y = \frac{31}{6}$
20. $-4x + 3y = -20$, $2x + 6y = -15$
21. $9x - 12 = 4y$, $3x + 2y = 3$
22. $\dfrac{x-y}{3} - \dfrac{y}{6} = \dfrac{2}{3}$, $22x + 9(y - 2x) = 8$

Verify that the determinant of the coefficients is zero for each of the following. Then decide whether the system is dependent or inconsistent.

23. $5x - 2y = 3$, $-15x + 6y = -4$
24. $2x - 3y = 5$, $10 - 4x = -6y$
25. $16x - 4y = 20$, $12x - 3y = 15$
26. $2x - 6y = -12$, $-3x + 9y = 18$
27. $3x = 5y - 10$, $6x - 10y = -25$
28. $10y = 2x - 4$, $x - 5y = 2$

When variables are used for some of the entries in the symbolism of a determinant, the determinant itself can be used to state equations. Solve for x.

29. $\begin{vmatrix} x & 2 \\ 5 & 3 \end{vmatrix} = 8$ 30. $\begin{vmatrix} 7 & 3 \\ 4 & x \end{vmatrix} = 15$ 31. $\begin{vmatrix} -2 & 4 \\ x & 3 \end{vmatrix} = -1$

Solve each system.

32. $\begin{vmatrix} x & y \\ 3 & 2 \end{vmatrix} = 2$, $\begin{vmatrix} x & -1 \\ y & 3 \end{vmatrix} = 14$
33. $\begin{vmatrix} x & y \\ 2 & 4 \end{vmatrix} = 5$, $\begin{vmatrix} 1 & y \\ -1 & x \end{vmatrix} = -\frac{1}{2}$
34. $\begin{vmatrix} 3 & x \\ 2 & y \end{vmatrix} = 13$, $\begin{vmatrix} 3 & 2 \\ y & x \end{vmatrix} = -12$

35. Show that if the rows and columns of a second-order determinant are interchanged, the value of the determinant remains the same.

36. Show that if one of the rows of $\begin{vmatrix} a_1 & b_1 \\ a_2 & b_2 \end{vmatrix}$ is a multiple of the other, then the determinant is zero. (*Hint:* Let $a_2 = ka_1$ and $b_2 = kb_1$.)

37. Use Exercises 35 and 36 to demonstrate that the determinant is zero if one column is a multiple of the other.

38. Show that if each element of a row (or column) of a second-order determinant is multiplied by the same number k, the value of the determinant is multiplied by k.

39. Make repeated use of the result in Exercise 38 to show the following:

$$\begin{vmatrix} 27 & 3 \\ 105 & -75 \end{vmatrix} = (45)\begin{vmatrix} 9 & 1 \\ 7 & -5 \end{vmatrix} \quad \text{or} \quad \begin{vmatrix} 27 & 3 \\ 105 & -75 \end{vmatrix} = (45)\begin{vmatrix} 3 & 1 \\ 7 & -15 \end{vmatrix}$$

Then evaluate each side to check.

*40. Prove:

$$\begin{vmatrix} a_1 + t_1 & b_1 \\ a_2 + t_2 & b_2 \end{vmatrix} = \begin{vmatrix} a_1 & b_1 \\ a_2 & b_2 \end{vmatrix} + \begin{vmatrix} t_1 & b_1 \\ t_2 & b_2 \end{vmatrix}$$

*41. Prove that if to each element of a row (or column) of a second-order determinant we add k times the corresponding element of another row (or column), then the value of the new determinant is the same as that of the original determinant.

42. **(a)** Evaluate $\begin{vmatrix} 3 & 5 \\ -6 & -1 \end{vmatrix}$ by definition.

(b) Evaluate the same determinant using the result of Exercise 41 by adding 2 times row one to row two.

(c) Evaluate the same determinant using the result of Exercise 41 adding -6 times column two to column one.

Use the results of Exercises 38 and 41 to evaluate each determinant.

43. $\begin{vmatrix} 12 & -42 \\ -6 & 27 \end{vmatrix}$

44. $\begin{vmatrix} 45 & 75 \\ 40 & -25 \end{vmatrix}$

10.5 SOLVING LINEAR SYSTEMS USING THIRD-ORDER DETERMINANTS

Just as a system of two linear equations in two variables can be solved using second-order determinants, so can a system of three linear equations in three variables be solved by using *third-order* determinants.

A **third-order determinant** may be defined in terms of second-order determinants as follows:

DEFINITION OF A THIRD-ORDER DETERMINANT

$$\begin{vmatrix} a_1 & b_1 & c_1 \\ a_2 & b_2 & c_2 \\ a_3 & b_3 & c_3 \end{vmatrix} = a_1\begin{vmatrix} b_2 & c_2 \\ b_3 & c_3 \end{vmatrix} - a_2\begin{vmatrix} b_1 & c_1 \\ b_3 & c_3 \end{vmatrix} + a_3\begin{vmatrix} b_1 & c_1 \\ b_2 & c_2 \end{vmatrix}$$

Note that the first term on the right is the product of a_1 times a second-order determinant. This determinant, also called the **minor of a_1**, can be found by eliminating the row and column that a_1 is in. Thus

$$\begin{vmatrix} a_1 & b_1 & c_1 \\ a_2 & b_2 & c_2 \\ a_3 & b_3 & c_3 \end{vmatrix} \longrightarrow \begin{vmatrix} b_2 & c_2 \\ b_3 & c_3 \end{vmatrix} \quad \text{minor of } a_1$$

Similar schemes can be used to obtain the minors for a_2 and a_3.

$$\begin{vmatrix} a_1 & b_1 & c_1 \\ a_2 & b_2 & c_2 \\ a_3 & b_3 & c_3 \end{vmatrix} \longrightarrow \begin{vmatrix} b_1 & c_1 \\ b_3 & c_3 \end{vmatrix} \quad \text{minor of } a_2$$

$$\begin{vmatrix} a_1 & b_1 & c_1 \\ a_2 & b_2 & c_2 \\ a_3 & b_3 & c_3 \end{vmatrix} \longrightarrow \begin{vmatrix} b_1 & c_1 \\ b_2 & c_2 \end{vmatrix} \quad \text{minor of } a_3$$

EXAMPLE 1 Evaluate: $\begin{vmatrix} 2 & -2 & 2 \\ 3 & 1 & 0 \\ 2 & -1 & 1 \end{vmatrix}$

Solution

$$\begin{aligned} \begin{vmatrix} 2 & -2 & 2 \\ 3 & 1 & 0 \\ 2 & -1 & 1 \end{vmatrix} &= 2\begin{vmatrix} 1 & 0 \\ -1 & 1 \end{vmatrix} - 3\begin{vmatrix} -2 & 2 \\ -1 & 1 \end{vmatrix} + 2\begin{vmatrix} -2 & 2 \\ 1 & 0 \end{vmatrix} \\ &= 2(1 - 0) - 3(-2 + 2) + 2(0 - 2) \\ &= 2 - 0 - 4 \\ &= -2 \end{aligned}$$

■

A third-order determinant can be evaluated and simplified as follows:

Simplified form of a third-order determinant

$$\begin{aligned} \begin{vmatrix} a_1 & b_1 & c_1 \\ a_2 & b_2 & c_2 \\ a_3 & b_3 & c_3 \end{vmatrix} &= a_1(b_2c_3 - b_3c_2) - a_2(b_1c_3 - b_3c_1) + a_3(b_1c_2 - b_2c_1) \\ &= a_1b_2c_3 + a_2b_3c_1 + a_3b_1c_2 - a_1b_3c_2 - a_2b_1c_3 - a_3b_2c_1 \end{aligned}$$

The given definition of a third-order determinant can be described as an *expansion by minors* along the first column. It turns out that there are six such expansions, one for each row and column, all giving the same result. Here is the expansion by minors along the first row.

Simplify this expansion to see that it agrees with the preceding simplification of the expansion along the first column.

$$\begin{vmatrix} a_1 & b_1 & c_1 \\ a_2 & b_2 & c_2 \\ a_3 & b_3 & c_3 \end{vmatrix} = +a_1\begin{vmatrix} b_2 & c_2 \\ b_3 & c_3 \end{vmatrix} - b_1\begin{vmatrix} a_2 & c_2 \\ a_3 & c_3 \end{vmatrix} + c_1\begin{vmatrix} a_2 & b_2 \\ a_3 & b_3 \end{vmatrix}$$

When the expansion by minors is done along the second column we have

$$\begin{aligned} \begin{vmatrix} a_1 & b_1 & c_1 \\ a_2 & b_2 & c_2 \\ a_3 & b_3 & c_3 \end{vmatrix} &= -b_1\begin{vmatrix} a_2 & c_2 \\ a_3 & c_3 \end{vmatrix} + b_2\begin{vmatrix} a_1 & c_1 \\ a_3 & c_3 \end{vmatrix} - b_3\begin{vmatrix} a_1 & c_1 \\ a_2 & c_2 \end{vmatrix} \\ &= -b_1(a_2c_3 - a_3c_2) + b_2(a_1c_3 - a_3c_1) - b_3(a_1c_2 - a_2c_1) \\ &= a_1b_2c_3 + a_2b_3c_1 + a_3b_1c_2 - a_1b_3c_2 - a_2b_1c_3 - a_3b_2c_1 \end{aligned}$$

For any expansion along a row or column keep the following display of signs in mind. It gives the signs preceding the row or column elements in the expansion.

Observe how the signs in the second column have been used in the preceding expansion.

$$\begin{vmatrix} + & - & + \\ - & + & - \\ + & - & + \end{vmatrix}$$

EXAMPLE 2 Evaluate $\begin{vmatrix} 2 & 1 & -3 \\ -4 & 0 & 2 \\ 5 & -1 & 6 \end{vmatrix}$ using each of these methods.

(a) Expand by minors along the third column.
(b) Expand by minors along the second row.

Solution

(a)
$$\begin{vmatrix} 2 & 1 & -3 \\ -4 & 0 & 2 \\ 5 & -1 & 6 \end{vmatrix} = -3\begin{vmatrix} -4 & 0 \\ 5 & -1 \end{vmatrix} - 2\begin{vmatrix} 2 & 1 \\ 5 & -1 \end{vmatrix} + 6\begin{vmatrix} 2 & 1 \\ -4 & 0 \end{vmatrix}$$
$$= -3(4 - 0) - 2(-2 - 5) + 6(0 + 4)$$
$$= -12 + 14 + 24$$
$$= 26$$

(b)
$$\begin{vmatrix} 2 & 1 & -3 \\ -4 & 0 & 2 \\ 5 & -1 & 6 \end{vmatrix} = -(-4)\begin{vmatrix} 1 & -3 \\ -1 & 6 \end{vmatrix} + 0\begin{vmatrix} 2 & -3 \\ 5 & 6 \end{vmatrix} - 2\begin{vmatrix} 2 & 1 \\ 5 & -1 \end{vmatrix}$$
$$= 4(3) + 0 - 2(-7)$$
$$= 26$$

■

Here is another procedure that can be used to evaluate a third-order determinant. Rewrite the first two columns at the right as shown. Follow the arrows pointing downward to get the three products having a plus sign, and the arrows pointing upward give the three products having a minus sign.

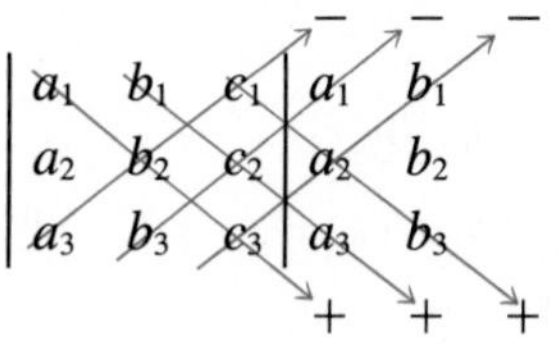

$$= a_1b_2c_3 + b_1c_2a_3 + c_1a_2b_3 - a_3b_2c_1 - b_3c_2a_1 - c_3a_2b_1$$

EXAMPLE 3 Evaluate by rewriting the first two columns: $\begin{vmatrix} -1 & 3 & 4 \\ 2 & 1 & 2 \\ 5 & 1 & 3 \end{vmatrix}$

Solution

$$\left|\begin{array}{rrr|rr} -1 & 3 & 4 & -1 & 3 \\ 2 & 1 & 2 & 2 & 1 \\ 5 & 1 & 3 & 5 & 1 \end{array}\right. = -3 + 30 + 8 - 20 - (-2) - 18 = -1$$

■

TEST YOUR UNDERSTANDING
Think Carefully

Evaluate each determinant using each of these methods:

(a) Expand by minors along one of the rows.
(b) Expand by minors along one of the columns.
(c) Rewrite the first two columns.

1. $\begin{vmatrix} 2 & 1 & 0 \\ -1 & 0 & 3 \\ 0 & 4 & -5 \end{vmatrix}$ **2.** $\begin{vmatrix} 3 & 6 & 9 \\ 0 & 0 & 0 \\ -2 & -4 & -6 \end{vmatrix}$ **3.** $\begin{vmatrix} 3 & -2 & 7 \\ 0 & 3 & -5 \\ 0 & 0 & 3 \end{vmatrix}$

4. $\begin{vmatrix} -1 & 4 & -2 \\ 6 & -6 & 1 \\ 3 & 3 & 2 \end{vmatrix}$ **5.** $\begin{vmatrix} 1 & -2 & 3 \\ -4 & 5 & -4 \\ 3 & -2 & 1 \end{vmatrix}$ **6.** $\begin{vmatrix} 2 & -1 & 9 \\ -7 & 3 & -4 \\ 2 & -1 & 9 \end{vmatrix}$

(Answers: Page 491)

Third-order determinants can be used to extend Cramer's rule for the solution of three linear equations in three variables. Consider this general system and assume it has just one solution for x, y, and z.

$$a_1x + b_1y + c_1z = d_1$$
$$a_2x + b_2y + c_2z = d_2$$
$$a_3x + b_3y + c_3z = d_3$$

By completing the tedious computations involved, it can be shown that the unique solution for this system is the following:

Note that D is the determinant of coefficients of the variables, in order. Then the numerators for the solutions for x, y, and z consist of the coefficients, but in each case the constants are used to replace the coefficients of the variable under consideration.

$$x = \frac{\begin{vmatrix} d_1 & b_1 & c_1 \\ d_2 & b_2 & c_2 \\ d_3 & b_3 & c_3 \end{vmatrix}}{D} \qquad y = \frac{\begin{vmatrix} a_1 & d_1 & c_1 \\ a_2 & d_2 & c_2 \\ a_3 & d_3 & c_3 \end{vmatrix}}{D} \qquad z = \frac{\begin{vmatrix} a_1 & b_1 & d_1 \\ a_2 & b_2 & d_2 \\ a_3 & b_3 & d_3 \end{vmatrix}}{D}$$

where

$$D = \begin{vmatrix} a_1 & b_1 & c_1 \\ a_2 & b_2 & c_2 \\ a_3 & b_3 & c_3 \end{vmatrix} \quad \text{and} \quad D \neq 0$$

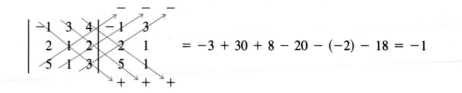

Now let the determinants in the numerators be denoted by D_x, D_y, and D_z:

$$D_x = \begin{vmatrix} d_1 & b_1 & c_1 \\ d_2 & b_2 & c_2 \\ d_3 & b_3 & c_3 \end{vmatrix} \qquad D_y = \begin{vmatrix} a_1 & d_1 & c_1 \\ a_2 & d_2 & c_2 \\ a_3 & d_3 & c_3 \end{vmatrix} \qquad D_z = \begin{vmatrix} a_1 & b_1 & d_1 \\ a_2 & b_2 & d_2 \\ a_3 & b_3 & d_3 \end{vmatrix}$$

Then the solution of the system in Cramer's rule can be stated in the following condensed form.

$$x = \frac{D_x}{D} \qquad y = \frac{D_y}{D} \qquad z = \frac{D_z}{D}, \qquad \text{where } D \neq 0$$

EXAMPLE 4 Use Cramer's rule to solve this system.

$$\begin{aligned} x + 2y + z &= 3 \\ 2x - y - z &= 4 \\ -x - y + 2z &= -5 \end{aligned}$$

Solution First we find D and note that $D \neq 0$.

$$D = \begin{vmatrix} 1 & 2 & 1 \\ 2 & -1 & -1 \\ -1 & -1 & 2 \end{vmatrix} = 1\begin{vmatrix} -1 & -1 \\ -1 & 2 \end{vmatrix} - 2\begin{vmatrix} 2 & 1 \\ -1 & 2 \end{vmatrix} + (-1)\begin{vmatrix} 2 & 1 \\ -1 & -1 \end{vmatrix}$$
$$= -12$$

Verify each of the following computations.

$$x = \frac{D_x}{D} = \frac{\begin{vmatrix} 3 & 2 & 1 \\ 4 & -1 & -1 \\ -5 & -1 & 2 \end{vmatrix}}{D} = \frac{-24}{-12} = 2$$

$$y = \frac{D_y}{D} = \frac{\begin{vmatrix} 1 & 3 & 1 \\ 2 & 4 & -1 \\ -1 & -5 & 2 \end{vmatrix}}{D} = \frac{-12}{-12} = 1$$

$$z = \frac{D_z}{D} = \frac{\begin{vmatrix} 1 & 2 & 3 \\ 2 & -1 & 4 \\ -1 & -1 & -5 \end{vmatrix}}{D} = \frac{12}{-12} = -1$$

Check this solution in the original system.

Thus $x = 2$, $y = 1$, and $z = -1$. ■

EXERCISES 10.5

Evaluate $\begin{vmatrix} 6 & -2 & -1 \\ 0 & -9 & 4 \\ -3 & 5 & 1 \end{vmatrix}$ using each of these methods.

(a) Expand by minors along the first column.
(b) Expand by minors along the third row.
(c) Rewrite the first two columns.

2. Evaluate $\begin{vmatrix} -8 & -1 & 0 \\ 4 & 7 & -5 \\ 3 & 0 & 2 \end{vmatrix}$ using each of these methods.

(a) Expand by minors along the second column.
(b) Expand by minors along the second row.
(c) Rewrite the first two columns.

Evaluate each determinant.

3. $\begin{vmatrix} 2 & 2 & -1 \\ -1 & 3 & -3 \\ 1 & 2 & 3 \end{vmatrix}$
4. $\begin{vmatrix} 2 & 0 & -1 \\ 3 & -2 & 1 \\ -3 & 0 & 4 \end{vmatrix}$
5. $\begin{vmatrix} 1 & -3 & 2 \\ -5 & 2 & 0 \\ 4 & -1 & 3 \end{vmatrix}$
6. $\begin{vmatrix} 1 & 2 & 3 \\ 4 & 5 & 6 \\ 7 & 8 & 9 \end{vmatrix}$
7. $\begin{vmatrix} 1 & 1 & 1 \\ -1 & 1 & 1 \\ -1 & -1 & 1 \end{vmatrix}$
8. $\begin{vmatrix} 1 & 1 & 4 \\ 2 & 2 & -5 \\ 3 & 3 & 6 \end{vmatrix}$

Solve for x.

9. $\begin{vmatrix} -1 & x & -1 \\ x & -3 & 0 \\ -3 & 5 & -1 \end{vmatrix} = 0$
10. $\begin{vmatrix} x & 5 & 2x \\ 2x & 0 & x^2 \\ 1 & -1 & 2 \end{vmatrix} = 0$
11. $\begin{vmatrix} 5-x & 0 & -2 \\ 4 & -1-x & 3 \\ 2 & 0 & 1-x \end{vmatrix} = 0$

Evaluate the following for this system:

$$\begin{aligned} 3x - y + 4z &= 2 \\ -5x + 3y - 7z &= 0 \\ 7x - 4y + 4z &= 12 \end{aligned}$$

12. D
13. D_x
14. $\dfrac{D_x}{D}$
15. $\dfrac{D_y}{D}$

Use Cramer's rule to solve each system.

16. $\begin{aligned} x + y + z &= 2 \\ x - y + 3z &= 12 \\ 2x + 5y + 2z &= -2 \end{aligned}$

17. $\begin{aligned} x + 2y + 3z &= 5 \\ 3x - y &= -3 \\ -4x + z &= 6 \end{aligned}$

18. $\begin{aligned} x - 8y - 2z &= 12 \\ -3x + 3y + z &= -10 \\ 4x + y + 5z &= 2 \end{aligned}$

19. $\begin{aligned} 2x + y &= 5 \\ 3x - 2z &= -7 \\ -3y + 8z &= -5 \end{aligned}$

20. $\begin{aligned} 4x - 2y - z &= 1 \\ 2x + y + 2z &= 9 \\ x - 3y - z &= \tfrac{3}{2} \end{aligned}$

21. $\begin{aligned} 6x + 3y - 4z &= 5 \\ \tfrac{3}{2}x + y - 4z &= 0 \\ 3x - y + 8z &= 5 \end{aligned}$

A general property of determinants is stated in each exercise. Prove this property for the indicated special case. (Note that in the symbol $|A|$, A represents the nine numbers of the determinant.)

22. *If $|A|$ contains a row of zeros or a column of zeros, then $|A| = 0$.* Prove this case:

$$\begin{vmatrix} a_1 & b_1 & c_1 \\ 0 & 0 & 0 \\ a_3 & b_3 & c_3 \end{vmatrix} = 0$$

23. *Interchanging the rows and columns in $|A|$ does not effect the value of the determinant.* Prove this case:

$$\begin{vmatrix} a_1 & b_1 & c_1 \\ a_2 & b_2 & c_2 \\ a_3 & b_3 & c_3 \end{vmatrix} = \begin{vmatrix} a_1 & a_2 & a_3 \\ b_1 & b_2 & b_3 \\ c_1 & c_2 & c_3 \end{vmatrix}$$

24. **(a)** *If one row in $|A|$ is a multiple of another row, then $|A| = 0$. Prove this case:*

$$\begin{vmatrix} a & b & c \\ ka & kb & kc \\ d & e & f \end{vmatrix} = 0$$

(b) Use the property in part (a) and the property in Exercise 23 to prove that *if one column in $|A|$ is a multiple of another column, then $|A| = 0$.*

25. *Interchanging any two rows or any two columns changes the sign of the determinant.* Prove this case:

$$\begin{vmatrix} c_1 & b_1 & a_1 \\ c_2 & b_2 & a_2 \\ c_3 & b_3 & a_3 \end{vmatrix} = -\begin{vmatrix} a_1 & b_1 & c_1 \\ a_2 & b_2 & c_2 \\ a_3 & b_3 & c_3 \end{vmatrix}$$

26. *If each element in a row or in a column in $|A|$ is multiplied by a number k the resulting determinant is k times the original.* Prove this case:

$$\begin{vmatrix} ka_1 & b_1 & c_1 \\ ka_2 & b_2 & c_2 \\ ka_3 & b_3 & c_3 \end{vmatrix} = k\begin{vmatrix} a_1 & b_1 & c_1 \\ a_2 & b_2 & c_2 \\ a_3 & b_3 & c_3 \end{vmatrix}$$

27. *Adding to a row (or column) in $|A|$ k times another row (or column), does not affect the value of the determinant.* Prove this case:

$$\begin{vmatrix} a_1 + kb_1 & b_1 & c_1 \\ a_2 + kb_2 & b_2 & c_2 \\ a_3 + kb_3 & b_3 & c_3 \end{vmatrix} = \begin{vmatrix} a_1 & b_1 & c_1 \\ a_2 & b_2 & c_2 \\ a_3 & b_3 & c_3 \end{vmatrix}$$

28. Use the property in Exercise 24(a) to evaluate

$$\begin{vmatrix} 3 & 5 & -5 \\ 1 & 2 & 3 \\ -4 & -8 & -12 \end{vmatrix}$$

29. Use the property in Exercise 24(b) to evaluate

$$\begin{vmatrix} 7 & 2 & -14 \\ 0 & 6 & 0 \\ -3 & 1 & 6 \end{vmatrix}$$

30. Use the property in Exercise 26 to explain each step.

(i) $$\begin{vmatrix} 8 & -10 & 2 \\ 4 & 25 & -1 \\ 2 & 10 & 0 \end{vmatrix} = 2\begin{vmatrix} 4 & -10 & 2 \\ 2 & 25 & -1 \\ 1 & 10 & 0 \end{vmatrix}$$

(ii) $$= 10\begin{vmatrix} 4 & -2 & 2 \\ 2 & 5 & -1 \\ 1 & 2 & 0 \end{vmatrix}$$

(iii) $$= 20\begin{vmatrix} 2 & -1 & 1 \\ 2 & 5 & -1 \\ 1 & 2 & 0 \end{vmatrix}$$

31. Evaluate the determinant given on the left side of (i) in Exercise 30, and also evaluate the result given in (iii).

32. You can verify that

$$\begin{vmatrix} 1 & 2 & 3 \\ -4 & -1 & 5 \\ 3 & 1 & 7 \end{vmatrix} = 71$$

It follows directly from this result that

$$\begin{vmatrix} 1 & -4 & 3 \\ 2 & -1 & 1 \\ 3 & 5 & 7 \end{vmatrix} = 71$$

Which of the preceding properties (see Exercises 22–27) justifies this conclusion without doing any further calculations?

33. Evaluate

$$\begin{vmatrix} 5 & -4 & 3 \\ -6 & 6 & 2 \\ -7 & 3 & 4 \end{vmatrix}$$

by first making repeated use of the property in Exercise 27 to obtain either a row or a column that has two zeros. First study this sample problem:

$$\begin{vmatrix} 3 & -6 & 4 \\ -5 & 2 & 7 \\ -1 & 4 & -3 \end{vmatrix} = \begin{vmatrix} 0 & 6 & -5 \\ -5 & 2 & 7 \\ -1 & 4 & -3 \end{vmatrix}$$ 3 times row 3 added to row 1 (Exercise 27)

$$= \begin{vmatrix} 0 & 6 & -5 \\ 0 & -18 & 22 \\ -1 & 4 & -3 \end{vmatrix}$$ −5 times row 3 added to row 2 (Exercise 27)

$$= (-1)\begin{vmatrix} 6 & -5 \\ -18 & 22 \end{vmatrix}$$ Expansion by minors along column 1

$$= (-6)\begin{vmatrix} 1 & -5 \\ -3 & 22 \end{vmatrix}$$ (Exercise 26)

$$= -6(22 - 15)$$

$$= -42$$

34. Show that

$$\begin{vmatrix} a^2 & b^2 & c^2 \\ a & b & c \\ 1 & 1 & 1 \end{vmatrix} = (a - b)(a - c)(b - c)$$

CHALLENGE
Think Creatively

The straight line through the two points (a, b) and (c, d), where $a \neq c$, is given by this equation:

$$\begin{vmatrix} 1 & 1 & 1 \\ x & a & c \\ y & b & d \end{vmatrix} = 0$$

Show how to convert this equation into the point-slope form for this line. (*Hint:* Apply the result in Exercise 27 to the columns in the determinant.)

10.6 SOLVING LINEAR SYSTEMS USING MATRICES

Methods of solving linear systems were studied in earlier sections of this chapter. In this section we develop another, more efficient method. We begin by forming the rectangular array of numbers, called a *matrix,* consisting of the coefficients and constants of the system. For example, the following linear system:

(1) $2x + 5y + 8z = 11$

(2) $x + 4y + 7z = 10$

(3) $3x + 6y + 12z = 15$

is replaced by this corresponding matrix:

Coefficients

$x \quad y \quad z$

$\downarrow \quad \downarrow \quad \downarrow$

$$\begin{array}{l} \text{Row 1} \\ \text{Row 2} \\ \text{Row 3} \end{array} \left[\begin{array}{ccc:c} 2 & 5 & 8 & 11 \\ 1 & 4 & 7 & 10 \\ 3 & 6 & 12 & 15 \end{array}\right]$$

The dashed vertical line serves as a reminder that the coefficients of the variables are to the left. The numbers to the right are the constants on the right-hand side of the equal signs in the given system.

Compare the steps in the two columns that follow. Note that the objective is to transform the given linear system into the form reached in step 5 below.

Working with the equations	Working with the matrices
Step 1 Write the system. $\begin{aligned} 2x + 5y + 8z &= 11 \\ x + 4y + 7z &= 10 \\ 3x + 6y + 12z &= 15 \end{aligned}$	Step 1 Write the corresponding matrix. $\begin{bmatrix} 2 & 5 & 8 & 11 \\ 1 & 4 & 7 & 10 \\ 3 & 6 & 12 & 15 \end{bmatrix}$
Step 2 Interchange the first two equations. $\begin{aligned} x + 4y + 7z &= 10 \\ 2x + 5y + 8z &= 11 \\ 3x + 6y + 12z &= 15 \end{aligned}$	Step 2 Interchange the first two rows. $\begin{bmatrix} 1 & 4 & 7 & 10 \\ 2 & 5 & 8 & 11 \\ 3 & 6 & 12 & 15 \end{bmatrix}$
Step 3 Add -2 times the first equation to the second, and -3 times the first to the third. $\begin{aligned} x + 4y + 7z &= 10 \\ -3y - 6z &= -9 \\ -6y - 9z &= -15 \end{aligned}$	Step 3 Add -2 time the first row to the second, and -3 times the first to the third. $\begin{bmatrix} 1 & 4 & 7 & 10 \\ 0 & -3 & -6 & -9 \\ 0 & -6 & -9 & -15 \end{bmatrix}$
Step 4 Multiply the second equation by $-\frac{1}{3}$. $\begin{aligned} x + 4y + 7z &= 10 \\ y + 2z &= 3 \\ -6y - 9z &= -15 \end{aligned}$	Step 4 Multiply row 2 by $-\frac{1}{3}$. $\begin{bmatrix} 1 & 4 & 7 & 10 \\ 0 & 1 & 2 & 3 \\ 0 & -6 & -9 & -15 \end{bmatrix}$
Step 5 Add 6 times the second equation to the third. $\begin{aligned} x + 4y + 7z &= 10 \\ y + 2z &= 3 \\ 3z &= 3 \end{aligned}$	Step 5 Add 6 times row 2 to row 3. $\begin{bmatrix} 1 & 4 & 7 & 10 \\ 0 & 1 & 2 & 3 \\ 0 & 0 & 3 & 3 \end{bmatrix}$

The operations on the equations in the left column produced **equivalent systems** of equations. Equivalent systems have the same solutions. Since each of these systems has its corresponding matrix produced by comparable operations on the rows of the matrices, we say that the matrices are **row-equivalent.**

In step 5, we reached a row-equivalent matrix whose corresponding linear system

When a different sequence of operations on the rows is used, it will most likely result in a different, but equivalent, triangular form.

$$\begin{aligned} x + 4y + 7z &= 10 \\ y + 2z &= 3 \\ 3z &= 3 \end{aligned}$$

is said to be in **triangular form**. From this form we solve for the variables using **back-substitution**. That is, we find z from the last equation, then substitute back into the second to find y, and finally substitute back into the first to find x. Thus

$$3z = 3 \longrightarrow z = 1$$
$$y + 2(1) = 3 \longrightarrow y = 1$$
$$x + 4(1) + 7(1) = 10 \longrightarrow x = -1$$

The solution is the ordered triple $(-1, 1, 1)$, which can be checked in the given system.

To summarize, our new matrix method for solving a linear system has two major parts:

Part A: Use the following **fundamental row operations** to transform the initial matrix corresponding to the linear system into a row-equivalent matrix as in step 5.

(1) Interchange two rows.
(2) Multiply a row by a nonzero constant.
(3) Add a multiple of a row to another row.

Part A can be completed by using a variety of row operations. The steps of a solution are therefore not unique, but the final solution will be the same.

Part B: Convert the matrix obtained in Part A back into a linear system, which will be equivalent to the original system, and solve for the variables by back-substitution.

EXAMPLE 1 Solve the linear system using row-equivalent matrices.

$$\begin{aligned} 2x + 14y - 4z &= -2 \\ -4x - 3y + z &= 8 \\ 3x - 5y + 6z &= 7 \end{aligned}$$

Solution Begin by writing the matrix corresponding to the linear system and apply the fundamental row operations.

The explanations pointing to the rows state what row operations were applied to the rows of the ***preceding matrix*** *to obtain the designated row.*

$$\left[\begin{array}{rrr|r} 2 & 14 & -4 & -2 \\ -4 & -3 & 1 & 8 \\ 3 & -5 & 6 & 7 \end{array}\right]$$ This is the matrix for the given system.

Getting a 1 in the circled position will make it easier to get zeros *below* this 1 in the next step using row operation (3).

$$\left[\begin{array}{rrr|r} \textcircled{1} & 7 & -2 & -1 \\ -4 & -3 & 1 & 8 \\ 3 & -5 & 6 & 7 \end{array}\right] \begin{array}{l} \longleftarrow \frac{1}{2} \times (\text{row } 1) \\ \\ \\ \end{array}$$

$$\left[\begin{array}{rrr|r} 1 & 7 & -2 & -1 \\ 0 & 25 & -7 & 4 \\ 0 & -26 & 12 & 10 \end{array}\right] \begin{array}{l} \\ \longleftarrow 4 \times (\text{row } 1) + \text{row } 2 \\ \longleftarrow -3 \times (\text{row } 1) + \text{row } 3 \end{array}$$

$$\left[\begin{array}{rrr|r} 1 & 7 & -2 & -1 \\ 0 & 25 & -7 & 4 \\ 0 & -1 & 5 & 14 \end{array}\right] \begin{array}{l} \\ \\ \longleftarrow \text{Row } 2 + \text{row } 3 \end{array}$$

Geting the -1 in the circled position will make it easier to get zero *below* this -1 in the next step.

$$\left[\begin{array}{ccc|c} 1 & 7 & -2 & -1 \\ 0 & \boxed{-1} & 5 & 14 \\ 0 & 25 & -7 & 4 \end{array}\right] \begin{array}{l} \\ \longleftarrow \text{Interchange rows 2} \\ \longleftarrow \text{and 3} \end{array}$$

$$\left[\begin{array}{ccc|c} 1 & 7 & -2 & 1 \\ 0 & -1 & 5 & 14 \\ 0 & 0 & 118 & 354 \end{array}\right] \begin{array}{l} \\ \\ \longleftarrow 25 \times (\text{row 2}) + \text{row 3} \end{array}$$

Now convert to the corresponding linear system and solve for the variables using back-substitution.

$$\begin{aligned} x + 7y - 2z &= -1 \\ -y + 5z &= 14 \\ 118z &= 354 \end{aligned}$$

$$118z = 354 \longrightarrow z = 3$$
$$-y + 5(3) = 14 \longrightarrow y = 1$$
$$x + 7(1) - 2(3) = -1 \longrightarrow x = -2$$

Thus the solution is the ordered triple $(-2, 1, 3)$. ■

This matrix procedure also reveals when a linear system has no solutions, that is, when it is an inconsistent system. In such a case we will obtain a row in a matrix of the form

$$0 \quad 0 \quad \cdots \quad 0 \mid p$$

where $p \neq 0$. But when this row is converted to an equation, we get the false statement $0 = p$. For example, solving the system

$$\begin{aligned} 3x - 6y &= 9 \\ -2x + 4y &= -8 \end{aligned}$$

we have

$$\left[\begin{array}{cc|c} 3 & -6 & 9 \\ -2 & 4 & -8 \end{array}\right]$$

$$\left[\begin{array}{cc|c} 1 & -2 & 3 \\ -2 & 4 & -8 \end{array}\right] \longleftarrow \tfrac{1}{3} \times (\text{row 1})$$

$$\left[\begin{array}{cc|c} 1 & -2 & 3 \\ 0 & 0 & -2 \end{array}\right] \longleftarrow 2 \times (\text{row 1}) + \text{row 2}$$

The last row gives the false equation $0 = -2$. Therefore, the system is inconsistent; there are no solutions.

The next example demonstrates how the matrix method can be used to solve a system that has infinitely many solutions, that is, a dependent system.

CAUTION
Note that there is no y-term in the third equation. You should think of this as 0y and record the 0 in the y-position in the third row of the matrix as shown.

EXAMPLE 2 Solve the system

$$\begin{aligned} x + 2y - z &= 1 \\ 2x - y + 3z &= 4 \\ 5x \quad\quad + 5z &= 9 \end{aligned}$$

Solution

$$\left[\begin{array}{rrr|r} 1 & 2 & -1 & 1 \\ 2 & -1 & 3 & 4 \\ 5 & 0 & 5 & 9 \end{array}\right]$$

$$\left[\begin{array}{rrr|r} 1 & 2 & -1 & 1 \\ 0 & -5 & 5 & 2 \\ 0 & -10 & 10 & 4 \end{array}\right] \begin{array}{l} \\ \longleftarrow -2 \times (\text{row } 1) + \text{row } 2 \\ \longleftarrow -5 \times (\text{row } 1) + \text{row } 3 \end{array}$$

$$\left[\begin{array}{rrrr} 1 & 2 & -1 & 1 \\ 0 & -5 & 5 & 2 \\ 0 & 0 & 0 & 0 \end{array}\right] \begin{array}{l} \\ \\ \longleftarrow -2 \times (\text{row } 2) + \text{row } 3 \end{array}$$

Since the last row contains all zeros, we have an equivalent linear system of two equations in three variables.

$$\begin{aligned} x + 2y - z &= 1 \\ -5y + 5z &= 2 \end{aligned}$$

This system has an *incomplete* triangular form.

Again we use back-substitution. First use the last equation to solve for y in terms of z to get $y = -\frac{2}{5} + z$. Now let $z = c$ represent any number, giving $y = -\frac{2}{5} + c$, and substitute back into the first equation.

$$x + 2y - z = x + 2(-\tfrac{2}{5} + c) - c = 1 \quad \longrightarrow \quad x = \tfrac{9}{5} - c$$

Find the specific solutions of the system for the values $c = 0$, $c = 1$, $c = \frac{2}{5}$, and $c = -2$.

The solutions are

$$(\tfrac{9}{5} - c, -\tfrac{2}{5} + c, c) \qquad \text{for any number } c$$

This result can be checked in the original system as follows.

$$\begin{aligned} x + 2y - z &= \tfrac{9}{5} - c + 2(-\tfrac{2}{5} + c) - c = 1 \\ 2x - y + 3z &= 2(\tfrac{9}{5} - c) - (-\tfrac{2}{5} + c) + 3c = 4 \\ 5x \quad\quad + 5z &= 5(\tfrac{9}{5} - c) + 5c = 9 \end{aligned}$$

■

The solutions for the system in Example 2 are stated in terms of $z = c$, where c (or z) represents any number. There are other ways in which these solutions can be stated. For example, when c is replaced by z in the given solutions we have

$$x = \tfrac{9}{5} - z \quad \text{or} \quad z = \tfrac{9}{5} - x$$

and

$$y = -\tfrac{2}{5} + z$$

Now let $x = d$ be any number. Then

$$z = \tfrac{9}{5} - x = \tfrac{9}{5} - d$$

and

$$y = -\tfrac{2}{5} + z = -\tfrac{2}{5} + (\tfrac{9}{5} - d) = \tfrac{7}{5} - d$$

You should also be able to obtain the solutions in this form $(\tfrac{7}{5} - e, e, \tfrac{2}{5} + e)$ *for any number* e.

Then the solutions can be stated in this form:

$$(d, \tfrac{7}{5} - d, \tfrac{9}{5} - d) \qquad \text{for any number } d$$

When using the matrix method of this section keep the following observations in mind.

1. If you reach a linear system in triangular form, as in Example 1, the system has a unique solution that can be found by back-substitution.
2. If you reach a linear system that has an incomplete triangular form, and there are no false equations, as in Example 2, then the system has many solutions that can be found by back-substitution.
3. If you reach a linear system having a false equation, then the system has no solutions.

EXERCISES 10.6

Use matrices and fundamental row operations to solve each system.

1. $\begin{aligned} x + 5y &= -9 \\ 4x - 3y &= -13 \end{aligned}$

2. $\begin{aligned} 4x - y &= 6 \\ 2x + 3y &= 10 \end{aligned}$

3. $\begin{aligned} 3x + 2y &= 18 \\ 6x + 5y &= 45 \end{aligned}$

4. $\begin{aligned} 2x + 4y &= 24 \\ -3x + 5y &= -25 \end{aligned}$

5. $\begin{aligned} 4x - 5y &= -2 \\ 16x + 2y &= 3 \end{aligned}$

6. $\begin{aligned} x &= y - 7 \\ 3y &= 2x + 16 \end{aligned}$

7. $\begin{aligned} 2x &= -8y + 2 \\ 4y &= x - 1 \end{aligned}$

8. $\begin{aligned} 2x - 5y &= 4 \\ -10x + 25y &= -20 \end{aligned}$

9. $\begin{aligned} 30x + 45y &= 60 \\ 4x + 6y &= 8 \end{aligned}$

10. $\begin{aligned} x - y &= 3 \\ -\tfrac{1}{3}x + \tfrac{1}{3}y &= 1 \end{aligned}$

11. $\begin{aligned} 2x &= 8 - 3y \\ 6x + 9y &= 14 \end{aligned}$

12. $\begin{aligned} -10x + 5y &= 8 \\ 15x - 10y &= -4 \end{aligned}$

13. $\begin{aligned} x - 2y + 3z &= -2 \\ -4x + 10y + 2z &= -2 \\ 3x + y + 10z &= 7 \end{aligned}$

14. $\begin{aligned} 2x + 4y + 8z &= 14 \\ 4x - 2y + 2z &= 6 \\ -5x + 3y - z &= -4 \end{aligned}$

15. $\begin{aligned} -x + 2y + 3z &= 11 \\ 2x - 3y &= -6 \\ 3x - 3y + 3z &= 3 \end{aligned}$

16. $\begin{aligned} -2x + y + 2z &= 14 \\ 5x + z &= -10 \\ x - 2y - 3z &= -14 \end{aligned}$

17. $\begin{aligned} x - 2z &= 5 \\ 3y + 4z &= -2 \\ -2x + 3y + 8z &= 4 \end{aligned}$

18. $\begin{aligned} 2x + y &= 3 \\ 4x + 5z &= 6 \\ -2y + 5z &= -4 \end{aligned}$

19.
$$\begin{aligned} x - 2y + z &= 1 \\ -6x + y + 2z &= -2 \\ -4x - 3y + 4z &= 0 \end{aligned}$$

20.
$$\begin{aligned} 4x - 3y + z &= 0 \\ -3x + y + 2z &= 0 \\ -2x - y + 5z &= 0 \end{aligned}$$

21.
$$\begin{aligned} w - x + 2y + 2z &= 0 \\ 2w - y - 3z &= 0 \\ 4x - 3y + z &= -2 \\ -3w + 2x + 4z &= 1 \end{aligned}$$

22.
$$\begin{aligned} 4w - 5x + 2z &= 0 \\ -2w + 10x + y - 3z &= 8 \\ 5x - 2y + 4z &= -16 \\ 6w + 3z &= 0 \end{aligned}$$

23.
$$\begin{aligned} v + 2w - x + 3z &= 4 \\ -w + 5x - y - z &= 3 \\ 3v + 6x + 2z &= 1 \\ 2v + 3w + 3x - y + 5z &= 10 \\ v + 9x - 2y + z &= 5 \end{aligned}$$

24.
$$\begin{aligned} w + 3y &= 10 \\ -x + 2y + 6z &= 2 \\ -3w - 2x - 4z &= 2 \\ 4x - y &= 8 \end{aligned}$$

Written Assignment: Continue using row operations after Step 5 on page 461 and reach the form

$$\left[\begin{array}{ccc|c} 1 & 0 & 0 & -1 \\ 0 & 1 & 0 & 1 \\ 0 & 0 & 1 & 1 \end{array}\right]$$

Describe the steps you have used to obtain this form and explain why no more needs to be done.

10.7 SYSTEMS OF LINEAR INEQUALITIES

The graphing of linear inequalities in two variables was introduced in Section 6.6. (You should review that material at this time.) There you learned that the graph of an inequality such as $3x - 5y < 10$ consists of all points in the plane on one side of the line $3x - 5y = 10$. To determine which side of the line is correct, two procedures were used.

Method I

(i) Draw the line $3x - 5y = 10$

(ii) Solve the inequality for y.

$$3x - 5y < 10$$
$$-5y < -3x + 10$$
$$y > \tfrac{3}{5}x - 2$$

(iii) Since $y > \frac{3}{5}x - 2$, the graph consists of all points above the line indicated by the shading.

Method II

(i) Draw the line $3x - 5y = 10$

(ii) Select a convenient point, such as (0, 0), on one side of the line and substitute the coordinates into the inequality

$$3(0) - 5(0) = 0 < 10$$

(iii) Since $0 < 10$ is a true statement, (0, 0) is on the correct side of the line, and the graph consists of all points on the same side of the line as the point (0, 0), indicated by the shading.

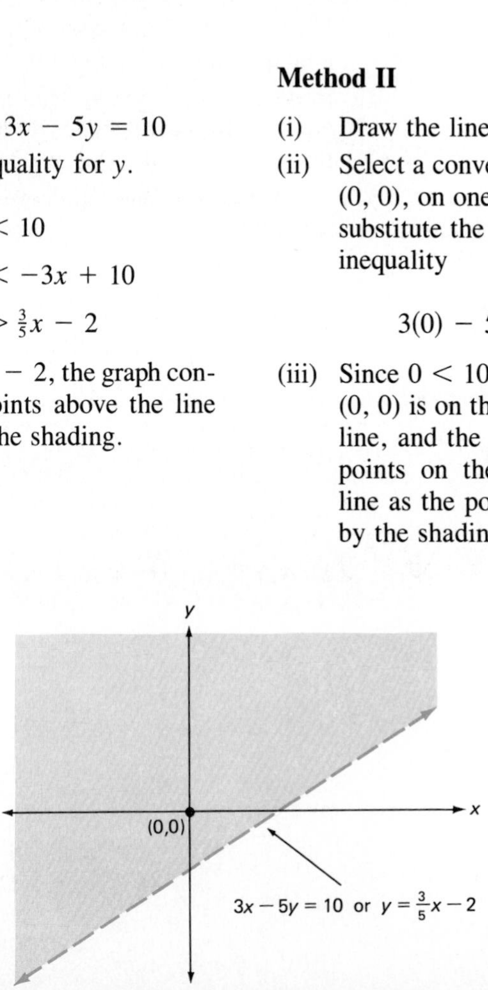

The dashed line means that the points on the line $3x - 5y = 10$ are not part of the graph. The graph of $3x - 5y \leq 10$ would include the points on the line, and a solid line would be used to indicate this.

The methods of graphing a linear inequality can be extended to graphing systems of linear inequalities. This is demonstrated in Example 1.

EXAMPLE 1 Graph the system of linear inequalities:

$$2x + y \leq 6$$
$$3x - 4y \geq 12$$

Solution Draw the lines $2x + y = 6$ and $3x - 4y = 12$. Shade the region $2x + y \leq 6$ vertically and the region $3x - 4y \geq 12$ horizontally. Since the coordinates of a point $P(x, y)$ must satisfy *both* conditions, the graph consists of all points shaded in both directions, as shown. The parts of the two lines above their point of intersection are dashed since they are not included.

Observe that the unshaded region consists of the points that satisfy this system:

$$2x + y > 6$$
$$3x - 4y < 12$$

What is the system for the part shaded only vertically? See Exercise 29.

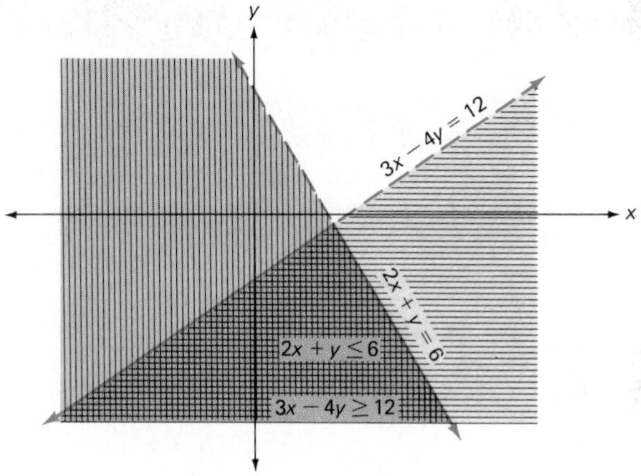

■

A system of linear inequalities may have more than two inequalities. Example 2 demonstrates how to graph a system of three linear inequalities.

EXAMPLE 2 Graph the system:

$$x + 3y \geq 12$$
$$-2x + y \leq 4$$
$$8x + 3y \leq 54$$

Solution Draw the three lines ℓ_1: $x + 3y = 12$; ℓ_2: $-2x + y = 4$; and ℓ_3: $8x + 3y = 54$. Solve each inequality for y to obtain the equivalent system

$$y \geq -\tfrac{1}{3}x + 4$$
$$y \leq 2x + 4$$
$$y \leq -\tfrac{8}{3}x + 18$$

The first inequality gives the points on and above ℓ_1, and the last two inequalities give the points on and below ℓ_2 and above ℓ_3. Since a point (x, y) belongs to the graph of the system when it satisfies all three inequalities, the graph is the triangular region including the sides of the triangle as shown.

As an alternative procedure draw the three lines and use a test point from each of the 7 regions determined by the lines (six regions are outside the triangle and one is inside). When a point from a region satisfies all 3 inequalities, then that region is part of the graph.

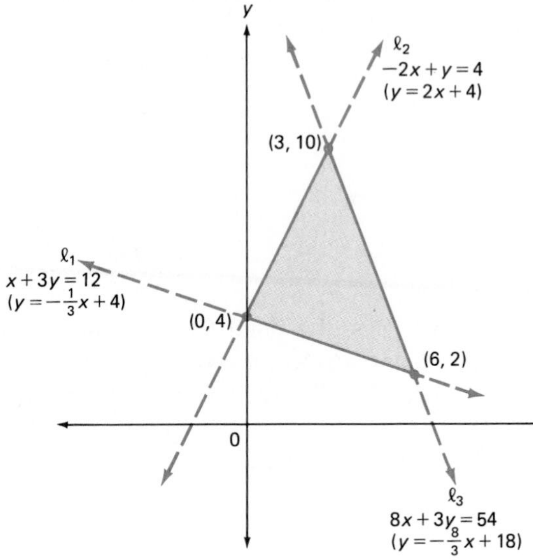

■

The final example shows how absolute values can be used to write and graph systems of linear inequalities.

EXAMPLE 3 Graph the system:

$$|x| \geq 3$$
$$|y| \leq 5$$

Solution The inequality $|x| \geq 3$ is equivalent to the compound inequality $x \leq -3$ or $x \geq 3$. Thus the points in the plane that satisfy $|x| \geq 3$ are all points to the left or on the vertical line $x = -3$, as well as all points to the right or on the line $x = 3$.

Similarly, since $|y| \leq 5$ is equivalent to $-5 \leq y \leq 5$, the points for this inequality are those on or between the horizontal lines $y = \pm 5$.

Since a point is on the graph of the given system provided *both* inequalities of the system are satisfied, the graph of the given system is the shaded region below.

Compound inequalities were introduced in Section 3.4.

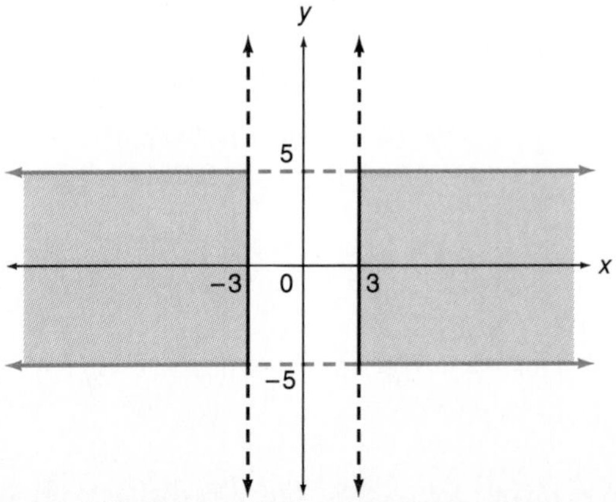

EXERCISES 10.7

Note: Exercises 1–12 review the graphing of linear inequalities introduced in Section 6.6.

Graph the regions on a plane satisfying the given inequalities.

1. $y \geq x - 2$
2. $y \leq x + 1$
3. $y > x$
4. $y \geq -\frac{1}{3}x - 2$
5. $x + y \leq 4$
6. $3x - y < 6$
7. $|x| \leq 3$
8. $|x| \geq 3$
9. $|x| < 2$
10. $|y| > 1$
11. $|y| \leq 2$
12. $|y| \geq 2$

Graph each system of inequalities.

13. $y \geq x + 2$, $y \leq -x + 1$
14. $y \geq -4x + 1$, $y \leq 4x + 1$
15. $y \leq -4x + 1$, $y \geq 4x + 1$
16. $2x - y + 1 \geq 0$, $x - 2y + 2 \leq 0$
17. $3x + y < 6$, $x > 1$
18. $2x - 3y > 6$, $y > -2$
19. $x + 2y \leq 10$, $3x + 2y \leq 18$, $x \geq 0, y \geq 0$
20. $x + 2y \geq 10$, $3x + 2y \leq 18$, $x \geq 0$
21. $x + 2y \leq 10$, $3x + 2y \geq 18$, $y \geq 0$
22. $x + 2y \geq 10$, $3x + 2y \geq 18$, $x \geq 0$, $y \geq 0$
23. $x - y \leq -1$, $2x + y \geq 7$, $4x - y \leq 11$
24. $x - y \geq -1$, $2x + y \geq 7$, $4x - y \leq 11$
25. $x - y \geq -1$, $2x + y \leq 7$, $4x - y \leq 11$, $x \geq 0$, $y \geq 0$
26. $|x| > 2$, $|y| > 1$
27. $|x - 2| \leq 1$, $|y + 1| \geq 2$
28. $|x| \leq 2$, $|y| \leq 1$
29. Find the system of linear inequalities whose graph is the region having only vertical shading in the figure for Example 1, page 467.
30. Find the system of linear inequalities whose graph is the region having only horizontal shading in the figure for Example 1, page 467.

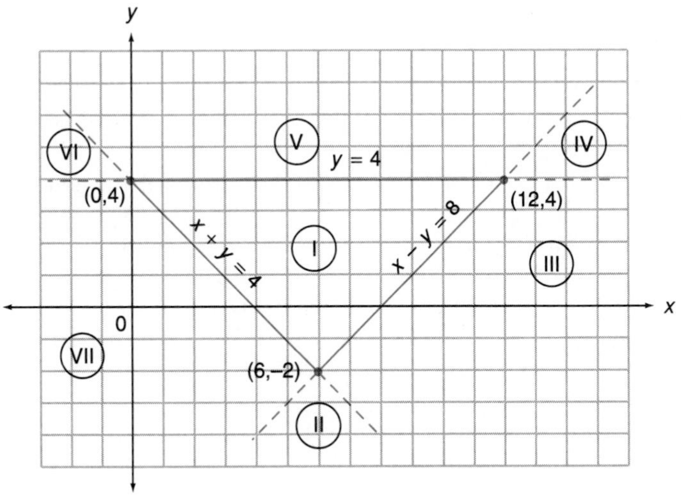

For Exercises 31–37 find the system of linear inequalities whose graph is indicated by the Roman numeral. Note, for example, that region I includes all three sides, whereas region III includes one side and excludes the other two.

31. I 32. II 33. III 34. IV 35. V 36. VI 37. VII

EXPLORATIONS
Think Critically

1. Both in terms of their graphs and their solutions, describe the difference between an inconsistent system and a dependent system of two linear equations in two variables.
2. Under what conditions does the following represent an inconsistent system?

$$ax + by = c$$
$$dx + ey = f$$

3. An infinite set of points (x, y) are on the line $-3x + 7y = 6$, and another infinite set of points (x, y) are on the line $11x - 7y = 10$. Why then, when solving the system of two equations, may we add to obtain $8x = 16$?
4. What must be true for the determinant of the coefficients of a linear system in two variables and two equations in order for the system to have a unique solution?
5. Consider the procedure for evaluating a 3 by 3 determinant by repeating the first two columns to the right of the determinant. Find similar procedures by repeating parts to the left, by repeating parts below, and by repeating parts above.
6. Back-substitution can be avoided when solving a linear system by continuing to apply fundamental row-operations until the form shown below is reached. From this form the solution is immediately available. Try this for Example 1, page 000, as well as for the system on page 000. Now adjust this procedure to solve Example 2, page 000.

$$\left|\begin{array}{ccc|c} 1 & 0 & 0 & a \\ 0 & 1 & 0 & b \\ 0 & 0 & 1 & c \end{array}\right|$$

7. Write the system of linear inequalities whose graph consists of all points inside and on the square whose vertices are on the coordinate axes, and whose diagonals have length d.

10.8 LINEAR PROGRAMMING

Linear programming is a mathematical procedure for solving problems related to the logistics of decision making. Applications of this procedure can be found in numerous areas, including business and industry, agriculture, the field of nutrition, and the military.

This introduction to linear programming will primarily be a geometric approach based on the graphing of linear systems in two variables. We begin with the following system and use it to develop some of the basic concepts.

$$x \geq 0$$
$$y \geq 0$$
$$\tfrac{3}{4}x + \tfrac{1}{2}y \leq 6$$
$$\tfrac{1}{2}x + y \leq 6$$

The graph of this system is a region R consisting of all points (x, y) that satisfy *each* of the inequalities in the system. First note that the conditions $x \geq 0$ and $y \geq 0$ require the points to be in quadrant I, or on the nonnegative parts of each axis. Now graph the lines $\frac{3}{4}x + \frac{1}{2}y = 6$, $\frac{1}{2}x + y = 6$ and determine their point of intersection (6, 3). Then, by the procedure used in Section 10.7, the (x, y) that satisfy both inequalities $\frac{3}{4}x + \frac{1}{2}y \leq 6$ and $\frac{1}{2}x + y \leq 6$ are found to be below or on these lines. The completed region R is shown on the next page.

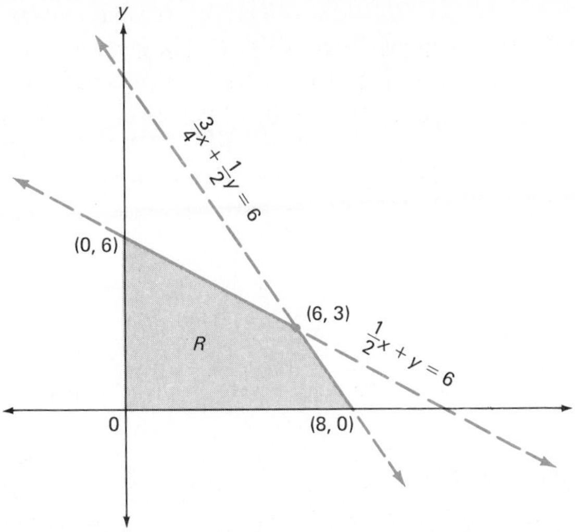

The region R is said to be a **convex set of points**, because any two points inside R can be joined by a line segment that is totally inside R. Roughly speaking, the boundary of R "bends outward." Here is a figure of a region that is *not* convex; note, in this figure, that points A and B cannot be connected by a line segment that is totally inside the region.

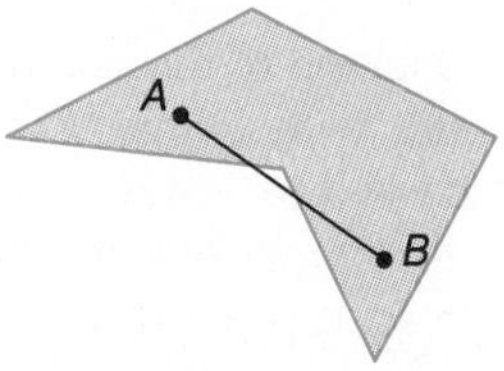

Draw the line ℓ_1: $x + y = 4$ in the same coordinate system as region R. (See the next figure.) All points (x, y) in R that are on this line have coordinates whose sum is 4. Now draw lines ℓ_2 and ℓ_3 parallel to ℓ_1, and also having equations $x + y = 2$ and $x + y = 8$, respectively. Obviously, all points in R on ℓ_2 or on ℓ_3 have coordinates whose sum is *not* 4. These three lines all have the form $x + y = k$, where k is some constant. There is an infinite number of such parallel lines, depending on the constant k. They all have slope -1. Some of these lines intersect R and others do not; we will be interested only in those that do.

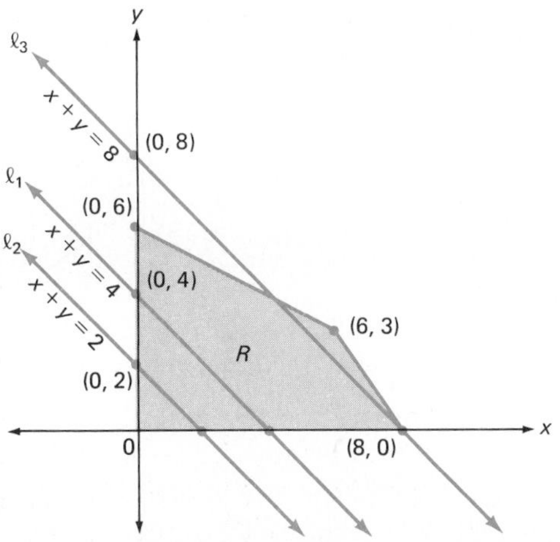

Can you guess which line of this form will intersect R and have the largest possible value k?

It is not difficult to see the answer, because all these lines are parallel; and the "higher" the line, the larger will be the value k. As a matter of fact, the line $x + y = k$ has k as its y-intercept. So the line we are looking for will be the line that intersects the y-axis as high up as possible, has slope -1, and still meets region R. Because of the shape of R (it is convex), this will be the line through (6, 3) with slope -1, as shown in the following figure.

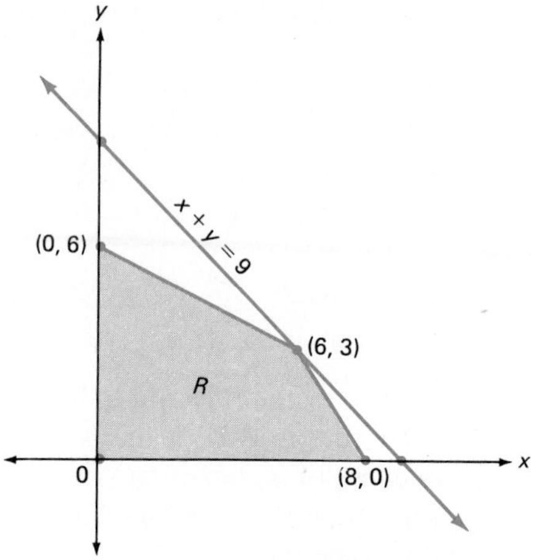

The equation of the line through (6, 3) with slope -1 is $x + y = 9$. Any other parallel line with a larger k-value will be higher and cannot intersect R; and those others that do will be lower and have a smaller k-value.

To sum up, we can say that of all the lines with form $x + y = k$ that intersect region R, the one that has the largest k-value is the line $x + y = 9$. Putting it another way, we can say that of all the points in R, the point (6, 3) is the point whose coordinates give the maximum value for the quantity $k = x + y$.

Suppose that we now look for the point in R that produces the maximum value for k, where $k = 2x + 3y$. First we draw a few parallel lines each with equation of the form $k = 2x + 3y$. Below are such lines for k taking on the values 6, 10, and 18.

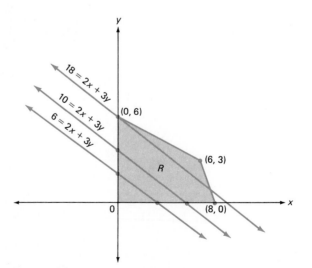

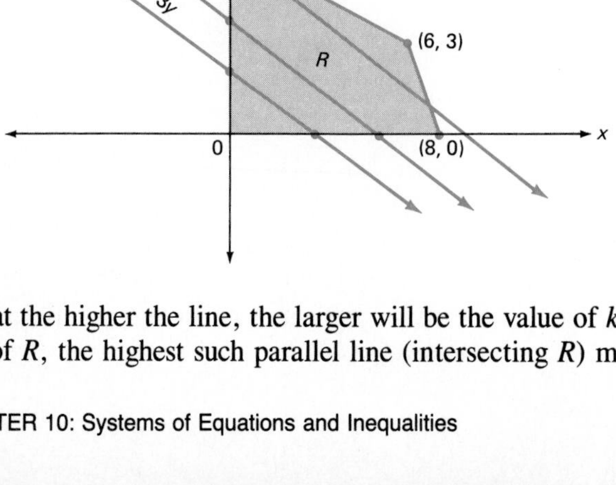

We see that the higher the line, the larger will be the value of k. Because of the convex shape of R, the highest such parallel line (intersecting R) must pass through the vertex (6, 3). For this line we get $2(6) + 3(3) = 21 = k$. This is the largest value of $k = 2x + 3y$ for the (x, y) in R.

The preceding observations should be convincing evidence for the following result:

Whenever we have a convex-shaped region R, then the point in R that produces the maximum value of a quantity of the form $k = ax + by$, where a and b are positive, will be at a vertex of R.

Because of this result it is no longer necessary to draw lines through R. All that needs to be done to find the maximum of $k = ax + by$ is to graph the region R, find the coordinates of all vertices on the boundary, and see which one produces the largest value for k.

EXAMPLE 1 Maximize the quantity $k = 4x + 5y$ for (x, y) in the region S given by this system:

$$\begin{aligned} -x + 2y &\leq 2 \\ 3x + 2y &\leq 10 \\ x + 6y &\geq 6 \\ 3x + 2y &\geq 6 \end{aligned}$$

Solution First graph the corresponding four lines and shade in the required region. Next find the four vertices of the region S by solving the appropriate pairs of equations. Since $k = 4x + 5y$ will be a maximum only at a vertex, the listing below shows that 18 is the maximum value.

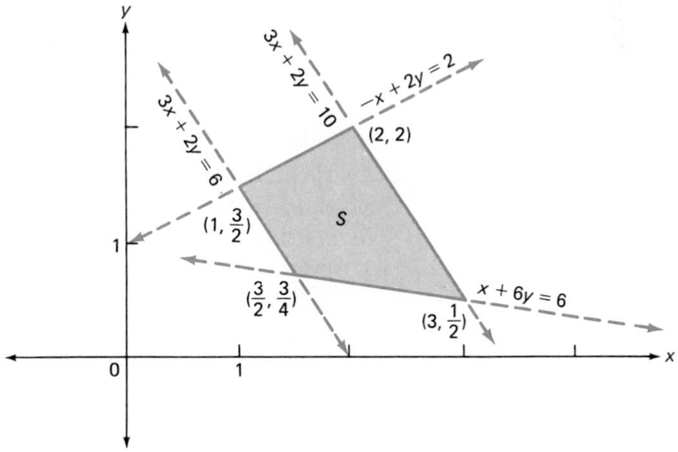

Vertex	$k = 4x + 5y$
$(1, \frac{3}{2})$	$4(1) + 5(\frac{3}{2}) = 11\frac{1}{2}$
$(\frac{3}{2}, \frac{3}{4})$	$4(\frac{3}{2}) + 5(\frac{3}{4}) = 9\frac{3}{4}$
$(2, 2)$	$4(2) + 5(2) = 18$
$(3, \frac{1}{2})$	$4(3) + 5(\frac{1}{2}) = 14\frac{1}{2}$

■

Instead of asking for the maximum of $k = 4x + 5y$, as in Example 1, it is also possible to find the minimum value of $k = 4x + 5y$ for the same region S. The earlier discussions explaining why a vertex of a convex region gives the maximum can be easily adjusted to show that a vertex will also give the minimum. Essentially, it becomes a matter of finding the line with correct slope, that has the lowest possible y-intercept, and that still intersects S. Because of this, the table in the solution to the preceding example shows that $9\frac{3}{4}$ is the minimum value of $k = 4x + 5y$. It occurs at the vertex $(\frac{3}{2}, \frac{3}{4})$.

TEST YOUR UNDERSTANDING
Think Carefully

Find the required values for the region R, where R is the graph of the system.

$$5x + 4y \geq 40$$
$$x + 4y \leq 40$$
$$7x + 4y \leq 112$$

1. Maximum of $k = x + y$
2. Minimum of $k = x + y$
3. Maximum of $k = x + 2y$
4. Minimum of $k = x + 2y$
5. Maximum of $k = x + 5y$
6. Minimum of $k = x + 5y$
7. Maximum of $k = 2x + y$
8. Minimum of $k = 2x + y$
9. Maximum of $k = 3x + 20y$
10. Minimum of $k = 3x + 20y$

(Answers: Page 491)

The preceding method of finding the maximum or minimum of a quantity $k = ax + by$, relative to a convex region, can be applied to a variety of applied situations. Here is a typical problem.

Before studying the solution, read this problem several times. Then list the given information and try to write the algebraic expression that needs to be maximized.

EXAMPLE 2 A store sells two kinds of bicycles, model A and model B. The store buys them unassembled from a wholesaler. Two employees are responsible for assembling the bicycles, and they are permitted to work no more than 6 hours each per week to do this job. Working together to assemble model A, employee I works $\frac{3}{4}$ hour and employee II works $\frac{1}{2}$ hour. Model B requires $\frac{1}{2}$ hour's work by employee I as well as 1 hour's work by employee II. There is a \$55 profit on each model A sold and \$48 on each model B. Because of the popularity of the sport, the store is able to sell as many bicycles as they decide to assemble. How many bicycles of each model should they assemble in order to get the maximum profit?

Solution Let x be the number of model A bicycles assembled per week and y the number of model B's per week. Then $55x$ is the profit earned for model A, and $48y$ is the profit for model B. The total profit p is given by $p = 55x + 48y$. It is this quantity we need to maximize.

Next we find the time each employee works. For employee I, $\frac{3}{4}x + \frac{1}{2}y$ will be the number of hours worked per week, because each model A (there are x of these) requires $\frac{3}{4}$ hour, and each model B (there are y of these) requires $\frac{1}{2}$ hour. But each employee works *no more than* 6 hours per week. Thus the total time for this worker satisfies

$$\frac{3}{4}x + \frac{1}{2}y \leq 6$$

By very similar reasoning, the total time for employee II is $\frac{1}{2}x + y$, which satisfies

$$\tfrac{1}{2}x + y \leq 6$$

It is also known that $x \geq 0$ and $y \geq 0$ because there cannot be a negative number of either model.

Collecting the preceding conditions, we have that x and y must satisfy this system:

$$\begin{aligned} x &\geq 0 \\ y &\geq 0 \\ \tfrac{3}{4}x + \tfrac{1}{2}y &\leq 6 \\ \tfrac{1}{2}x + y &\leq 6 \end{aligned}$$

We want to find the (x, y) for this system so that $p = 55x + 48y$ is a maximum.

The graph of this system is the same region R given at the beginning of this section. The points (x, y) in R represent all the possibilities because for all such points the numbers x and y satisfy *all* the conditions of the stated problem, and any (x, y) not in R cannot be a possibility because the x- and y-values would not satisfy *all* the stated conditions.

The table below shows that the vertex (6, 3) gives the answer.

Vertex	$p = 55x + 48y$
(0, 0)	0
(0, 6)	288
(6, 3)	474
(8, 0)	440

Therefore, a maximum weekly profit is realized by assembling 6 bicycles of model A and 3 of model B. ■

Problems such as this are called **linear programming** problems. The inequalities that give the region R are sometimes called the **constraints.** The region R is called the set of **feasible points,** and (6, 3) is referred to as the **optimal point.**

EXERCISES 10.8

1. **(a)** Graph the region given by this system of constraints.

$$x + 3y \leq 6, \qquad x \geq 0, \qquad y \geq 0$$

(b) Find all the vertices.
(c) Find the maximum and minimum of $p = x + y$.
(d) Find the maximum and minimum of $q = 6x + 10y$.
(e) Find the maximum and minimum of $r = 2x + 9y$.

Follow the instructions in Exercise 1 for each of these systems.

2. $x + 4y \le 72$, $5x + 4y \le 120$, $x \ge 0$, $y \ge 0$

3. $x + 6y \le 96$, $4x + 5y \le 118$, $3x + y \le 72$, $x \ge 0$, $y \ge 0$

4. $x - 2y \ge -10$, $2x + 7y \le 57$, $5x + 6y \le 85$, $5x + 2y \le 75$, $x \ge 0$, $y \ge 0$

5. $y - x \le 3$, $x + 4y \le 22$, $2x + y \le 16$, $x - 3y \le 1$, $x + 4y \ge 8$, $x \ge 0$

Find the maximum and minimum values of p, q, and r for the given constraints.

6. $y \ge 0$, $8x + 7y \le 56$, $8x + 3y \ge 24$; $p = 2x + y$, $q = 5x + 4y$, $r = 3x + 3y$

7. $x \ge 0$, $6x + 5y \le 60$, $2x + 5y \ge 20$; $p = x + 3y$, $q = 4x + 3y$, $r = \frac{1}{2}x + \frac{1}{4}y$

8. $x \ge 4$, $y \ge 2$, $x + 2y \le 20$; $p = x + 3y$, $q = 2x + 3y$, $r = 2x + 5y$

9. $x \le 7$, $0 \le y \le 5$, $2x + y \ge 8$; $p = 2x + 2y$, $q = 3x + y$, $r = x + \frac{1}{5}y$

The regions for the systems in Exercises 10 and 11 are "open-ended"; that is, their borders do not form a closed polygon and the graph extends endlessly. These regions are still convex in the sense discussed in the text.

10. Graph the region given by the system of constraints.

$$4x - y \ge 20, \qquad 2x - 3y \ge 0, \qquad x - 4y \ge -20, \qquad y \ge 0$$

and evaluate the maximum and minimum (when they exist) of the following:
(a) $p = x + y$ **(b)** $q = 6x + 10y$ **(c)** $r = 2x + y$

11. Follow the instructions of Exercise 10 for this system.

$$5x + 2y \ge 22, \quad 6x + 7y \ge 54, \quad 2x + 11y \ge 44, \quad 4x - 5y \le 34, \quad x \ge 0$$

12. A publisher prints and sells both hardcover and paperback copies of the same book. Two machines, M_1 and M_2, are needed jointly to manufacture these books. To produce one hardcover copy, machine M_1 works $\frac{1}{6}$ hour and machine M_2 works $\frac{1}{12}$ hour. For a paperback copy, machines M_1 and M_2 work $\frac{1}{15}$ hour and $\frac{1}{10}$ hour, respectively. Each machine may be operated no more than 12 hours per day. If the profit is \$2 on a hardcover copy and \$1 on a paperback copy, how many of each type should be made per day to earn the maximum profit?

13. A manufacturer produces two models of a certain product; model I and model II. There is a \$5 profit on model I and an \$8 profit on model II. Three machines, M_1, M_2, and M_3, are used jointly to manufacture these models. The number of hours that each machine operates to produce 1 unit of each model is given in the table:

	Model I	Model II
Machine M_1	$1\frac{1}{2}$	1
Machine M_2	$\frac{3}{4}$	$1\frac{1}{2}$
Machine M_3	$1\frac{1}{3}$	$1\frac{1}{3}$

No machine is in operation more than 12 hours per day.
(a) If x is the number of model I made per day, and y the number of model II per day, show that x and y satisfy the following constraints.

$$x \ge 0, \quad y \ge 0, \quad \tfrac{3}{2}x + y \le 12, \quad \tfrac{3}{4}x + \tfrac{3}{2}y \le 12, \quad \tfrac{4}{3}x + \tfrac{4}{3}y \le 12$$

(b) Express the daily profit p in terms of x and y.

(c) Graph the feasible region given by the constraints in part (a) and find the coordinates of the vertices.

(d) What is the maximum profit, and how many of each model are produced daily to realize it?

(e) Find the maximum profit possible for the constraints stated in part (a) if the unit profits are \$8 and \$5 for models I and II, respectively.

14. A farmer buys two varieties of animal feed. Type A contains 8 ounces of corn and 4 ounces of oats per pound; type B contains 6 ounces of corn and 8 ounces of oats per pound. The farmer wants to combine the two feeds so that the resulting mixture has at least 60 pounds of corn and at least 50 pounds of oats. Feed A costs him 5¢ per pound and feed B costs 6¢ per pound. How many pounds of each type should the farmer buy to minimize the cost?

15. An appliance manufacturer makes two kinds of refrigerators, model A that earns \$100 profit and model B that earns \$120 profit. Each month the manufacturer can produce up to 600 units of model A and up to 500 units of model B. If there are only enough man-hours available to produce no more than a total of 900 refrigerators per month, how many of each kind should be produced to obtain the maximum profit?

16. Two dog foods A and B each contain three types of ingredients I_1, I_2, I_3. The number of ounces of these ingredients in each pound of a dog food is given in the table.

	I_1	I_2	I_3
A	8	3	1
B	8	1	3

A mixture of the two dog foods is to be formed to contain at least 1600 ounces of I_1, at least 300 ounces of I_2, and at least 360 ounces of I_3. If a pound of dog food A costs 8¢ and a pound of dog food B costs 6¢, how many pounds of each dog food should be used so that the resulting mixture meets all the requirements at the least cost?

10.9 SOLVING NONLINEAR SYSTEMS

A straight line will intersect a parabola or a circle twice, or once, or not at all. Two parabolas of the form $y = ax^2 + bx + c$ can intersect at most two times; the same is true for two circles. A circle and a parabola can intersect at most four times. These diagrams illustrate some of these possibilities.

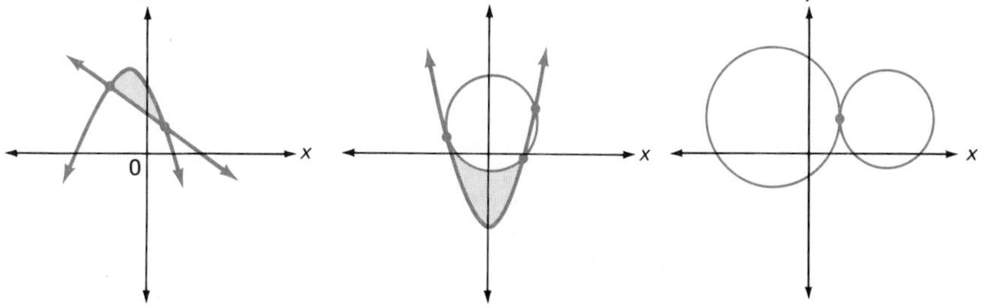

In each of the examples that follow, at least one of the two equations will not be linear. Thus we will be learning how to solve certain types of *nonlinear systems*. The underlying strategy in solving such systems will be the same as it was for linear systems, namely, first eliminate one of the two variables so as to obtain an equation in one unknown.

A Parabola and a Line

EXAMPLE 1 Solve the system of two equations and graph:

$$y = x^2$$
$$y = -2x + 8$$

Solution Let (x, y) represent the points of intersection. Since these x- and y-values are the same in both equations, we may set the two values for y equal to each other and solve for x.

$$x^2 = -2x + 8$$
$$x^2 + 2x - 8 = 0$$
$$(x + 4)(x - 2) = 0$$
$$x = -4 \quad \text{or} \quad x = 2$$

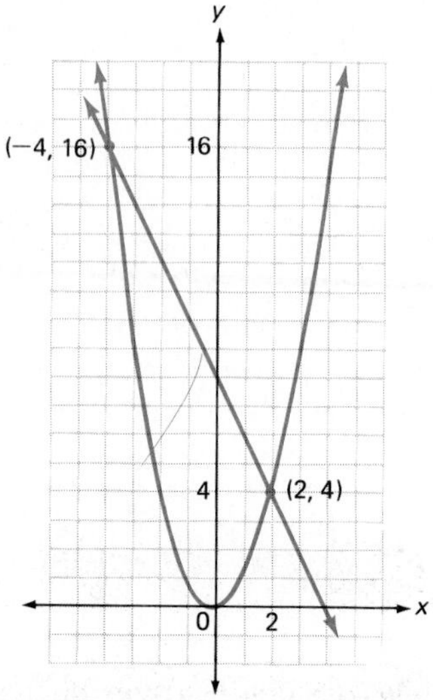

To find the corresponding y-values, either of the original equations may be used. Using $y = -2x + 8$, we have:

For $x = -4$; $\quad y = -2(-4) + 8 = 16$

For $x = 2$: $\quad y = -2(2) + 8 = 4$

The solution of the system consists of the two ordered pairs $(-4, 16)$ and $(2, 4)$. The other equation can be used as a check of these results. ■

Two Parabolas

EXAMPLE 2 Solve the system and graph:

$$y = x^2 - 2$$
$$y = -2x^2 + 6x + 7$$

Solution Set the two values for y equal to each other and solve for x.

$$x^2 - 2 = -2x^2 + 6x + 7$$
$$3x^2 - 6x - 9 = 0$$
$$x^2 - 2x - 3 = 0$$
$$(x + 1)(x - 3) = 0$$
$$x = -1 \quad \text{or} \quad x = 3$$

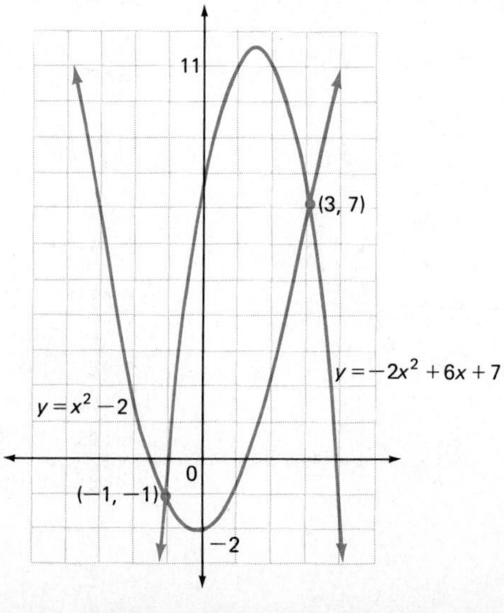

Use $y = x^2 - 2$ to solve for y.

Check these points in the second equation of the given system.

$$y = (-1)^2 - 2 = -1 \qquad y = 3^2 - 2 = 7$$

The points of intersection are $(-1, -1)$ and $(3, 7)$. ■

A Circle and a Parabola

EXAMPLE 3 Solve the system and graph.

$$x^2 + y^2 - 8y = -7$$
$$y - x^2 = 1$$

Solution Solve the second equation for x^2.

$$x^2 = y - 1$$

Substitute into the first equation and solve for y.

Note: Alternative methods can be used to solve this example (as well as others). Another easy way begins by adding the given equations. Try it. We may also solve $y - x^2 = 1$ for y and substitute into the first equation. You will find that the latter method is more difficult. With practice you will learn how to find the easier methods.

$$(y - 1) + y^2 - 8y = -7$$
$$y^2 - 7y + 6 = 0$$
$$(y - 1)(y - 6) = 0$$
$$y = 1 \quad \text{or} \quad y = 6$$

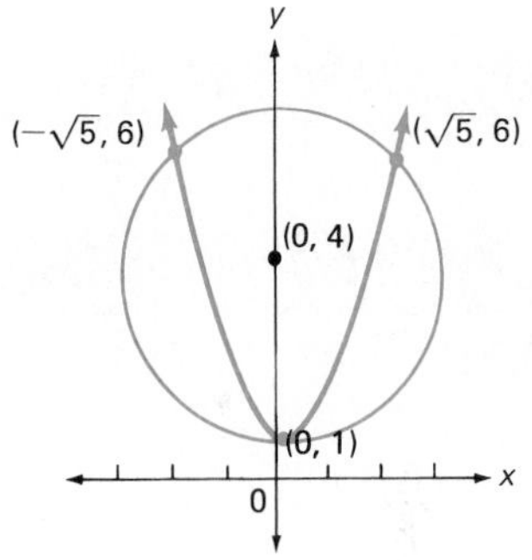

Use $x^2 = y - 1$ to solve for x.

$$\text{For } y = 1: \quad x^2 = 1 - 1 = 0 \qquad x = 0$$
$$\text{For } y = 6: \quad x^2 = 6 - 1 = 5 \qquad x = \pm\sqrt{5}$$

Check these points in the given system.

The points of intersection are $(0, 1)$, $(\sqrt{5}, 6)$, and $(-\sqrt{5}, 6)$. ■

A Circle and an Ellipse

EXAMPLE 4 Solve the system and graph.

$$x^2 + y^2 = 9$$
$$\frac{x^2}{25} + \frac{y^2}{16} = 1$$

Solution You should recognize the equations as those of a circle and an ellipse. First rewrite the second equation in this form:

$$16x^2 + 25y^2 = 400$$

Then solve the first equation for either x^2 or y^2, say y^2, and substitute into the second equation:

$$
\begin{aligned}
y^2 &= 9 - x^2 \\
16x^2 + 25(9 - x^2) &= 400 \\
16x^2 + 225 - 25x^2 &= 400 \\
-9x^2 &= 175 \\
x^2 &= -\frac{175}{9} \\
x &= \pm\sqrt{-\frac{175}{9}}
\end{aligned}
$$

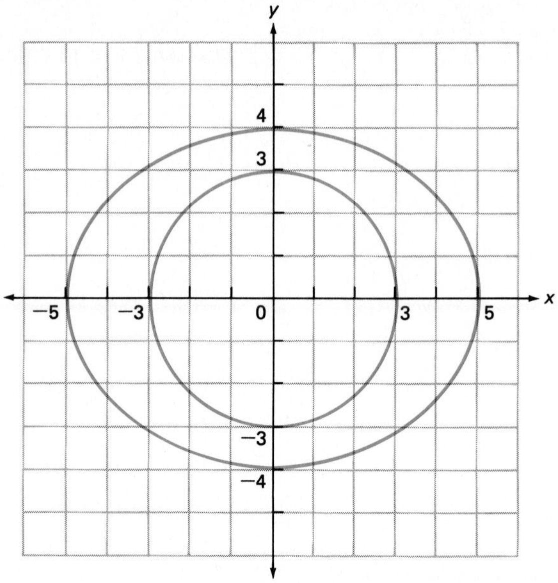

Our solution produces the square root of a negative number, which is imaginary. Thus there are no real solutions and the two curves do not intersect. ■

An Ellipse and a Hyperbola

EXAMPLE 5 Solve the system and graph.

$$
\begin{aligned}
x^2 - y^2 &= 1 \\
9x^2 + y^2 &= 9
\end{aligned}
$$

Solution Add the two equations:

$$
\begin{aligned}
x^2 - y^2 &= 1 \\
\underline{9x^2 + y^2} &\underline{= 9} \\
10x^2 &= 10 \\
x^2 &= 1 \\
x &= \pm 1 \quad \text{and} \quad y = 0
\end{aligned}
$$

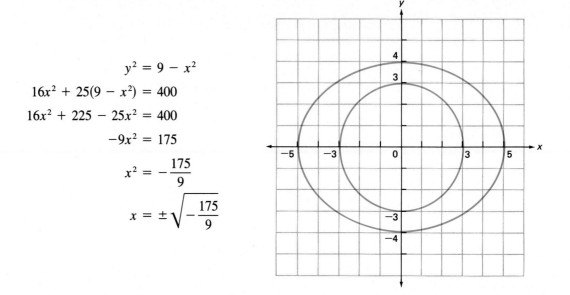

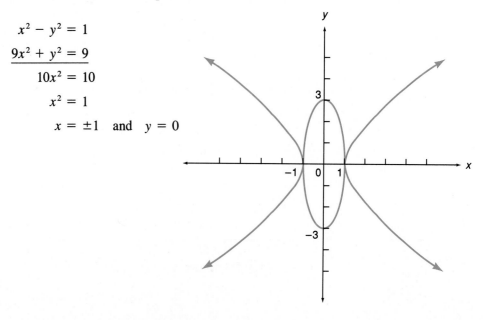

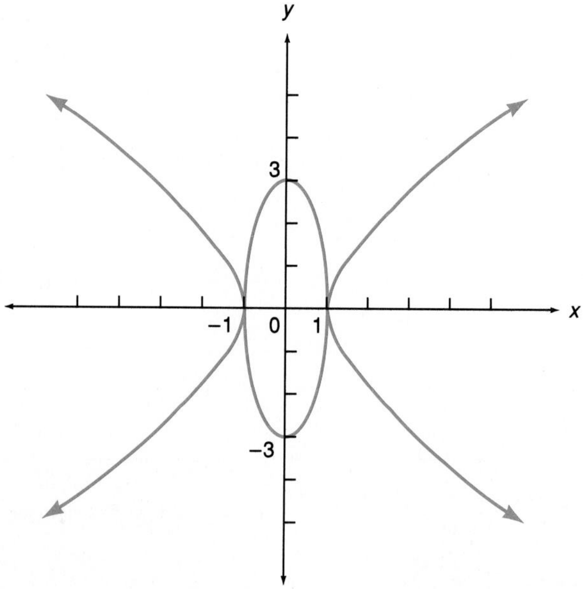

The points of intersection are located at $(\pm 1, 0)$. ■

The Cube Root and a Line

EXAMPLE 6 Solve the system and graph:

$$y = \sqrt[3]{x}$$
$$y = \tfrac{1}{4}x$$

Solution For the points of intersection the x-and y-values are the same in both equations. Thus, for such points, we set the y-values equal to one another.

$$\tfrac{1}{4}x = \sqrt[3]{x} \quad \longleftarrow \quad x^{1/3} \text{ can be used in place of } \sqrt[3]{x}$$

Cube both sides and solve for x.

$$\tfrac{1}{64}x^3 = x$$
$$x^3 = 64x$$
$$x^3 - 64x = 0$$
$$x(x^2 - 64) = 0 \qquad \text{(factoring out } x\text{)}$$
$$x(x + 8)(x - 8) = 0$$
$$x = 0 \quad \text{or} \quad x = -8 \quad \text{or} \quad x = 8$$

CAUTION
A common error is to take $x^3 - 64x = 0$ and divide through by x to get $x^2 - 64 = 0$. This step produces the roots ± 8. The root 0 has been lost because we divided by x, and 0 is the number for which the factor x in $x(x^2 - 64)$ is zero. You may always divide by a nonzero expression and get an equivalent form of the equation. But when you divide by a variable quantity there is the danger of losing some roots, those for which the divisor is 0.

Substitute these values into either of the given equations to obtain the corresponding y-values. The remaining equation can be used for checking. The solutions are $(-8, -2)$, $(0, 0)$, and $(8, 2)$.

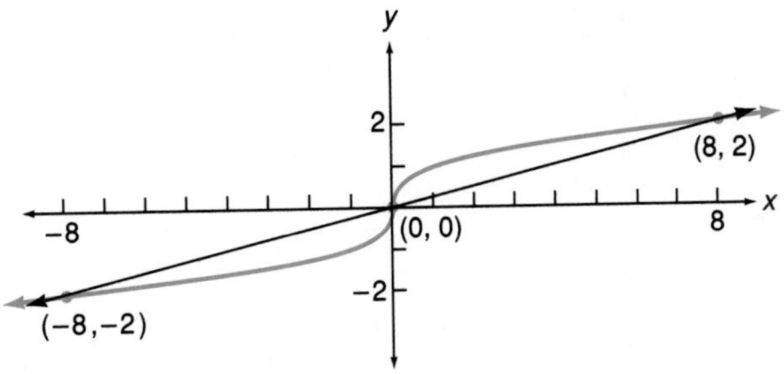

The final example illustrates how to graph a system of nonlinear inequalities.

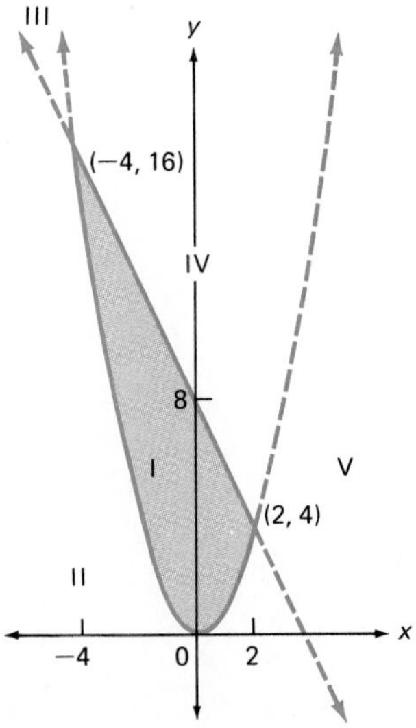

EXAMPLE 7 Graph the system of inequalities

$$y \geq x^2$$
$$y \leq -2x + 8$$

Solution First draw the graph of the equations $y = x^2$ and $y = -2x + 8$ (see Example 1). Now use test points from the numbered regions determined by the two graphs. If a test point satisfies both inequalities, then the region it comes from is part of the graph of the system. Otherwise, the region is not part of the graph. Using $(0, 4)$ from region I gives true statements:

$$4 \geq 0^2 = 0$$
$$4 \leq -2(0) + 8 = 8$$

Therefore, region I is included. Using (3, 4) from region V gives

$$4 \geq 3^2 = 9$$

$$4 \leq -2(3) + 8 = 2$$

Since (3, 4) does not satisfy *both* inequalities region V is excluded. You should verify that regions II, III, IV are also excluded. The graph of the system is the shaded part as shown. Note that the boundaries of region I are included since the given system uses the inequality symbols ≤ and ≥. ■

Another way to obtain the graph in Example 7 is to observe that $y \geq x^2$ calls for all points on or above the parabola $y = x^2$, and $y \leq -2x + 8$ calls for all points on or below the line $y = -2x + 8$.

EXERCISES 10.9

Solve each system and graph.

1. $y = -x^2 - 4x + 1$
   $y = 2x + 10$
2. $3x - 4y = -5$
   $(x + 3)^2 + (y + 1)^2 = 25$
3. $(x + 4)^2 + (y - 1)^2 = 16$
   $(x + 4)^2 + (y - 3)^2 = 4$
4. $y = x^2 - 6x + 9$
   $(x - 3)^2 + (y - 9)^2 = 9$
   (*Hint*: Factor $x^2 - 6x + 9$ and substitute.)
5. $y = x^2 + 6x + 6$
   $y = -x^2 - 6x + 6$
6. $y = \frac{1}{3}(x - 3)^2 - 3$
   $(x - 3)^2 + (y + 2)^2 = 1$
   (*Hint*: Solve the first equation for $(x - 3)^2$ and substitute into the second.)

Solve each system.

7. $y = (x + 1)^2$
   $y = (x - 1)^2$
8. $x^2 + y^2 = 9$
   $y = x^2 - 3$
9. $y = x^2$
   $y = -x^2 + 8x - 16$
10. $y = x^2$
    $y = x^2 - 8x + 24$
11. $7x + 3y = 42$
    $y = -3x^2 - 12x - 15$
12. $y + 2x = 1$
    $x^2 + 4x = 6 - 2y$
13. $y - x = 0$
    $(x - 2)^2 + (y + 5)^2 = 25$
14. $y - 2x = 0$
    $(x - 2)^2 + (y + 5)^2 = 25$
15. $x - 2y^2 = 0$
    $x^2 - y^2 = 3$
16. $x^2 + y^2 = 25$
    $2x^2 + y^2 = 34$
17. $4x^2 + y^2 = 4$
    $2x - y = 2$
18. $4x^2 + y^2 = 4$
    $x + y = 3$
19. $4x^2 - 9y^2 = 36$
    $9x^2 + 4y^2 = 36$
20. $2x^2 - y^2 = 1$
    $y^2 - x^2 = 3$
21. $2x^2 + y^2 = 11$
    $x^2 - 2y^2 = -2$
22. $x^2 - 2y^2 = 8$
    $3x + 4y = 4$
23. $y = -x^2 + 2x$
    $x^2 - 2x + y^2 - 2y = 0$
24. $x^2 + 4x + y^2 - 4y = -4$
    $(x - 2)^2 + (y - 2)^2 = 4$

*25. $(x - 1)^2 + y^2 = 1$
    $x^2 + (y - 1)^2 = 1$

26. $4x + 3y = 25$
    $x^2 + y^2 = 25$

*27. $y = \frac{1}{3}(x - 3)^2 - 3$
    $x^2 - 6x + y^2 + 2y = -6$

28. $y = x^2 - 6x + 9$
    $(x - 3)^2 + (y - 2)^2 = 58$
29. $\dfrac{x^2}{4} - y^2 = 1$
    $y^2 - \dfrac{x^2}{8} = 1$
30. $x^2 - 4y^2 = 4$
    $x^2 + y^2 = 24$

Solve the system and graph.

31. $2x + 3y = 7$
    $y = \dfrac{1}{x}$
32. $y = x^2 - 2x - 4$
    $y = -\dfrac{8}{x}$
33. $y = \sqrt{x}$
    $y = \frac{1}{2}x$
34. $y = \dfrac{2}{\sqrt{x}}$
    $x + 3y = 7$

Graph each system.

35. $y \geq -x^2 - 4x + 1$
$y \leq 2x + 10$
(See Exercise 1.)

36. $3x - 4y \geq -5$
$(x + 3)^2 + (y + 1)^2 \leq 25$
(See Exercise 2.)

37. $(x + 4)^2 + (y - 1)^2 \leq 16$
$(x + 4)^2 + (y - 3)^2 \geq 4$
(See Exercise 3.)

38. $y \geq x^2 - 6x + 9$
$(x - 3)^2 + (y - 9)^2 \geq 9$
(See Exercise 4.)

39. $y \geq x^2 + 6x + 6$
$y \leq -x^2 - 6x + 6$
(See Exercise 5.)

40. $y \geq \frac{1}{3}(x - 3)^2 - 3$
$(x - 3)^2 + (y + 2)^2 \geq 1$
(See Exercise 6.)

41. $9x^2 + 25y^2 \leq 225$
$x^2 + 4y^2 \geq 16$

42. $\frac{x^2}{16} + \frac{y^2}{9} \leq 1$
$\frac{x^2}{9} + \frac{y^2}{16} \leq 1$

43. $\frac{x^2}{16} + \frac{y^2}{9} \leq 1$
$\frac{x^2}{9} + \frac{y^2}{16} \geq 1$

CHAPTER 10 SUMMARY

Review these key terms and concepts so that you are able to define or describe them. A clear understanding of these terms will be very helpful when reviewing the developments of this chapter.

The basic strategy for solving linear systems is to systematically reduce the number of variables and equations until one equation in one variable is obtained. Solve for this variable and use back-substitution to solve for the remaining variables.

Methods of Solving Linear Systems

Using Substitution

For a system of two equations in two variables solve one equation for one of the variables in terms of the other, and then substitute into the other equations.

Using Multiplication-Addition

Use multiplcation to modify two of the equations so that the coefficients of the same variable in each equation are opposites. Then add to eliminate that variable.

Using Determinants

Second order:

$$\begin{vmatrix} a & b \\ c & d \end{vmatrix} = ad - bc$$

Third order:

$$\begin{vmatrix} a_1 & b_1 & c_1 \\ a_2 & b_2 & c_2 \\ a_3 & b_3 & c_3 \end{vmatrix} = a_1\begin{vmatrix} b_2 & c_2 \\ b_3 & c_3 \end{vmatrix} - a_2\begin{vmatrix} b_1 & c_1 \\ b_3 & c_3 \end{vmatrix} + a_3\begin{vmatrix} b_1 & c_1 \\ b_2 & c_2 \end{vmatrix}$$

Cramer's rule for a 2-by-2 system:

The system of equations $a_1x + b_1y = c_1 \quad a_2x + b_2y = c_2$

has the unique solution $x = \dfrac{\begin{vmatrix} c_1 & b_1 \\ c_2 & b_2 \end{vmatrix}}{\begin{vmatrix} a_1 & b_1 \\ a_2 & b_2 \end{vmatrix}}$ and $y = \dfrac{\begin{vmatrix} a_1 & c_1 \\ a_2 & c_2 \end{vmatrix}}{\begin{vmatrix} a_1 & b_1 \\ a_2 & b_2 \end{vmatrix}}$

provided that $a_1b_2 - a_2b_1 \neq 0$.

Fundamental Row Operations:

(1) Interchange two rows.

(2) Multiply a row by a nonzero constant.

(3) Add a multiple of a row to another row.

Using Matrices

Begin with the matrix of coefficients and constants of the given system. Apply fundamental row operations to obtain a row-equivalent matrix that gives a linear system in triangular form. Then use back-substitution.

To graph a system of linear inequalities, first draw the related straight lines. One way to find the correct graph of the system is to use test points from the various regions determined by the lines. Substitute the coordinates of these points into the system to determine which region is the correct graph.

A *linear programming* problem calls for finding the maximum or minimum value of a quantity $ax + by$ subject to *constraints*. The constraints form a system of linear inequalities in two variables whose graph is the *feasible* region. The point that gives the solution will be at a vertex of the feasible region, found by substituting the coordinates of all vertices into $ax + by$.

A nonlinear system of two equations in two variables can be solved by using substitution and/or addition-multiplication procedures to eliminate one of the variables.

REVIEW EXERCISES

The solutions to the following exercises can be found within the text of Chapter 10. Try to answer each question before referring to the text.

Section 10.1

1. Find the point of intersection of the lines given by $2x - y = -4$ and $x - 2y = -5$.
2. Solve by the substitution method.

$$4x - 3y = 5$$
$$2x - y = 2$$

Solve by the multiplication-additon method.

3. $2x + 3y = 12$
   $3x + 2y = 12$
4. $\frac{1}{3}x - \frac{2}{5}y = 4$
   $7x + 3y = 27$

Decide if each of the systems is consistent, or inconsistent, or dependent.

5. $39x - 91y = -28$
   $6x - 14y = 7$
6. $y = -\frac{2}{3}x + 5$
   $2x + 3y = 15$

Section 10.2

7. A field goal in basketball is worth 2 or 3 points and a free throw is worth 1 point. In a recent game the school basketball team scored 85 points. If there were twice as many

8. The sum of two numbers is 101. Five times the smaller is one more than the larger. Find the numbers.

9. There is a two-digit number that equals four times the sum of its digits. Also, if 27 is added to the number, then its digits would be reversed. What is the number?

10. For her participation in a recent "walk-for-poverty," Ellen collected \$2 per mile for a total of \$52. She recorded that for a certain time she walked at the rate of 3 miles per hour (mph) and the rest at 4 mph. Afterward she mentioned that it was too bad that she did not have the energy to reverse the rates. For if she could have walked 4 mph for the same time that she actually walked 3 mph, and vice versa, she would have collected a total of \$60. How long did her walk take?

11. A grocer sells Brazilian coffee at \$5.00 per pound and Colombian coffee at \$8.50 per pound. How many pounds of each should he mix in order to have a blend of 50 pounds that he can sell at \$7.10 per pound?

12. It costs the Roller King Company \$8 to produce one skateboard. In addition, there is a \$200 daily fixed cost for building maintenance.
   (a) Find the total daily cost for producing x skateboards per day.
   (b) Find the total daily revenue if the company sells x skateboards per day for \$16 each.
   (c) Find the daily break-even point. That is, find the coordinates of the point at which the cost equals the revenue.

Section 10.3

13. $$\begin{aligned} 2x - 5y + z &= -10 \\ x + 2y + 3z &= 26 \\ -3x - 4y + 2z &= 5 \end{aligned}$$

14. $$\begin{aligned} 3x - 2y + z &= 1 \\ x - y - z &= 2 \\ 6x - 4y + 2z &= 3 \end{aligned}$$

15. A veterinarian wants to control the diet of an animal so that on a monthly basis the animal consumes (besides hay, grass, and water) 60 pounds of oats, 75 pounds of corn, and 55 pounds of soybeans. The veterinarian has three feeds available, each consisting of oats, corn, and soybeans, as shown in the table. How many pounds of each feed should be used to obtain the desired mix?

	Oats	Corn	Soybeans
1 lb of feed *A*	6 oz	5 oz	5 oz
1 lb of feed *B*	6 oz	6 oz	4 oz
1 lb of feed *C*	4 oz	7 oz	5 oz

Section 10.4

Evaluate each determinant.

16. $\begin{vmatrix} -20 & 8 \\ 10 & 4 \end{vmatrix}$

17. $\begin{vmatrix} 10 & -20 \\ 8 & 4 \end{vmatrix}$

Use determinants to solve each system.

18. $$\begin{aligned} 5x - 9y &= 7 \\ -8x + 10y &= 2 \end{aligned}$$

19. $$\begin{aligned} 3x &= 2y + 22 \\ 2(x + y) &= x - 2y - 2 \end{aligned}$$

Section 10.5

Evaluate each determinant.

20. $\begin{vmatrix} 2 & -2 & 2 \\ 3 & 1 & 0 \\ 2 & -1 & 1 \end{vmatrix}$

21. $\begin{vmatrix} 2 & 1 & -3 \\ -4 & 0 & 2 \\ 5 & -1 & 6 \end{vmatrix}$

22. $\begin{vmatrix} -1 & 3 & 4 \\ 2 & 1 & 2 \\ 5 & 1 & 3 \end{vmatrix}$

23. Use Cramer's rule to solve this system.

$$\begin{aligned} x + 2y + z &= 3 \\ 2x - y - z &= 4 \\ -x - y + 2z &= -5 \end{aligned}$$

Section 10.6

Use matrices to solve each system.

24. $\begin{aligned} 2x + 5y + 8z &= 11 \\ x + 4y + 7z &= 10 \\ 3x + 6y + 12z &= 15 \end{aligned}$

25. $\begin{aligned} 2x + 14y - 4z &= -2 \\ -4x - 3y + z &= 8 \\ 3x - 5y + 6z &= 7 \end{aligned}$

26. $\begin{aligned} x + 2y - z &= 1 \\ 2x - y + 3z &= 4 \\ 5x + 5z &= 9 \end{aligned}$

27. $\begin{aligned} 3x - 6y &= 9 \\ -2x + 4y &= -8 \end{aligned}$

Section 10.7

28. Graph the region satisfying the linear inequality $3x - 5y < 10$.

Graph the system of inequalities.

29. $\begin{aligned} 2x + y &\le 6 \\ 3x - 4y &\ge 12 \end{aligned}$

30. $\begin{aligned} x + 3y &\ge 12 \\ -2x + y &\le 4 \\ 8x + 3y &\le 54 \end{aligned}$

31. Graph the system: $\begin{aligned} |x| &\ge 3 \\ |y| &\le 5 \end{aligned}$

Section 10.8

32. Maximize the quantity $k = 4x + 5y$ for (x, y) in the region S given by this system:

$$-x + 2y \le 2, \qquad 3x + 2y \le 10, \qquad x + 6y \ge 6, \qquad 3x + 2y \ge 6$$

33. Find the minimum of $k = 4x + 5y$ for the system in Exercise 32.
34. A store sells two kinds of bicycles, model A and model B. The store buys them unassembled from a wholesaler. Two employees are responsible for assembling the bicycles, and they are permitted to work no more than 6 hours each per week to do this job. Working together to assemble model A, employee I works $\frac{3}{4}$ hour and employee II works $\frac{1}{2}$ hour. Model B requires $\frac{1}{2}$ hour's work by employee I as well as 1 hour's work by employee II. There is a \$55 profit on each model A sold and \$48 on each model B. Because of the popularity of the sport, the store is able to sell as many bicycles as they decide to assemble. How many bicycles of each model should they assemble in order to get the maximum profit?

Section 10.9

Solve each system and graph.

35. $\begin{aligned} y &= x^2 \\ y &= -2x + 8 \end{aligned}$

36. $\begin{aligned} y &= x^2 - 2 \\ y &= -2x^2 + 6x + 7 \end{aligned}$

37. $\begin{aligned} x^2 + y^2 - 8y &= -7 \\ y - x^2 &= 1 \end{aligned}$

38. $\begin{aligned} x^2 + y^2 &= 9 \\ \frac{x^2}{25} + \frac{y^2}{16} &= 1 \end{aligned}$

39. $\begin{aligned} x^2 - y^2 &= 1 \\ 9x^2 + y^2 &= 9 \end{aligned}$

40. $\begin{aligned} y &= \sqrt[3]{x} \\ y &= \tfrac{1}{4}x \end{aligned}$

41. Graph the system of inequalities:

$$\begin{aligned} y &\ge x^2 \\ y &\le -2x + 8 \end{aligned}$$

CHAPTER 10 TEST: STANDARD ANSWER

Use these questions to test your knowledge of the basic skills and concepts of Chapter 10. Then check your answers with those given at the back of the book.

Solve each system.

1. $\begin{aligned} 3x - 2y &= 8 \\ x + 2y &= 4 \end{aligned}$

2. $\begin{aligned} 3x + 2y + z &= 6 \\ -2x - y + 2z &= 10 \\ -5x + 4y - z &= 8 \end{aligned}$

3. Find the point of intersection of the lines given by $2x + 3y = 5$ and $3x - y = -9$.

For problems 4, 5, and 6, decide if the system is consistent, inconsistent, or dependent.

4. $\begin{aligned} x + 3y &= 5 \\ 6x + 4y &= 2 \end{aligned}$

5. $\begin{aligned} -8x + 4y &= 28 \\ 6x - 3y &= 15 \end{aligned}$

6. $\begin{aligned} 6x - 3y &= -4 \\ -x + \tfrac{1}{2}y &= \tfrac{2}{3} \end{aligned}$

Evaluate each determinant.

7. (a) $\begin{vmatrix} 2 & -3 \\ 5 & 9 \end{vmatrix}$ (b) $\begin{vmatrix} 7 & 4 \\ -\frac{1}{2} & -\frac{2}{7} \end{vmatrix}$

8. $\begin{vmatrix} 2 & 0 & -3 \\ -3 & 4 & 1 \\ 2 & -3 & 5 \end{vmatrix}$

9. There were 2010 paid admissions to a basketball game. Children's tickets cost 75¢ and adults paid \$1.50. If the receipts totaled \$2781.00, how many children and how many adults attended the game?

10. The difference between two numbers is 23. Three times the smaller number is 6 more than twice the larger number. Find the two numbers.

11. It costs a company \$2.50 to produce a certain item that it sells for \$3.10. In addition to the production cost per item, the company has a daily overhead cost of \$450. Find the daily break-even point.

12. A nutritionist wants to combine three foods so that the resulting mixture will contain 550 units of ingredient *A*, 300 units of ingredient *B*, and 350 units of ingredient *C*. The units of each ingredient per ounce of food is given in the table. How many ounces of each food should be combined to obtain the required mixture?

	A	B	C
1 oz of food I	25 units	20 units	15 units
1 oz of food II	35 units	15 units	25 units
1 oz of food III	40 units	20 units	20 units

Use Cramer's rule to solve each system.

13. $\begin{aligned} 4x - 2y &= 15 \\ 3x + 2y &= -8 \end{aligned}$

14. $\begin{aligned} 2x + y - z &= -3 \\ x - 2y + z &= 8 \\ 3x - y - 2z &= -1 \end{aligned}$

15. $\begin{aligned} x + y - 2z &= 5 \\ 2x - y - 2z &= 2 \\ -x - 2y + 4z &= -9 \end{aligned}$

Solve the linear system using row-equivalent matrices.

16. $\begin{aligned} 2x + 5y &= 4 \\ 4x - 3y &= -18 \end{aligned}$

17. $\begin{aligned} -6x + 3y &= 9 \\ 10x - 5y &= -12 \end{aligned}$

18. $\begin{aligned} 2x - y + 2z &= -3 \\ x + 4y - 3z &= 18 \\ -4x + 2y - 3z &= 0 \end{aligned}$

19. Graph this system of inequalities:

$$x + y \le 2 \qquad 2x - y \ge -2$$

20. Graph the system

$$x \ge 0, \qquad y \ge 0, \qquad x + 2y \le 8, \qquad x + y \le 5, \qquad 2x + y \le 8$$

21. Find the maximum value of $k = 3x + 2y$ for the region in Question 20.
22. A manufacturer of television sets makes two kinds of 19-inch color TV sets, model A that earns \$80 profit and model B that earns \$100 profit. Each month the manufacturer can produce up to 400 units of model A and up to 300 units of model B, but no more than 600 sets altogether. How many of each model should be produced to obtain the maximum profit?

Solve each system.

23. $2x^2 - 3y^2 = 15$, $3x^2 + 2y^2 = 29$
24. $y^2 - 2x^2 = 16$, $y - x = 2$
25. $y = x^2 + 2x + 1$, $y = x^2 - 4x - 4$

CHAPTER 10 TEST: MULTIPLE CHOICE

1. Which of the following statements is true for the system shown below?

$$2x - y = 4$$
$$x + 3y = 7$$

(a) The graph consists of two parallel lines.
(b) The graph consists of a single line.
(c) The graph consists of two lines that intersect in the first quadrant.
(d) The graph consists of two lines that intersect in the fourth quadrant.
(e) None of the preceding.

2. Which of the following represents a dependent system?
(a) $2x - 3y = 8$, $2x - 3y = 5$ **(b)** $2x - 3y = 8$, $4x - 6y = 5$ **(c)** $2x - 3y = 8$, $x - \frac{3}{2}y = 4$
(d) $2x - 3y = 8$, $x - \frac{3}{2}y = 8$ **(e)** $2x - 3y = 8$, $-4x + 6y = -4$

3. A toy manufacturer can produce a rag doll at a cost of \$3 and can sell it for \$9. The daily fixed cost for the production of these dolls is \$150. What is the daily break-even point?
(a) (12.5, 112.5) **(b)** (12.5, 187.5) **(c)** (25, 125) **(d)** (25, 225)
(e) None of the preceding

4. Which of the following is a permissible fundamental row operation when using matrices to solve a linear system.
I. Interchange two rows.
II. Multiply a row by a nonzero constant.
III. Add a multiple of a row to another row.
(a) Only I **(b)** Only II **(c)** Only III **(d)** I, II, and III
(e) None of the preceding

5. Which of the following is equal to this determinant? $\begin{vmatrix} 2 & 5 \\ 7 & 3 \end{vmatrix}$

(a) $\begin{vmatrix} 5 & 7 \\ 3 & 2 \end{vmatrix}$ **(b)** $\begin{vmatrix} 7 & 3 \\ 2 & 5 \end{vmatrix}$ **(c)** $\begin{vmatrix} 3 & 7 \\ 5 & 2 \end{vmatrix}$ **(d)** $\begin{vmatrix} 2 & 5 \\ 3 & 7 \end{vmatrix}$ **(e)** $\begin{vmatrix} -3 & -5 \\ 7 & 2 \end{vmatrix}$

6. Evaluate: $\begin{vmatrix} 1 & -2 & 3 \\ -1 & 0 & 1 \\ -2 & 3 & -1 \end{vmatrix}$

(a) 8 **(b)** -14 **(c)** -6 **(d)** 0 **(e)** None of the preceding

7. Which system of equations can be used to solve this problem?

Wendy has 31 bills consisting of fives, f, tens, t, and twenties, n. The total value is \$210 and she has one more five than the total number of the other bills. How many of each type bill does she have?

(a) $\begin{aligned} f + t + n &= 210 \\ 5f + 10t + 20n &= 31 \\ f &= n + t + 1 \end{aligned}$ **(b)** $\begin{aligned} 5f + 10t + 20n &= 210 \\ f + t + n &= 31 \\ f - t - n &= 1 \end{aligned}$

(c) $\begin{aligned} 5f + 10t + 20n &= 210 \\ f + t + n &= 31 \\ f - t + n &= 1 \end{aligned}$ **(d)** $\begin{aligned} f + t + n &= 31 \\ 5f + 10t + 20n &= 210 \\ f + 1 &= t + n \end{aligned}$

(e) None of the preceding

8. A jogger ran 15 miles. The first part of the 15 miles took x hours running at 8 mph, and the final part took y hours running at 10 mph. If the jogger had reversed the rates and ran at 10 mph and 8 mph for the same times that were actually run at 8 mph and 10 mph respectively, then an additional 1.5 miles would have been run. Which system of equations can be used to find the times for each part of the run?

(a) $\begin{aligned} 8x + 10y &= 16.5 \\ 10x + 8y &= 15 \end{aligned}$ **(b)** $\begin{aligned} 8x + 10y &= 15 \\ 10x - 8y &= 13.5 \end{aligned}$ **(c)** $\begin{aligned} 8x + 10y &= 15 \\ 10x + 8y &= 16.5 \end{aligned}$

(d) $\begin{aligned} \frac{x}{8} + \frac{y}{10} &= 15 \\ \frac{x}{10} - \frac{y}{8} &= 13.5 \end{aligned}$ **(e)** None of the preceding

9. Which of the following is one of the three numbers in the solution of this system?

$$\begin{aligned} 2x - 3y + z &= 11 \\ 3x + y - z &= 9 \\ -x + 3y - 2z &= -1 \end{aligned}$$

(a) -3 **(b)** -2 **(c)** 5 **(d)** -8 **(e)** None of the preceding

10. When Cramer's rule is used to solve for y for the system shown below, which of the following becomes the numerator of the fraction?

$$\begin{aligned} 2x + y - 3z &= 5 \\ -x + 2y + z &= 1 \\ 3x - y + 2z &= -3 \end{aligned}$$

(a) $\begin{vmatrix} 2 & 1 & -3 \\ -1 & 2 & 1 \\ 3 & -1 & 2 \end{vmatrix}$ **(b)** $\begin{vmatrix} 2 & -3 & 5 \\ -1 & 1 & 1 \\ 3 & 2 & -3 \end{vmatrix}$ **(c)** $\begin{vmatrix} 5 & 2 & -3 \\ 1 & -1 & 1 \\ -3 & 3 & 2 \end{vmatrix}$

(d) $\begin{vmatrix} 5 & 2 & 1 \\ 1 & -1 & 2 \\ -3 & 3 & -1 \end{vmatrix}$ **(e)** None of the preceding

11. Which system of equations can be used to solve this problem?

The units' digit of a two-digit number is one more than twice the tens' digit. The sum of the digits is 10. What is the number? (Use u for the units' digit and t for the tens' digit.)

(a) $\begin{aligned} u + 1 &= 2t \\ t + u &= 10 \end{aligned}$ **(b)** $\begin{aligned} u &= 2t + 1 \\ t + u &= 10 \end{aligned}$ **(c)** $\begin{aligned} t &= 2u + 1 \\ t + u &= 10 \end{aligned}$

(d) $\begin{aligned} u &= 2(t + 1) \\ t + u &= 10 \end{aligned}$ **(e)** $\begin{aligned} 2u &= t + 1 \\ t &= 10 - u \end{aligned}$

12. Suppose a linear system of 3 equations in 3 variables is being solved by making use of row-equivalent matrices. What does it mean when we arrive at the following matrix?

$$\left[\begin{array}{ccc|c} 1 & -3 & 6 & 4 \\ 0 & 4 & 3 & -1 \\ 0 & 0 & 0 & 0 \end{array}\right]$$

(a) $(4, -1, 0)$ is a solution of the original system.
(b) The original system of equations has more than one solution.
(c) The original system of equations has exactly one solution.
(d) The original system of equations has no solutions.
(e) None of the preceding

13. Which response below is true regarding the graph of this system?

$$\begin{aligned} 2x + y &\le 6 \\ x + y &\ge 4 \\ y &\ge 0 \\ x &\ge 0 \end{aligned}$$

(a) The graph consists of all points inside and on the border of just one triangle.
(b) The graph consists of all points inside and on the border of two triangles.
(c) The graph is a four-sided figure.
(d) All points on the positive part of the y-axis are in the graph.
(e) None of the preceding

14. What is the maximum value of $k = 4x + 3y$ for the region given by the system shown below?

$$\begin{aligned} x &\ge 0 \\ y &\ge 0 \\ x + 2y &\le 22 \\ 3x + 2y &\le 30 \end{aligned}$$

(a) 88 **(b)** 44 **(c)** 43 **(d)** 33 **(e)** None of the preceding

15. The number of solutions of the given system is which of the following?

$$y = x^2 - 2x$$
$$y = x^2 + 4x - 1$$

(a) None **(b)** One **(c)** Two **(d)** Three **(e)** None of the preceding

ANSWERS TO THE TEST YOUR UNDERSTANDING EXERCISES

Page 431

1. $(1, 2)$ 2. $(-3, 4)$ 3. $(2, -1)$ 4. $(\frac{1}{5}, \frac{1}{5})$ 5. $(-4, 1)$ 6. $(35, 50)$
7. $(-1, 2)$ 8. $(\frac{1}{2}, -3)$ 9. $(\frac{3}{2}, \frac{5}{7})$

Page 444

1. $(1, 2, 3)$ 2. No solution 3. $(-3, 0, 2)$ 4. $(\frac{1}{2}, -3, 4)$

Page 448

1. -2 2. -2 3. 2 4. -10 5. 5 6. 5
7. 20 8. 2 9. -4 10. 1 11. -4 12. 0

Page 455

1. −29	2. 0	3. 27	4. −93	5. −8	6. 0

Page 474

1. 19	2. 1	3. 26	4. −34	5. 50	6. −139
7. 37	8. 10	9. 200	10. −592		

CHAPTER 11

SEQUENCES AND SERIES

11.1 SEQUENCES

The same equation can be used to define a variety of functions by changing the domain. For example, below are the graphs of three functions all of whose range values are given by the equation $y = x^2$ for the indicated domains.

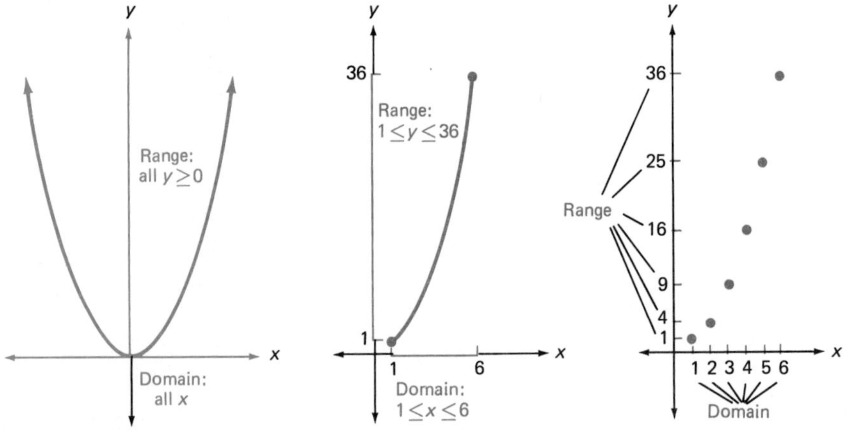

The type of function that is studied in this chapter is illustrated by the preceding graph at the right, where the domain consists of the consecutive positive integers 1, 2, 3, 4, 5, 6. This kind of function is called a **sequence**.

DEFINITION OF A SEQUENCE

A sequence is a function whose domain is a set of consecutive positive integers.

a_n is read "a-sub-n" and has the same meaning as the functional notation $a(n)$, that is, "a of n."

Instead of using the variable x, letters, such as n, k, i are normally used for the domain variable of a sequence. Frequently, sequences (functions) will be denoted by the lowercase letter s and the range values by a_n, which are also called the **terms** of the sequence.

Sequences are often given by stating their **general** or ***n*th terms**. Thus the general term of the sequence, previously given by $y = x^2$, becomes $a_n = n^2$.

EXAMPLE 1 Find the range values of the sequence given by $a_n = \frac{1}{n}$ for the domain $\{1, 2, 3, 4, 5\}$.

Solution The range values are:

This is an example of a finite sequence since the domain is finite. That is, the domain is a set of integers having a last element.

$$a_1 = \frac{1}{1} = 1 \qquad a_4 = \frac{1}{4}$$

$$a_2 = \frac{1}{2} \qquad a_5 = \frac{1}{5}$$

$$a_3 = \frac{1}{3} \qquad a_6 = \frac{1}{6}$$

■

EXAMPLE 2 List the first six terms of the sequence given by $b_k = \frac{(-1)^k}{k^2}$.

Solution

$$b_1 = \frac{(-1)^1}{1^2} = -1 \qquad b_4 = \frac{(-1)^4}{4^2} = \frac{1}{16}$$

$$b_2 = \frac{(-1)^2}{2^2} = \frac{1}{4} \qquad b_5 = \frac{(-1)^5}{5^2} = -\frac{1}{25}$$

$$b_3 = \frac{(-1)^3}{3^2} = -\frac{1}{9} \qquad b_6 = \frac{(-1)^6}{6^2} = \frac{1}{36}$$

■

TEST YOUR UNDERSTANDING
Think Carefully

Write the first five terms of the given sequence.

1. $a_n = 2n + 1$
2. $a_n = -2n$
3. $a_n = -2n + 2$
4. $b_k = \frac{(-1)^k}{k}$
5. $b_k = \frac{1}{k^2}$
6. $b_k = \frac{-3}{k(k+1)}$
7. $c_n = \frac{3}{n(2n-1)}$
8. $c_n = \left(\frac{1}{3}\right)^n$
9. $c_n = 1 - (-1)^n$

(Answers Page 524)

Sometimes a sequence is given by a verbal description. If, for example, we ask for the increasing sequence of odd integers beginning with -3, then this implies the **infinite sequence** whose first few terms are

$$-3, -1, 1, \ldots$$

A sequence can also be given by presenting a listing of its first few terms, possibly including the general term. Thus the preceding sequence is

$$-3, -1, 1, \ldots, 2n - 5, \ldots$$

EXAMPLE 3 Find the tenth term of the sequence

*This is an example of an **infinite** sequence since the domain is infinite. That is, the domain consists of all the positive integers.*

$$-3, 4, \frac{5}{3}, \ldots, \frac{n+2}{2n-3}, \ldots$$

Solution Since the first term -3 is obtained by letting $n = 1$ in the general term $\dfrac{n+2}{2n-3}$, the tenth term is

$$a_{10} = \frac{10 + 2}{2(10) - 3} = \frac{12}{17}$$ ■

EXAMPLE 4 Write the first four terms of the sequence given by $a_n = \left(1 + \dfrac{1}{n}\right)^n$. Round off to two decimal places when appropriate.

Solution

$$a_1 = (1 + \tfrac{1}{1})^1 = 2$$
$$a_2 = (1 + \tfrac{1}{2})^2 = (\tfrac{3}{2})^2 = \tfrac{9}{4} = 2.25$$
$$a_3 = (1 + \tfrac{1}{3})^3 = (\tfrac{4}{3})^3 = \tfrac{64}{27} = 2.37$$
$$a_4 = (1 + \tfrac{1}{4})^4 = (\tfrac{5}{4})^4 = \tfrac{625}{256} = 2.44$$ ■

The terms of the sequence in Example 4 are getting successively larger. But the increase from term to term is getting smaller. That is, the differences between successive terms are decreasing:

$$a_2 - a_1 = 0.25$$
$$a_3 - a_2 = 0.12$$
$$a_4 - a_3 = 0.07$$

If more terms of $a_n = \left(1 + \dfrac{1}{n}\right)^n$ were computed, you would see that while the terms keep on increasing, the amount by which each new term increases keeps getting smaller.

Use a calculator to verify these table entries to four decimal places.

n	$a_n = \left(1 + \frac{1}{n}\right)^n$
10	2.5937
50	2.6916
100	2.7048
500	2.7156
1000	2.7169
5000	2.7180
10,000	2.7181

It turns out that no matter how large n is, the value of $\left(1 + \frac{1}{n}\right)^n$ is never more than 2.72. In fact, the larger the n that is taken, the closer $\left(1 + \frac{1}{n}\right)^n$ gets to the irrational value $e = 2.71828 \ldots$. This is the number that was introduced in Chapter 9 in reference to natural logarithms and exponential functions.

EXERCISES 11.1

The domain of the sequence in each exercise consists of the integers 1, 2, 3, 4, 5. Write the corresponding range values.

1. $a_n = 2n - 1$
2. $a_n = 10 - n^2$
3. $a_k = (-1)^k$
4. $b_k = -\frac{6}{k}$
5. $b_i = 8(-\frac{1}{2})^i$
6. $b_i = (\frac{1}{2})^{i-3}$

Write the first four terms of the sequence given by the formula in each exercise.

7. $c_k = (-1)^k k^2$
8. $c_j = 3(\frac{1}{10})^{j-1}$
9. $c_j = 3(\frac{1}{10})^j$
10. $a_j = 3(\frac{1}{10})^{j+1}$
11. $a_j = 3(\frac{1}{10})^{2j}$
12. $a_n = \frac{(-1)^{n+1}}{n + 3}$
13. $a_n = \frac{1}{n} - \frac{1}{n + 1}$
14. $a_n = \frac{n^2 - 4}{n + 2}$
15. $a_k = (2k - 10)^2$
16. $a_k = 1 + (-1)^k$
17. $a_n = -2 + (n - 1)(3)$
18. $a_n = a_1 + (n - 1)(d)$
19. $b_i = \frac{i - 1}{i + 1}$
20. $b_i = 64^{1/i}$
21. $b_n = \left(1 + \frac{1}{n}\right)^{n-1}$
22. $u_n = \frac{1}{2^n}$
23. $u_n = -2(\frac{3}{4})^{n-1}$
24. $u_k = a_1 r^{k-1}$
25. $x_k = \frac{k}{2^k}$
26. $x_n = \frac{(-1)^n}{n} + n$
27. $x_k = \frac{k}{k + 1} - \frac{k + 1}{k}$
28. $y_n = \left(1 + \frac{1}{n + 1}\right)^n$
29. $y_n = 4$
30. $y_n = \frac{n}{(n + 1)(n + 2)}$
31. Find the sixth term of $1, 2, 5, \ldots, \frac{1}{2}(1 + 3^{n-1}), \ldots$.
32. Find the ninth and tenth terms of $0, 4, 0, \ldots, \frac{2^n + (-2)^n}{n}, \ldots$.
33. Find the seventh term of $a_k = 3(0.1)^{k-1}$.
34. Find the twentieth term of $a_n = (-1)^{n-1}$.
35. Find the twelfth term of $a_i = i$.
36. Find the twelfth term of $a_i = (i - 1)^2$.
37. Find the twelfth term of $a_i = (1 - i)^3$.
38. Find the hundredth term of $a_n = \frac{n + 1}{n^2 + 5n + 4}$.
39. Write the first four terms of the sequence of even increasing integers beginning with 4.
40. Write the first four terms of the sequence of decreasing odd integers beginning with 3.
41. Write the first five positive multiples of 5 and find the formula for the nth term.
42. Write the first five powers of 5 and find the formula for the nth term.
43. Write the first five powers of -5 and find the formula for the nth term.
44. Write the first five terms of the sequence of reciprocals of the negative integers and find the formula for the nth term.

45. The numbers 1, 3, 6, and 10 are called **triangular numbers** because they correspond to the number of dots in the triangular arrays. Find the next three triangular numbers.

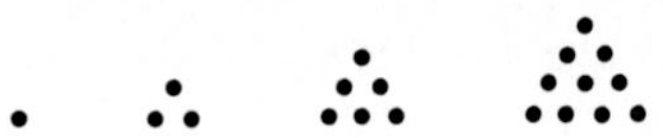

46. Write the first eight terms of $s_n = \dfrac{1 + (-1)^{n+1}}{2i^{n-1}}$. (See page 207 for the powers of $i = \sqrt{-1}$.)

47. Write the first five terms of $a_n = \dfrac{3^n}{2^n + 1}$.

48. Write the first seven terms of $a_n = n!$ where $n!$ is read as "n factorial" and is defined by

$$n! = n(n - 1) \cdot (n - 2) \cdots 3 \cdot 2 \cdot 1$$

49. Write the first four terms of

$$a_n = \frac{3 \cdot 5 \cdots (2n - 1)(2n + 1)}{2 \cdot 4 \cdots (2n - 2)(2n)}$$

50. When an investment earns **simple interest** it means the interest is earned only on the original investment. For example, if P dollars are invested in a bank that pays simple interest at the annual rate of r percent, then the interest for the first year is Pr, and the amount in the bank at the end of the year is $P + Pr$. For the second year, the interest is again Pr; the amount now would be $(P + Pr) + Pr = P + 2Pr$.
   (a) What is the amount after n years?
   (b) What is the amount in the bank if an investment of \$750 has been earning simple interest for 5 years at the annual rate of 12%?
   (c) If the amount in the bank is \$5395 after 12 years, what was the original investment P if it has been earning simple interest at the annual rate of $12\frac{1}{2}$%?

CHALLENGE
Think Creatively

The number of dots in the square arrays of dots correspond to the first four square numbers 1, 4, 9, 16.

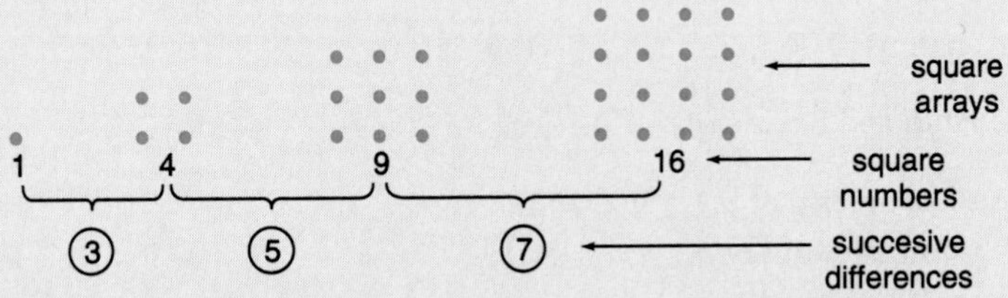

The circled numbers are the differences between successive square numbers. They represent the number of dots that need to be added to the preceding array to obtain the next array. Since these differences 3, 5, 7 increase by 2, the next difference is 9. Then the next square number is 16 + 9 = 25.

Now consider the pentagonal numbers. The first two and their arrays of dots are

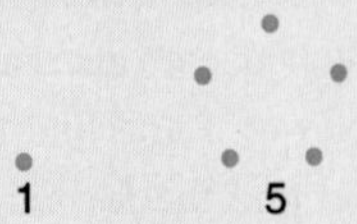

Construct the arrays for the next two pentagonal numbers. Then, using differences as demonstrated for the square numbers, find the fifth and sixth pentagonal numbers.

11.2 SUMS OF FINITE SEQUENCES

How long would it take you to add up the integers from 1 to 1000? Here is a quick way. List the sequence displaying the first few and last few terms.

$$1, 2, 3, \ldots, 998, 999, 1000$$

Add them in pairs, the first and last, the second and second from last, and so on.

It is told that Carl Friedrich Gauss (1777–1855) discovered how to compute such sums when he was 10 years old. He became one of the greatest mathematicians of all time.

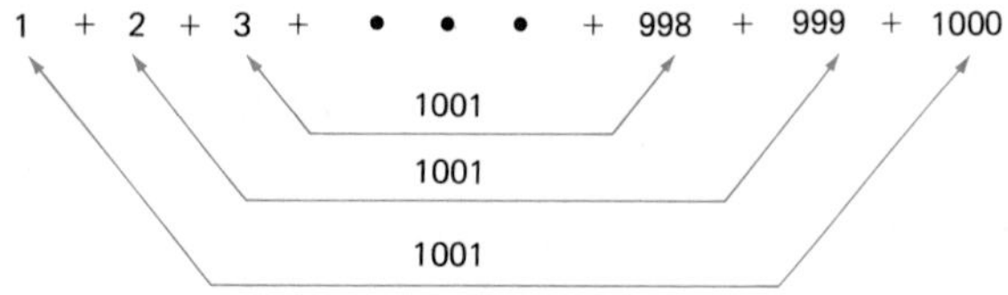

Since there are 500 such pairs to be added, the total is

$$500(1001) = 500{,}500$$

For any finite sequence we can add up all its terms and say that we have found the *sum of the sequence*. The sum of a sequence is called a **series**. For example, the sequence

$$1, \quad 3, \quad 5, \quad 7, \quad 9, \quad 11$$

can be associated with the series

$$1 + 3 + 5 + 7 + 9 + 11$$

The sum of the terms in this series can easily be found, by adding, to be 36. As another example, the sequence $a_n = \dfrac{1}{n}$ for $n = 1, 2, 3, 4, 5$ has the sum

$$1 + \frac{1}{2} + \frac{1}{3} + \frac{1}{4} + \frac{1}{5} = \frac{60 + 30 + 20 + 15 + 12}{60} = \frac{137}{60}$$

EXAMPLE 1 Find the sum of the first seven terms of $a_k = 2k$.

Solution

$$a_1 + a_2 + a_3 + a_4 + a_5 + a_6 + a_7$$
$$= 2 + 4 + 6 + 8 + 10 + 12 + 14 = 56$$

■

There is a very handy notational device available for expressing the sum of a sequence. The Greek letter Σ (capital sigma) is used for this purpose. Referring to Example 1, the sum of the seven terms is expressed by the symbol $\Sigma_{k=1}^{7} a_k$; that is,

Just think of sigma as a command to add.

$$\sum_{k=1}^{7} a_k = a_1 + a_2 + a_3 + a_4 + a_5 + a_6 + a_7$$

Add the terms a_k for consecutive values of k, starting with $k = 1$ up to and including

The letter k that appears on the bottom of the sigma, and takes on the values from 1 to 7, is called the ***index of summation****. The letter n is used as the index of summation in Example 2.*

$k = 7$. With this symbolism, the question in Example 1 can now be stated by asking for the value of $\Sigma_{k=1}^{7} a_k$, where $a_k = 2k$, or by asking for the value of $\Sigma_{k=1}^{7} 2k$.

EXAMPLE 2 Find $\sum_{n=1}^{5} b_n$, where $b_n = \dfrac{n}{n+1}$.

Solution

$$\sum_{n=1}^{5} b_n = b_1 + b_2 + b_3 + b_4 + b_5$$

Now replace each b_n by its numerical value:

$$\sum_{n=1}^{5} b_n = \frac{1}{2} + \frac{2}{3} + \frac{3}{4} + \frac{4}{5} + \frac{5}{6}$$

$$= \frac{30 + 40 + 45 + 48 + 50}{60}$$

$$= \frac{213}{60} = \frac{71}{20}$$

■

EXAMPLE 3 Evaluate $\sum_{k=1}^{4} (2k + 1)$.

Solution It is understood here that we are to find the sum of the first four terms of the sequence whose general term is $a_k = 2k + 1$.

$$\sum_{k=1}^{4} (2k + 1) = (2 \cdot 1 + 1) + (2 \cdot 2 + 1) + (2 \cdot 3 + 1) + (2 \cdot 4 + 1)$$

$$= 3 + 5 + 7 + 9$$

$$= 24$$

■

EXAMPLE 4 Find $\sum_{i=1}^{5} x_i$, where $x_i = (-1)^i(i + 1)$.

Solution

$$\sum_{i=1}^{5} x_i = x_1 + x_2 + x_3 + x_4 + x_5$$

$$= (-1)^1(1 + 1) + (-1)^2(2 + 1) + (-1)^3(3 + 1) + (-1)^4(4 + 1) + (-1)^5(5 + 1)$$

$$= -2 + 3 - 4 + 5 - 6$$

$$= -4$$

■

TEST YOUR UNDERSTANDING
Think Carefully

Evaluate each of the following.

1. $\sum_{k=1}^{5} (4k)$ **2.** $\sum_{k=1}^{5} (2k - 1)$ **3.** $\sum_{k=1}^{5} (k^2 - k)$

4. $\sum_{n=1}^{6} (-1)^n$ **5.** $\sum_{n=1}^{4} (2n^2 - n)$ **6.** $\sum_{n=1}^{4} 2(-\frac{1}{2})^{n+1}$

(Answers Page 524)

When the general term of a given series can be found, then the series can be rewritten using the sigma notation. For example, since the series

$$3 + 6 + 9 + 12 + 15 + 18 + 21$$

Note: In some of the exercises the summation begins with values of i, n, or k other than 1. For example, see Exercises 8, 22, 24, and 25.

is the sum of the first seven multiples of 3, it can be written as $\Sigma_{k=1}^{7} 3k$.

EXERCISES 11.2

Find the sum of the first five terms of the sequence given by the formula in each exercise.

1. $a_n = 3n$ **2.** $a_k = (-1)^k \frac{1}{k}$ **3.** $a_i = i^2$

4. $b_i = i^3$ **5.** $b_k = \frac{3}{10^k}$ **6.** $b_n = -6 + 2(n - 1)$

7. Find $\sum_{n=1}^{8} t_n$ where $t_n = 2^n$.

8. Find $\sum_{n=0}^{8} x_n$ where $x_n = \frac{1}{2^n}$.

9. Find $\sum_{k=1}^{20} y_k$ where $y_k = 3$.

Find each of the following sums for n = 7.

10. $2 + 4 + \cdots + 2n$ **11.** $2 + 4 + \cdots + 2^n$

12. $-7 + 2 + \cdots + (9n - 16)$ **13.** $3 + \frac{3}{2} + \cdots + 3(\frac{1}{2})^{n-1}$

Evaluate each of the following.

14. $\sum_{k=1}^{6} (5k)$ **15.** $5\left(\sum_{k=1}^{6} k\right)$ **16.** $\sum_{n=1}^{4} (n^2 + n)$

17. $\sum_{n=1}^{4} n^2 + \sum_{n=1}^{4} n$ **18.** $\sum_{i=1}^{8} (i - 2i^2)$ **19.** $\sum_{k=1}^{4} \frac{k}{2^k}$

20. $\sum_{k=1}^{7} (-1)^k$ **21.** $\sum_{k=1}^{8} (-1)^k$ **22.** $\sum_{k=3}^{7} (2k - 5)$

23. $\sum_{j=1}^{6} [-3 + (j - 1)5]$ **24.** $\sum_{k=-3}^{3} 10^k$ **25.** $\sum_{k=-3}^{3} \frac{1}{10^k}$

26. $\sum_{k=1}^{5} 4(-\frac{1}{2})^{k-1}$ **27.** $\sum_{i=1}^{4} (-1)^i 3^i$ **28.** $\sum_{n=1}^{3} \left(\frac{n+1}{n} - \frac{n}{n+1}\right)$

29. $\sum_{n=1}^{3} \frac{n+1}{n} - \sum_{n=1}^{3} \frac{n}{n+1}$ **30.** $\sum_{k=1}^{8} \frac{1 + (-1)^k}{2}$ **31.** $\sum_{k=1}^{3} (0.1)^{2k}$

Rewrite each series using sigma notation.

32. $4 + 8 + 12 + 16 + 20 + 24$ **33.** $5 + 10 + 15 + 20 + \cdots + 50$

34. $-4 - 2 + 0 + 2 + 4 + 6 + 8$ **35.** $-9 - 6 - 3 + 0 + 3 + \cdots + 24$

36. Read the discussion at the beginning of this section, where we found the sum of the first 1000 positive integers, and find a formula for the sum of the first n positive integers for n even.

37. **(a)** Find $\Sigma_{k=1}^{n} (2k - 1)$ for each of the following values of n: 2, 3, 4, 5, 6.

(b) On the basis of the results in part (a), find a formula for the sum of the first n odd numbers.

38. The sequence 1, 1, 2, 3, 5, 8, 13 . . . is called the *Fibonacci sequence*. Its first two terms are ones, and each term thereafter is computed by adding the preceding two terms.

(a) Write the next seven terms of this sequence.

(b) Let $u_1, u_2, u_3, \ldots, u_n, \ldots$ be the Fibonacci sequence. Evaluate $S_n = \sum_{k=1}^{n} u_k$ for these values of n: 1, 2, 3, 4, 5, 6, 7, 8.

***(c)** Note that $u_1 = u_3 - u_2$, $u_2 = u_4 - u_3$, $u_3 = u_5 - u_4$, and so on. Use this form for the first n numbers to derive a formula for the sum of the first n Fibonacci numbers.

***39.** Let a_n be a sequence with $a_1 = 2$ and $a_m + a_n = a_{m+n}$, where m and n are any positive integers. Show that $a_n = 2n$ for any n.

40. Show that $\sum_{k=1}^{9} \log_{10} \frac{k+1}{k} = 1$. $\left(\textit{Hint: } \log_{10} \frac{a}{b} = \log_{10} a - \log_{10} b.\right)$

***41.** Prove: $\sum_{k=1}^{n} a_k + \sum_{k=1}^{n} b_k = \sum_{k=1}^{n} (a_k + b_k)$.

***42.** Prove: $\sum_{k=1}^{n} ca_k = c \sum_{k=1}^{n} a_k$, c a constant.

***43.** Prove: $\sum_{k=1}^{n} (a_k + c) = \left(\sum_{k=1}^{n} a_k\right) + nc$, c a constant.

***44. (a)** Evaluate $\sum_{k=1}^{10} \frac{1}{k(k+1)}$ using the result $\frac{1}{k(k+1)} = \frac{1}{k} - \frac{1}{k+1}$

(b) Use the identity given in part (a) to prove that

$$\frac{1}{1 \cdot 2} + \frac{1}{2 \cdot 3} + \frac{1}{3 \cdot 4} + \cdots + \frac{1}{98 \cdot 99} + \frac{1}{99 \cdot 100} = \frac{99}{100}$$

45. Use the identity $\frac{2}{k(k+2)} = \frac{1}{k} - \frac{1}{k+2}$ to show that $\sum_{k=1}^{10} \frac{2}{k(k+2)} = \frac{175}{132}$.

11.3 ARITHMETIC SEQUENCES AND SERIES

Here are the first five terms of the sequence whose general term is $a_k = 7k - 2$:

$$5, \quad 12, \quad 19, \quad 26, \quad 33$$

Do you notice any special pattern? It does not take long to observe that each term, after the first, is 7 more than the preceding term. This sequence is an example of an ***arithmetic sequence***.

DEFINITION OF AN ARITHMETIC SEQUENCE

A sequence is said to be *arithmetic* if each term, after the first, is obtained from the preceding term by adding a common value.

An arithmetic sequence is also referred to as an ***arithmetic progression****.*

Let us consider the first four terms of two different arithmetic sequences:

$$2, \quad 4, \quad 6, \quad 8, \ldots$$

$$-\tfrac{1}{2}, \quad -1, \quad -\tfrac{3}{2}, \quad -2, \ldots$$

For the first sequence, the common value (or difference) that is added to each term to get the next is 2. Thus it is easy to see that 10, 12, and 14 are the next three terms. You might guess that the nth term is $a_n = 2n$.

The second sequence has the common difference $-\frac{1}{2}$. This can be found by subtracting the first term from the second, or the second from the third, and so forth. The nth term is $a_n = -\frac{1}{2}n$.

Unlike the preceding sequences, it is not always easy to see what the nth term of a specific arithmetic sequence is. Therefore, we will now develope a general form that makes it possible to write the nth term of any such sequence.

Can you find the nth term of the arithmetic sequence 11, 2, −7, −16, . . . ?

Let a_n be the nth term of an arithmetic sequence, and let d be the **common difference**. Then the first four terms are:

$$a_1$$

$$a_2 = a_1 + d$$

$$a_3 = a_2 + d = (a_1 + d) + d = a_1 + 2d$$

$$a_4 = a_3 + d = (a_1 + 2d) + d = a_1 + 3d$$

The pattern is clear. Without further computation we see that

$$a_5 = a_1 + 4d \quad \text{and} \quad a_6 = a_1 + 5d$$

Since the coefficient of d is always 1 less than the number of the term, the nth term is given as follows.

This formula says that the nth term of an arithmetic sequence is completely identified by its first term a and its common difference d. Also note that $a_n = a_{n-1} + d$ *and* $d = a_n - a_{n-1}$.

GENERAL TERM OF AN ARITHMETIC SEQUENCE

The nth term of an arithmetic sequence is

$$a_n = a_1 + (n - 1)d$$

where a_1 is the first term and d is the common difference.

To illustrate the use of this result, consider the arithmetic sequence

$$11, \quad 2, \quad -7, \quad -16, \ldots$$

We can use the formula, with $a_1 = 11$ and $d = a_2 - a_1 = 2 - 11 = -9$, to find the nth term:

$$\begin{aligned} a_n &= 11 + (n - 1)(-9) \\ &= -9n + 20 \end{aligned}$$

TEST YOUR UNDERSTANDING
Think Carefully

Each of the following gives the first few terms of an arithmetic sequence. Find the nth term in each case.

1. 5, 10, 15, . . .
2. 6, 2, −2, . . .
3. $\frac{1}{10}, \frac{1}{5}, \frac{3}{10}, \ldots$
4. −5, −13, −21, . . .
5. 1, 2, 3, . . .
6. −3, −2, −1, . . .

Find the nth term a_n of the arithmetic sequence with the given values for the first term and the common difference.

7. $a_1 = \frac{2}{3}; d = \frac{2}{3}$
8. $a_1 = 53; d = -12$
9. $a_1 = 0; d = \frac{1}{5}$
10. $a_1 = 2; d = 1$

(Answers Page 525)

EXAMPLE 1 The first term of an arithmetic sequence is -15 and the fifth term is 13. Find the fortieth term.

Solution Since $a_5 = 13$ use $n = 5$ in the formula $a_n = a_1 + (n - 1)d$ to solve for d.

$$a_5 = a_1 + (5 - 1)d$$
$$13 = -15 + 4d$$
$$28 = 4d$$
$$7 = d$$

Then $a_{40} = -15 + (39)7 = 258$. ■

Adding the terms of a finite sequence may not be much work when the number of terms to be added is small. When many terms are to be added, however, the amount of time and effort needed can be overwhelming. For example, to add the first 10,000 terms of the arithmetic sequence beginning with

$$246, \quad 261, \quad 276, \ldots$$

would call for an enormous effort, unless some shortcut could be found. Fortunately, there is an easy way available to find such sums. This method (in disguise) was already used in the question at the start of Section 11.2. Let us look at the general situation. Let S_n denote the sum of the first n terms of the arithmetic sequence given by $a_k = a_1 + (k - 1)d$:

The sum of an arithmetic sequence is called an ***arithmetic series****.*

$$S_n = \sum_{k=1}^{n} [a_1 + (k - 1)d]$$
$$= a_1 + [a_1 + d] + [a_1 + 2d] + \cdots + [a_1 + (n - 1)d]$$

Put this sum in reverse order and write the two equalities together as follows:

$$\begin{array}{ccccccccc} S_n = & a_1 & + & [a_1 + d] & + \cdots + & [a_1 + (n-2)d] & + & [a_1 + (n - 1)d] \\ & \updownarrow & & \updownarrow & & \updownarrow & & \updownarrow \\ S_n = & [a_1 + (n - 1)d] & + & [a_1 + (n - 2)d] & + \cdots + & [a_1 + d] & + & a_1 \end{array}$$

Now add to get

$$2S_n = [2a_1 + (n - 1)d] + [2a_1 + (n - 1)d] + \cdots + [2a_1 + (n - 1)d] + [2a_1 + (n - 1)d]$$

On the right-hand side of this equation there are n terms, each of the form $2a_1 + (n - 1)d$. Therefore,

$$2S_n = n[2a_1 + (n - 1)d]$$

Divide by 2 to solve for S_n:

$$S_n = \frac{n}{2}[2a_1 + (n - 1)d]$$

Returning to the sigma notation, we can summarize our results this way:

ARITHMETIC SERIES

$$\sum_{k=1}^{n} [a_1 + (k - 1)d] = \frac{n}{2}[2a_1 + (n - 1)d]$$

EXAMPLE 2 Find S_{20} for the arithmetic sequence whose first term is $a_1 = 3$ and whose common difference is $d = 5$.

Solution Substituting $a_1 = 3$, $d = 5$, and $n = 20$ into the formula for S_n, we have

$$\begin{aligned} S_{20} &= \frac{20}{2}[2(3) + (20 - 1)5] \\ &= 10(6 + 95) \\ &= 1010 \end{aligned}$$

■

EXAMPLE 3 Find the sum of the first 10,000 terms of the arithmetic sequence beginning with 246, 261, 276, ...

Solution Since $a = 246$ and $d = 15$,

$$\begin{aligned} S_{10,000} &= \frac{10,000}{2}[2(246) + (10,000 - 1)15] \\ &= 5000(150,477) \\ &= 752,385,000 \end{aligned}$$

■

EXAMPLE 4 Find the sum of the first n positive integers.

Solution First observe that the problem calls for the sum of the sequence $a_k = k$ for $k = 1, 2, \ldots, n$. This is an arithmetic sequence with $a_1 = 1$ and $d = 1$. Therefore,

$$\sum_{k=1}^{n} k = \frac{n}{2}[2(1) + (n - 1)1] = \frac{n(n + 1)}{2}$$

■

With the result of Example 4 we are able to check the answer for the sum of the first 1000 positive integers, found at the beginning of Section 11.2, as follows:

$$\sum_{k=1}^{1000} k = \frac{1000(1001)}{2} = 500,500$$

The form $a_k = a_1 + (k - 1)d$ for the general term of an arithmetic sequence easily converts to $a_k = dk + (a_1 - d)$. It is this latter form that is ordinarily used when the general term of a *specific* arithmetic sequence is given. For example, we would usually begin with the form $a_k = 3k + 5$ instead of $a_k = 8 + (k - 1)3$. The important thing to notice in the form $a_k = dk + (a_1 - d)$ is that the common difference is the coefficient of k.

EXAMPLE 5 Evaluate: $\sum_{k=1}^{50} (-6k + 10)$

Solution First note that $a_k = -6k + 10$ is an arithmetic sequence with $d = -6$ and with $a_1 = 4$.

$$\sum_{k=1}^{50} (-6k + 10) = \tfrac{50}{2}[2(4) + (50 - 1)(-6)]$$

$$= -7150$$

■

EXERCISES 11.3

Each of the following gives the first two terms of an arithmetic sequence. Write the next three terms; find the nth term; and find the sum of the first 20 terms.

1. $1, 3, \ldots$
2. $2, 4, \ldots$
3. $2, -4, \ldots$
4. $1, -3, \ldots$
5. $\frac{15}{2}, 8, \ldots$
6. $-\frac{4}{3}, -\frac{11}{3}, \ldots$
7. $\frac{2}{5}, -\frac{1}{5}, \ldots$
8. $-\frac{1}{2}, \frac{1}{4}, \ldots$
9. $50, 100, \ldots$
10. $-27, -2, \ldots$
11. $-10, 10, \ldots$
12. $225, 163, \ldots$

Find the indicated sum by using ordinary addition; also find the sum by using the formula for the sum of an arithmetic sequence.

13. $5 + 10 + 15 + 20 + 25 + 30 + 35 + 40 + 45 + 50 + 55 + 60 + 65$
14. $-33 - 25 - 17 - 9 - 1 + 7 + 15 + 23 + 31 + 39$
15. $\frac{3}{4} + 1 + \frac{5}{4} + \frac{3}{2} + \frac{7}{4} + 2 + \frac{9}{4} + \frac{5}{2} + \frac{11}{4}$
16. $128 + 71 + 14 - 43 - 100 - 157$
17. Find a_{30} for the arithmetic sequence having $a_1 = -30$ and $a_{10} = 69$.
18. Find a_{51} for the arithmetic sequence having $a_1 = 9$ and $a_8 = -19$.

Find S_{100} for the arithmetic sequence with the given values for a_1 and d.

19. $a_1 = 3; d = 3$
20. $a_1 = 1; d = 8$
21. $a_1 = -91; d = 21$
22. $a_1 = -7; d = -10$
23. $a_1 = \frac{1}{7}; d = 5$
24. $a_1 = \frac{2}{5}; d = -4$
25. $a_1 = 725; d = 100$
26. $a_1 = 0.1; d = 10$

27. Find S_{28} for the sequence $-8, 8, \ldots, 16n - 24, \ldots$.
28. Find S_{25} for the sequence $96, 100, \ldots, 4n + 92, \ldots$.
29. Find the sum of the first 50 positive multiples of 12.
30. **(a)** Find the sum of the first 100 positive even numbers.
    (b) Find the sum of the first n positive even numbers.
31. **(a)** Find the sum of the first 100 positive odd numbers.
    (b) Find the sum of the first n positive odd numbers.

Evaluate the series in each exercise.

32. $\sum_{k=1}^{12} [3 + (k - 1)9]$
33. $\sum_{k=1}^{9} [-6 + (k - 1)\frac{1}{2}]$
34. $\sum_{k=1}^{20} (4k - 15)$
35. $\sum_{k=1}^{30} (10k - 1)$
36. $\sum_{k=1}^{40} (-\frac{1}{3}k + 2)$
37. $\sum_{k=1}^{49} (\frac{3}{4}k - \frac{1}{2})$
38. $\sum_{k=1}^{20} 5k$
39. $\sum_{k=1}^{n} 5k$

40. Find u such that 7, u, 19 is an arithmetic sequence.
41. Find u such that -7, u, $\frac{5}{2}$ is an arithmetic sequence.
42. Find the twenty-third term of the arithmetic sequence 6, -4, ...
43. Find the thirty-fifth term of the arithmetic sequence $-\frac{2}{3}$, $-\frac{1}{5}$, ...
44. An object is dropped from an airplane and falls 32 feet during the first second. During each successive second it falls 48 feet more than in the preceding second. How many feet does it travel during the first 10 seconds? How far does it fall during the tenth second?
45. Suppose you save \$10 one week and that each week thereafter you save 50¢ more than the preceding week. How much will you have saved by the end of 1 year?
46. A pyramid of blocks has 26 blocks in the bottom row and 2 fewer blocks in each successive row thereafter. How many blocks are there in the pyramid?
*47. Evaluate $\Sigma_{n=6}^{20} (5n - 3)$.
*48. Find the sum of all the even numbers between 33 and 427.
*49. If $\Sigma_{k=1}^{30} [a_1 + (k - 1)d] = -5865$ and $\Sigma_{k=1}^{20} [a_1 + (k - 1)d] = -2610$, find a_1 and d.
50. Listing the first few terms of a sequence like 2, 4, 6, . . . , without stating its general term or describing just what kind of sequence it is makes it impossible to predict the next term. Show that both sequences $t_n = 2n$ and $s_n = 2n + (n - 1)(n - 2)(n - 3)$ produce these first three terms but that their fourth terms are different.
*51. Find u and v such that 3, u, v, 10 is an arithmetic sequence.
*52. Show that the sum of an arithmetic sequence of n terms is n times the average of the first and last terms.

Use the result in Exercise 52 to find S_{80} for the given arithmetic sequences.

53. $a_k = 3k - 8$ 54. $a_k = \frac{1}{2}k + 10$ 55. $a_k = -5k$
56. What is the connection between arithmetic sequences and linear functions?
57. The function f defined by $f(x) = 3x + 7$ is a linear function. Evaluate the series $\Sigma_{k=1}^{16} f_k$, where $f_k = f(k)$ is the arithmetic sequence associated with f.

Written Assignment: Explain, with examples, the difference between an arithmetic sequence and an arithmetic series.

11.4 GEOMETRIC SEQUENCES AND SERIES

Suppose that a ball is dropped from a height of 4 feet and bounces straight up and down, always bouncing up exactly one-half the distance it just came down. How far will the ball have traveled if you catch it after it reaches the top of the fifth bounce? The following figure will help you to answer this question. For the sake of clarity, the bounces have been separated in the figure.

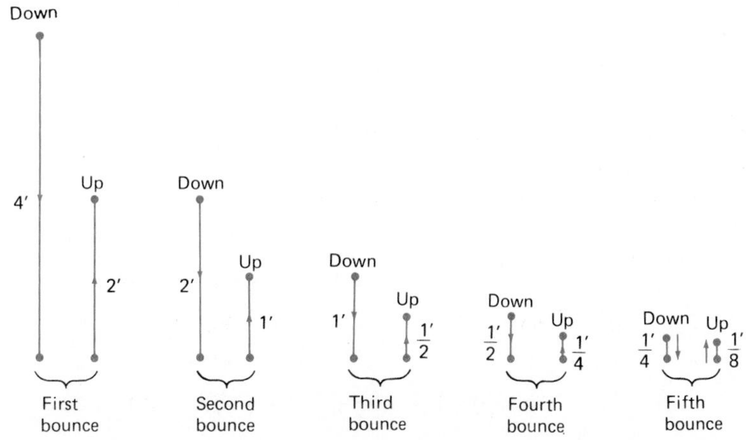

From this diagram we can determine how far the ball has traveled on each bounce. On the first bounce it goes 4 feet down and 2 feet up, for a total of 6 feet; on the second bounce the total distance is $2 + 1 = 3$ feet; and so on. These distances form the following sequence of five terms (one for each bounce):

$$6, \quad 3, \quad \frac{3}{2}, \quad \frac{3}{4}, \quad \frac{3}{8}$$

This sequence has the special property that, after the first term, each successive term can be obtained by multiplying the preceding term by $\frac{1}{2}$; that is, the second term, 3, is half the first, 6, and so on. This is an example of a ***geometric sequence***. Later we will develop a formula for finding the sum of such a sequence; in the meantime, we can find the total distance the ball has traveled during the five bounces by adding the first five terms as follows:

$$6 + 3 + \frac{3}{2} + \frac{3}{4} + \frac{3}{8} = \frac{48 + 24 + 12 + 6 + 3}{8} = 11\frac{5}{8}$$

A geometric sequence is also referred to as a ***geometric progression.***

DEFINITION OF A GEOMETRIC SEQUENCE

A sequence is said to be *geometric* if each term, after the first, is obtained by multiplying the preceding term by a common value.

Here are the first four terms of a geometric sequence.

$$2, \quad -4, \quad 8, \quad -16, \ldots$$

By inspection you can determine that the common multiplier for this sequence is -2. To find the nth term we first derive the formula for the nth term of *any* geometric sequence.

Let a_n be the nth term of a geometric sequence, and let a_1 be its first term. The common multiplier, which is also called the **common ratio**, is denoted by r. Here are the first five terms:

$$\begin{aligned} &a_1 \\ &a_2 = a_1 r \\ &a_3 = a_2 r = (a_1 r)r = a_1 r^2 \\ &a_4 = a_3 r = (a_1 r^2)r = a_1 r^3 \\ &a_5 = a_4 r = (a_1 r^3)r = a_1 r^4 \end{aligned}$$

Notice that the exponent of r is 1 less than the number of the term. This observation allows us to write the nth term as follows:

This formula says that the nth term of a geometric sequence is completely determined by its first term a_1 and common ratio r.

THE GENERAL TERM OF A GEOMETRIC SEQUENCE

The nth term of a geometric sequence is

$$a_n = a_1 r^{n-1}$$

where a_1 is the first term and r is the common ratio.

With this result, the first four terms and the nth terms of the geometric sequence given previously are as follows.

$$2, \quad -4, \quad 8, \quad -16, \ldots, 2(-2)^{n-1} \qquad (r = -2)$$

Here are two more illustrations:

$$1, \quad \tfrac{1}{3}, \quad \tfrac{1}{9}, \quad \tfrac{1}{27}, \ldots, 1(\tfrac{1}{3})^{n-1} \qquad (r = \tfrac{1}{3})$$

$$5, \quad -5, \quad 5, \quad -5, \ldots, 5(-1)^{n-1} \qquad (r = -1)$$

You can substitute the values $n = 1, 2, 3$, and 4 into the forms for the nth terms and see that the given first four terms are obtained in each case.

EXAMPLE 1 Find the hundredth term of the geometric sequence having $r = \frac{1}{2}$ and $a_1 = \frac{1}{2}$.

Solution The nth term of this sequence is given by

$$a_n = \frac{1}{2}\left(\frac{1}{2}\right)^{n-1} = \frac{1}{2}\left(\frac{1}{2^{n-1}}\right) = \frac{1}{2^n}$$

Thus $a_{100} = \dfrac{1}{2^{100}}$. ■

The reason r is called the common ratio of a geometric sequence $a_n = a_1 r^{n-1}$ is that for each n the ratio of the $(n + 1)$st term to the nth term equals r. Thus

$$\frac{a_{n+1}}{a_n} = \frac{a_1 r^n}{a_1 r^{n-1}} = r$$

EXAMPLE 2 Find the nth term of the geometric sequence beginning with

$$6, 9, \tfrac{27}{2}, \ldots$$

and find the seventh term.

Solution First find r.

r can also be found using a_2 and a_3.

$$\frac{s_3}{s_2} = \frac{\frac{27}{2}}{9} = \frac{27}{2 \cdot 9} = \frac{3}{2}$$

$$r = \frac{a_2}{a_1} = \frac{9}{6} = \frac{3}{2}$$

Then the nth term is

$$a_n = 6\left(\frac{3}{2}\right)^{n-1}$$

Let $n = 7$ to get

$$a_7 = 6\left(\frac{3}{2}\right)^{7-1} = 6\left(\frac{3}{2}\right)^6$$

$$= (3 \cdot 2) \cdot \frac{3^6}{2^6} = \frac{3^7}{2^5} = \frac{2187}{32}$$ ■

EXAMPLE 3 Write the kth term of the geometric sequence $a_k = (\frac{1}{2})^{2k}$ in the form $a_1 r^{k-1}$ and find the value of a_1 and r.

Solution

Note: The first term a_1 can also be found by simply computing s_1 in the given formula for a_k. Then find a_2 and evaluate $r = \frac{a_2}{a_1}$.

$$a_k = (\tfrac{1}{2})^{2k} = [(\tfrac{1}{2})^2]^k = (\tfrac{1}{4})^k$$

$$= \tfrac{1}{4}(\tfrac{1}{4})^{k-1} \longleftarrow \text{this is now in the form } a_1 r^{n-1}$$

Then $a_1 = \frac{1}{4}$ and $r = \frac{1}{4}$. ■

TEST YOUR UNDERSTANDING
Think Carefully

Write the first five terms of the geometric sequences with the given general term. Also write the nth term in the form ar^{n-1} and find r.

1. $a_n = (\frac{1}{2})^{n-1}$ **2.** $a_n = (\frac{1}{2})^{n+1}$
3. $a_n = (-\frac{1}{2})^n$ **4.** $a_n = (-\frac{1}{3})^{3n}$

Find r and the nth term of the geometric sequence with the given first two terms.

5. $\frac{1}{5}, 2$ **6.** $27, -12$

(Answers Page 525)

EXAMPLE 4 A geometric sequence consisting of positive numbers has $a_1 = 18$ and $a_5 = \dfrac{32}{9}$. Find r.

Solution Use $n = 5$ in $a_n = a_1 r^{n-1}$:

$$\frac{32}{9} = 18r^4$$

$$r^4 = \frac{32}{9 \cdot 18} = \frac{16}{81}$$

$$r = \pm\sqrt[4]{\frac{16}{81}} = \pm\frac{2}{3}$$

Since the terms are positive, $r = \dfrac{2}{3}$. ■

Check this result by writing $a_1 = 18$ and finding a_2, a_3, a_4, and a_5 using $a_n = a_{n-1}r$.

Let us return to the original problem of this section. We found that the total distance the ball traveled was $11\frac{5}{8}$ feet. This is the sum of the first five terms of the geometric sequence whose nth term is $6(\frac{1}{2})^{n-1}$. Adding these five terms was easy. But what about adding the first 100 terms? There is a formula for the sum of a geometric sequence that will enable us to find such answers efficiently.

The sum of a geometric sequence is called a **geometric series**. Just as with arithmetic series, there is a formula for finding such sums. To discover this formula, let $a_k = a_1 r^{k-1}$ be a geometric sequence and denote the sum of the first n terms by $S_n = \sum_{k=1}^{n} a_1 r^{k-1}$. Then

$$S_n = a_1 + a_1 r + a_1 r^2 + \cdots + a_1 r^{n-2} + a_1 r^{n-1}$$

Multiplying this equation by r gives

$$rS_n = a_1r + a_1r^2 + \cdots + a_1r^{n-1} + a_1r^n$$

Now consider these two equations:

$$\begin{aligned} S_n &= a_1 + a_1r + a_1r^2 + \cdots + a_1r^{n-2} + a_1r^{n-1} \\ &\qquad\quad \updownarrow \qquad \updownarrow \qquad\qquad\quad \updownarrow \qquad\quad \updownarrow \\ rS_n &= \qquad\quad a_1r + a_1r^2 + \cdots + a_1r^{n-2} + a_1r^{n-1} + a_1r^n \end{aligned}$$

Subtract and factor:

$$S_n - rS_n = a_1 - a_1r^n$$

$$(1 - r)S_n = a_1(1 - r^n)$$

Divide by $1 - r$ to solve for S_n:

Here $r \neq 1$. However, when $r = 1$, $a_k = a_1r^{k-1} = a_1$, which is also an arithmetic sequence having $d = 0$.

$$S_n = \frac{a_1(1 - r^n)}{1 - r}$$

Returning to sigma notation, we can summarize our results this way:

GEOMETRIC SERIES

$$\sum_{k=1}^{n} a_1r^{k-1} = \frac{a_1(1 - r^n)}{1 - r}$$

The preceding formula can be used to verify the earlier result for the bouncing ball:

$$\begin{aligned} \sum_{k=1}^{5} 6\left(\frac{1}{2}\right)^{k-1} &= \frac{6[1 - (\frac{1}{2})^5]}{1 - \frac{1}{2}} \\ &= \frac{6(1 - \frac{1}{32})}{\frac{1}{2}} \\ &= \frac{93}{8} = 11\frac{5}{8} \end{aligned}$$

EXAMPLE 5 Find the sum of the first 100 terms of the geometric sequence given by $a_k = 6(\frac{1}{2})^{k-1}$ and show that the answer is very close to 12.

Solution

$$\begin{aligned} S_{100} &= \frac{6\left(1 - \frac{1}{2^{100}}\right)}{1 - \frac{1}{2}} \\ &= 12\left(1 - \frac{1}{2^{100}}\right) \end{aligned}$$

Next observe that the fraction $\frac{1}{2^{100}}$ is so small that $1 - \frac{1}{2^{100}}$ is very nearly equal to 1, and therefore S_{100} is very close to 12. ■

Another way to find a_1 and r is to write the first few terms as follows:

$$\frac{3}{100} + \frac{3}{1000} + \frac{3}{10{,}000} + \cdots = \frac{3}{100} + \frac{3}{100}\left(\frac{1}{10}\right) + \frac{3}{100}\left(\frac{1}{10}\right)^2 + \cdots$$

Then $a_1 = 0.03$ and $r = 0.1$.

EXAMPLE 6 Evaluate: $\sum_{k=1}^{8} 3\left(\frac{1}{10}\right)^{k+1}$

Solution

$$3\left(\frac{1}{10}\right)^{k+1} = \frac{3}{100}\left(\frac{1}{10}\right)^{k-1}$$

Then $a_1 = 0.03$, $r = 0.1$, and

$$S_8 = \frac{0.03[1 - (0.1)^8]}{1 - 0.1} = \frac{0.03(1 - 0.00000001)}{0.9} = 0.033333333$$

■

Geometric sequences have many applications as illustrated by Examples 7 and 8. You will find others in the exercises at the end of this section.

EXAMPLE 7 Suppose that you save \$128 in January and that each month thereafter you only manage to save half of what you saved the previous month. How much do you save in the tenth month, and what are your total savings after 10 months?

Solution The amounts saved each month form a geometric sequence with $a_1 = 128$ and $r = \frac{1}{2}$. Then $a_n = 128(\frac{1}{2})^{n-1}$ and

$$a_{10} = 128\left(\frac{1}{2}\right)^9 = \frac{2^7}{2^9} = \frac{1}{4} = 0.25$$

This means that you saved 25¢ in the tenth month. Your total savings are:

$$S_{10} = \frac{128\left(1 - \frac{1}{2^{10}}\right)}{1 - \frac{1}{2}} = 256\left(1 - \frac{1}{2^{10}}\right) = 256 - \frac{256}{2^{10}} = 256 - \frac{2^8}{2^{10}} = 255.75$$

The total savings is \$255.75. ■

EXAMPLE 8 A roll of wire is 625 feet long. If $\frac{1}{5}$ of the wire is cut off repeatedly, what is the general term of the sequence for the length of wire remaining? Use a calculator and the general term to determine the length of wire remaining after 7 cuts.

Solution Since $\frac{1}{5}$ is cut off, $\frac{4}{5} = 0.8$ must remain. Thus, $625(0.8) = 500$ feet remain after one cut, $625(0.8)(0.8) = 500(0.8) = 400$ feet remain after two cuts, and after n cuts are made, the length of wire remaining is $625(0.8)^n$ feet. Using a calculator, we have

$$625(0.8)^7 = 131.072$$

Therefore, to the nearest tenth of a foot, 131.1 feet of wire remains after 7 cuts.

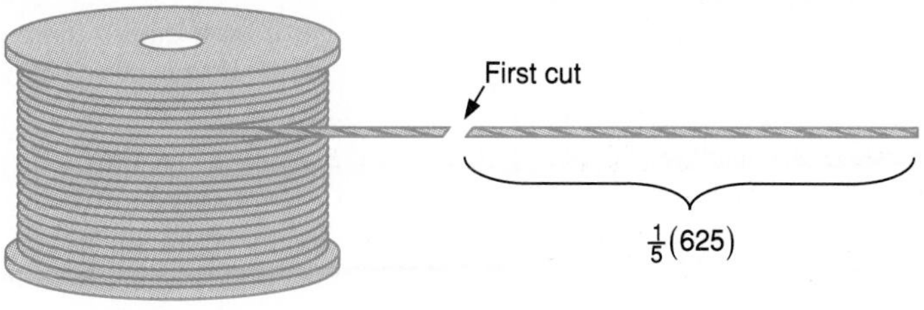

■

EXERCISES 11.4

The first three terms of a geometric sequence are given. Write the next three terms and also find the formula for the nth term.

1. 2, 4, 8, . . .
2. 2, −4, 8, . . .
3. 1, 3, 9, . . .
4. 2, −2, 2, . . .
5. $-3, 1, -\frac{1}{3}, \ldots$
6. 100, 10, 1, . . .
7. −1, −5, −25, . . .
8. 12, −6, 3, . . .
9. $-6, -4, -\frac{8}{3}, \ldots$
10. −64, 16, −4, . . .
11. $\frac{1}{1000}, \frac{1}{10}, 10, \ldots$
12. $\frac{27}{8}, \frac{3}{2}, \frac{2}{3}, \ldots$

Find the sum of the first six terms of the indicated sequence by using ordinary addition and also by using the formula for a geometric series.

13. The sequence in Exercise 1.
14. The sequence in Exercise 5.
15. The sequence in Exercise 9.
16. Find the tenth term of the geometric sequence 2, 4, 8, ...
17. Find the fourteenth term of the geometric sequence $\frac{1}{8}, \frac{1}{4}, \frac{1}{2}, \ldots$.
18. Find the fifteenth term of the geometric sequence $\frac{1}{100{,}000}, \frac{1}{10{,}000}, \frac{1}{1000}, \ldots$.
19. What is the one-hundred-first term of the geometric sequence having $a_1 = 3$ and $r = -1$?
20. For the geometric sequence with $a_1 = 100$ and $r = \frac{1}{10}$, use the formula $a_n = a_1 r^{n-1}$ to find which term is equal to $\frac{1}{10^{10}}$.
21. Find r for the geometric series having $a_1 = 20$ and $a_6 = -\frac{5}{8}$.
22. Find r for the geometric series having $a_1 = -25$ and $a_5 = -3.24$.

Evaluate the series.

23. $\sum_{k=1}^{10} 2^{k-1}$
24. $\sum_{j=1}^{10} 2^{j+2}$
25. $\sum_{k=1}^{n} 2^{k-1}$
26. $\sum_{k=1}^{8} 3(\frac{1}{10})^{k-1}$
27. $\sum_{k=1}^{5} 3^{k-4}$
28. $\sum_{k=1}^{6} (-3)^{k-2}$
29. $\sum_{j=1}^{5} (\frac{2}{3})^{j-2}$
30. $\sum_{k=1}^{8} 16(\frac{1}{2})^{k+2}$
31. $\sum_{k=1}^{8} 16(-\frac{1}{2})^{k+2}$
32. Find $u > 0$ such that 2, u, 98 forms a geometric sequence.
33. Find $u < 0$ such that $\frac{1}{7}$, u, $\frac{25}{63}$ forms a geometric sequence.
34. Find a sequence whose first term is 5 that is both geometric and arithmetic. What are r and d?
35. Suppose someone offered you a job that pays 1¢ the first day, 2¢ the second day, 4¢ the third day, etc., each day earning double of what was earned the preceding day. How many dollars would you earn for 30 days work?
36. Suppose that the amount you save in any given month is twice the amount you saved in the previous month. How much will you have saved at the end of 1 year if you save \$1 in January? How much if you saved 25¢ in January?

37. A certain bacterial culture doubles in number every day. If there were 1000 bacteria at the end of the first day, how many will there be after 10 days? How many after n days?

38. A radioactive substance is decaying so that at the end of each month there is only one-third as much as there was at the beginning of the month. If there were 75 grams of the substance at the beginning of the year, how much is left at midyear?

39. Suppose that an automobile depreciates 10% in value each year for the first 5 years. What is it worth after 5 years if its original cost was \$5280? (*Hint:* Use $a_1 = 5280$ and $n = 6$.)

40. In the compound-interest formula $A_t = P(1 + r)^t$, developed on page 406, P is the initial investment, r is the annual interest rate, and t is the number of years during which the interest has been compounded annually to obtain the total value A_t. Explain how this formula may be viewed as the general term of a geometric sequence.

41. A sum of \$800 is invested at 11% interest compounded annually.
   (a) What is the amount after n years?
   (b) What is the amount after 5 years?

42. How much money must be invested at the interest rate of 12%, compounded annually, so that after 3 years the amount is \$1000?

43. Find the amount of money that an investment of \$1500 earns at the interest rate of 8% compounded annually for 5 years.

44. **(a)** If $\frac{3}{5}$ of the wire in Example 8, page 510, is cut off repeatedly, what is the general form of the sequence for the length of the remaining wire?
   (b) What length remains after six cuts have been made? Give the answer to the nearest tenth of a foot.
   (c) What is the general form for the total length of wire that has been cut off after n cuts have been made?

45. **(a)** A set of containers are decreasing in size so that the second container is $\frac{1}{2}$ the volume of the first, the third is $\frac{1}{2}$ the volume of the second, etc. If the first container is empty and the other five are filled with water, can all the full ones be emptied into the first without the water spilling over? Explain.

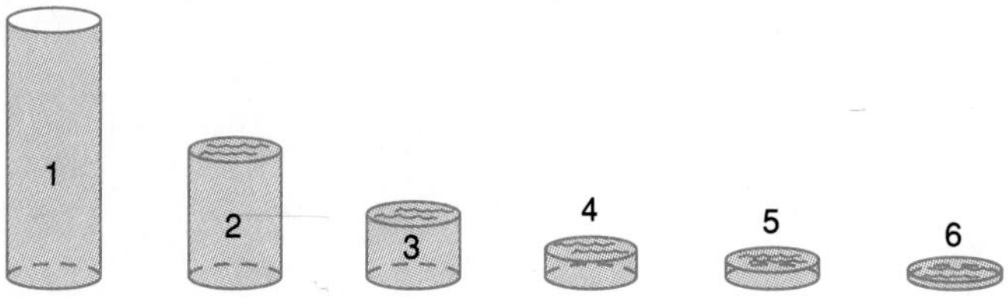

   (b) Answer the question in part (a) assuming that each container, after the first, is $\frac{2}{3}$ the volume of the one preceding it.

Written Assignment: Explain, with examples, the difference between an arithmetic sequence and a geometric sequence.

CHALLENGE
Think Creatively

Suppose you snap your fingers, wait 1 minute, and snap them again. Then you snap them after 2 more minutes, then again after 4 more minutes, again after 8 more minutes, etc., each time waiting twice as long as you waited for the preceding snap. First *guess* how many times you would snap your fingers if you continued this process for one year. Now use a calculator to compute how long it would take you to snap your fingers (a) 10 times, (b) 15 times, and (c) 20 times.

11.5 INFINITE GEOMETRIC SERIES

In decimal form the fraction $\frac{3}{4}$ becomes 0.75, which means $\frac{75}{100}$. This can also be written as $\frac{7}{10} + \frac{5}{100}$. What about $\frac{1}{3}$? As a decimal we can write

$$\tfrac{1}{3} = 0.333 \cdots$$

where the dots mean that the 3 repeats endlessly. We can express this decimal as the sum of fractions whose denominators are powers of 10:

$$\frac{1}{3} = \frac{3}{10} + \frac{3}{100} + \frac{3}{1000} + \cdots$$

*The sum of an infinite sequence is an **infinite series.***

The numbers being added here are the terms of the *infinite geometric sequence* with first term $a_1 = \frac{3}{10}$ and common ratio $r = \frac{1}{10}$. Thus the nth term is

$$\begin{aligned} a_1 r^{n-1} = \frac{3}{10}\left(\frac{1}{10}\right)^{n-1} &= 3\left(\frac{1}{10}\right)\left(\frac{1}{10}\right)^{n-1} \\ &= 3\left(\frac{1}{10}\right)^n \\ &= \frac{3}{10^n} \end{aligned}$$

The sum of the first n terms is found by using the formula

$$S_n = \sum_{k=1}^{n} a_1 r^{k-1} = \frac{a_1(1 - r^n)}{1 - r}$$

Here are some cases:

$$S_1 = \frac{\frac{3}{10}\left(1 - \frac{1}{10}\right)}{1 - \frac{1}{10}} = \frac{1}{3}\left(1 - \frac{1}{10}\right) = 0.3$$

$$S_2 = \frac{\frac{3}{10}\left(1 - \frac{1}{10^2}\right)}{1 - \frac{1}{10}} = \frac{1}{3}\left(1 - \frac{1}{10^2}\right) = 0.33$$

$$S_{10} = \frac{\frac{3}{10}\left(1 - \frac{1}{10^{10}}\right)}{1 - \frac{1}{10}} = \frac{1}{3}\left(1 - \frac{1}{10^{10}}\right) = 0.\underbrace{3333333333}_{10 \text{ places}}$$

$$S_n = \frac{\frac{3}{10}\left(1 - \frac{1}{10^n}\right)}{1 - \frac{1}{10}} = \frac{1}{3}\left(1 - \frac{1}{10^n}\right) = 0.\underbrace{333 \ldots 3}_{n \text{ places}}$$

You can see that as more and more terms are added, the closer and closer the answer gets to $\frac{1}{3}$. This can be seen by studying the form for the sum of the first n terms:

$$S_n = \frac{1}{3}\left(1 - \frac{1}{10^n}\right)$$

It is clear that the bigger n is, the closer $\frac{1}{10^n}$ is to zero, the closer $1 - \frac{1}{10^n}$ is to 1 and, finally, the closer S_n is to $\frac{1}{3}$. Although it is true that S_n is never exactly equal to $\frac{1}{3}$, for very large n the difference between S_n and $\frac{1}{3}$ is very small. Saying this another way:

By taking n large enough, S_n can be made as close to $\frac{1}{3}$ as we like.

This is what we mean when we say that the sum of all the terms is $\frac{1}{3}$.

$$\frac{3}{10} + \frac{3}{10^2} + \frac{3}{10^3} + \cdots + \frac{3}{10^n} + \cdots = \frac{1}{3}$$

The summation symbol, Σ, can also be used here after an adjustment in notation is made. Traditionally, the symbol ∞ has been used to suggest an infinite number of objects. So we use this and make the transition from the sum of a finite number of terms

$$S_n = \sum_{k=1}^{n} \frac{3}{10^k} = \frac{3}{10} + \frac{3}{10^2} + \cdots + \frac{3}{10^n} = \frac{1}{3}\left(1 - \frac{1}{10^n}\right)$$

to the sum of an infinite number of terms:

In calculus the symbol S_∞ is replaced by $\lim_{n\to\infty} S_n = \frac{1}{3}$, which is read as "the limit of S_n as n gets arbitrarily large is $\frac{1}{3}$."

$$S_\infty = \sum_{k=1}^{\infty} \frac{3}{10^k} = \frac{3}{10} + \frac{3}{10^2} + \cdots + \frac{3}{10^n} + \cdots = \frac{1}{3}$$

Not every geometric sequence produces an infinite geometric series that has a finite sum. For instance, the sequence

$$2, 4, 8, \ldots, 2^n, \ldots$$

is geometric, but the corresponding geometric series

$$2 + 4 + 8 + \cdots + 2^n + \cdots$$

cannot have a finite sum.

By now you might suspect that the common ratio r determines whether or not an infinite geometric sequence can be added. This turns out to be true. To see how this works, the general case will be considered next.

Let

$$a_1, a_1r, a_1r^2, \ldots, a_1r^{n-1}, \ldots$$

be an infinite geometric sequence.

Then the sum of the first n terms is

$$S_n = \frac{a_1(1 - r^n)}{1 - r}$$

Rewrite in this form:

$$S_n = \frac{a_1}{1 - r}(1 - r^n)$$

Use a calculator to verify the powers of $r = 0.9$ and $r = 1.1$ to the indicated decimal places.

$(0.9)^1 = 0.9$
$(0.9)^{10} = 0.35$
$(0.9)^{20} = 0.12$
$(0.9)^{40} = 0.015$
$(0.9)^{80} = 0.0002$
$(0.9)^{100} = 0.00003$
↓
getting close to 0

$(1.1)^1 = 1.1$
$(1.1)^5 = 1.6$
$(1.1)^{10} = 2.6$
$(1.1)^{20} = 6.7$
$(1.1)^{50} = 117.4$
$(1.1)^{100} = 13780.6$
↓
getting very large

At this point the importance of r^n becomes clear. If, as n gets larger, r^n gets very large, then the infinite geometric series will not have a finite sum. But if r^n gets arbitrarily close to zero as n gets larger, then $1 - r^n$ gets close to 1 and S_n gets closer and closer to $\frac{a_1}{1 - r}$.

The values of r for which r^n gets arbitrarily close to zero are precisely those values between -1 and 1; that is, $|r| < 1$. For instance, $\frac{3}{5}$, $-\frac{1}{10}$, and 0.09 are values of r for which r^n gets close to zero; and 1.01, -2, and $\frac{3}{2}$ are values for which the series does not have a finite sum.

To sum up, we have the following useful result:

SUM OF AN INFINITE GEOMETRIC SERIES

If $|r| < 1$, then $\sum_{k=1}^{\infty} a_1 r^{k-1} = \frac{a_1}{1 - r}$. For other values of r the series has no finite sum.

EXAMPLE 1 Find the sum of the infinite geometric series

$$27 + 3 + \frac{1}{3} + \cdots$$

Solution Since $r = \frac{3}{27} = \frac{1}{9}$ and $a_1 = 27$, the preceding result gives

$$\begin{aligned} 27 + 3 + \frac{1}{3} + \cdots &= \frac{27}{1 - \frac{1}{9}} \\ &= \frac{27}{\frac{8}{9}} \\ &= \frac{243}{8} \end{aligned}$$

■

EXAMPLE 2 Why does the infinite geometric series $\sum_{k=1}^{\infty} 5(\frac{4}{3})^{k-1}$ have no finite sum?

Solution The series has no finite sum because the common ratio $r = \frac{4}{3}$ is not between -1 and 1. ■

Another way to find a_1 is to let $k = 1$ in $\frac{7}{10^{k+1}}$:

$$a_1 = \frac{7}{10^2} = \frac{7}{100}$$

Also, r can be found by taking the ratio of the second term to the first term.

$$r = \frac{\frac{7}{10^3}}{\frac{7}{10^2}} = \frac{1}{10}$$

EXAMPLE 3 Find: $\sum_{k=1}^{\infty} \frac{7}{10^{k+1}}$.

Solution Since $\frac{7}{10^{k+1}} = 7\left(\frac{1}{10^{k+1}}\right) = 7\left(\frac{1}{10^2}\right)\frac{1}{10^{k-1}} = \frac{7}{100}\left(\frac{1}{10}\right)^{k-1}$, it follows that $a_1 = \frac{7}{100}$ and $r = \frac{1}{10}$. Therefore, by the formula for the sum of an infinite geometric series we have

$$S_\infty = \sum_{k=1}^{\infty} \frac{7}{10^{k+1}} = \frac{\frac{7}{100}}{1 - \frac{1}{10}} = \frac{7}{100 - 10} = \frac{7}{90}$$ ■

TEST YOUR UNDERSTANDING Think Carefully

Find the common ratio r, and then find the sum if the given infinite geometric series has one.

1. $10 + 1 + \frac{1}{10} + \cdots$
2. $\frac{1}{64} + \frac{1}{16} + \frac{1}{4} + \cdots$
3. $36 - 6 + 1 - \cdots$
4. $-16 - 4 - 1 - \cdots$
5. $\sum_{k=1}^{\infty} (\frac{4}{3})^{k-1}$
6. $\sum_{k=1}^{\infty} 3(0.01)^k$
7. $\sum_{i=1}^{\infty} (-1)^i 3^i$
8. $\sum_{n=1}^{\infty} 100(-\frac{9}{10})^{n+1}$
9. $101 - 102.01 + 103.0301 - \cdots$

(Answers Page 525)

The introduction to this section indicated how the endless repeating decimal 0.333. . . can be regarded as an infinite geometric series. The next example illustrates how such decimal fractions can be written in the rational form $\frac{a}{b}$ (the ratio of two integers) by using the formula for the sum of an infinite geometric series.

Compare the method shown in Example 4 with that developed in Exercise 61 of Section 1.5.

EXAMPLE 4 Express the repeating decimal 0.242424. . . in rational form.

Solution First write

$$0.242424\ldots = \frac{24}{100} + \frac{24}{10{,}000} + \frac{24}{1{,}000{,}000} + \cdots$$

$$= \frac{24}{10^2} + \frac{24}{10^4} + \frac{24}{10^6} + \cdots + \frac{24}{10^{2k}} + \cdots$$

$$0.242424\ldots = \frac{24}{100} + \frac{24}{100}\left(\frac{1}{100}\right) + \frac{24}{100}\left(\frac{1}{100}\right)^2 + \cdots + \frac{24}{100}\left(\frac{1}{100}\right)^{k-1} + \cdots$$

Observe that

$$\frac{24}{10^{2k}} = 24\left(\frac{1}{10^{2k}}\right) = 24\left(\frac{1}{10^2}\right)^k = 24\left(\frac{1}{100}\right)^k = \frac{24}{100}\left(\frac{1}{100}\right)^{k-1}$$

Then $a_1 = \frac{24}{100}$, $r = \frac{1}{100}$, and

$$0.242424\ldots = \sum_{k=1}^{\infty} \frac{24}{100}\left(\frac{1}{100}\right)^{k-1} = \frac{\frac{24}{100}}{1 - \frac{1}{100}} = \frac{24}{99} = \frac{8}{33}$$

Check this result by dividing 33 into 8. ■

It seems as if the horse cannot finish the race this way. But read on to see that there is really no contradiction with this interpretation.

EXAMPLE 5 A racehorse running at the constant rate of 30 miles per hour will finish a 1-mile race in 2 minutes. Now consider the race broken down into the following parts: before the racehorse can finish the 1-mile race it must first reach the halfway mark; having done that, the horse must next reach the quarter pole; then it must reach the eighth pole; and so on. That is, it must always cover half the distance remaining before it can cover the whole distance. Show that the sum of the infinite number of time intervals is also 2 minutes.

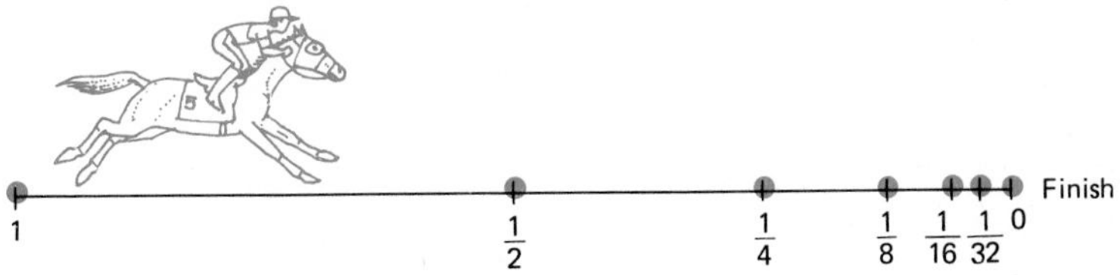

$T = \frac{D}{R}\left(\textit{time} = \frac{\textit{distance}}{\textit{rate}}\right)$

Solution For the first $\frac{1}{2}$ mile the time will be $\frac{\frac{1}{2}}{\frac{1}{2}} = 1$ minute; for the next $\frac{1}{4}$ mile the time will be $\frac{\frac{1}{4}}{\frac{1}{2}} = \frac{1}{2}$ minute; for the next $\frac{1}{8}$ mile the time will be $\frac{\frac{1}{8}}{\frac{1}{2}} = \frac{1}{4}$ minute; and for the nth distance, which is $\frac{1}{2^n}$ miles, the time will be $\frac{\frac{1}{2^n}}{\frac{1}{2}} = \frac{1}{2^{n-1}}$.

Thus the total time is given by this series:

$$\sum_{k=1}^{\infty} \frac{1}{2^{k-1}} = 1 + \frac{1}{2} + \frac{1}{4} + \cdots + \frac{1}{2^{n-1}} + \cdots$$

This is an infinite geometric series having $a_1 = 1$ and $r = \frac{1}{2}$. Thus

$$\sum_{k=1}^{\infty} 1\left(\frac{1}{2}\right)^{k-1} = \frac{1}{1 - \frac{1}{2}} = 2$$

which is the same result as before. ■

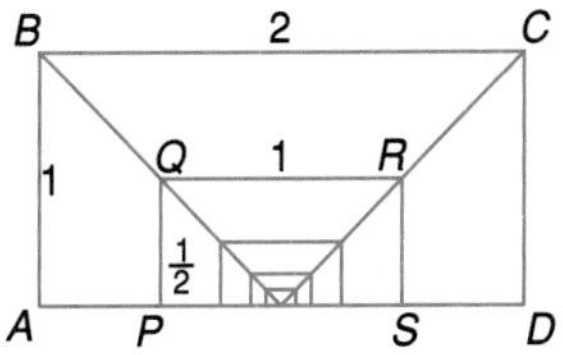

EXAMPLE 6 Rectangle $ABCD$ has dimensions 1 by 2. The next rectangle $PQRS$ has dimensions $\frac{1}{2}$ by 1. In like manner, each inner rectangle has dimensions half the size of the preceding rectangle. If this sequence of rectangles continues endlessly, what is the sum of the areas of all the rectangles?

Solution The area of rectangle $ABCD$ is $1 \cdot 2$, the area of rectangle $PQRS$ is $\frac{1}{2} \cdot 1$, the next has area $\frac{1}{4} \cdot \frac{1}{2}$, and so on. The sum of all the areas is this infinite geometric series.

$$\begin{aligned} &1 \cdot 2 + \tfrac{1}{2} \cdot 1 + \tfrac{1}{4} \cdot \tfrac{1}{2} + \tfrac{1}{8} \cdot \tfrac{1}{4} + \cdots \\ &= 2 + \tfrac{1}{2} + \tfrac{1}{8} + \tfrac{1}{32} + \cdots \\ &= 2 + \tfrac{1}{2} + (\tfrac{1}{2})^3 + (\tfrac{1}{2})^5 + \cdots \end{aligned}$$

Since $a_1 = 2$ and $r = \frac{1}{4}$, the sum equals

$$\frac{a_1}{1 - r} = \frac{2}{1 - \frac{1}{4}} = \frac{8}{3} = 2\frac{2}{3}$$

■

EXERCISES 11.5

Find the sum, if it exists, of each infinite geometric series.

1. $2 + 1 + \frac{1}{2} + \cdots$
2. $8 + 4 + 2 + \cdots$
3. $25 + 5 + 1 + \cdots$
4. $1 + \frac{4}{3} + \frac{16}{9} + \cdots$
5. $1 - \frac{1}{2} + \frac{1}{4} - \cdots$
6. $100 - 1 + \frac{1}{100} - \cdots$
7. $1 + 0.1 + 0.01 + \cdots$
8. $52 + 0.52 + 0.0052 + \cdots$
9. $-2 - \frac{1}{4} - \frac{1}{32} - \cdots$
10. $-729 + 81 - 9 + \cdots$

*Explain what is **wrong** with each of these statements.*

11. $\sum_{n=1}^{\infty} \left(\frac{1}{2}\right)^{n+1} = \frac{1}{1 - \frac{1}{2}}$
12. $\sum_{n=1}^{\infty} (-1)^{n-1} = \frac{1}{1-(-1)} = \frac{1}{2}$
13. $\sum_{n=1}^{\infty} \left(-\frac{1}{3}\right)^{n-1} = \frac{1}{1 - \frac{1}{3}}$
14. $\sum_{n=1}^{\infty} 10(1.01)^{n-1} = \frac{10}{1 - 1.01}$

Decide whether or not the given infinite geometric series has a sum. If it does, find it using $S_\infty = \dfrac{a_1}{1 - r}$.

15. $\sum_{k=1}^{\infty} (\frac{1}{3})^{k-1}$
16. $\sum_{k=1}^{\infty} (\frac{1}{3})^{k}$
17. $\sum_{k=1}^{\infty} (\frac{1}{3})^{k+1}$
18. $\sum_{n=1}^{\infty} \frac{1}{2^{n+1}}$
19. $\sum_{n=1}^{\infty} \frac{1}{2^{n-2}}$
20. $\sum_{k=1}^{\infty} (\frac{1}{10})^{k-1}$
21. $\sum_{k=1}^{\infty} 2(0.1)^{k-1}$
22. $\sum_{k=1}^{\infty} (-\frac{1}{2})^{k-1}$
23. $\sum_{n=1}^{\infty} (\frac{3}{2})^{n-1}$
24. $\sum_{n=1}^{\infty} (-\frac{1}{3})^{n+2}$
25. $\sum_{k=1}^{\infty} (0.7)^{k-1}$
26. $\sum_{k=1}^{\infty} 5(0.7)^{k}$
27. $\sum_{k=1}^{\infty} 5(1.01)^{k}$
28. $\sum_{k=1}^{\infty} (\frac{1}{10})^{k-4}$
29. $\sum_{k=1}^{\infty} 10(\frac{2}{3})^{k-1}$
30. $\sum_{k=1}^{\infty} (-1)^{k}$
31. $\sum_{k=1}^{\infty} (0.45)^{k-1}$
32. $\sum_{k=1}^{\infty} (-0.9)^{k+1}$
33. $\sum_{n=1}^{\infty} 7(-\frac{3}{4})^{n-1}$
34. $\sum_{k=1}^{\infty} (0.1)^{2k}$
35. $\sum_{k=1}^{\infty} (-\frac{2}{5})^{2k}$

Find a rational form for each of the following repeating decimals in a manner similar to that in Example 4. Check your answers.

36. 0.444 . . .
37. 0.777 . . .
38. 7.777 . . .
39. 0.131313 . . .
40. 13.131313 . . .
41. 0.0131313 . . .
42. 0.050505 . . .
43. 0.999 . . .
44. 0.125125125 . . .
45. Suppose that a 1-mile distance a racehorse must run is divided into an infinite number of parts, obtained by always considering $\frac{2}{3}$ of the remaining distance to be covered. Then the lengths of these parts form the sequence

$$\frac{2}{3}, \frac{2}{9}, \frac{2}{27}, \ldots, \frac{2}{3^n}, \ldots$$

(a) Find the sequence of times corresponding to these distances. (Assume that the horse is moving at a rate of $\frac{1}{2}$ mile per minute.)

(b) Show that the sum of the times in part (a) is 2 minutes.

46. A certain ball always rebounds $\frac{1}{3}$ of the distance it falls. If the ball is dropped from a height of 9 feet, how far does it travel before coming to rest? (See the similar situation at the beginning of Section 11.4)
47. A substance initially weighing 64 grams is decaying at a rate such that after 4 hours there

are only 32 grams left. In another 2 hours only 16 grams remain; in another 1 hour after that only 8 grams remain; and so on so that the time intervals and amounts remaining form geometric sequences. How long does it take altogether until nothing of the substance is left?

48. After it is set in motion, each swing in either direction of a particular pendulum is 40% as long as the preceding swing. What is the total distance that the end of the pendulum travels before coming to rest if the first swing is 30 inches long?

*49. Assume that a racehorse takes 1 minute to go the first $\frac{1}{2}$ mile of a 1-mile race. After that, the horse's speed is no longer constant: for the next $\frac{1}{4}$ mile it takes $\frac{2}{3}$ minute; for the next $\frac{1}{8}$ mile it takes $\frac{4}{9}$ minute; for the next $\frac{1}{16}$ mile it takes $\frac{40}{81}$ minute; and so on, so that the time intervals form a geometric sequence. Why can't the horse finish the race?

50. **(a)** $ABCD$ is a square whose sides measure 8 units. $PQRS$ is a square whose sides are $\frac{1}{2}$ the length of the sides of square $ABCD$. The next square has sides $\frac{1}{2}$ the length as those of square $PQRS$. In like manner each inner square has sides $\frac{1}{2}$ the length of the preceding square. If this sequence of squares continues endlessly, what is the sum of all the areas of the squares in the sequence?

(b) What is the sum of all the perimeters?

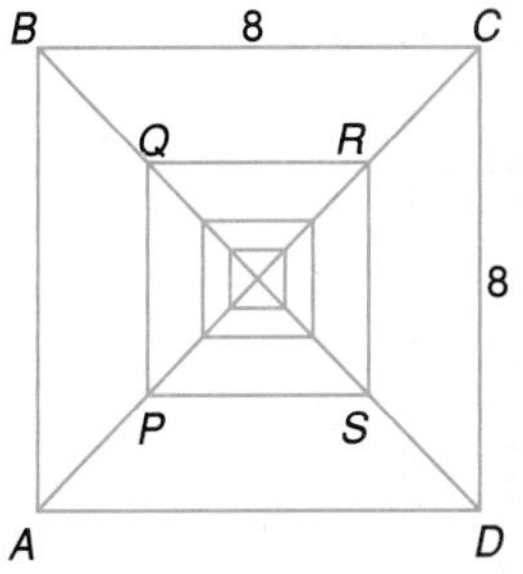

51. ABC is an isosceles right triangle with right angle at C. P_1 is the midpoint of the hypotenuse AB so that CP_1 divides triangle ABC into two congruent triangles. P_2 is the midpoint of BC so that P_1P_2 divides triangle CBP_1 into two congruent triangles. This process continues endlessly.

(a) If $AC = CB = 4$, what do you expect the sum of the area of all the triangles labeled 1, 2, 3, . . . to be equal to?

(b) Verify the result in part (a) by using an infinite geometric series.

(c) Find the sum of all the triangles labeled with odd numbers and also find the sum of all the triangles labeled with even numbers. What is the sum of the two sums?

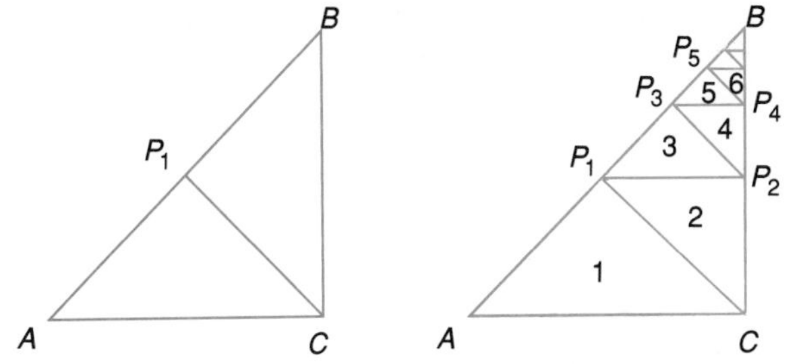

52. The largest circle has radius $A_1B = 1$. The next circle has radius $A_2B = \frac{1}{2}A_1B$, the one after that has radius $A_3B = \frac{1}{2}A_2B$, and so on. If these circles continue endlessly in this manner, what is the sum of the areas of all the circles?

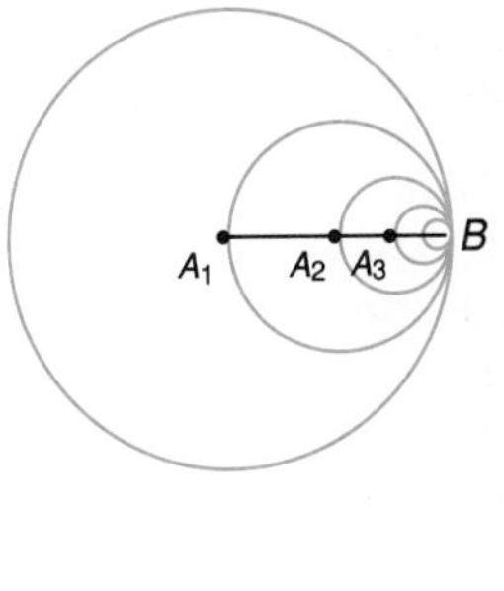

53. Triangle AB_1C_1 has a right angle at C_1. $AC_1 = 9$ and $B_1C_1 = 3$. Points $C_2, C_3, C_4, \ldots$ are chosen so that $AC_2 = \frac{2}{3}AC_1$, $AC_3 = \frac{2}{3}AC_2$, and so on. Find the sum of the areas of all the right triangles labeled AB_kC_k for $k = 1, 2, 3, \ldots$.

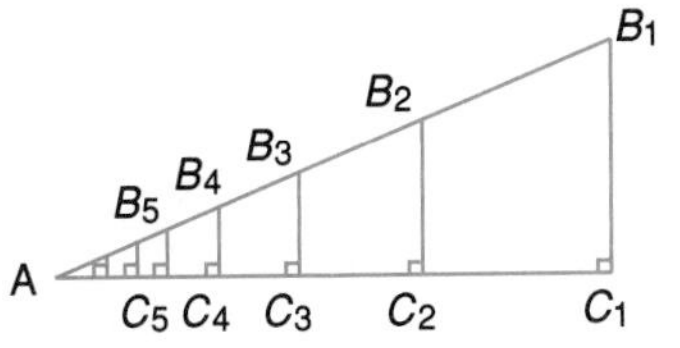

Written Assignment: Explain, in your own words, the meaning of an infinite geometric series. Under what conditions will such a series have a finite sum?

CHALLENGE
Think Creatively

Use the identity $\frac{1}{k(k+1)} = \frac{1}{k} - \frac{1}{k+1}$ to find the sum $\sum_{k=1}^{\infty} \frac{1}{k(k+1)}$. (*Hint:* Consider the sum $S_n = \sum_{k=1}^{n} \frac{1}{k(k+1)}$ as n gets arbitrarily large.)

EXPLORATIONS
Think Critically

1. Infinite sequences may or may not have an infinite number of distinct terms. Give examples of general terms for infinite sequences having the following number of distinct terms: (a) one, (b) two, (c) three.

2. If the addend $2k - 1$ in $\Sigma_{k=1}^{n} (2k - 1)$ is changed to $2k + 1$, then the value of the summation will remain the same provided that the proper change is made with the index of summation. Make this adjustment by completing the equation $\Sigma_{k=1}^{n} (2k - 1) = \Sigma\, (2k + 1)$. Now complete the equality $\Sigma_{k=3}^{n+1} (k + 1)^2 = \Sigma_{k=1}^{n-1}$ ______, in which the index of summation has been modified, by adjusting the addend.

3. In Exercises 30(b) and 31(b) on page 504 the sum of the first n odd integers and the sum of the first n even integers are called for. When the two answers are added, the result represents the sum of which consecutive integers beginning with 1? Use the formula for an arithmetic series to verify your result.

4. Which is larger, the sum of the first n powers of 2 starting with 2^0, or the single power 2^n? Justify your answer.

5. Observe that $\Sigma_{k=1}^{2n} [2^{k-1} + (-2)^{k-1}] = \Sigma_{k=1}^{2n} 2^{k-1} + \Sigma_{k=1}^{2n} (-2)^{k-1}$. Therefore, the series at the left can be evaluated by evaluating each series at the right and adding the results. Do this computation and use the answer to show that the series on the left can be written in the form $\Sigma_{k=1}^{n} a_1 r^{k-1}$.

6. Decide if the series $\Sigma_{k=1}^{\infty} \frac{3^{k+1}}{5^{k-1}}$ is a geometric series having ratio r, where $-1 < r < 1$. If it is, then find the sum. If it isn't, give a reason.

CHAPTER 11 SUMMARY

Review these key terms so that you are able to define or describe them. A clear understanding of these terms will be very helpful when reviewing the developments of this chapter.

A **sequence** is a function whose domain is a set of consecutive positive integers. For an **infinite sequence**, the domain consists of all the positive integers. The Greek letter Σ is used to express the sum of a sequence. For example, if we have a sequence of five terms, a_1, a_2, a_3, a_4, a_5, then the sum of this sequence is given by:

$$\sum_{k=1}^{5} a_k = a_1 + a_2 + a_3 + a_4 + a_5$$

For an **arithmetic sequence**, each term after the first is obtained from the preceding term by adding a common value. The nth term of an arithmetic sequence is

$$a_n = a_1 + (n - 1)d$$

where a_1 is the first term and d is the common difference. The sum of an arithmetic sequence is an **arithmetic series**:

$$\sum_{k=1}^{n} [a_1 + (k - 1)d] = \frac{n}{2}[2a_1 + n - 1)d]$$

For a **geometric sequence**, each term after the first is obtained by multiplying the preceding term by a common value. The nth term of a geometric series is

$$a_n = a_1 r^{n-1}$$

where a_1 is the first term and r is the common ratio. The sum of a geometric sequence is a **geometric series**:

$$\sum_{k=1}^{n} a_1 r^{k-1} = \frac{a_1(1 - r^n)}{1 - r}$$

The sum of an **infinite geometric series**: If $|r| < 1$, then $\sum_{k=1}^{\infty} a_1 r^{k-1} = \frac{a_1}{1 - r}$.

For other values of r, the series has no finite sum.

REVIEW EXERCISES

The solutions to the following exercises can be found within the text of Chapter 11. Try to answer each question before referring to the text.

Section 11.1

1. State the definition of a sequence.
2. Find the range values of the sequence $a_n = \frac{1}{n}$ for the domain {1, 2, 3, 4, 5,}.
3. List the first six terms of the sequence $s_k = \frac{(-1)^k}{k^2}$.
4. Find the tenth term of the sequence $a_n = \frac{n + 2}{2n - 3}$.
5. Write the first four terms of the sequence $a_n = \left(1 + \frac{1}{n}\right)^n$ and round off the terms to two decimal places.

Section 11.2

6. Find the sum of the first seven terms of $a_k = 2k$.
7. Find $\sum_{n=1}^{5} a_n$, where $a_n = \frac{n}{n + 1}$.
8. Evaluate $\sum_{k=1}^{4} (2k + 1)$.
9. Evaluate $\sum_{i=1}^{5} (-1)^i(i + 1)$.
10. Rewrite the series using the sigma notation: $3 + 6 + 9 + 12 + 15 + 18 + 21$.

Section 11.3

11. State the definition of an arithmetic sequence.
12. What is the nth term of an arithmetic sequence whose first term is a_1 and whose common difference is d?
13. Find the nth term of the arithmetic sequence 11, 2, −7, ...
14. Write the formula for the sum S_n of the arithmetic sequence $a_k = a_1 + (k - 1)d$.
15. The first term of an arithmetic sequence is −15 and the fifth term is 13. Find the fortieth term.
16. Find S_{20} for the arithmetic sequence with $a_1 = 3$ and $d = 5$.

17. Find the sum of the first 10,000 terms of the arithmetic sequence 246, 261, 276, ...
18. Find the sum of the first n positive integers.
19. Evaluate $\sum_{k=1}^{50} (-6k + 10)$.

Section 11.4

20. What is a geometric sequence?

Write the nth term of the geometric sequence.

21. 2, −4, 8, . . . 22. 5, −5, 5, . . . 23. 6, 9, $\frac{27}{2}$, . . .
24. Find the hundredth term of the geometric sequence having $r = \frac{1}{2}$ and $a_1 = \frac{1}{2}$.
25. Write the kth term of the geometric sequence $a_k = (\frac{1}{2})^{2k}$ in the form $a_1 r^{k-1}$ and find the values of a_1 and r.
26. A geometric sequence consisting of positive numbers has $a_1 = 18$ and $a_5 = \frac{32}{9}$. Find r.
27. Write the formula for the sum S_n of a geometric sequence $a_k = a_1 r^{k-1}$.
28. Find the sum of the first 100 terms of the geometric sequence $a_k = 6(\frac{1}{2})^{k-1}$.
29. Evaluate $\sum_{k=1}^{8} 3\left(\frac{1}{10}\right)^{k+1}$.
30. Suppose that you save \$128 in January and that each month thereafter you only manage to save half of what you saved the previous month. How much do you save in the tenth month? What are your total savings after 10 months?
31. A roll of wire is 625 feet long. If $\frac{1}{5}$ of the wire is cut off repeatedly, what is the general term of the sequence for the length of wire remaining? Use a calculator and the general term to determine the length of wire remaining after 7 cuts.

Section 11.5

32. For which values of r does $\sum_{k=1}^{\infty} a_1 r^{k-1} = \frac{a_1}{1 - r}$?
33. Find the sum of the infinite geometric series $27 + 3 + \frac{1}{3} + \cdots$.
34. Why does the infinite series $\sum_{k=1}^{\infty} 5(\frac{4}{3})^{k-1}$ have no finite sum?
35. Evaluate $\sum_{k=1}^{\infty} \frac{7}{10^{k+1}}$.
36. Express the repeating decimal 0.242424 . . . in rational form (the ratio of two integers).
37. A racehorse running at the constant rate of 30 miles per hour will finish a 1-mile race in 2 minutes. Now consider the race broken down into the following parts. Before the racehorse can finish the 1-mile race it must first reach the halfway mark; having done that, the horse must next reach the quarter pole; then it must reach the eighth pole; and so on. That is, it must always cover half the distance before it can cover the whole distance. Show that the sum of the infinite number of time intervals is also 2 minutes.

CHAPTER 11 TEST: STANDARD ANSWER

Use these questions to test your knowledge of the basic skills and concepts of Chapter 11. Then check your answers with those given at the back of the book.

1. Find the first four terms of the sequence given by $a_n = \frac{n^2}{6 - 5n}$.
2. Find the tenth term of the sequence in Question 1.

3. Write the first four terms of the sequence given by $a_n = (-1)^n + n$.

4. Find the tenth term of $a_i = \dfrac{(1 - i)^4}{(-3)^{i-1}}$.

5. Find the hundredth term of $a_n = \dfrac{n + 2}{3n^2 + 6n}$

6. Write the first five positive multiples of 9 and find a formula for the nth term.

In Questions 7 and 8, an arithmetic sequence has $a_1 = -3$ and $d = \frac{1}{2}$.

7. Find the forty-ninth term.

8. What is the sum of the first 20 terms?

9. Find the sum of the first 100 positive odd integers.

10. Evaluate: $\sum_{k=1}^{50} (-5k + 20)$

11. Rewrite the arithmetic series using sigma notation:
$-12 - 8 - 4 + 0 + 4 + \cdots + 36$

12. Write the next three terms of the geometric sequence $-768, 192, -48, \ldots$.

13. Write the nth term of the sequence in Question 12.

14. Find $\sum_{n=1}^{5} a_n$, where $a_n = \dfrac{(-2)^n}{n}$.

15. Find the sum of the first 50 positive multiples of 7.

16. Suppose the cost of a new car is \$9500 and that it depreciates 11% per year. Write the general term of the sequence that gives the value of the car at the end of the nth year.

17. Use the formula for the sum of a finite geometric sequence to show that

$$\sum_{k=1}^{4} 8\left(\tfrac{1}{2}\right)^k = \tfrac{15}{2}$$

18. Evaluate $\sum_{j=1}^{101} (4j - 50)$.

19. Evaluate $\sum_{k=1}^{8} 12\left(\frac{1}{2}\right)^{k-1}$.

Decide whether each of the given infinite geometric series has a sum. Find the sum if it exists; otherwise, give a reason why there is no sum.

20. $\sum_{k=1}^{\infty} 8\left(\frac{3}{4}\right)^{k+1}$

21. $1 + \frac{3}{2} + \frac{9}{4} + \cdots$

22. $0.06 - 0.009 + 0.00135 - \cdots$

23. Change the repeating decimal 0.363636 . . . into rational form.

24. Suppose you save \$10 one week and that each week thereafter you save 10¢ more than the week before. How much will you have saved after 1 year?

25. An object is moving along a straight line such that each minute it travels one-third as far as it did during the preceding minute. How far will the object have moved before coming to rest if it moves 24 feet during the first minute?

CHAPTER 11 TEST: MULTIPLE CHOICE

1. What is the tenth term of the sequence $1, \dfrac{1}{3}, \ldots, \dfrac{n + 2}{2n^2 + 3n - 2}, \ldots$?

(a) $\dfrac{1}{7}$ **(b)** $\dfrac{1}{19}$ **(c)** $\dfrac{3}{107}$ **(d)** $\dfrac{2}{201}$ **(e)** None of the preceding

2. List the first four terms of the sequence given by the formula $a_n = 1 + (-1)^n$.

(a) 0, 2, 0, 2 **(b)** 2, 0, 2, 0 **(c)** 0, −2, 0, −2 **(d)** −2, 0, −2, 0 **(e)** None of the preceding

3. Find the sum of the first fifteen terms of the sequence $s_k = -2k$.
(a) 180 (b) -30 (c) -240 (d) -480 (e) None of the preceding

4. Evaluate: $3[\sum_{n=1}^{4} (n^2 - n)]$.
(a) 20 (b) 30 (c) 60 (d) 90 (e) None of the preceding

5. Find $\sum_{n=1}^{4} a_n$ where $a_n = 3^{n-1}(n-1)$.
(a) 39 (b) 102 (c) 103 (d) 306 (e) None of the preceding

6. What is the fiftieth term of the arithmetic sequence having $a_1 = -2$ and $d = 5$?
(a) 243 (b) 245 (c) 248 (d) 252 (e) None of the preceding

7. Find S_{20} for the arithmetic sequence whose first term is $a_1 = 2$ and whose common difference is $d = -3$.
(a) 640 (b) 610 (c) -690 (d) -530 (e) None of the preceding

8. Find $\sum_{k=1}^{50} (-2k + 3)$.
(a) 2550 (b) 2500 (c) -2300 (d) -2400 (e) None of the preceding

9. A slow leak in a water pipe develops in such a way that the first day of the leak one ounce of water drips out. Each day thereafter the amount of water lost is one-half ounce more than the day before. How many ounces of water will leak out in 60 days?
(a) $30\frac{1}{2}$ (b) 915 (c) 945 (d) 960 (e) None of the preceding

10. What is the hundredth term of the geometric sequence having $r = \frac{1}{2}$ and $a_1 = -\frac{1}{2}$?
(a) 2^{-100} (b) $-\frac{1}{4^{99}}$ (c) $-\frac{1}{2^{100}}$ (d) $-\frac{1}{2^{98}}$
(e) None of the preceding

11. Find $\sum_{k=1}^{100} 3\left(\frac{1}{3}\right)^k$.
(a) $\frac{3}{2}\left(1 - \frac{1}{3^{100}}\right)$ (b) $1 - \frac{1}{3^{100}}$ (c) $\frac{2}{3}\left(1 - \frac{1}{3^{100}}\right)$ (d) $1 - \left(\frac{1}{3}\right)^{99}$
(e) None of the preceding

12. Suppose you save \$512 in January and then each month thereafter you save only half as much as you saved the preceding month. How much money will you have saved after one year?
(a) \$1023 (b) \$1023.25 (c) \$1023.50 (d) 1023.75 (e) None of the preceding

13. Find the sum of the infinite geometric sequence $-27, -9, -3, \ldots$.
(a) $\frac{81}{4}$ (b) $-\frac{81}{4}$ (c) $\frac{81}{2}$ (d) -81 (e) None of the preceding

14. Compute: $\sum_{n=1}^{\infty} 100(\frac{7}{100})^{n+1}$.
(a) $\frac{49}{93}$ (b) $\frac{490}{3}$ (c) $\frac{1000}{3}$ (d) No finite sum (e) None of the preceding

15. Compute: $\sum_{n=1}^{\infty} (-1)^n 2^n$.
(a) $-\frac{1}{2}$ (b) $\frac{1}{2}$ (c) 2 (d) No finite sum (e) None of the preceding

ANSWERS TO THE TEST YOUR UNDERSTANDING EXERCISES

Page 493

1. 3, 5, 7, 9, 11
2. $-2, -4, -6, -8, -10$
3. $0, -2, -4, -6, -8$
4. $-1, \frac{1}{2}, -\frac{1}{3}, \frac{1}{4}, -\frac{1}{5}$
5. $1, \frac{1}{4}, \frac{1}{9}, \frac{1}{16}, \frac{1}{25}$
6. $-\frac{3}{2}, -\frac{1}{2}, -\frac{1}{4}, -\frac{3}{20}, -\frac{1}{10}$
7. $3, \frac{1}{2}, \frac{1}{5}, \frac{3}{28}, \frac{1}{15}$
8. $\frac{1}{3}, \frac{1}{9}, \frac{1}{27}, \frac{1}{81}, \frac{1}{243}$
9. 2, 0, 2, 0, 2

Page 498

1. 60 2. 25 3. 40 4. 0 5. 50 6. $\frac{5}{16}$

Page 501

1. $5n$
2. $-4n + 10$
3. $\frac{1}{10}n$
4. $-8n + 3$
5. n
6. $n - 4$
7. $\frac{2}{3}n$
8. $-12n + 65$
9. $\frac{1}{5}n - \frac{1}{5}$
10. $n + 1$

Page 508

1. $1, \frac{1}{2}, \frac{1}{4}, \frac{1}{8}, \frac{1}{16}; 1(\frac{1}{2})^{n-1}; r = \frac{1}{2}$
2. $\frac{1}{4}, \frac{1}{8}, \frac{1}{16}, \frac{1}{32}, \frac{1}{64}; \frac{1}{4}(\frac{1}{2})^{n-1}; r = \frac{1}{2}$
3. $-\frac{1}{2}, \frac{1}{4}, -\frac{1}{8}, \frac{1}{16}, -\frac{1}{32}; -\frac{1}{2}(-\frac{1}{2})^{n-1}; r = -\frac{1}{2}$
4. $-\frac{1}{27}, \frac{1}{27^2}, -\frac{1}{27^3}, \frac{1}{27^4}, -\frac{1}{27^5}; -\frac{1}{27}\left(-\frac{1}{27}\right)^{n-1}; r = -\frac{1}{27}$
5. $r = 10; \frac{1}{5}(10)^{n-1}$
6. $r = -\frac{4}{9}; 27(-\frac{4}{9})^{n-1}$

Page 576

1. $r = \frac{1}{10}; S_\infty = 11\frac{1}{9}$
2. $r = 4$; no finite sum.
3. $r = -\frac{1}{6}; S_\infty = 30\frac{6}{7}$
4. $r = \frac{1}{4}; S_\infty = -21\frac{1}{3}$
5. $r = \frac{4}{3}$; no finite sum.
6. $r = 0.01; S_\infty = \frac{1}{33}$
7. $r = -3$; no finite sum.
8. $r = -\frac{9}{10}; S_\infty = \frac{810}{19}$
9. $r = -1.01$; no finite sum.

CHAPTER

12 PERMUTATIONS, COMBINATIONS, PROBABILITY

12.1 PERMUTATIONS

You will be able to answer this question using the concept of combinations in Section 12.2. See Exercise 15, page 538.

Suppose that there are 30 students in your class, and you all decide to become acquainted by shaking hands. Each person shakes hands with every other person. How many handshakes take place? Although this problem may not be an important or realistic one, it does suggest types of problems we may solve using counting procedures that will be studied in this chapter.

As a start, let us assume that you are planning a trip that will consist of visiting three cities, A, B, and C. You have your choice as to the order in which you are to visit the cities. How many different trips are possible? A trip begins with a stop at any one of the three cities, the second stop will be at any one of the remaining two cities, and the trip is completed by stopping at the remaining city. One way to answer this question is to sketch the possible trips, one of which is shown in the margin. However, it can get clumsy or tedious to find all trips using such diagrams, especially if more cities are involved.

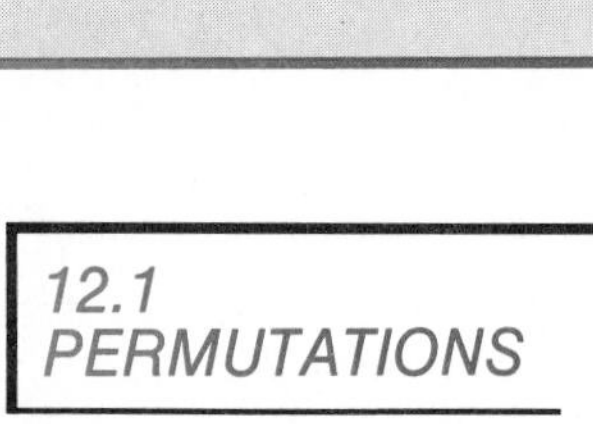

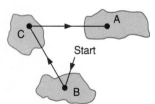

This map shows the trip where the first city is B, then C, and then A.

A better way to obtain the solution is to draw a *tree diagram* that illustrates all possible routes. From the diagram we can read the six possible trips. The arrangement ABC means that the trip begins with city A, goes to B, and ends at C. The other arrangements have similar interpretations.

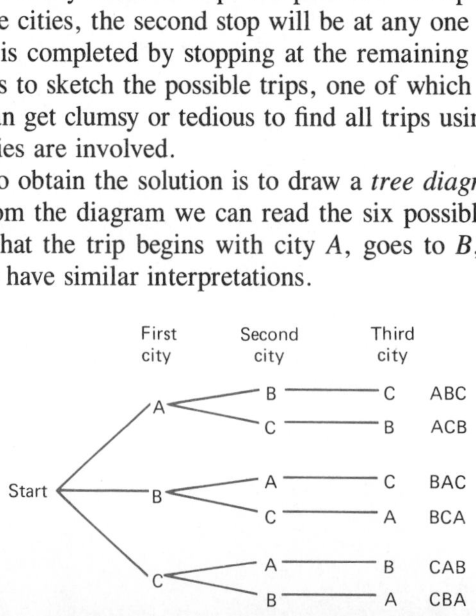

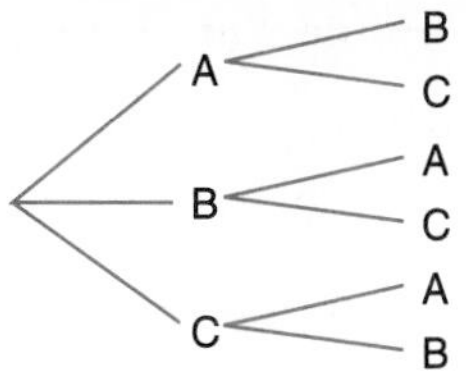

$3 \cdot 2 = 6$

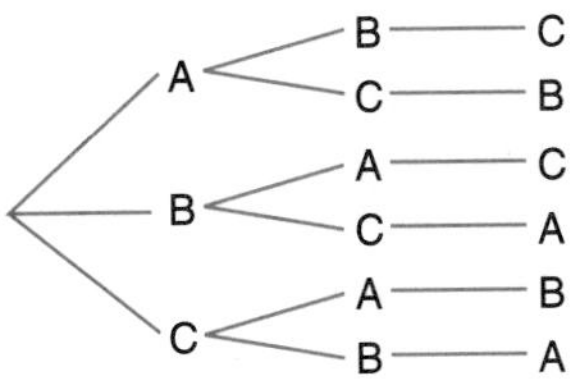

$3 \cdot 2 \cdot 1 = 6$

Tree diagrams can be useful in solving such counting problems when the number of possibilities is relatively small. However, when large numbers of possibilities are involved such tree diagrams are not practical. A more efficient method is needed. There is such a method whose underlying idea can be observed using the preceding tree diagram.

After the start, the tree branches into 3 parts, A, B, and C. Then, from each of these 3 points there are 2 new branches, giving 6 paths thus far. In other words, using the concept of multiplication, we have 3 groups, each containing 2 paths for a total of $3 \cdot 2 = 6$ possibilities for the first two cities. Then, the last choice consists of only the 1 remaining city. So now we have $3 \cdot 2 = 6$ groups, each of which contains 1 path, for a final total of $(3 \cdot 2) \cdot 1 = 6$ possible trips.

When the preceding observations are generalized we obtain the following important principle of counting.

FUNDAMENTAL PRINCIPLE OF COUNTING

Suppose that a first task can be completed in m_1 ways, a second task in m_2 ways, and so on, until we reach the kth task that can be done in m_k ways; then the total number of ways in which these tasks can be completed together is the product

$$m_1 m_2 \cdots m_k$$

The following three examples illustrate the use of this counting principle. Because of the large number of possibilities involved, these examples also demonstrate the advantage this principle has over the construction of tree diagrams.

We must assume here that no person may hold two offices at the same time.

EXAMPLE 1 A club consists of 15 boys and 20 girls. They wish to elect officers consisting of a girl as president and a boy as vice president. They also wish to elect a treasurer and a secretary who may be of either sex. How many sets of officers are possible?

Solution There are 20 choices for president and 15 choices for vice president. Thereafter, since two club members have been chosen, and the remaining positions can be filled by either a boy or girl, 33 members are left for the post of treasurer, and then 32 choices for secretary. Then by the fundamental principle of counting, the total number of choices is

$$20 \cdot 15 \cdot 33 \cdot 32 = 316{,}800$$

■

EXAMPLE 2

(a) How many three-digit whole numbers can be formed if zero is not an acceptable digit in the hundreds place and repetitions of digits are allowed?

(b) How many if repetitions are not allowed?

Solution

(a) Imagine that you must place a digit in each of three positions, as in the following display.

____ ____ ____ Digits to use:

↑

0 cannot be used here 0, 1, 2, 3, 4, 5, 6, 7, 8, 9

A tree diagram for part (b) would begin this way:

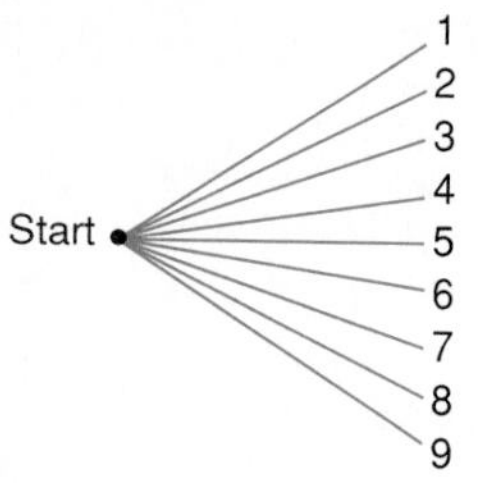

If the tree were continued what nine numbers would follow the 4? What eight numbers would follow the path 47?

There are 9 choices for the first position, and 10 for each of the others. By the fundamental principle of counting, the solution is

$$9 \cdot 10 \cdot 10 = 900.$$

(b) There are 9 choices for the hundreds place since zero is not allowed here. Once a choice is made, there are still 9 choices available for the tens digit since zero is permissible here. Finally, there are only 8 choices available for the units digits. The final solution is the product $9 \cdot 9 \cdot 8 = 648$. ■

In our first illustration where the six trips to the three cities were listed, the order of the elements A, B, C was crucial; trip ABC is different from trip BAC. We say that each of the six arrangements is a *permutation* of three objects taken three at a time. In general, for n elements (n a positive integer) and r a positive integer, $r \leq n$, we have this definition.

When considering permutations, the order of the elements is important. Thus 213 and 312 are two different permutations of the digits 1, 2, 3.

DEFINITION OF PERMUTATION

A **permutation** of n elements taken r at a time is an ordered arrangement, without repetitions, of r of the n elements. The number of permutations of n elements taken r at a time is denoted by ${}_nP_r$.

EXAMPLE 3 How many three-letter "words" composed from the 26 letters of the alphabet are possible? No duplication of letters is permitted.

Solution Since duplication of letters is *not* permitted, once a letter is chosen, it may not be selected again. Therefore, the first letter may be any one of the 26 letters of the alphabet, the second may be any one of the remaining 25, and the third is chosen from the remaining 24. Thus the total number of different "words" is $26 \cdot 25 \cdot 24 = 15{,}600$. Since we have taken 3 elements out of 26, without repetitions and in all possible orders, we may say that there are 15,600 permutations; that is

Note that a "word" here is to be interpreted as any collection of three letters. That is, "nlm" is considered to be a word in this example.

$${}_{26}P_3 = 26 \cdot 25 \cdot 24 = 15{,}600$$ ■

Note that ${}_{26}P_3 = 26 \cdot 25 \cdot 24$ has three factors beginning with 26 and with each successive factor decreasing by 1. In general, for ${}_nP_r$ there will be r factors beginning with n, as follows.

$${}_nP_r = n(n-1)(n-2)(n-3)\cdots[n-(r-1)]$$

Thus

$${}_nP_r = n(n-1)(n-2)(n-3)\cdots(n-r+1)$$

When applying this formula, begin with n and use a total of r factors successively decreasing by 1 as shown:

$n = 20$

$${}_{20}P_{8} = 20 \cdot 19 \cdot 18 \cdot 17 \cdot 16 \cdot 15 \cdot 14 \cdot 13$$

$r = 8$ factors successively decreasing by 1

A specific application of this formula occurs when $n = r$. In this case we have the permutation of n elements taken n at a time, and the product has n factors.

$$_nP_n = n(n-1)(n-2)(n-3)\cdots 3\cdot 2\cdot 1$$

In the notation n! the ! is not used as a typical exclamation mark. Rather, it means to multiply all the positive integers from n down to 1 as shown.

We may abbreviate this formula by using **factorial notation.**

$$_nP_n = n!$$

For example:

$$_3P_3 = 3! = 3\cdot 2\cdot 1 = 6$$

$$_4P_4 = 4! = 4\cdot 3\cdot 2\cdot 1 = 24$$

$$_5P_5 = 5! = 5\cdot 4\cdot 3\cdot 2\cdot 1 = 120$$

For future consistency we find it convenient to define 0! as equal to 1. That is

$$0! = 1$$

A tree diagram would begin in this way:

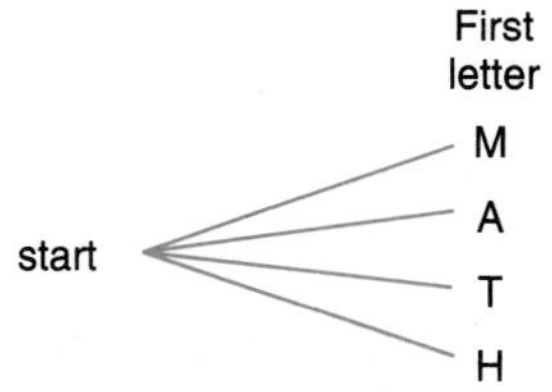

EXAMPLE 4 How many different ways can the four letters of the word MATH be arranged using each letter only once in each arrangement?

Solution Here we have the permutations of four elements taken four at a time. Thus $_4P_4 = 4! = 4\cdot 3\cdot 2\cdot 1 = 24$. This includes such arrangements as MATH, AMTH, TMAH, and HMAT. Can you list all 24 possibilities? Complete the tree diagram in the margin to help you find all possible cases. ■

TEST YOUR UNDERSTANDING
Think Carefully

Evaluate.

1. $_{10}P_4$ **2.** $_8P_3$ **3.** $_6P_6$ **4.** $\frac{10!}{8!}$ **5.** $\frac{12!}{9!\,3!}$

6. How many four-letter "words" from the 26 letters of the alphabet are possible? No duplication of letters is permitted.
7. Answer Exercise 6 if duplications are permitted.
8. How many different ways can the letters of the word EAT be arranged using each letter once in each arrangement? List all the possibilities.
9. Draw a tree diagram that shows all the three-digit whole numbers that can be formed using the digits 2, 5, 8 so that each of the digits is used once in each number.
10. How many four-digit whole numbers greater than 5000 can be formed using each of the digits 3, 5, 6, 8 once in each number?

(Answers: Page 559)

You have seen that $n!$ consists of n factors, beginning with n and successively decreasing to 1. At times it will be useful to display only some of the specific factors in $n!$ For example:

$$n! = n(n-1)! = n(n-1)(n-2)! = n(n-1)(n-2)(n-3)!$$

In particular,

$$5! = 5 \cdot 4! = 5 \cdot 4 \cdot 3! = 5 \cdot 4 \cdot 3 \cdot 2!$$

We can use the formula for ${}_nP_n$ to obtain a different form for ${}_nP_r$. To do so we write the formula for ${}_nP_r$ and then multiply numerator and denominator by $(n - r)!$ as follows.

$${}_nP_r = \frac{n(n-1)(n-2)\cdots[n-(r-1)]}{1} \cdot \frac{(n-r)!}{(n-r)!}$$

Now observe, after multiplying, that the numerator can be written as $n!$ to produce a useful formula for the permutation of n elements taken r at a time.

When applying these formulas there are three things to keep in mind. First, the n elements or objects must all be different. Second, no element is repeated more than once in a permutation. Third, order of the elements is important.

PERMUTATION FORMULAS

For n distinct elements taken r at a time, where $1 \le r \le n$:

$${}_nP_r = n(n-1)(n-2)\cdots(n-r+1)$$

$${}_nP_r = \frac{n!}{(n-r)!}$$

Illustrations:

Using the first formula, ${}_7P_4 = 7 \cdot 6 \cdot 5 \cdot 4 = 840$

Using the second formula, ${}_7P_4 = \frac{7!}{3!} = \frac{7 \cdot 6 \cdot 5 \cdot 4 \cdot \cancel{3!}}{\cancel{3!}} = 840$

EXAMPLE 5 A club contains 10 members. They wish to elect officers consisting of a president, vice-president, and secretary-treasurer. How many sets of officers are possible?

Solution We may think of the three offices to be filled in terms of a first office (president), a second office (vice-president), and a third office (secretary-treasurer). Therefore we need to select 3 out of 10 members and arrange them in all possible orders; we need to find the permutations of 10 elements taken 3 at a time.

$${}_{10}P_3 = \frac{10!}{7!} = \frac{10 \cdot 9 \cdot 8 \cdot \cancel{7!}}{\cancel{7!}} = 720$$

■

This example makes use of both permutations and the fundamental counting principle.

EXAMPLE 6 A family of 5 consisting of the parents and 3 children are going to be arranged in a row by a photographer. If the parents are to be next to each other, how many arrangements are possible?

Solution Suppose the parents occupy the first two positions and the children the last three. In this case we have the following:

$$\begin{pmatrix} \text{Parents occupy} \\ \text{these two positions} \\ \text{in } {}_2P_2 \text{ ways} \\ \underline{2} \cdot \underline{1} \end{pmatrix} \cdot \begin{pmatrix} \text{Children occupy} \\ \text{these three positions} \\ \text{in } {}_3P_3 \text{ ways} \\ \underline{3} \cdot \underline{2} \cdot \underline{1} \end{pmatrix}$$

These are the 4 adjacent positions for the parents.

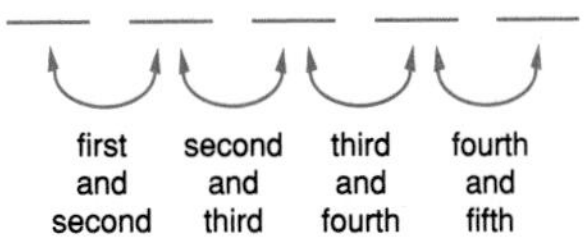

Since the parents can occupy the first two positions in 2 ways and the children occupy the remaining three positions in 6 ways, the fundamental counting principle gives $2 \cdot 6 = 12$ ways. But there are 4 adjacent positions for the parents possible, and each of these gives 12 arrangements. Therefore, the total number of arrangement is $4(12) = 48$. ■

Permutations can also be formed using collections of objects not all of which are distinct from one another. For example, in the word ALL the two L's are not distinguishable. However, it is easy to see that the letters A, L, L produce the 3 *distinguishable permutations* listed below at the left. If subscripts are used to distinguish the L's we get the $3! = 6$ permutations of the three letters A, L_1, L_2.

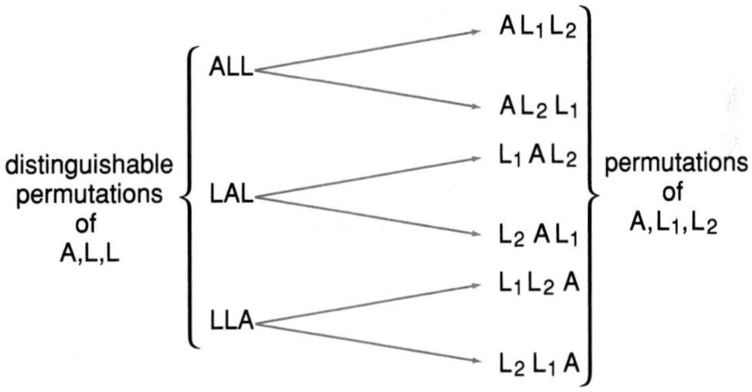

You can see that each permutation at the left produces 2 at the right by interchanging the positions of the L's, but keeping A fixed. If, for the moment, we assumed that we didn't know there were three permutations at the left, and used x for this unknown number, then the connection between the number of permutations in the two columns is given by $2x = 6$. Since $2! = 2$ is the number of ways that L_1, L_2 can occupy two positions, we have

$$2!\,x = 3! \quad \text{or} \quad x = \frac{3!}{2!} = 3$$

This type of reasoning can be extended to produce the following result:

> The number of distinguishable permutations of n objects of which n_1 are of one kind, n_2 are of another kind, . . . , n_k are of another kind is given by
>
> $$\frac{n!}{n_1!\,n_2!\cdots n_k!}$$

EXAMPLE 7 Find the number of distinguishable permutations using all of the letters in each word.
(a) BOOK **(b)** REFERRED **(c)** BEGINNING

These are the permutations for part (a):

BKOO	*OBKO*
BOKO	*OKBO*
BOOK	*OBOK*
KBOO	*OKOB*
KOBO	*OOBK*
KOOB	*OOKB*

Solution

(a) Use $n_1 = 2$, since there are two O's, and $n = 4$ to get $\dfrac{4!}{2!} = 12$

(b) Use $n_1 = 3$, $n_2 = 3$, and $n = 8$ to get $\dfrac{8!}{3!\,3!} = 1120$

(c) Use $n_1 = 2$, $n_2 = 2$, $n_3 = 3$, and $n = 9$ to get

$$\frac{9!}{2!\,2!\,3!} = \frac{9\cdot 8\cdot 7\cdot 6\cdot 5\cdot 4\cdot \cancel{3!}}{2\cdot 2\cdot \cancel{3!}}$$

$$= 9\cdot 8\cdot 7\cdot 6\cdot 5 = 15{,}120$$

■

EXERCISES 12.1

Evaluate.

1. $\dfrac{7!}{6!}$ 2. $\dfrac{12!}{10!}$ 3. $\dfrac{12!}{2!\,10!}$ 4. $\dfrac{15!}{10!\,5!}$

5. ${}_5P_4$ 6. ${}_5P_5$ 7. ${}_4P_1$ 8. ${}_8P_5$

9. Write ${}_nP_{n-3}$ in factorial notation.
10. Show that ${}_nP_n = {}_nP_{n-1}$.
11. How many ways can the manager of a baseball team select a pitcher and a catcher for a game if there are 5 pitchers and 3 catchers on the team?
12. How many different outfits can Laura wear if she is able to match any one of five blouses, four skirts, and three pairs of shoes?

Consider three-letter "words" to be formed by using the vowels a, e, i, o, and u.

13. How many different three-letter words can be formed if repetitions are not allowed?
14. How many different words can be formed if repetitions are allowed?
15. How many different words without repetitions can be formed whose middle letter is *o*?
16. How many different words without repetitions can be formed whose first letter is *e*?
17. How many different words without repetitions can be formed whose letters at the ends are *u* and *i*?
18. If repetitions are allowed, how many different words can be formed whose middle letter is *a*?
*19. How many different words can be formed containing the letter *a* and two other letters?
*20. How many different words can be formed containing the letters *a* and *e* so that these letters are not next to each other?

For Exercises 21–26, consider three-digit numbers to be formed using the digits 1, 2, 3, 4, 5, 6, 7, 8, and 9. Also assume that repetition of digits is not allowed unless specified otherwise.

21. How many three-digit whole numbers can be formed?
22. How many three-digit whole numbers can be formed if repetition of digits is allowed?
23. How many three-digit whole numbers can be formed that are even?
24. How many three-digit whole numbers can be formed that are divisible by 5?
25. How many three-digit whole numbers can be formed that are greater than 600?
26. How many three-digit whole numbers can be formed that are less than 400 and are divisible by 5?
27. **(a)** In how many different ways can the letters of STUDY be arranged using each letter only once in each arrangement?

(b) How many arrangements are there if the S and T are in the first two positions?

(c) How many arrangements are there if S and T are next to each other?

(d) How many arrangements are there if the S and T are not next to each other?

28. A class consists of 20 members. In how many different ways can the class select a set of officers consisting of a president, a vice-president, a secretary, and treasurer?

29. A baseball team consists of nine players. How many different batting orders are possible? How many are possible if the pitcher bats last? (Use a calculator for your computations.)

30. (a) Each question in a multiple-choice exam has the four choices indicated by the letters a, b, c, and d. If there are eight questions, how many ways can the test be answered? (*Hint:* Use the fundamental counting principle.)

(b) How many ways are there if no two consecutive answers can have the same answer?

31. When people are seated at a circular table, we consider only their positions relative to each other and are not concerned with the particular seat that a person occupies. How many arrangements are there for seven people to seat themselves around a circular table? (*Hint:* Consider one person's position as fixed.)

32. Review Exercise 31 and conjecture a formula for the number of different permutations of n distinct objects placed around a circle.

33. A license plate is formed by listing two letters of the alphabet followed by three digits. How many different license plates are possible:

(a) If repetitions of letters and digits are not allowed?

(b) If repetitions are allowed?

34. Write an expression that gives the number of arrangements of n objects taken r at a time if repetitions are allowed.

35. Solve for n: **(a)** ${}_nP_1 = 10$; **(b)** ${}_nP_2 = 12$.

36. Show that $2({}_nP_{n-2}) = {}_nP_{n-1}$.

37. A pair of dice is rolled. Each die has six faces on each of which is one of the numbers 1 through 6. One possible outcome is (3, 5), where the first digit shows the outcome of one die and the second digit shows the outcome of the other die. How many different outcomes are possible? List all possible outcomes as pairs of numbers (x, y).

38. To avoid electronic detection, a ship can send coded messages to neighboring ships by displaying a sequence of signal flags having different shapes. If twelve different shaped flags are available, how many messages can be displayed using a four-flag sequence?

39. (a) A social security number is a sequence of nine digits. How many different social security numbers are possible?

(b) How many are there in which no digits repeat?

(c) How many are there in which there are some repetitions of digits?

*(d) How many are there in which a digit appears exactly three times in succession and no other digits repeat?

40. A local telephone number consists of seven digits, the first three of which are called the telephone exchange, such as in

$$\underbrace{627}_{\text{(telephone exchange)}}\text{–}4195$$

(a) For a given exchange, how many different telephone numbers are possible?

(b) Suppose a city has 73,500 telephones. What is the minimum number of exchanges needed to accommodate the city's phones?

41. A student has room for six books on a shelf near her study area. The books consist of a dictionary and textbooks in the areas of chemistry, English, history, mathematics, and philosophy.

(a) In how many ways can the books be arranged on the shelf?

(b) In how many ways can they be arranged if the dictionary is put into the first position?

*__(c)__ How many arrangements are possible in which the mathematics and philosophy books are next to each other?

42. Solve for n:

(a) ${}_nP_6 = 15({}_nP_5)$ **(b)** ${}_nP_6 = 90({}_nP_4)$

*43. Solve for r: ${}_{12}P_r = 8({}_{12}P_{r-1})$

How many distinguishable permutations can be formed using all the letters in each word?

44. **(a)** SEVEN **(b)** INNING **(c)** ORDERED

45. **(a)** DELEGATE **(b)** COLLEGE **(c)** STATEMENTS

46. List the distinguishable permutations using all the letters in ERIE. (*Hint:* use a tree diagram.)

CHALLENGE
Think Creatively

How many ways can you choose the letters to form the word PYRAMID in the given diagram if

(a) each letter in the word PYRAMID can be any one of that particular letter listed?

(b) the letter being chosen, other than P, is below and to the immediate left or right of the preceding letter?

(c) The letters are chosen according to part (b) except that the D must be the middle D in the last row?

P
Y Y
R R R
A A A A
M M M M M
I I I I I I
D D D D D D D

12.2 COMBINATIONS

A permutation may be regarded as an *ordered* collection of elements. For example, a visit to cities A, B, and C in the order ABC is different from a visit in the order ACB. On the other hand, there are times when we need to consider situations where the order of the elements is not essential. For instance, suppose that on a mathematics test you are given the choice of answering any three of five given questions denoted by Q_1, Q_2, Q_3, Q_4, Q_5. If you choose questions Q_2, Q_3, Q_4, it makes no difference in which order you answer the three questions, so this is not a permutation. Rather, we say that the triple Q_2, Q_3, Q_4 is a *combination* of 3 things taken out of 5, according to the following definition, in which n and r are integers with $0 \le r \le n$.

Recall that a set A is a subset of a set B, provided that each element in A is also in B.

DEFINITION OF COMBINATION

A **combination** is a subset of r distinct elements selected out of n elements without regard to order. The number of combinations of n elements taken r at a time is denoted by either of these symbols:

$${}_nC_r \quad \text{or} \quad \binom{n}{r}$$

As an example, suppose that a class of 10 members wishes to elect a committee consisting of three of its members. These three members are not to be designated as holding any special office. Then a committee consisting of members David, Ellen, and Robert is the same regardless of the order in which they are selected. In other words, using D, E, and R as abbreviations for their names, the combination {D, E, R} gives rise to these six permutations:

DER   DRE   EDR   ERD   RDE   RED

Each combination of three members in this illustration actually gives rise to $3! = 6$ permutations. To find the number of possible committees we need to find the combinations of 10 elements taken 3 at a time. The *number* of such combinations is expressed as ${}_{10}C_3$ and is read as "the number of combinations of 10 elements taken 3 at a time." Since each of these combinations produces $3! = 6$ permutations, it follows that

The 120 combinations give rise to 720 permutations.

$$3!({}_{10}C_3) = {}_{10}P_3 \quad \text{or} \quad {}_{10}C_3 = \frac{{}_{10}P_3}{3!} = \frac{10 \cdot 9 \cdot 8}{3 \cdot 2 \cdot 1} = 120$$

In general, using ${}_nP_r = \dfrac{n!}{(n-r)!}$, we obtain

$$ {}_nC_r = \frac{{}_nP_r}{r!} = \frac{\dfrac{n!}{(n-r)!}}{r!} = \frac{n}{r!(n-r)!}$$

*When applying these formulas there are three things to keep in mind. First, the n elements or objects must all be different. Second, no element is used more than once in a combination. Third, order of the elements is **not** important.*

COMBINATION FORMULAS

For n elements taken r at a time, where $0 \le r \le n$:

$$ {}_nC_r = \frac{{}_nP_r}{r!}$$

$$ {}_nC_r = \frac{n!}{r!(n-r)!}$$

As noted, the symbol $\binom{n}{r}$ can be used in place of ${}_nC_r$. Thus $\binom{n}{0} = \dfrac{n!}{0!\,n!} = 1$; also $\binom{n}{n} = \dfrac{n!}{n!\,0!} = 1$. Can you prove that $\binom{n}{r} = \binom{n}{n-r}$?

Recall that we have defined $0! = 1$. Do you see here why this definition was made?

Illustrations:

(a) $$\binom{10}{2} = \frac{10!}{2!\,8!} = \frac{10 \cdot 9 \cdot \not{8!}}{2 \cdot \not{8!}} = 45$$

(b) $$ {}_{10}C_9 = \frac{10!}{9!\,1!} = \frac{10 \cdot \not{9!}}{\not{9!}} = 10$$

The question of order is the essential ingredient that determines whether a problem involves permutations or combinations. Examples 1 and 2 are illustrations of this distinction.

EXAMPLE 1 Using the digits 1 through 9, how many different four-digit whole numbers can be formed if repetition of digits is not allowed?

Solution Order is important here; thus 4923 is a different number from 9432. Therefore we need to find the *permutations* of nine elements taken four at a time.

$$ {}_9P_4 = \frac{9!}{5!} = 9 \cdot 8 \cdot 7 \cdot 6 = 3024 $$

■

EXAMPLE 2 A student has a penny, a nickel, a dime, a quarter, and a half-dollar and wishes to leave a tip consisting of exactly three coins. How many different amounts as tips are possible?

Here are four of the 10 *possible tips:*

$1 + 5 + 10 = 16$

$1 + 5 + 25 = 31$

$1 + 5 + 50 = 56$

$1 + 10 + 25 = 36$

Complete the list of 10 *possibilities.*

Solution Order is not important here; a tip of 5¢ + 10¢ + 25¢ is the same as one of 25¢ + 10¢ + 5¢. Therefore, we need to find the *combinations* of five things taken three at a time.

$$ \binom{5}{3} = {}_5C_3 = \frac{5!}{3!\,2!} = \frac{5 \cdot 4}{2} = 10 $$

■

The next example illustrates how the fundamental principle of counting is used in a problem involving combinations.

EXAMPLE 3 A class consists of 10 boys and 15 girls. How many committees of five can be selected if each committee is to consist of two boys and three girls?

Solution The order is not essential here since the committee members do not hold any special offices. Thus the problem involves combinations.

To select two boys: $\quad {}_{10}C_2 = \dfrac{10!}{2!\,8!} = 45$

To select three girls: $\quad {}_{15}C_3 = \dfrac{15!}{3!\,12!} = 455$

Since there are 45 pairs of boys possible, and since each of these pairs can be matched with any of the possible 455 triples of girls, the fundamental principle of counting gives

$$ 45 \cdot 455 = 20{,}475 $$

as the total number of committees that can be formed. ■

In the following example an ordinary deck of playing cards consisting of 52 different cards is used. These are divided into four suits: spades, hearts, diamonds, and clubs. There are 13 cards in each suit: from 1 (ace) through 10, jack, queen, and king.

EXAMPLE 4 A "poker hand" consists of 5 cards. How many different hands can be dealt from a deck of 52 cards?

If available, you should use a calculator to solve problems with extensive computations such as shown here. First simplify the fraction.

Solution The order of the 5 cards dealt is not important, so that this becomes a problem involving combinations rather than permutations. We wish to find the number of combinations of 52 elements taken 5 at a time.

$$_{52}C_5 = \frac{52!}{5!\,47!} = \frac{52 \cdot 51 \cdot 50 \cdot 49 \cdot 48 \cdot \cancel{47!}}{5 \cdot 4 \cdot 3 \cdot 2 \cdot 1 \cdot \cancel{47!}} = 2{,}598{,}960$$ ■

TEST YOUR UNDERSTANDING Think Carefully

Evaluate.

1. $_{10}C_4$ **2.** $_5C_5$ **3.** $_8C_0$ **4.** $\binom{12}{3}$ **5.** $\binom{8}{5}$

6. Show that $_{10}C_3 = {}_{10}C_7$.

7. How many different ways can a committee of 4 be selected from a group of 12 students?

8. A supermarket carries 6 brands of canned peas and 8 brands of canned corn. A shopper wants to try 2 different brands of peas and 3 different brands of corn. How many ways can the shopper select the 5 items?

9. How many lines are determined by eight points in a plane if no three points are on the same line?

10. How many triangles can be formed using five points in a plane no three of which are on the same line?

(Answers: Page 560)

Sometimes finding the number of possibilities can be done in more than one way. Example 5 illustrates a situation in which the number of possibilities can be found either by using the fundamental counting principle or by using combinations.

EXAMPLE 5 An ice cream parlor advertises that you may have your choice of five different toppings, and you may choose none, one, two, three, four, or all five toppings. How many choices are there in all?

Solution Here are two ways to approach this problem. From one point of view you may consider yourself on a cafeteria line with five stations. At each one you have two choices, to accept the topping or not to accept it. Thus, by the fundamental counting principle the total number of choices is $2 \cdot 2 \cdot 2 \cdot 2 \cdot 2 = 32$. From another point of view, the solution is the number of different ways that we can select none, one, two, three, four, or five elements from a total of five possibilities; that is,

Since, for example, choosing one topping or choosing two toppings are not done together, we add the number of possibilities rather than multiply.

$$\binom{5}{0} + \binom{5}{1} + \binom{5}{2} + \binom{5}{3} + \binom{5}{4} + \binom{5}{5}$$

Show that this sum is also equal to 32. ■

EXAMPLE 6 How many subsets containing 3 elements does set $S = \{a, b, c, d, e\}$ have? List these subsets.

Solution Since a subset is a combination, the number of required subsets is given by

$$_5C_3 = \frac{5!}{3!\,2!} = 10$$

The subsets are

How many times does each of the 5 elements of S appear in the 10 subsets?

$$\{a, b, c\},\ \{a, b, d\},\ \{a, b, e\},\ \{a, c, d\},\ \{a, c, e\}$$
$$\{a, d, e\},\ \{b, c, d\},\ \{b, c, e\},\ \{b, d, e\},\ \{c, d, e\}$$

■

EXAMPLE 7 A dinner party of 10 people arrives at a restaurant that has only two tables available. One table seats 6 and the other 4. If the seating arrangement at either table is not taken into account, how many ways can the 10 people divide themselves to be seated at these two tables?

Solution Each time 6 people sit at the table for 6, the remaining 4 will be at the other table. Therefore, it is necessary to find the number of ways we can select subsets of 6 out of 10 people. Thus

If the seating arrangements at each table are taken into account, how many ways can the people be seated? (See Exercise 41, page 540)

$$\binom{10}{6} = \frac{10!}{6!\,4!} = \frac{10 \cdot 9 \cdot 8 \cdot 7 \cdot \cancel{6!}}{\cancel{6!} \cdot 4 \cdot 3 \cdot 2 \cdot 1} = 210$$

The dinner party can split into the two tables in 210 ways. ■

EXERCISES 12.2

Evaluate.

1. $_5C_2$
2. $_{10}C_1$
3. $_{10}C_0$
4. $_4C_3$
5. $\binom{15}{15}$
6. $\binom{30}{3}$
7. $\binom{30}{27}$
8. $\binom{n}{3}$
9. A class consists of 20 members. In how many different ways can the class select **(a)** a committee of 4? **(b)** a set of 4 officers?
10. On a test a student must select 8 questions out of a total of 10. In how many different ways can this be done?
11. How many different ways can six people be split up into two teams of three each?
12. How many straight lines are determined by five points, no three of which are collinear?
13. There are 15 women on a basketball team. In how many different ways can a coach field a team of 5 players?
14. Answer Exercise 13 if two of the players can only play center and the others can play any of the remaining positions. (Assume that exactly one center is in a game at one time.)
15. Answer the question stated at the beginning of Section 12.1: How many handshakes take place when each person in a group of 30 shakes hands with every other person?
16. Box A contains 8 balls and box B contains 10 balls. In how many different ways can 5 balls be selected from these boxes if 2 are to be taken from box A and 3 from box B?

17. A class consists of 12 women and 10 men. A committee is to be selected consisting of 3 women and 4 men. How many different committees are possible?

18. Solve for n: **(a)** ${}_nC_1 = 6$; **(b)** ${}_nC_2 = 6$.

19. Convert to fraction form and simplify: **(a)** ${}_nC_{n-1}$; **(b)** ${}_nC_{n-2}$.

20. Prove: $\binom{n}{r} = \binom{n}{n-r}$.

21. Evaluate ${}_nC_4$ given that ${}_nP_4 = 1680$.

22. Solve for n: $5\binom{n}{2} = 2\binom{n+2}{2}$

23. Consider this expression: $\binom{n}{0} + \binom{n}{1} + \binom{n}{2} + \cdots + \binom{n}{n-1} + \binom{n}{n}$. Evaluate for:

(a) $n = 2$ **(b)** $n = 3$ **(c)** $n = 4$ **(d)** $n = 5$

24. Use the results of Exercise 23 and conjecture the value of the expression for any positive integer n.

*25. Explain the equality in Exercise 20 in terms of subsets.

26. Give subset interpretations of the results ${}_nC_0 = 1$ and and ${}_nC_n = 1$.

27. Interpret the result in Exercise 24 in terms of subsets of a set.

28. Ten points are marked on a circle. How many different triangles do these points determine so that the vertices of each triangle are marked points on the circle?

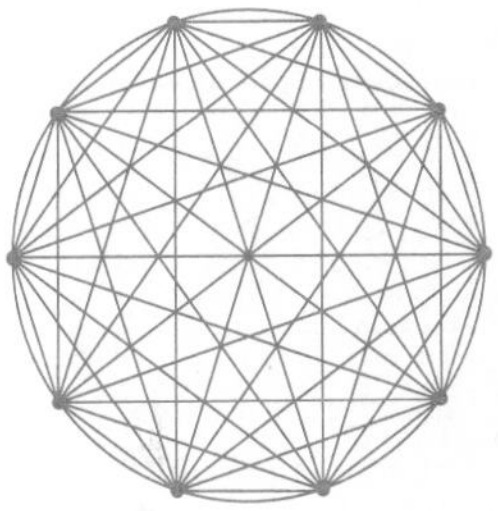

29. How many ways can 5-card hands be selected out of a deck of 52 cards so that all 5 cards are in the same suit?

30. How many ways can 5-card hands be selected out of a deck of 52 cards such that 4 of the cards have the same face value? (Four 10's and some fifth card is one such hand.)

31. How many ways can 5-card hands be selected out of a deck of 52 cards so that the 5 cards consist of a pair and three of a kind? (Two kings and three 7's is one such hand.)

32. A student wants to form a schedule consisting of 2 mathematics courses, 2 history courses, and 1 art course. The student can make these selections from 6 mathematics courses, 10 history courses, and 5 art courses. Assuming that there are no time conflicts, how many ways can the student select the 5 courses?

33. Suppose that in the U.S. Senate 25 Republicans and 19 Democrats are eligible for membership on a new committee. If this new committee is to consist of 9 senators, how many committees would be possible if the committee contained:

(a) 5 Republicans and 4 Democrats?

(b) 5 Democrats and 4 Republicans?

34. In Exercise 33, how many committees are possible if the chairperson of the committee is an eligible Republican senator appointed by the Vice-President, and the rest of the committee is evenly divided between Democrats and Republicans?

35. A basketball squad has 12 players consisting of 3 centers, 5 forwards, and 4 guards. How many ways can the coach field a team having 1 center, 2 forwards, and 2 guards?

36. The Green Lawn Tennis Club has scheduled a round-robin tennis tournament in which each player plays one match against every other player. If 12 players are signed up for the tournament, how many matches have been scheduled?

37. A college club has 18 members of whom 10 are men and 8 are women. The chairperson of the club is one of the men. A committee of 5 club members is to be formed that must include the chairperson of the club.

(a) How many committees can be formed consisting of 2 women and 3 men?

***(b)** How many committees can be formed containing no less than 1 woman?

38. Solve for n: ${}_nC_6 = 3({}_{n-2}C_4)$

*39. Suppose that a 9-member committee is to vote on an amendment. How many different ways can the votes be cast so that the amendment passes by a simple majority? (A simple majority means 5 or more yes votes.)

*40. In Exercise 39, how many different favorable ways can the votes be cast if the amendment needs at least a $\frac{2}{3}$ majority?

*41. If the dinner tables in Example 7 of this section are round, how many seating arrangements are possible if the arrangement at each table is taken into account?

Written Assignment: Use specific examples of your own to explain the distinction between a permutation and a combination of n elements taken r at a time.

CHALLENGE
Think Creatively

Prove: $\binom{n}{r-1} + \binom{n}{r} = \binom{n+1}{r}$

12.3 THE BINOMIAL EXPANSION

The factored form of the trinomial square $a^2 + 2ab + b^2$ is $(a + b)^2$. Turning this around, we say that the *expanded form* of $(a + b)^2$ is $a^2 + 2ab + b^2$. And if $(a + b)^2$ is multiplied by $a + b$, we get the expansion of $(a + b)^3$. Here is a list of the expansions of the first five powers of the binomial $a + b$.

You can verify these results by multiplying the expansion in each row by $a + b$ to get the expansion in the next row.

$$(a + b)^1 = a + b$$

$$(a + b)^2 = a^2 + 2ab + b^2$$

$$(a + b)^3 = a^3 + 3a^2b + 3ab^2 + b^3$$

$$(a + b)^4 = a^4 + 4a^3b + 6a^2b^2 + 4ab^3 + b^4$$

$$(a + b)^5 = a^5 + 5a^4b + 10a^3b^2 + 10a^2b^3 + 5ab^4 + b^5$$

Our objective here is to learn how to find such expansions directly without having to multiply. That is, we want to be able to expand $(a + b)^n$, especially for larger values of n, without having to multiply $a + b$ by itself repeatedly.

Let n represent a positive integer. As seen in the preceding display, each expansion begins with a^n and ends with b^n. Moreover, each expansion, has $n + 1$ terms that are all preceded by plus signs. Now look at the case for $n = 5$. Replace the first term a^5 by a^5b^0 and use a^0b^5 in place of b^5. Then

$$(a + b)^5 = a^5b^0 + 5a^4b + 10a^3b^2 + 10a^2b^3 + 5ab^4 + a^0b^5$$

In this form it becomes clear that (from left to right) the exponents of a successively decrease by 1, beginning with 5 and ending with zero. At the same time, the exponents of b increase from zero to 5. Also note that the sum of the exponents for each term is 5. Verify that similar patterns also hold for the other cases shown.

Using the preceding observations we would *expect* the expansion of $(a + b)^6$ to have seven terms that, except for the unknown coefficients, look like this:

$$a^6 + \_\_a^5b + \_\_a^4b^2 + \_\_a^3b^3 + \_\_a^2b^4 + \_\_ab^5 + b^6$$

Our list of expansions reveals that the second coefficient, as well as the coefficient of the next to the last term, is the number n. Filling in these coefficients for the case $n = 6$ gives

$$a^6 + 6a^5b + \_\_a^4b^2 + \_\_a^3b^3 + \_\_a^2b^4 + 6ab^5 + b^6$$

To get the remaining coefficients we return to the case $n = 5$ and learn how such coefficients can be generated. Look at the second and third terms.

$$\underbrace{\textcircled{5}a^{\textcircled{4}}b}_{\text{2nd term}} \qquad \underbrace{10a^3b^{\textcircled{2}}}_{\text{3rd term}}$$

If the exponent 4 of a in the *second* term is multiplied by the coefficient 5 of the *second* term and then divided by the exponent 2 of b in the *third* term, the result is 10, the coefficient of the third term.

coefficient of 2nd term → 5; exponent of a in 2nd term → (4)

$$\text{coefficient of third term} = \frac{5(4)}{2} = 10$$

↑ exponent of b in 3rd term (the denominator 2)

Verify that this procedure works for the next coefficient.

On the basis of the evidence we expect the missing coefficients for the case $n = 6$ to be obtainable in the same way. Here are the computations:

Use $\textcircled{6}\ a^{\textcircled{5}}b + \_\_a^4b^{\textcircled{2}}$: 3rd coefficient $= \dfrac{6(5)}{2} = 15$

Use $\textcircled{15}\ a^{\textcircled{4}}b^2 + \_\_a^3b^{\textcircled{3}}$: 4th coefficient $= \dfrac{15(4)}{3} = 20$

Use $\textcircled{20}\ a^{\textcircled{3}}b^3 + \_\_a^2b^{\textcircled{4}}$: 5th coefficient $= \dfrac{20(3)}{4} = 15$

You can verify that this equality is correct by multiplying the expansion for $(a + b)^5$ by $a + b$.

Finally, we may write the following expansion:

$$(a + b)^6 = a^6 + 6a^5b + 15a^4b^2 + 20a^3b^3 + 15a^2b^4 + 6ab^5 + b^6$$

More labor can be saved by observing the symmetry in the expansions of $(a + b)^n$. For instance, when $n = 6$ the coefficients around the middle term are symmetric. Similarly, when $n = 5$ the coefficients around the two middle terms are symmetric.

To get an expansion of the binomial $a - b$, write $a - b = a + (-b)$ and substitute into the previous form. For example, with $n = 6$,

$$\begin{aligned}(a - b)^6 = [a + (-b)]^6 &= a^6 + 6a^5(-b) + 15a^4(-b)^2 + 20a^3(-b)^3 \\ &\quad + 15a^2(-b)^4 + 6a(-b)^5 + (-b)^6 \\ &= a^6 - 6a^5b + 15a^4b^2 - 20a^3b^3 + 15a^2b^4 - 6ab^5 + b^6\end{aligned}$$

This result indicates that the expansion of $(a - b)^n$ is the same as the expansion of $(a + b)^n$ except that the signs alternate, beginning with plus.

EXAMPLE 1 Expand: **(a)** $(x + 2)^7$ **(b)** $(x - 2)^7$

Solution

(a) Let x and 2 play the roles of a and b in $(a + b)^7$, respectively.

$$(x + 2)^7 = x^7 + 7x^6 2 + \underline{\quad} x^5 2^2 + \underline{\quad} x^4 2^3 + \underline{\quad} x^3 2^4 + \underline{\qquad} x^2 2^5 + 7x 2^6 + 2^7$$

Now find the missing coefficients as follows:

$$\text{3rd coefficient} = \frac{7(6)}{2} = 21 = \text{6th coefficient}$$

$$\text{4th coefficient} = \frac{21(5)}{3} = 35 = \text{5th coefficient}$$

The completed expansion may now be given as follows:

$$\begin{aligned}(x + 2)^7 &= x^7 + 7x^6 2 + 21x^5 2^2 + 35x^4 2^3 + 35x^3 2^4 + 21x^2 2^5 + 7x 2^6 + 2^7 \\ &= x^7 + 14x^6 + 84x^5 + 280x^4 + 560x^3 + 672x^2 + 448x + 128\end{aligned}$$

(b) The expansion of $(x - 2)^7$ may be obtained from the expansion of $(x + 2)^7$ by alternating the signs. Thus,

$$(x - 2)^7 = x^7 - 14x^6 + 84x^5 - 280x^4 + 560x^3 - 672x^2 + 448x - 128 \quad ■$$

The preceding work can be used to obtain the expansion of the general form $(a + b)^n$. Begin by writing the variable parts of the first few terms.

$$a^n + \underline{\quad} a^{n-1}b^1 + \underline{\quad} a^{n-2}b^2 + \underline{\quad} a^{n-3}b^3 + \cdots$$

As before, to get the second coefficient multiply 1 by n and divide by 1.

$$a^n + \frac{n}{1}a^{n-1}b^1 + \underline{\quad} a^{n-2}b^2 + \underline{\quad} a^{n-3}b^3 + \cdots$$

To get the third coefficient multiply $\frac{n}{1}$ by $n - 1$ and divide by 2.

$$a^n + \frac{n}{1}a^{n-1}b^1 + \frac{n(n-1)}{1 \cdot 2}a^{n-2}b^2 + \underline{\quad} a^{n-3}b^3 + \cdots$$

The next coefficient is $\frac{n(n-1)}{1 \cdot 2}$ times $n - 2$ divided by 3, and we now have

$$a^n + \frac{n}{1}a^{n-1}b^1 + \frac{n(n-1)}{1 \cdot 2}a^{n-2}b^2 + \frac{n(n-1)(n-2)}{1 \cdot 2 \cdot 3}a^{n-3}b^3 + \cdots$$

Proceeding in this manner and noting the symmetry of the coefficients, we obtain the following result.

BINOMIAL FORMULA

$$(a + b)^n = a^n + \frac{n}{1}a^{n-1}b + \frac{n(n-1)}{1\cdot 2}a^{n-2}b^2$$
$$+ \frac{n(n-1)(n-2)}{1\cdot 2\cdot 3}a^{n-3}b^3 + \cdots + \frac{n}{1}ab^{n-1} + b^n$$

The term having the factor b^r, is the $(r + 1)$st term and can be written as

$$\frac{n(n-1)(n-2)\cdots(n-r+1)}{r!}a^{n-r}b^r$$

When expanding a binomial, you may find it easier to follow the steps that precede the statement of this formula rather than substitute directly into it.

EXAMPLE 2 Use the binomial formula to write the expansion of $(x + 2y)^4$.

Solution Use the formula with $a = x$, $b = 2y$ and $n = 4$. Then simplify.

$$(x + 2y)^4 = x^4 + \frac{4}{1}x^3(2y) + \frac{4\cdot 3}{1\cdot 2}x^2(2y)^2 + \frac{4\cdot 3\cdot 2}{1\cdot 2\cdot 3}x(2y)^3 + (2y)^4$$
$$= x^4 + 8x^3y + 24x^2y^2 + 32xy^3 + 16y^4$$

■

EXAMPLE 3 Expand $(2x - y)^5$ and simplify.

Solution Use $a = 2x$, $b = y$, and $n = 5$ in the binomial formula with alternating signs.

$$(2x - y)^5 = (2x)^5 - \frac{5}{1}(2x)^4y + \frac{5\cdot 4}{1\cdot 2}(2x)^3y^2 - \frac{5\cdot 4\cdot 3}{1\cdot 2\cdot 3}(2x)^2y^3$$
$$+ \frac{5\cdot 4\cdot 3\cdot 2}{1\cdot 2\cdot 3\cdot 4}(2x)y^4 - \frac{5\cdot 4\cdot 3\cdot 2\cdot 1}{1\cdot 2\cdot 3\cdot 4\cdot 5}y^5$$
$$= 32x^5 - 80x^4y + 80x^3y^2 - 40x^2y^3 + 10xy^4 - y^5$$

■

EXAMPLE 4 Evaluate 2^7 by expanding $(1 + 1)^7$.

Solution Since all powers of 1 equal 1, the expansion of $(1 + 1)^7$ is the sum of the coefficients in the expansion of $(a + b)^7$. Thus

$$2^7 = (1 + 1)^7 = 1 + 7 + 21 + 35 + 35 + 21 + 7 + 1 = 128$$

■

The coefficients in the expansion of $(a + b)^7$ were found in Example 1.

Another way to develop the binomial formula is to make use of our knowledge of combinations. Let us consider the expansion of $(a + b)^5$ again from a different point of view.

$$(a + b)^5 = \underbrace{(a + b)(a + b)(a + b)(a + b)(a + b)}_{\text{5 factors}}$$

To expand $(a + b)^5$, consider each term in the expansion as follows.

First term: Multiply all the a's together to obtain a^5.

Second term: We need to combine all terms of the form a^4b. How are these terms formed in the multiplication process that gives the expansion? One of these terms is formed by multiplying the a's in the first four factors times the b in the last factor.

$$(a + b)(a + b)(a + b)(a + b)(a + \boxed{b})$$

Product $= a^4b$

Another of these terms is formed like this:

$$(a + b)(a + b)(a + b)(a + \boxed{b})(a + b)$$

Product $= a^4b$

Now you can see that the number of such terms is the same as the number of ways we can select just one of the b's from the five factors.

This can be done in five ways, which can be expressed as ${}_5C_1$ or as $\binom{5}{1}$, the coefficient of a^4b.

Third term: Search for all terms of the form a^3b^2. The number of ways of selecting two b's from the five factors is ${}_5C_2$ or $\binom{5}{2}$.

Fourth term: The number of terms of the form a^2b^3 is the number of ways of selecting three b's from the five factors is ${}_5C_3$ or $\binom{5}{3}$.

Fifth term: The number of ways of selecting four b's from the five factors is ${}_5C_4$ or $\binom{5}{4}$, the coefficient of ab^4.

Sixth term: Multiply all the b's together to obtain b^5.

Thus we may write the expansion of $(a + b)^5$ in this form:

$$a^5 + \binom{5}{1}a^4b + \binom{5}{2}a^3b^2 + \binom{5}{3}a^2b^3 + \binom{5}{4}ab^4 + b^5$$

For consistency of form, we may write the coefficient of a^5 as $\binom{5}{0}$, and that of b^5 as $\binom{5}{5}$. In each case note that $\binom{5}{0} = \binom{5}{5} = 1$.

A similar argument can be made for each of the terms of the expansion of $(a + b)^n$. For example, to find the coefficient of the term that has the factors $a^{n-r}b^r$, we need to find the number of different ways of selecting r b's from n factors. This can be expressed as ${}_nC_r$ or $\binom{n}{r}$. Now we are ready to generalize and write this second form of the *binomial formula:*

This expansion can be written in sigma notation as

$$(a+b)^n = \sum_{r=0}^{n} \binom{n}{r} a^{n-r} b^r$$

$$(a+b)^n = \binom{n}{0}a^n b^0 + \binom{n}{1}a^{n-1}b^1 + \binom{n}{2}a^{n-2}b^2 + \cdots + \binom{n}{r}a^{n-r}b^r + \cdots + \binom{n}{n-1}a^1 b^{n-1} + \binom{n}{n}a^0 b^n$$

The numbers $\binom{n}{r}$ are referred to as ***binomial coefficients.***

Observe that the $(r+1)$st term here and in the formula on page 543 are the same:

$$\binom{n}{r}a^{n-r}b^r = \frac{{}_nP_r}{r!}a^{n-r}b^r = \frac{n(n-1)(n-2)\cdots(n-r+1)}{r!}a^{n-r}b^r$$

In the examples that follow you will also need to recall that in each term of the expansion the sum of the exponents is equal to n.

EXAMPLE 5 Find the sixth term in the expansion of $(a+b)^8$.

Solution Note that for any term in the expansion the exponent r of b is one less than the number of the term. Then, since we need the sixth term, $r = 5$. Also, since the sum of the exponents is 8, the sixth term is

$$\binom{8}{5} = \frac{8!}{5!\,3!} = 56$$

$$\binom{8}{5}a^3b^5 = 56a^3b^5$$ ■

Example 5 can also be solved by following the method used in the following example.

EXAMPLE 6 Find the fourth term in the expansion of $(x-2y)^{10}$.

Solution Use the general term $\binom{n}{r}a^{n-r}b^r$, which is the $(r+1)$st term. Then $r+1=4$ and $r=3$, $n=10$, and $n-r=7$. Then the fourth term is

Note that in Example 6 we may think of $(x-2y)^{10}$ as $[x+(-2y)]^{10}$ so that we may apply the binomial formula for $(a+b)^n$.

$$\binom{10}{3}x^7(-2y)^3 = 120x^7(-8y^3) = -960x^7y^3$$ ■

EXERCISES 12.3

Expand and simplify.

1. $(x+1)^5$
2. $(x-1)^6$
3. $(x+1)^7$
4. $(x-1)^8$
5. $(a-b)^4$
6. $(3x-2)^4$
7. $(3x-y)^5$
8. $(x+y)^8$
9. $(a^2+1)^5$
10. $(2+h)^9$
11. $(1-h)^{10}$
12. $(-2+x)^7$
13. $\left(\frac{1}{2}-a\right)^4$
14. $\left(\frac{x}{2}+\frac{2}{y}\right)^5$
15. $\left(\frac{1}{x}-x^2\right)^6$
16. $\left(2a-\frac{1}{a^2}\right)^6$

Simplify.

17. $\dfrac{(c + h)^3 - c^3}{h}$ 18. $\dfrac{(3 + h)^4 - 81}{h}$

19. Evaluate 2^{10} by expanding $(1 + 1)^{10}$.
20. Write the first five terms in the expansion of $(x + 1)^{15}$. What are the last five terms?
21. Write the first five terms and the last five terms in the expansion of $(c + h)^{20}$.
22. Write the first four terms and the last four terms in the expansion of $(a - 1)^{30}$.
23. Study this triangular array of numbers and discover the connection with the expansions of $(a + b)^n$, where $n = 1, 2, 3, 4, 5, 6$.

1 1
1 2 1
1 3 3 1
1 4 6 4 1
1 5 10 10 5 1
1 6 15 20 15 6 1

This triangular array of numbers is called **Pascal's triangle,** named after French mathematician Blaise Pascal (1623–1662). However, the triangle appeared in Chinese writings as early as 1303. How many properties of the triangle can you find? For example, find the sum of the entries in each row.

In Exercise 23 you learned that the nth row in Pascal's triangle contains the coefficients in the expansion of $(a + b)^n$. Use this result to expand the following.

24. $(x + 1)^5$ 25. $(a + 2)^6$ 26. $(2x - 3)^3$ 27. $(3p + 2q)^4$

28. Discover how the 6th row of the triangle in Exercise 23 can be obtained from the 5th row by studying the connection between the 4th and 5th rows indicated by this scheme.

1 + 4 + 6 + 4 + 1
1 5 10 10 5 1

29. Using the result of Exercise 28, write the 7th, 8th, 9th, and 10th rows of the triangle.
30. Use the 9th row found in Exercise 29 to expand $(x + h)^9$.
31. Use the 10th row found in Exercise 29 to expand $(x - h)^{10}$.
32. Why does the sum of all the numbers in one line of Pascal's triangle equal twice the sum of the numbers in the preceding line?
33. Find the sixth term in the expansion of $(a + 2b)^{10}$.
34. Find the fifth term in the expansion of $(2x - y)^8$.
35. Find the fourth term in the expansion of $\left(\dfrac{1}{x} + \sqrt{x}\right)^7$.
36. Find the eighth term in the expansion of $\left(\dfrac{a}{2} + \dfrac{b^2}{3}\right)^{10}$.
37. Find the term that contains x^5 in the expansion of $(2x + 3y)^8$.
38. Find the term that contains y^{10} in the expansion of $(x - 2y^2)^8$.
39. Write the middle term of the expansion of $\left(3a - \dfrac{b}{2}\right)^{10}$.
40. Write the last three terms of the expansion of $(a^2 - 2b^3)^7$.
41. Evaluate $(2.1)^4$ by expanding $(2 + 0.1)^4$.
42. Evaluate $(1.9)^4$ by expanding $(2 - 0.1)^4$.
43. Evaluate $(3.98)^3$ by expanding an appropriate binomial.
44. Evaluate $(1.2)^5$ by expanding an appropriate binomial.

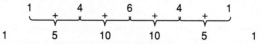

Written Assignment: Explain what Pascal's triangle has to do with the binomial expansion (see Exercise 23).

EXPLORATIONS Think Critically

1. True or False: The permutation formula ${}_nP_r = n(n - 1)(n - 2) \cdots (n - r + 1)$ is based directly on the fundamental principle of counting. Justify your answer.
2. From Exercise 27, page 539, the number of subsets of a set of n elements is 2^n. Arrive at this result by making use of the fundamental principle of counting. (*Hint:* Use a tree diagram applied to a set of three elements and generalize for n elements.)
3. In order to win a certain state lottery, a player must have chosen the same six integers that are selected at random from the integers 1 through 40 by the lottery agency. Are the number of groups of six numbers a matter of permutations or combinations? How many such groups are there?
4. What can be done to the formula ${}_nP_r = \dfrac{n!}{(n-r)!}$ in order to produce the formula for ${}_{n+1}P_r$? On the basis of the preceding observation explain why ${}_{n+1}P_r \geq {}_nP_r$.
5. For any positive integer n evaluate the expression

$$\binom{n}{0} - \binom{n}{1} + \binom{n}{2} - \binom{n}{3} + \cdots + (-1)^n\binom{n}{n}$$

On the basis of the preceding result, what can you conclude about the number of subsets formed from a set of n elements that have an even number of elements and the number of subsets that have an odd number of elements?

12.4 PROBABILITY

Concepts of probability are encountered frequently in daily life, such as a weather forecaster's statement that there is "a 20% chance of rain." Actually, the formal study of probability started in the seventeenth century when two famous mathematicians, Pascal and Fermat, considered the following problem that was posed to them by a gambler. Two people are involved in a game of chance and are forced to quit before either one has won. The number of points needed to win the game is known, and the number of points that each player has at the time is known. The problem was to determine how the stakes should be divided.

From this beginning mathematicians developed the theory of probability that has had far-reaching effects in many fields of endeavor. In this section we explore some basic aspects of probability, and the counting procedures studied earlier in this chapter will be useful in solving a variety of probability questions.

Let us begin by considering the situation of tossing two coins. What is the probability that both coins will be heads? We can approach this problem by forming a list of all possible outcomes. (A tree diagram is helpful in identifying all possibilities.)

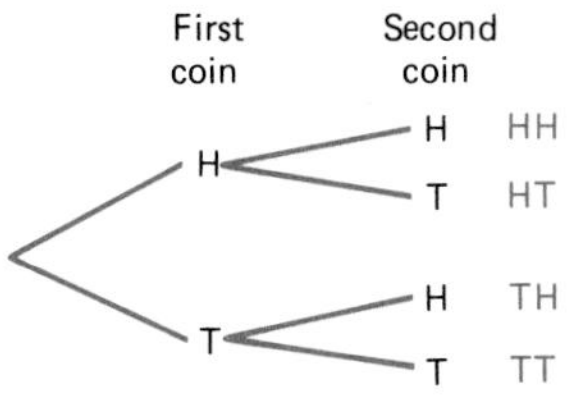

First coin	Second coin
Heads	Heads
Heads	Tails
Tails	Heads
Tails	Tails

Event	Probability
2 heads	$\frac{1}{4}$
1 head	$\frac{2}{4}$
0 heads	$\frac{1}{4}$

$P(\text{HH}) = \frac{1}{4}$ is read as "the probability of two heads is $\frac{1}{4}$."

Notice that there are four possible outcomes {HH, HT, TH, TT} and that only one of these gives the required two heads. Thus we say that the probability that both coins will be heads is $\frac{1}{4}$. In symbols we may write $P(\text{HH}) = \frac{1}{4}$. This reflects what happens *in the long run*. If we continue to toss two coins repeatedly, we would expect that on the average one out of four tosses will show two heads. Of course, there may be times when we toss double heads several times in a row; but if the experiment were to be repeated 1000 times, we can expect to have *about* 250 cases of double heads.

Let us assume that an experiment can have n different outcomes, each *equally likely*, and that s of these outcomes produce the event E. Then the **probability** that the event E will occur, $P(E)$, is given as

***Note:** This definition assumes that all events are equally likely to occur; that is, each event has the same chance of happening as any other event.*

DEFINITION OF PROBABILITY

$$P(E) = \frac{s}{n} = \frac{\text{number of successful outcomes}}{\text{total number of outcomes}}$$

EXAMPLE 1 A die is tossed. What is the probability of tossing a 5?

Solution There are six possible outcomes {1, 2, 3, 4, 5, 6}. Each has the same chance of occurring as the others. There is only one way to succeed, namely by tossing a 5. Thus $P(5) = \dfrac{s}{n} = \dfrac{1}{6}$. ■

EXAMPLE 2 To win the jackpot of a state lottery, the six numbers chosen by a person, from the numbers 1 through 54, must be the same as the six numbers selected at random by the state lottery system. On each lottery ticket purchased, there are two separate selections for the six numbers. What is the probability of winning the jackpot with one lottery ticket?

Solution Since the order of the six numbers chosen does not matter, the number of ways of selecting six numbers out of 54 is ${}_{54}C_6$.

$$\begin{aligned} {}_{54}C_6 &= \frac{54 \cdot 53 \cdot 52 \cdot 51 \cdot 50 \cdot 49 \cdot \cancel{48!}}{6 \cdot 5 \cdot 4 \cdot 3 \cdot 2 \cdot 1 \cdot \cancel{48!}} \\ &= 25{,}827{,}165 \end{aligned}$$

Then, since there are two chances on one ticket,

The chances of winning the jackpot is about 1 in 13 million; an extremely small possibility.

$$P(\text{jackpot}) = \frac{2}{25{,}827{,}165} \approx \frac{1}{12{,}913{,}583}$$ ■

EXAMPLE 3 A die is tossed. What is the probability of tossing a 7?

We can phrase the result of Example 3 in another way: $P(7 \text{ cannot occur}) = 1$.

Solution None of the outcomes is 7. Thus $P(7) = \frac{0}{6} = 0$. ■

From Example 3 we see that the probability of an event that *cannot* occur is 0. Furthermore, the probability for an event that will *always* occur is 1. For example, the probability of tossing a number less than 7 on a single toss of a die is $\frac{6}{6} = 1$ since all numbers are less than 7. This leads to the following observation for the probability that an event E will occur:

$$0 \le P(E) \le 1$$

As an extension of this idea, we note that every event will either occur or fail to occur. That is, $P(E) + P(\text{not } E) = 1$. Therefore,

$$P(\text{not } E) = 1 - P(E)$$

EXAMPLE 4 Two cards are drawn simultaneously from a deck of playing cards. What is the probability that both cards are not spades?

The advantage of the formula $P(\text{not } E) = 1 - P(E)$ is demonstrated by this example. It is much more difficult to solve otherwise. Try to explain this solution:

$$P(\text{not 2 spades}) = \frac{{}_{39}C_2 + {}_{39}C_1 \cdot {}_{13}C_1}{{}_{52}C_2} = \frac{16}{17}.$$

Solution Two cards can be selected out of 52 in ${}_{52}C_2$ ways. Also, since there are 13 spades, ${}_{13}C_2$ is the number of ways of selecting 2 spades. Then

$$P(\text{2 spades}) = \frac{{}_{13}C_2}{{}_{52}C_2} = \frac{1}{17}$$

Now use the formula $P(\text{not } E) = 1 - P(E)$.

$$P(\text{not 2 spades}) = 1 - P(\text{2 spades}) = 1 - \frac{1}{17} = \frac{16}{17}$$ ■

We need to be careful when adding probabilities since we may only do this when events are *mutually exclusive,* that is, when they cannot both happen at the same time. For example, consider the probability of drawing a king or a queen when a single card is drawn from a deck of cards.

$$P(\text{king}) = \frac{4}{52} \qquad P(\text{queen}) = \frac{4}{52}$$

$$P(\text{king or queen}) = \frac{4}{52} + \frac{4}{52} = \frac{8}{52} = \frac{2}{13}$$

This seems to agree with our intuition since there are 8 cards in a deck of 52 cards that meet the necessary conditions. These conditions are mutually exclusive because a card cannot be a king and a queen at the same time. Note, however, the difference in the conditions of Example 5.

For mutually exclusive events E or F,

$$P(E \text{ or } F) = P(E) + P(F)$$

EXAMPLE 5 A single card is drawn from a deck of cards. What is the probability that the card is either a queen or a spade?

*In Example 5 the events are **not** mutually exclusive because it is possible for a card to be both a queen and a spade.*

Solution The probability of drawing a queen is $\frac{4}{52}$, and the probability of drawing a spade is $\frac{13}{52}$. However there is one card that is being counted twice in these probabilities, namely the queen of spades. Thus we account for this as follows:

$$P(\text{queen or spade}) = \frac{13}{52} + \frac{4}{52} - \frac{1}{52} = \frac{16}{52} = \frac{4}{13}$$ ■

In general, for events E and F,

$$P(E \text{ or } F) = P(E) + P(F) - P(E \text{ and } F)$$

TEST YOUR UNDERSTANDING
Think Carefully

A pair of coins is tossed. Find the probability of each outcome.

1. Both are tails.
2. One is heads and one is tails.
3. At least one is heads.
4. Both are heads or both are tails.

A single die is tossed. Find the probability of each outcome.

5. An even number comes up.
6. A number less than 5 comes up.
7. An even number or a number greater than 3 comes up.

A single card is drawn from a deck of cards. Find the probability of each outcome.

8. A red card is drawn.
9. A heart or a spade is drawn.
10. An ace or a heart is drawn.
11. A picture card is drawn.
12. A picture card or an ace is drawn.

(Answers: Page 560)

We consider the picture cards to be the jacks, queens, and kings.

Sometimes a probability example can be solved in a variety of ways. For example, consider the probability of drawing two aces when two cards are selected from a deck of cards. In all such cases, unless stated otherwise, we shall assume that a card is drawn and *not* replaced in the deck before the second card is drawn.

$$\text{Probability that the first card is an ace} = \tfrac{4}{52}.$$

Assume that an ace was drawn. Then there are only 51 cards left in the deck, of which 3 are aces. Thus the probability that the second card is an ace $= \frac{3}{51}$.

We now make use of the following principle. Suppose that $P(E)$ represents the probability that an event E will occur, and $P(F \text{ given } E)$ is the probability that event F occurs after E has occurred. Then

$$P(E \text{ and } F) = P(E) \times P(F \text{ given } E)$$

Thus to find the probability that both cards are aces, we must multiply:

$$\frac{4}{52} \cdot \frac{3}{51} = \frac{1}{221}$$

This example can also be expressed through the use of combinations. Thus the total number of ways to select two cards from the deck is ${}_{52}C_2$. Furthermore, the number of ways of selecting two aces from the four aces in a deck of cards is ${}_4C_2$. Therefore,

Show that $\frac{{}_4C_2}{{}_{52}C_2} = \frac{1}{221}$.

$$\text{Probability of selecting two aces} = \frac{{}_4C_2}{{}_{52}C_2} = \frac{1}{221}$$

When each of two events can occur so that neither one affects the occurrence of the other, we say that the events are **independent.** For example, if E is the event that a head will come up when tossing a coin, and F is the event that a 4 comes up when rolling a die, then neither outcome affects the other and therefore they are independent events. Also, since $P(E) = \frac{1}{2}$ and $P(F) = \frac{1}{6}$, the probability of both events occurring is found by multiplying. Thus

This can also be found by noting that there are two outcomes when a coin is tossed and 6 when a die is rolled. Then there are $2 \cdot 6 = 12$ ways for both events to occur together. Since only 1 of these 12 consists of heads on the coin and 4 on the die, $P(E \text{ and } F) = \frac{1}{12}$.

$$P(E \text{ and } F) = \frac{1}{2} \cdot \frac{1}{6} = \frac{1}{12}$$

In general,

> For independent events E and F,
>
> $$P(E \text{ and } F) = P(E) \cdot P(F)$$

EXAMPLE 6 A single card is drawn from a deck of cards. It is then replaced and a second card is drawn. What is the probability that both cards are aces?

Solution Since the card is replaced after the first drawing, the two events are independent; neither outcome depends on the other. In each case the probability is $\frac{4}{52}$. The probability that both are aces is $\frac{4}{52} \cdot \frac{4}{52} = \frac{1}{169}$. ■

The final two illustrative examples demonstrate the use of combinations in conjunction with the fundamental counting principle to solve probability problems.

*In Example 7 we assume that we are to draw **exactly** two aces. A more difficult problem is to find the probability of drawing **at least** two aces. Can you solve that problem?*

EXAMPLE 7 Five cards are dealt from a deck of 52 cards. What is the probability that exactly two of the cards are aces?

Solution The number of ways of selecting two aces from the four aces in the deck is ${}_4C_2$. However, we also need to select three additional cards from the remaining 48 cards in the deck that are *not* aces; this can be done in ${}_{48}C_3$ ways. Thus, by the fundamental counting principle, the total number of ways of drawing two aces (and three other cards) is the product $({}_4C_2) \cdot ({}_{48}C_3)$.

$$\text{Probability of drawing two aces in five cards} = \frac{({}_4C_2) \cdot ({}_{48}C_3)}{{}_{52}C_5}$$

Show that this ratio reduces to $\frac{2162}{54{,}145} \approx 0.04$ ■

If the members are selected ***at random,*** *then each one has an equally likely chance of being selected. For example, a random selection could involve having all 18 names placed in a hat and 4 names drawn as in a lottery.*

EXAMPLE 8 A class consists of 10 men and 8 women. Four members are to be selected *at random* to represent the class. What is the probability that the selection will consist of two men and two women?

Solution Number of ways of selecting two men $= {}_{10}C_2$.
Number of ways of selecting two women $= {}_{8}C_2$.
Number of ways of selecting four members of the class $= {}_{18}C_4$.
Thus the probability that the committee will consist of two men and two women is given as

$$\frac{({}_{10}C_2)\cdot({}_{8}C_2)}{{}_{18}C_4} = \frac{7}{17}$$ ■

EXERCISES 12.4

Use the tree diagram below, showing the results of tossing three coins, to find the probability of each outcome.

1. All three coins are heads.
2. Exactly one coin is heads.
3. At least one coin is heads.
4. None of the coins is heads.
5. At most one coin is heads.
6. At least two coins are heads.

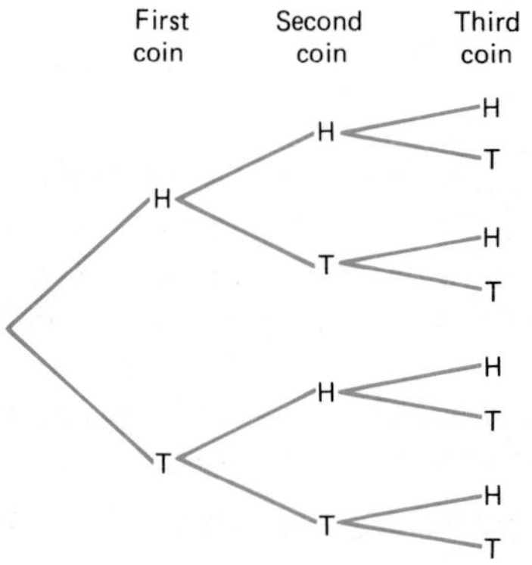

Assume that after spinning the pointer, its chance of stopping in any one region is just as likely as its chance of stopping in any of the others, and that it does not stop on a line. Find the probabilities.

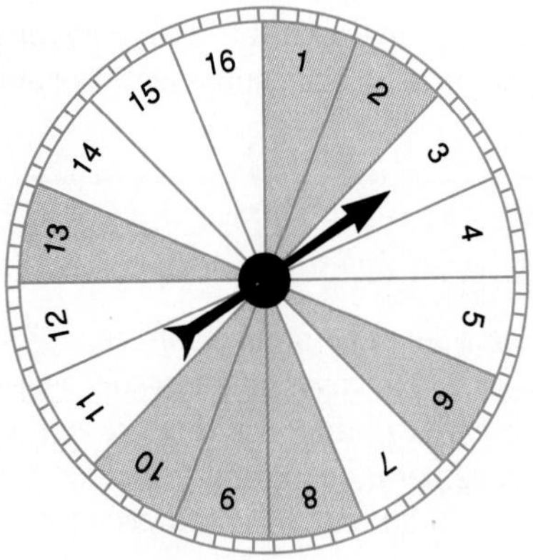

7. P(odd number)
8. P(multiple of 3)
9. P(white)
10. P(7 or 11)
11. P(even and red)
12. P(no more than 7)
13. P(prime)
14. P(red and between 3 and 10)

There are 36 different ways that a pair of dice can land, as shown in the diagram. Each outcome is shown as a pair of numbers showing the outcome on each die. The first number in each pair shows the outcome of tossing one die; the other number is the outcome of the second die. Use this diagram to find the probability of each outcome when a pair of dice are tossed.

(1, 1)	(1, 2)	(1, 3)	(1, 4)	(1, 5)	(1, 6)
(2, 1)	(2, 2)	(2, 3)	(2, 4)	(2, 5)	(2, 6)
(3, 1)	(3, 2)	(3, 3)	(3, 4)	(3, 5)	(3, 6)
(4, 1)	(4, 2)	(4, 3)	(4, 4)	(4, 5)	(4, 6)
(5, 1)	(5, 2)	(5, 3)	(5, 4)	(5, 5)	(5, 6)
(6, 1)	(6, 2)	(6, 3)	(6, 4)	(6, 5)	(6, 6)

15. Both dice show the same number.
16. The sum is 11.
17. The sum is 7.
18. The sum is 7 or 11.
19. The sum is 2, 3, or 12.
20. The sum is 6 or 8.
21. The sum is an odd number.
22. The sum is not 7.

Two cards are drawn from a deck of 52 playing cards without replacement. Find the probability of each outcome.

23. Both cards are red.
24. Both cards are spades.
25. Both cards are the ace of hearts.
*26. Both cards are the same suit.

(*Hint for Exercise* 26: A successful outcome is to have both cards spades *or* both hearts *or* both diamonds *or* both clubs. The sum of these probabilities gives the solution.)

Two cards are drawn from a deck of 52 playing cards with the first card replaced before the second card is drawn. Find the probability of each outcome.

27. Both cards are black or both are red.
28. Both cards are hearts.
29. Both cards are the ace of hearts.
30. Both cards are picture cards.
31. Neither card is an ace.
32. Neither card is a spade.
33. The first card is an ace and the second card is a king.
34. The first card is an ace and the second card is not an ace.

A bag of marbles contains 8 red marbles and 5 green marbles. Three marbles are drawn at random at the same time. Find the probability of each outcome.

35. All are red.
36. All are green.
37. Two are red and one is green.
38. One is red and two are green.
39. A student takes a true-false test consisting of 10 questions by just guessing at each answer. Find the probability that the student's score will be:
    (a) 100% **(b)** 0% **(c)** 80% or better
40. A die is tossed three times in succession. Find the probability that:
    (a) All three tosses show 5. **(b)** Exactly one of the tosses shows 5.
    (c) At least one of the tosses shows 5.

In Exercises 41 and 42 use $P(\text{not } E) = 1 - P(E)$.

41. Two cards are drawn simultaneously from a deck of playing cards. Find P(not two red cards).
*42. Three cards are drawn simultaneously from a deck of playing cards. Find P(not three hearts).

43. Five cards are dealt from a deck of 52 cards. Find the probability of obtaining:
   (a) Four aces
   (b) Four of a kind (that is, four aces or four twos or four threes, etc.).
   (c) A flush (that is, all five cards of the same suit).
   *(d) A royal straight flush (10, jack, queen, king, ace, all in the same suit.)

*44. You are given 10 red marbles and 10 green marbles to distribute into two boxes. You will then be blindfolded and asked to select one of the boxes and then draw one marble from that box. You win \$10,000 if the marble you select is red. Finally, you are allowed to distribute the 20 marbles into the two boxes in any way that you wish before you begin to make the selection. Try to determine what is the best strategy for distributing the marbles; that is, how many of each color should you place in each box?

Mathematical expectation is defined as the product of the probability that an event will occur and the amount to be received if it does occur. For example, if you are to receive \$12 if you toss a 6 on one throw of a die, your expectation is $\frac{1}{6}(\$12) = \$2$. *What is your expectation for each of the events described in Exercises 45–48?*

45. You toss a coin twice and receive \$20 if both tosses show heads.

46. You select a card from a deck of cards and receive \$26 if it is an ace or a king.

47. You toss two coins. If both show heads, you receive \$20; if only one coin shows heads, you receive \$10; and you receive nothing if both coins are tails.

*48. You throw a pair of dice. If the sum of the outcomes is 7, you receive \$30; if the sum is more than 7, you receive \$12; and you pay \$24 (you receive minus \$24) if the sum is less than 7.

The odds in favor of an event occurring is the ratio of the probability that it will occur to the probability that it will not occur. For example, the odds in favor of tossing two heads in two tosses of a coin are $\frac{1/4}{3/4} = \frac{1}{3}$ *or 1 to 3. The odds against this event are* 3 *to* 1. *Exercises 49–56 involve computation of odds.*

49. A card is drawn from a deck of cards. What are the odds in favor of obtaining an ace? What are the odds against this event occurring?

50. What are the odds in favor of tossing three successive heads with a coin?

51. What are the odds against tossing a 7 or an 11 in a single throw of a pair of dice?

52. Two cards are drawn simultaneously from a deck of cards. What are the odds in favor of both cards being hearts?

53. What are the odds for getting 2 heads when tossing 3 coins?

54. Suppose you have four playing cards in your hand whose face values are 5, 6, 7, 8. What are the odds of selecting a fifth card from the remaining 48 that has a face value of either 4 or 9?

*55. Show that if the odds for an event E are a to b, then $P(E) = \frac{a}{a+b}$.

*56. Suppose you are given 1 to 3 odds that event E occurs and 3 to 5 odds that event F occurs. If E and F cannot occur together, what are the odds that E or F will occur?

Written Assignment: Explain, in your own words and with an example, what is meant by $P(E)$, the probability that an event E will occur.

CHALLENGE
Think Creatively

Three cards are placed in a hat. One is red on each side, one is green on each side, and the third is red on one side and green on the other side. A card is drawn at random and placed on the table. Assume that the color showing is red. What is the probability that the other side is also red? Note that the answer, surprisingly, is *not* $\frac{1}{2}$. (*Hint:* Consider making a list of all possibilities using R_1, R_2, and R_3 for the red sides and G_1, G_2, and G_3 for the green sides.)

CHAPTER 12 SUMMARY

Review these key terms and concepts so that you are able to define or describe them. A clear understanding of these will be very helpful when reviewing the developments of this chapter.

Fundamental Principle of Counting

If a first task can be completed in m_1 ways, a second task in m_2 ways, and the kth task in m_k ways, then the total number of ways in which these tasks can be completed together is the product $m_1 m_2 \cdots m_k$.

Factorial Notation

$$n! = n(n-1)(n-2)(n-3)\cdots 3\cdot 2\cdot 1, \qquad 0! = 1$$

A **permutation** of n elements taken r at a time is an ordered arrangement, without repetition, of r of the n elements. The number of such permutations is denoted by ${}_nP_r$:

$${}_nP_r = n(n-1)(n-2)\cdots(n-r+1) = \frac{n!}{(n-r)!}$$

A **combination** is a subset of r distinct elements selected out of n elements without regard to order. The number of such combinations is denoted by ${}_nC_r$ or $\binom{n}{r}$:

$${}_nC_r = \frac{{}_nP_r}{r!} = \frac{n!}{r!(n-r)!}$$

Binomial Formula:

$$(a+b)^n = \binom{n}{0}a^n b^0 + \binom{n}{1}a^{n-1}b^1 + \binom{n}{2}a^{n-2}b^2 + \cdots \binom{n}{r}a^{n-r}b^r + \cdots + \binom{n}{n-1}a^1 b^{n-1} + \binom{n}{n}a^0 b^n = \sum_{r=0}^{n}\binom{n}{r}a^{n-r}b^r$$

Each coefficient $\binom{n}{r}$ is called a **binomial coefficient** and is the same as ${}_nC_r$.

The **probability** that an event E will occur is

$$P(E) = \frac{s}{n} = \frac{\text{number of successful outcomes}}{\text{total number of outcomes}}$$

For an event E, $0 \le P(E) \le 1$ and $P(\text{not } E) = 1 - P(E)$.
For events E and F, $P(E \text{ or } F) = P(E) + P(F) - P(E \text{ and } F)$.
For **mutually exclusive events** E or F, $P(E \text{ or } F) = P(E) + P(F)$.
For **independent events** E and F, $P(E \text{ and } F) = P(E)\cdot P(F)$.

REVIEW EXERCISES

The solutions to the following exercises can be found within the text of Chapter 12. Try to answer each question before referring to the text.

Section 12.1

1. You are planning a trip that will consist of visiting three cities, A, B, and C. You have your choice as to the order in which you are to visit the cities. List all possible different trips.
2. A club consists of 15 boys and 20 girls. They wish to elect officers consisting of a girl as president and a boy as vice-president. They also wish to elect a treasurer and a secretary who may be of either sex. How many sets of officers are possible?
3. How many three-digit whole numbers can be formed if zero is not an acceptable digit in the hundreds place and repetitions are allowed?
4. Repeat Exercise 3 if repetitions are not allowed.
5. How many three-letter "words" composed from the 26 letters of the alphabet are possible? No duplication of letters is permitted.
6. How many different ways can the four letters of the word MATH be arranged using each letter only once in each arrangement?
7. Evaluate ${}_7P_4$ by using each of the formulas for ${}_nP_r$.
8. A club contains 10 members. They wish to elect officers consisting of a president, vice-president, and secretary-treasurer. How many sets of officers are possible?
9. A family of 5 consisting of the parents and 3 children are going to be arranged in a row by a photographer. If the parents are to be next to each other, how many arrangements are possible?

Section 12.2

10. Evaluate:

    (a) $\binom{10}{2}$ **(b)** ${}_{10}C_9$.
11. Using the digits 1 through 9, how many different four-digit whole numbers can be formed if repetition of digits is not allowed?
12. A student has a penny, a nickel, a dime, a quarter, and a half-dollar and wishes to leave a tip consisting of exactly three coins. How many different amounts as tips are possible?
13. A class consists of 10 boys and 15 girls. How many committees of five can be selected if each committee is to consist of two boys and three girls?
14. A "poker hand" consists of 5 cards. How many different hands can be dealt from a deck of 52 cards?
15. An ice cream parlor advertises that you may have your choice of five different toppings, and you may choose none, one, two, three, four, or all five toppings. How many choices are there in all?
16. How many subsets containing 3 elements does set $S = \{a, b, c, d, e\}$ have? List these subsets.
17. How many ways can a dinner party of 10 people divide themselves into two tables if one table seats 6 and the other 4, and arrangements at each table are not taken into account.

Section 12.3

18. Write the expansion of $(x + 2)^7$ and of $(x - 2)^7$.
19. Write the expansion of $(x + 2y)^4$.
20. Expand $(2x - y)^5$ and simplify.

21. Evaluate 2^7 by expanding $(1 + 1)^7$. 22. Expand: $(a + b)^5$.

23. Write the $(r + 1)$st term in the expansion of $(a + b)^n$.

24. Find the sixth term in the expansion of $(a + b)^8$.

25. Find the fourth term in the expansion of $(x - 2y)^{10}$.

Section 12.4

26. A die is tossed.
    (a) What is the probability of tossing a 5?
    (b) What is the probability of tossing a 7?

27. The jackpot in a state lottery can be won if the six numbers, from 1 through 54, chosen by a person, match the six numbers selected at random by the lottery system. On each lottery ticket two separate selections of the six numbers are made. What is the probability of winning the jackpot with one lottery ticket?

28. Two cards are drawn simultaneously from a deck of cards. What is the probability that both cards are not spades?

29. A single card is drawn from a deck of cards. What is the probability that the card is either a queen or a king?

30. A single card is drawn from a deck of cards. What is probability that the card is either a queen or a spade?

31. What is the probability of drawing two aces when two cards are selected simultaneously from a deck of cards?

32. A single card is drawn from a deck of cards. It is then replaced and a second card is drawn. What is the probability that both cards are aces?

33. Five cards are dealt from a deck of 52 cards. What is the probability that exactly two of the cards are aces?

34. A class consists of 10 men and 8 women. Four members are to be selected at random to represent the class. What is the probability that the selection will consist of two men and two women?

CHAPTER 12 TEST: STANDARD ANSWER

Use these questions to test your knowledge of the basic skills and concepts of Chapter 12. Then check your answers with those given at the back of the book.

1. Evaluate: **(a)** ${}_{10}P_3$ **(b)** ${}_{10}C_3$

2. How many different ways can the letters of the word TODAY be arranged so that each arrangement uses each of the letters once?

3. How many three-digit whole numbers can be formed if zero is not an acceptable hundreds digit?
   (a) Repetitions are allowed. **(b)** Repetitions are not allowed.

4. Write ${}_nP_{n-5}$ in factorial notation.

5. How many three-digit odd whole numbers can be formed if repetition of digits is allowed?

6. In how many different ways can a class of 15 students select a committee of three students?

7. A class consists of 12 boys and 10 girls. How many committees of four can be selected if each committee is to consist of two boys and two girls?

8. How many different four-letter "words" can be formed from the letters m, o, n, d, a, y if no letter may be used more than once?

9. Evaluate: **(a)** $\binom{10}{5}$ **(b)** $\binom{n}{4}$

10. Solve for n: ${}_nC_2 = 55$

11. A student takes a five-question true–false test just by guessing at each answer.
    (a) How many different sets of answers are possible?
    (b) What is the probability of obtaining a score of 100% on the test?
    (c) What is the probability that the score will be 0%?

12. A single die is tossed. What is the probability that it will show an even number or a number less than 3?

13. A pair of dice is tossed. What is the probability that the sum will be:
    (a) 12 **(c)** Not 12 **(c)** Less than 5?

14. Two cards are drawn from a deck of 52 playing cards. What is the probability that they are both aces or both kings if:
    (a) The first card is replaced before the second card is drawn?
    (b) No replacements are made?

15. Five cards are dealt from a deck of 52 playing cards. What is the probability that exactly four of the cards are picture cards?

16. A box contains 20 red chips and 15 green chips. Five chips are selected from the box at random. What is the probability that exactly four of the chips will be red?

17. Four coins are tossed. Find the probability that:
    (a) All four coins will land showing heads.
    (b) None of the coins will show heads.

18. Two integers from 0 through 10 are selected simultaneously and at random. What is the probability that they will both be odd?

19. Box A contains 5 green and 3 red chips. Box B contains 4 green and 6 red chips. You are to select one chip from each box. What is the probability that both chips will be red?

20. In Exercise 19, suppose that you are blindfolded and told to select just one chip from one of the boxes. What is the probability that the chip you select will be red?

21. Expand: $(x - 2y)^5$.

22. Write the first four terms of the expansion of $\left(\dfrac{1}{2a} - b\right)^{10}$.

23. Write the seventh term in the expansion of $(3a + b)^{11}$.

24. Write the middle term in the expansion of $(2x - y)^{16}$.

25. Evaluate $(3.1)^4$ by expanding $(3 + 0.1)^4$.

CHAPTER 12 TEST: MULTIPLE CHOICE

1. How many four-digit whole numbers can be formed if zero is not an acceptable digit in the thousands place and repetition of digits is not allowed?
   (a) 3360 **(b)** 4032 **(c)** 4536 **(d)** 5040 **(e)** None of the preceding

2. Which of the following are correct?

   I. ${}_nP_r = \dfrac{n!}{(n-r)!}$ **II.** ${}_nP_n = n!$ **III.** ${}_nC_r = \dfrac{n!}{r!(n-r)!}$

   (a) Only I **(b)** Only II **(c)** Only III **(d)** I, II, and III **(e)** None of the preceding

3. How many different ways can the five letters of the word EIGHT be arranged using each letter only once in each arrangement?
   (a) ${}_8P_5$ **(b)** ${}_5C_5$ **(c)** $5!$ **(d)** 5^5 **(e)** None of the preceding

4. A group consists of 5 boys and 6 girls. How many committees of five can be selected if each committee is to consist of 2 boys and 3 girls?

(a) 150 **(b)** 200 **(c)** 1800 **(d)** 2400 **(e)** None of the preceding

5. Solve for n: ${}_nP_2 = 56$.

(a) 7 **(b)** 8 **(c)** 14 **(d)** Not enough information is given **(e)** None of the preceding

6. Which of the following are true?

I. ${}_{10}C_3 = {}_{10}C_7$ **II.** ${}_{10}C_3 = \dfrac{{}_{10}P_3}{3!}$ **III.** ${}_{10}P_3 = {}_{10}P_7$

(a) Only I **(b)** Only II **(c)** Only III **(d)** I, II, and III **(e)** None of the preceding

7. How many triangles can be formed using six points in a plane no three of which are on the same straight line?

(a) 10 **(b)** 12 **(c)** 18 **(d)** 20 **(e)** None of the preceding

8. What is the coefficient of x^3 in the expansion of $(2 + x)^5$?

(a) 40 **(b)** 20 **(c)** 10 **(d)** 80 **(e)** None of the preceeding

9. What is the fifth term in the expansion of $(3a - b)^6$?

(a) $-135a^2b^4$ **(b)** $135a^2b^4$ **(c)** $-540a^3b^3$ **(d)** $-18ab^5$ **(e)** None of the preceding

10. What is the middle term in the expansion of $(x - 2)^6$?

(a) $-20x^3$ **(b)** $60x^4$ **(c)** $-160x^3$ **(d)** $240x^2$ **(e)** None of the preceding

11. Which of the following expressions represents the ninth term in the expansion of $(a + b)^n$?

(a) $\binom{n}{8}a^{n-8}b^8$ **(b)** $\binom{n}{8}a^8b^{n-8}$ **(c)** $\binom{n}{9}a^{n-9}b^9$ **(d)** $\binom{n}{9}a^9b^{n-9}$

(e) None of the preceding

12. A coin is tossed three times. What is the probability that not all three tosses are the same?

(a) $\frac{1}{8}$ **(b)** $\frac{3}{8}$ **(c)** $\frac{1}{4}$ **(d)** $\frac{3}{4}$ **(e)** None of the preceding

13. Five cards are dealt from a deck of 52 cards. Which of the following shows the probability that four aces will be dealt?

(a) $\dfrac{{}_4C_4}{{}_{12}C_5}$ **(b)** $\dfrac{{}_{52}C_4}{{}_{52}C_5}$ **(c)** $\dfrac{({}_4C_4)({}_{48}C_1)}{{}_{52}C_5}$ **(d)** $\dfrac{({}_4C_4)({}_{52}C_1)}{{}_{52}C_5}$ **(e)** None of the preceding.

14. Two cards are drawn from a deck of 52 cards with the first card replaced before the second card is drawn. What is the probability that neither card is a spade?

(a) $\frac{9}{16}$ **(b)** $\frac{3}{4}$ **(c)** $\frac{1}{16}$ **(d)** $\frac{19}{34}$ **(e)** None of the preceding

15. A pair of dice is tossed. What is the probability that the sum of the faces showing on top is 10?

(a) $\frac{2}{9}$ **(b)** $\frac{1}{12}$ **(c)** $\frac{1}{9}$ **(d)** $\frac{1}{6}$ **(e)** None of the preceding

ANSWERS TO THE TEST YOUR UNDERSTANDING EXERCISES

Page 529

1. 5040 2. 336 3. 720 4. 90 5. 220

6. 358,800 7. 456,976 8. 6; EAT, ETA, AET, ATE, TEA, TAE

9.

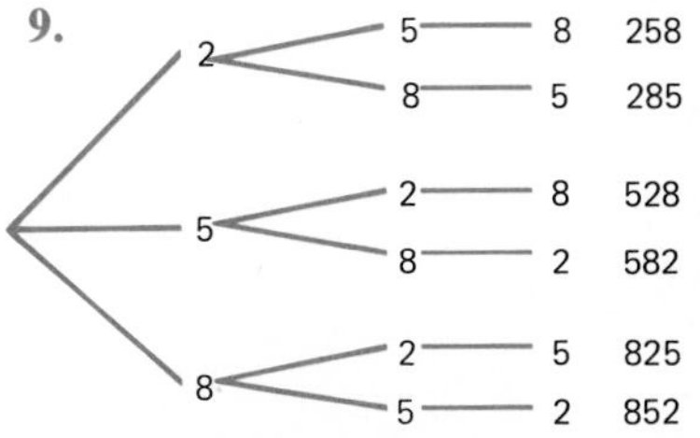

10. 18.

Page 537

1. 210
2. 1
3. 1
4. 220
5. 56
6. ${}_{10}C_3 = \dfrac{10!}{3!\,7!} = \dfrac{10!}{7!\,3!} = {}_{10}C_7$
7. 495
8. 840
9. 28
10. 10

Page 550

1. $\frac{1}{4}$ 2. $\frac{1}{2}$ 3. $\frac{3}{4}$ 4. $\frac{1}{2}$ 5. $\frac{1}{2}$ 6. $\frac{2}{3}$
7. $\frac{2}{3}$ 8. $\frac{1}{2}$ 9. $\frac{1}{2}$ 10. $\frac{4}{13}$ 11. $\frac{3}{13}$ 12. $\frac{4}{13}$

CUMULATIVE REVIEW QUESTIONS: CHAPTERS 10–12

Solve each system using multiplication–addition (subtraction) or substitution methods.

1. $\begin{aligned} -3x - 7y &= 8 \\ 10x + 14y &= 20 \end{aligned}$
2. $\begin{aligned} 10x + 2y &= -2 \\ 3y + 3 &= 9x \end{aligned}$
3. $\begin{aligned} \tfrac{1}{2}x - y &= 4 \\ 2x + \tfrac{1}{2}y &= 7 \end{aligned}$
4. Decide if the system is consistent, inconsistent, or dependent:

$$\begin{aligned} 8x - 6y &= 5 \\ 4x - 3y &= 2 \end{aligned}$$

5. Janice has \$365 consisting of five- and ten-dollar bills. The number of five-dollar bills is one more than seven times the number of ten-dollar bills. How many of each type of bill does she have?
6. One investment earns 8% interest per year and another 9%. If the total of the two investments is \$7700, and the total annual interest earned is \$658, how much is invested at each rate?
7. A custom tailor can make shirts for \$24 per shirt and sells them for \$42 each. If the daily overhead expenses are \$126, what is the break-even point?

Solve each system using multiplication–addition (subtraction) or substitution methods.

8. $\begin{aligned} x + y + z &= 0 \\ 2x + 4y - z &= 13 \\ 3x - y + 2z &= -5 \end{aligned}$
9. $\begin{aligned} x + y - 2z &= 11 \\ 2x - y + z &= -3 \\ 5x - 2y - z &= 2 \end{aligned}$

10. A field goal in basketball is worth either 2 or 3 points, and a free throw is worth 1 point. A team scored a total of 76 points of which the number of 2-point field goals was 14 more than the combined number of free throws and 3-point field goals. If also the points scored by 2-point field goals was equal to the number of free throws plus 5 times the points scored by 3-point field goals, how many of each were there?

Evaluate the determinants.

11. **(a)** $\begin{vmatrix} 3 & -1 \\ -2 & 5 \end{vmatrix}$ **(b)** $\begin{vmatrix} 2 & -3 \\ 3 & -2 \end{vmatrix}$ **(c)** $\begin{vmatrix} -5 & -4 \\ -1 & 3 \end{vmatrix}$

12. $\begin{vmatrix} 2 & -3 & 5 \\ 3 & 2 & -1 \\ -5 & 4 & 6 \end{vmatrix}$
13. $\begin{vmatrix} -4 & 1 & 7 \\ 6 & -3 & 2 \\ 0 & -2 & 5 \end{vmatrix}$

Solve each system using Cramer's rule.

14. $\begin{aligned} 2x + 3y &= -10 \\ 3x - y &= -4 \end{aligned}$

15. $\begin{aligned} 2x + 3y &= 10 \\ -4x + 3y &= 7 \end{aligned}$

16. $\begin{aligned} 3x + y + z &= 4 \\ 2x - 3y + 2z &= 2 \\ 5x + 2y - z &= -6 \end{aligned}$

17. Solve for x: $\begin{vmatrix} x-4 & x & 0 \\ 2 & x+5 & 0 \\ -3 & 6 & x \end{vmatrix} = 0$

Use matrices and fundamental row operations to solve each system.

18. $\begin{aligned} 3x + 4y &= 16 \\ -6x - 3y &= 3 \end{aligned}$

19. $\begin{aligned} 2x - 6y &= 10 \\ -3x + 9y &= -15 \end{aligned}$

20. $\begin{aligned} 2x - 3y + z &= -4 \\ 4x + 8y - 2z &= -6 \\ x + 2y + 3z &= 9 \end{aligned}$

21. Graph the system:

$$\begin{aligned} x + y &\le 4 \\ -2x + y &\le 4 \\ x - y &\le 0 \end{aligned}$$

22. Find the maximum and minimum values of $p = 2x + 5y$ for the following constraints:

$$x + 2y \ge 2, \qquad x - 2y \le 2, \qquad x - y \ge -1, \qquad 2x + 3y \le 18$$

Solve each system.

23. $\begin{aligned} x^2 - 9y^2 &= 9 \\ 18y^2 - x^2 &= 18 \end{aligned}$

24. $\begin{aligned} x^2 + (y-4)^2 &= 16 \\ y &= 8 - x^2 \end{aligned}$

25. Write the first four terms of each sequence

 (a) $s_n = \dfrac{(-1)^n}{n^2}$ **(b)** $s_n = (2n - 1) + (n - 1)(n - 2)(n - 3)$

 (c) $a_n = \left(1 - \dfrac{1}{n}\right)^n$

26. Write the first five even multiples of $\frac{1}{6}$ and write the formula for the nth term.

27. Evaluate $\sum_{n=0}^{5} s_n$ where $s_n = \dfrac{3}{n+1}$.

28. Evaluate $\sum_{k=1}^{6} (-1)^{k-1}(k + 3)$.

29. For the arithmetic sequence with first term $a_1 = \frac{2}{3}$ and common difference $d = -3$, write the nth term and find the sum S_{20}.

30. Evaluate the arithmetic series $\Sigma_{k=1}^{50}(-3k + 5)$.

31. A promotional display of canned peaches in a supermarket is in the form of a pyramid. A sign that gives the discount price for the peaches is standing on the top row consisting of 3 cans of peaches. The second row has 4 cans, the third row has 5, etc. If the pyramid has 20 rows, how many cans are there in the pyramid?

32. Find the nth term of the geometric sequence with first terms $a_1 = 128$ and $r = -\frac{1}{2}$. What is the 17th term?

33. Find the nth term of the geometric sequence $16, -12, 9, \ldots$.

Evaluate each geometric series.

34. $\sum_{k=1}^{10} 2^{k+1}$ 35. $\sum_{k=1}^{5} 9\left(-\frac{1}{3}\right)^{k-1}$

36. Suppose you save 1¢ the first day, 2¢ the second day, etc., each day saving twice as much as the preceding day. How many dollars have you saved after 20 days?
37. A roll of string is 1024 feet long. If $\frac{1}{4}$ of the string is cut off repeatedly, what is the general term of the sequence for the remaining length of string? How much string remains after 5 cuts?
38. Find the sum of each infinite geometric series if it exists.
    (a) $9 + 3 + 1 + \cdots$ **(b)** $2 - 3 + \frac{9}{2} - \cdots$
39. Evaluate each geometric series.
    (a) $\sum_{k=1}^{\infty} \left(\frac{2}{3}\right)^k$ **(b)** $\sum_{k=1}^{\infty} 15(-0.2)^{k-1}$
40. A ball always rebounds $\frac{3}{4}$ the distance it falls. If the ball is dropped from a height of 20 feet, how far does it travel before coming to rest?
41. **(a)** How many different four-letter "words" can be formed from the letters in the word *compute* if no letter is used more than once in any word?
    (b) How many of the four-letter words in part (a) begin with a vowel?
    (c) How many of the four-letter words in part (a) begin and end with a vowel?
42. **(a)** How many 3-digit even numbers can be formed from 1, 2, 3, 4, 5, 6, 7, 8, 9?
    (b) How many of the 3-digit even numbers in part (a) are more than 700?
43. Evaluate:
    (a) ${}_9P_4$ **(b)** ${}_9C_4$ **(c)** ${}_6P_6$ **(d)** $\binom{38}{35}$
44. The membership of a club consists of 14 women and 12 men. Find how many committees can be formed consisting of 6 club members if
    (a) all 6 are women.
    (b) 3 are men and 3 are women.
    (c) 2 particular club members must be on the committee, and the remaining committee members can be either men or women.
45. Expand and simplify: $(2x - 3)^5$
46. Find the 12th term in the expansion of $(2a - 1)^{15}$.
47. Evaluate $(3.1)^4$ by expanding an appropriate binomial.
48. **(a)** A card is selected at random from an ordinary deck of playing cards. Find the probability that the card is either a 7 or a red card.
    (b) If two cards are drawn at random, without replacement, find P(two clubs).
49. Three coins are tossed. Find the probability of each outcome.
    (a) Two heads and one tail.
    (b) All three are either heads or tails.
    (c) At least one coin is a tail.
50. A pair of dice are rolled. Find the probability that the sum of the outcomes on the dice is
    (a) 8 **(b)** 4 or 10 **(c)** less than 6

TABLES

I *Square Roots and Cube Roots*
II *Exponential Functions*
III *Natural Logarithms (Base e)*
IV *Four-Place Common Logarithms (Base 10)*

Table I begins on page 564
Table II begins on page 565.
Table III begins on page 566.
Table IV begins on page 567.

TABLE I: SQUARE ROOTS AND CUBE ROOTS

N	$\sqrt{N}$	$\sqrt[3]{N}$	N	$\sqrt{N}$	$\sqrt[3]{N}$	N	$\sqrt{N}$	$\sqrt[3]{N}$	N	$\sqrt{N}$	$\sqrt[3]{N}$
1	1.000	1.000	51	7.141	3.708	101	10.050	4.657	151	12.288	5.325
2	1.414	1.260	52	7.211	3.733	102	10.100	4.672	152	12.329	5.337
3	1.732	1.442	53	7.280	3.756	103	10.149	4.688	153	12.369	5.348
4	2.000	1.587	54	7.348	3.780	104	10.198	4.703	154	12.410	5.360
5	2.236	1.710	55	7.416	3.803	105	10.247	4.718	155	12.450	5.372
6	2.449	1.817	56	7.483	3.826	106	10.296	4.733	156	12.490	5.383
7	2.646	1.913	57	7.550	3.849	107	10.344	4.747	157	12.530	5.395
8	2.828	2.000	58	7.616	3.871	108	10.392	4.762	158	12.570	5.406
9	3.000	2.080	59	7.681	3.893	109	10.440	4.777	159	12.610	5.418
10	3.162	2.154	60	7.746	3.915	110	10.488	4.791	160	12.649	5.429
11	3.317	2.224	61	7.810	3.936	111	10.536	4.806	161	12.689	5.440
12	3.464	2.289	62	7.874	3.958	112	10.583	4.820	162	12.728	5.451
13	3.606	2.351	63	7.937	3.979	113	10.630	4.835	163	12.767	5.463
14	3.742	2.410	64	8.000	4.000	114	10.677	4.849	164	12.806	5.474
15	3.873	2.466	65	8.062	4.021	115	10.724	4.863	165	12.845	5.485
16	4.000	2.520	66	8.124	4.041	116	10.770	4.877	166	12.884	5.496
17	4.123	2.571	67	8.185	4.062	117	10.817	4.891	167	12.923	5.507
18	4.243	2.621	68	8.246	4.082	118	10.863	4.905	168	12.961	5.518
19	4.359	2.668	69	8.307	4.102	119	10.909	4.919	169	13.000	5.529
20	4.472	2.714	70	8.367	4.121	120	10.954	4.932	170	13.038	5.540
21	4.583	2.759	71	8.426	4.141	121	11.000	4.946	171	13.077	5.550
22	4.690	2.802	72	8.485	4.160	122	11.045	4.960	172	13.115	5.561
23	4.796	2.844	73	8.544	4.179	123	11.091	4.973	173	13.153	5.572
24	4.899	2.884	74	8.602	4.198	124	11.136	4.987	174	13.191	5.583
25	5.000	2.924	75	8.660	4.217	125	11.180	5.000	175	13.229	5.593
26	5.099	2.962	76	8.718	4.236	126	11.225	5.013	176	13.267	5.604
27	5.196	3.000	77	8.775	4.254	127	11.269	5.027	177	13.304	5.615
28	5.292	3.037	78	8.832	4.273	128	11.314	5.040	178	13.342	5.625
29	5.385	3.072	79	8.888	4.291	129	11.358	5.053	179	13.379	5.636
30	5.477	3.107	80	8.944	4.309	130	11.402	5.066	180	13.416	5.646
31	5.568	3.141	81	9.000	4.327	131	11.446	5.079	181	13.454	5.657
32	5.657	3.175	82	9.055	4.344	132	11.489	5.092	182	13.491	5.667
33	5.745	3.208	83	9.110	4.362	133	11.533	5.104	183	13.528	5.677
34	5.831	3.240	84	9.165	4.380	134	11.576	5.117	184	13.565	5.688
35	5.916	3.271	85	9.220	4.397	135	11.619	5.130	185	13.601	5.698
36	6.000	3.302	86	9.274	4.414	136	11.662	5.143	186	13.638	5.708
37	6.083	3.332	87	9.327	4.431	137	11.705	5.155	187	13.675	5.718
38	6.164	3.362	88	9.381	4.448	138	11.747	5.168	188	13.711	5.729
39	6.245	3.391	89	9.434	4.465	139	11.790	5.180	189	13.748	5.739
40	6.325	3.420	90	9.487	4.481	140	11.832	5.192	190	13.784	5.749
41	6.403	3.448	91	9.539	4.498	141	11.874	5.205	191	13.820	5.759
42	6.481	3.476	92	9.592	4.514	142	11.916	5.217	192	13.856	5.769
43	6.557	3.503	93	9.644	4.531	143	11.958	5.229	193	13.892	5.779
44	6.633	3.530	94	9.695	4.547	144	12.000	5.241	194	13.928	5.789
45	6.708	3.557	95	9.747	4.563	145	12.042	5.254	195	13.964	5.799
46	6.782	3.583	96	9.798	4.579	146	12.083	5.266	196	14.000	5.809
47	6.856	3.609	97	9.849	4.595	147	12.124	5.278	197	14.036	5.819
48	6.928	3.634	98	9.899	4.610	148	12.166	5.290	198	14.071	5.828
49	7.000	3.659	99	9.950	4.626	149	12.207	5.031	199	14.107	5.838
50	7.071	3.684	100	10.000	4.642	150	12.247	5.313	200	14.142	5.848

TABLE II: EXPONENTIAL FUNCTIONS

x	e^x	e^{-x}	x	e^x	e^{-x}
0.0	1.00	1.000	3.1	22.2	0.045
0.1	1.11	0.905	3.2	24.5	0.041
0.2	1.22	0.819	3.3	27.1	0.037
0.3	1.35	0.741	3.4	30.0	0.033
0.4	1.49	0.670	3.5	33.1	0.030
0.5	1.65	0.607	3.6	36.6	0.027
0.6	1.82	0.549	3.7	40.4	0.025
0.7	2.01	0.497	3.8	44.7	0.022
0.8	2.23	0.449	3.9	49.4	0.020
0.9	2.46	0.407	4.0	54.6	0.018
1.0	2.72	0.368	4.1	60.3	0.017
1.1	3.00	0.333	4.2	66.7	0.015
1.2	3.32	0.301	4.3	73.7	0.014
1.3	3.67	0.273	4.4	81.5	0.012
1.4	4.06	0.247	4.5	90.0	0.011
1.5	4.48	0.223	4.6	99.5	0.010
1.6	4.95	0.202	4.7	110	0.0091
1.7	5.47	0.183	4.8	122	0.0082
1.8	6.05	0.165	4.9	134	0.0074
1.9	6.69	0.150	5.0	148	0.0067
2.0	7.39	0.135	5.5	245	0.0041
2.1	8.17	0.122	6.0	403	0.0025
2.2	9.02	0.111	6.5	665	0.0015
2.3	9.97	0.100	7.0	1097	0.00091
2.4	11.0	0.091	7.5	1808	0.00055
2.5	12.2	0.082	8.0	2981	0.00034
2.6	13.5	0.074	8.5	4915	0.00020
2.7	14.9	0.067	9.0	8103	0.00012
2.8	16.4	0.061	9.5	13360	0.00075
2.9	18.2	0.055	10.0	22026	0.000045
3.0	20.1	0.050			

TABLE III: NATURAL LOGARITHMS (BASE *e*)

x	$\ln x$	x	$\ln x$	x	$\ln x$
0.0		3.4	1.224	6.8	1.917
0.1	−2.303	3.5	1.253	6.9	1.932
0.2	−1.609	3.6	1.281	7.0	1.946
0.3	−1.204	3.7	1.308	7.1	1.960
0.4	−0.916	3.8	1.335	7.2	1.974
0.5	−0.693	3.9	1.361	7.3	1.988
0.6	−0.511	4.0	1.386	7.4	2.001
0.7	−0.357	4.1	1.411	7.5	2.015
0.8	−0.223	4.2	1.435	7.6	2.028
0.9	−0.105	4.3	1.459	7.7	2.041
1.0	0.000	4.4	1.482	7.8	2.054
1.1	0.095	4.5	1.504	7.9	2.067
1.2	0.182	4.6	1.526	8.0	2.079
1.3	0.262	4.7	1.548	8.1	2.092
1.4	0.336	4.8	1.569	8.2	2.104
1.5	0.405	4.9	1.589	8.3	2.116
1.6	0.470	5.0	1.609	8.4	2.128
1.7	0.531	5.1	1.629	8.5	2.140
1.8	0.588	5.2	1.649	8.6	2.152
1.9	0.642	5.3	1.668	8.7	2.163
2.0	0.693	5.4	1.686	8.8	2.175
2.1	0.742	5.5	1.705	8.9	2.186
2.2	0.788	5.6	1.723	9.0	2.197
2.3	0.833	5.7	1.740	9.1	2.208
2.4	0.875	5.8	1.758	9.2	2.219
2.5	0.916	5.9	1.775	9.3	2.230
2.6	0.956	6.0	1.792	9.4	2.241
2.7	0.993	6.1	1.808	9.5	2.251
2.8	1.030	6.2	1.825	9.6	2.262
2.9	1.065	6.3	1.841	9.7	2.272
3.0	1.099	6.4	1.856	9.8	2.282
3.1	1.131	6.5	1.872	9.9	2.293
3.2	1.163	6.6	1.887	10.0	2.303
3.3	1.194	6.7	1.902		

TABLE IV: FOUR-PLACE COMMON LOGARITHMS (BASE 10)

N	0	1	2	3	4	5	6	7	8	9
1.0	.0000	.0043	.0086	.0128	.0170	.0212	.0253	.0294	.0334	.0374
1.1	.0414	.0453	.0492	.0531	.0569	.0607	.0645	.0682	.0719	.0755
1.2	.0792	.0828	.0864	.0899	.0934	.0969	.1004	.1038	.1072	.1106
1.3	.1139	.1173	.1206	.1239	.1271	.1303	.1335	.1367	.1399	.1430
1.4	.1461	.1492	.1523	.1553	.1584	.1614	.1644	.1673	.1703	.1732
1.5	.1761	.1790	.1818	.1847	.1875	.1903	.1931	.1959	.1987	.2014
1.6	.2041	.2068	.2095	.2122	.2148	.2175	.2201	.2227	.2253	.2279
1.7	.2304	.2330	.2355	.2380	.2405	.2430	.2455	.2480	.2504	.2529
1.8	.2553	.2577	.2601	.2625	.2648	.2672	.2695	.2718	.2742	.2765
1.9	.2788	.2810	.2833	.2856	.2878	.2900	.2923	.2945	.2967	.2989
2.0	.3010	.3032	.3054	.3075	.3096	.3118	.3139	.3160	.3181	.3201
2.1	.3222	.3243	.3263	.3284	.3304	.3324	.3345	.3365	.3385	.3404
2.2	.3424	.3444	.3464	.3483	.3502	.3522	.3541	.3560	.3579	.3598
2.3	.3617	.3636	.3655	.3674	.3692	.3711	.3729	.3747	.3766	.3784
2.4	.3802	.3820	.3838	.3856	.3874	.3892	.3909	.3927	.3945	.3962
2.5	.3979	.3997	.4014	.4031	.4048	.4065	.4082	.4099	.4116	.4133
2.6	.4150	.4166	.4183	.4200	.4216	.4232	.4249	.4265	.4281	.4298
2.7	.4314	.4330	.4346	.4362	.4378	.4393	.4409	.4425	.4440	.4456
2.8	.4472	.4487	.4502	.4518	.4533	.4548	.4564	.4579	.4594	.4609
2.9	.4624	.4639	.4654	.4669	.4683	.4698	.4713	.4728	.4742	.4757
3.0	.4771	.4786	.4800	.4814	.4829	.4843	.4857	.4871	.4886	.4900
3.1	.4914	.4928	.4942	.4955	.4969	.4983	.4997	.5011	.5024	.5038
3.2	.5051	.5065	.5079	.5092	.5105	.5119	.5132	.5145	.5159	.5172
3.3	.5185	.5198	.5211	.5224	.5237	.5250	.5263	.5276	.5289	.5302
3.4	.5315	.5328	.5340	.5353	.5366	.5378	.5391	.5403	.5416	.5428
3.5	.5441	.5453	.5465	.5478	.5490	.5502	.5514	.5527	.5539	.5551
3.6	.5563	.5575	.5587	.5599	.5611	.5623	.5635	.5647	.5658	.5670
3.7	.5682	.5694	.5705	.5717	.5729	.5740	.5752	.5763	.5775	.5786
3.8	.5798	.5809	.5821	.5832	.5843	.5855	.5866	.5877	.5888	.5899
3.9	.5911	.5922	.5933	.5944	.5955	.5966	.5977	.5988	.5999	.6010
4.0	.6021	.6031	.6042	.6053	.6064	.6075	.6085	.6096	.6107	.6117
4.1	.6128	.6138	.6149	.6160	.6170	.6180	.6191	.6201	.6212	.6222
4.2	.6232	.6243	.6253	.6263	.6274	.6284	.6294	.6304	.6314	.6325
4.3	.6335	.6345	.6355	.6365	.6375	.6385	.6395	.6405	.6415	.6425
4.4	.6435	.6444	.6454	.6464	.6474	.6484	.6493	.6503	.6513	.6522
4.5	.6532	.6542	.6551	.6561	.6571	.6580	.6590	.6599	.6609	.6618
4.6	.6628	.6637	.6646	.6656	.6665	.6675	.6684	.6693	.6702	.6712
4.7	.6721	.6730	.6739	.6749	.6758	.6767	.6776	.6785	.6794	.6803
4.8	.6812	.6821	.6830	.6839	.6848	.6857	.6866	.6875	.6884	.6893
4.9	.6902	.6911	.6920	.6928	.6937	.6946	.6955	.6964	.6972	.6981
5.0	.6990	.6998	.7007	.7016	.7024	.7033	.7042	.7050	.7059	.7067
5.1	.7076	.7084	.7093	.7101	.7110	.7118	.7126	.7135	.7143	.7152
5.2	.7160	.7168	.7177	.7185	.7193	.7202	.7210	.7218	.7226	.7235
5.3	.7243	.7251	.7259	.7267	.7275	.7284	.7292	.7300	.7308	.7316
5.4	.7324	.7332	.7340	.7348	.7356	.7364	.7372	.7380	.7388	.7396
N	0	1	2	3	4	5	6	7	8	9

TABLE IV: CONTINUED

N	0	1	2	3	4	5	6	7	8	9
5.5	.7404	.7412	.7419	.7427	.7435	.7443	.7451	.7459	.7466	.7474
5.6	.7482	.7490	.7497	.7505	.7513	.7520	.7528	.7536	.7543	.7551
5.7	.7559	.7566	.7574	.7582	.7589	.7597	.7604	.7612	.7619	.7627
5.8	.7634	.7642	.7649	.7657	.7664	.7672	.7679	.7686	.7694	.7701
5.9	.7709	.7716	.7723	.7731	.7738	.7745	.7752	.7760	.7767	.7774
6.0	.7782	.7789	.7796	.7803	.7810	.7818	.7825	.7832	.7839	.7846
6.1	.7853	.7860	.7868	.7875	.7882	.7889	.7896	.7903	.7910	.7917
6.2	.7924	.7931	.7938	.7945	.7952	.7959	.7966	.7973	.7980	.7987
6.3	.7993	.8000	.8007	.8014	.8021	.8028	.8035	.8041	.8048	.8055
6.4	.8062	.8069	.8075	.8082	.8089	.8096	.8102	.8109	.8116	.8122
6.5	.8129	.8136	.8142	.8149	.8156	.8162	.8169	.8176	.8182	.8189
6.6	.8195	.8202	.8209	.8215	.8222	.8228	.8235	.8241	.8248	.8254
6.7	.8261	.8267	.8274	.8280	.8287	.8293	.8299	.8306	.8312	.8319
6.8	.8325	.8331	.8338	.8344	.8351	.8357	.8363	.8370	.8376	.8382
6.9	.8388	.8395	.8401	.8407	.8414	.8420	.8426	.8432	.8439	.8445
7.0	.8451	.8457	.8463	.8470	.8476	.8482	.8488	.8494	.8500	.8506
7.1	.8513	.8519	.8525	.8531	.8537	.8543	.8549	.8555	.8561	.8567
7.2	.8573	.8579	.8585	.8591	.8597	.8603	.8609	.8615	.8621	.8627
7.3	.8633	.8639	.8645	.8651	.8657	.8663	.8669	.8675	.8681	.8686
7.4	.8692	.8698	.8704	.8710	.8716	.8722	.8727	.8733	.8739	.8745
7.5	.8751	.8756	.8762	.8768	.8774	.8779	.8785	.8791	.8797	.8802
7.6	.8808	.8814	.8820	.8825	.8831	.8837	.8842	.8848	.8854	.8859
7.7	.8865	.8871	.8876	.8882	.8887	.8893	.8899	.8904	.8910	.8915
7.8	.8921	.8927	.8932	.8938	.8943	.8949	.8954	.8960	.8965	.8971
7.9	.8976	.8982	.8987	.8993	.8998	.9004	.9009	.9015	.9020	.9025
8.0	.9031	.9036	.9042	.9047	.9053	.9058	.9063	.9069	.9074	.9079
8.1	.9085	.9090	.9096	.9101	.9106	.9112	.9117	.9122	.9128	.9133
8.2	.9138	.9143	.9149	.9154	.9159	.9165	.9170	.9175	.9180	.9186
8.3	.9191	.9196	.9201	.9206	.9212	.9217	.9222	.9227	.9232	.9238
8.4	.9243	.9248	.9253	.9258	.9263	.9269	.9274	.9279	.9284	.9289
8.5	.9294	.9299	.9304	.9309	.9315	.9320	.9325	.9330	.9335	.9340
8.6	.9345	.9350	.9355	.9360	.9365	.9370	.9375	.9380	.9385	.9390
8.7	.9395	.9400	.9405	.9410	.9415	.9420	.9425	.9430	.9435	.9440
8.8	.9445	.9450	.9455	.9460	.9465	.9469	.9474	.9479	.9484	.9489
8.9	.9494	.9499	.9504	.9509	.9513	.9518	.9523	.9528	.9533	.9538
9.0	.9542	.9547	.9552	.9557	.9562	.9566	.9571	.9576	.9581	.9586
9.1	.9590	.9595	.9600	.9605	.9609	.9614	.9619	.9624	.9628	.9633
9.2	.9638	.9643	.9647	.9652	.9657	.9661	.9666	.9671	.9675	.9680
9.3	.9685	.9689	.9694	.9699	.9703	.9708	.9713	.9717	.9722	.9727
9.4	.9731	.9736	.9741	.9745	.9750	.9754	.9759	.9763	.9768	.9773
9.5	.9777	.9782	.9786	.9791	.9795	.9800	.9805	.9809	.9814	.9818
9.6	.9823	.9827	.9832	.9836	.9841	.9845	.9850	.9854	.9859	.9863
9.7	.9868	.9872	.9877	.9881	.9886	.9890	.9894	.9899	.9903	.9908
9.8	.9912	.9917	.9921	.9926	.9930	.9934	.9939	.9943	.9948	.9952
9.9	.9956	.9961	.9965	.9969	.9974	.9978	.9983	.9987	.9991	.9996
N	0	1	2	3	4	5	6	7	8	9

Answers to Odd-Numbered Exercises, Chapter Tests, and Cumulative Review Questions

CHAPTER 1: THE SET OF REAL NUMBERS

1.1 Sets of Numbers (Page 5)

1. {1} **3.** {2, 4, 6, . . .} **5.** {101, 102, 103, . . .} **7.** {1, 3, 5, . . .} **9.** {0} **11.** True **13.** True **15.** True **17.** True **19.** False; for example, there is no integer between 2 and 3. **21.**

$$\begin{matrix} a & b & c \\ \updownarrow & \updownarrow & \updownarrow \\ m & n & p \end{matrix}$$

(One of six possible solutions)

23.

$$\begin{matrix} 0 & 1 & 2 & 3 & 4 \\ \updownarrow & \updownarrow & \updownarrow & \updownarrow & \updownarrow \\ -2 & -1 & 0 & 1 & 2 \end{matrix}$$

25. ∅, {1}, {2}, {1,2} **27.** **(a)** 4; **(b)** 8; **(c)**. 16; **(d)** 32

29. There is no integer, for example, between 4 and 5 **31.**

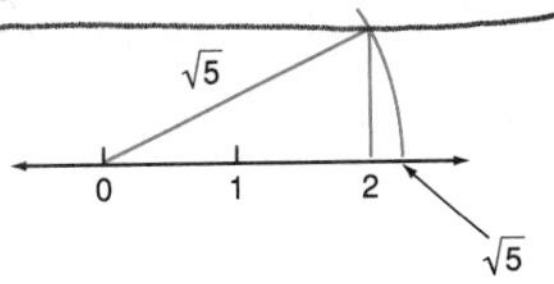

1.2 Properties of the Real Numbers (Page 10)

1. Closure property for addition **3.** Inverse property for addition **5.** Commutative property for multiplication **7.** Commutative property for addition **9.** Identity property for addition **11.** Additive inverse **13.** Multiplication property of zero **15.** Distributive property **17.** 3 **19.** 5 **21.** 7 **23.** −9 **25.** If $ab = 0$, then $a = 0$ or $b = 0$; $x + 2 = 0$ and $x = -2$ **27.** **(i)** Commutative property for multiplication; **(ii)** distributive property; **(iii)** same as (i); **(d)** same as (ii) **29.** True **31.** False; $7 \div 3$ is not an integer. **33.** False; $8 - 2 \neq 2 - 8$ **35.** True **37.** True **39.** No; for example $2^3 \neq 3^2$ **41.** $(\sqrt{3})(\sqrt{3}) = 3$, a rational number

1.3 Operations with Integers: Addition and Subtraction (Page 15)

1. +14 3. 0 5. +7 7. −2 9. −10 11. +30 13. −8 15. 0 17. +39 19. +4 21. −10 23. +13 25. −27 27. +2 29. +213 31. −7 33. −427 35. +2473 37. +45 39. −15 41. 0 43. −12 45. 12 47. −5 49. 5 51. 1 53. 19 55. −4 57. −10 59. 1

61. (i) Definition of subtraction; (ii) distributive property; (iii) the opposite of a product is the product of one number times the opposite of the other; (iv) definition of subtraction

63. Either x and y are both negative, or if they are of opposite signs, the one with the larger absolute value is negative

65. **(a)** $|x| - |y|$; **(b)** $-(|y| - |x|)$

1.4 Operations with Integers: Multiplication and Division (Page 20)

1. −40 3. +64 5. +40 7. +52 9. +96 11. −20 13. −128 15. −128 17. +1296 19. +4 21. −4 23. +7 25. 0 27. −6 29. +8 31. −12 33. +12 35. −11 37. +6 39. −42 41. −27 43. −60 45. +2 47. 15 49. 15 51. 0 53. 5 55. 5 57. −8 59. 2 61. −54 63. −16 65. 11 67. 51 69. 0 71. $\frac{5}{4}$ 73. 23 75. No. $a \div bc \neq (a \div b)(a \div c)$; for example, $12 \div (2)(4) \neq (12 \div 2)(12 \div 4)$

77. Let $\frac{0}{0} = x$, where x is some real number. Then the definition of division gives $0 \cdot x = 0$. Since any number x will work, the answer to $\frac{0}{0}$ is not unique. Therefore, $\frac{0}{0}$ is undefined.

1.5 Operations with Real Numbers (Page 25)

1. −14.35 3. −21.34 5. +0.21 7. +6.84 9. −8.66 11. −11.41 13. −0.6375 15. +17.8308 17. −2.9718 19. +145 21. −45 23. −7.5 25. $-\frac{17}{12}$ 27. $+\frac{3}{2}$ 29. $-\frac{11}{12}$ 31. $-\frac{1}{6}$ 33. $-\frac{1}{2}$ 35. $-3\frac{1}{2}$ 37. $-\frac{14}{3}$ 39. +35 41. 7.646 43. −0.247 45. −1.536 47. 3.082 49. −1.463 51. −0.982

53. **(a)** 4444544444; **(b)** 1222212222 55. **(a)** 0.23; $0.\overline{23}$; 0.23233; 0.232332333. . . **(b)** 0.07; 0.07007; 0.070070007; $0.\overline{07}$ 57. $0.\overline{714285}$ 59. 0.203125 61. **(a)** $\frac{5}{11}$; **(b)** $\frac{37}{99}$; **(c)** $\frac{26}{111}$

CHAPTER 1 TEST: STANDARD ANSWER (Page 29)

1.

	Whole numbers	Integers	Rational numbers	Irrational numbers	Negative integers
−6		✓	✓		✓
0.231			✓		
$\sqrt{5}$				✓	
$\frac{2}{3}$			✓		
1991	✓	✓	✓		

2. {0, 1, 2, 3, 4} 3. {4, 5, 6, 7} 4. {−1, −2, −3, −4, −5} 5. {−2, −1, 0, 1} 6. {4, 5, 6, . . .} 7. False 8. True 9. False 10. True 11. False 12. True 13. Commutative for multiplication 14. Commutative for addition 15. Distributive 16. Associative for multiplication 17. Identity property for addition 18. Multiplicative inverse 19. **(a)** −4 **(b)** +5 20. **(a)** −6 **(b)** −15 21. **(a)** −21 **(b)** +54 22. **(a)** −8 **(b)** +9 23. **(a)** +14 **(b)** −21 24. **(a)** −4 **(b)** +4 25. **(a)** −48 **(b)** +7

CHAPTER 1 TEST: MULTIPLE CHOICE (Page 30)

1. (d) 2. (c) 3. (b) 4. (c) 5. (b) 6. (c) 7. (c) 8. (e) 9. (c) 10. (d) 11. (d) 12. (e) 13. (a) 14. (a) 15. (d)

CHAPTER 2: FUNDAMENTAL OPERATIONS

2.1 Integral Exponents (Page 35)

1. False; 3^6 3. False; 2^7 5. True 7. True 9. False; $2 \cdot 3^4$ 11. False; 1 13. 100,000 15. -72 17. $\frac{1}{64}$ 19. 9 21. $-\frac{27}{64}$ 23. 2 25. x^6 27. x^4y^6 29. $\frac{x^4}{y^6}$ 31. a^6 33. 1 35. 1 37. x^6y^2 39. $\frac{x^8}{y^6}$ 41. $\frac{4x^6}{9y^4}$ 43. x^8y^{10} 45. $(x - 2y)^2$ 47. x^{4a} 49. $\frac{(10^6)(10^4)}{10^7} = 1000$

51. For $\frac{x^3y}{y^4}$: $\frac{(-8)(3)}{81} = -\frac{8}{27}$; for $\frac{x^3}{y^3}$: $\frac{(-2)^3}{(3)^3} = -\frac{8}{27}$ 53. For $\frac{(x^2y)^4}{(xy)^2}$: $\frac{(\frac{1}{4} \cdot 4)^4}{(-\frac{1}{2} \cdot 4)^2} = \frac{1}{4}$; for x^6y^2: $\left(-\frac{1}{2}\right)^6(4)^2 = \frac{16}{64} = \frac{1}{4}$

55. For $\left(\frac{3xy^2}{2x^2y^3}\right)^2$: $\left(\frac{3(2.3)(1.5)^2}{2(2.3)^2(1.5)^3}\right)^2 = \left(\frac{15.525}{35.7075}\right)^2 = (0.43478)^2 = 0.189$; for $\frac{9}{4x^2y^2}$: $\frac{9}{4(2.3)^2(1.5)^2} = \frac{9}{47.61} = 0.189$

57. 5 grams; $640(\frac{1}{2})^n$ grams 59. 32 feet; $243(\frac{2}{3})^n$ feet 61. \$1331 63. $(2^3)^4 = 2^{12} \neq 2^{81} = 2^{(3^4)}$

2.2 Negative Exponents and Scientific Notation (Page 44)

1. $\frac{1}{2}$ 3. $\frac{1}{8}$ 5. $-\frac{1}{8}$ 7. 1 9. $\frac{1}{64}$ 11. 100 13. $\frac{16}{9}$ 15. $\frac{1}{12}$ 17. $\frac{a^2}{b^3}$ 19. $\frac{1}{x^3}$ 21. $\frac{1}{8x^3}$ 23. x^2 25. x^5 27. $\frac{b^2}{a^2}$ 29. x^2 31. x^{15} 33. x^{12} 35. x^5y^6 37. x^6y^3 39. $(x - 2y)^4$ 41. x^3 43. $\frac{1}{x^6}$ 45. $\frac{27y^6}{x^6}$ 47. $\frac{2y^3}{x^4}$ 49. $\frac{y^{10}}{x^{12}}$ 51. $\frac{x^9}{2y^6}$ 53. $\frac{6b^2d^4}{a^3c^2}$ 55. $\frac{4}{x^6y^6}$ 57. $4x^{24}y^4$ 59. $-\frac{b^{10}}{a^{31}}$ 61. $(s + t)^6$ 63. 9 65. 8 67. -3 69. 4.68×10^3 71. 9.2×10^{-1} 73. 7.583×10^6 75. 2.5×10^1 77. 5.55×10^{-7} 79. 2.024×10^2 81. 78,900 83. 3000 85. 0.174 87. 17.4 89. 0.0906 91. 10^2 93. 10^0 95. 10^{17} 97. 2000 99. 0.02 101. 0.000000000216 103. 500 seconds

2.3 Addition and Subtraction of Polynomials (Page 49)

1. $8x^2 - 2x + 7$ 3. $x^3 + 2x^2 + x + 4$ 5. $3x^2 - 11x - 8$ 7. $x^3 - x^2 + 12x + 2$ 9. $4x^2 + 17x + 8$ 11. $3x^3 + x^2 - 10x + 6$ 13. $2x^3 - 8x^2 + 14x - 6$ 15. $11x + 3y$ 17. $5xy + 5$ 19. $x + 4y$ 21. $6x + 6y$ 23. $5x + 2$ 25. $x^2y - xy^2 + xy$ 27. -2 29. $a + 5b - 9c$ 31. $-5b$ 33. $6x + y$ 35. $3c + 5$ 37. $-2x^4 - 2x^3 + 5x^2 + 3x + 10$ 39. $-2x^3 - 2x^2 + 8x + 3$ 41. $5x + 3y + 6z$ 43. $a^2 - ab + 2b + 3b^2$ 45. $4x^2 + 5x - 3$ 47. $-a + 2b + 6$ 49. $7a - 4b^2 - 9b$ 51. $-2a - 2c$ 53. **(a)** 5 **(b)** 9 55. **(a)** -3 **(b)** -93 57. **(a)** -50.35 **(b)** -20.3345 59. **(a)** $5.8109 + 17.4518 = 23.2627$ **(b)** $3x^2 + 2x - 1$; 23.2627

2.4 Multiplication of Polynomials (Page 54)

1. $-20x^5$ 3. $-16x^6$ 5. $4x^4$ 7. $6a^4b^4$ 9. $-4a^4b^7$ 11. $-15x^7$ 13. $-6a^4x^4$ 15. $9x^6y^5$ 17. $-2a^4 + 6a$ 19. $-2x^4 - 10x^3 + 2x$ 21. $2x^5 - 5x^4 + 3x^3$ 23. $2a^4b - 6a^3b^2 + 2a^2b^3$ 25. $x^3 + 4x^2 + 7x + 4$ 27. $x^3 + 4x^2 - 7x + 2$ 29. $x^2 + 2x + 1$ 31. $6x^2 + 13x + 6$ 33. $6x^3 - 7x^2 + 12x - 5$ 35. $x^4 - 2x^3 - 4x^2 - 5x - 2$ 37. $x^6 - 64$ 39. $-8x^5 + 36x^3 + 44x^2 + 24x$ 41. **(a)** $(7.55)(-4.5775) = -34.560125$ **(b)** $3x^3 + 11x^2 - 11x - 35$; -34.560125 43. **(a)** $-3x^3 - 7x^2$ **(b)** 244.67924

2.5 The Product of Binomials (Page 58)

1. $2x^2 + 5x + 3$ 3. $x^2 + 10x + 24$ 5. $x^2 - 2x - 24$ 7. $x^2 - 6x + 9$ 9. $6x^2 + 5x + 1$ 11. $16x^2 - 20x - 50$ 13. $20x^2 - 42x + 18$ 15. $20x^2 + 18x - 18$ 17. $x^2 - 4x + 4$ 19. $4a^2 - 4ab + b^2$ 21. $x^2 + x + \frac{1}{4}$ 23. $x^2 - \frac{7}{2}x - 2$ 25. $270x^2 + 135x - 810$ 27. $-6x^2 - 3x + 18$ 29. $-6x^2 + 3x + 18$

31. $\frac{1}{4}x^2 - 16$ 33. $\frac{1}{4}x^2 - \frac{1}{3}x + \frac{1}{9}$ 35. $63 - x - 12x^2$ 37. $a^2x^2 + 2abx + b^2$ 39. $a^2 - 4ab + 4b^2$
41. $9x^2 + 24xy + 16y^2$ 43. $4x^2 - 2x + \frac{1}{4}$ 45. $a^2 - 2ab + b^2$ 47. $x^4 + 4x^3 + 6x^2 + 4x + 1$
49. $a^4 - 4a^3b + 6a^2b^2 - 4ab^3 + b^4$ 51. $\frac{1}{4}x^4 - 4x^2 + 16$ 53. $\frac{1}{8}x^3 - \frac{3}{4}x^2 + \frac{3}{2}x - 1$ 55. $-3a^2 - 4ab - 10b^2$
57. $8x^3 - 14x^2 + 13x - 70$ 59. $3x^3 - 6x^2 + 3x$ 61. $x^3 + 10x^2 - 100x - 1000$ 63. $c^4 - 9d^4$
65. $x^6 + 4x^3 + 4$ 67. $p^6 - 2p^3q^2 + q^4$ 69. $a^{2n} - 16$
71. $(x + 1)^2(x - 1)^2 = [(x + 1)(x - 1)]^2 = (x^2 - 1)^2 = x^4 - 2x^2 + 1$
$(x + 1)^2(x - 1)^2 = (x^2 + 2x + 1)(x^2 - 2x + 1) = x^4 - 2x^2 + 1$

2.6 Introduction to Factoring (Page 64)

1. $5(x - 1)$ 3. $7(x + 2)$ 5. $a(x + y)$ 7. $2x(x + 2)$ 9. $2x^2(1 - 2x)$ 11. $2ab(b - 2a)$
13. $4x(2x^2 + x - 1)$ 15. $-3xy(4x^2 - 3xy + 2y^2)$ 17. $x(a + b - c)$ 19. $(x + 3)(x - 3)$ 21. $(2 + x)(2 - x)$
23. $(x + 10)(x - 10)$ 25. $(2x + 3)(2x - 3)$ 27. $(8a + b)(8a - b)$ 29. $(a + 11b)(a - 11b)$ 31. 1599
33. 6391 35. 9984 37. 14,391 39. $(x - 2)(x^2 + 2x + 4)$ 41. $(x - 5)(x^2 + 5x + 25)$
43. $(6 - a)(36 + 6a + a^2)$ 45. $(2x + 1)(4x^2 - 2x + 1)$ 47. $(3a - 2)(9a^2 + 6a + 4)$
49. $(2 - 3a)(4 + 6a + 9a^2)$ 51. $(2x + 7y)(4x^2 - 14xy + 49y^2)$ 53. $(1 - 9b)(1 + 9b + 81b^2)$
55. $(4a^2 + b^2)(2a + b)(2a - b)$ 57. $(a^4 + b^4)(a^2 + b^2)(a + b)(a - b)$
59. $(a^8 + 1)(a^4 + 1)(a^2 + 1)(a + 1)(a - 1)$ 61. $a(x + y)(x^2 - xy + y^2)$ 63. $3(3x - y)(9x^2 + 3xy + y^2)$
65. $5a(a + 5)(a - 5)$ 67. $5(1 + 4x^2)(1 + 2x)(1 - 2x)$ 69. $2d(5c - d)(25c^2 + 5cd + d^2)$
71. $2p^2q(4p + q)(16p^2 - 4pq + q^2)$
73. **(a)** $x^4 - y^4 = (x - y)(x + y)(x^2 + y^2) = (x - y)(x^3 + x^2y + xy^2 + y^3)$
(b) $x^6 - y^6 = (x - y)(x^5 + x^4y + x^3y^2 + x^2y^3 + xy^4 + y^5)$
(c) $x^5 - y^5 = (x - y)(x^4 + x^3y + x^2y^2 + xy^3 + y^4)$
$x^7 - y^7 = (x - y)(x^6 + x^5y + x^4y^2 + x^3y^3 + x^2y^4 + xy^5 + y^6)$
$x^8 - y^8 = (x - y)(x^7 + x^6y + x^5y^2 + x^4y^3 + x^3y^4 + x^2y^5 + xy^6 + y^7)$
75. $x^n + y^n = (x + y)(x^{n-1} - x^{n-2}y + x^{n-3}y^2 - \cdots + x^2y^{n-3} - xy^{n-2} + y^{n-1})$ 77. $(\frac{1}{3}x - 4)(\frac{1}{3}x + 4)$
79. $\left(\frac{c}{6} - \frac{7}{d}\right)\left(\frac{c}{6} + \frac{7}{d}\right)$
81. **(a)** $y^2 - 4x^2 = (y - 2x)(y + 2x)$; $(12.8 - 4.8)(12.8 + 4.8) = (8)(17.6) = 140.8$
(b) $(y - 2x)^2$ is the area of the square determined by the inner vertices of the corner squares, and $4x(y - 2x)$ is the area of the remaining four rectangular strips. Also, factoring out $y - 2x$ gives $(y - 2x)(y - 2x + 4x) = (y - 2x)(y + 2x)$.

2.7 Factoring Trinomials (Page 71)

1. $(x + 2)^2$ 3. $(a - 7)^2$ 5. $(1 + b)^2$ 7. $4(a + 1)^2$ 9. $9(x - y)^2$ 11. $(x + 2)(x + 3)$ 13. $(x + 17)(x + 3)$
15. $(5a - 1)(4a - 1)$ 17. $(x + 2)(x + 18)$ 19. $(3x + 1)^2$ 21. $(5a - 1)^2$ 23. $(3x + 2)(x + 6)$
25. $(7x + 1)(2x + 5)$ 27. $(8x - 1)(x - 1)$ 29. $(2a + 5)^2$ 31. $(b + 15)(b + 3)$ 33. $2(2x - 3)(2x - 1)$
35. $(9t - 2)(2t - 7)$ 37. $(5x + 1)(3x - 2)$ 39. $(3a + 7)(2a - 3)$ 41. $(2x - 1)(2x + 3)$
43. $2(b - 3)(6b + 1)$ 45. $(8a + 3b)(3a + 2b)$ 47. $5(x + 1)(x + 4)$ 49. $a(2x + 1)^2$ 51. Not factorable
53. $2(3x - 5)(x + 2)$ 55. Not factorable 57. $2(x - 8)(x + 7)$ 59. $2(b + 2)(b + 4)$ 61. $ab(a - b)^2$
63. $8(2x - 1)(x - 1)$ 65. $25(a + b)^2$ 67. $-1(10x + 3)(3x - 1)$ 69. $-3xy(2x - y)^2$ 71. $\frac{1}{4}(x + 8)(x - 4)$
73. $\frac{1}{12}(2x - 1)(2x + 3)$

2.8 Using Various Factoring Methods (Page 75)

1. $(y - 2)(x + 3)$ 3. $(2h - 3)(7 - 3h)$ 5. $(2a + b)(a - b)(a + b)$ 7. $a(x - 1)(a - b)(a + b)$
9. $(a + 2)(a - b)$ 11. $(x + 1)(y + 1)$ 13. $(x + y)(a + b)$ 15. $(2 - y^2)(1 + x)$ 17. $(ax - by)(x^2 + y^2)$
19. $2xy(2x - 3)(4x + 7)$ 21. $5a(2b - a)(4b^2 + 2ab + a^2)$ 23. $(x - y)(a + b)(a^2 - ab + b^2)$
25. $3(3 - 2b)(9 + 6b + 4b^2)$ 27. $7(a + 5)(a - b)$ 29. $(a - 1)^2(a^2 + a + 1)^2$ 31. $3xy(x^2 + 2y)(2x^2 - 5y)$
33. $(a - b)^2(a^2 + ab + b^2)^2$ 35. $(2c - d)(c + 2d^2)(c^2 - 2cd^2 + 4d^4)$
37. $[(x + 5) + 7][(x + 5) - 2] = (x + 12)(x + 3)$ 39. $a(a + 1)^3$

CHAPTER 2 TEST: STANDARD ANSWER (Page 78)

1. **(a)** False **(b)** True **(c)** True **(d)** False **(e)** False 2. -200 3. $1\frac{9}{16}$ 4. -32 5. $9a^6$ 6. $4a^8$ 7. $\frac{a^4}{b^4}$ 8. ab^5 9. 1 10. $\frac{4x^8}{y^7}$ 11. $\frac{4y^7}{3x^6}$ 12. **(a)** 2.37×10^4; **(b)** 5.08×10^{-3} 13. **(a)** 563,000 **(b)** 0.000000806 14. $3x^2 + 2x - 2$ 15. $4x^2 + 5x - 16$ 16. $-2x^5 - 2x^4 + 6x^3 + 4x^2$ 17. $15x^2 - x - 6$ 18. $2x^3 - 3x^2 + x + 15$ 19. $x^2 + 3x + 7$ 20. $(7x + 2y)(7x - 2y)$ 21. $(2x + 3)(4x^2 - 6x + 9)$ 22. $5xy^2(3x^2 - 2xy^2 + y)$ 23. $(a - b)(a^2 + 7)$ 24. $2(2a + b)(3a - 2b)$ 25. $2x(4x - y)(16x^2 + 4xy + y^2)$

CHAPTER 2 TEST: MULTIPLE CHOICE (Page 79)

1. (c) 2. (e) 3. (b) 4. (a) 5. (b) 6. (d) 7. (c) 8. (d) 9. (c) 10. (d) 11. (a) 12. (b) 13. (e) 14. (b) 15. (c)

CHAPTER 3: INTRODUCTION TO EQUATIONS AND INEQUALITIES

3.1 Introduction to Equations (Page 86)

1. $x = 4$ 3. $x = -4$ 5. $x = -4$ 7. $x = -8$ 9. $x = \frac{9}{2}$ 11. $x = -9$ 13. $x = -20$ 15. $x = -32$ 17. $x = -25$ 19. $x = \frac{20}{3}$ 21. $x = 15$ 23. $x = -10$ 25. $x = -\frac{22}{3}$ 27. $x = 5$ 29. $x = 3$ 31. $\{-\frac{11}{2}\}$ 33. $\{3\}$ 35. $\{5\}$ 37. $\{-2\}$ 39. $\{3\}$ 41. $\{-4\}$ 43. $\{-21\}$ 45. $\{-\frac{1}{2}\}$ 47. $\frac{P - 2\ell}{2}$ 49. $\frac{N - u}{10}$ 51. $\frac{C}{2\pi}$ 53. $\frac{w - 7}{4}$ 55. **(a)** $\frac{2A - ha}{h}$ **(b)** 5.48 cm 57. 50°F 59. 33.8°F 61. 0°C 63. Inconsistent, Ø 65. Consistent, $x = 4$ 67. Identity, all real numbers 69. -6 71. $h = \frac{200}{\ell w}$

3.2 Applications: Problem Solving (Page 93)

1. 8 3. $w = 7$, $\ell = 21$ 5. 36, 38 7. Robert is 2, Ellen is 12 9. 18 11. **(a)** \$90 **(b)** \$270 **(c)** \$600 13. 10 at 15¢, 13 at 20¢, 5 at 25¢ 15. $3\frac{1}{2}$ hours 17. 24 19. Ben is 10 and Bob is 30 21. \$7000 at 9%; \$9700 at 12% 23. \$6562.50 25. \$7300 at 9%; \$5500 at 8% 27. 12, 14, 16, 18, 20 29. Let $x + (x + 2) + (x + 4) = 180$; then $3x = 174$ and $x = 58$; therefore, the integers must be even. 31. $w = 5$, $\ell = 14$ 33. $x, x + 2, 3x + 6, 3x + 15, 6x + 30, x + 5, (x + 5) - x = 5$

3.3 Introduction to Statements of Inequality (Page 101)

1. $\{x \mid x > 12\}$ 3. $\{x \mid x < 13\}$ 5. $\{x \mid x \geq 4\}$ 7. $\{x \mid x \leq -3\}$ 9. $\{x \mid x < 10\}$ 11. $\{x \mid x > -17\}$ 13. $\{x \mid x \geq -18\}$ 15. **(a)** $\{0, 1, 2, 3, 4, 5\}$ **(b)** $\{1, 2, 3, 4, 5\}$ 17. $\{x \mid x > 5\}$ 19. $\{x \mid x \geq -7\}$ 21. $\{x \mid x > -10\}$ 23. $\{x \mid x > 36\}$ 25. $\{x \mid x \leq -10\}$ 27. $\{x \mid x > -18\}$ 29. $\{x \mid x \leq 4\}$ 31. $\{x \mid x > \frac{5}{14}\}$

33. $\{x \mid x \geq 4\}$ 

35. $\{x \mid x \geq -9\}$ 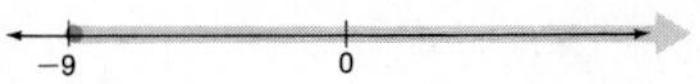

37. $\{x \mid x < -8\}$ 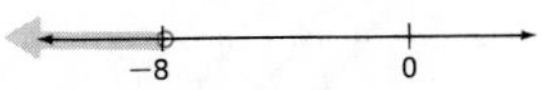

39. $\{x \mid x > -11\}$ 

41. $\{x \mid x > -\frac{7}{2}\}$ 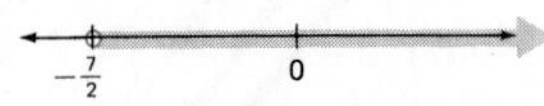

43. $\{x \mid x < -4\}$ 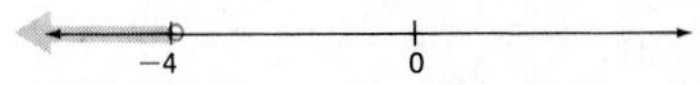

45. $\{x \mid x > 32\}$ 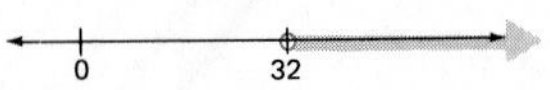

47. $\{x \mid x > -65\}$ 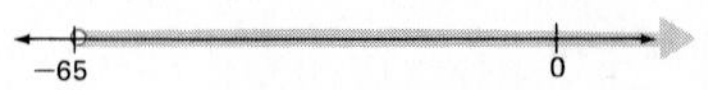

49. $\{x \mid x > -25\}$ 

51. $\left\{x \mid x > \dfrac{c-b}{a}\right\}$ 53. $\{x \mid x \geq 1\}$ 55. $\{x \mid x > 9\}$ 57. $\{x \mid x > 3\}$ 59. $\{x \mid x > 1\}$ 61. $\{x \mid x < 2\}$
63. $\{x \mid x < 8\}$ 65. After $\frac{2}{3}$ hour 67. $2\frac{1}{3}$ inches 69. 94; 74
71. \$18 for entertainment, \$27 for transportation, and \$108 for food 73. 22
75. Divide both sides by ab, which is positive;

$$\frac{a}{ab} < \frac{b}{ab}, \text{ giving } \frac{1}{b} < \frac{1}{a} \text{ or } \frac{1}{a} > \frac{1}{b}$$

3.4 Compound Inequalities (Page 107)

1. 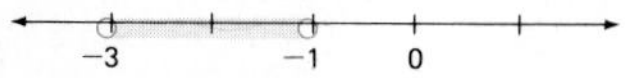
3. 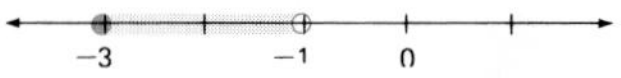
5. 
7. 

9. $[-5, 2]$ 11. $[-6, 0)$ 13. $(-10, 10)$ 15. $(-\infty, 5)$ 17. $[-2, \infty)$ 19. $(-\infty, -1]$
21. $-1 \leq x < 3$; $[-1, 3)$ 23. $-1 < x < 3$; $(-1, 3)$ 25. $x < 1$; $(-\infty, 1)$ 27. 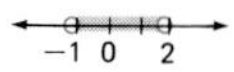
29. 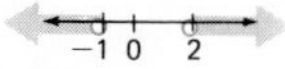 31. 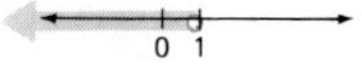 33. $-2 \leq x \leq 3$ 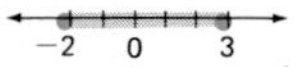
35. $-2 < x < 5$ 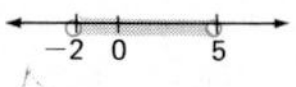 37. $1 < x < 2$ 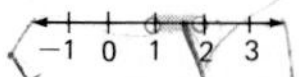 39. $x < -3$ or $x > -1$ 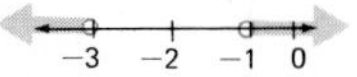
41. $x > -2$ and $x < 1$ 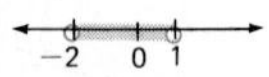 43. $-1 < n < 4$ 45. $n \leq -\frac{1}{2}$ or $n \geq \frac{1}{2}$ 47. $-\frac{4}{3} \leq n \leq \frac{5}{3}$
49. $n \geq 3$ or $n \leq -5$ 51. $n \geq -4$ and $n \leq 1$ 53. 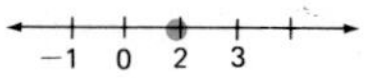 55. 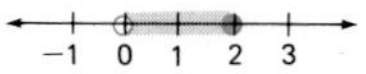
57. 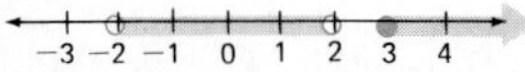 59. $2 \leq x \leq 7$ or $x > 10$ 61. At least 83 but less than 93
63. Between 77°F and 86°F 65. **(a)** $8 \leq g \leq 68$ **(b)** $2 \leq h \leq 16$
67. Between \$67.50 and \$82.50, and between \$82.50 and \$97.50.
69. Between \$86 and \$108 for 12 and between \$64 and \$86 for the other 8.

3.5 Inequalities and Absolute Value (Page 113)

1. $x = -3$, $x = 3$ 3. $x = -2$, $x = 4$ 5. $x = 2$, $x = 4$ 7. $x = -13$, $x = 13$ 9. $x = -2$, $x = 5$ 11. 
13. 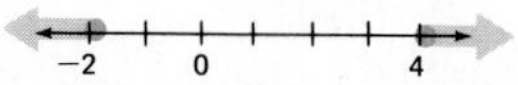 15. 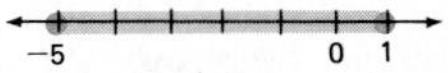
17. 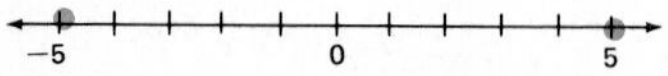 19. 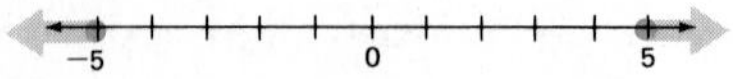
21. 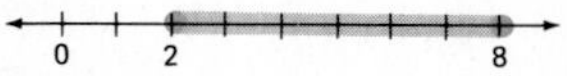 23. 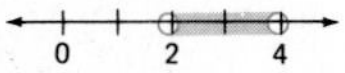 25. 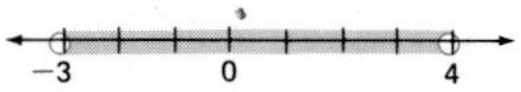
27. 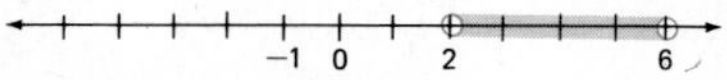 29. 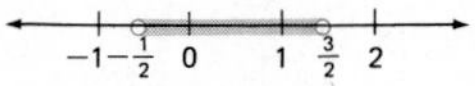
31. 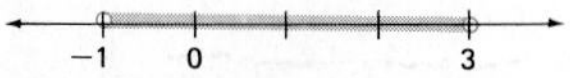 33. 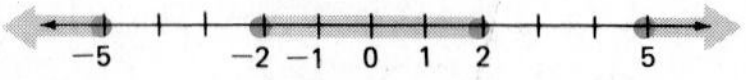
35. 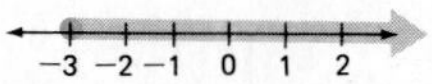 37. 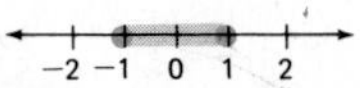 39. 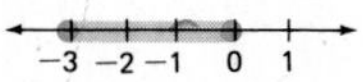
41. 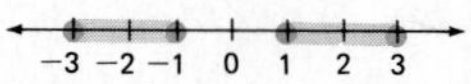 43. 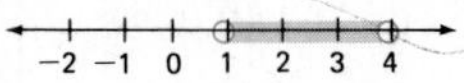 45. 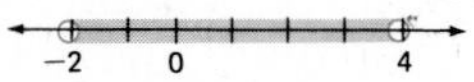

47. **(a)** $-4 \leq x \leq 12$ **(b)** $-4 \leq x \leq -2$
49. The graph of $|x - 3| \leq 2$ can be obtained by shifting the graph of $|x| \leq 2$ three units to the right; similarly, the graph of $|x - 3| \geq 2$ can be obtained by shifting the graph of $|x| \geq 2$ three units to the right.
51. **(a)** $|x - 1| < 3$ **(b)** $|x - 1| \geq 3$ 53. **(a)** $|x - \frac{7}{2}| < \frac{3}{2}$ **(b)** $|x - \frac{7}{2}| > \frac{3}{2}$
55. $|x - 6| < \frac{2}{5}$; $-\frac{2}{5} < x - 6 < \frac{2}{5}$; $-\frac{1}{5} < \frac{1}{2}x - 3 < \frac{1}{5}$; $|\frac{1}{2}x - 3| < \frac{1}{5}$

57. **(a)** $|7| \cdot |5| = 7 \cdot 5 = 35$ and $|7 \cdot 5| = |35| = 35$
$|-7| \cdot |5| = 7 \cdot 5 = 35$ and $|-7 \cdot 5| = |-35| = 35$
$|-7| \cdot |-5| = 7 \cdot 5 = 35$ and $|(-7)(-5)| = |35| = 35$
(b) (i) $|x| = -x$ since $x < 0$ and $|y| = y$ since $y > 0$.
(ii) Opposite of a product (see page 9)
(iii) $xy < 0$

59. **(a)** $|5 - 2| \geq ||5| - |2||; 3 \geq 3$
(b) $|5 - (-2)| \geq ||5| - |-2||; 7 \geq 3$
(c) $|-5 - 2| \geq ||-5| - |2||; 7 \geq 3$
(d) $|-5 - (-2)| \geq ||-5| - |-2||; 3 \geq 3$

61. $x > -2$ 63. All $x \neq 3$

3.6 Introduction to Quadratic Equations (Page 118)

1. $x = 2$ 3. $x = 0, x = 10$ 5. $x = -\frac{1}{5}, x = \frac{3}{2}$ 7. $x = 4$ 9. $x = -2, x = 18$ 11. $x = -7, x = 4$
13. $x = 0, x = 5$ 15. $x = -3, x = 3$ 17. 8 19. 3 by 3 21. -11, 11
23. There is no real number x whose square is -1. 25. 23 by 14 27. -3, 0, 2
29. 7 centimeters by 16 centimeters 31. 6 boxes, 24 cans per box 33. 7, 9, 11 or -5, -3, -1
35. $\ell = 12, w = 2, h = 4$

CHAPTER 3 TEST: STANDARD ANSWER (Page 122)

1. $x = 9$ 2. $x = -4$ 3. $x = -\frac{7}{2}$ 4. $x = 4$ 5. $x > -5$ 6. $x < 2$ 7. $x \geq -\frac{17}{2}$ 8. $x \leq -\frac{11}{3}$

9. 

10. 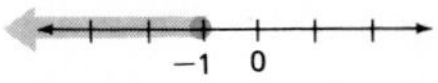

11. 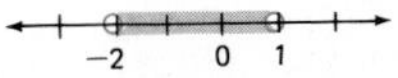

12. 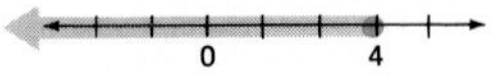

13. 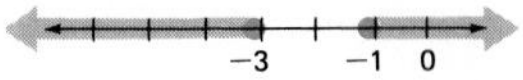

14. 

15. $x \leq 2$ and $x \geq -3$ 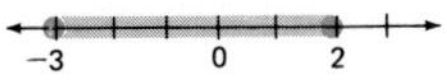

16. $x < -1$ or $x > 2$ 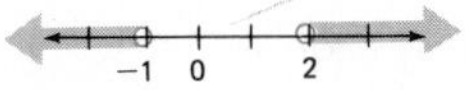

17. **(a)** 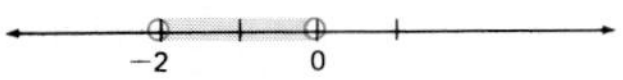

(b) 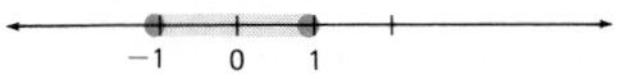

(c) 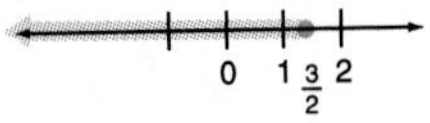

18. **(a)** $x = -5$ or $x = \frac{2}{3}$ **(b)** $x = 3$ or $x = -3$ 19. $h = \dfrac{S - 2\pi r^2}{2\pi r}$ 20. **(a)** $|x - 3| < 2$ **(b)** $|x - 5| > 2$
21. $w = 7$ inches, $\ell = 19$ inches 22. \$4600 23. After $\frac{1}{4}$ hour 24. 2 inches 25. **(a)** 130 **(b)** 146

CHAPTER 3 TEST: MULTIPLE CHOICE (Page 122)

1. (b) 2. (c) 3. (a) 4. (e) 5. (c) 6. (b) 7. (c) 8. (b) 9. (d) 10. (a) 11. (e) 12. (c) 13. (a)
14. (d) 15. (c)

ANSWERS TO CUMULATIVE REVIEW FOR CHAPTERS 1–3 (Page 125)

1. +32 2. +25 3. -7 4. +7 5. -5 6. -6 7. Commutative property for addition
8. Commutative property for multiplication 9. Distributive property 10. Associative property for addition
11. True 12. True 13. False 14. False 15. True 16. y^5 17. x^5 18. $\dfrac{b^9}{a^{11}}$ 19. $-32xy$ 20. $\dfrac{y^3}{x^3}$ 21. x^2y

22. If $a \times b = 0$, then $a = 0$ or $b = 0$ or both. 23. (a) 3.05×10^6 (b) 5.7×10^{-5}
24. (a) 472,000 (b) 0.000108 25. $8x^3 - 3x^2 - 5x + 1$ 26. $-x^3 - 8x^2 + 14x - 9$ 27. $15x^5 - 5x^4 + 25x^2$
28. $2x^3 - 5x^2 + 5x - 6$ 29. $15x^2 + 29xy - 14y^2$ 30. $4a^4 + 4a^2b^2 + b^4$ 31. $2x(x - 5)(x + 5)$
32. Cannot be factored 33. $(2x + 3)(x + 1)$ 34. $xy(x - 4)(x^2 + 4x + 16)$ 35. $x(3x - 7)(x - 2)$
36. $(x - 2)(x^3 + 3)$ 37. $x = -13$ 38. $x = -54$ 39. $\dfrac{b^2}{a - 2}$ 40. width = 5 cm, length = 13 cm 41. \$3500

42. $x \leq -2$ 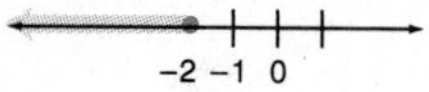

43. $-4 \leq x \leq 2$ 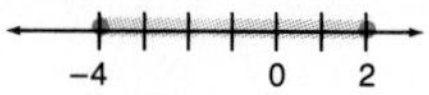

44. $x < \frac{1}{3}$ or $x > 1$ 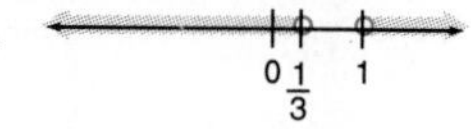

45. $x \leq -2$ and $x > -4$ 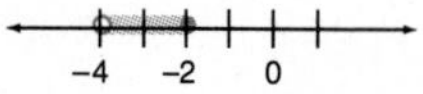

46. $x = 0$ or $x = 5$ 47. $x = -4$ or $x = \frac{3}{4}$

48. $x = -1$ or $x = \frac{3}{2}$ 49. (a) 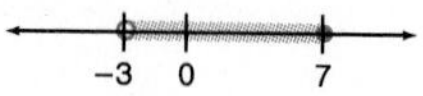

(b) 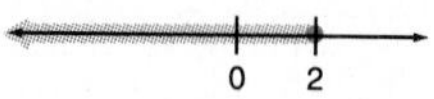

50. 47

CHAPTER 4: FUNDAMENTAL OPERATIONS WITH RATIONAL EXPRESSIONS

4.1 Simplifying Rational Expressions (Page 131)

1. $x = 0$ 3. $x = -y$ 5. $x = 3$ or $x = -3$ 7. none 9. $\frac{3}{5}$ 11. $\frac{3}{2}$ 13. $\frac{5}{3}$ 15. $\dfrac{2x}{3z}$ 17. $-8n$ 19. $\dfrac{3x}{4}$

21. $-\dfrac{2a}{3d}$ 23. $\dfrac{a}{2b^2}$ 25. -1 27. -1 29. $\dfrac{2x - 3}{3x - 2}$ 31. x 33. $\dfrac{1}{n - 1}$ 35. $\dfrac{n + 1}{n^2 + 1}$ 37. $\dfrac{x + 1}{x - 1}$ 39. $\dfrac{a - 2}{a + 1}$

41. $-\dfrac{1}{a^2}$ 43. -1 45. (a) $\frac{1}{8}$ (b) $-\frac{5}{12}$ 47. (a) 0 (b) undefined 49. (a) undefined (b) 0.8 51. $\dfrac{c - 3}{c + 3}$

53. $-\dfrac{x + 1}{x - 1}$ 55. $\dfrac{2n + 1}{1 - n}$ 57. $\dfrac{3n + 2}{n + 2}$ 59. $(x - 3)(x + 1)$ 61. $\dfrac{-5}{x^2 - 3x + 9}$ 63. $\dfrac{x + y}{x + 2}$

65. $\dfrac{a - 4b}{a^2 - 4ab + 16b^2}$ 67. $-(2 + x)$ 69. $\dfrac{b}{a^n}$ 71. $\dfrac{a^n - b^n}{a^n + b^n}$

4.2 Multiplication and Division of Rational Expressions (Page 135)

1. $\frac{15}{56}$ 3. 16 5. $\frac{40}{57}$ 7. $\dfrac{3x^2}{2y^2}$ 9. $\dfrac{3}{8x^2}$ 11. $\dfrac{5x}{2}$ 13. $\dfrac{8x}{9}$ 15. $\dfrac{3b^2x}{2a^2}$ 17. xy 19. $\dfrac{a^2}{b}$ 21. $\dfrac{2a^4}{b^3c^2}$ 23. $\dfrac{2}{a^2}$

25. $\dfrac{x^2 + 1}{3(x + 1)}$ 27. $\dfrac{a}{2(a - 5)}$ 29. $x(x + 1)$ 31. 1 33. $\dfrac{1}{x + 2}$ 35. $\dfrac{x + 2}{x + 1}$ 37. $\dfrac{t - 6}{t(3t + 1)}$ 39. $\dfrac{n - 2}{n - 3}$

41. $\dfrac{a - 2b}{a + 2b}$ 43. $\dfrac{x + 1}{2x + 1}$ 45. $\dfrac{a - b}{a + b}$ 47. $\dfrac{1}{n}$ 49. $\dfrac{x + 2y}{2x + y}$ 51. $-\dfrac{2}{y^2(y + 1)}$ 53. -0.25

4.3 Addition and Subtraction of Rational Expressions (Page 141)

1. $\frac{7}{5}$ 3. $\frac{17}{10}$ 5. $\frac{13}{12}$ 7. $\dfrac{4x}{5}$ 9. $\dfrac{5x}{a}$ 11. x 13. $\dfrac{2x + 5}{4}$ 15. $a - 1$ 17. $x + 1$ 19. $\dfrac{4a + 3b}{a}$ 21. 1

23. $\dfrac{11x + 8y}{15}$ 25. $\dfrac{23}{6x}$ 27. $\dfrac{3x - 1}{3x}$ 29. $\dfrac{5x}{a + b}$ 31. $x + 1$ 33. $\dfrac{5y^2 - y}{y^2 - 1}$ 35. $\dfrac{3a - a^2}{a^2 - 1}$ 37. $\dfrac{3a^2 + 4b - a}{a^3b^2}$

39. $a + 3$ 41. $\dfrac{3x}{x + 1}$ 43. $\dfrac{2x - 9}{(x + 2)(x + 3)}$ 45. $\dfrac{2x}{x + 1}$ 47. $x - 8$ 49. $\dfrac{-x^2 + 11x - 14}{(x + 2)(x - 1)(x - 3)}$ 51. $\frac{11}{12}$

53. $\dfrac{(2x - 3)(x + 2)}{x(x + 3)(x - 2)(x - 1)}$ 55. $\dfrac{a + 11}{a^2 - 4}$ 57. $\dfrac{5}{x - 5}$ 59. $\dfrac{1}{y + 1}$

61. $\dfrac{3x(x - 2) + x(x^2 - 4)}{(x^2 - 4)(x - 2)} = \dfrac{x(x + 5)}{x^2 - 4}$; $\dfrac{3x}{x^2 - 4} + \dfrac{x(x + 2)}{x^2 - 4} = \dfrac{x(x + 5)}{x^2 - 4}$

63. (a) 2.5327103 (b) 4.4354067

4.4 Complex Fractions (Page 146)

1. 1 3. $\frac{20}{19}$ 5. 30 7. $\frac{a}{bc}$ 9. $\frac{ad}{bc}$ 11. $\frac{x}{15y^2}$ 13. $\frac{1}{6x^2y}$ 15. $\frac{x+3}{2}$ 17. $\frac{3}{5}$ 19. $\frac{15}{2}$ 21. $\frac{2}{3}$ 23. $-\frac{64}{5}$

25. $\frac{2x+3}{2x-1}$ 27. $-\frac{1}{2x}$ 29. $-\frac{3+x}{9x^2}$ 31. $\frac{a^2b^2}{a+b}$ 33. $-\frac{1}{2(2+h)}$ 35. $-\frac{x+5}{x^2y}$ 37. $2a$ 39. $\frac{x(x-3)}{x-5}$

41. $\frac{t+3}{t-3}$ 43. $\frac{b^2-a^2}{a^2b^2}$ 45. $\frac{b(b-a^2)}{a(b^2-a)}$ 47. 2 49. $\frac{3x+2}{2x+1}$

51. **(a)** $\dfrac{\frac{a+b+c}{3}+d}{2} = \dfrac{a+b+c+3d}{6}$ **(b)** $\dfrac{6\left(\frac{a+b+c}{3}\right)+4d}{10} = \dfrac{a+b+c+2d}{5}$

4.5 Division of Polynomials (Page 152)

1. $3a^2-5a$ 3. $3x+1$ 5. $3x^2+2x+1$ 7. a^3+2a+3 9. $4-2xy+x^2y^2$ 11. $y+x^2y^2$
13. $a+1-ab$ 15. $3-4a^2x^4$ 17. $x+3$ 19. $2x+3$ 21. $5x-8, r=9$ 23. $12x-39, r=124$
25. x^2-5x+2 27. $x^2, r=7$ 29. $2x+3$ 31. $x^2+7x+7, r=22$ 33. $3(x-y)+4$
35. $-5(a+b)+1$ 37. $3y-2$ 39. a^2+3a+9 41. x^3+2x^2+4x+8 43. $a^2+3ab+9b^2$
45. $x^4+2x^3+4x^2+8x+16$ 47. $4x+3, r=7x-7$ 49. $x+2$ 51. **(a)** -4; **(b)** -4
53. **(a)** -55; **(b)** -55

4.6 Equations with Rational Expressions (Page 156)

1. $x=20$ 3. $x=7$ 5. $x=4$ 7. $x=\frac{17}{4}$ 9. $x=2$ 11. $x=9$ 13. $x=-4$ 15. $x=12$ 17. $x=-2$

19. $x=-\frac{5}{2}$, $x=2$ 21. $x=1$, $x=2$ 23. $x=20$ 25. $x=-3$, $x=3$ 27. $x=-\frac{1}{2}$, $x=3$ 29. $\emptyset$ 31. $x=-3$ 33. $x=\frac{ab}{c}$

35. $x=\frac{ay}{b+y}$ 37. $x=\frac{a-2b}{2c}$ 39. $m=\frac{2gK}{v^2}$ 41. $s=a+(n-1)d$ 43. $m=\frac{fp}{p-f}$

45. $R=\frac{V+\frac{1}{3}\pi h^3}{\pi h^2}=\frac{3V+\pi h^3}{3\pi h^2}$ 47. $\ell=\frac{S-2hw}{2w+2h}$ 49. $s=\frac{S-\pi r^2}{\pi r}$

4.7 Applications of Fractional Equations (Page 162)

1. $x=\frac{15}{2}$ 3. $x=-3$, $x=3$ 5. $x=-\frac{5}{2}$, $x=\frac{5}{2}$ 7. $x=6$ 9. $\frac{2}{3},\frac{3}{2}$ 11. 5 13. 2 15. 3 17. $1\frac{5}{7}$ hours 19. $\frac{2}{9},\frac{4}{9}$

21. 20, 21, 22 23. 60 feet 25. 170 feet 27. 8000 29. 7 31. $\frac{2}{3}$ 33. 4, 8 35. $2\frac{2}{3}$ hours

37. $\frac{a}{b}+1=\frac{c}{d}+1;\ \frac{a}{b}+\frac{b}{b}=\frac{c}{d}+\frac{d}{d};\ \frac{a+b}{b}=\frac{c+d}{d}$ 39. 92.31 ohms 41. 10 centimeters

43. 110 paperbacks and 55 hardcover copies

4.8 Variation (Page 169)

1. $P=4s;\ k=4$ 3. $A=5l;\ k=5$ 5. $z=kxy^3$ 7. $z=\frac{kx}{y^3}$ 9. $w=\frac{kx^2}{yz}$ 11. $\frac{1}{2}$ 13. $\frac{1}{5}$ 15. $\frac{2}{5}$ 17. $\frac{243}{8}$

19. 27 feet 21. 4 pounds 23. 75 lbs/in^2 25. 784 feet 27. $\frac{32\pi}{3}$ cubic inches 29. 5.12 ohms; 512 ohms

31. **(a)** $ky(\frac{25}{4}-y^2)$ **(b)** $\frac{1}{2}kx^2\sqrt{25-4x^2}$ **(c)** $5.52k$

CHAPTER 4 TEST: STANDARD ANSWER (Page 173)

1. $\frac{ac^3}{2b^2}$ 2. $\frac{3x}{2(x+2)^2}$ 3. $-\frac{x}{x+2}$ 4. $\frac{x+1}{2x-1}$ 5. $\frac{x+1}{x+2}$ 6. 2 7. $-\frac{2}{x}$ 8. $\frac{3x(x-1)}{(x+3)(x-3)}$ 9. $\frac{x^2+4}{(x+2)^2}$
10. $\frac{x+1}{x-1}$ 11. $\frac{x-3}{x-2}$ 12. $4ab^2 - 2 + 3a^2b^4$ 13. **(a)** $\frac{2}{x}$ **(b)** $-\frac{1}{3x}$ 14. $\frac{x(y^2+x)}{y(y+x^2)}$ 15. $\frac{3(x^2+y^2)}{y(x+y)}$
16. $2x^2 + 5x - 3$ 17. $x^2 + 2x - 3, r = -x + 5$ 18. $x = -1$ 19. $x = 6, x = 10$ 20. Ø 21. $x = 2$
22. $\frac{7}{13}$ 23. 90 minutes 24. 24, 25 25. $z = 2$

CHAPTER 4 TEST: MULTIPLE CHOICE (Page 174)

1. (e) 2. (c) 3. (d) 4. (a) 5. (b) 6. (a) 7. (b) 8. (a) 9. (e) 10. (c) 11. (b) 12. (b) 13. (d)
14. (e) 15. (c)

CHAPTER 5: RADICALS, COMPLEX NUMBERS, AND THE QUADRATIC FORMULA

5.1 Radicals (Page 182)

1. 9 3. 3 5. 3 7. $\frac{1}{5}$ 9. Not real since $-4 < 0$ and $n = 2$ is even 11. $4x$
13. Not real since $-81 < 0$ and $n = 4$ is even 15. 0.1 17. -1 19. -4 21. -0.2 23. -3
25. $3\sqrt{5} = 6.708$ 27. $3\sqrt[3]{10} = 6.462$ (by Table I) 29. $2\sqrt{14}$ 31. $4\sqrt{5x}$ 33. $2x\sqrt[3]{10x}$ 35. $3x^2\sqrt[3]{3}$
37. $-2\sqrt[4]{5}$ 39. $5y\sqrt{2y}$ 41. $\frac{8}{5}$ 43. $-2\sqrt[3]{2}$ 45. $3\sqrt{2}$ 47. $5\sqrt{2}$ 49. $3t\sqrt{10}$ 51. 85 53. $18x^2\sqrt{5}$
55. -3 57. 5 59. $-\sqrt{2}$ 61. $\frac{1}{2}\sqrt[3]{4}$ 63. $7\sqrt{2}$ 65. 0.3 67. $-\frac{3}{s}$ 69. $35\sqrt{3}$ 71. $4\sqrt[3]{35}$ 73. 2
75. $4\sqrt{5}$ 77. $4x\sqrt{2}$ 79. $\frac{\sqrt{2}}{6}$ 81. $\frac{\sqrt{3}}{2}$ 83. $4\sqrt[3]{4}$ 85. $2\sqrt[4]{5}$ 87. $\frac{9\sqrt{3+x}}{3+x}$ 89. $\frac{2\sqrt[3]{b}}{b^3}$
91. **(a)** 25 centimeters **(b)** 18.87 centimeters
93. $ab = x^ny^n = (xy)^n$, thus $\sqrt[n]{ab} = xy$; but $x = \sqrt[n]{a}$ and $y = \sqrt[n]{b}$, thus $\sqrt[n]{ab} = \sqrt[n]{a} \cdot \sqrt[n]{b}$
95. **(a)** $\sqrt[4]{\sqrt{256}} = \sqrt[4]{16} = 2$; $\sqrt{\sqrt[4]{256}} = \sqrt{4} = 2$; $\sqrt[8]{256} = 2$
(b) $\sqrt[3]{\sqrt{729}} = \sqrt[3]{27} = 3$; $\sqrt{\sqrt[3]{729}} = \sqrt{9} = 3$; $\sqrt[6]{729} = 3$
(c) $\sqrt[3]{\sqrt[3]{-512}} = \sqrt[3]{-8} = -2$; $\sqrt[9]{-512} = -2$;
$\sqrt[m]{\sqrt[n]{a}} = \sqrt[mn]{a}$

5.2 Combining Radicals (Page 189)

1. 8 3. $\frac{7}{2}$ 5. $\frac{1}{4}$ 7. $|x|$ 9. $2|s|$ 11. $b \neq 0$ 13. $x \leq 1$ 15. $x \geq 0, y \neq 0$ 17. $x \geq 0$ 19. $4\sqrt{2}$
21. 12 23. $-\sqrt{3}$ 25. $10\sqrt{2}$ 27. $17\sqrt{5}$ 29. $10\sqrt{2}$ 31. $6\sqrt[3]{2}$ 33. $7\sqrt{2}$ 35. $14\sqrt{2}$ 37. $\frac{\sqrt{3}}{9}$
39. $2\sqrt[3]{2}$ 41. $3\sqrt[3]{7x}$ 43. $|x|\sqrt{2}$ 45. $5\sqrt[4]{2}$ 47. $5\sqrt{6} - 3\sqrt{2}$ 49. $5\sqrt{10}$ 51. $7\sqrt{2x}$ 53. $\frac{33\sqrt{2}}{2}$ 55. 0
57. $12|x|$ 59. $|x|\sqrt{y} + 12|x|\sqrt{2y}$ 61. $7a\sqrt{5a}$ 63. $\frac{\sqrt[3]{4x}}{x}$ 65. $2\sqrt{3} - 6$ 67. $6\sqrt{2}$ 69. $9 - 14\sqrt{6}$
71. $6x^2 - 5x\sqrt{y} + y$ 73. $x - 4 + \frac{4}{x}$ 75. $6(\sqrt{5} + \sqrt{3})$ 77. $-2(\sqrt{2} + 3)$ 79. $\frac{x + 2\sqrt{xy} + y}{x - y}$
81. $8 - 3\sqrt{7}$ 83. $\sqrt{x} + \sqrt{y}$ 85. $\frac{1}{3 - \sqrt{5}}$ 87. $\frac{1}{3\sqrt{7} - 4\sqrt{5}}$ 89. 4
91. $-1 + \sqrt{2} = .4142$; $(0.4142)^2 + 2(0.4142) - 1 = -0.0000348$

5.3 Rational Exponents (Page 196)

1. $\sqrt{3}$ 3. $\sqrt{8}$ 5. $\sqrt[4]{15}$ 7. $\sqrt[4]{5^3}$ 9. $\dfrac{1}{\sqrt[3]{7^2}}$ 11. $(\sqrt[3]{\tfrac{4}{5}})^2 = \sqrt[3]{\tfrac{16}{25}}$ 13. $11^{1/2}$ 15. $9^{1/4}$ 17. $(-20)^{1/5}$
19. $(-7)^{2/3}$ 21. $4^{-2/3}$ 23. $(\frac{1}{10})^{1/2}$ 25. 11 27. -5 29. $\frac{1}{4}$ 31. $\frac{1}{16}$ 33. $\frac{1}{16}$ 35. $\frac{4}{9}$ 37. 1 39. -9 41. $\frac{1}{32}$
43. 36 45. $\frac{1}{9}$ 47. $\dfrac{a^2c^6}{b}$ 49. $\dfrac{4a^2}{b^6}$ 51. $\dfrac{1}{a^3b^6}$ 53. 1 55. $\dfrac{1}{a^2b^{2/3}}$ 57. $\dfrac{4}{a^2b^4}$ 59. x 61. $\dfrac{1}{xy^n}$ 63. $\frac{5}{6}$ 65. $\frac{8}{9}$
67. $-\frac{1}{4}\sqrt[3]{1001}$ 69. $10x^{1/3} + 4$ 71. $4a - 1$ 73. $b + 2 + \dfrac{1}{b}$ 75. $\dfrac{3}{y^{1/2}} - \dfrac{13}{y^{1/4}} - 10$ 77. $1 + y^2$
79. $y^{-1/2} - y^{3/2} = \dfrac{1}{y^{1/2}} - y^{3/2} = \dfrac{1 - y^2}{y^{1/2}} = \dfrac{1 - y^2}{\sqrt{y}}$ 81. $y^{-3/2} + y^{1/2} = \dfrac{1}{y^{3/2}} + y^{1/2} = \dfrac{1 + y^2}{y^{3/2}} = \dfrac{1 + y^2}{\sqrt{y^3}}$
83. $b^{2/3} + \dfrac{2}{3}b^{-1/3}(b - 10) = b^{2/3} + \dfrac{2b - 20}{3b^{1/3}} = \dfrac{3b + 2b - 20}{3b^{1/3}} = \dfrac{5b - 20}{3\sqrt[3]{b}}$ 85. $\dfrac{a^{3/2} + 1}{a^3 - 1}$ 87. $\dfrac{a + 1}{a - 1}$
89. $\sqrt[4]{-16}$ is not a real number.

5.4 Equations With Radicals or Rational Exponents (Page 201)

1. 17 3. 9 5. 9 7. 10 9. $-5, 5$ 11. $\emptyset$ 13. $\frac{9}{2}$ 15. $\frac{1}{3}$ 17. 9 19. 16 21. 2 23. 2 25. 10
27. $\frac{17}{3}$ 29. $\frac{5}{16}$ 31. 0, 8 33. 1 35. 9 37. 4 39. $-1, 7$ 41. 9 43. 7 45. $A = 4\pi r^2$; 16π
47. $h = \sqrt{s^2 - r^2}$; 14.72

5.5 Complex Numbers (Page 208)

1. True 3. True 5. False 7. $5 + 2i$ 9. $-5 + 0i$ 11. $4i$ 13. $12i$ 15. $\frac{3}{4}i$ 17. $-\sqrt{5}i$ 19. -27
21. $-\sqrt{6}$ 23. $15i$ 25. $12i$ 27. $5i\sqrt{2}$ 29. $(3 - \sqrt{3})i$ 31. $10 + 7i$ 33. $5 - 3i$ 35. $10 + 2i$
37. $-10 + 6i$ 39. $13i$ 41. $23 + 14i$ 43. $5 - 3i$ 45. $\frac{13}{5} + \frac{1}{5}i$ 47. $\frac{4}{5} - \frac{3}{5}i$ 49. $4 - 3i$ 51. $\frac{53}{13} - \frac{21}{13}i$
53. $-3i$ 55. $-2i$ 57. 4 59. $\frac{3}{13} - \frac{2}{13}i$ 61. **(a)** $(\sqrt{2}i)^2 + 2 = 2i^2 + 2 = -2 + 2 = 0$ **(b)** $x = -\sqrt{2}i$
63. $[(3 + i)(3 - i)](4 + 3i) = (9 - i^2)(4 + 3i) = 10(4 + 3i) = 40 + 30i$
$(3 + i)[(3 - i)(4 + 3i)] = (3 + i)(15 + 5i) = 40 + 30i$
65. $\dfrac{ac + bd}{c^2 + d^2} + \dfrac{bc - ad}{c^2 + d^2}i$ 67. $(x + i)(x - i)$ 69. $3(x + 5i)(x - 5i)$

5.6 Solving Quadratic Equations by Completing the Square (Page 213)

1. $x = \pm 6$ 3. $x = 0$ 5. $x = \pm 2\sqrt{3}$ 7. $x = 3$, $x = -1$ 9. $x = -3 \pm \sqrt{5}i$ 11. $x = \dfrac{-1 \pm \sqrt{6}}{2}$ 13. 9 15. 16
17. $\frac{1}{4}$ 19. $x = -3$, $x = 1$ 21. $x = -7$, $x = 1$ 23. $x = 1$, $x = 3$ 25. $x = 2 \pm \sqrt{2}i$ 27. $x = -1 \pm \sqrt{3}$ 29. $x = \dfrac{1 \pm \sqrt{13}}{2}$
31. $x = \dfrac{-1 \pm i}{2}$ 33. $x = -1$, $x = \frac{2}{3}$ 35. $x = \dfrac{3 \pm \sqrt{6}}{3}$ 37. $x = \dfrac{3 \pm \sqrt{11}}{2}$ 39. $x = -2 \pm \sqrt{11}$ 41. $x = -3 \pm \sqrt{19}$

5.7 The Quadratic Formula (Page 217)

1. $x = -2$, $x = 5$ 3. $x = -2$, $x = \frac{1}{2}$ 5. $x = 3$ 7. $x = \pm\frac{3}{2}$ 9. $x = \pm\dfrac{\sqrt{6}}{2} \approx \pm 1.22$ 11. $x = 3 + \sqrt{3} \approx 4.73$, $x = 3 - \sqrt{3} \approx 1.27$
13. $x = 1 + \sqrt{5} \approx 3.24$, $x = 1 - \sqrt{5} \approx -1.24$ 15. $x = 1 \pm i$ 17. $x = 3 \pm \sqrt{5}i$ 19. $x = \dfrac{3 + \sqrt{17}}{4} \approx 1.78$, $x = \dfrac{3 - \sqrt{17}}{4} \approx -0.28$ 21. $x = \frac{1}{2}$, $x = 2$

23. $x = \dfrac{-1 \pm \sqrt{17}i}{6}$ 25. (a) 27. (b) 29. (d) 31. (c) 33. (c) 35. (d) 37. $b = \pm 6$ 39. $b = \pm 12$

41. $k > -4$ 43. $k > -\frac{1}{4}; k \neq 0$ 45. $t > 9$

47. $\dfrac{-b + \sqrt{b^2 - 4ac}}{2a} + \dfrac{-b - \sqrt{b^2 - 4ac}}{2a} = \dfrac{-2b}{2a} = -\dfrac{b}{a};$

$\left(\dfrac{-b + \sqrt{b^2 + 4ac}}{2a}\right)\left(\dfrac{-b - \sqrt{b^2 - 4ac}}{2a}\right) = \dfrac{b^2 - (b^2 - 4ac)}{4a^2} = \dfrac{4ac}{4a^2} = \dfrac{c}{a}$

49. Sum $= -\frac{5}{6}$; product $= -\frac{2}{3}$ 51. Sum $= \frac{26}{3}$; product $= \frac{35}{3}$ 53. Sum $= -3$; product $= \frac{9}{2}$ 55. $x = -\frac{3}{2}$, $x = \frac{1}{4}$

57. $x = -1$, $x = 5$ 59. $x = -\sqrt{3} \pm 1$ 61. $x = -i$, $x = 2i$ 63. $x = \pm 2$, $x = \pm 1$ 65. $x = -4$, $x = -1, x = 2$ 67. $x = 0$, $x = \frac{4}{3}$ 69. $x = -3$, $x = \pm 2$

5.8 Applications of Quadratic Equations (Page 222)

1. 14, 15 3. $2 + \sqrt{3} \approx 3.73$; $2 - \sqrt{3} \approx 0.27$ 5. 5, 8 7. 2 feet 9. Width = 3 centimeters; length = 5 centimeters 11. $\dfrac{5\sqrt{2}}{2} \approx 3.54$ seconds

13. 8, 10 or $-8, -10$ 15. **(a)** 3 seconds; 5 seconds **(b)** 8 seconds 17. $\dfrac{8 + \sqrt{14}}{2} \approx 5.87$ seconds; $\dfrac{8 - \sqrt{14}}{2} \approx 2.13$ seconds 19. $n = 15$ 21. 7, 9

23. $\dfrac{-1 + \sqrt{33}}{4} \approx 1.19$; $\dfrac{-1 - \sqrt{33}}{4} \approx -1.69$ 25. 8 centimeters 27. 20 mph; 1.5 hours downstream; 2.25 hours upstream 29. 1.2 hours

31. 6 hours for José; 10 hours for brother 33. $V = 4 \cdot 4 \cdot 7 = 112$ cubic inches 35. **(a)** 116 ft **(b)** 197 ft **(c)** 262 ft

CHAPTER 5 TEST: STANDARD ANSWER (Page 225)

1. **(a)** -8 **(b)** $\dfrac{1}{9}$ 2. $\dfrac{a^{1/6}}{b^{1/6}}$ 3. $\dfrac{a^{7/6}}{b^{3/4}}$ 4. **(a)** $4\sqrt{3}$ **(b)** $3\sqrt{5}$ **(c)** $-3x^2\sqrt[3]{3}$ 5. $10\sqrt{2}$ 6. $6\sqrt{3}$ 7. $5|x|$

8. $4|x|y\sqrt{2y}$ 9. $27 - \sqrt{7}$ 10. $x = 64$ 11. $x = 9$ 12. 5 13. **(a)** $12\sqrt{3}i$ **(b)** $-6\sqrt{2}$ 14. $9 - 7i$

15. $1 + i$ 16. $x = 2, x = 6$ 17. $x = \dfrac{-3 \pm \sqrt{17}}{2}$ 18. $x = \dfrac{3 \pm \sqrt{33}}{2}$ 19. $x = \dfrac{-1 \pm \sqrt{7}}{3}$

20. $x = -\dfrac{5}{2}, x = 1$ 21. $2 \pm 2i$ 22. **(a)** 41; two irrational numbers **(b)** -23; two imaginary numbers

23. $k > 2$ 24. 11 and 13 or -11 and -13 25. 2 seconds, 8 seconds

CHAPTER 5 TEST: MULTIPLE CHOICE (Page 229)

1. (a) 2. (b) 3. (a) 4. (b) 5. (e) 6. (c) 7. (b) 8. (d) 9. (d) 10. (b) 11. (c) 12. (b) 13. (a) 14. (c) 15. (d)

CHAPTER 6: GRAPHING LINEAR EQUATIONS AND INEQUALITIES; THE FUNCTION CONCEPT

6.1 The Cartesian Coordinate System (Page 235)

1. $A(-4, 0)$; $B(-2, 2)$; $C(0, 4)$; $D(2, 2)$; $E(4, 0)$; $F(2, -2)$; $G(0, -4)$; $H(-2, -2)$

3. $A(-2, -2)$; $B(-\frac{3}{2}, -2)$; $C(-1, -2)$; $D(-\frac{1}{2}, -2)$; $E(0, -2)$; $F(\frac{1}{2}, -2)$

5. 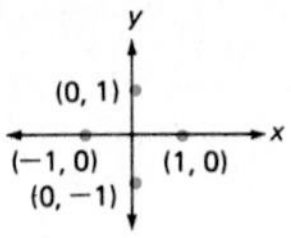

7. 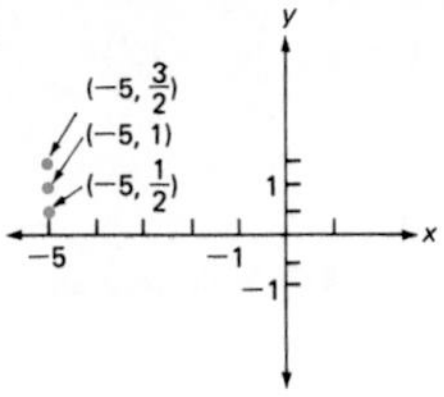

9. 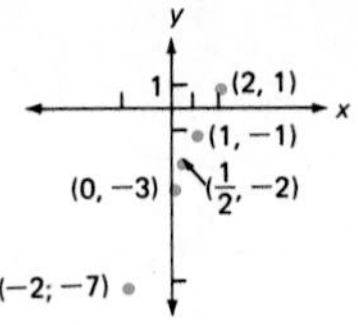

11. 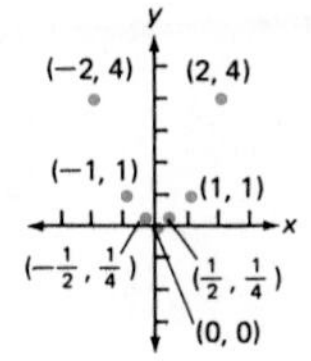

13. 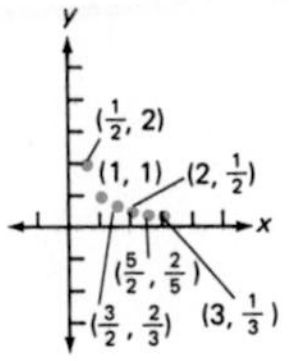

15. 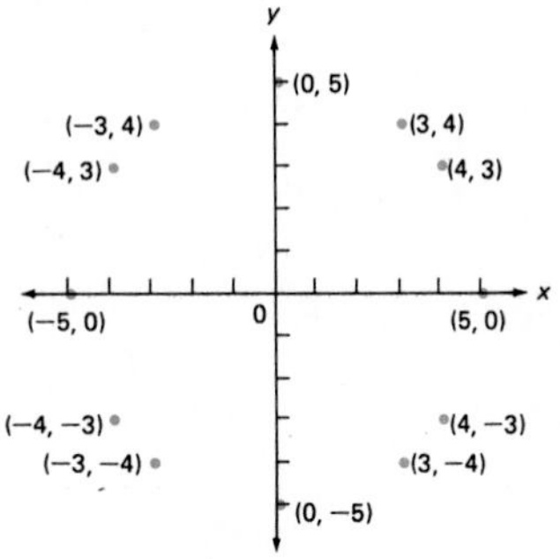

17. $y = -x^2$ 19. $C(6, 3)$ 21. **(a)** $(11, 4)$ **(b)** $(-5, 4)$

6.2 Graphing Linear Equations (Page 241)

1.

x	−3	−2	−1	0	1	2
y	−5	−4	−3	−2	−1	0

3.

x	−2	−1	0	1	2
y	−8	−6	−4	−2	0

5. $x = -2$ 7. $y = 2$

9. $y = x$

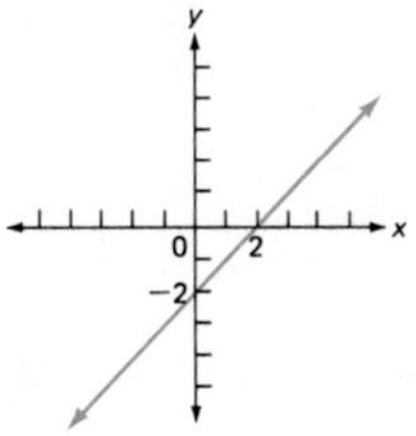

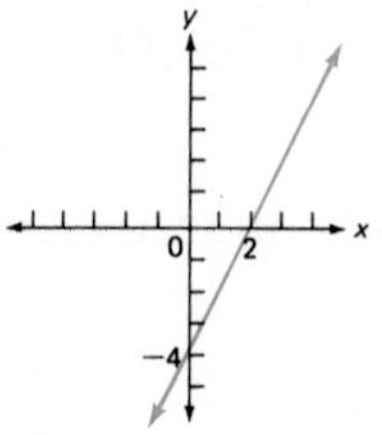

11. 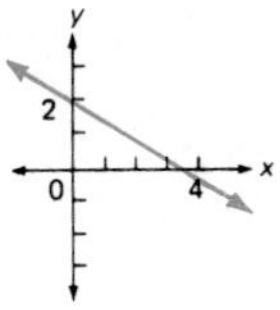

13. 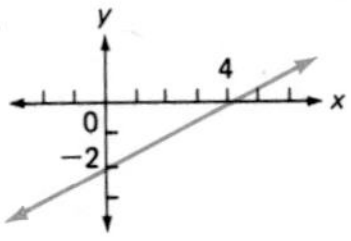

15. 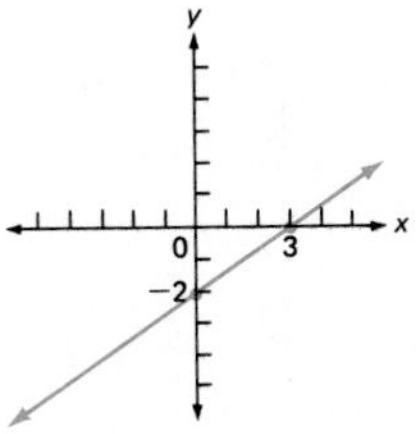

17. 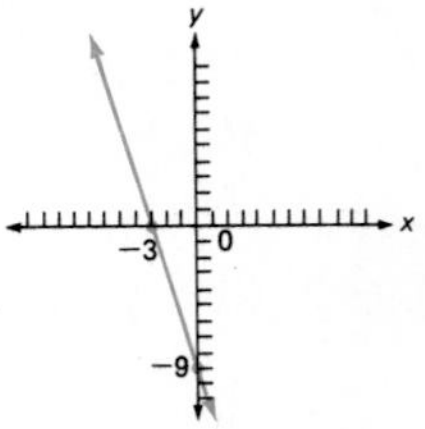

19. 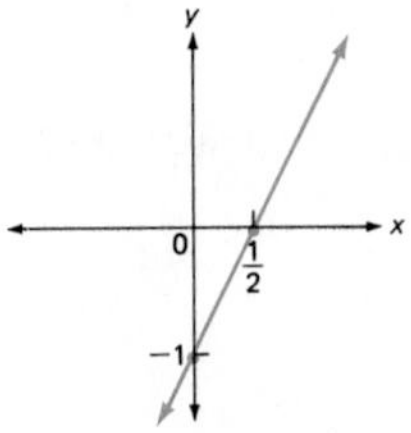

21. 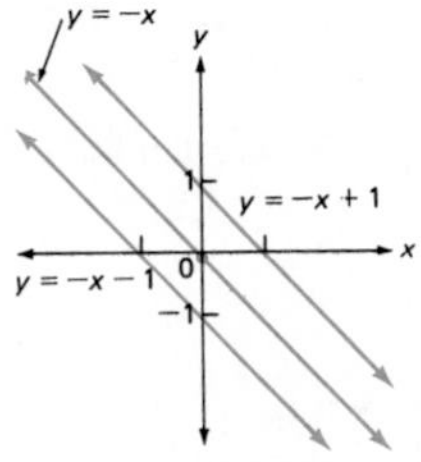

23. 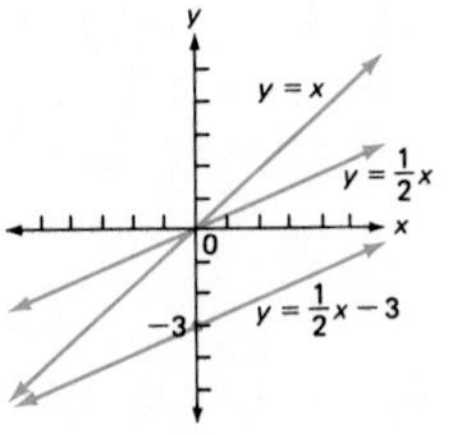

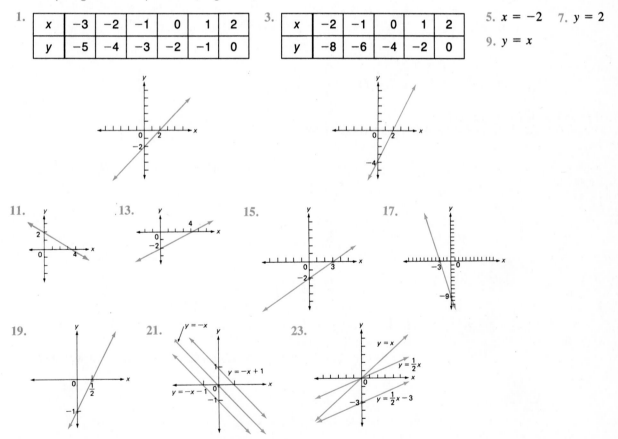

25. **(a)** Horizontal line **(b)** Vertical line **(c)** A nonhorizontal, nonvertical line through the origin **(d)** The entire plane

6.3 The Slope of a Line (Page 247)

1. Each quotient gives the same slope, $-\frac{1}{2}$. **(a)** $\dfrac{3-1}{-2-2}$ **(b)** $\dfrac{2-0}{0-4}$ **(c)** $\dfrac{1-0}{2-4}$ **(d)** $\dfrac{3-(-1)}{-2-6}$ **(e)** $\dfrac{2-(-1)}{0-6}$ **(f)** $\dfrac{1-(-1)}{2-6}$

3. $\frac{1}{2}$ 5. -1 7. $-\frac{3}{2}$ 9. 0 11.

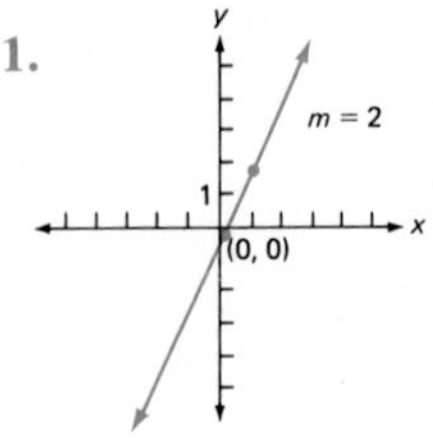

13.

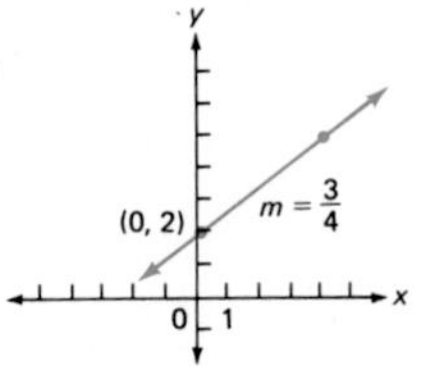

15.

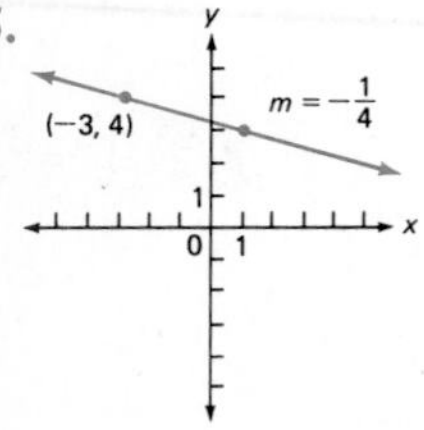

17.

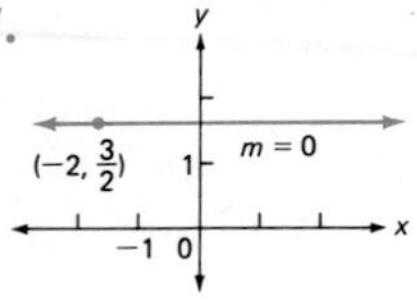

19.

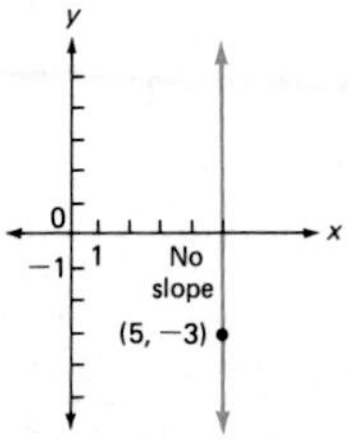

21.

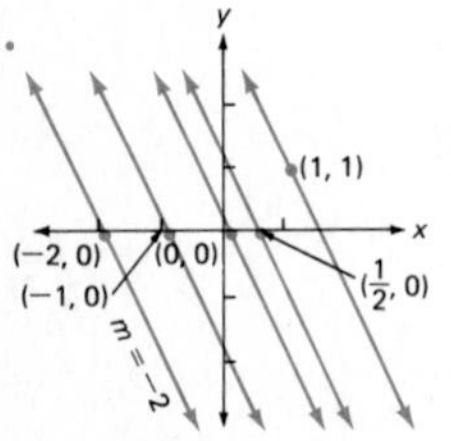

23.

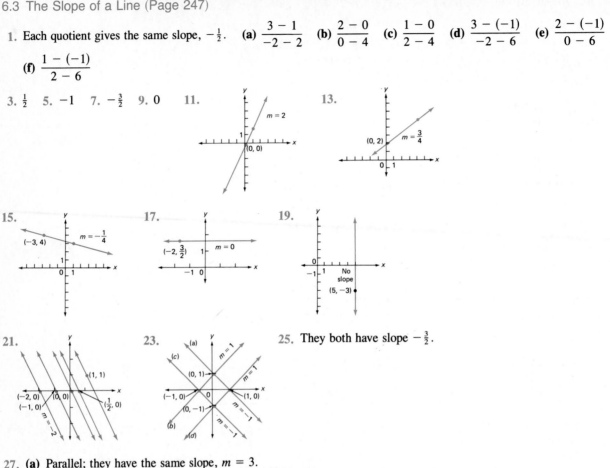

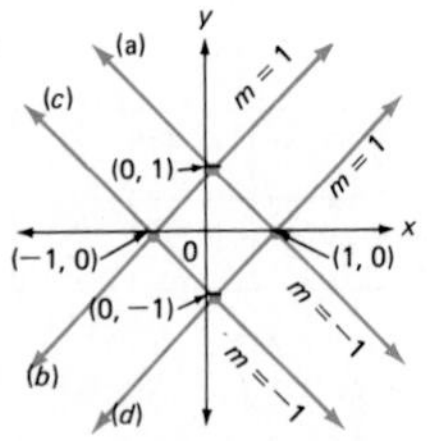

25. They both have slope $-\frac{3}{2}$.

27. **(a)** Parallel; they have the same slope, $m = 3$.
(b) Perpendicular; the slopes are negative reciprocals of one another.

29. $t = 4$

31. Since ℓ_2 has slope $\frac{5}{2}$, start at $(-2, -6)$, move 2 units right and then 5 units upward to reach $(0, -1)$ on ℓ_2. The y-intercept is -1.

33. \$135; 1625

6.4 Algebraic Forms of a Line (Page 254)

1. $y = -3x + 4$, $m = -3$, $b = 4$
3. $y = -x + 5$, $m = -1$, $b = 5$
5. $y = 2x - \frac{1}{3}$, $m = 2$, $b = -\frac{1}{3}$
7. $y = \frac{5}{3}$, $m = 0$, $b = \frac{5}{3}$
9. $y = x + 3$, $m = 1$, $b = 3$
11. $y = -\frac{1}{2}x + 3$, $m = -\frac{1}{2}$, $b = 3$

13. $y = 2x + 3$ 15. $y = x + 1$ 17. $y = 5$ 19. $y = \frac{1}{2}x + 3$ 21. $y = \frac{1}{4}x - 2$ 23. $y - 3 = x - 2$
25. $y - 3 = 4(x + 2)$ 27. $y + 2 = \frac{1}{2}(x + 1)$ 29. $y + 2 = -2(x + 2)$ 31. $x + y = 5$ 33. $x - y = 3$
35. $4x + 3y = 0$ 37. $x + 3y = -8$ 39. $x + y = 0$ 41. $y = -\frac{1}{3}x + \frac{25}{3}$ 43. $y = -\frac{1}{2}x - \frac{3}{2}$
45. $y = -\frac{2}{3}x - \frac{1}{3}$ 47. $t = \frac{34}{3}$ 49. $y = -x + 3$; $y = \frac{1}{2}x + \frac{3}{2}$ 51. They are negative reciprocals of each other.

53.
$$y - y_1 = m(x - x_1)$$
$$y - y_1 = \frac{y_2 - y_1}{x_2 - x_1}(x - x_1)$$
$$\frac{y - y_1}{y_2 - y_1} = \frac{x - x_1}{x_2 - x_1}$$

6.5 The Distance and Midpoint Formulas (Page 259)

1. 5; $(\frac{3}{2}, 2)$ **3.** 13; $(\frac{11}{2}, 2)$ **5.** 12; $(4, \frac{5}{4})$ **7.** $\sqrt{20}$; $(-1, 6)$ **9.** 20; $(2\sqrt{2}, -3\sqrt{2})$

11. **(a)** $(QR)^2 = 10 = 5 + 5 = (PQ)^2 + (PR)^2$ **(b)** $(PR)^2 = 226 = 113 + 113 = (PQ)^2 + (QR)^2$

13. Length $PQ = \sqrt{17}$ = length RS; slope $PQ = 4$ = slope RS

Length $SQ = \sqrt{53}$ = length RP; slope $SQ = \dfrac{-7}{2}$ = slope RP

15. $A(\frac{5}{2}, 6)$ = midpoint RS
$B(4, \frac{9}{2})$ = midpoint SQ
$C(\frac{9}{2}, -1)$ = midpoint QP
$D(3, \frac{1}{2})$ = midpoint PR

Length $AB = \sqrt{\dfrac{18}{4}}$ = length DC; slope $AB = -1$ = slope DC

Length $AD = \sqrt{\dfrac{122}{4}}$ = length BC; slope $AD = -11$ = slope BC

17. $(6, -2)$ **19.** $(3, \frac{13}{4}), (3, -\frac{11}{4})$

21. The slope of the line joining the midpoints is

$$\frac{\dfrac{y_1 + y_2}{2} - \dfrac{y_1}{2}}{\dfrac{x_1 + x_2}{2} - \dfrac{x_1}{2}} = \frac{y_2}{x_2} = \text{slope of } DF$$

The length of this segment is

$$\sqrt{\left(\frac{x_1 + x_2}{2} - \frac{x_1}{2}\right)^2 + \left(\frac{y_1 + y_2}{2} - \frac{y_1}{2}\right)^2} = \sqrt{\left(\frac{x_2}{2}\right)^2 + \left(\frac{y_2}{2}\right)^2} = \frac{1}{2}\sqrt{(x_2)^2 + (y_2)^2}, \text{ which is } \tfrac{1}{2} \text{ of length } DF$$

23. **(a)** By construction, each of the four triangles inside the square is congruent to the given right triangle, and therefore they are congruent to each other. Thus the sides labeled c are congruent. Also, since $\theta + \phi = 90°$ in the given triangle, it now follows that each angle of the inside quadrilateral is 90°. Therefore, this quadrilateral is a square.

(b) $4(\frac{1}{2}ab) + c^2 = 2ab + c^2$

(c) We have $(a + b)^2 = a^2 + 2ab + b^2$. Then $a^2 + 2ab + b^2 = 2ab + c^2$, by (b); therefore, $a^2 + b^2 = c^2$.

6.6 Linear Inequalities (Page 266)

1. $y \leq 2x$ **3.** $y \leq 1$

5.

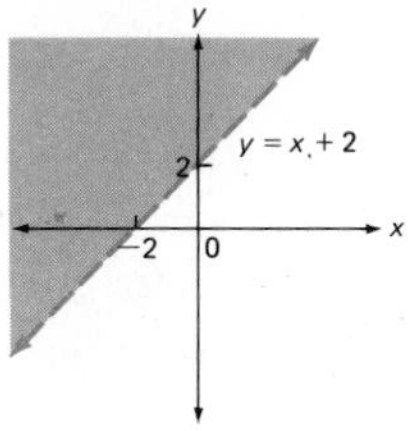

7.

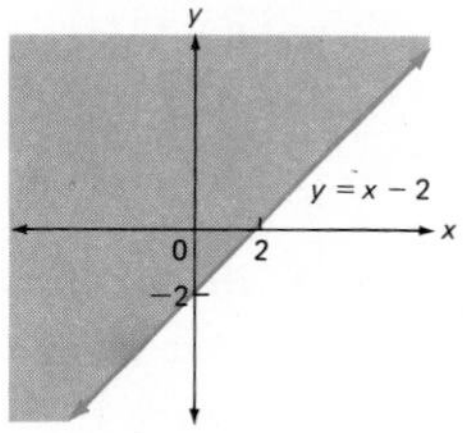

9.

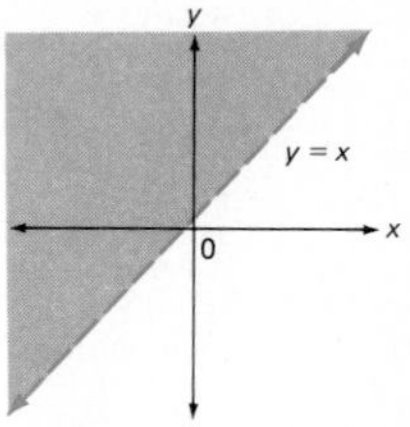

11.

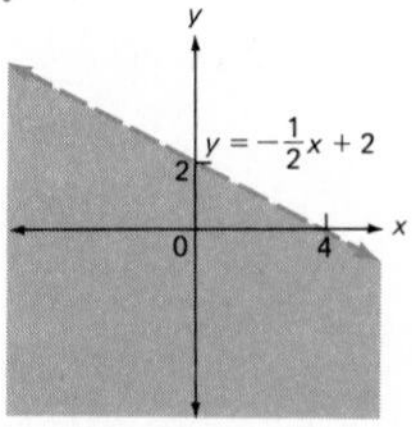

13.

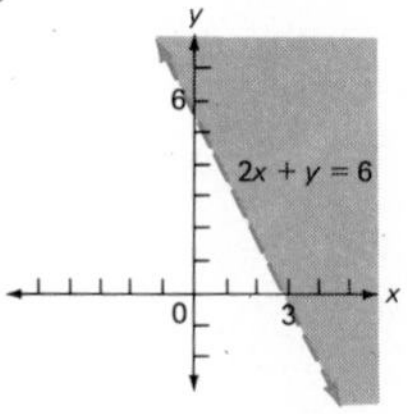

15.

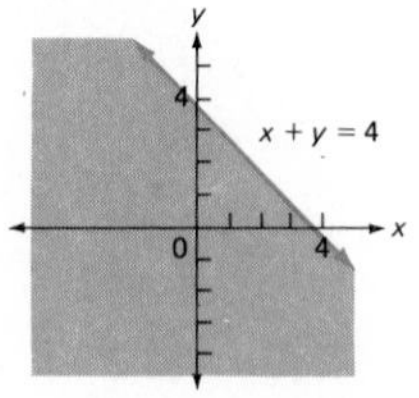

17.

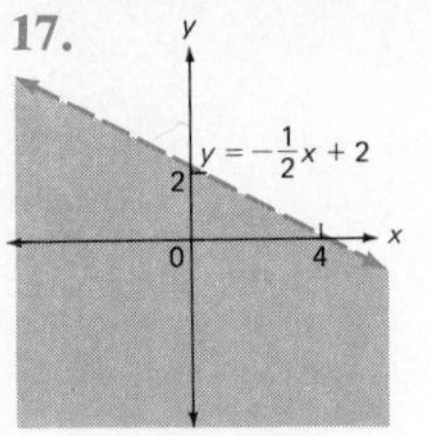

19.

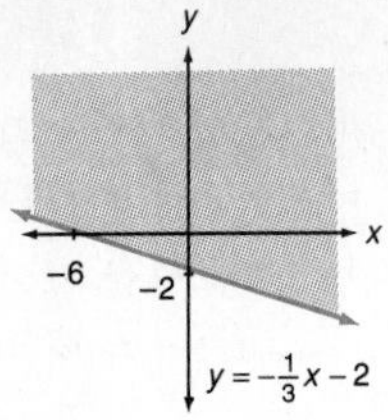

21.

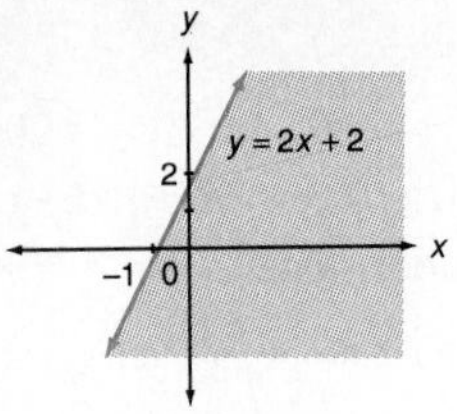

23.

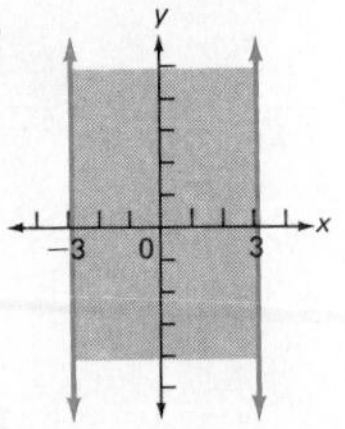

25.

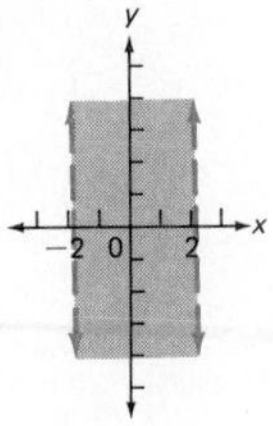

27.

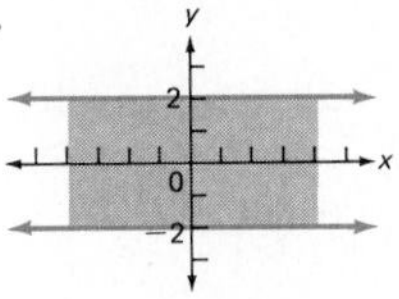

29.

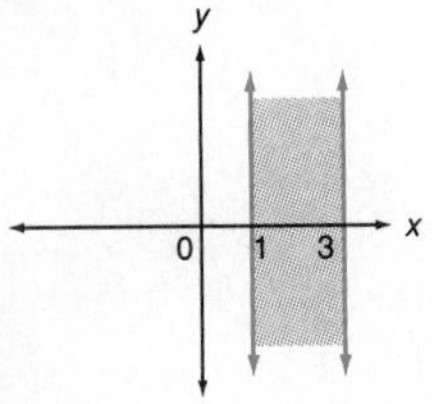

31.

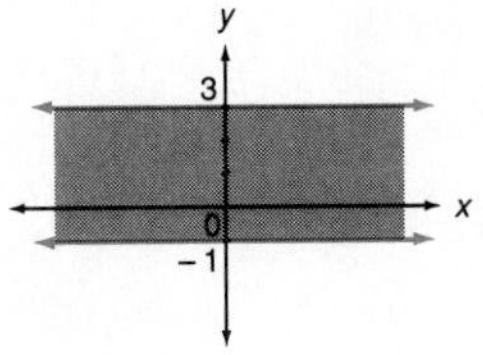

6.7 Introduction to the Function Concept (Page 271)

1. Function: all reals **3.** Function: all $x > 0$ **5.** Not a function **7.** Function: all $x \neq -1$ **9.** Not a function
11. Function: all $x > 4$

13.

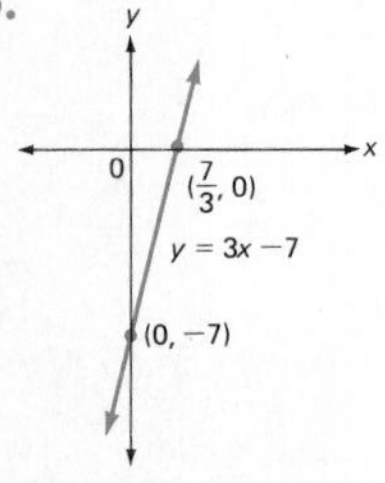

15.

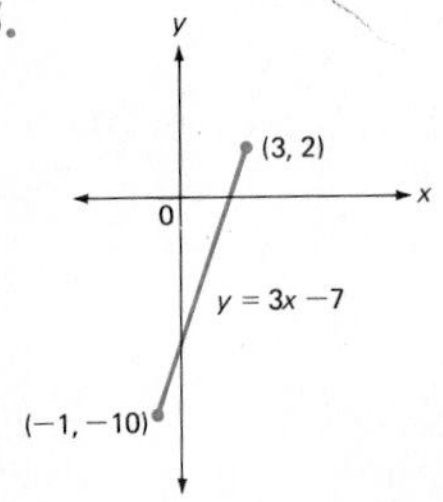

17.

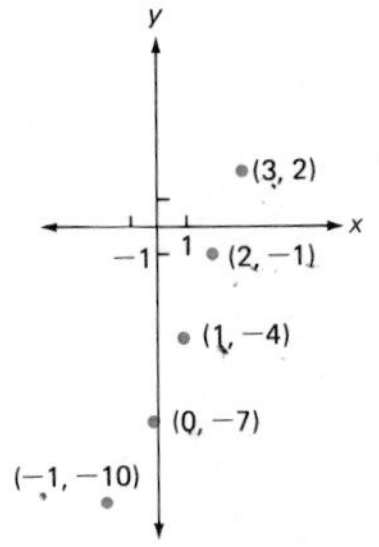

19. $\{y \mid y \leq 0\}$ **21.** **(a)** $\{x \mid 1 \leq x \leq 4\}$ **(b)** $\{y \mid 1 \leq y \leq 3\}$ **(c)** $y = \frac{2}{3}x + \frac{1}{3}$
23. For some values of x, two different values of y fit the equation; for example, if $x = 6$, then $y = \pm 2$. **25.** Yes
27. Yes **29.** No **31.** Yes

33.

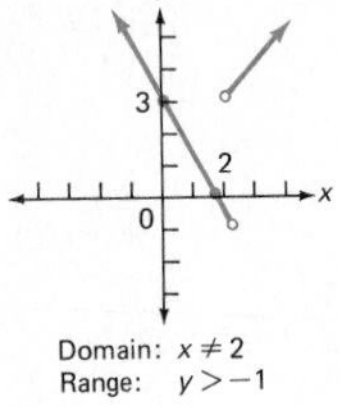

35.

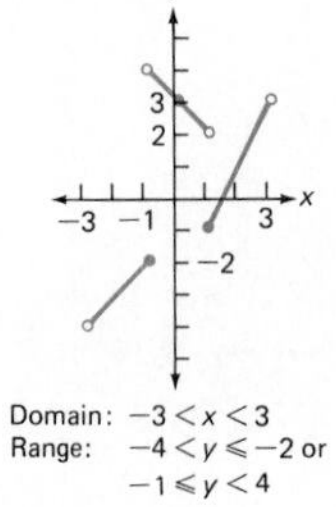

37.

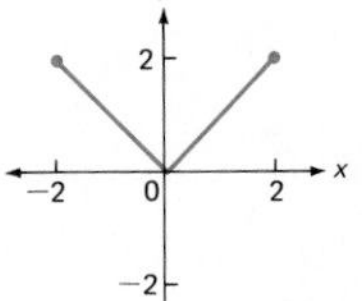

6.8 The Function Notation (Page 276)

1. **(a)** -3 **(b)** -1 **(c)** 0 **(d)** 1 **(e)** 3 **3.** **(a)** 12 **(b)** 6 **(c)** $\frac{15}{4}$ **(d)** 2 **(e)** 0
5. **(a)** $\frac{15}{2}$ **(b)** 0 **(c)** $-\frac{9}{4}$ **(d)** $-\frac{13}{2}$ **(e)** -30 **7.** **(a)** Does not exist **(b)** -1 **(c)** $-\frac{4}{3}$ **(d)** Does not exist **(e)** $\frac{1}{3}$
9. **(a)** 2 **(b)** $\sqrt{8}$ **(c)** $\sqrt{10}$ **(d)** $\sqrt{12}$ **(e)** 4 **11.** -43.9327; 9.2085; $-330{,}618$
13. 4.2942; 2.0298; 14.1421 **15.** True **17.** False; 6 **19.** True **21.** False; $-x^2 + 16$ **23.** True
25. **(a)** $l(2) + l(3) = 6 + 9 = 15 = l(5)$ **(b)** $4l(7) = (4)(21) = 84 = l(28)$
(c) $-2l(3) + 4l(-2) = (-18) + (-24) = -42 = l(-14)$
27. **(a)** $3x + 9$ **(b)** $-3x + 24$ **(c)** $3x - 6$ **(d)** $-x + 18$
29. **(a)** $16x^2 - 12x + 1$ **(b)** $x^2 + 3x + 1$ **(c)** $x^2 + 5x + 5$ **(d)** $x^4 - 3x^2 + 1$ **31.** **(a)** $2u$ **(b)** $\dfrac{u}{2}$ **(c)** u
33. **(a)** $f(1) = -1, g(1) = 5, f(1) + g(1) = 4$ **(b)** $s(x) = 5x - 1$ **(c)** $s(1) = 4$
35. **(a)** $f(1) = -1, g(1) = 5, f(1) \cdot g(1) = -5$ **(b)** $p(x) = 6x^2 - 5x - 6$ **(c)** $p(1) = -5$
37. **(a)** $(x - 5) + \dfrac{3}{x + 5}$ **(b)** $(x - 5) - \dfrac{3}{x + 5}$ **(c)** $\dfrac{3}{x + 5} - (x - 5)$ **(d)** $\dfrac{3(x - 5)}{x + 5}$ **(e)** $\dfrac{x^2 - 25}{3}$ **(f)** $\dfrac{3}{x^2 - 25}$
39. **(a)** $\dfrac{x - 2}{x} + \dfrac{x}{x^2 + 2x + 4}$ **(b)** $\dfrac{x - 2}{x} - \dfrac{x}{x^2 + 2x + 4}$ **(c)** $\dfrac{x}{x^2 + 2x + 4} - \dfrac{x - 2}{x}$ **(d)** $\dfrac{x - 2}{x^2 + 2x + 4}$
(e) $\dfrac{(x - 2)(x^2 + 2x + 4)}{x^2}$ **(f)** $\dfrac{x^2}{(x - 2)(x^2 + 2x + 4)}$ **41.** **(a)** 11; 2 **(b)** $\frac{4}{5}$; 5 **(c)** $5a + 1$ **(d)** a **(e)** $\frac{1}{5}$; 2
43. **(a)** $p(2) = 2$; $q(2) = 7$; $p(2) - q(2) = -5$ **(b)** $d(x) = 2x^3 - x^2 - 9x + 1$; $d(2) = -5$ **45.** $x^3 + 4x^2 - 3$
47. **(a)** $h^2 + 4h + 3$ **(b)** $\dfrac{(h^2 + 4h + 3) - 3}{h} = \dfrac{h^2 + 4h}{h} = h + 4$
49. If $f(x) = mx$, then $f(a) = ma$, $f(b) = mb$, and $f(a) + f(b) = ma + mb$. Also, $f(a + b) = m(a + b) = ma + mb$.

CHAPTER 6 TEST: STANDARD ANSWER (Page 281)

1. **(a)** True **(b)** False **(c)** True **(d)** False **(e)** True **2.** $2x + 5y = 19$ **3.**

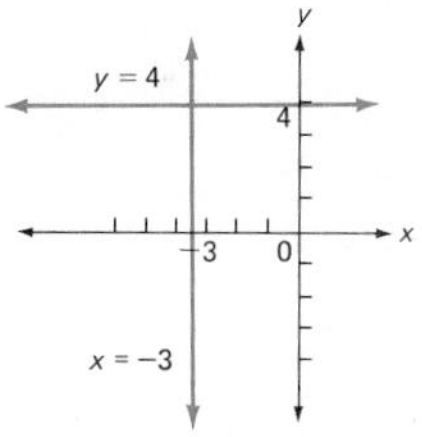

4. $-\frac{2}{7}$ **5.**

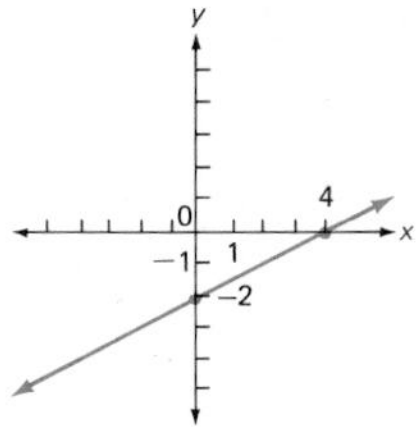

6. $y = -\frac{3}{4}x$ **7.** $y = -\frac{3}{2}x$ **8.** $y + 5 = -3(x - 2)$ **9.** $y + 4 = -\frac{1}{3}(x + 15)$ **10.** $y - 5 = 3(x - 8)$
11. **(a)** $(-6, -7)$ **(b)** $(-\frac{7}{2}, \frac{1}{2})$ **12.** **(a)** 360 **(b)** 250 **(c)** 610 **13.** $(PR)^2 = 610 = 360 + 250 = (PQ)^2 + (QR)^2$
14. Slope $PQ = -\dfrac{1}{3} = -\dfrac{1}{\text{slope } QR}$ **15.** Domain; $\{x \mid -15 \le x \le 8\}$; range: $\{y \mid -4 \le y \le 5\}$
16. **(a)** $y - 2 = -1(x + 1)$ **(b)** $y = -x + 1$ **(c)** $x + y = 1$ **17.** $(4, 1)$; $y = 2x - 7$
18. Midpoints are located at $K(2, 1)$, $L(5, 3)$, $M(6, 1)$, and $N(3, -1)$. Slope of KL = slope of $MN = \frac{2}{3}$; slope of KN = slope of $LM = -2$.

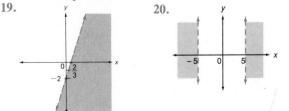

19.

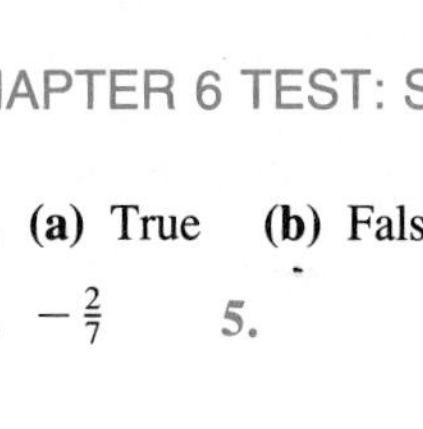

20.

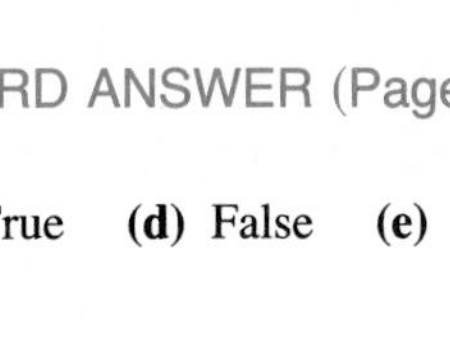

21. For some x there is more than one corresponding y-value; for example, for $x = 0$, $y = 1$ or $y = -1$.
22. **(a)** $x \geq 3$ **(b)** all real numbers

23.
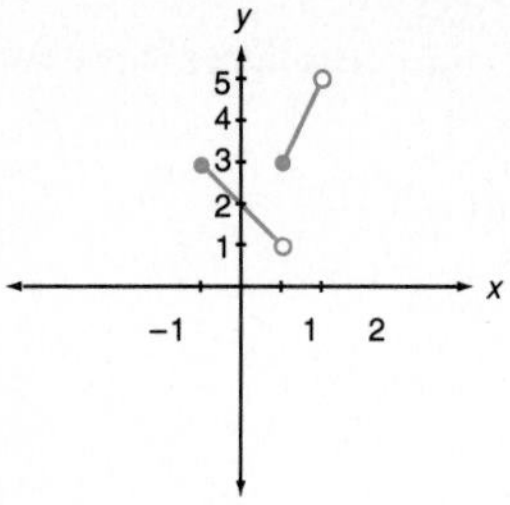

Domain: $-1 \leq x < 2$
Range: $1 < y < 5$

24. **(a)** x^2 **(b)** $\dfrac{1}{x^2 + 4x + 4}$ 25. **(a)** $x^4 - x^3 + x^2 - 7x + 5$ **(b)** -15

CHAPTER 6 TEST: MULTIPLE CHOICE (Page 282)

1. (c) 2. (a) 3. (d) 4. (d) 5. (e) 6. (c) 7. (b) 8. (b) 9. (e) 10. (c) 11. (e) 12. (a)
13. (a) 14. (b) 15. (a)

ANSWERS TO CUMULATIVE REVIEW FOR CHAPTERS 4–6 (Page 285)

1. **(a)** -1 **(b)** $\dfrac{2x - 1}{2x + 1}$ 2. 2 3. $\dfrac{-2x}{x^2 - 4}$ 4. $\dfrac{a^2 - a + 8}{12a^2}$ 5. $\dfrac{x - y}{x + y}$ 6. $3b^2 - 2a^2b^4 + 1$ 7 **7.** $\dfrac{a^2 + b^2}{(a - b)^2}$

8. $\dfrac{x - 3}{4x - 3}$ 9. $-(3x + 2)$ 10. $-\dfrac{1}{4x}$ 11. $\dfrac{y(y^2 + x^3)}{x(y^3 + x^2)}$ 12. $x = 3$ 13. $x = 5$ 14. 15, 16 15. $2\frac{2}{9}$ hr

16. $\frac{3}{2}$ 17. 75π cm³ 18. **(a)** $\frac{1}{9}$ **(b)** 8 19. $\dfrac{b^{1/6}}{a^{1/6}}$ 20. $ab^{1/6}$ 21. **(a)** $4\sqrt{3}$ **(b)** $\sqrt{3}$ 22. $5\sqrt{5}$ 23. $|a|b\sqrt{3b}$

24. $21\sqrt{2}i$ 25. $-10\sqrt{3}$ 26. $17 + 7i$ 27. $\dfrac{1}{2} + \dfrac{1}{2}i$ 28. $x = 3$ or $x = -1/2$ 29. $x = \frac{5}{3}$ 30. $2 \pm 2\sqrt{3}$

31. $\dfrac{3 \pm i}{2}$ 32. two irrational roots 33. two imaginary roots 34. $k < \frac{1}{3}$ 35. $\dfrac{3 \pm \sqrt{5}}{2}$ 36. At $t = 2$ and $t = 4$

37. $\frac{3}{5}$ 38. $y = -x - 1$ 39. $y + 2 = -\frac{3}{4}(x - 4)$ 40. $y + 2 = \frac{2}{3}(x - 3)$ 41. $y = -\frac{1}{2}x$

42.
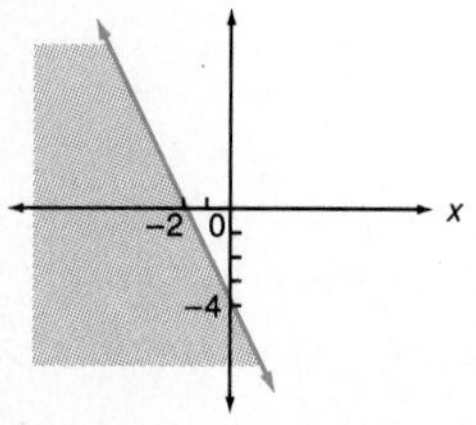

43.
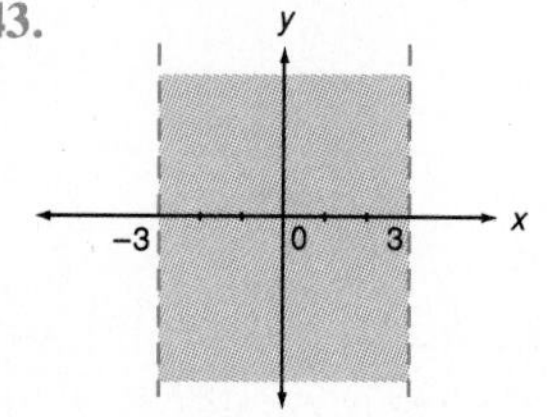

44. **(a)** $x \leq 2$ **(b)** $x \neq \pm 2$ 45. **(a)** $(4, -4)$ **(b)** $2\sqrt{2}$ 46. Lengths $AB = CD = \sqrt{13}$; slope AB = slope $CD = \frac{3}{2}$

47. **(a)** x^3 **(b)** $\dfrac{1}{(x + 1)^3}$ 48. 5 49. **(a)** $-x^3 + 2x^2 - 3x - 1$ **(b)** -5

50. Domain: $-2 \leq x < 3$
Range: $-4 \leq y < -1$; $0 < y \leq 2$

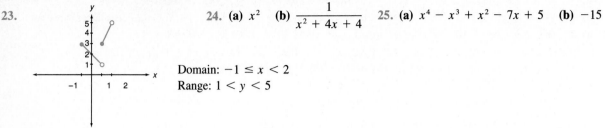

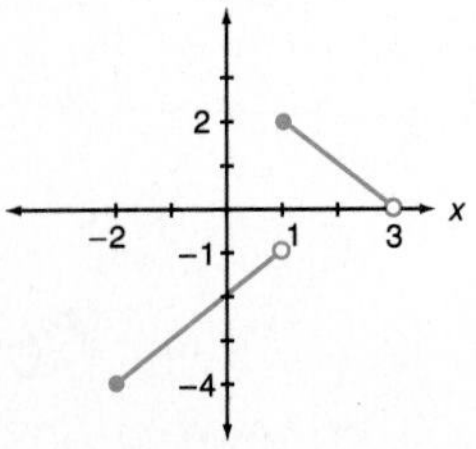

7.1 Graphing Quadratic Functions (Page 294)

1.
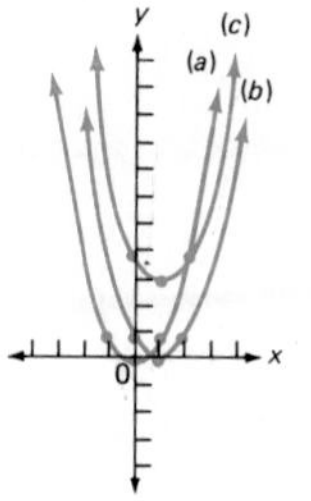

3.
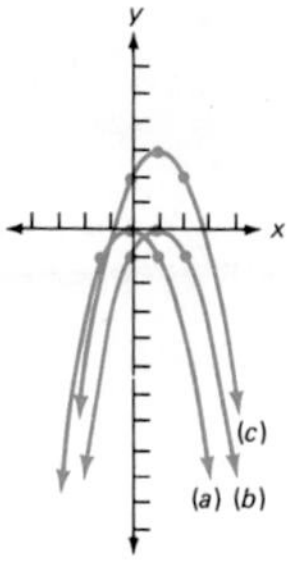

5.
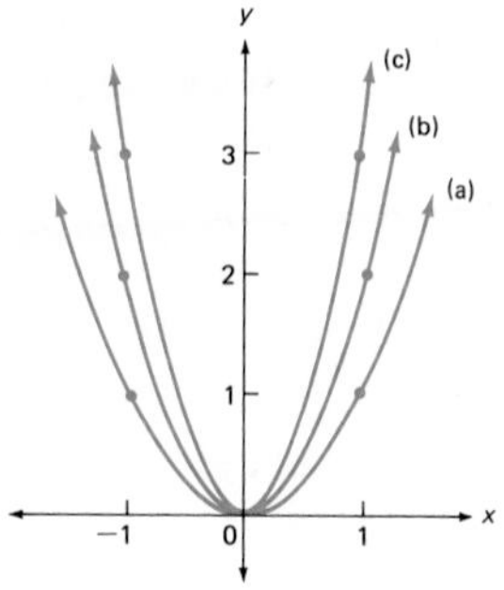

7.
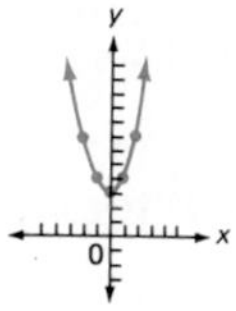

9.
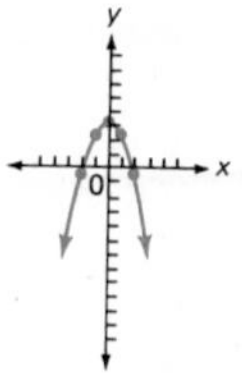

11.
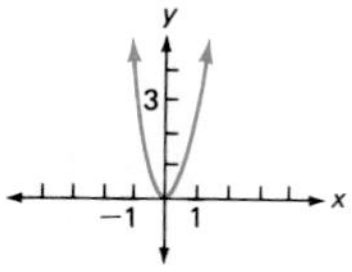

13.
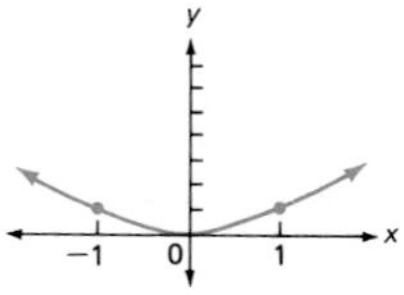

15.
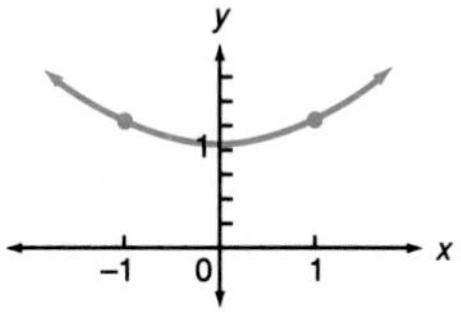

17.
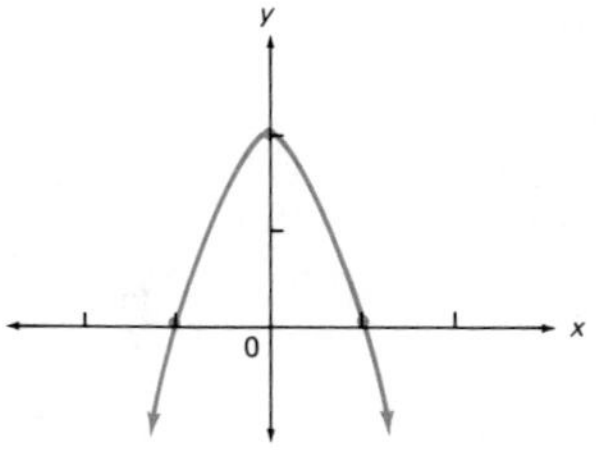

19. Decreasing on $(-\infty, 1]$; increasing on $[1, \infty)$; concave up

21. Increasing on $(-\infty, -1]$; decreasing on $[-1, \infty)$; concave down

23. Decreasing on $(-\infty, 3]$; increasing on $[3, \infty)$; concave up

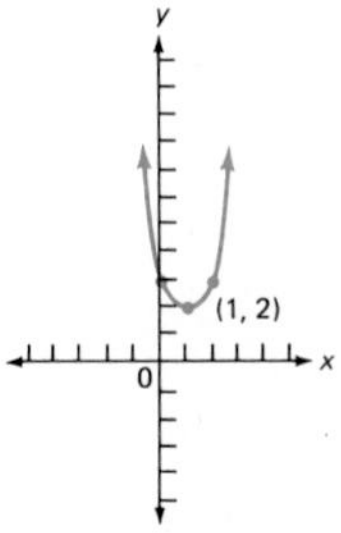

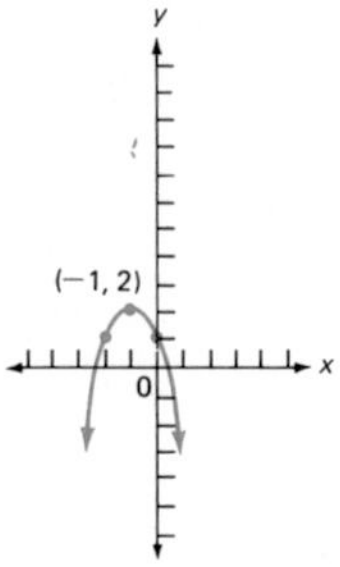

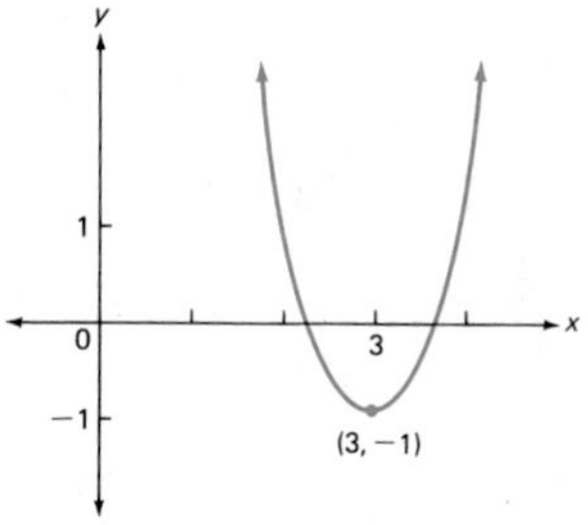

25. **(a)** $(3, 5)$; **(b)** $x = 3$; **(c)** Set of real numbers; **(d)** $y \geq 5$

27. **(a)** $(3, 5)$; **(b)** $x = 3$; **(c)** Set of real numbers; **(d)** $y \leq 5$

29. **(a)** $(-1, -3)$; **(b)** $x = -1$; **(c)** Set of real numbers; **(d)** $y \geq -3$

31. **(a)** $(1, 2)$; **(b)** $x = 1$; **(c)** Set of real numbers; **(d)** $y \leq 2$

33. **(a)** $(-2, -4)$; **(b)** $x = -2$; **(c)** Set of real numbers; **(d)** $y \geq -4$

35. $a = -2$ **37.** $k = 3$

39.
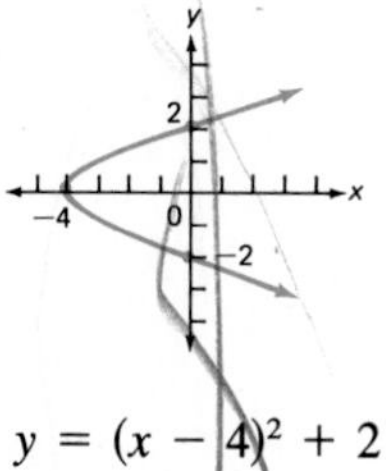

41.
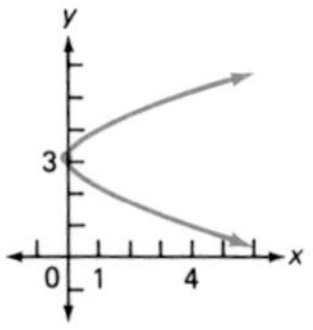

43.
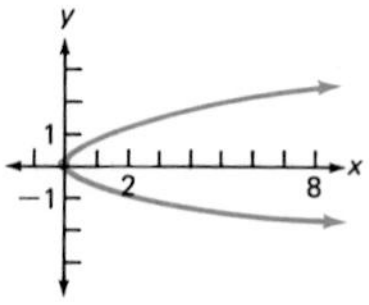

45.
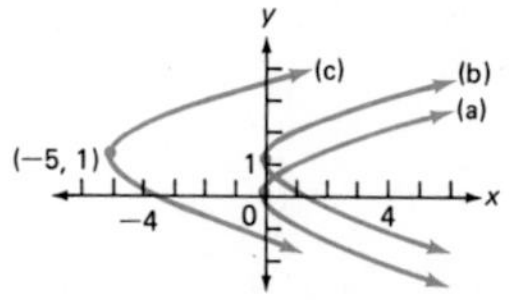

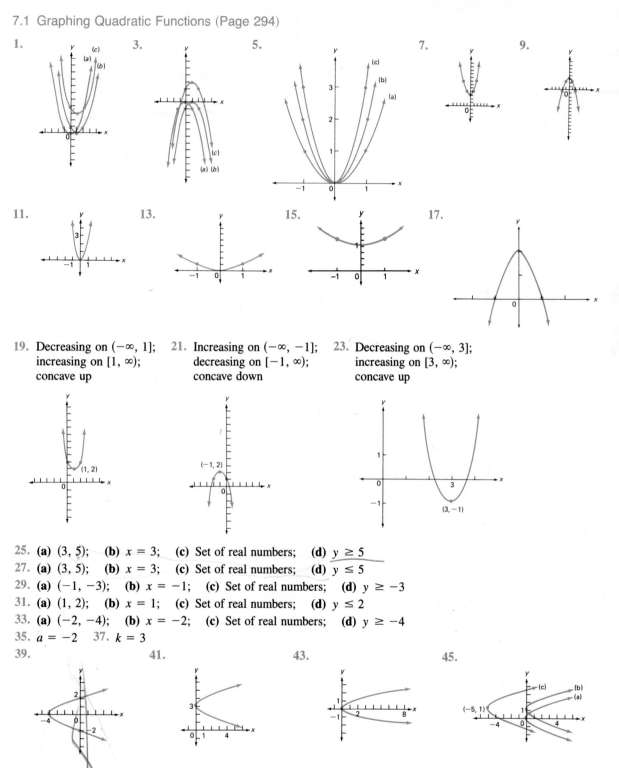

47. $y = (x - 4)^2 + 2$ **49.** $x = (y - 2)^2 - 5$ **51.** $y = \frac{1}{2}(x - 3)^2 - 2$

7.2 Applying the Standard Form $y = a(x - h)^2 + k$ (Page 301)

1. $y = (x + 1)^2 - 6$ 3. $y = (x + 3)^2 - 11$ 5. $y = (x - 2)^2 - 3$ 7. $y = (x - \frac{3}{2})^2 + \frac{7}{4}$ 9. $y = (x - \frac{5}{2})^2 - \frac{33}{4}$
11. $y = 2(x - 1)^2 + 1$ 13. $y = 3(x - 1)^2 + 2$ 15. $y = -\frac{1}{3}(x + 9)^2 + 32$
17. $y = (x + 1)^2 - 2$ **(a)** $(-1, -2)$ **(b)** $x = -1$ **(c)** -1 **(d)** $-1 \pm \sqrt{2}$
19. $y = -(x - 2)^2 + 3$ **(a)** $(2, 3)$ **(b)** $x = 2$ **(c)** -1 **(d)** $2 \pm \sqrt{3}$
21. $y = 3(x + 1)^2 - 6$ **(a)** $(-1, -6)$ **(b)** $x = -1$ **(c)** -3 **(d)** $-1 \pm \sqrt{2}$
23. $y = -(x - 1)^2$ **(a)** $(1, 0)$ **(b)** $x = 1$ **(c)** -1 **(d)** 1
25. $y = 2(x + \frac{1}{2})^2 + \frac{1}{2}$ **(a)** $(-\frac{1}{2}, \frac{1}{2})$ **(b)** $x = -\frac{1}{2}$ **(c)** 1 **(d)** None
27. $y = (x + \frac{3}{2})^2 - \frac{5}{4}$ **(a)** $(-\frac{3}{2}, -\frac{5}{4})$ **(b)** $x = -\frac{3}{2}$ **(c)** 1 **(d)** $\dfrac{-3 \pm \sqrt{5}}{2}$
29. **(a)** $(1, 4)$ **(b)** 3 **(c)** $-1, 3$
31. **(a)** $(-\frac{3}{4}, \frac{27}{4})$ **(b)** 9 **(c)** None
33. **(a)** $(3, -5)$ **(b)** 4 **(c)** $3 \pm \sqrt{5}$
35. $b = \pm 4\sqrt{3}$ 37. $b = \pm 3$ 39. Minimum $= -32$ at $x = -9$ 41. Maximum $= -\frac{3}{2}$ at $x = 3$
43. Minimum $= -25$ at $x = 5$ 45. 60; \$600 47. $\frac{1}{8}$ and $\frac{1}{8}$; $p = \frac{1}{64}$ 49. 15 and 5 51. 25 ft by 25 ft; $A = 625$ ft^2
53. $x = 25$ ft, $y = 50$ ft for each rectangle 55. 15 cm by 15 cm 57. 256 ft 59. 8 sec 61. \$15
63. Five additional trees for a total of 70 65. $(\frac{3}{2}, \frac{9}{2})$; $\frac{9}{2}$ is a minimum value 67. $(-\frac{1}{3}, \frac{19}{18})$; $\frac{19}{18}$ is a maximum value

7.3 Quadratic Inequalities (Page 308)

1. $x < -1$ or $x > 3$ 3. $-2 \le x \le 4$ 5. $x < -5$ or $x > 5$ 7. $x \le -4$ or $x \ge 4$ 9. $x < -3$ or $x > 0$
11. $x \le 0$ or $x \ge \frac{1}{2}$ 13. $x < -\frac{1}{2}$ or $x > \frac{1}{2}$ 15. $0 \le x \le \frac{3}{8}$ 17. $x < -5$ or $x > 2$ 19. $-4 < x < 3$
21. $x \le -3$ or $x \ge 5$ 23. $-2 < x < \frac{1}{2}$ 25. $x \le -\frac{1}{2}$ or $x \ge 3$ 27. $x < -5$ or $x > 6$ 29. $\frac{1}{3} \le x \le 1$
31. $x \le -\frac{1}{2}$ or $x \ge \frac{1}{3}$ 33. $x < -2 - \sqrt{6}$ or $x > -2 + \sqrt{6}$ 35. $\dfrac{-3 - \sqrt{13}}{2} < x < \dfrac{-3 + \sqrt{13}}{2}$
37. $x < \dfrac{-1 - \sqrt{17}}{4}$ or $x > \dfrac{-1 + \sqrt{17}}{4}$ 39. $\emptyset$ 41. $x < -\frac{3}{2}$ or $x > \frac{2}{3}$ 43. All real numbers x
45. $x < b$ or $x > a$ 47. $2 < x < 3$ 49. $x < 0$ or $x > 6$ 51. $-\frac{1}{2} \le x < 1$ 53. $x < -2$ or $x > 3$
55. $x \le -\frac{5}{4}$ or $x > \frac{3}{2}$ 57. $2 < x < 6$ 59. $x < -1$ or $x > 1$ 61. $x < 0$ or $x > 3$ 63. $-\frac{3}{4} \le x < 0$
65. $-9 < x < 0$ 67. $x < -\frac{5}{2}$ or $x > 1$

69. 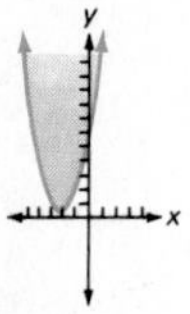

71. 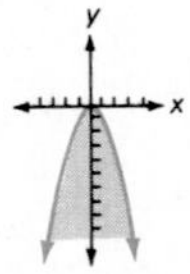

73. 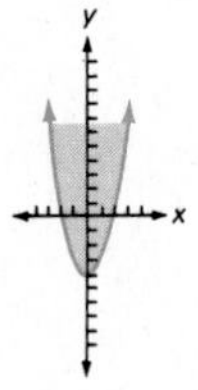

75. 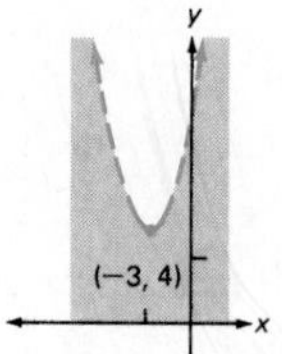

77. 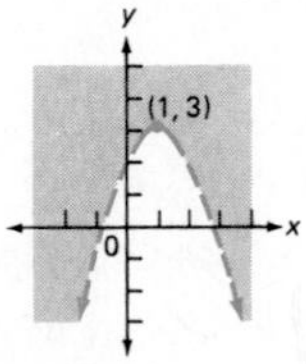

7.4 Conic Sections: The Circle (Page 315)

1. 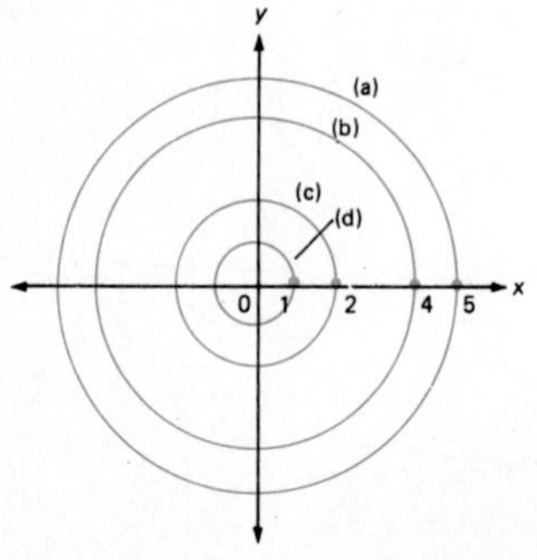

3. $x^2 + y^2 = \frac{9}{4}$ 5. 6 7. $\frac{1}{2}$ 9. $\sqrt{5}$ 11. $(-4, 3)$; $(-4, -3)$

13. (0, 5); (0, −5) **15.** (8, −6); (−8, −6) **17.** (6, 0); (−6, 0) **19.** $y - 4 = 2(x + 8)$ **21.** $y + 12 = -\frac{5}{12}(x + 5)$

23.

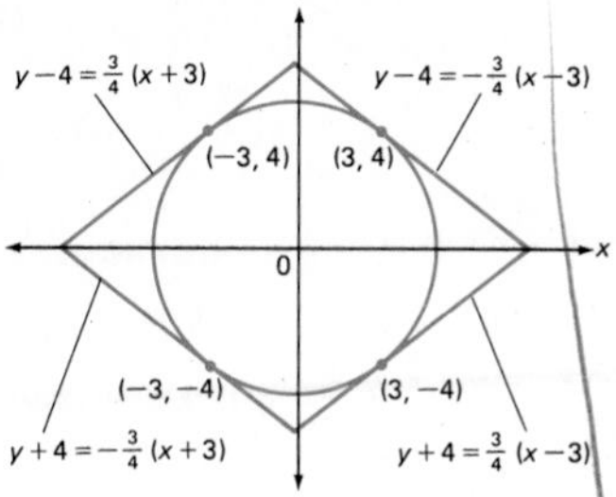

25.

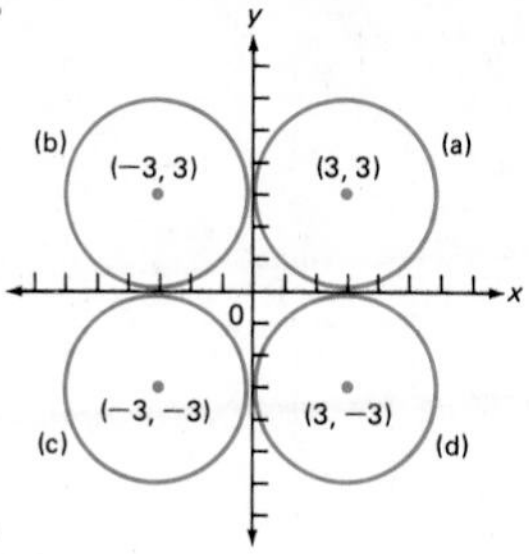

27. $(x + 2)^2 + (y + 3)^2 = 9$

29. $(x - 2)^2 + (y - 3)^2 = 13$ **31.** $x^2 + (y + 5)^2 = 25$ **33.** $(x - 2)^2 + y^2 = 25$; (2, 0); 5
35. $(x - 1)^2 + (y - 3)^2 = 1$; (1, 3); 1 **37.** $(x - 2)^2 + (y - 5)^2 = 1$; $C(2, 5)$, $r = 1$
39. $(x - 4)^2 + (y - 0)^2 = 2$; $C(4, 0)$, $r = \sqrt{2}$ **41.** $(x - 10)^2 + (y + 10)^2 = 100$; $C(10, -10)$, $r = 10$
43. $(x + \frac{3}{4})^2 + (y - 1)^2 = 9$; $C(-\frac{3}{4}, 1)$, $r = 3$

45.

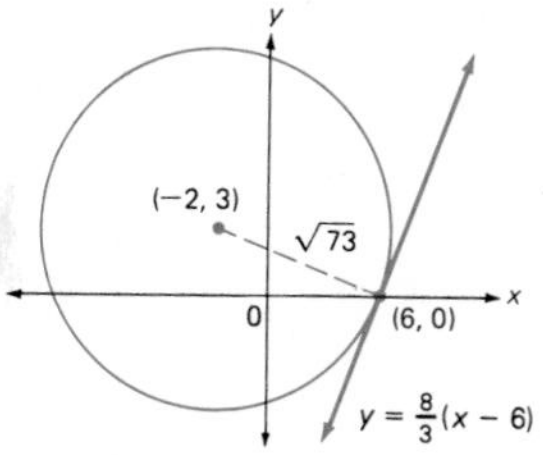

47.

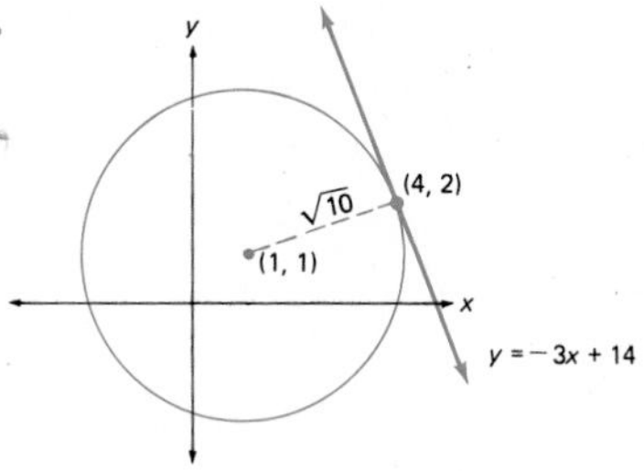

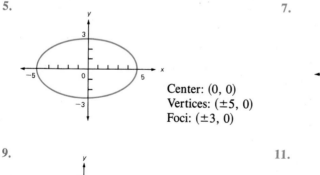

49. $x^2 + (y - 4)^2 = 49$; $C(0, 4)$, $r = 7$ **51.** $x + 2y = 20$ **53.** $5x + 12y = 26$

7.5 Conic Sections: The Ellipse (Page 320)

1. $\frac{x^2}{9} + \frac{y^2}{4} = 1$; $(\pm\sqrt{5}, 0)$ **3.** $\frac{x^2}{9} + \frac{y^2}{25} = 1$; $(0, \pm 4)$

5.

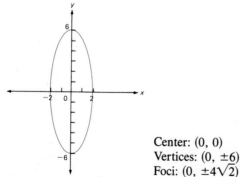

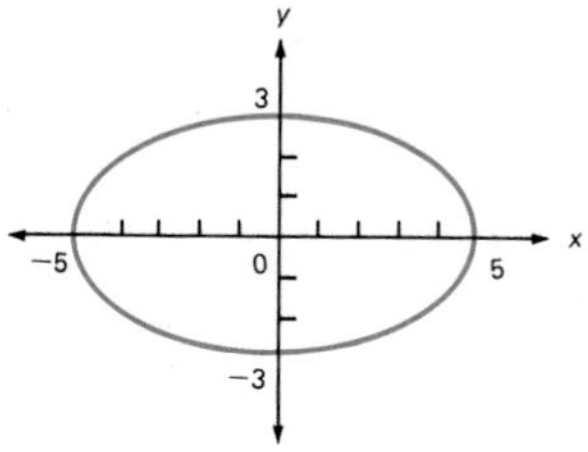

Center: (0, 0)
Vertices: (±5, 0)
Foci: (±3, 0)

7.

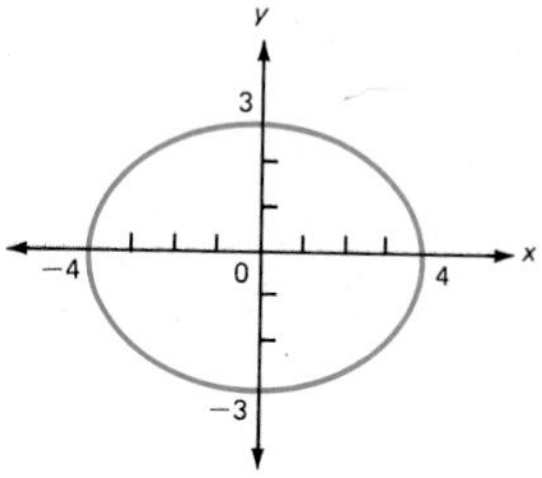

Center: (0, 0)
Vertices: (±4, 0)
Foci: $(\pm\sqrt{7}, 0)$

9.

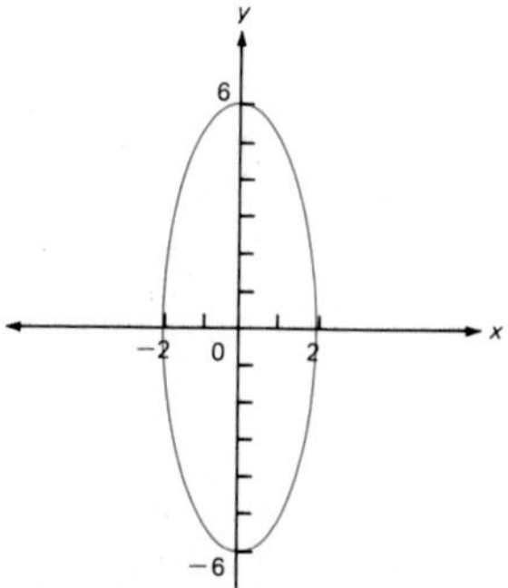

Center: (0, 0)
Vertices: (0, ±6)
Foci: $(0, \pm 4\sqrt{2})$

11.

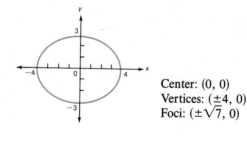

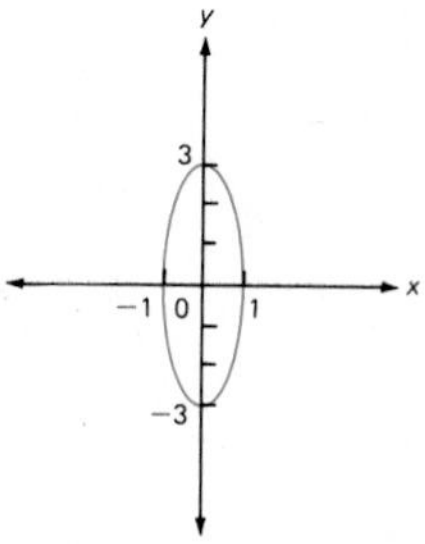

Center: (0, 0)
Vertices: (0, ±3)
Foci: $(0, \pm 2\sqrt{2})$

13.

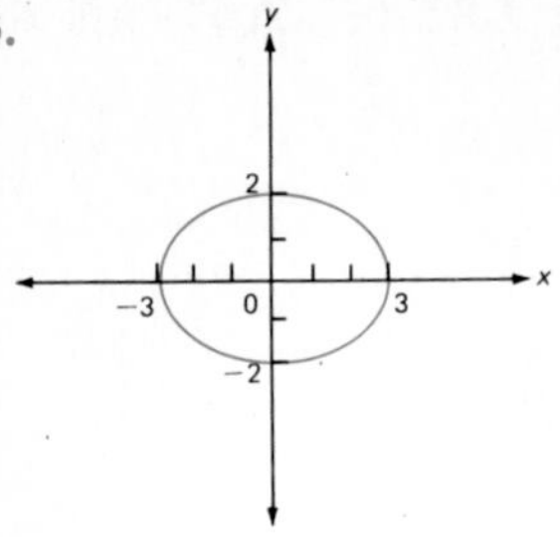

Center: (0, 0)
Vertices: $(\pm 3, 0)$
Foci: $(\pm\sqrt{5}, 0)$

15.

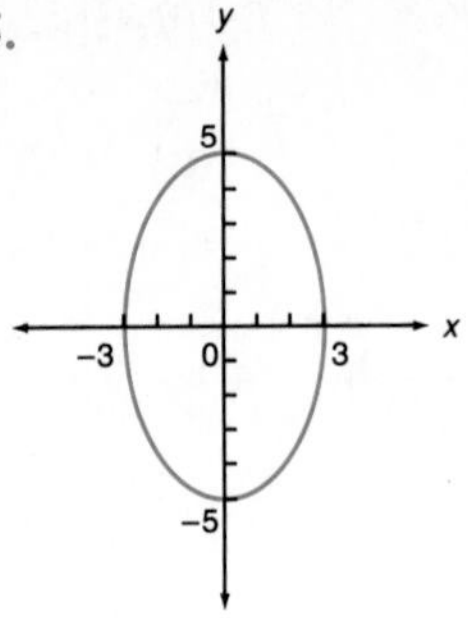

Center: (0, 0)
Vertices: $(0, \pm 5)$
Foci: $(0, \pm 4)$

17. $\dfrac{x^2}{64} + \dfrac{y^2}{16} = 1$ 19. $\dfrac{x^2}{25} + \dfrac{y^2}{9} = 1$ 21. $\dfrac{x^2}{16} + \dfrac{y^2}{25} = 1$

23. **(a)** $\dfrac{x^2}{b^2} + \dfrac{y^2}{(4355)^2}$ where $b^2 = (4355)^2 - (225)^2$ **(b)** 4349 miles; $\dfrac{x^2}{(4349)^2} + \dfrac{y^2}{(4355)^2} = 1$

25. **(a)** $y = \frac{2}{3}\sqrt{900 - 400} = \frac{20}{3}\sqrt{5}$ **(b)** 14.9 ft

27.
$$\begin{aligned}(x + 4)^2 + y^2 &= 100 - 20\sqrt{(x - 4)^2 + y^2} + (x - 4)^2 + y^2\\ 16x - 100 &= -20\sqrt{(x - 4)^2 + y^2}\\ 4x - 25 &= -5\sqrt{(x - 4)^2 + y^2}\\ 16x^2 - 200x + 625 &= 25(x^2 - 8x + 16 + y^2)\\ 225 &= 9x^2 + 25y^2\end{aligned}$$

29.

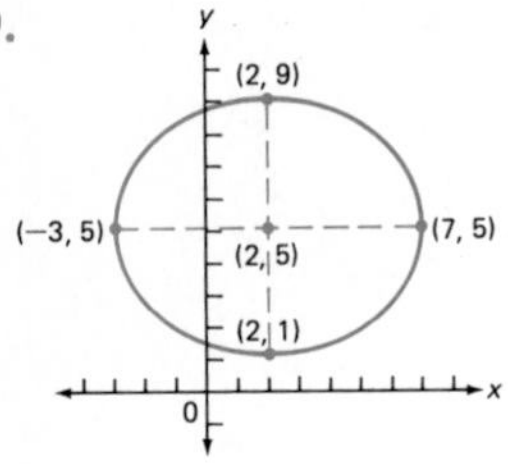

31.

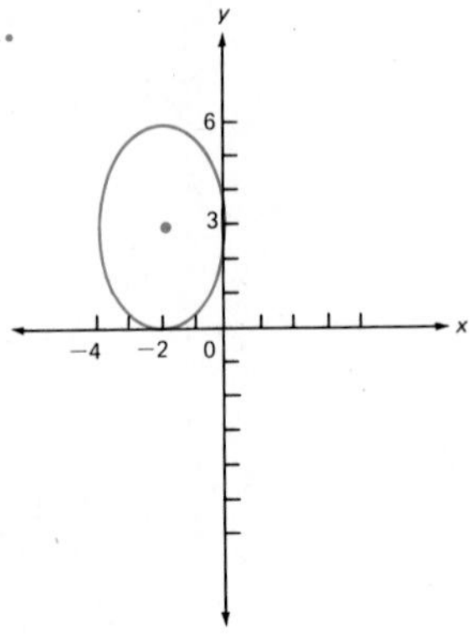

33. $\dfrac{(x - 2)^2}{16} + \dfrac{(y + 3)^2}{36} = 1$

7.6 Conic Sections: The Hyperbola (Page 326)

1. $\dfrac{x^2}{16} - \dfrac{y^2}{4} = 1$; $y = \pm\frac{1}{2}x$, $(\pm 2\sqrt{5}, 0)$ 3. $\dfrac{y^2}{25} - \dfrac{x^2}{64} = 1$; $y = \pm\frac{5}{8}x$, $(0, \pm\sqrt{89})$

5.

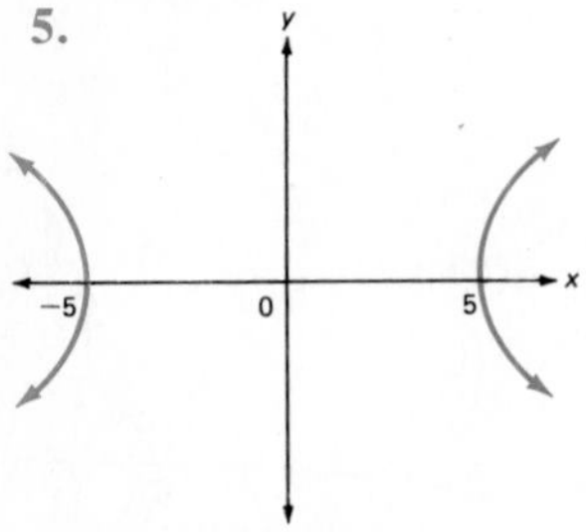

Center: (0, 0)
Vertices: $(\pm 5, 0)$
Foci: $(\pm\sqrt{34}, 0)$
Asymptotes: $y = \pm\frac{3}{5}x$

7.

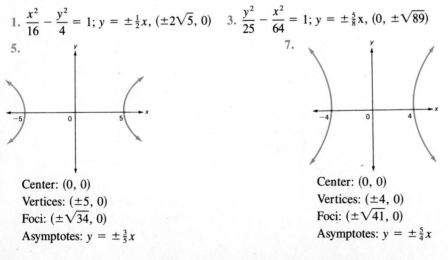

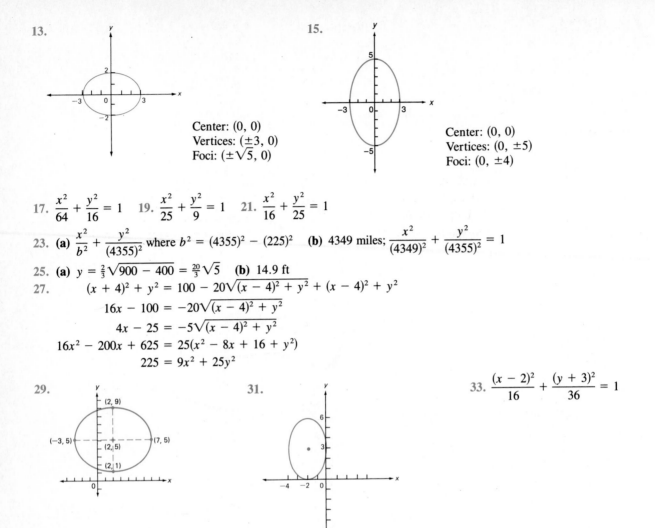

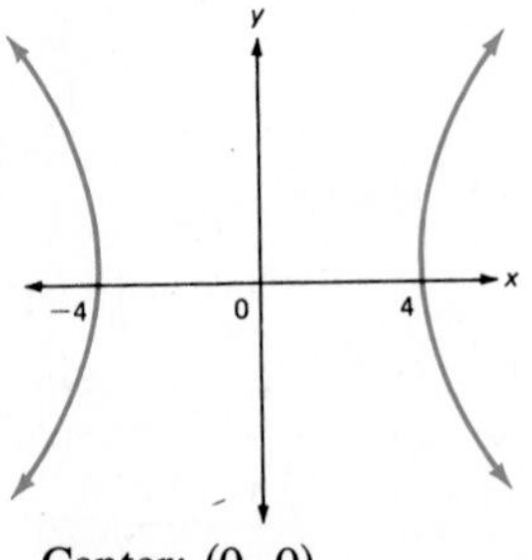

Center: (0, 0)
Vertices: $(\pm 4, 0)$
Foci: $(\pm\sqrt{41}, 0)$
Asymptotes: $y = \pm\frac{5}{4}x$

9.

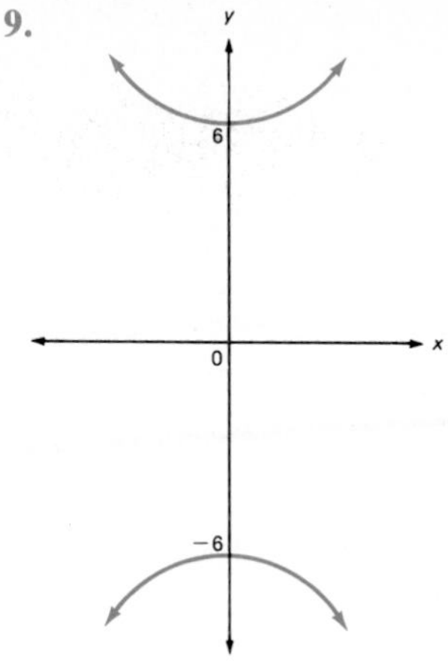

Center: (0, 0)
Vertices: (0, ±6)
Foci: (0, $\pm 3\sqrt{5}$)
Asymptotes: $y = \pm 2x$

11.

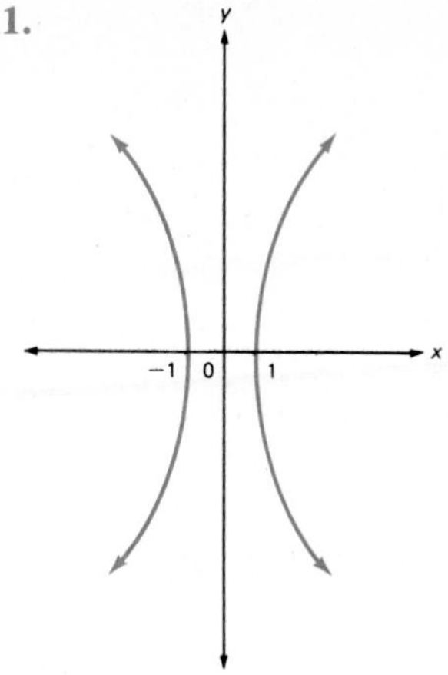

Center: (0, 0)
Vertices: (±1, 0)
Foci: ($\pm\sqrt{10}$, 0)
Asymptotes: $y = \pm 3x$

13.

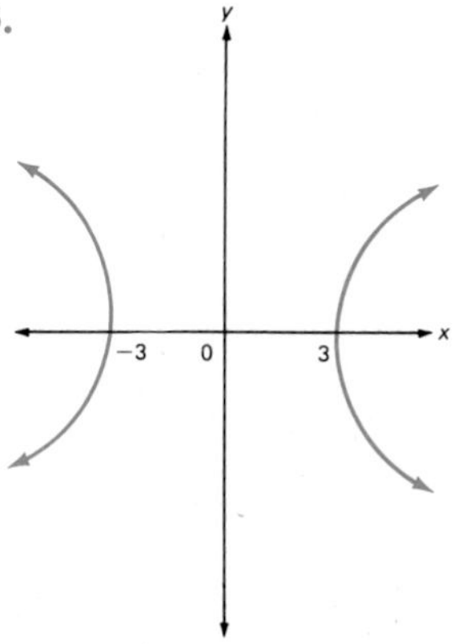

Center: (0, 0)
Vertices: (±3, 0)
Foci: ($\pm\sqrt{13}$, 0)
Asymptotes: $y = \pm \frac{2}{3}x$

15.

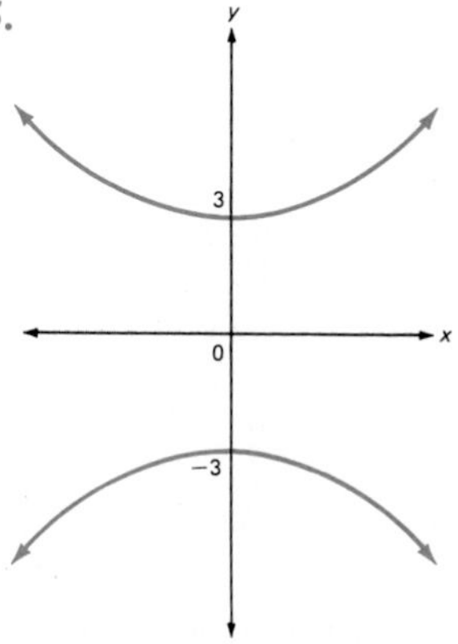

Center: (0, 0)
Vertices: (0, ±3)
Foci: (0, $\pm\sqrt{34}$)
Asymptotes: $y = \pm \frac{3}{5}x$

17. $\dfrac{x^2}{16} - \dfrac{y^2}{20} = 1$ 19. $\dfrac{y^2}{18} - \dfrac{x^2}{7} = 1$ 21. $\dfrac{x^2}{16} - \dfrac{y^2}{4} = 1$ 23. $\dfrac{y^2}{64} - \dfrac{x^2}{225} = 1$

25.

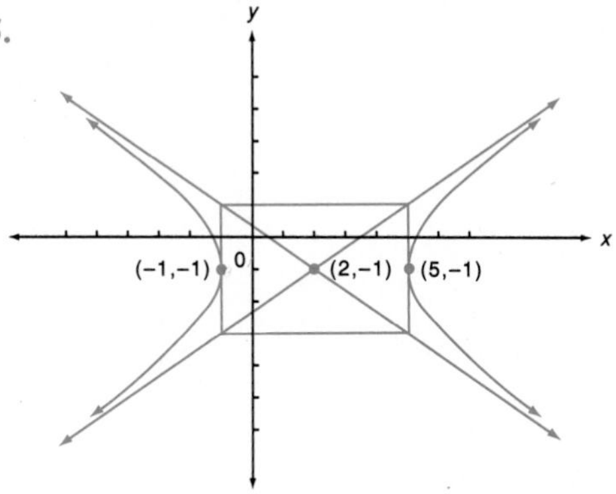

27.

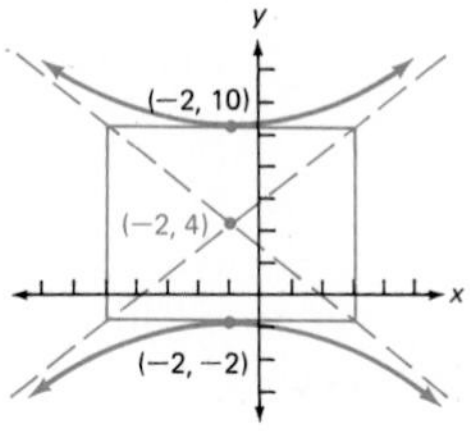

29. $\dfrac{(y-3)^2}{9} - \dfrac{(x+2)^2}{7} = 1$

7.7 Conic Sections: The Parabola (Page 331)

1. $x^2 = 8y$; $y = -2$ 3. $y^2 = -9x$; $F(-\frac{9}{4}, 0)$, $x = \frac{9}{4}$ 5. Focus: $(0, \frac{1}{8})$; directrix: $y = -\frac{1}{8}$
7. Focus: $(0, \frac{1}{16})$; directrix: $y = -\frac{1}{16}$ 9. Focus: $(0, -\frac{1}{16})$; directrix: $y = \frac{1}{16}$ 11. Focus: $(\frac{1}{2}, 0)$; directrix $x = -\frac{1}{2}$
13. Focus: $(-\frac{1}{12}, 0)$; directrix $x = \frac{1}{12}$ 15. Focus: $(-\frac{3}{8}, 0)$; directrix $x = \frac{3}{8}$
17. $x^2 = -12y$

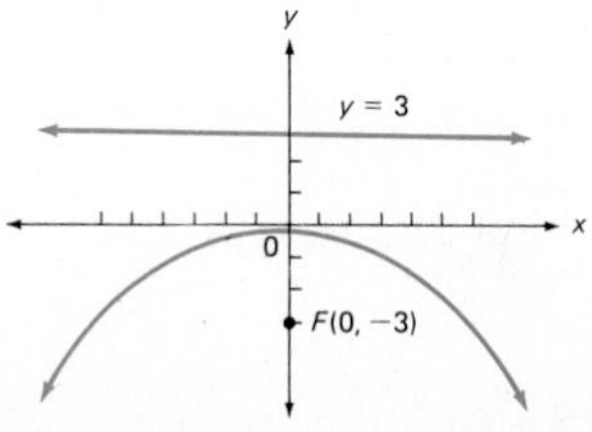

19. $x^2 = \frac{8}{3}y$

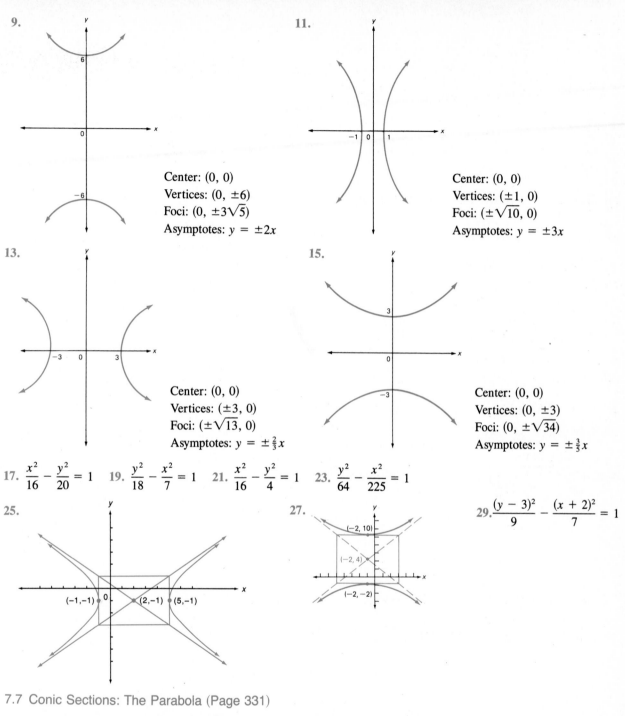

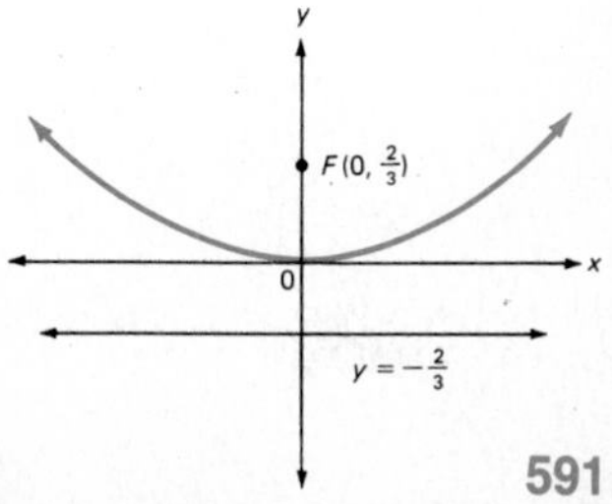

21. $y^2 = -3x$

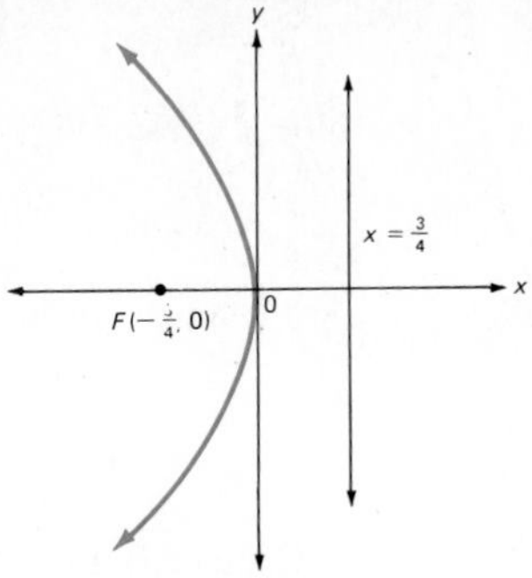

23. $y^2 = 3x$

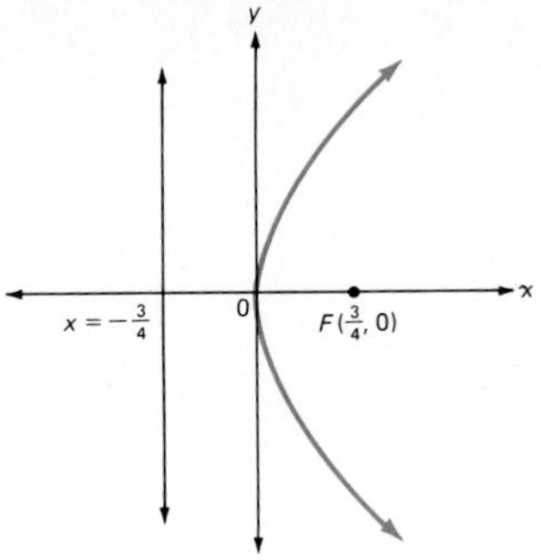

25. $\frac{9}{8}$ ft; 8π ft 27. 90 ft 29. $V(-5, 2)$; axis: $y = 2$; $F(-4, 2)$; directrix: $x = -6$ 31. $(y - 2)^2 = 8(x + 5)$

CHAPTER 7 TEST: STANDARD ANSWER (Page 336)

1.

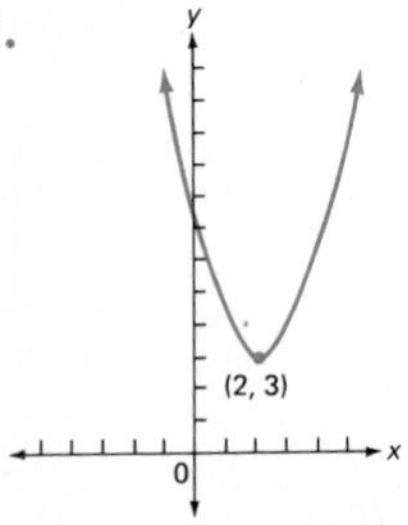

2.

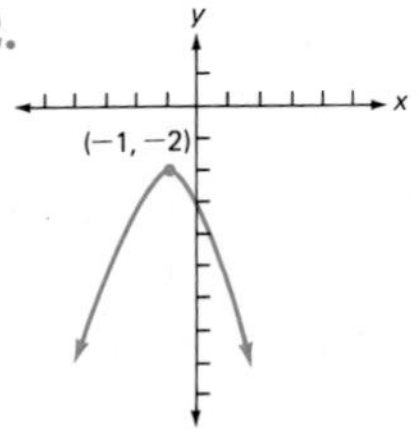

3. **(a)** $(1, -5)$ **(b)** $x = 1$ **(c)** All real numbers **(d)** All real numbers greater than or equal to -5
4. Increasing on $(-\infty, 4]$; Decreasing on $[4, \infty)$; Concave down on $(-\infty, \infty)$ 5. $y = (x + 2)^2 - 13$
6. $y = 2(x - 2)^2 - 5$ 7. 5; two irrational numbers 8. 169; two rational numbers 9. $2 - \sqrt{11}$, $2 + \sqrt{11}$
10. 9 ft by 9 ft 11. 64 ft 12. $x \le -2$ or $x \ge 8$ 13. $\frac{3 - \sqrt{5}}{4} < x < \frac{3 + \sqrt{5}}{4}$ 14. $x < -\frac{1}{2}$ or $x > 4$

15. **(a)**

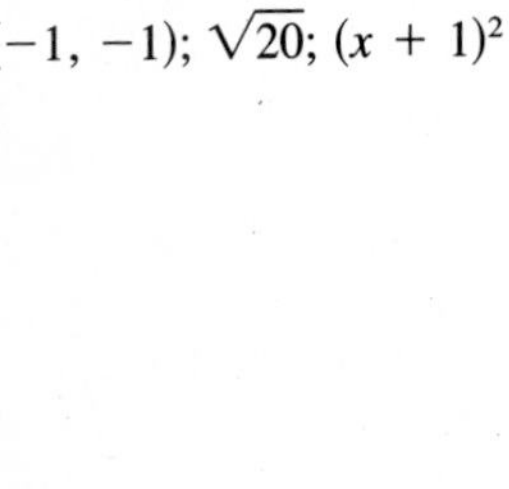

(b) $y = -\frac{3}{4}x + \frac{9}{2}$ 16. $(x + 5)^2 + y^2 = 10$; $C(-5, 0)$, $r = \sqrt{10}$

17. $(-1, -1)$; $\sqrt{20}$; $(x + 1)^2 + (y + 1)^2 = 20$ 18. Ellipse 19. Hyperbola 20. $\frac{x^2}{16} + \frac{y^2}{9} = 1$

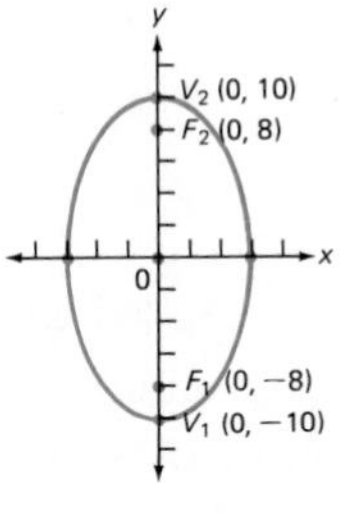

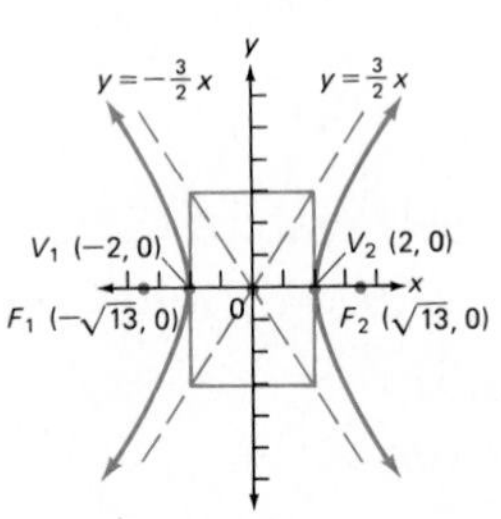

21. $\frac{x^2}{36} - \frac{y^2}{28} = 1$ 22. $\frac{y^2}{25} - \frac{x^2}{144} = 1$ 23. Focus: $(0, -2)$; directrix: $y = 2$ 24. $y^2 = -\frac{8}{3}x$; focus: $(-\frac{2}{3}, 0)$
25. $\frac{x^2}{b^2} + \frac{y^2}{6000^2} = 1$, where $b^2 = (6000)^2 - (1000)^2 \approx (5916)^2$

CHAPTER 7 TEST: MULTIPLE CHOICE (Page 337)

1. (c) **2.** (e) **3.** (a) **4.** (b) **5.** (a) **6.** (b) **7.** (c) **8.** (a) **9.** (c) **10.** (c) **11.** (d) **12.** (c) **13.** (e) **14.** (b) **15.** (a)

CHAPTER 8: GRAPHING FUNCTIONS: ROOTS OF POLYNOMIAL FUNCTIONS

8.1 Graphing Polynomial Functions (Page 345)

1.

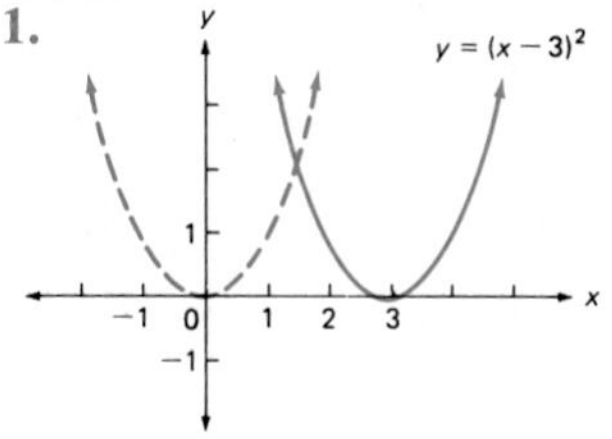

For $y = (x - 3)^2$:
Domain: all reals
Range: $y \geq 0$
Decreasing on $(-\infty, 3]$
Increasing on $[3, \infty)$
Concave up on $(-\infty, \infty)$

3.

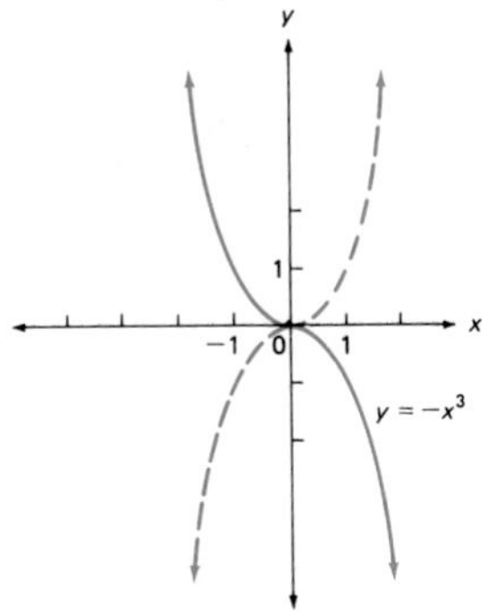

For $y = -x^3$:
Domain: all reals
Range: all reals
Decreasing on $(-\infty, \infty)$
Concave up on $(-\infty, 0)$
Concave down on $(0, \infty)$

5.

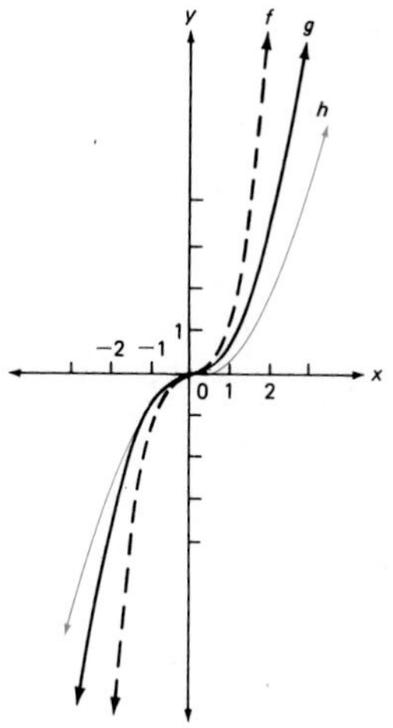

For $h(x) = \frac{1}{4}x^3$:
Domain: all reals
Range: all reals
Increasing on $(-\infty, \infty)$
Concave down on $(-\infty, 0)$
Concave up on $(0, \infty)$

7.

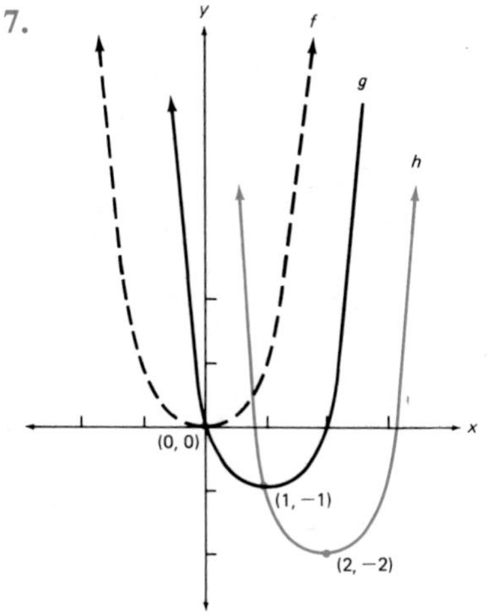

For $h(x) = (x - 2)^4 - 2$
Domain: all reals
Range: $y \geq -2$
Decreasing on $(-\infty, 2]$
Increasing on $[2, \infty)$
Concave up on $(-\infty, \infty)$

9.

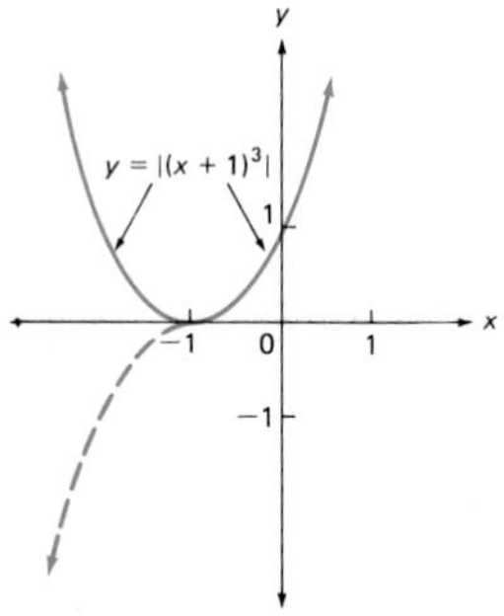

11.

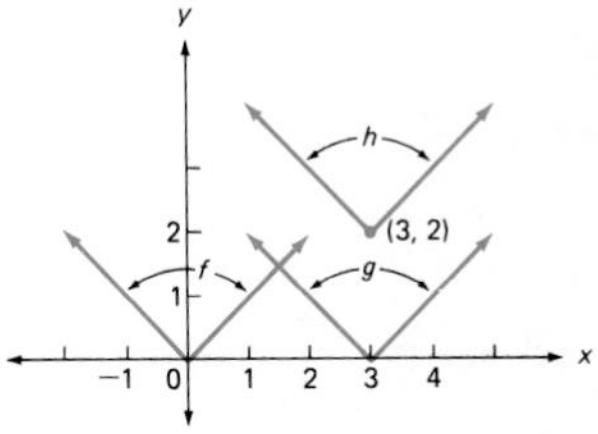

For f: x-intercept $= 0$
y-intercept $= 0$
For g: x-intercept $= 3$
y-intercept $= 3$
For h: y-intercept $= 5$

13.

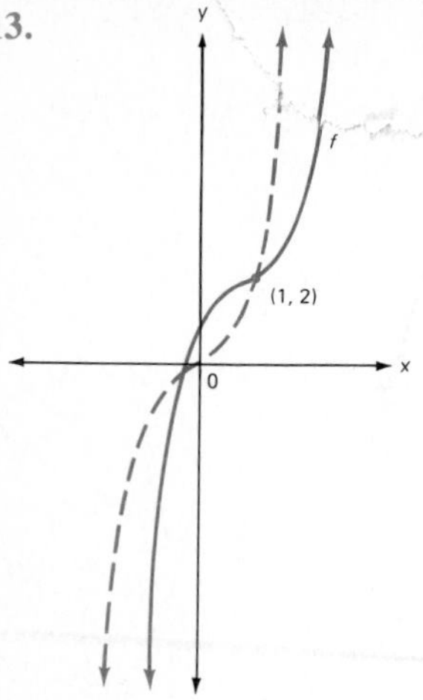

Translate the graph of $y = x^3$ one unit to the right and 2 units up.

15.

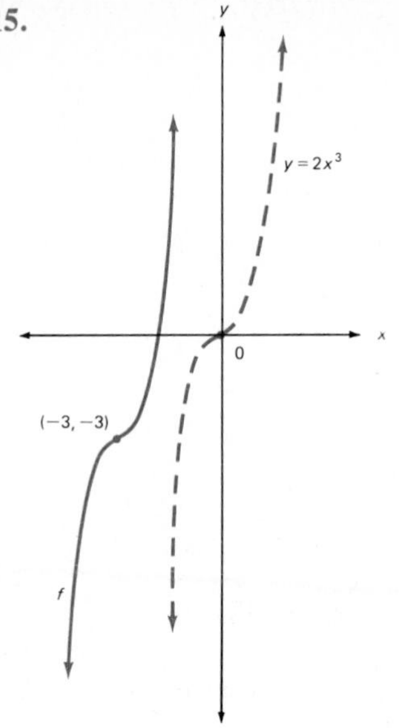

First sketch $y = 2x^3$ by multiplying the ordinates of $y = x^3$ by 2. Then translate $y = 2x^3$ three units left, and 3 units down.

17.

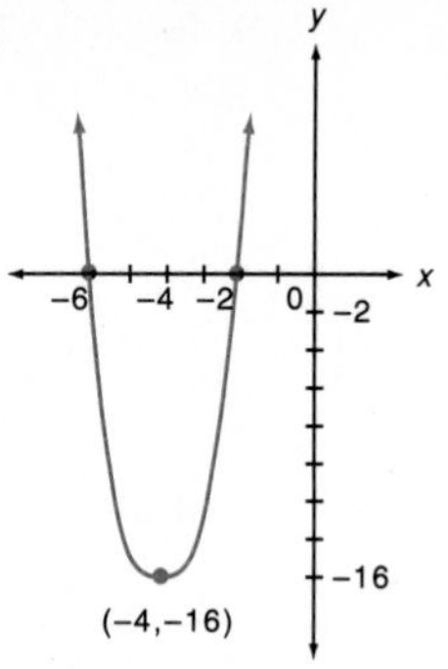

x-intercepts: -2, -6
y-intercept: 240

19. $y = (x - 3)^4 + 2$ **21.** $y = |x + \frac{3}{4}|$ **23.** $y = |x^4 - 4|$ **25.** $y = |x^3 - 1|$

27.

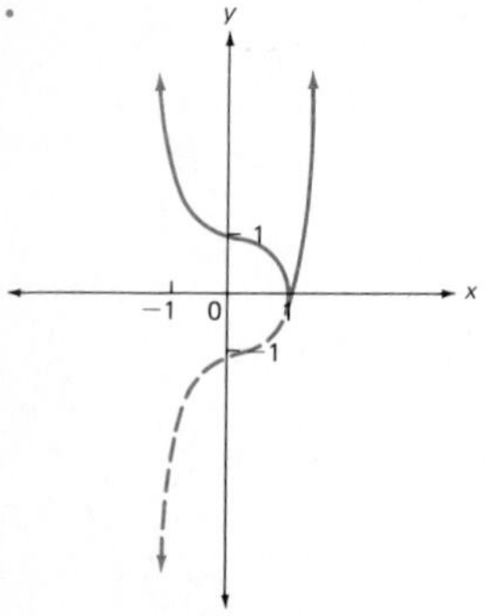

29.

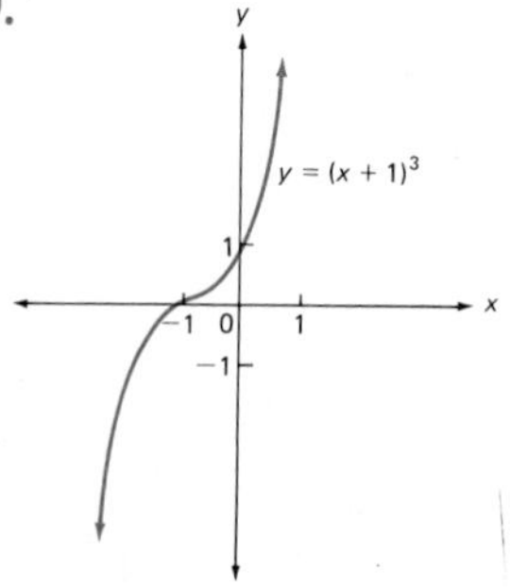

31.

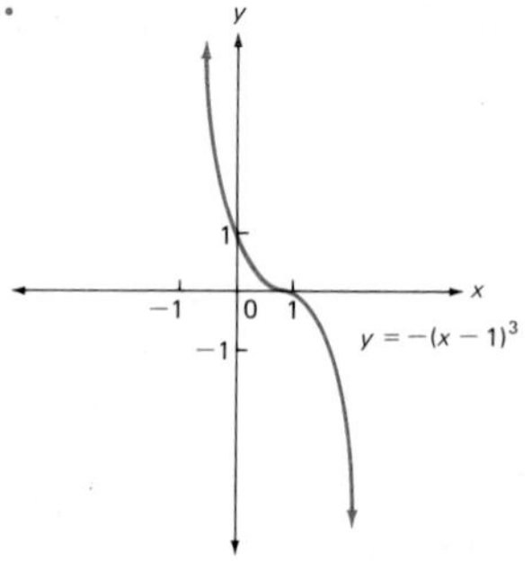

33. $x^2 + 3x + 9$ **35.** $4 + 6h + 4h^2 + h^3$

8.2 Graphing Rational Functions (Page 352)

1.

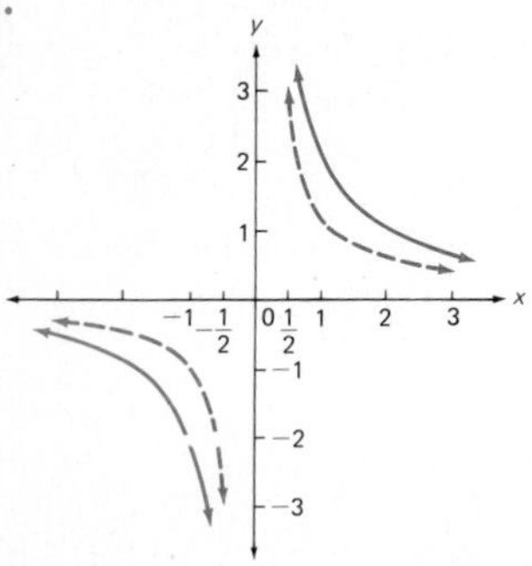

3.

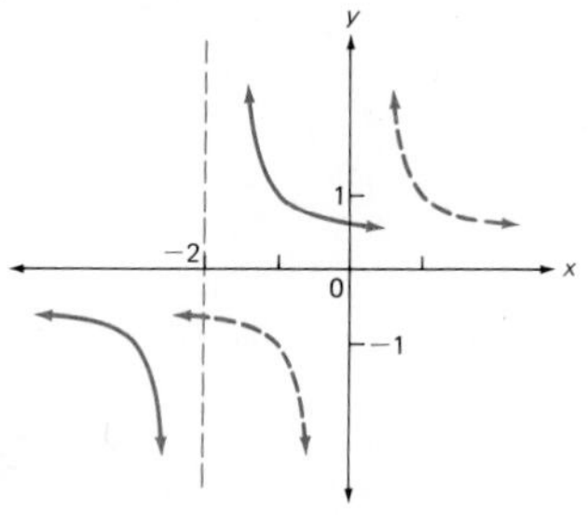

5.

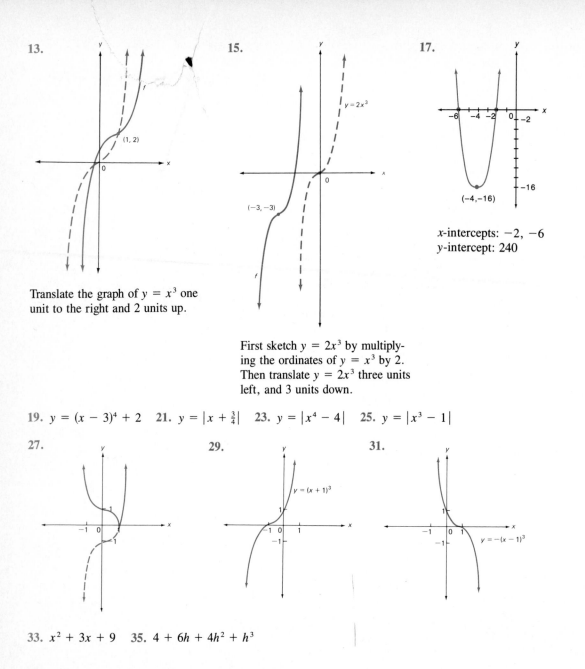

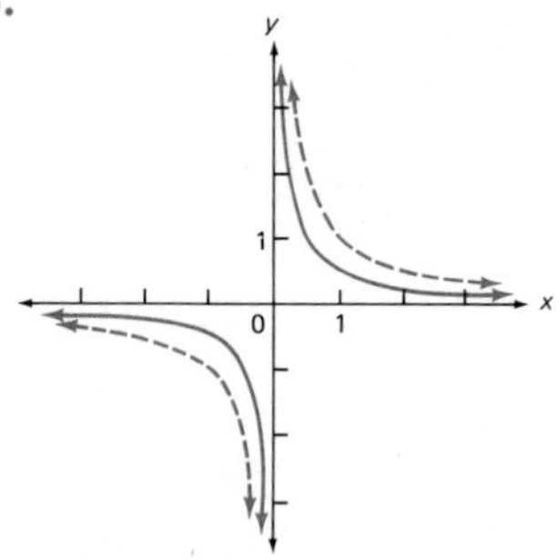

7.

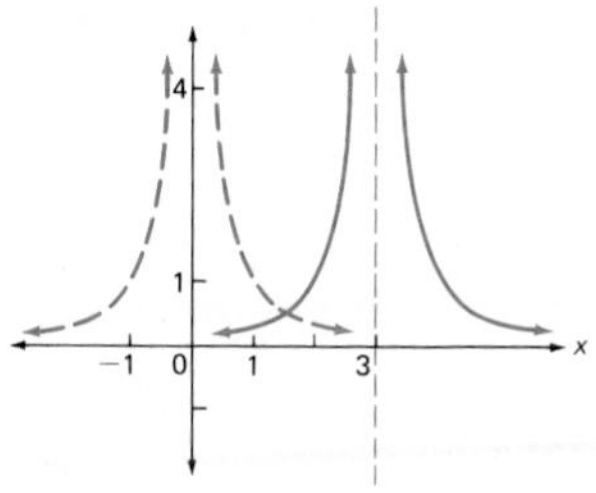

9. Asymptotes: $x = 0$, $y = 2$

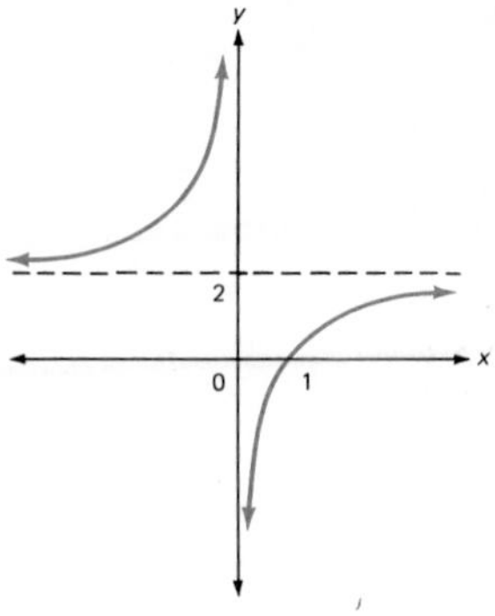

Domain: all $x \neq 0$
Range: all $y \neq 2$
Increasing and concave up on $(-\infty, 0)$
Increasing and concave down on $(0, \infty)$

11. Asymptotes $x = -4$, $y = -2$

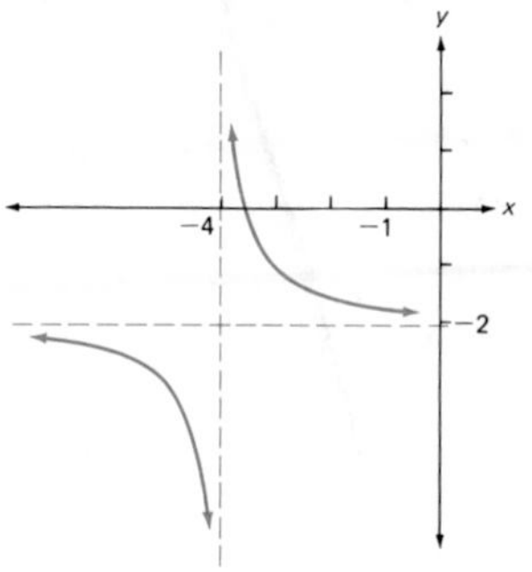

Domain: All $x \neq -4$
Range: all $y \neq -2$
Decreasing and concave down on $(-\infty, -4)$
Decreasing and concave up on $(-4, \infty)$

13. Asymptotes: $x = 2$, $y = 1$

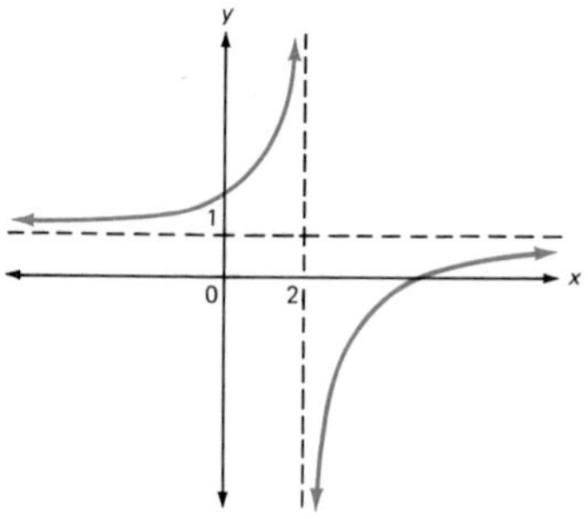

Domain: all $x \neq 2$
Range: All $y \neq 1$
Increasing and concave up on $(-\infty, 2)$
Increasing and concave down on $(2, \infty)$

15. Asymptotes: $x = -1$, $y = -2$

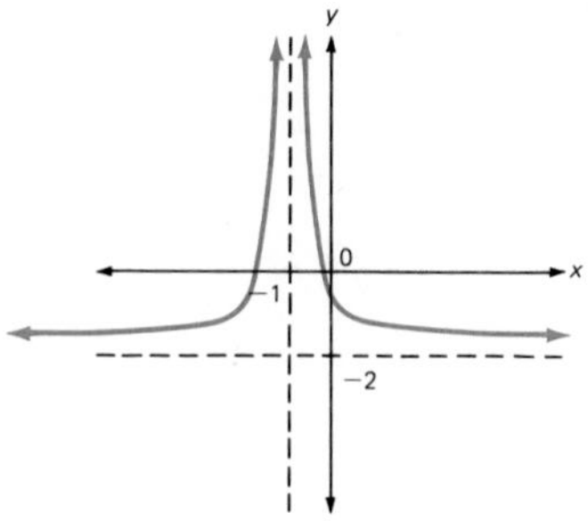

Domain: All $x \neq -1$
Range: All $y > -2$
Increasing and concave up on $(-\infty, -1)$
Decreasing and concave up on $(-1, \infty)$

17. Asymptotes: $x = 2$, $y = 0$

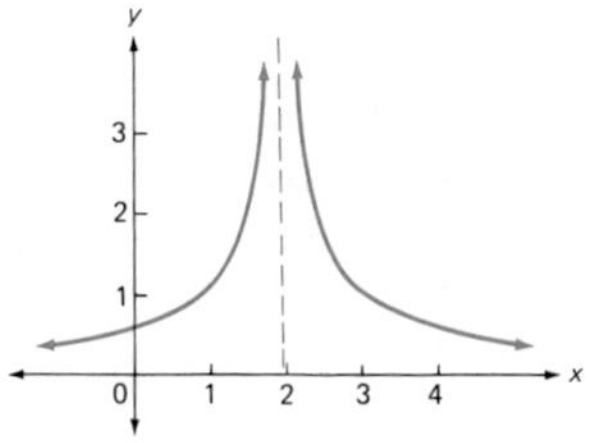

Domain: All $x \neq 2$
Range: All $y > 0$
Increasing and concave up on $(-\infty, 2)$
Decreasing and concave up on $(2, \infty)$

19. Asymptotes: $y = 0$; $x = 0$

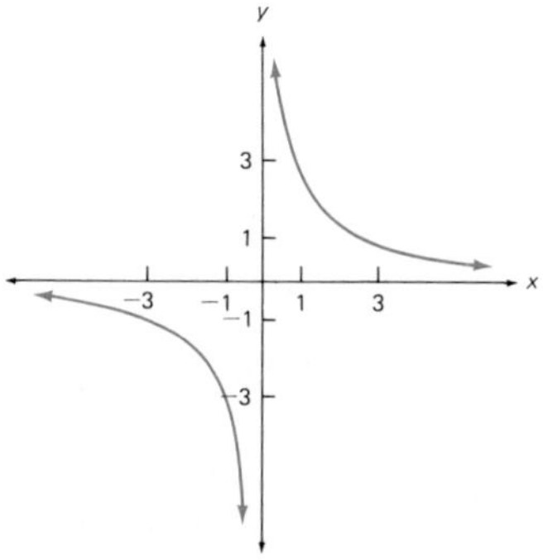

21. Asymptotes: $y = 0$, $x = 1$

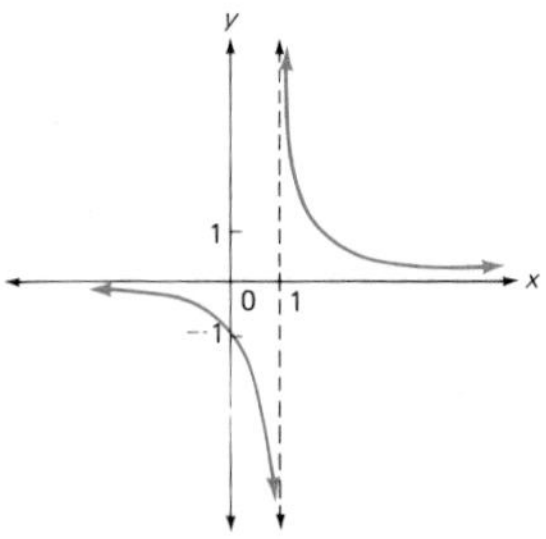

23. $f(x) = x + 3$, $x \neq 3$

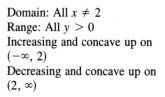

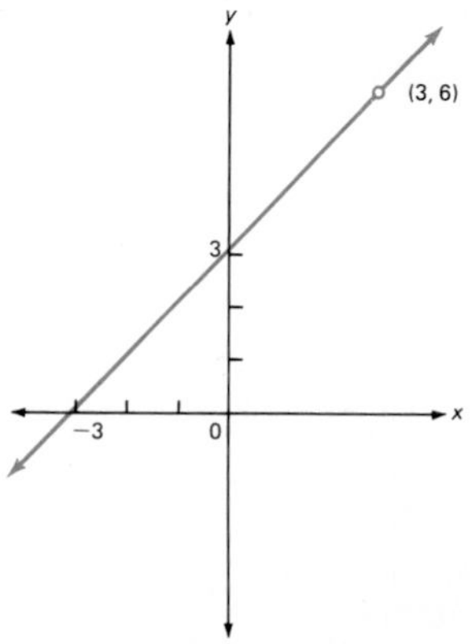

25. $f(x) = x + 2, x \neq 3$

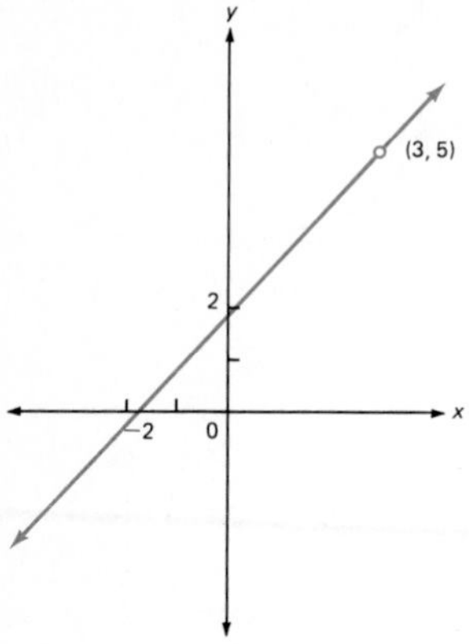

27. $f(x) = \dfrac{1}{x - 1}, x \neq -1$
Asymptotes: $x = 1, y = 0$

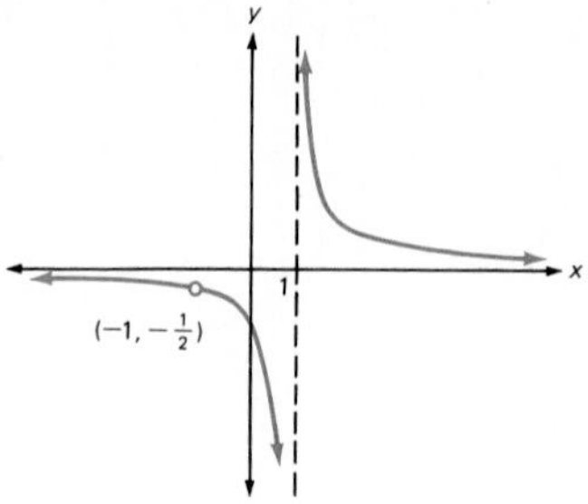

29. $f(x) = x^2 + 2x + 4, x \neq 2$

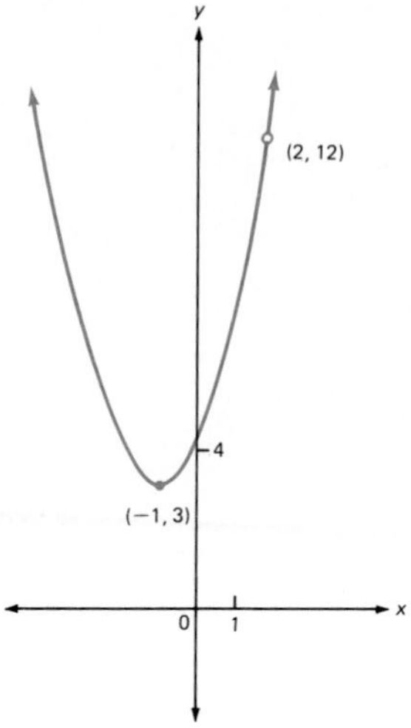

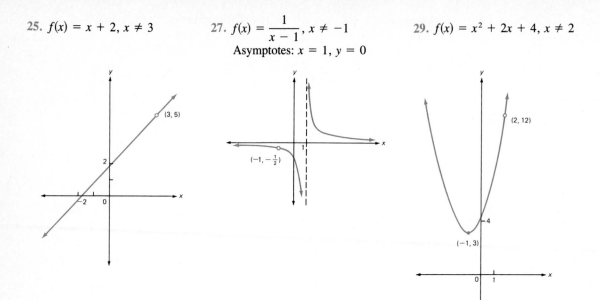

31.

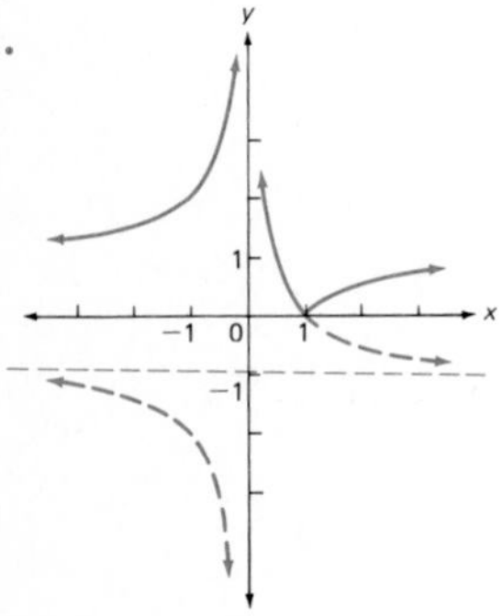

33. $-\dfrac{1}{3x}$

8.3 Graphing Radical Functions (Page 356)

1.

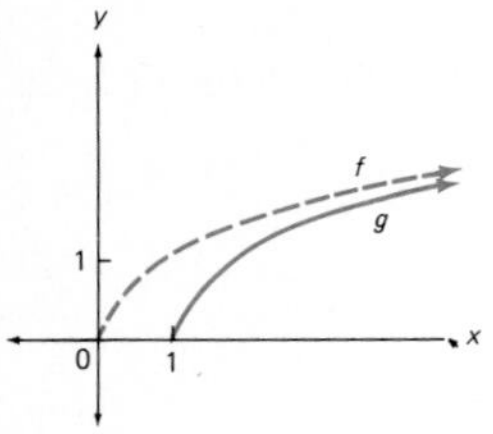

3.

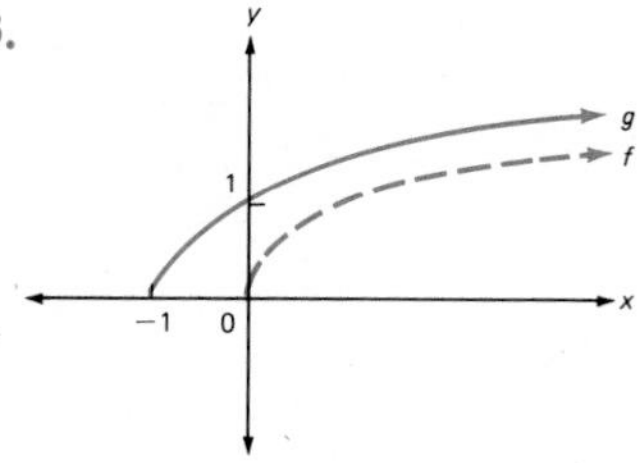

5.

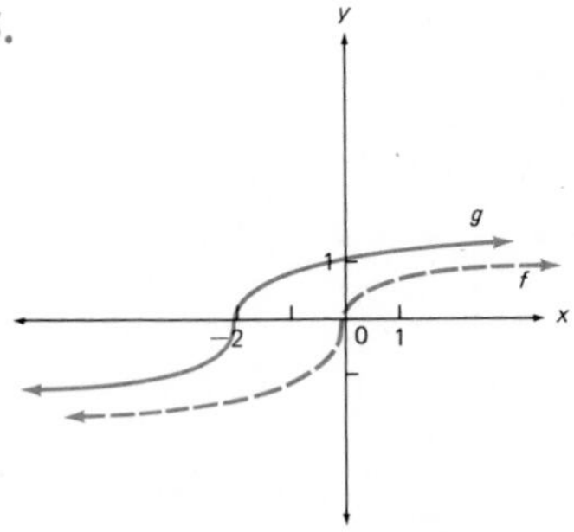

7.

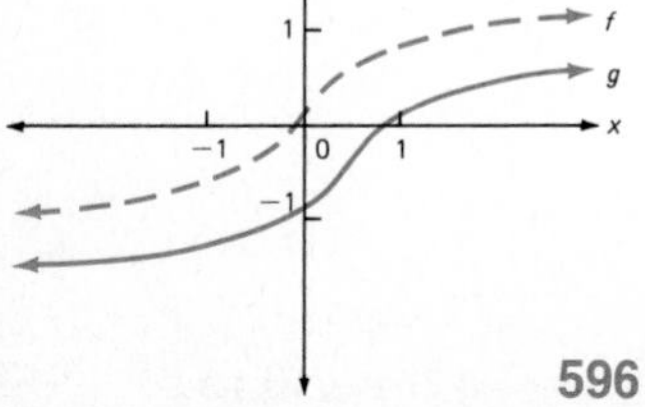

9. Domain: $x \geq -2$
Range: $y \geq 0$

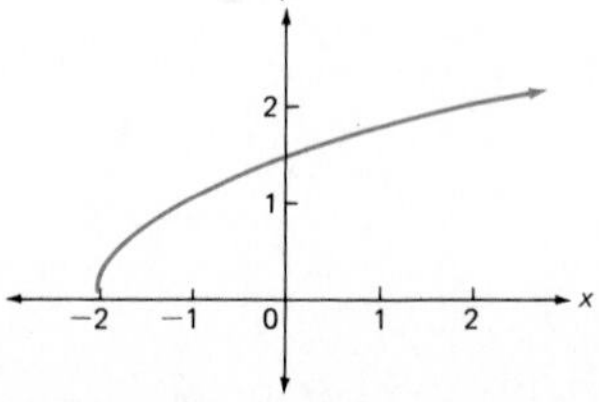

Increasing and concave down on $(2, \infty)$

11. Domain: $x \geq 3$; Range: $y \geq -1$

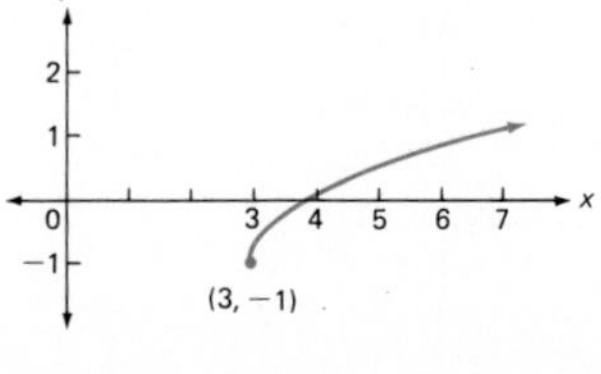

Increasing and concave down on $(3, \infty)$

13. Domain: $x \le 0$
Range: $y \ge 0$

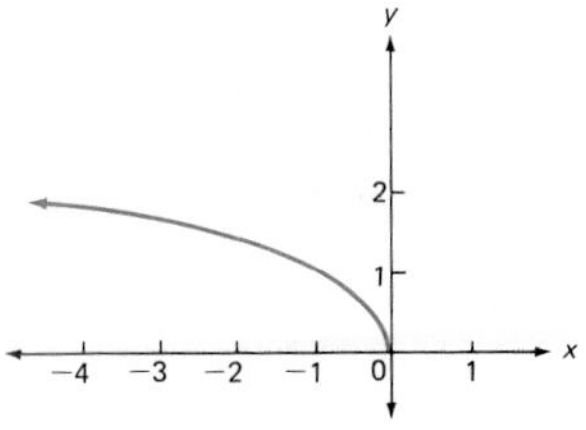

Decreasing and concave down on $(-\infty, 0)$

15. Domain: all real x
Range: all real x

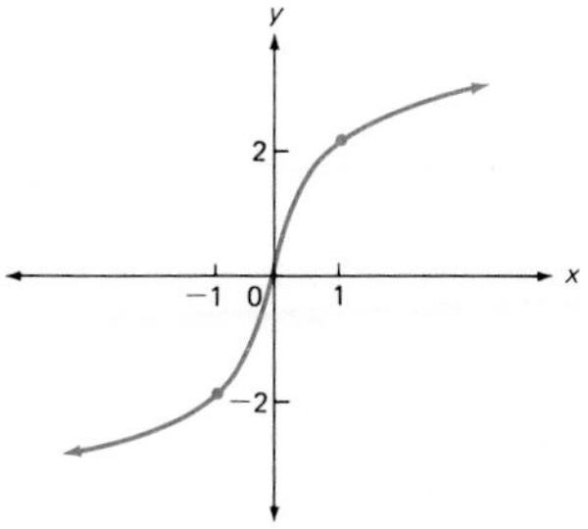

Increasing on $(-\infty, \infty)$; concave up on $(-\infty, 0)$; concave down on $(0, \infty)$

17. Domain: all real x
Range: all real y

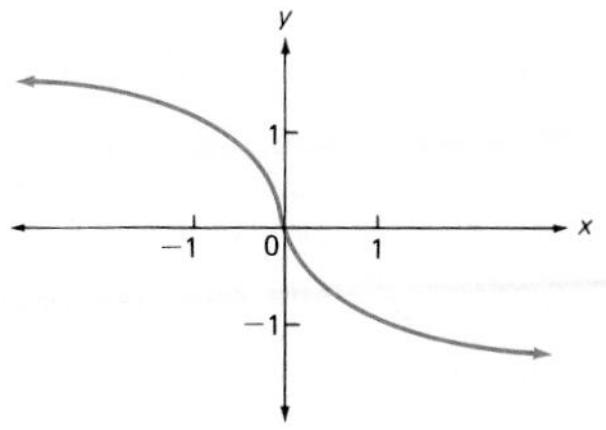

Decreasing on $(-\infty, \infty)$; concave down on $(-\infty, 0)$; concave up on $(0, \infty)$

19. Domain: $x > 0$; Range: $y > -1$
Asymptotes: $x = 0$, $y = -1$

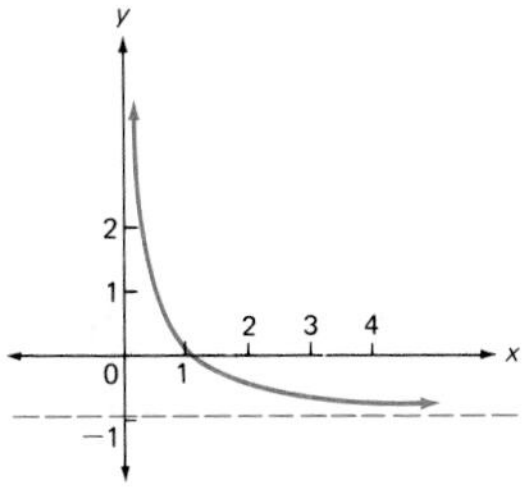

Decreasing and concave up on $(0, \infty)$

21. **(a)** $f(-x) = \dfrac{1}{\sqrt[3]{-x}} = \dfrac{1}{-\sqrt[3]{x}} = -\dfrac{1}{\sqrt[3]{x}} = -f(x)$

(b) all $x \neq 0$

(c)

x	$\frac{1}{27}$	$\frac{1}{8}$	1	8
y	3	2	1	$\frac{1}{2}$

(d) $x = 0$; $y = 0$

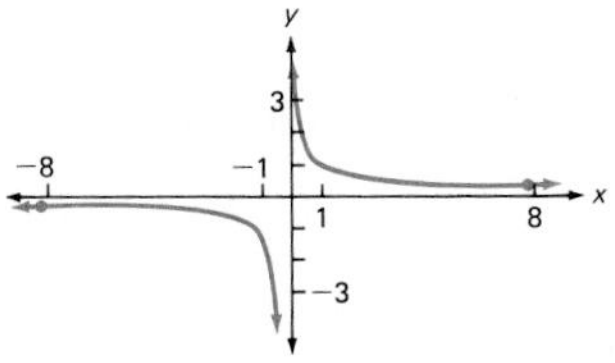

23. $y = \sqrt[4]{x}$ is equivalent to $y^4 = x$ for $x \ge 0$.

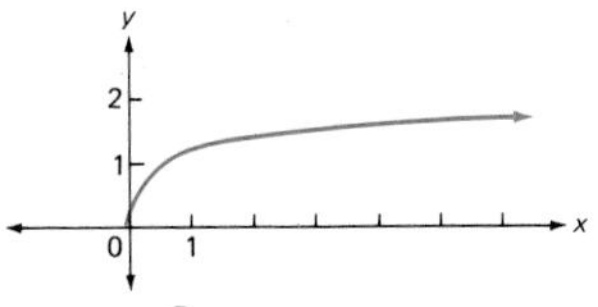

25.

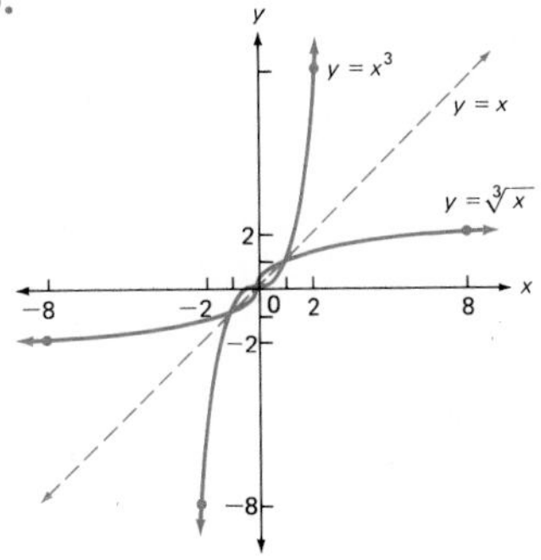

27.

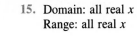

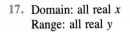

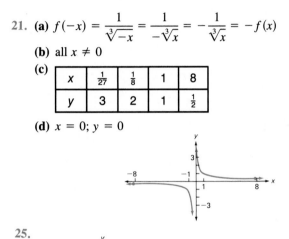

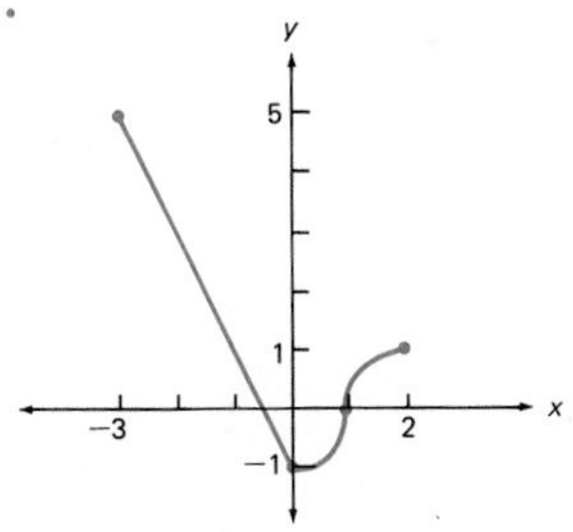

29. $$\frac{\sqrt{4+h} - 2}{h} = \frac{(\sqrt{4+h} - 2)(\sqrt{4+h} + 2)}{h(\sqrt{4+h} + 2)}$$
$$= \frac{4 + h - 4}{h(\sqrt{4+h} + 2)} = \frac{1}{\sqrt{4+h} + 2}$$

31. **(a)** $d(x) = \sqrt{25 + x^2} + \sqrt{100 + (20 - x)^2}$; **(b)** $t(x) = \dfrac{\sqrt{25 + x^2}}{12} + \dfrac{\sqrt{100 + (20 - x)^2}}{10}$; **(c)** 2.4

8.4 Synthetic Division (Page 361)

1. $x^2 + x - 2; r = 0$ 3. $2x^2 - x - 2; r = 9$ 5. $x^2 + 7x + 7; r = 22$ 7. $x^3 - 5x^2 + 17x - 36; r = 73$
9. $2x^3 + 2x^2 - x + 3; r = 1$ 11. $x^2 + 3x + 9; r = 0$ 13. $x^2 - 3x + 9; r = 0$ 15. $x^3 + 2x^2 + 4x + 8; r = 0$
17. $x^3 - 2x^2 + 4x - 8; r = 32$ 19. $x^3 + \frac{1}{2}x^2 + \frac{5}{6}x + \frac{7}{12}; r = \frac{47}{60}$

8.5 The Remainder and Factor Theorems (Page 364)

1. 8 3. 59 5. 0 7. 1 9. 67 11. −4

For Exercises 13–23, use synthetic division to obtain $p(c) = 0$, showing $x - c$ to be a factor of $p(x)$. The remaining factors of $p(x)$ are obtained by factoring the quotient obtained in the synthetic division.

13. $(x + 1)(x + 2)(x + 3)$ 15. $(x - 2)(x + 3)(x + 4)$ 17. $-(x + 2)(x - 3)(x + 1)$
19. $(x - 5)(3x + 4)(2x - 1)$ 21. $(x + 2)^2(x + 1)(x - 1)$ 23. $x^2(x + 3)^2(x + 1)(x - 1)$ 25. 2, 3 27. 2
29. $(x + 2)(x + 1)(x - 1)(x^2 - x + 3)$

8.6 The Rational Root Theorem (Page 370)

1. −3(double root), 5 3. $-\frac{2}{3}, -5, 5$ 5. −1(triple root), 0 7. $-5, -\sqrt{3}, -1, \sqrt{3}$ 9. −3, 3
11. $-\sqrt{3}, 0, \sqrt{3}, 2, 3$ 13. $-1, \dfrac{1 - \sqrt{7}}{2}, \dfrac{1 + \sqrt{7}}{2}$ 15. $\dfrac{2}{3}, 4$ 17. $-(x + 4)^2(x - 5)$ 19. $3(x + 2)(2x - 1)(x^2 + 1)$
21. By the rational root theorem the only possible rational roots are $c = \pm 1, \pm 2, \pm 4, \pm 8, \pm 16$. Synthetic division can be used to show that $p(c) \neq 0$ for each value c. Thus, there can be no rational roots.
23. $-1, 2, 1 - 2i, 1 + 2i$ 25. $2, -\dfrac{1}{6} \pm \dfrac{\sqrt{47}}{6}i$ 27. 1(triple root), $3 + i, 3 - i$

29.

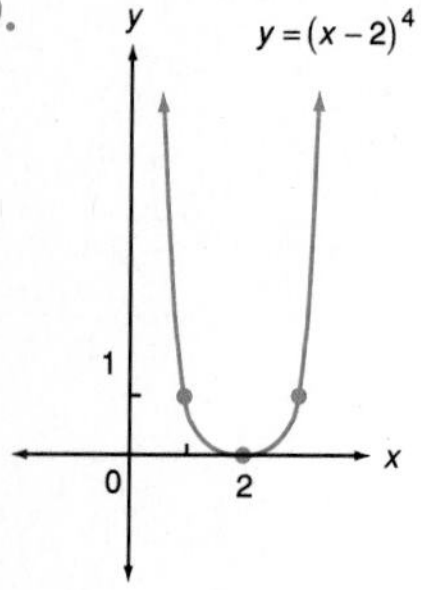

31.

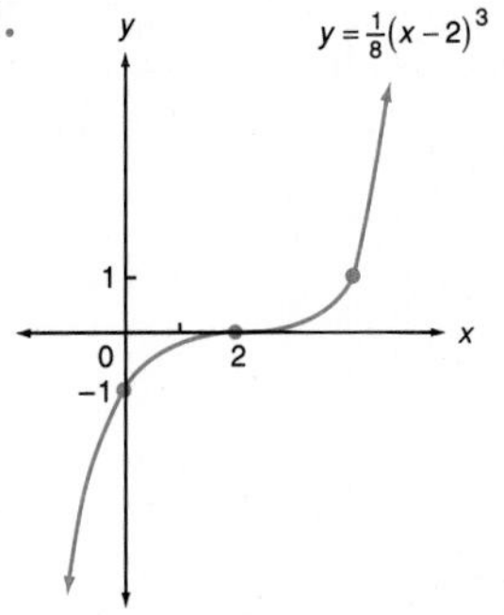

CHAPTER 8 TEST: STANDARD ANSWER (Page 374)

1. No asymptotes

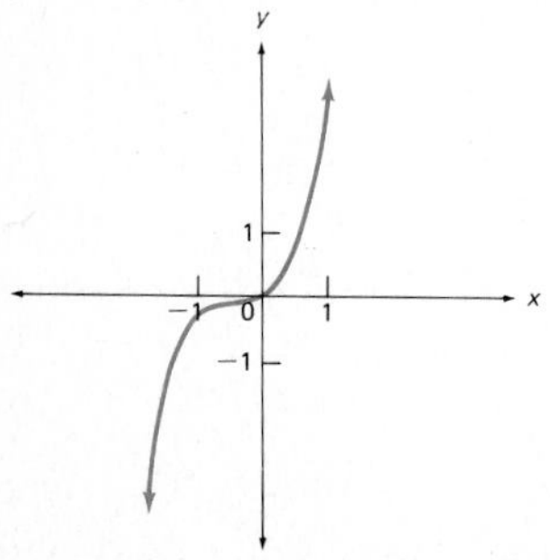

2. No asymptotes

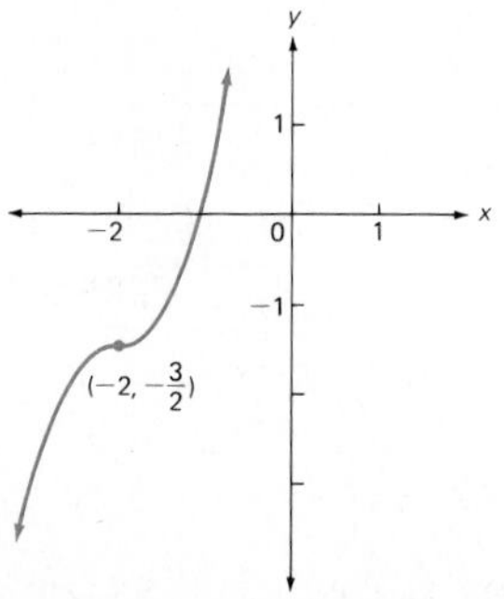

3. Asymptotes: $x = 2, y = 0$

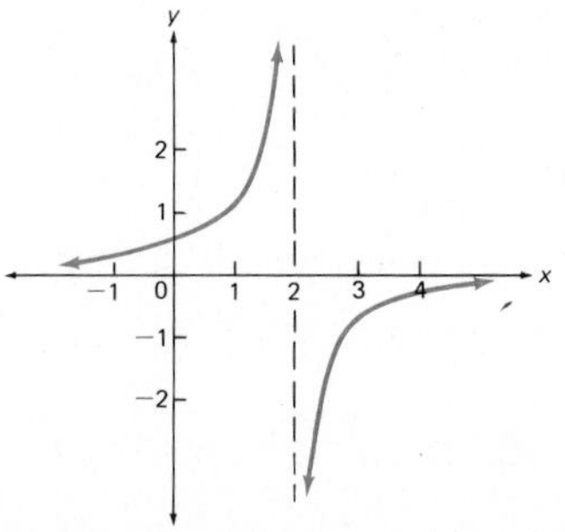

4. Domain: all $x \neq 2$; range: all $y \neq 0$; increasing and concave up on $(-\infty, 2)$; increasing and concave down on $(2, \infty)$.
5. Domain: all real x; range: all $y \geq -16$; x-intercepts: ± 2; y-intercept: -16

6. Domain: all reals

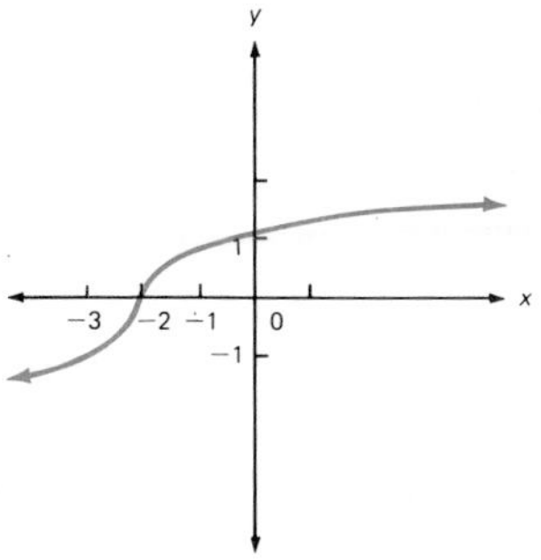

7. Domain: all $x > 0$
Asymptotes: $y = 2$, $x = 0$

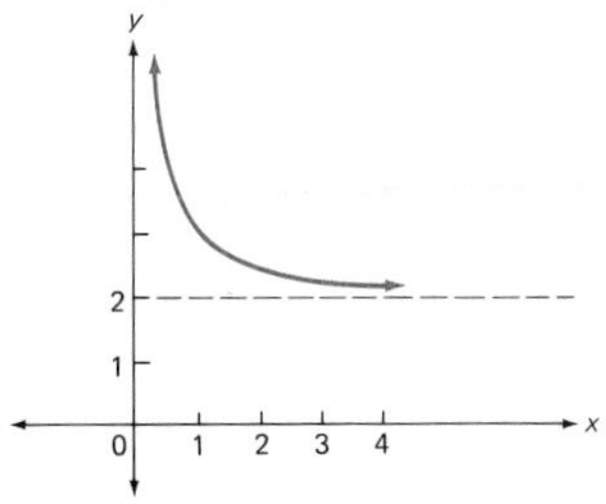

8. Domain: all reals

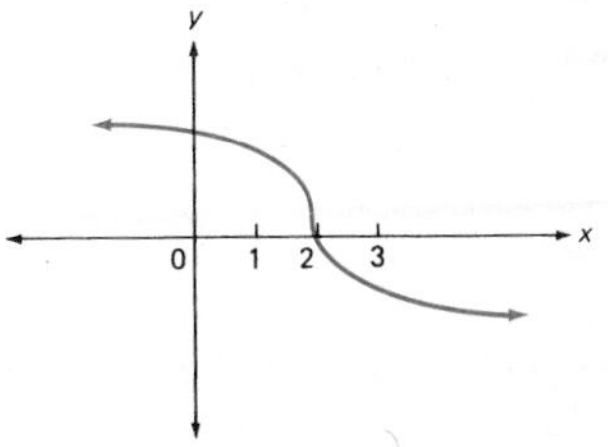

9. Range: all $y > 2$; decreasing and concave up on $(0, \infty)$ 10. $x = -3, y = 0$ 11. $x = 0, y = 3$
12. Translate graph of f 5 units right and 2 units up.
13. Translate graph of f 4 units left and 3 units down.
14. Reflect graph of f through the x-axis and translate 4 units down.
15. Reflect the negative part of $f(x) = x^3$ through the x-axis and translate 1 unit down.
16. Multiply the ordinates of g by $\frac{1}{2}$ and translate 3 units down.
17. $q(x) = 5x^2 - 2x - 7, r = -8$ 18. $q(x) = 2x^4 - x^3 + 3x^2 - 10x + 9, r = -20$
19. $q(x) = x^4 - 2x^3 + 4x^2 - 8x + 16, r = 0$
20. **(a)** When $p(x)$ is divided by $x - \frac{1}{3}$, the remainder is $\frac{2}{3}$. Then, by the remainder theorem, we have $p(\frac{1}{3}) = \frac{2}{3}$.
(b) Since $p(\frac{1}{3}) \neq 0$, the factor theorem says that $x - \frac{1}{3}$ is not a factor of $p(x)$.
21. Use synthetic division and the factor theorem to obtain $p(x) = (x + 4)(2x^2 - 5x - 3)$. Factor the quadratic to obtain $p(x) = (x + 4)(2x + 1)(x - 3)$.
22. Use synthetic division, the factor theorem, and factoring by grouping to obtain

$$\begin{aligned} p(x) &= (x - 2)(x^3 - 2x^2 + 3x - 6) = (x - 2)[x^2(x - 2) + 3(x - 2)] = (x - 2)(x^2 + 3)(x - 2) \\ &= (x - 2)^2(x^2 + 3) \end{aligned}$$

23. $(x + 3)^2(x^2 - x + 1)$ 24. $-1, -3, 2$ 25. $\frac{1}{2}, -3, -5$

CHAPTER 8 TEST: MULTIPLE CHOICE (Page 375)

1. (d) 2. (b) 3. (a) 4. (c) 5. (e) 6. (c) 7. (d) 8. (a) 9. (e) 10. (d) 11. (c) 12. (e) 13. (c) 14. (b) 15. (a)

CHAPTER 9: EXPONENTIAL AND LOGARITHMIC FUNCTIONS

9.1 Inverse Functions (Page 383)

1. **(a)** 2, 1 **(b)** $6x + 1$ **(c)** 1 3. **(a)** 6 **(b)** 8 **(c)** x 5. **(a)** $-\dfrac{215}{216}$ **(b)** $-\dfrac{511}{512}$ **(c)** $\dfrac{1}{x^3} - 1$
7. **(a)** 24 **(b)** $14 + 6\sqrt{6}$ **(c)** $x + 6\sqrt{x - 2} + 6$ 9. $(f \circ g)(x) = -8x + 7$, $(g \circ f)(x) = -8x + 19$
11. $(f \circ g)(x) = \dfrac{1}{2x^2} - 1$, $(g \circ f)(x) = \dfrac{1}{4x^2 - 2}$ 13. $(f \circ g)(x) = (x - 1)^2$; $(g \circ f)(x) = x^2 - 1$
15. $(f \circ g)(x) = \dfrac{x + 3}{3 - x}$; $(g \circ f)(x) = 4 - \dfrac{6}{x}$ 17. $(f \circ g)(x) = x^2$; $(g \circ f)x = x^2 + 2x$
19. $(f \circ g)(x) = 2$; $(g \circ f)(x) = 4$ 21. Not one-to-one 23. One-to-one 25. Not one-to-one

27. $(f \circ g)(x) = \frac{1}{3}(3x + 9) - 3 = x$
$(g \circ f)(x) = 3(\frac{1}{3}x - 3) + 9 = x$

29. $(f \circ g)(x) = (\sqrt[3]{x - 1} + 1)^3 = x$
$(g \circ f)(x) = \sqrt[3]{(x + 1)^3} - 1 = x$

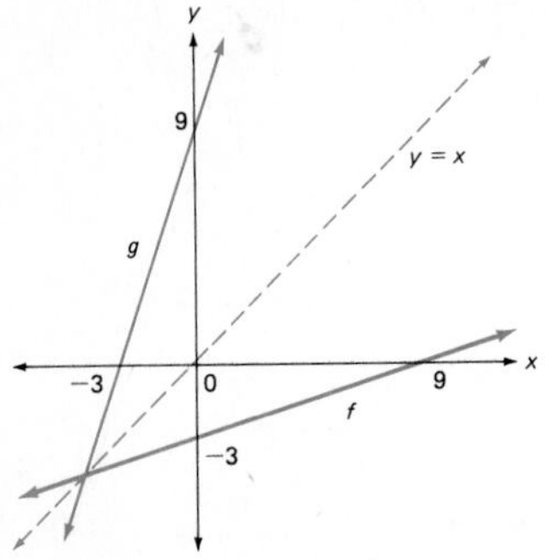

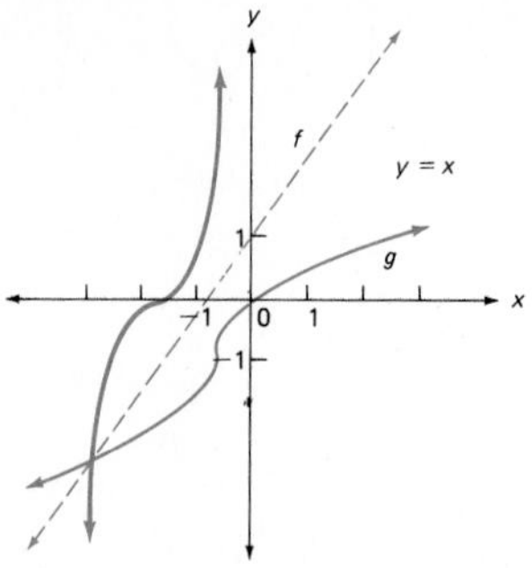

31. $g(x) = \sqrt[3]{x} + 5$ 33. $g(x) = \frac{3}{2}x + \frac{3}{2}$ 35. $g(x) = \sqrt[5]{x} + 1$

37. $f^{-1}(x) = 2 + \dfrac{2}{x}; f(f^{-1}(x)) = f\left(2 + \dfrac{2}{x}\right) = \dfrac{2}{\left(2 + \dfrac{2}{x}\right) - 2} = \dfrac{2}{\dfrac{2}{x}} = x$

$$f^{-1}(f(x)) = f^{-1}\left(\frac{2}{x - 2}\right) = 2 + \frac{2}{\dfrac{2}{x - 2}} = 2 + (x - 2) = x$$

39. $f^{-1}(x) = \dfrac{3}{x} - 2; f(f^{-1}(x)) = f\left(\dfrac{3}{x} - 2\right) = \dfrac{3}{\left(\dfrac{3}{x} - 2\right) + 2} = \dfrac{3}{\dfrac{3}{x}} = x$

$$f^{-1}(f(x)) = f^{-1}\left(\frac{3}{x + 2}\right) = \frac{3}{\dfrac{3}{x + 2}} - 2 = (x + 2) - 2 = x$$

41. $f^{-1}(x) = x^{-1/5}; f(f^{-1}(x)) = f(x^{-1/5}) = (x^{-1/5})^{-5} = x; f^{-1}(f(x)) = f^{-1}(x^{-5}) = (x^{-5})^{-1/5} = x$

43. $(f \circ f)(x) = f(f(x)) = f\left(\dfrac{1}{x}\right) = \dfrac{1}{\dfrac{1}{x}} = x$

45. $(f \circ f)(x) = f(f(x)) = f\left(\dfrac{x}{x - 1}\right) = \dfrac{\dfrac{x}{x - 1}}{\dfrac{x}{x - 1} - 1} = \dfrac{x}{x - (x - 1)} = x$

47. $g(x) = \sqrt{x} - 1$ 49. $g(x) = \dfrac{1}{x^2}$ 51. $y = mx + k$, where $m = -1$

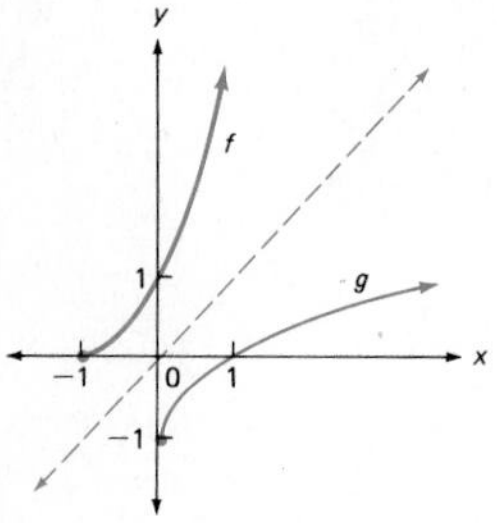

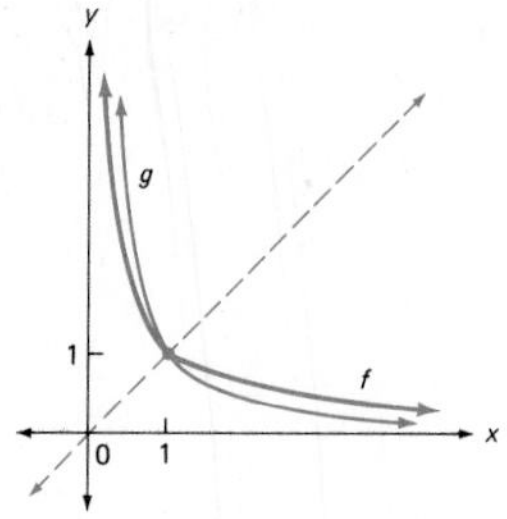

53. **(a)** $\dfrac{1}{2\sqrt[3]{x} - 1}$ **(b)** $\dfrac{2}{\sqrt[3]{x}} - 1$ **(c)** $\dfrac{1}{\sqrt[3]{2x - 1}}$ 55. **(a)** x^2 **(b)** x^2 **(c)** x^2

Note: For Exercises 1–7 the x-axis is the horizontal asymptote.

1.

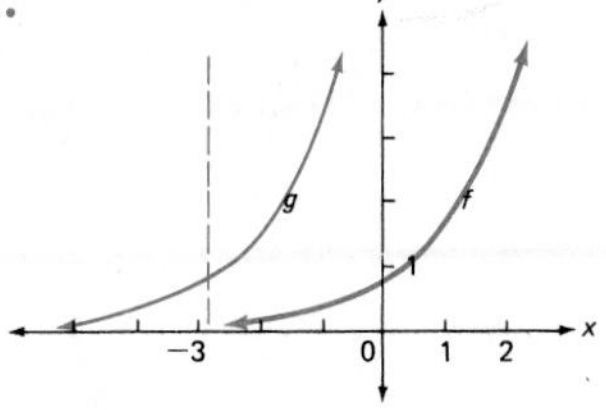

3.

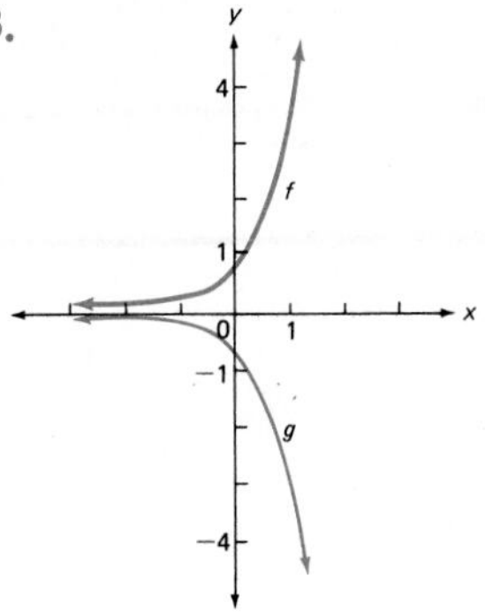

5.

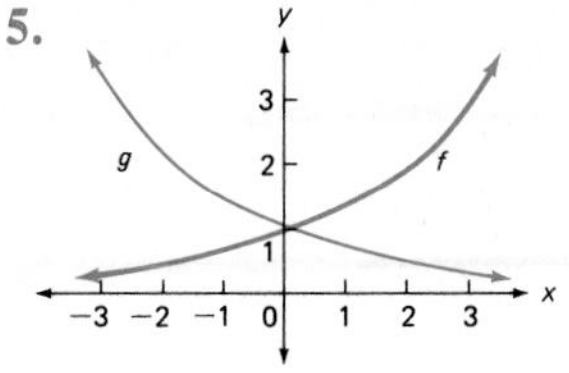

7.

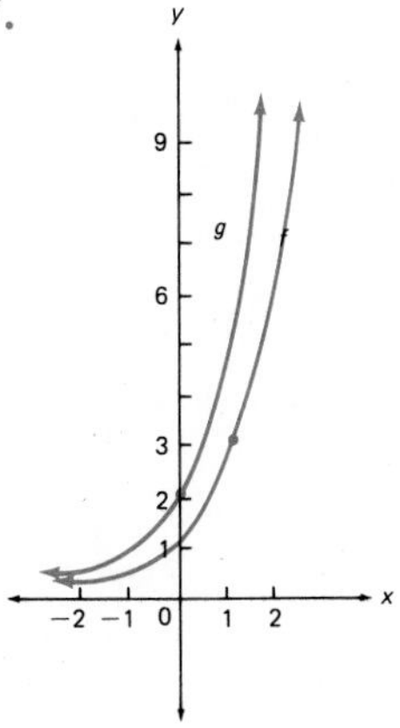

9.

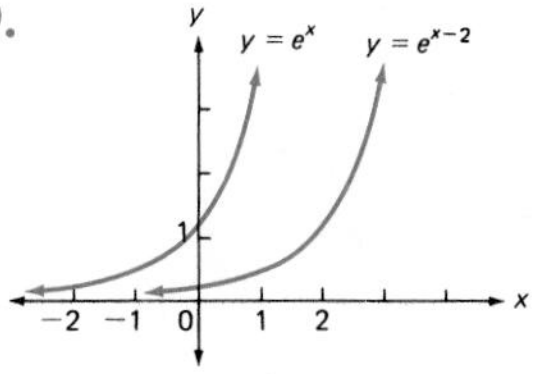

11.

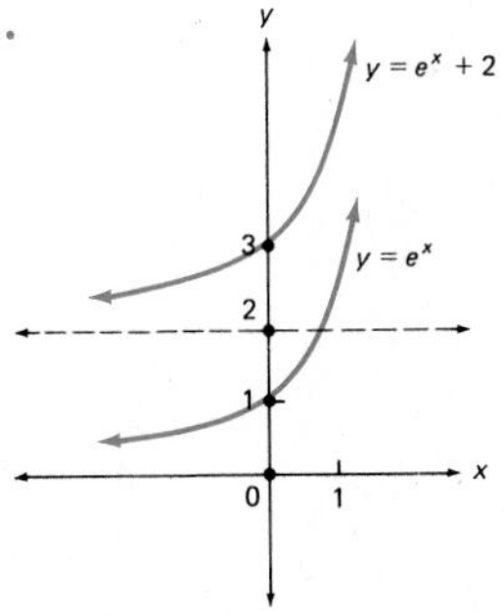

13.

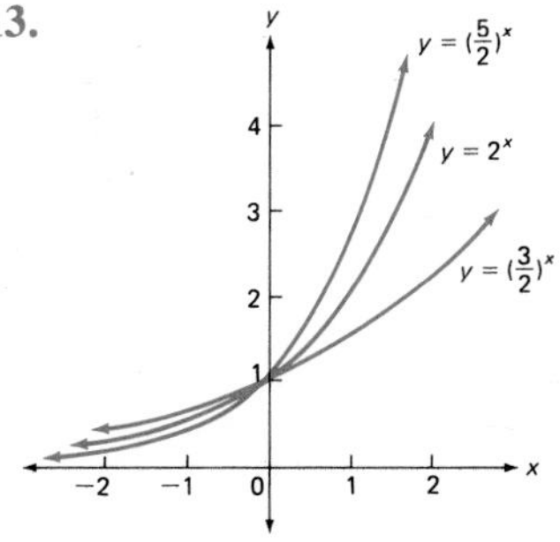

15. 6 **17.** ± 3 **19.** 1 **21.** $-2; 1$ **23.** -5 **25.** $\frac{1}{2}$ **27.** $\frac{3}{2}$ **29.** $-\frac{1}{2}$

31. $\frac{2}{3}$ **33.** $-\frac{1}{2}; \frac{3}{2}$ **35.**

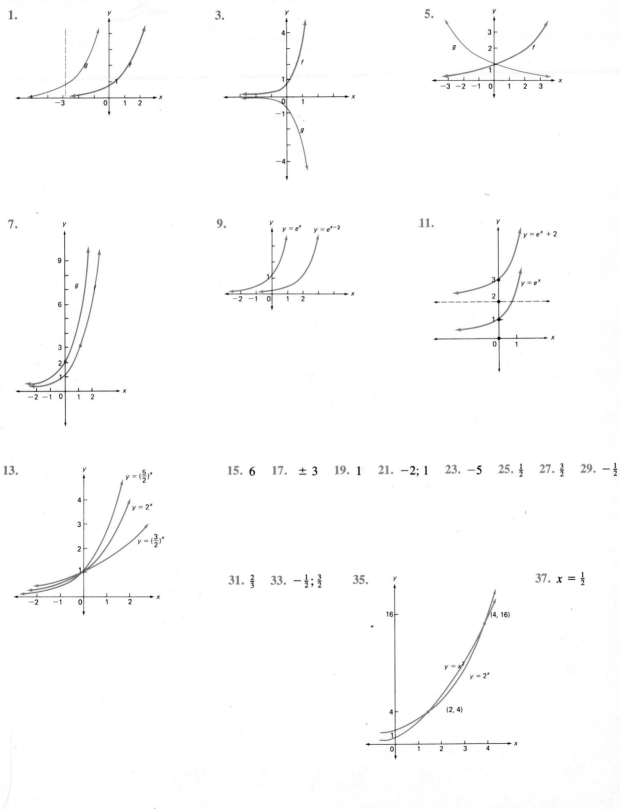

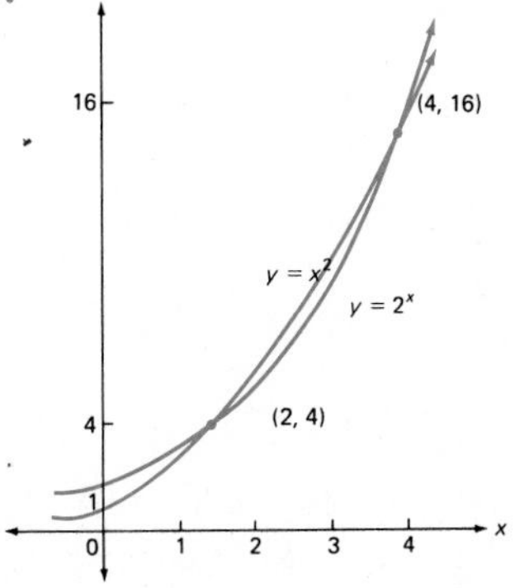

37. $x = \frac{1}{2}$

9.3 Logarithmic Functions (Page 395)

1. $g(x) = \log_4 x$

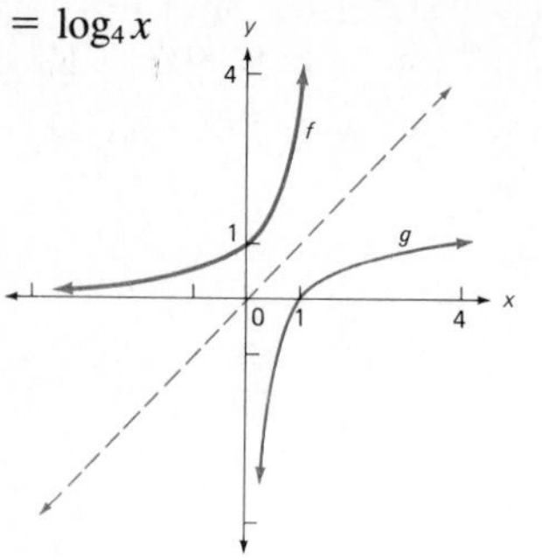

3. $g(x) = \log_{1/3} x$

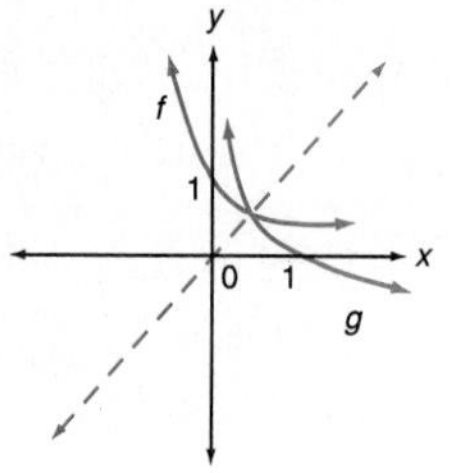

5. Shift 2 units left; $x > -2$; $x = -2$ 7. Shift 2 units upward; $x > 0$; $x = 0$.

9. Domain: all $x > 0$

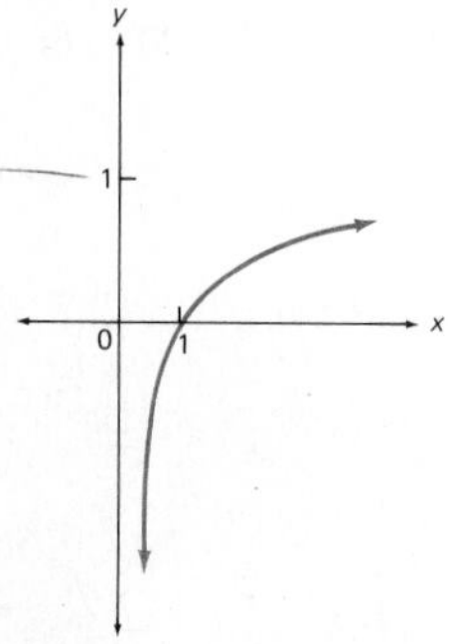

11. Domain: all $x > 0$

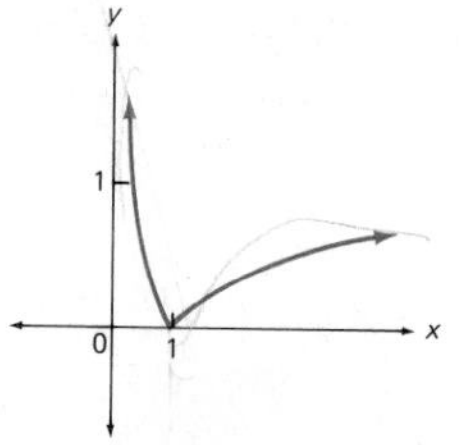

13. Domain: all $x \neq 0$

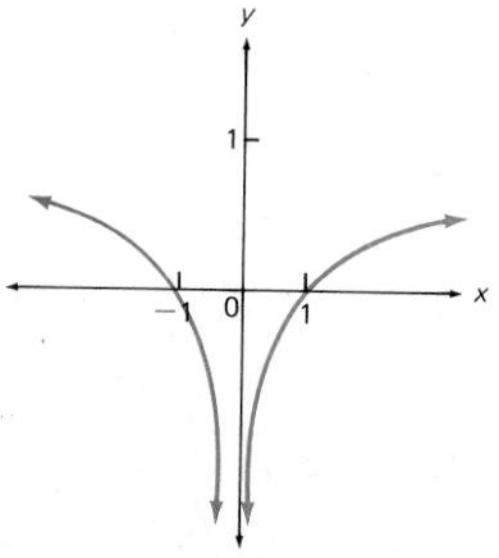

15.

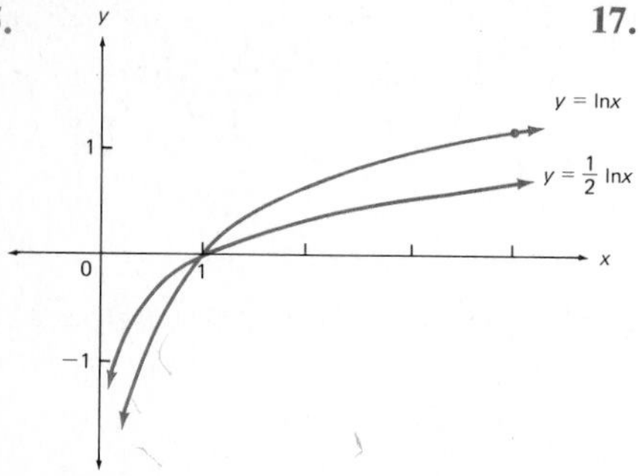

17.

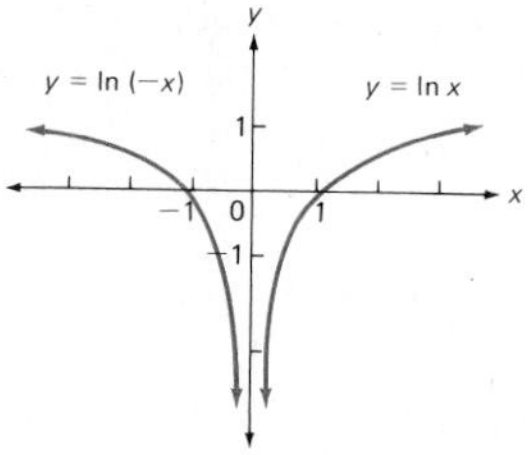

19. $\sqrt{x}$ 21. $\dfrac{1}{x^2}$ 23. $\log_2 256 = 8$ 25. $\log_{1/3} 3 = -1$

27. $\log_{17} 1 = 0$ 29. $10^{-4} = 0.0001$ 31. $(\sqrt{2})^2 = 2$ 33. $12^{-3} = \frac{1}{1728}$ 35. 4 37. -3 39. $\frac{1}{216}$ 41. 5 43. 4 45. $\frac{1}{3}$ 47. $\frac{2}{3}$ 49. 9 51. -2 53. $-\frac{1}{3}$ 55. $x = -100 \ln 27$ 57. $x = \frac{1}{3}$ 59. $x = 8$ 61. 2

63. $g(x) = \log_2 x - 1$;
$(f \circ g)(x) = 2^{(\log_2 x - 1)+1} = 2^{\log_2 x} = x$;
$(g \circ f)(x) = \log_2(2^{x+1}) - 1 = (x + 1) - 1 = x$

9.4 The Laws of Logarithms (Page 400)

1. $\log_b 3 + \log_b x - \log_b (x + 1)$ 3. $\frac{1}{2}\log_b (x^2 - 1) - \log_b x = \frac{1}{2}\log_b (x + 1) + \frac{1}{2}\log_b (x - 1) - \log_b x$

5. $-2 \log_b x$ 7. $\ln(x - 1) + 2 \ln (x + 3) - \frac{1}{2}\ln (x^2 + 2)$ 9. $\ln \sqrt{x}(x^2 + 5)$ 11. $\log_b \dfrac{x + 1}{x + 2}$

13. $\log_b \sqrt{\dfrac{x^2 - 1}{x^2 + 1}}$ 15. $\log_b \dfrac{x^3}{2(x + 5)}$ 17. $\log_b 27 + \log_b 3 = \log_b 81$ (Law 1)
$\log_b 243 - \log_b 3 = \log_b 81$ (Law 2)

19. $-2 \log_b \frac{4}{9} = \log_b (\frac{4}{9})^{-2}$ (Law 3) $= \log_b \frac{81}{16}$ 21. **(a)** 0.6020; **(b)** 0.9030; **(c)** -0.3010

23. **(a)** 1.6811; **(b)** -0.1761; **(c)** 2.0970 25. **(a)** 0.2330; **(b)** 1.9515; **(c)** 1.4771 27. 20 29. $\frac{1}{20}$ 31. 17
33. 8 35. 2 37. 7 39. 1.01 41. 5 43. 3 45. 0, -2 47. Since $f(x) = 1 + \ln x$, shift 1 unit upward.
49. Since $f(x) = \ln(x - 1)$, shift 1 unit to the right. 51. $x = \dfrac{e^2}{1 + e^2}$

53. Let $r = \log_b M$ and $s = \log_b N$. Then $b^r = M$ and $b^s = N$. Divide: $\dfrac{M}{N} = \dfrac{b^r}{b^s} = b^{r-s}$

Convert to log form and substitute: $\log_b \dfrac{M}{N} = r - s = \log_b M - \log_b N$

55. -1; 0 57. 8

9.5 Exponential Growth and Decay (Page 408)

te: The answers to some of these exercises will vary slightly depending on whether the solution was obtained using the es or a calculator)

1. $\dfrac{1.792}{0.693} = 2.586$ 3. $\dfrac{2.079}{-1.609} = -1.292$

5. $\ln 15 = \ln(3)(5) = \ln 3 + \ln 5 = 1.099 + 1.609 = 2.708$; $\dfrac{2.708}{2(1.099)} = 1.232$

7. $\ln 100 = \ln 10^2 = 2 \ln 10$; $\dfrac{2 \ln 10}{-4 \ln 10} = -0.5$

9. 2010 11. 27 13. $\frac{1}{2} \ln 100$ 15. $\frac{1}{4} \ln \frac{1}{3}$ 17. 667,000 19. 1.83 days 21. 17.33 years
23. **(a)** $\frac{1}{5} \ln \frac{4}{5}$; **(b)** 6.4 grams; **(c)** 15.5 years 25. 93.2 seconds 27. 18.47 years 29. 4200 years
31. 115,000 years 33. **(a)** \$15,605 **(b)** \$15,657 **(c)** \$15,677 **(d)** \$15,682 **(e)** \$15,683
35. **(a)** \$13,655 **(b)** \$13,686 **(c)** \$13,699 **(d)** \$13,702 **(e)** \$13,703 37. 7.7 years; 5.8 years 39. 13.86%
41. 8.69 years 43. \$2914 45. \$11,009

9.6 Common Logarithms and Applications (Page 415)

(These answers were found using Table IV. If a calculator is used for the common logarithms, then some of these answers will be slightly different.)

1. 2.6599 3. 9.6599 − 10 5. 1.8627 7. 3.72 9. 68.1 11. 0.14 13. 43,000,000 15. 2770
17. 0.000125 19. 1.22 21. 0.00887 23. 6.58 25. \$1.21 per gallon 27. 2.27 years 29. \$4050
31. 0.23 gal 33. 7220 cubic centimeters using $\frac{4}{3} = 1.33$ 35. 2.15 seconds
37. For $1 \le x < 10$ we get $\log 1 \le \log x < \log 10$ because $f(x) = \log x$ is an increasing function. Substituting $0 = \log 1$ and $1 = \log 10$ into the preceding inequality gives $0 \le \log x < 1$.
39. **(a)** 2.825; **(b)** 9.273; **(c)** 1.378; **(d)** 5.489; **(e)** 4.495; **(f)** 7.497; **(g)** 1.022; **(h)** 9.508; **(i)** 2.263

CHAPTER 9 TEST: STANDARD ANSWER (Page 421)

1. $(f \circ g)(x) = \dfrac{1}{1 - x}$; $(g \circ f)(x) = \dfrac{1}{\sqrt{1 - x^2}}$ 2. $f^{-1}(x) = \dfrac{x + 2}{3}$

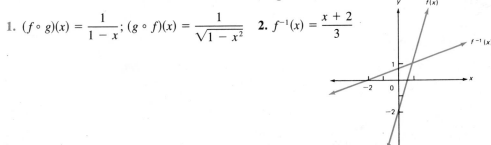

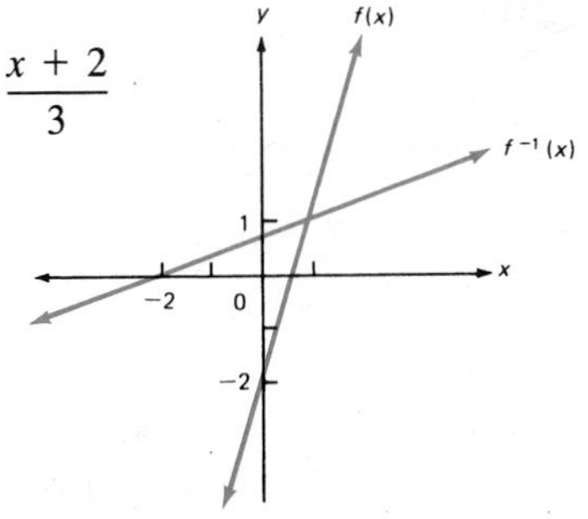

3. $g(x) = (x+1)^3$; $(f \circ g)(x) = f(g(x)) = f((x+1)^3) = \sqrt[3]{(x+1)^3} - 1 = x$;
$(g \circ f)(x) = g(f(x)) = g(\sqrt[3]{x} - 1) = (\sqrt[3]{x} - 1 + 1)^3 = (\sqrt[3]{x})^3 = x$
4. (i) a; (ii) c; (iii) e; (iv) f; (v) d; (vi) b
5. **(a)** $5^3 = 125$ **(b)** $9^{-2} = \frac{1}{81}$ 6. **(a)** $\log_{16} 8 = \frac{3}{4}$ **(b)** $\log_7 \frac{1}{49} = -2$ 7. **(a)** $b = \frac{2}{3}$ **(b)** -2
8. **(a)** $x = \frac{1}{2}$ **(b)** $x = 9$ 9. $x = \pm\sqrt{7}$ 10. $x = -7$ 11. $x = \dfrac{\ln 321}{2\ln 9}$ 12. Domain: all real x; $y = -4$
13. Domain: $x > 4$; $x = 4$ 14. $\log_b x + 10\log_b(x^2+1)$ 15. $3\ln x - \ln(x+1) - \frac{1}{2}\ln(x^2+2)$ 16. $\log_7 10x^2$
17. $\log_b \dfrac{\sqrt[3]{x}}{(x+2)^2}$ 18. $x = \dfrac{10}{3}$ 19. $x = 5$ 20.

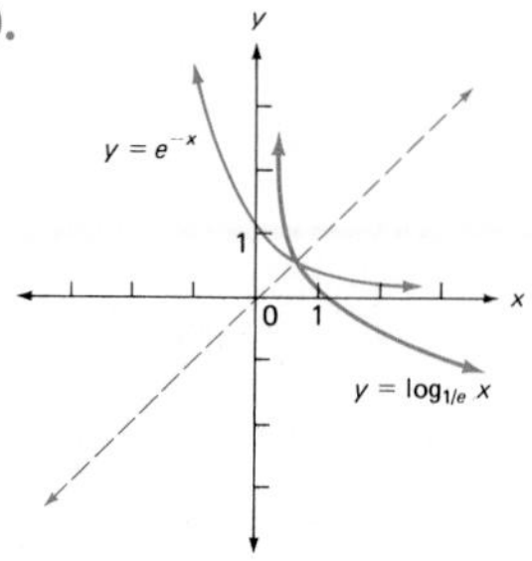

21. $25\ln 2$ 22. $2000(1.02)^{24}$

23. $12,060 (using Table 2), $12,083 (using a calculator) 24. 0.11 25. $x = \dfrac{e-2}{e+2}$

CHAPTER 9 TEST: MULTIPLE CHOICE (Page 422)

1. (d) 2. (b) 3. (d) 4. (a) 5. (c) 6. (e) 7. (a) 8. (c) 9. (a) 10. (b) 11. (a) 12. (a) 13. (b)
14. (c) 15. (d)

ANSWERS TO CUMULATIVE REVIEW FOR CHAPTERS 7–9 (Page 425)

1. **(a)** Translate $y = x^2$ three units right and five units down. x-intercepts $= 3 \pm \sqrt{5}$; y-intercept $= 4$
(b) Reflect $y = x^2$ through the x-axis and shift (translate) $\frac{1}{2}$ unit up. x-intercepts $= \pm\dfrac{\sqrt{2}}{2}$; y-intercept $= \frac{1}{2}$.
2. **(a)** (i) $(0, -3)$ (ii) $x = 0$ (iii) Domain: all reals, range: $y \geq -3$ (iv) decreasing on $(-\infty, 0]$, increasing on $[0, \infty)$ (v) concave up on $(-\infty, \infty)$
(b) (i) $(-1, 4)$ (ii) $x = -1$ (iii) Domain: all reals, range: $y \leq 4$ (iv) increasing on $(-\infty, -1]$, decreasing on $[-1, \infty)$ (v) concave down on $(-\infty, \infty)$.
3. 1, 5 4. -7; none 5. $y = 2(x-3)^2 - 23$; -23 is minimum; range: $y \geq -23$. 6. 16 cm by 16 cm
7. $-5 < x < \frac{1}{2}$ 8. $x < 0$ or $x > 2$ 9. $(x+7)^2 + (y-4)^2 = 81$ 10. $C(3, -5)$, $r = 6$
11. $y + 5 = \frac{12}{5}(x - 12)$ 12. $\dfrac{x^2}{81} + \dfrac{y^2}{45} = 1$ 13. Foci: $(0, \pm 4\sqrt{2})$; Vertices: $(0, \pm 6)$; $2b = 4$
14. $\dfrac{x^2}{425} + \dfrac{y^2}{441} = 1$ 15. $\dfrac{x^2}{25} - \dfrac{y^2}{39} = 1$ 16. $\dfrac{x^2}{9} - \dfrac{y^2}{4} = 1$
17. Foci: $(0, \pm 5)$; Vertices: $(0, \pm 3)$; $y = \pm\frac{3}{4}x$ 18. $F(0, \frac{1}{2})$; $y = -\frac{1}{2}$ 19. $y^2 = -\frac{20}{3}x$; $F(-\frac{5}{3}, 0)$
20. **(a)** Translate $y = x^3$ two units left and four units down.
(b) Translate $y = x^3$ five units right and reflect the negative part through the x-axis.
21. Decreasing in $(-\infty, 5]$; increasing on $[5, \infty)$; x-intercept $= 5$; y-intercept $= 125$
22. $y = \dfrac{1}{x+3} - 4$; $x = -3$, $y = -4$

23. Domain: $x \neq 2$; range: $y > 1$; concave up and increasing on $(-\infty, 2)$, concave up and decreasing on $(2, \infty)$

24. $y = \sqrt{x - 2} + 5$

25.

	Domain	Intervals Where Increasing	Intervals Where Decreasing	Asymptotes
(a)	$x > -2$	none	$(-2, \infty)$	$x = -2, y = 0$
(b)	$x > 0$	none	$(0, \infty)$	$x = 0, y = 2$
(c)	all reals	$(-\infty, \infty)$	none	none

26. Symmetric around y-axis: (a), (c), (e), (i), (j), (k)
Symmetric through origin: (b), (d), (g), (h), (l)

27. $q(x) = 2x^4 + x^3 + 2x^2 + 11x + 10$, $r = 26$

28. Substitute -3 into $p(x)$ to obtain $p(-3) = r$, or use synthetic division to divide $p(x)$ by $x + 3$ and obtain $r = p(-3)$.

29. Since $p(1) = 1 \neq 0$, $x - 1$ is not a factor of $p(x)$.

30. By synthetic division $p(-3) = 3d - 12$. Then $p(-3) = 0$ when $d = 4$ and $x - 3$ is a factor of $p(x)$.

31. $(x + 1)(x + 4)(x - 2)$ 32. $\frac{1}{2}, 2, i, -i$ 33. $-3, \frac{1}{3}, 5$ 34. **(a)** $-\frac{4}{3}, 4$ **(b)** $\dfrac{2(x + 1)}{1 - 4x}$ **(c)** $\dfrac{3x - 5}{2x}$

35. $f^{-1}(x) = \dfrac{2}{3 - x}$

36. **(a)** Domain: all reals; range: $y > 0$ **(b)** Increasing on $(-\infty, \infty)$ **(c)** concave up on $(-\infty, \infty)$
(d) no x-intercepts; y-intercept: 1 **(e)** $y = 0$

37. $y = \dfrac{3}{2^{x-3}}$ 38. $x = -1, 3$ 39. $g(x) = \log_5 x$; domain: $x > 0$ 40. **(a)** $x = -2, 3$ **(b)** 10

41. **(a)** 9 **(b)** $\frac{2}{3}$ 42. $\frac{1}{2}(1 + \ln 25)$ 43. $3 \log_b x + \frac{1}{2}\log(x^2 + 1) - \log_b(5x - 1)$ 44. 10 45. 9

46. Since $y = -\ln x$, reflect $y = \ln x$ through the x-axis. 47. **(a)** $\dfrac{\ln 12}{2 \ln 3}$ **(b)** 1.13 48. 2.88 hrs

49. **(a)** \$5867 **(b)** \$5868 50. 0.000148

CHAPTER 10: SYSTEMS OF EQUATIONS AND INEQUALITIES

10.1 Systems of Linear Equations in Two Variables (Page 433)

1. $(1, -1)$ 3. $(2, -2)$ 5. $(5, 1)$ 7. $(-2, -6)$ 9. $(4, -10)$ 11. $(3, -2)$ 13. $(-7, 3)$ 15. $(\frac{3}{4}, \frac{2}{3})$
17. $(\frac{17}{10}, \frac{4}{5})$ 19. $(3, 2)$ 21. $(\frac{1}{2}, \frac{1}{3})$ 23. $(5, 4)$ 25. $(-1, 2)$ 27. $(4, 4)$ 29. $(-5, 2)$ 31. $(\frac{36}{11}, -\frac{6}{11})$
33. $(-16, -4)$ 35. $(0, 1)$ 37. $(1, \frac{1}{2})$ 39. $(2, -1)$ 41. $(\frac{17}{18}, \frac{7}{18})$ 43. Inconsistent 45. Inconsistent
47. Consistent 49. Consistent 51. Dependent 53. Inconsistent 55. $(-20, -20)$ 57. $a = 4, b = -7$
59. $(\frac{1}{11}, \frac{1}{7})$ 61. $(\frac{3}{2}, \frac{1}{9})$ 63. $(\frac{88}{21}, \frac{22}{9})$

10.2 Applications of Linear Systems (Page 439)

1. 13, 43 3. $l = 21, w = 9$ 5. 26, 37 7. 8 pounds of potatoes and $1\frac{1}{2}$ pounds of string beans
9. 10 fours and 8 fives 11. 47 13. \$5200 tuition; \$3200 room and board
15. 15 hours work-study; $8\frac{1}{2}$ hours babysitting 17. $2\frac{3}{4}$ hours at 5 kph; $1\frac{3}{4}$ hours at 3 kph; total time = $4\frac{1}{2}$ hours
19. Speed of plane = 410 miles per hour; wind velocity = 10 miles per hour 21. 59 23. 3 miles per hour; 6 miles
25. $-43, 9$ 27. 550 at 25¢ and 360 at 45¢ 29. \$2800 at 8%; \$3200 at $7\frac{1}{2}$% 31. 16 milliliters of each
33. 6 miles by car; 72 miles by train 35. An infinite number of answers

37. Yes, the clerk was wrong. If there were common unit prices, say x = cost per orange and y = cost per tangerine, then

$$6x + 12y = 144$$
$$2x + 4y = 47$$

which is an inconsistent system. (*Note:* The smaller bag is a better buy since

$$2(\text{oranges}) + 4(\text{tangerines}) = \tfrac{1}{3}(6 \text{ oranges} + 12 \text{ tangerines}) = \tfrac{1}{3}(1.44) = 0.48$$

39. \$120 41. 90 43. (90, 1800) 45. 115

10.3 Systems of Linear Equations in Three Variables With Applications (Page 445)

1. $(-1, 0, 2)$ 3. $(1, 2, -3)$ 5. $(3, -2, -4)$ 7. $(\frac{1}{2}, -2, 5)$ 9. $(1, 0, 3)$ 11. Inconsistent; no solution 13. $(4, -3, 2)$ 15. 25 ones, 30 fives, 40 tens 17. 40°, 50°, 90° 19. 15 pounds peanuts, 10 pounds pecans, 25 pounds brazil nuts 21. 10 grams of A, 15 grams of B, 20 grams of C

10.4 Solving Linear Systems Using Second-Order Determinants (Page 451)

1. 17 3. 94 5. -60 7. 0 9. $(-1, 2)$ 11. $(3, 2)$ 13. $(1, -1)$ 15. $(12, 24)$ 17. $(5, 5)$ 19. $(-5, -7)$ 21. $(\frac{6}{5}, -\frac{3}{10})$ 23. Inconsistent 25. Dependent 27. Inconsistent 29. 6 31. $-\frac{5}{4}$ 33. $(\frac{2}{3}, -\frac{7}{6})$

35. $\begin{vmatrix} a_1 & b_1 \\ a_2 & b_2 \end{vmatrix} = a_1b_2 - a_2b_1 = a_1b_2 - b_1a_2 = \begin{vmatrix} a_1 & a_2 \\ b_1 & b_2 \end{vmatrix}$

37. Let $b_1 = ka_1$, $b_2 = ka_2$; then $\begin{vmatrix} a_1 & b_1 \\ a_2 & b_2 \end{vmatrix} = \begin{vmatrix} a_1 & ka_1 \\ a_2 & ka_2 \end{vmatrix} = \begin{vmatrix} a_1 & a_2 \\ ka_1 & ka_2 \end{vmatrix}$ (by Exercise 35)

$= 0$ (by Exercise 36)

39. $\begin{vmatrix} 27 & 3 \\ 105 & -75 \end{vmatrix} = 3\begin{vmatrix} 9 & 1 \\ 105 & -75 \end{vmatrix} = 45\begin{vmatrix} 9 & 1 \\ 7 & -5 \end{vmatrix} = -2340$

$\begin{vmatrix} 27 & 3 \\ 105 & -75 \end{vmatrix} = 3\begin{vmatrix} 9 & 1 \\ 105 & -75 \end{vmatrix} = 9\begin{vmatrix} 3 & 1 \\ 33 & -75 \end{vmatrix} = 45\begin{vmatrix} 3 & 1 \\ 7 & -15 \end{vmatrix} = -2340$

41. $\begin{vmatrix} a_1 + kb_1 & b_1 \\ a_2 + kb_2 & b_2 \end{vmatrix} = \begin{vmatrix} a_1 & b_1 \\ a_2 & b_2 \end{vmatrix} + \begin{vmatrix} kb_1 & b_1 \\ kb_2 & b_2 \end{vmatrix}$ (by Exercise 40)

$= \begin{vmatrix} a_1 & b_1 \\ a_2 & b_2 \end{vmatrix} + 0$ (by Exercise 37)

$= \begin{vmatrix} a_1 & b_1 \\ a_2 & b_2 \end{vmatrix}$

43. (Sample Solution)

$$\begin{vmatrix} 12 & -42 \\ -6 & 27 \end{vmatrix} \underset{\text{Exercise 38}}{=} 6\begin{vmatrix} 2 & -42 \\ -1 & 27 \end{vmatrix} \underset{\text{Exercise 38}}{=} 12\begin{vmatrix} 1 & -21 \\ -1 & 27 \end{vmatrix} \underset{\text{Exercise 41}}{=} 12\begin{vmatrix} 1 & -21 \\ 0 & 6 \end{vmatrix} = 12(6) = 72$$

10.5 Solving Linear Systems Using Third-Order Determinants (Page 457)

1. $|A| = -123$ 3. 35 5. -45 7. 4 9. $x = 2, x = 3$ 11. $x = -1, x = 3$ (3 is a double root) 13. -92 15. 2 17. $(-1, 0, 2)$ 19. $(-1, 7, 2)$ 21. $(\frac{2}{3}, 1, \frac{1}{2})$

23. $\begin{vmatrix} a_1 & b_1 & c_1 \\ a_2 & b_2 & c_2 \\ a_3 & b_3 & c_3 \end{vmatrix} = a_1b_2c_3 + a_2b_3c_1 + a_3b_1c_2 - a_1b_3c_2 - a_2b_1c_3 - a_3b_2c_1 = \begin{vmatrix} a_1 & a_2 & a_3 \\ b_1 & b_2 & b_3 \\ c_1 & c_2 & c_3 \end{vmatrix}$

25. $\begin{vmatrix} c_1 & b_1 & a_1 \\ c_2 & b_2 & a_2 \\ c_3 & b_3 & a_3 \end{vmatrix} = c_1b_2a_3 + c_2b_3a_1 + c_3b_1a_2 - c_1b_3a_2 - c_2b_1a_3 - c_3b_2a_1$

$$= a_3b_2c_1 + a_1b_3c_2 + a_2b_1c_3 - a_2b_3c_1 - a_3b_1c_2 - a_1b_2c_3$$
$$= a_1b_3c_2 + a_2b_1c_3 + a_3b_2c_1 - a_1b_2c_3 - a_2b_3c_1 - a_3b_1c_2$$
$$= -(a_1b_2c_3 + a_2b_3c_1 + a_3b_1c_2 - a_1b_3c_2 - a_2b_1c_3 - a_3b_2c_1)$$
$$= -\begin{vmatrix} a_1 & b_1 & c_1 \\ a_2 & b_2 & c_2 \\ a_3 & b_3 & c_3 \end{vmatrix}$$

27. $\begin{vmatrix} a_1 + kb_1 & b_1 & c_1 \\ a_2 + kb_2 & b_2 & c_2 \\ a_3 + kb_3 & b_3 & c_3 \end{vmatrix} = (a_1 + kb_1)\begin{vmatrix} b_2 & c_2 \\ b_3 & c_3 \end{vmatrix} - (a_2 + kb_2)\begin{vmatrix} b_1 & c_1 \\ b_3 & c_3 \end{vmatrix} + (a_3 + kb_3)\begin{vmatrix} b_1 & c_1 \\ b_2 & c_2 \end{vmatrix}$

$$= \left(a_1\begin{vmatrix} b_2 & c_2 \\ b_3 & c_3 \end{vmatrix} - a_2\begin{vmatrix} b_1 & c_1 \\ b_3 & c_3 \end{vmatrix} + a_3\begin{vmatrix} b_1 & c_1 \\ b_2 & c_2 \end{vmatrix}\right) + kb_1\begin{vmatrix} b_2 & c_2 \\ b_3 & c_3 \end{vmatrix}$$
$$- kb_2\begin{vmatrix} b_1 & c_1 \\ b_3 & c_3 \end{vmatrix} + kb_3\begin{vmatrix} b_1 & c_1 \\ b_2 & c_2 \end{vmatrix}$$
$$= \begin{vmatrix} a_1 & b_1 & c_1 \\ a_2 & b_2 & c_2 \\ a_3 & b_3 & c_3 \end{vmatrix} + k(b_1b_2c_3 - b_1b_3c_2 - b_1b_2c_3 + b_2b_3c_1 + b_1b_3c_2 - b_2b_3c_1)$$
$$= \begin{vmatrix} a_1 & b_1 & c_1 \\ a_2 & b_2 & c_2 \\ a_3 & b_3 & c_3 \end{vmatrix}$$

29. 0, since third column is -2 times first column. 31. $80 = 20(4)$

33. (Sample Solution)

$\begin{vmatrix} 5 & -4 & 3 \\ -6 & 6 & 2 \\ -7 & 3 & 4 \end{vmatrix} = \begin{vmatrix} 14 & -4 & 3 \\ 0 & 6 & 2 \\ 5 & 3 & 4 \end{vmatrix}$ 3 times third column added to first column (Exercise 27)

$= \begin{vmatrix} 14 & -13 & 3 \\ 0 & 0 & 2 \\ 5 & -9 & 4 \end{vmatrix}$ -3 times third column added to second column (Exercise 27)

$= -2\begin{vmatrix} 14 & -13 \\ 5 & -9 \end{vmatrix}$ Expansion by minors along row 2

$= 2\begin{vmatrix} 14 & 13 \\ 5 & 9 \end{vmatrix}$ (Exercise 26)

$= 2(126 - 65)$

$= 122$

10.6 Solving Linear Systems Using Matrices (Page 465)

1. $(-4, -1)$ 3. $(0, 9)$ 5. $(\frac{1}{8}, \frac{1}{2})$ 7. $(1, 0)$ 9. $(2 - \frac{3}{2}c, c)$ for all c, or $(d, \frac{4}{3} - \frac{2}{3}d)$ for all d
11. No solutions 13. $(5, 2, -1)$ 15. $(3, 4, 2)$ 17. No solutions
19. $(\frac{3}{11} + \frac{5}{11}c, -\frac{4}{11} + \frac{8}{11}c, c)$ for all c, or $(d, -\frac{4}{5} + \frac{8}{5}d, -\frac{3}{5} + \frac{11}{5}d)$ for all d, or $(\frac{1}{2} + \frac{5}{8}e, e, \frac{1}{2} + \frac{11}{8}e)$ for all e
21. $(-1, 3, 4, -2)$ 23. No solutions

10.7 Systems of Linear Inequalities (Page 469)

1.

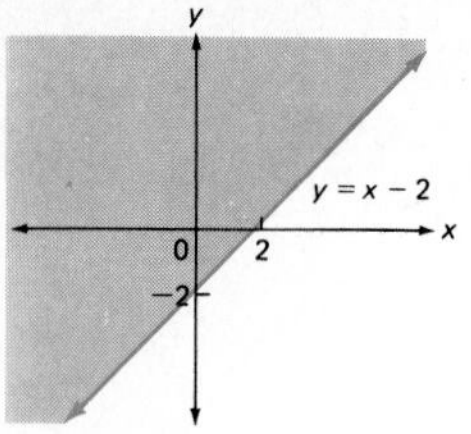

3.

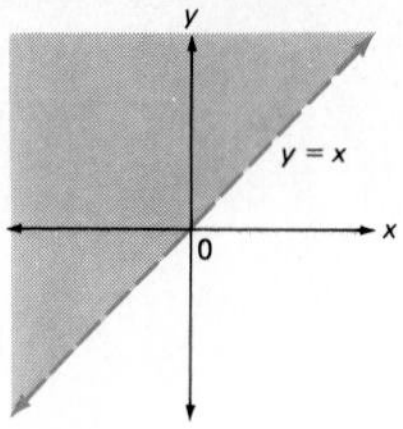

5.

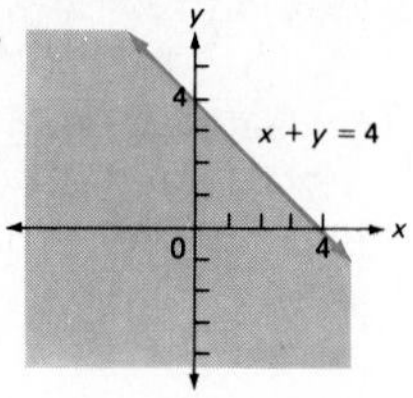

7.

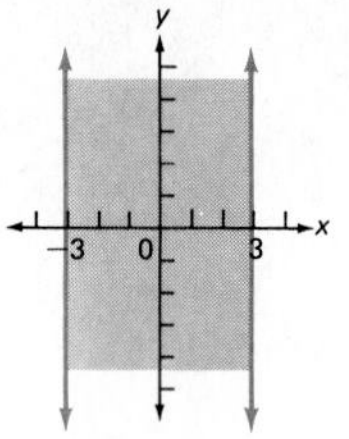

9.

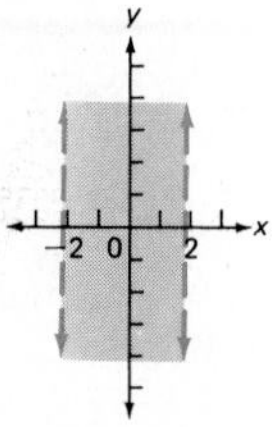

11.

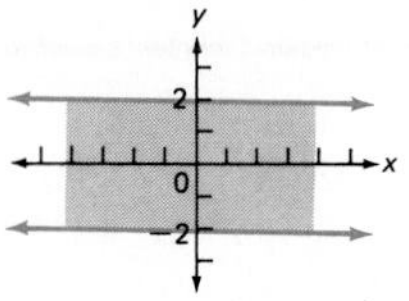

13.

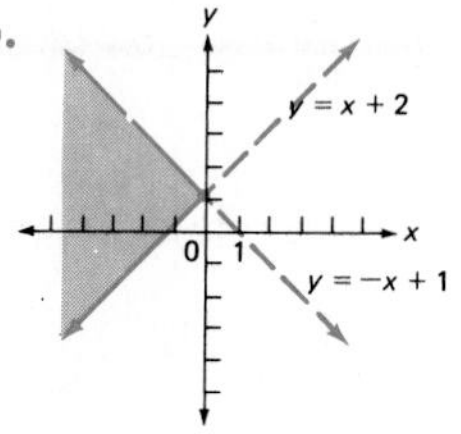

15.

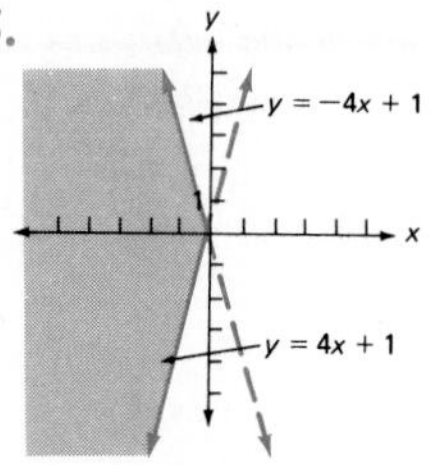

17.

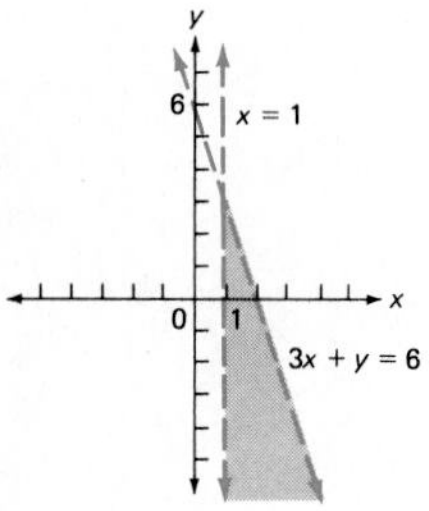

19.

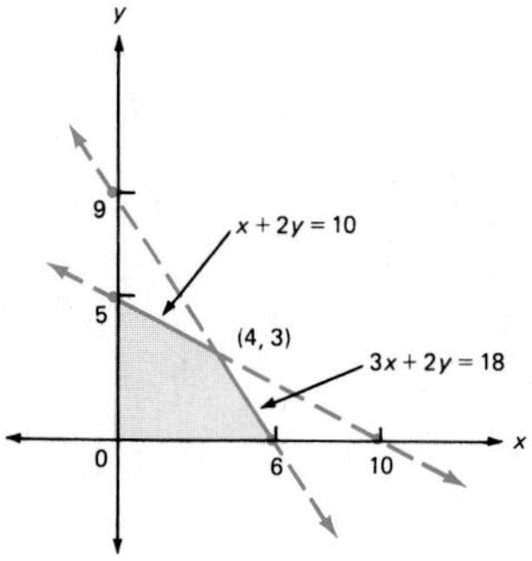

21.

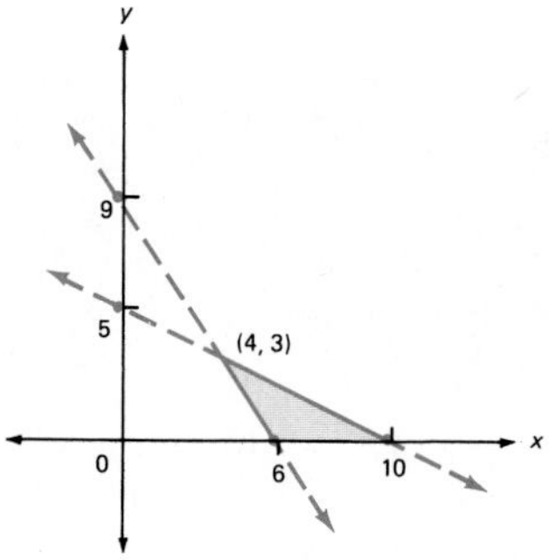

23.

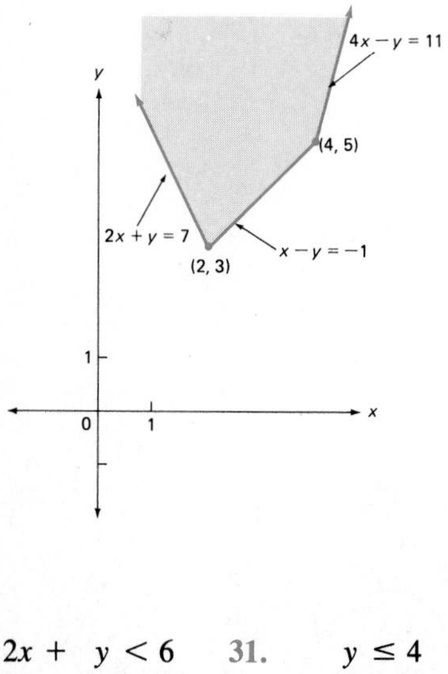

25.

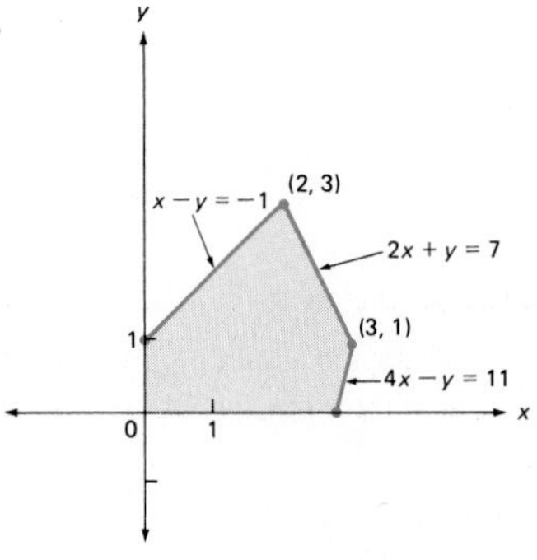

27.

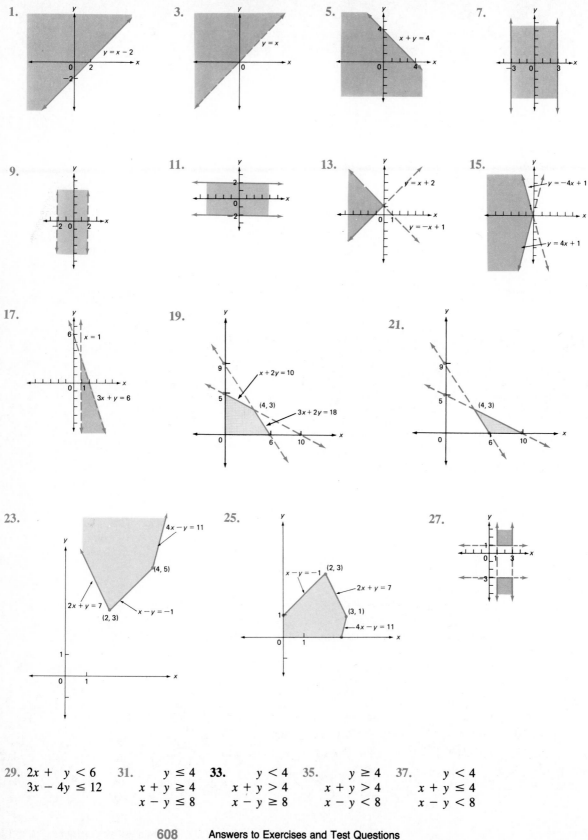

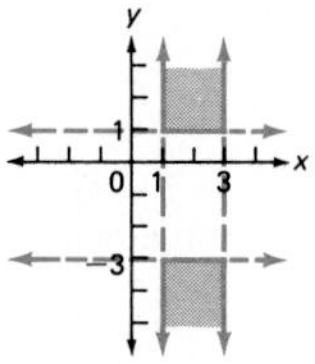

29. $2x + y < 6$
$3x - 4y \le 12$

31. $y \le 4$
$x + y \ge 4$
$x - y \le 8$

33. $y < 4$
$x + y > 4$
$x - y \ge 8$

35. $y \ge 4$
$x + y > 4$
$x - y < 8$

37. $y < 4$
$x + y \le 4$
$x - y < 8$

1. **(a)**

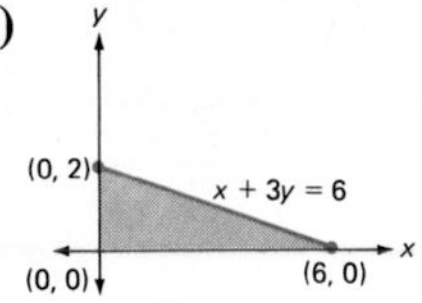

(b) (0, 0), (6, 0), (0, 2) **(c)** Maximum = 6; minimum = 0

(d) Maximum = 36, minimum = 0 **(e)** Maximum = 18; minimum = 0

3. **(a)**

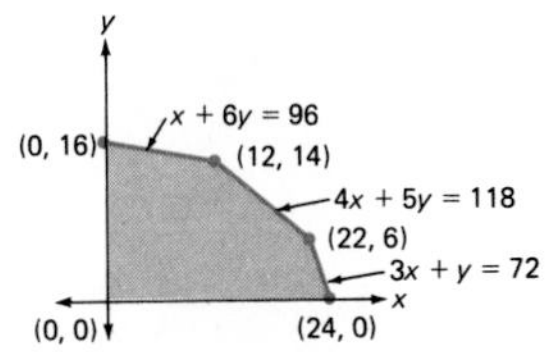

(b) (0, 0); (24, 0); (22, 6); (12, 14); (0, 16) **(c)** Maximum = 28; minimum = 0

(d) Maximum = 212; minimum = 0; **(e)** Maximum = 150; minimum = 0

5. **(a)**

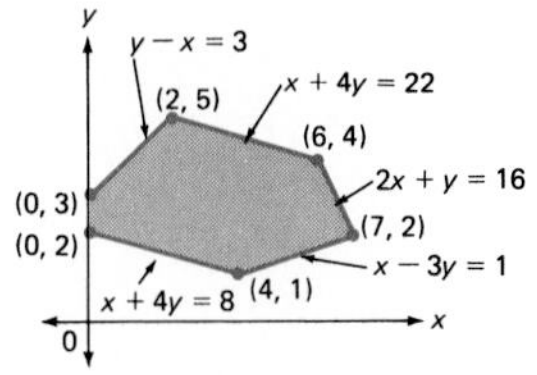

(b) (0, 2), (4, 1), (7, 2), (6, 4), (2, 5), (0, 3) **(c)** Maximum = 10; minimum = 2

(d) Maximum = 76; minimum = 20 **(e)** Maximum = 49; minimum = 17

7. For $p = x + 3y$, max = 36 at (0, 12); min = 10 at (10, 0)
For $q = 4x + 3y$, max = 40 at (10, 0); min = 12 at (0, 4)
For $r = \frac{1}{2}x + \frac{1}{4}y$, max = 5 at (10, 0); min = 1 at (0, 4)

9. For $p = 2x + 2y$, max = 24 at (7, 5); min = 8 at (4, 0)
For $q = 3x + y$, max = 26 at (7, 5); min = 9.5 at $(\frac{3}{2}, 5)$
For $r = x + \frac{1}{5}y$, max = 8 at (7, 5); min = $\frac{5}{2}$ at $(\frac{3}{2}, 5)$

11.

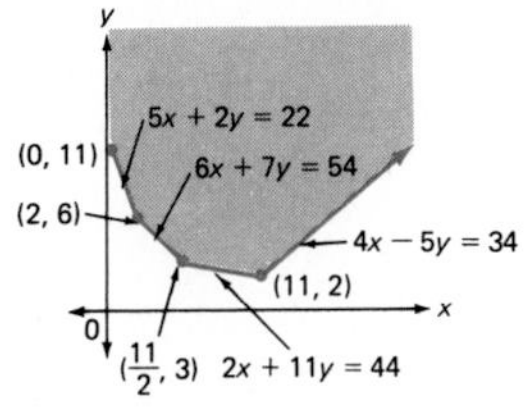

(a) No maximum value; minimum = 8 **(b)** No maximum value; minimum = 63

(c) No maximum value; minimum = 10

13. **(a)** $x \geq 0$, $y \geq 0$ because a negative number of either model is not possible; $\frac{3}{2}x + y$ is the amount of time that machine M_1 works per day, and $\frac{3}{2}x + y \leq 12$ says that M_1 works at most 12 hours daily. The remaining inequalities are the contraints for machines M_2 and M_3; the explanations are similar, as for M_1.

(b) $p = 5x + 8y$

(c)

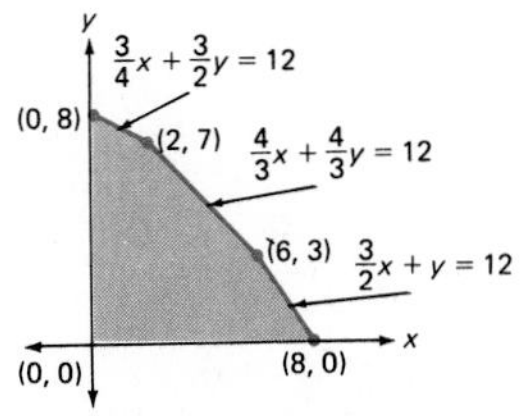

(d) Maximum = \$66 when $x = 2$ and $y = 7$ **(e)** Maximum = \$64 when $x = 8$ and $y = 0$

15. 400 of model A and 500 of model B. Profit \$100,000.

1.

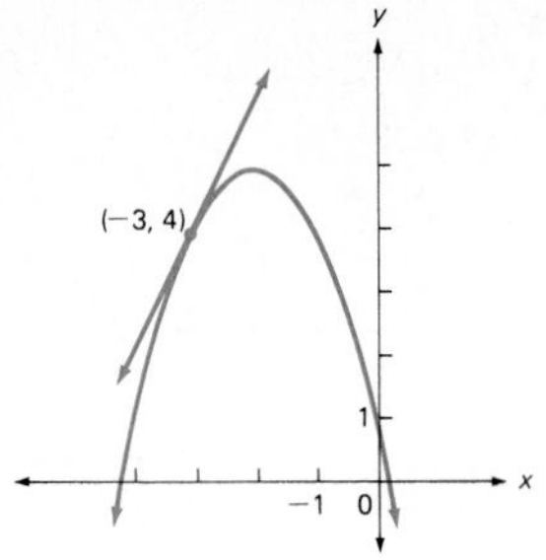

3.

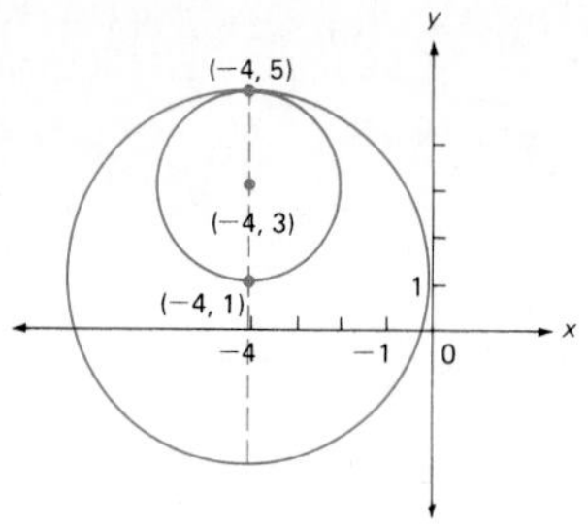

5.

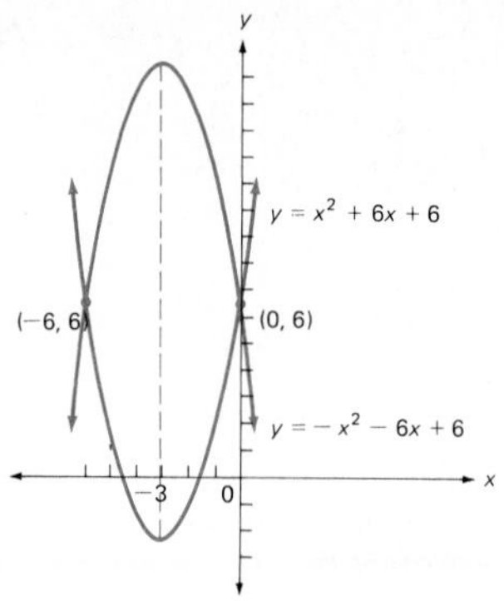

7. (0, 1) **9.** No solutions **11.** No solutions **13.** $(-1, -1); (-2, -2)$
15. $(2, 1), (2, -1)$ **17.** $(1, 0), (0, -2)$ **19.** No solutions **21.** $(2, \sqrt{3}), (2, -\sqrt{3}), (-2, \sqrt{3}), (-2, -\sqrt{3})$
23. $(0, 0); (2, 0)$ **25.** $(0, 0); (1, 1)$ **27.** $(3, -3); (3 + \sqrt{3}, -2); (3 - \sqrt{3}, -2)$
29. $(4, \sqrt{3}), (4, -\sqrt{3}), (-4, \sqrt{3}), (-4, -\sqrt{3})$

31.

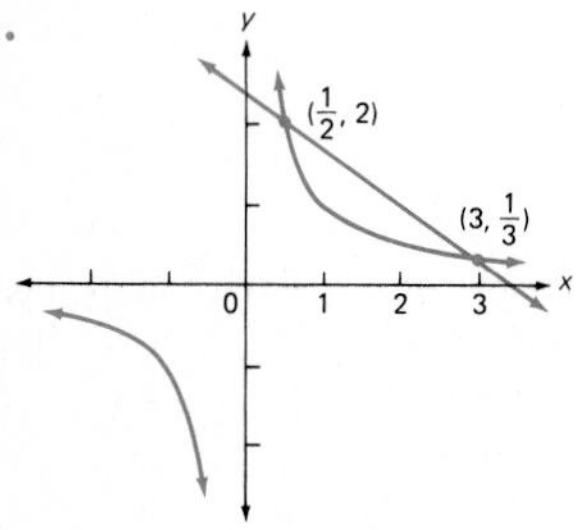

33.

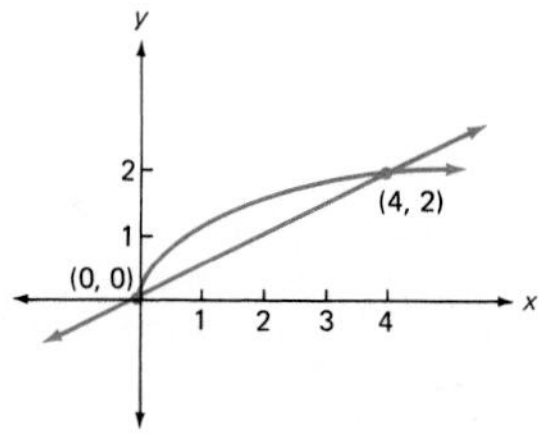

35.

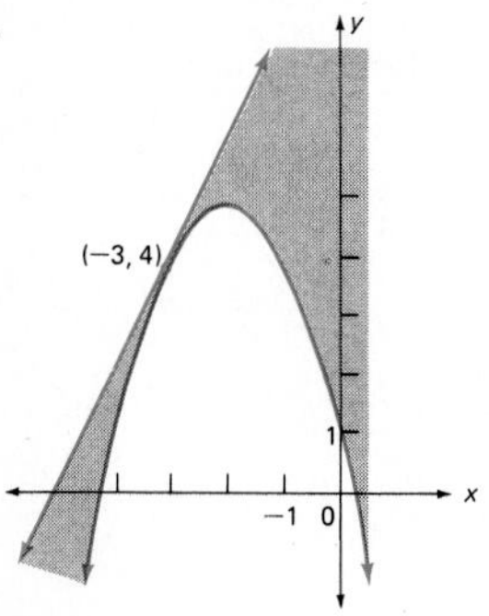

37.

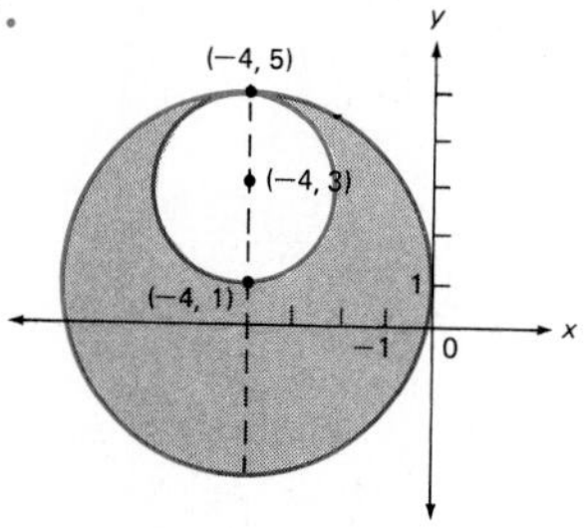

39.

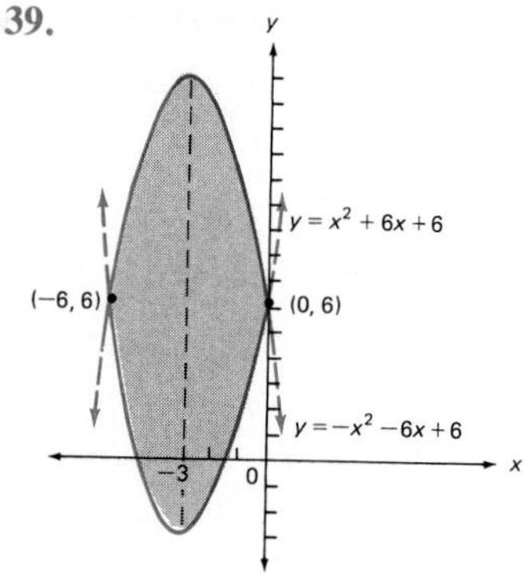

41.

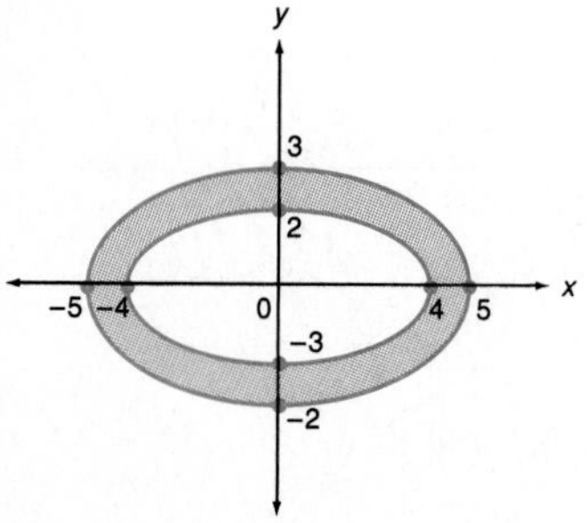

43.

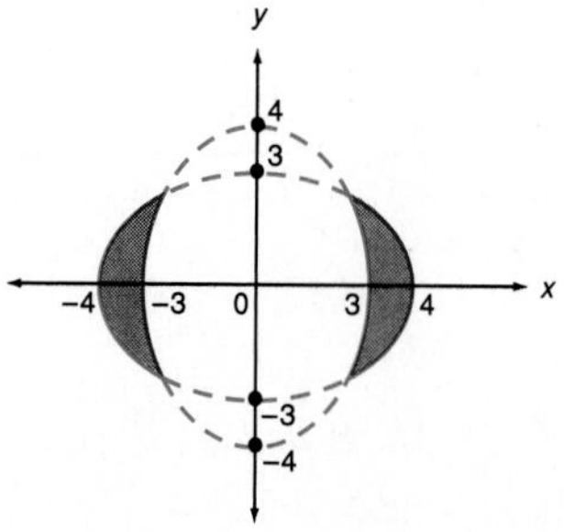

CHAPTER 10 TEST: STANDARD ANSWER (Page 487)

1. $(3, \frac{1}{2})$ 2. $(-1, 2, 5)$ 3. $(-2, 3)$ 4. Consistent 5. Inconsistent 6. Dependent 7. **(a)** 33 **(b)** 0 8. 43 9. 312 children; 1698 adults 10. 52, 75 11. (750, 2325) 12. 6 oz. of I; 8 oz. of II; 3 oz. of III 13. $(1, -\frac{11}{2})$ 14. $(1, -2, 3)$ 15. $(1, 2, -1)$ 16. $(-3, 2)$ 17. No solution 18. $(4, -1, -6)$

19.

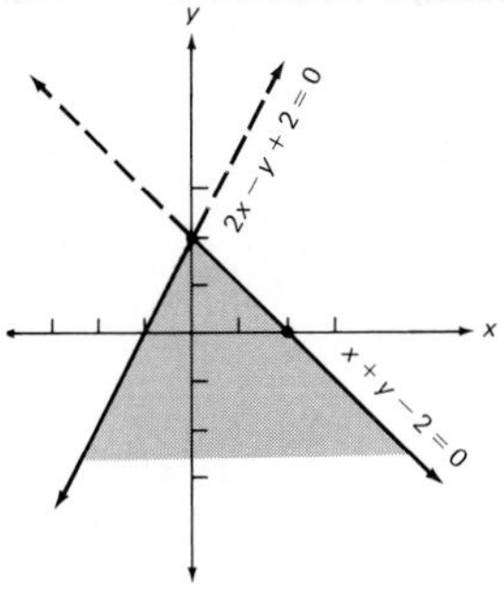

20.

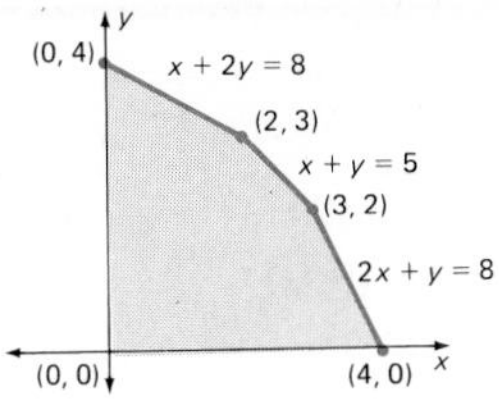

21. 13

22. 300 of model A, 300 of model B 23. $(3, 1), (3, -1), (-3, 1), (-3, -1)$ 24. No solution 25. $(-\frac{5}{6}, \frac{1}{36})$

CHAPTER 10 TEST: MULTIPLE CHOICE (Page 488)

1. (c) 2. (c) 3. (d) 4. (d) 5. (c) 6. (c) 7. (b) 8. (c) 9. (d) 10. (e) 11. (b) 12. (b) 13. (a) 14. (c) 15. (b)

CHAPTER 11: SEQUENCES AND SERIES

11.1 Sequences (Page 495)

1. $a_1 = 1, a_2 = 3, a_3 = 5, a_4 = 7, a_5 = 9$ 3. $a_1 = -1, a_2 = 1, a_3 = -1, a_4 = 1, a_5 = -1$ 5. $b_1 = -4, b_2 = 2, b_3 = -1, b_4 = \frac{1}{2}, b_5 = -\frac{1}{4}$ 7. $-1, 4, -9, 16$ 9. $\frac{3}{10}, \frac{3}{100}, \frac{3}{1000}, \frac{3}{10{,}000}$ 11. $\frac{3}{100}, \frac{3}{10{,}000}, \frac{3}{1{,}000{,}000}, \frac{3}{100{,}000{,}000}$ 13. $\frac{1}{2}, \frac{1}{6}, \frac{1}{12}, \frac{1}{20}$ 15. 64, 36, 16, 4 17. $-2, 1, 4, 7$ 19. $0, \frac{1}{3}, \frac{1}{2}, \frac{3}{5}$ 21. $1, \frac{3}{2}, \frac{16}{9}, \frac{125}{64}$ 23. $-2, -\frac{3}{2}, -\frac{9}{8}, -\frac{27}{32}$ 25. $\frac{1}{2}, \frac{1}{2}, \frac{3}{8}, \frac{1}{4}$ 27. $-\frac{3}{2}, -\frac{5}{6}, -\frac{7}{12}, -\frac{9}{20}$ 29. 4, 4, 4, 4 31. 122 33. 0.000003 35. 12 37. -1331 39. 4, 6, 8, 10 41. 5, 10, 15, 20, 25; $s_n = 5n$ 43. $-5, 25, -125, 625, -3125$; $s_n = (-5)^n$ 45. 15, 21, 28 47. $1, \frac{9}{5}, 3, \frac{81}{17}, \frac{81}{11}$ 49. $\frac{3}{2}, \frac{15}{8}, \frac{35}{16}, \frac{315}{128}$

11.2 Sums of Finite Sequences (Page 499)

1. 45 3. 55 5. 0.33333 7. 510 9. 60 11. 254 13. $\frac{381}{64}$ 15. 105 17. 40 19. $\frac{13}{8}$ 21. 0 23. 57 25. 1111.111 27. 60 29. $\frac{35}{12}$ 31. 0.010101 33. $\sum_{k=1}^{10} 5k$ 35. $\sum_{k=1}^{12} 3(k-4)$ 37. **(a)** 4, 9, 16, 25, 36; **(b)** n^2

39. $a_n = a_1 + a_{n-1} = a_1 + (a_1 + a_{n-2}) = \cdots = \overbrace{a_1 + a_1 + a_1 + \cdots + a_1}^{n \text{ terms}} = 2 + 2 + 2 + \cdots + 2 = 2n$

41. $\sum_{k=1}^{n} a_k + \sum_{k=1}^{n} b_k = (a_1 + a_2 + \cdots + a_n) + (b_1 + b_2 + \cdots + b_n) = (a_1 + b_1) + (a_2 + b_2) + \cdots + (a_n + b_n)$
$= \sum_{k=1}^{n} (a_k + b_k)$

43. $\sum_{k=1}^{n} (a_k + c) = (a_1 + c) + (a_2 + c) + \cdots + (a_n + c) = (a_1 + a_2 + \cdots + a_n) + (c + c + \cdots + c)$
$= \left(\sum_{k=1}^{n} a_k\right) + nc$

45. $\sum_{k=1}^{10} \frac{2}{k(k+2)} = \sum_{k=1}^{10}\left(\frac{1}{k} - \frac{1}{k+2}\right)$
$= (1 - \frac{1}{3}) + (\frac{1}{2} - \frac{1}{4}) + (\frac{1}{3} - \frac{1}{5}) + (\frac{1}{4} - \frac{1}{6}) + (\frac{1}{5} - \frac{1}{7}) + (\frac{1}{6} - \frac{1}{8}) + (\frac{1}{7} - \frac{1}{9}) + (\frac{1}{8} - \frac{1}{10}) + (\frac{1}{9} - \frac{1}{11}) + (\frac{1}{10} - \frac{1}{12})$
$= 1 + \frac{1}{2} - \frac{1}{11} - \frac{1}{12} = \frac{264 + 132 - 24 - 22}{2\cdot 11\cdot 12} = \frac{350}{2\cdot 11\cdot 12} = \frac{175}{132}$

11.3 Arithmetic Sequences and Series (Page 504)

1. 5, 7, 9; $2n - 1$; 400 **3.** −10, −16, −22; $-6n + 8$; −1100 **5.** $\frac{17}{2}$, 9, $\frac{19}{2}$; $\frac{1}{2}n + 7$; 245 **7.** $-\frac{4}{5}$, $-\frac{7}{5}$, −2; $-\frac{3}{5}n + 1$; −106 **9.** 150, 200, 250; $50n$; 10,500 **11.** 30, 50, 70; $20n - 30$; 3600 **13.** 455 **15.** $\frac{63}{4}$ **17.** 289 **19.** 15,150 **21.** 94,850 **23.** $\frac{173{,}350}{7}$ **25.** 567,500 **27.** 5824 **29.** 15,300 **31.** **(a)** 10,000; **(b)** n^2 **33.** −36 **35.** 4620 **37.** $\frac{3577}{4}$ **39.** $\frac{5}{2}n(n+1)$ **41.** $-\frac{9}{4}$ **43.** $\frac{228}{15} = \frac{76}{5}$ **45.** $1183 **47.** 930 **49.** $a_1 = -7$; $d = -13$ **51.** $u = \frac{16}{3}$, $v = \frac{23}{3}$ **53.** 9080 **55.** −16,200 **57.** 520

11.4 Geometric Sequences and Series (Page 511)

1. 16, 32, 64; 2^n **3.** 27, 81, 243; 3^{n-1} **5.** $\frac{1}{9}$, $-\frac{1}{27}$, $\frac{1}{81}$; $-3(-\frac{1}{3})^{n-1}$ **7.** −125, −625, −3125; -5^{n-1} **9.** $-\frac{16}{9}$, $-\frac{32}{27}$, $-\frac{64}{81}$; $-6(\frac{2}{3})^{n-1}$ **11.** 1000, 100,000, 10,000,000; $\frac{1}{1000}(100)^{n-1}$ **13.** 126 **15.** $-\frac{1330}{81}$ **17.** 1024 **19.** 3 **21.** $-\frac{1}{2}$ **23.** 1023 **25.** $2^n - 1$ **27.** $\frac{121}{27}$ **29.** $\frac{211}{54}$ **31.** $-\frac{85}{64}$ **33.** $-\frac{5}{21}$ **35.** $10,737,418 **37.** 512,000; $1000(2^{n-1})$ **39.** $3117.79 **41.** **(a)** $\$800(1.11)^n$; **(b)** $1350 **43.** $703.99

45. **(a)** If the volume of the first container is V, then
$\frac{1}{2}V + (\frac{1}{2})^2 V + (\frac{1}{2})^3 V + (\frac{1}{2})^4 V + (\frac{1}{2})^5 V = \frac{31}{32}V$ is the sum of the volumes of the other five. Since $\frac{31}{32}V < V$, the answer is yes.
(b) $\sum_{k=1}^{5} (\frac{2}{3})^k V = \frac{422}{243}V > V$; therefore no.

11.5 Infinite Geometric Series (Page 518)

1. 4 **3.** $\frac{125}{4}$ **5.** $\frac{2}{3}$ **7.** $\frac{10}{9}$ **9.** $-\frac{16}{7}$ **11.** The numerator at the right should be $\frac{1}{4}$, not 1. **13.** The denominator at the right should be $1 - (-\frac{1}{3})$ since $r = -\frac{1}{3}$, not $\frac{1}{3}$. **15.** $\frac{3}{2}$ **17.** $\frac{1}{6}$ **19.** 4 **21.** $\frac{20}{9}$ **23.** No finite sum **25.** $\frac{10}{3}$ **27.** No finite sum **29.** 30 **31.** $\frac{20}{11}$ **33.** 4 **35.** $\frac{4}{21}$ **37.** $\frac{7}{9}$ **39.** $\frac{13}{99}$ **41.** $\frac{13}{990}$ **43.** 1

45. **(a)** $\frac{4}{3}, \frac{4}{9}, \frac{4}{27}, \ldots, \frac{4}{3^n}, \ldots$; **(b)** $\sum_{n=1}^{\infty} \frac{4}{3^n} = \frac{\frac{4}{3}}{1 - \frac{1}{3}} = 2$ **47.** 8 hours

49. The time for the last $\frac{1}{2}$ mile would have to be $\sum_{n=1}^{\infty} \frac{2}{5}(\frac{10}{9})^{n-1}$, which is not a finite sum, since $\frac{10}{9} > 1$.

51. **(a)** $\frac{1}{2}(AC)(CB) = \frac{1}{2}(4)(4) = 8$ **(b)** $4 + 2 + 1 + \frac{1}{2} + \cdots = \frac{4}{1 - \frac{1}{2}} = 8$

(c) For the odd-numbered triangles:

$$4 + 1 + \frac{1}{4} + \cdots = \frac{4}{1 - \frac{1}{4}} = \frac{16}{3}$$

For the even-numbered triangles:

$$2 + \frac{1}{2} + \frac{1}{8} + \cdots = \frac{2}{1 - \frac{1}{4}} = \frac{8}{3}$$

$$\frac{16}{3} + \frac{8}{3} = 8$$

53. $\frac{1}{2}(9)(3) + \frac{1}{2}(6)(2) + \frac{1}{2}(4)\left(\frac{4}{3}\right) + \cdots = \frac{27}{2} + 6 + \frac{8}{3} + \cdots = \frac{\frac{27}{2}}{1 - \frac{4}{9}} = \frac{243}{10}$

CHAPTER 11 TEST: STANDARD ANSWER (Page 522)

1. $1, -1, -1, -\frac{8}{7}$ **2.** $-\frac{25}{11}$ **3.** 0, 3, 2, 5 **4.** $-\frac{1}{3}$ **5.** $\frac{1}{300}$ **6.** 9, 18, 27, 36, 45; $a_n = 9n$ **7.** 21 **8.** 35 **9.** 10,000 **10.** -5375 **11.** $\sum_{n=1}^{15} (4n - 16)$ **12.** $12, -3, \frac{3}{4}$ **13.** $-768(-\frac{1}{4})^{n-1}$ **14.** $-\frac{76}{15}$ **15.** 8925 **16.** $9500(0.89)^n$

17. $\sum_{k=1}^{4} 8\left(\frac{1}{2}\right)^k = \frac{4\left(1 - \frac{1}{2^4}\right)}{1 - \frac{1}{2}} = \frac{15}{2}$ **18.** 15,554 **19.** $24\left(1 - \frac{1}{2^8}\right) = \frac{765}{32}$ **20.** 18 **21.** No finite sum since $r = \frac{3}{2} > 1$ **22.** $\frac{6}{115}$ **23.** $\frac{4}{11}$ **24.** \$652.60 **25.** 36 feet

CHAPTER 11 TEST: MULTIPLE CHOICE (Page 523)

1. (b) **2.** (a) **3.** (c) **4.** (c) **5.** (b) **6.** (a) **7.** (d) **8.** (d) **9.** (c) **10.** (c) **11.** (a) **12.** (d) **13.** (e) **14.** (a) **15.** (d)

CHAPTER 12: PERMUTATIONS, COMBINATIONS AND PROBABILITY

12.1 Permutations (Page 532)

1. 7 **3.** 66 **5.** 120 **7.** 4 **9.** $\frac{n!}{[n - (n - 3)]!} = \frac{n!}{3!}$ **11.** 15 **13.** 60 **15.** 12 **17.** 6 **19.** 36 **21.** 504 **23.** 224 **25.** 224 **27.** **(a)** 120 **(b)** 12 **(c)** 48 **(d)** 72 **29.** 362,880; 40,320 **31.** 720 **33.** 468,000; 676,000 **35.** **(a)** 10 **(b)** 4

37. 36;

(1, 1)	(2, 1)	(3, 1)	(4, 1)	(5, 1)	(6, 1)
(1, 2)	(2, 2)	(3, 2)	(4, 2)	(5, 2)	(6, 2)
(1, 3)	(2, 3)	(3, 3)	(4, 3)	(5, 3)	(6, 3)
(1, 4)	(2, 4)	(3, 4)	(4, 4)	(5, 4)	(6, 4)
(1, 5)	(2, 5)	(3, 5)	(4, 5)	(5, 5)	(6, 5)
(1, 6)	(2, 6)	(3, 6)	(4, 6)	(5, 6)	(6, 6)

39. **(a)** 1,000,000,000 **(b)** 3,628,800 **(c)** 996,371,200 **(d)** $10 \cdot 7(9 \cdot 8 \cdot 7 \cdot 6 \cdot 5 \cdot 4) = 4{,}233{,}600$ **41.** **(a)** 720 **(b)** 120 **(c)** 240 **43.** 5 **45.** **(a)** 6720; **(b)** 1260; **(c)** 151,200

12.2 Combinations (Page 538)

1. 10 **3.** 1 **5.** 1 **7.** 4060 **9.** **(a)** 4845 **(b)** 116,280 **11.** 20 **13.** 3003 **15.** 435 **17.** 46,200 **19.** **(a)** n **(b)** $\frac{n(n-1)}{2}$ **21.** 70 **23.** **(a)** 2^2 **(b)** 2^3 **(c)** 2^4 **(d)** 2^5

25. Each time a subset of r elements is chosen out of n elements there are $n - r$ elements left over. Likewise, when $n - r$ elements are chosen out of n elements there are $n - (n - r) = r$ elements left over. Therefore, there must be the same number of subsets of size r, $\binom{n}{r}$, as there are subsets of size $\binom{n}{n-r}$.

27. A set of n elements has a total of 2^n subsets of all possible sizes, including the empty set and the set itself.

29. $4\binom{13}{5} = 5148$ **31.** $13\binom{4}{2}12\binom{4}{3} = 3744$ **33.** **(a)** 205,931,880 **(b)** 147,094,200 **35.** 180 **37.** **(a)** 1008 **(b)** 2254 **39.** 256 **41.** $210 \times 5! \times 3! = 151{,}200$

12.3 The Binomial Expansion (Page 545)

1. $x^5 + 5x^4 + 10x^3 + 10x^2 + 5x + 1$ **3.** $x^7 + 7x^6 + 21x^5 + 35x^4 + 35x^3 + 21x^2 + 7x + 1$ **5.** $a^4 - 4a^3b + 6a^2b^2 - 4ab^3 + b^4$ **7.** $243x^5 - 405x^4y + 270x^3y^2 - 90x^2y^3 + 15xy^4 - y^5$

9. $a^{10} + 5a^8 + 10a^6 + 10a^4 + 5a^2 + 1$

11. $1 - 10h + 45h^2 - 120h^3 + 210h^4 - 252h^5 + 210h^6 - 120h^7 + 45h^8 - 10h^9 + h^{10}$

13. $\frac{1}{16} - \frac{1}{2}a + \frac{3}{2}a^2 - 2a^3 + a^4$ **15.** $\frac{1}{x^6} - \frac{6}{x^3} + 15 - 20x^3 + 15x^6 - 6x^9 + x^{12}$ **17.** $3c^2 + 3ch + h^2$

19. $(1 + 1)^{10} = 1 + 10 + 45 + 120 + 210 + 252 + 210 + 120 + 45 + 10 + 1 = 1024$

21. $c^{20} + 20c^{19}h + 190c^{18}h^2 + 1140c^{17}h^3 + 4845c^{16}h^4 + \cdots + 4845c^4h^{16} + 1140c^3h^{17} + 190c^2h^{18} + 20ch^{19} + h^{20}$

23. The nth row of the triangle contains the coefficients in the expansion of $(a + b)^n$.

25. $a^6 + 12a^5 + 60a^4 + 160a^3 + 240a^2 + 192a + 64$ **27.** $81p^4 + 216p^3q + 216p^2q^2 + 96pq^3 + 16q^4$

29.

```
         1   7  21  35  35  21   7   1
       1   8  28  56  70  56  28   8   1
     1   9  36  84 126 126  84  36   9   1
   1  10  45 120 210 252 210 120  45  10   1
```

31. $x^{10} - 10x^9h + 45x^8h^2 - 120x^7h^3 + 210x^6h^4 - 252x^5h^5 + 210x^4h^6 - 120x^3h^7 + 45x^2h^8 - 10xh^9 + h^{10}$

33. $8064a^5b^5$ **35.** $35x^{-5/2}$ **37.** $48{,}384x^5y^3$ **39.** $-\frac{15309}{8}a^5b^5$ **41.** 19.4481 **43.** 63.044792

12.4 Probability (Page 552)

1. $\frac{1}{8}$ **3.** $\frac{7}{8}$ **5.** $\frac{1}{2}$ **7.** $\frac{1}{2}$ **9.** $\frac{9}{16}$ **11.** $\frac{1}{4}$ **13.** $\frac{3}{8}$ **15.** $\frac{1}{6}$ **17.** $\frac{1}{6}$ **19.** $\frac{1}{9}$ **21.** $\frac{1}{2}$ **23.** $\frac{25}{102}$ **25.** 0 **27.** $\frac{1}{2}$
29. $\frac{1}{2704}$ **31.** $\frac{144}{169}$ **33.** $\frac{1}{169}$ **35.** $\frac{28}{143}$ **37.** $\frac{70}{143}$ **39.** **(a)** $\frac{1}{1024}$ **(b)** $\frac{1}{1024}$ **(c)** $\frac{7}{128}$ **41.** $\frac{77}{102}$

43. **(a)** $\frac{48}{\binom{52}{5}} = 0.0000185$ **(b)** $\frac{13 \cdot 48}{\binom{52}{5}} = 0.0002401$ **(c)** $\frac{4\binom{13}{5}}{\binom{52}{5}} = 0.0019808$ **(d)** $\frac{4}{\binom{52}{5}} = 0.0000015$

45. \$5 **47.** \$10 **49.** 1 to 12; 12 to 1 **51.** 7 to 2 **53.** 3 to 5

55. Let $x = P(E)$. Then $P(\text{not } E) = 1 - x$ and we have the odds for $E = \frac{a}{b} = \frac{x}{1 - x}$. Then

$$
\begin{aligned}
a(1 - x) &= bx \\
a - ax &= bx \\
a &= (a + b)x \\
\frac{a}{a + b} &= x \quad \text{or} \quad P(E) = \frac{a}{a + b}
\end{aligned}
$$

CHAPTER 12 TEST: STANDARD ANSWER (Page 557)

1. **(a)** 720 **(b)** 120 **2.** 120 **3.** **(a)** 900 **(b)** 648 **4.** $\frac{n!}{[n - (n - 5)]!} = \frac{n!}{5!}$ **5.** 450 **6.** 455 **7.** 2970

8. 360 **9.** **(a)** 252 **(b)** $\frac{n(n - 1)(n - 2)(n - 3)}{24}$ **10.** 11 **11.** **(a)** 32 **(b)** $\frac{1}{32}$ **(c)** $\frac{1}{32}$

12. $\frac{2}{3}$ **13.** **(a)** $\frac{1}{36}$ **(b)** $\frac{35}{36}$ **(c)** $\frac{1}{6}$ **14.** **(a)** $\frac{2}{169}$ **(b)** $\frac{2}{221}$ **15.** $\frac{\binom{12}{4} \cdot 40}{\binom{52}{5}} = 0.0076184$

16. $\frac{\binom{20}{4} \cdot 15}{\binom{35}{5}} = 0.2238689$ **17.** **(a)** $\frac{1}{16}$ **(b)** $\frac{1}{16}$ **18.** $\frac{2}{11}$ **19.** $\frac{9}{40}$ **20.** $\frac{39}{80}$

21. $x^5 - 10x^4y + 40x^3y^2 - 80x^2y^3 + 80xy^4 - 32y^5$ **22.** $\frac{1}{1024a^{10}} - \frac{5b}{256a^9} + \frac{45b^2}{256a^8} - \frac{15b^3}{16a^7}$

23. $\binom{11}{6}(3a)^5b^6 = 112{,}266a^5b^6$ **24.** $\binom{16}{8}(2x)^8(-y)^8 = 3{,}294{,}720x^8y^8$

25. $3^4 + 4(3^3)(.1) + 6(3^2)(.1)^2 + 4(3)(.1)^3 + (.1)^4 = 92.3521$

CHAPTER 12 TEST: MULTIPLE CHOICE (Page 558)

1. (c) 2. (d) 3. (c) 4. (b) 5. (b) 6. (e) 7. (d) 8. (a) 9. (b) 10. (c) 11. (a) 12. (d) 13. (c) 14. (a) 15. (b)

ANSWERS TO CUMULATIVE REVIEW QUESTIONS FOR CHAPTERS 10–12 (Page 560)

1. (9, −5) 2. (0, −1) 3. (4, −2) 4. Inconsistent 5. 57 fives and 8 tens 6. \$3500 at 8%; \$4200 at 9% 7. (7, 294) 8. (1, 2, −3) 9. (2, 5, −2) 10. 28 two pointers, 3 three pointers, 11 free throws 11. **(a)** 13 **(b)** 5 **(c)** −19 12. 181 13. −70 14. (−2, −2) 15. $(\frac{1}{2}, 3)$ 16. (−1, 2, 5) 17. −4, 0, 5 18. (−4, 7) 19. $(5 + 3c, c)$ for all c, or $(d, \frac{1}{3}d - \frac{5}{3})$ for all d 20. (−2, 1, 3) 21.

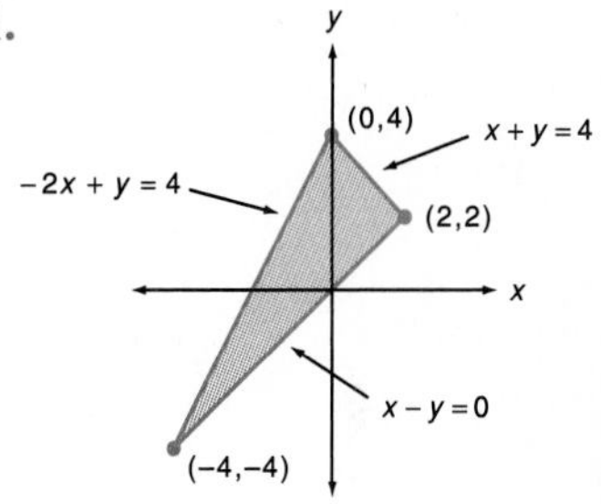

22. maximum: 26 minimum: 4 23. $(6, \sqrt{3}), (6, -\sqrt{3}), (-6, \sqrt{3}), (-6, -\sqrt{3})$ 24. $(\sqrt{7}, 1), (-\sqrt{7}, 1), (0, 8)$ 25. **(a)** $-1, \frac{1}{4}, -\frac{1}{9}, \frac{1}{16}$ **(b)** 1, 3, 5, 13 **(c)** $0, \frac{1}{4}, \frac{8}{27}, \frac{81}{256}$ 26. $\frac{1}{3}, \frac{2}{3}, 1, \frac{4}{3}, \frac{5}{3}; \frac{n}{3}$ 27. $\frac{147}{20}$ 28. −3 29. $\frac{11}{3} - 3n; S_{20} = -\frac{1670}{3}$ 30. −3575 31. 250 32. $128(-\frac{1}{2})^{n-1}; \frac{1}{512}$ 33. $16(-\frac{3}{4})^{n-1}$ 34. 4092 35. $\frac{61}{9}$ 36. \$10,485.75 37. $1024(\frac{3}{4})^n$; 243 ft 38. **(a)** $\frac{27}{2}$ **(b)** No finite sum 39. **(a)** 2 **(b)** 12.5 40. 224 ft 41. **(a)** 840 **(b)** 360 **(c)** 120 42. **(a)** 324 **(b)** 108 43. **(a)** 3024 **(b)** 126 **(c)** 720 **(d)** 8436 44. **(a)** 3003 **(b)** 80,080 **(c)** 10,626 45. $32x^5 - 240x^4 + 720x^3 - 1080x^2 + 810x - 243$ 46. $-21{,}840a^4$ 47. $(3 + 0.1)^4 = 92.3521$ 48. **(a)** $\frac{7}{13}$ **(b)** $\frac{1}{17}$ 49. **(a)** $\frac{3}{8}$ **(b)** $\frac{1}{4}$ **(c)** $\frac{7}{8}$ 50. **(a)** $\frac{5}{36}$ **(b)** $\frac{1}{6}$ **(c)** $\frac{5}{18}$

INDEX

F

G

H

I

J

L

M

N

SPECIAL FACTORING FORMULAS

Difference of squares: $a^2 - b^2 = (a - b)(a + b)$

Difference of cubes: $a^3 - b^3 = (a - b)(a^2 + ab + b^2)$

Sum of cubes: $a^3 + b^3 = (a + b)(a^2 - ab + b^2)$

Trinomial square: $a^2 + 2ab + b^2 = (a + b)^2$

Trinomial square: $a^2 - 2ab + b^2 = (a - b)^2$

PROPERTIES OF FRACTIONS

$\frac{ac}{bc} = \frac{a}{b}$ $\quad \frac{a}{b} \cdot \frac{c}{d} = \frac{ac}{bd}$ $\quad \frac{a}{b} \div \frac{c}{d} = \frac{ad}{bc}$

$\frac{a}{c} + \frac{b}{c} = \frac{a+b}{c}$ $\quad \frac{a}{c} - \frac{b}{c} = \frac{a-b}{c}$ $\quad$ If $\frac{a}{b} = \frac{c}{d}$, then $ad = bc$.

VARIATION

Direct: $y = kx$ $\quad$ Indirect: $y = \frac{k}{x}$ $\quad$ Joint: $y = kxy$

EXPONENTS

$b^0 = 1, \quad b^{-n} = \frac{1}{b^n}$

$b^m b^n = b^{m+n}$

$\frac{b^m}{b^n} = b^{m-n}$

$(b^m)^n = b^{mn}$

$(ab)^m = a^m b^m$

$\left(\frac{a}{b}\right)^m = \frac{a^m}{b^m}$

RADICALS

$a^{1/n} = \sqrt[n]{a}$

$a^{m/n} = (\sqrt[n]{a})^m = \sqrt[n]{a^m}$

$\sqrt[n]{a} \cdot \sqrt[n]{b} = \sqrt[n]{ab}$

$\frac{\sqrt[n]{a}}{\sqrt[n]{b}} = \sqrt[n]{\frac{a}{b}}$

$\sqrt[m]{\sqrt[n]{a}} = \sqrt[mn]{a}$

LOGARITHMS

$\log_b x = y$ means $b^y = x$

$\log_b MN = \log_b M + \log_b N$

$\log_b \frac{M}{N} = \log_b M - \log_b N$

$\log_b (N^b) = k \log_b N$

$\log_b (b^x) = x$ and $b^{\log_b x} = x$

INEQUALITIES

Transitive property: If $a < b$ and $b < c$ then $a < c$.

Addition property: If $a < b$ then $a + c < b + c$.

Multiplication properties: If $a < b$ and $c > 0$ then $ac < bc$.

If $a < b$ and $c < 0$ then $ac > bc$.